Ireneo Peral Alonso, Fernando Soria de Diego
Elliptic and Parabolic Equations Involving the Hardy-Leray Potential

De Gruyter Series in Nonlinear Analysis and Applications

Volume 38

Ireneo Peral Alonso, Fernando Soria de Diego

Elliptic and Parabolic Equations Involving the Hardy-Leray Potential

DE GRUYTER

Mathematics Subject Classification 2010
35A15, 35B25, 35B30, 35B33, 35B40, 35J58, 35B99, 35C15, 35D05, 35D10, 35D30, 35D35, 35J20, 35J25, 35J58, 35J70, 35K05, 35K15, 35K55, 35K57, 35K58, 35R09, 35R11, 45K05, 46E30, 46E35, 47G20

Authors
Prof. Dr. Ireneo Peral Alonso
Universidad Autonoma de Madrid
Departamento de Matemáticas
Calle de Tomás y Valiente 7
Ciudad Univ. de Cantoblanco
28049 Madrid
Spain
ireneo.peral@uam.es

Prof. Dr. Fernando Soria de Diego
Universidad Autonoma de Madrid
Departamento de Matemáticas
Calle de Tomás y Valiente 7
Ciudad Univ. de Cantoblanco
28049 Madrid
Spain
fernando.soria@uam.es

ISBN 978-3-11-060346-0
e-ISBN (PDF) 978-3-11-060627-0
e-ISBN (EPUB) 978-3-11-060560-0
ISSN 0941-813X

Library of Congress Control Number: 2020950485

Bibliographic information published by the Deutsche Nationalbibliothek
The Deutsche Nationalbibliothek lists this publication in the Deutsche Nationalbibliografie; detailed bibliographic data are available on the Internet at http://dnb.dnb.de.

Typesetting: VTeX UAB, Lithuania
Printing and binding: CPI books GmbH, Leck

www.degruyter.com

To our families

This book is dedicated to Magdalena and Angelines, for their love, support and perseverance; to our daughters and son, Irene and Magdalena and Irene, Eva and Alberto; to their spouses, David and Carlos and Carlos, Carlos and Irene; and to five important persons, our grandchildren, Iria, Oscar, Laura and Jaime and Emma.

Preface

This monograph contains part of the work developed by the first author during the last 25 years in connection with the study of solutions to certain elliptic and parabolic partial differential equations associated to the Hardy–Leray potential. So, the guest star of this story is nothing but

$$V(x) = \frac{1}{|x|^2}, \quad x \in \mathbb{R}^N,\ N \geq 3,$$

a potential function that is closely related to the *uncertainty principle* of Heisenberg in *quantum mechanics*.

The starting point of the theory is Hardy's inequality in one dimension, the only one properly attributed to Hardy. Our interest in this field came when reading the seminal work of Jean Leray [225] on the Navier–Stokes equations, where one can find the proof of a *similar* inequality to that by Hardy, but this time in dimension three, the natural dimension for these equations.

In the paper [257] this inequality was used to study the stability of the extremal solution of a problem related to a reaction combustion model. The idea behind it was to establish a blow-up result. But the first application of what is now called the Hardy–Leray inequality can be seen in the paper by Jean Leray about the Navier–Stokes equations quoted above. The Hardy–Leray inequality also appears as a tool in the study of the Schrödinger equation

$$iu_t + \Delta u = V(x,t)u, \quad V(x,t) = -\frac{(Zq - Q)}{4\pi|x|} - \frac{Q}{4\pi|x - v_0 t|}, \quad Z \in \mathbb{N},\ t > 0.$$

This describes the wave function of an electron under the action of an α-particle with charge Q that, at time $t = 0$, leaves the nucleus of an atom with velocity $v_0 \in \mathbb{R}^3$ (see [133, Chapter XVIII, page 752–758]).

The peculiar behavior of the Hardy–Leray potential is motivated by the fact that it has the same homogeneity as the Dirichlet integral of the Laplacian when rescaling the space variables. In this sense, a similar behavior is observed for

$$V(x) = \frac{1}{|x|^p}, \quad x \in \mathbb{R}^N \setminus \{0\},\ N > p > 1,$$

with respect to the p-Laplacian, i. e., the operator $\Delta_p(u) = \operatorname{div}(|\nabla u|^{p-2}\nabla u)$. See the paper [179], where we began to study some nontrivial extensions of the Hardy–Leray inequality. In this monograph we do not follow this direction, but suggest the reader to search for the information about the subsequent results in the bibliography.

If we consider

$$V(x) = \frac{1}{|x|^{2s}}, \quad x \in \mathbb{R}^N \setminus \{0\},\ N > 2s,\ 0 < s < 1,$$

https://doi.org/10.1515/9783110606270-201

we will find the same behavior of the homogeneity with respect to the so-called *fractional Laplacian* that was introduced my M. Riesz in [272]. In recent years this operator has attracted the attention of mathematicians and is appearing in a large number of problems. The definition and properties of the fractional Laplacian will be explained in Chapter 8.

The main goals of this work are to develop the key ingredients to study how the singular potentials

$$V(x) = \frac{1}{|x|^2} \quad \text{and} \quad V_s(x) = \frac{1}{|x|^{2s}}, \; 0 < s < 1, \; N > 2s,$$

affect some problems modeled by partial differential equations and integro-differential equations, respectively. It is in this last part in which the cooperation of the authors has been definitively used in a deep way in order to complete the total landscape of *both the local and the nonlocal setting.*

The Hardy–Leray potentials, $V(x)$ and $V_s(x)$, have something fascinating: they complicate the computations and are difficult to handle, but at the same time they provide a very rich set of properties. In this sense V and V_s are *both a monster and a friend*, as we will try to show in this monograph. We are really in the presence of an old classic theme, especially in the case of V, but with many new results in recent times for both potentials.

An important feature about $V(x)$ is its dimensionless character. This is due to the fact that *it corresponds, modulo constants, to the Fourier transform of the fundamental solution of the Laplace equation in any dimension greater than or equal to three.*

That is, in the sense of the Fourier transform defined on $\mathcal{S}'(\mathbb{R}^N)$, the dual of the Schwartz class, we find

$$\mathcal{F}\left(\frac{a_N}{|x|^{N-2}}\right)(\xi) = \frac{b_N}{|\xi|^2}.$$

The same property clearly holds for the potential V_s. See Chapters 8 and 9 to find the details.

There are many reasons, coming particularly from physics, about the importance of studying partial differential equations associated to the Hardy–Leray potential. From a purely point of view, we must say that the potential given by the reciprocal of the square function is a perfect *laboratory* to study *borderline properties or critical behavior* in many situations, like the existence of solutions to nonlinear partial differential equations, certain eigenvalue problems, instantaneous and complete blow up and many others.

The same can be said for the fractional Hardy potential, V_s, for which the framework is the *fractional Laplacian* that in the last years has received a lot of attention of researchers due to the analytical interest and the large amount of applications. The reader, again, can find many interesting references about this in the bibliography.

The first seven chapters of the monograph are devoted to the classical Hardy–Leray potential $V(x) = \frac{1}{|x|^2}$ in dimension $N \geq 3$.

In Chapter 1 we describe several settings in which the Hardy–Leray potential appears in a natural way. In quantum mechanics the Hardy–Leray inequality represents, perhaps, the most elementary way to express the uncertainty principle and provides as well a threshold for many associated results. From the analytical point of view, observing that $V \notin L^{\frac{N}{2}}_{loc}$, although $V(x) \in L^p_{loc}$ for $p < \frac{N}{2}$, it gives a good example for the spectral theory of the Laplacian in bounded domains.

We will see how the Hardy potential appears in the study of the stability of the singular solution of a reaction of combustion by Frank-Kamenetskii and by Gelfand and collaborators. The Hardy–Leray inequality also appears in [99] to study the stability of the Simmons cones (see [289]). In Section 1.3 we introduce a very useful tool, the Picone inequality.

Chapter 2 deals with the elementary properties of the Hardy–Leray potential, the proof of the Hardy–Leray inequality in $L^2(\mathbb{R}^N)$ and, in general, in $L^p(\mathbb{R}^N)$, $1 < p < N$. In particular the optimal constant is evaluated and its optimality is proved as well as its nonattainability. To this end, we will also present some results on classical symmetrization. The last section is devoted to obtaining the Hardy–Leray inequality with a remainder term, which depends on a weighted L^2-norm of the gradient.

It is well known, by now, what is the summability of the solution to the problem

$$-\Delta u = f, \quad \text{in } \Omega,\ u = 0,\ \text{on } \partial\Omega,$$

in terms of the summability of the datum f. Here we assume that $\Omega \subset \mathbb{R}^N$ is a bounded domain and $0 \in \Omega$. This can be obtained either as a byproduct of the Calderón–Zygmund theory or as a particular case of the results by Stampacchia about the summability of the solution of the equation in divergence form with measurable bounded coefficients in terms of the datum. Chapter 3 has as an objective to study the effect of a linear term involving the Hardy–Leray potential, that is, to analyze the problem

$$-\Delta u - \lambda \frac{u}{|x|^2} = f, \quad \text{in } \Omega,\ u = 0,\ \text{on } \partial\Omega,$$

where $0 < \lambda < \Lambda_{N,2}$ and $f \in L^m(\Omega)$, $m \geq 1$. Summarizing we can say that if we consider the function

$$\lambda(m) = \frac{N(m-1)(N-2m)}{m^2},$$

then, for $0 < \lambda < \lambda(m)$, the summability is exactly the same as that in the unperturbed problem. Moreover, these results are optimal. For instance, if $f \in L^\infty(\Omega)$ the solution is unbounded and the problem is not well posed in $L^1(\Omega)$, the opposite of what happens in the unperturbed problem. A necessary and sufficient condition for existence in weighted L^1-spaces, among other results, is obtained in Section 3.4.

In the paper by Brezis, Dupaigne and Tesei [89] and by Dupaigne [145], the effect of the Hardy–Leray potential was studied in some semilinear elliptic problems of the type

$$-\Delta u - \lambda \frac{u}{|x|^2} = u^p + f, \quad u \geq 0 \text{ in } \Omega,\ 0 \in \Omega,\ u = 0 \text{ on } \partial\Omega.$$

Chapter 4 collects these results and some additional topics of elliptic equations with measure data, potential theory and a proof of the *mountain pass theorem* by Ambrosetti and Rabinowitz. We start computing the critical exponent for the existence and then we give the existence result for the Dirichlet problem for elliptic equations in divergence form and with measure data. We also present a motivation of the concept of *Newtonian capacity* and a proof, based on arguments by I. Ekeland [147], of the *mountain pass theorem*. Chapter 4 includes the proofs of existence under the critical exponent, $p_+(\lambda)$, and the nonexistence result over the critical exponent. To understand better the nature of the nonexistence result we prove the so-called *complete blow-up*.

Following a path to better understanding of the effect of the Hardy–Leray potential, we devote Chapter 5 to studying the behavior of the heat equation with a reaction term of power type. An inspiring result for the linear heat equation perturbed by the Hardy–Leray potential is due to P. Baras and J. Goldstein in [41]. All along the monograph one will find different approaches to this type of result but in different settings. The threshold power for the existence turns out to be the same as the one we have found for the stationary equation in Chapter 4. Some previous results about the existence of weak solution to the heat equation are introduced in order to make the exposition more systematic. The proof of nonexistence if $p \geq p_+(\lambda)$ is more involved than in the elliptic case. Moreover the proof of the *instantaneous and complete blow-up* gives precise meaning to the nonexistence phenomenon. The Cauchy problem is also analyzed and one of the new facts that are worth to emphasize is the shift to the right of the *Fujita exponent* of the heat equation.

We explore in Chapter 6 some quasilinear problems, that is, those with a nonlinear term which is a power of the gradient. These kinds of problems appear in several contexts related to Hamilton–Jacobi equations, some growth models, as the stationary part of the Kardan–Parisi–Zhang model, some flame propagation, etc. See [213] for some other physical motivations.

These quasilinear problems have an extra degree of difficulty according to the nonlinear gradient term. Because of this, some compactness results have to be used, even more when there is a perturbation by the Hardy–Leray potential. Some analytical tools that will be used along the chapter are discussed as a starting point. Then we will find the critical power of the gradient, $q_+(\lambda)$, in order to have existence and we prove nonexistence if $q \geq q_+(\lambda)$ as well as complete blow-up. The arguments in the quasilinear case are more involved than in the sublinear case. Finally the existence result is obtained for $0 < \lambda < \Lambda_{N,2}$ and $1 < q < q_+(\lambda)$ by using the compactness arguments explained in the complementary analytical results at the beginning of Chapter 6.

In Chapter 7 we study the parabolic problem associated to the nonlinear elliptic problem analyzed in Chapter 6. As in the sublinear case, the introduction of a linear term with the Hardy–Leray potential gives the same threshold to the existence as in the elliptic case, $q_+(\lambda)$. Section 7.2 introduces some important analytical tools that will be essential to obtain the results in the chapter. These tools could be useful in other contexts. The nonexistence and instantaneous and complete blow-up are obtained if $p \geq q_+(\lambda)$. We emphasize here that the Caffarelli–Kohn–Nirenberg inequalities play an important role to prove this nonexistence result. Also, the existence of solution for $p \leq q_+(\lambda)$ is studied in detail giving precise results on the regularity of the solutions.

The results for the Cauchy problem are really surprising. It is well known that for the quasilinear problem associated to $\lambda = 0$, that is, without the additional term of the Hardy potential, there is no Fujita type exponent; see for instance [187]. However the presence of the term depending on the Hardy–Leray potential does produce a Fujita exponent.

In the last chapters of this work, Chapters 8 to 13, we will consider the *relativistic uncertainty principle*; that is, the corresponding fractional Hardy inequality and its consequences in problems with fractional diffusion. Among all the possibilities to define a fractional power, $(-\Delta)^s$, $0 < s < 1$, of the positive operator given by the Laplacian $-\Delta$, we will use the singular kernel that coincides with that coming from the Fourier transform. This is the point of view of potential theory that was introduced by Marcel Riesz in [272], where he suggested to compute the kernel via an analytical continuation in the complex plane of the exponents. This idea has been well established in the literature; in particular , it can be seen in the book by L. Landkof [223]. Nevertheless, we perform the detailed calculation to obtain the kernel in Chapter 8, where we will also study the Dirichlet problem for a related operator to the fractional Laplacian. The fractional Laplacian is in nature a nonlocal operator and, hence, the estimates are often different from the local case. We have tried to present the techniques and functional framework in this area, many of which come from Fourier analysis and potential theory. We place special emphasis on the regularity of the solution in terms of the regularity of the source term. Related to this, a fractional Picone inequality is proved as a tool that will be used in some subsequent chapters.

Chapter 9 represents a central point of the monograph. Here we give a direct proof of the Hardy inequality in this fractional setting. Our proof is inspired by the article of W. Beckner [54]. We give a general result which includes even the cases when $s > 1$ and with dimension $N > 2s$. Moreover, we find the value of the optimal constant $\Lambda_{N,2s}$ and prove that it is not attained. Looking at the Euler–Lagrange equation associated to the Hardy inequality we look at the problems

$$(-\Delta)^s u - \lambda \frac{u}{|x|^{2s}} = 0,$$

where $0 < \lambda < \Lambda_{N,2s}$, and we study the local behavior of the solution near 0. This result in the case $s = 1$, as we have seen, is elementary; in fact, it reduces to the integration of a linear differential equation of Euler type. The case $0 < s < 1$ is more involved and is obtained in detail in the corresponding section of Chapter 9. We complete the results with the *ground state representation* and the Hardy inequality with remainder term, which is another way to prove that the optimal constant is not attained. Chapters 8 and 9 give a self-contained approach to the main results with respect to the fractional Laplacian and the proof of the corresponding Hardy inequality.

The summability of the solutions of the Dirichlet problem for the fractional Laplacian, studied in Chapter 8, is deeply modified by the presence of a linear term involving the Hardy fractional potential. In Chapter 10 such effect is analyzed. The proofs of the results involve new functional settings such as the weighted Sobolev inequalities that are closely related with the equation obtained by the ground state representation. This new functional setting has an interest in itself. In order to obtain the results, the Harnack inequality for the positive solutions of the equation is obtained as a result of the ground state representation. A necessary and sufficient condition for weighted L^1-spaces is also obtained as a consequence of the Hardy inequality. In the last section of Chapter 10 we list some open questions.

The effect of the fractional Hardy potential in semilinear problems with fractional diffusion is studied in Chapter 11. Some important tools are presented in order to obtain an almost self-contained discussion. We concentrate in the first part of the chapter on the so-called *concave-convex problem*. This choice is made in order to complete the situation in Chapter 4 for the classical, local case. First, for each $\lambda \in (0, \Lambda_{N,2s})$ we find a critical exponent, $p(\lambda, s)$, such that if $0 < p < p(\lambda, s)$ we are able to find a positive minimal solution. The critical exponent gives, in turn, that the critical Sobolev exponent $p^*_{N,s}$ is less than $p(\lambda, s)$. Then, if $p \le p^*_{N,s}$ we study the existence of a second positive solution. To see that $p(\lambda, s)$ is the barrier for existence, we prove that if $p \ge p(\lambda, s)$ there is no positive solution to the problem. Also in this case, the nonexistence gives a complete blow-up behavior. To finish the chapter we give some results with nonlinearities that are singular at the boundary.

The last chapters, Chapters 12 and 13, are respectively dedicated to the fractional heat equation and to the effect of the Hardy potential on the existence and regularity of the solution to the evolution equation with fractional diffusion. In Chapter 12 we provide a quite complete exposition of the results of existence and regularity with different kinds of data. We would like to describe here a couple of interesting results related to the Cauchy problem. The first one is a detailed proof of the known asymptotic behavior at infinity of the fundamental solution of the fractional heat equation. See [262] for dimension $N = 1$ and [63] for $N \ge 1$. Our proof is inspired in the *probabilistic world* (see for instance [84] and the references therein), but translated to pure analytical language. In fact, the proof that we present uses Bernstein's theorem, which is based on the *Helly selection principle*. Since these results are in some way hidden in the literature we prove both by completeness. The second point is the uniqueness theorem for positive solutions of the Cauchy problem in the spirit of the classical D. Widder theorem for the heat equation. Widder's uniqueness result shows the coherence of the model with the principles of thermodynamics.

We finish the monograph with Chapter 13, where the starting point is to prove the Harnack inequality for the resulting weighted parabolic equation of the ground state representation. Observe that this is an equation with singular coefficients, which makes the proof much more difficult. As a consequence of this Harnack inequality and some extra estimates, we are able to prove the exact behavior of the solution near the pole of the Hardy potential. The first consequence of this provides a complete profile of the existence results in terms of the spectral parameter $\lambda > 0$ for the linear fractional heat equation. In this way, we obtain necessary and sufficient conditions for the existence, a result that can be seen as the Baras–Goldstein result in the fractional case. Nevertheless, the arguments given here are slightly different from those in the proof of the Baras–Goldstein result for the heat equation. We also study the effect of the Hardy potential on a semilinear fractional heat equation. The critical power is the same as in the elliptic case, as the reader will find in detail in Section 13.5.

One may ask also what happens to the similar problems that contain a power of the norm of the gradient as in the local case. At this point, we can only say, "do not worry," they are not here. We think that a monograph must be of finite length and should be finished at some point. However, the interested reader can still see some results for the elliptic case along this direction in [26].

As the reader may realize, the problems studied in this monograph require a large variety of techniques, which could be useful in many other different problems. We have explained in detail most of the techniques that we have used. In those cases where we could not do this, we have given precise references for anybody to complete the arguments in their totality if necessary. We hope that the contents of the monograph will be useful to any reader, especially for doctoral students working in some areas of analysis and, more specifically, for those interested in the area of partial differential equations.

This monograph is the result of many years of work and academic discussions with colleagues and doctoral students. We are deeply thankful to all our collaborators and colleagues for their continuous support and advice. In particular we point out that the paper with J. L. Vázquez [257] represented a first application of the Hardy–Leray inequality. We would like to thank Juan Luis Vázquez for this collaboration. A bit later, a starting piece of work was introduced in the article written in collaboration with Jesús García Azorero [179]. We deeply thank Jesús for this long scientific journey in which we have enjoyed mathematics and many more things together.

We are also thankful to our *maestros*, A. Ambrosetti, L. Boccardo, L. Caffarelli and Y. Meyer,[1] for their friendship and advice. We have learned from them the best flavor of mathematics in many and diverse areas.

1 During the proofreading one of our referents passed away, Antonio Ambrosetti. We will always remember him

Our special thanks go to Boumediene Abdellaoui and Ana Primo for the work we have shared for more than 15 years, enjoying mathematics in a friendly atmosphere. We thank Luigi Orsina for his friendly collaboration. Special thanks also go to Enrico Valdinoci, Begoña Barrios, Serena Dipierro, Tommaso Leonori, María Medina, Luigi Montoro and Dino Sciunzi for allowing us to join the "fractional" adventure. Young collaborators, as Pablo Ochoa and Leandro del Pezzo, gave us the energy to follow. Thanks go to Magdalena Walias, among many other things, for her collaboration and support. To Juanjo Manfredi and Julio D. Rossi we must thank many helpful discussions. We thank also David Arcoya and David Ruiz, the Granada team, and Andrea Dall'Aglio and Daniela Giachetti for their collaboration and friendly hospitality.

Fernando Charro, Eduardo Colorado, Carlos Escudero, Veronica Felli, Fausto Ferrari, Filippo Gazzola, Anna Mercaldo and Susana Merchán, have been close collaborators with whom we have worked on other topics that, still, have influenced this monograph; we thank them deeply.

The Department of Mathematics of the UAM has been our scientific home for almost our entire professional life; the Institution and its members deserve our recognition for giving us their support during the shared years.

The final work of corrections and adjustments was done during a confinement period in Madrid, due to the Covid-19 pandemic. We hope that this critical situation will help us to reflect and improve our human living conditions at all levels throughout the World.

With our best wishes for the readers,

Madrid May 31, 2020.
The Authors

Contents

1 Motivation of the Hardy–Leray potential

The aim of this chapter is to explain the motivation and basic facts about the Hardy–Leray potential,

$$V(x) = \frac{1}{|x|^2}.$$

The space dimension must be $N \geq 3$ for integrability reasons. In lower dimensions the setting is different: either we lose homogeneity, or we assume some particular behavior of the elements of the functional space close to the pole of the potential.

1.1 Introduction

The one dimensional Hardy inequality is the continuous translation of the following discrete inequality.

The discrete Hardy inequality. Let $\{a_n\}_{n\in\mathbb{N}}$ be a sequence of nonnegative real numbers which is not identically zero; then, for every real number $p > 1$, one has

$$\sum_{n=1}^{\infty}\left(\frac{a_1 + a_2 + \cdots + a_n}{n}\right)^p < \left(\frac{p}{p-1}\right)\sum_{n=1}^{\infty} a_n^p.$$

From this, Hardy's inequality in its continuous version can be formulated as follows.

If f is a positive, integrable function, then

$$\int_0^{\infty}\left(\frac{1}{x}\int_0^x f(s)\,ds\right)^p < \left(\frac{p}{p-1}\right)\int_0^{\infty} f(s)^p\,ds. \tag{1.1.1}$$

Moreover the equality is only true for $f \equiv 0$.

We are talking here about nothing more than real analysis and pure mathematics, with more frequency on the Hilbert space setting, that is, when $p = 2$.

Inequality (1.1.1) appeared without a proof for the first time in the paper by Hardy [195]. A proof was later presented in [196]. J. Leray established (see (1.14) in [225]) the following basic inequality in three dimensions.

Let u be a function with first derivatives $u_{x_i} \in L^2(\mathbb{R}^3)$, $i = 1, 2, 3$. Then

$$\int_{\mathbb{R}^3} \frac{u^2}{|x|^2}\,dx \leq 4\int_{\mathbb{R}^3} |\nabla u|^2\,dx. \tag{1.1.2}$$

We will call this the Hardy–Leray inequality.

The proof of Leray relies on a representation formula with respect to the fundamental solution of the stationary Stokes equations. We skip this proof and send the

https://doi.org/10.1515/9783110606270-001

reader to the original paper, which is a wonderful milestone in the mathematics of the twentieth century. In contrast to this, we will prove the inequality for $N \geq 3$ in an elementary way; see Theorem 2.2.1. Before going into this, note that the one dimensional Hardy inequality kills a bit the singularity of the potential by forcing the function to be equal to zero at the pole.

1.2 Where does the monster appear?

We present next some different problems in which the Hardy–Leray potential appears or is hidden behind some of the results. All the problems that we will consider could be described in some sense as classical and critical-supercritical.

1.2.1 The Hardy–Leray potential in quantum mechanics

From the beginning of quantum mechanics the presence of singular potentials appeared in different models. In particular, the Hardy potential is a fundamental object for quantum physicists who study the Schrödinger equation. Such kinds of singular potentials provide a new problematic in order to study the energy eigenvalues and eigenfunctions. We will follow the paper of K. M. Case [115] and the book by Moot and Massey [247]. Here the Hardy–Leray potential appears when studying the scattering of a beam of atomic particles in a central vector field in which the potential energy at distance r of the nucleus is $\frac{\gamma}{r^2}$. This model seems, historically, to be the first one where certain *spectral difficulties* had to be overcome.

Consider the stationary Schrödinger equation in three dimensions,

$$\Delta\psi + \{k^2 - U(r)\}\psi = 0, \tag{1.2.1}$$

where k depends of the mass m and the $\hbar$, the Planck constant. Solving the equation in polar coordinates, it is easy to see that

$$\phi = P_n(\cos\theta)f_n(r)$$

is a solution if f_n is, in turn, a solution to

$$\frac{1}{r^2}\frac{d}{dr}\left(r^2\frac{df}{dr}\right) + \left(k^2 - U(r) - \frac{n(n+1)}{r^2}\right)f = 0 \tag{1.2.2}$$

and $P_n(\cos\theta)$ is the nth Legendre function (see [315] page 302). Then the *superposition principle* helps to find general solutions.

In particular, if $U(r) = -\frac{\lambda}{r^2}$, equation (1.2.2) reads as

$$f''(r) + \frac{2}{r}f'(r) + \left(k^2 - \frac{n(n+1)+\lambda}{r^2}\right)f = 0. \tag{1.2.3}$$

It is a classical result that the solutions to equation (1.2.3) are

$$\frac{1}{r^{\frac{1}{2}}} J_{\nu+\frac{1}{2}}(kr),$$

where $J_{\nu+\frac{1}{2}}$ is the Bessel function of index $\nu + \frac{1}{2}$ and ν is a solution to the algebraic equation

$$\nu(\nu+1) = n(n+1) + \lambda, \quad \text{i. e.,} \quad \nu = \frac{1}{2}\{-1 + (1 + 4n + 4n^2 + 4\lambda)^{\frac{1}{2}}\}.$$

Now our wave function must be finite at the origin; therefore, since

$$\frac{1}{r^{\frac{1}{2}}} J_{\nu+\frac{1}{2}} \approx r^{\nu}, \quad \text{we need } \nu \geq 0.$$

In the case of a *repulsive field*, that is, if $\lambda > 0$, the condition is satisfied only for a root of the equation, and this fact means that the bounded solution is unique up to a multiplicative constant.

More interesting is the case of *attractive field*, that is, $\lambda < 0$. We have two possibilities:

1. If $-\frac{1}{4} < \lambda < 0$, then for $n = 0$ both solutions are unbounded at the origin, but there is one solution with a lower order of singularity at the origin that any other one. By choosing this solution, which in fact is locally of finite energy, it is possible to perform calculations and, as is explained in [247], *a formula for the scattered intensity may be obtained.* We will find many situations in which this behavior is crucial. Note that $\frac{1}{4}$ is a *magic quantity* that is related to inequality (1.1.2).
2. If $\lambda < -\frac{1}{4}$, then all nontrivial solutions to the equations behave like

$$r^{-\frac{1}{2}} e^{\pm i\alpha \log r}$$

near the origin. That is, in this case infinitely many oscillations happen in any neighborhood of the origin. As is explained in [247] in this case *there is no solution to the scattering problem.* We will find a lot of anomalies for $\lambda < -\frac{1}{4}$. The physicist tells us that in this unbounded interval there is no way do choose a convenient solution.

We call the attention of the reader about the existence of a constant that is a borderline in the behavior of the radial equation (1.2.3). We will elucidate this point below.

Nevertheless, in [115], a method to select such a solution with ideas from Von Neumann is developed. Such idea is a suitable understanding of the orthogonality of the solutions.

To give another idea of the peculiar behavior of the Hardy–Leray potential in quantum mechanics, let us describe an observation by C. Fefferman, in his paper [159].

We try to study the negative eigenvalues of the operator

$$L \equiv -\Delta + V(x) \quad \text{in } \mathbb{R}^N,\ N \geq 3,$$

where V has a suitable local summability condition. Fefferman refers to a theorem by Cwikel–Lieb–Rosenblum, which gives a precise estimate of the number of negatives eigenvalues of L below the level μ, $N(L,\mu)$; see [128], [229], [276] and the paper [166] for some fractional related results. Such theorem establishes that the number of eigenvalues of L below the level μ, $N(L,\mu)$, is controlled by a constant depending only on the dimension multiplied by the volume in the *phase space*, that is, in $\mathbb{R}^N \times \mathbb{R}^N$, of the set

$$A(L,\mu) = \{(x,\xi) \in \mathbb{R}^N \times \mathbb{R}^N \mid |\xi|^2 + V(x) < \mu\}.$$

Namely,

$$N(L,\mu) \leq C_N |A(L,\mu)|.$$

As a consequence, if for fixed μ, $|A(L,\mu)| < C_N^{-1}$, then $L \geq \mu$ in the sense of the operators.

If we now specialize the result for $L \equiv -\Delta - \frac{\lambda}{|x|^2}$, it is clear that $|A(L,\mu)| = +\infty$, so it could be thought that L is not bounded from below.

However, one of the more relevant classical results related to the *Hardy potential* is inequality (2.2.2), which contradicts the idea, given by the above argument, that $L \equiv -\Delta - \frac{\lambda}{|x|^2}$ is unbounded from below.

The conclusion is that the Hardy–Leray potential has not the summability required in the theorem of Cwikel–Lieb–Rosenblum; see in particular inequality (1.1) in [166] quantifying Cwikel's inequality.

1.2.2 More quantum mechanics: the uncertainty principle

Hardy's inequality (1.1.2) is the first mathematical formulation of the *Heisenberg uncertainty principle*, which, roughly speaking, says that the position and the momentum of a particle cannot be approximated with simultaneous accuracy. Indeed, we can read the Hardy inequality by saying that if u is concentrated close to the origin, that is, we have a very good approximation of the position of the particle, then the momentum has to be big.

Note that if we have an inequality of the form

$$C \int_{\mathbb{R}^N} |x|^\sigma u^2(x)\,dx \leq \int_{\mathbb{R}^N} |\nabla u|^2\,dx, \quad \text{for all } u \in C_0^\infty(\mathbb{R}^N),$$

then necessarily $\sigma = -2$. It is sufficient to check the inequality on the family of rescaled functions $\phi_\mu(x) = \phi(\mu x)$ to find the inverse square potential. That means that the inverse square potential has the same dimensionality of the gradient and this is one of the main characteristics to be important. To see more details and forms of the Heisenberg principle the reader can see the nice lecture notes by R. Frank [165].

1.2.3 Frank-Kamenetskii model in combustion: a Gelfand problem

The semilinear reaction–diffusion equation

$$u_t - \Delta u = \lambda e^u, \quad \lambda > 0, \tag{1.2.4}$$

and its stationary counterpart

$$\Delta u + \lambda e^u = 0 \tag{1.2.5}$$

appear in several contexts. For instance, equation (1.2.5) appears in astrophysics in the description of a ball of isothermal gas in gravitational equilibrium. This was already proposed by Lord Kelvin. Radially symmetric stationary solutions have been studied by Emden (1896), Fowler (1926) and Chandrasekar and Hopf in the 1930s; see Chandrasekar's book [118]. In combustion the evolution model (1.1) was proposed by Frank-Kamenetskii [169] and is known as the solid fuel ignition model. A more recent approach could be seen in [53].

In the form

$$\Delta u + K(x)e^u = 0$$

and space dimension 2, the stationary equation is found in Riemannian geometry in the search for certain Riemannian metrics with given Gaussian curvature; see the paper by Kazdan and Warner [211].

Some pioneering works in the 1950s and 1960s are [182] (in particular, Section 15, by G. I. Barenblatt) and [321].

Important mathematical contributions are due to Fujita [170], who studied the evolution problem with zero boundary data as well. An important contribution is due to Joseph and Lundgren [204], which describes the bifurcation diagram of equation (1.2.5) posed in a ball B with zero boundary data.

Now we will try to study the relation of the above equation with exponential reaction term and the Hardy–Leray potential. To answer this question let us recall the main results of the bifurcation analysis in [204].

We will consider the Dirichlet boundary problem,

$$\begin{aligned} \Delta u + \lambda e^u &= 0, \quad x \in B(0,1), \\ u &= 0, \quad \|x\| = 1. \end{aligned} \tag{1.2.6}$$

If λ_1 is the principal eigenvalue for the Laplacian in $B(0,1)$ and ϕ_1 is a positive eigenfunction, we have

$$\begin{aligned}\lambda_1 \int_{B(0,1)} u\phi_1\,dx &= \int_{B(0,1)} u(-\Delta\phi_1)\,dx\\ &= \int_{B(0,1)} (-\Delta u)\phi_1\,dx = \lambda \int_{B(0,1)} \phi_1 e^u\,dx \ge \lambda \int_{B(0,1)} \phi_1 u\,dx,\end{aligned}$$

which implies that if u is a solution of (1.2.6), then $\lambda < \lambda_1$. By using a monotony argument we are able to prove that there exists an interval $[0,\lambda^*]$ such that for $\lambda \in [0,\lambda^*]$, problem (1.2.6) has a minimal solution.

Moreover it is known that:

1. If $N \ge 10$, the regular minimal solutions for $\lambda \in [0,\lambda^*)$ converge to a solution u^* (*a priori* singular) when λ goes to λ^*.
2. If $N < 10$, the limit u^* is a regular solution.

See [176], [240], [241] and [204].

Writing equation (1.2.5) in radial coordinates we find

$$-u''(r) - \frac{N-1}{r}u'(r) = \lambda e^u, \quad r > 0. \tag{1.2.7}$$

By the classical change of variables given in [204] (see also [182]),

$$\begin{cases} s = \log r,\\ v = u_s,\\ w = -\lambda e^{2s} e^{u(s)}, \end{cases}$$

we obtain the dynamical system

$$\begin{cases} \dfrac{dv}{ds} = w - (N-2)v,\\ \dfrac{dw}{ds} = (2+v)w. \end{cases} \tag{1.2.8}$$

It is easy to check that the autonomous dynamical system (1.2.8) has two isolated critical points, i. e.,

$$P_1 = (0,0) \quad \text{and} \quad P_2 = -2(1, N-2).$$

Since $u > 0$, we are interested in $w < 0$. By analyzing carefully the linearized system we find that P_1 is an unstable hyperbolic point, independently of the dimension N. The stable manifold is the v-axis and the unstable manifold is tangent to $w = Nv$. The behavior of P_2 depends on the dimension, precisely,

$$P_2 \text{ is } \begin{cases} \text{a stable nodus if } N \ge 10,\\ \text{a stable focus if } 3 \le N \le 9. \end{cases}$$

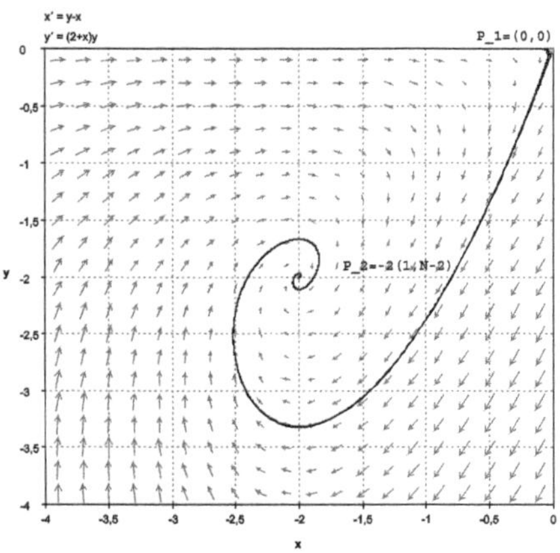

Dimension $N = 3$

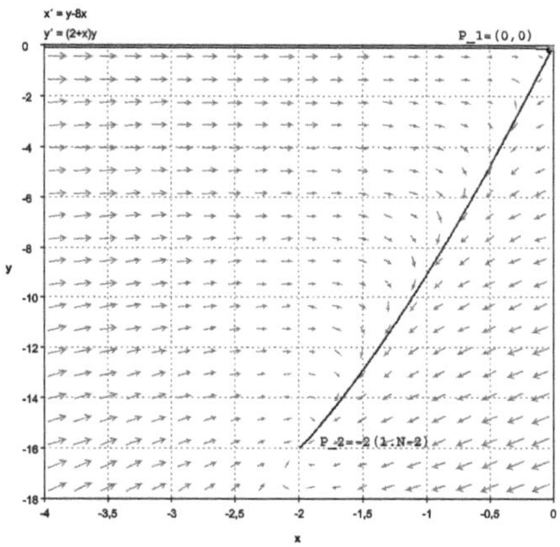

Dimension $N = 10$

Equation (1.2.6) is invariant with respect to the scaling

$$(x, u) \to (\alpha x, u(\alpha x) + 2\log\alpha),$$

and then we find a radial singular self-similar solution to (1.2.6), that is,

$$u(x) = \log\left(\frac{1}{|x|^2}\right), \quad \tilde{\lambda} = 2(N-2).$$

The direction field shows that the unstable manifold starting out from P_1 remains bounded, and therefore is a stable manifold for P_2. In fact, the direction field implies that either the manifold remains bounded or $\lim_{s\to+\infty} w(s) = -\infty$.

But in this case, we have a trajectory corresponding to a radial solution such that if we fix $\lambda > \lambda^*$, for some s_1, $w(s_1) = -\lambda$. By the invariance of the system under translations in the variable s, we can assume $s_1 = 0$, and then $w(0) = -\lambda = -\lambda e^{u(0)}$. Moreover, $\lim_{s\to-\infty} w(s) = 0 = \lim_{s\to-\infty} v(s)$. Then, in the r variable $u(1) = 0$ and $u(r)$ remains bounded in a neighborhood of the origin. Hence problem (1.2.6) has a solution for $\lambda > \lambda^*$ and this is a contradiction. This argument implies that:
1. There exists an heteroclinic orbit, φ, joining the critical points P_1 and P_2.
2. Given λ, problem (1.2.6) has a solution for each intersection of φ with the straight line $w = -\lambda$.

We need the following results to understand the behavior of problem (1.2.7) in terms of the dynamical system (1.2.8).

Lemma 1.2.1. *Let u be a radial solution of* (1.2.7) *and* (v, w) *the corresponding trajectory of* (1.2.8).

Then u is a regular solution of (1.2.7) *(i. e., $\lim_{s\to-\infty} u(s) = A < \infty$) if and only if*

$$\lim_{s\to-\infty} (v(s), w(s)) = (0, 0).$$

Proof. It is clear that if u is a regular solution, then $\lim_{s\to-\infty}(v(s), w(s)) = (0, 0)$.

Conversely, since the unstable manifold starting from the origin is tangent to the line $w = Nv$, if

$$\lim_{s\to-\infty}(v(s), w(s)) = (0, 0),$$

then

$$\lim_{s\to-\infty}\frac{w(s)}{v(s)} = N,$$

and in this way, there exists some negative constant s_0 such that if $s < s_0$, then $\frac{1}{2N} < \frac{v}{w} < \frac{2}{N}$. Hence,

$$2w + \left(\frac{1}{2N}\right)w^2 < \frac{dw}{ds} < 2w + \left(\frac{2}{N}\right)w^2.$$

Therefore, for $-w = y^{-1}$ we get

$$\left(\frac{1}{2N}\right) - 2y < \frac{dy}{ds} < \left(\frac{2}{N}\right) - 2y.$$

On the other hand, since $\lim_{s\to-\infty} w(s) = 0$ and $w \le 0$, $\lim_{s\to-\infty} y(s) = +\infty$ and then for s small,

$$\int_s^{s_0} \frac{y'(s)}{(\frac{2}{N}) - 2y}\,ds > (s_0 - s).$$

Finally, we get

$$y(s) > \frac{1}{2}\left(\left(\frac{2}{N}\right) + \left(2y(s_o) - \left(\frac{2}{N}\right)e^{2(s_0-s)}\right)\right).$$

But from this inequality it follows easily that $u(s) < K$. □

The next step is to identify the self-similar solution to (1.2.7) in the dynamical system (1.2.8).

Lemma 1.2.2. *The unique trajectory of* (1.2.8) *corresponding to a solution of* (1.2.7) *such that*

$$\lim_{s\to-\infty} u(s) = +\infty \quad \textit{and} \quad e^u \in L^1_{\mathrm{loc}}$$

is the critical point P_2.

Proof. Assume that $\lim_{s\to-\infty} u(s) = +\infty$. We organize the proof in several cases.

(A) We have $\lim_{s\to-\infty} v(s) = v_0$.

1. If $v_0 > -2$, then there exists some negative constant $s_0 < 0$ such that if $s < s_0$, $(2 + v) > \delta > 0$. Taking into account the sign of w,

$$\frac{dw}{ds} < \delta w, \quad \text{for } s < s_0.$$

Then, by integrating from s to s_0, we get $\lim_{s\to-\infty} w(s) = 0$. Therefore, $\lim_{s\to-\infty} \frac{dv}{ds} = -(N-2)v_0$, and this implies that $v_0 = 0$.
By Lemma 1.2.1 we reach a contradiction with the hypothesis that u is a singular solution.

2. If $v_0 < -2$, in the same way as before we obtain

$$\frac{dw}{ds} > -\delta w,$$

and then $\lim_{s\to-\infty} w(s) = -\infty$. Hence, $\lim_{s\to-\infty} \frac{dv}{ds} = -\infty$, in contradiction with the hypothesis $v(s) \to v_0$.

3. If $v_0 = -2$, it is possible to prove that $\lim_{s\to-\infty} w(s) = -2(N-2)$ (it is clear if we assume that the limit exists; we will prove in (C) below that this the case).
Then, as P_2 is an attractor, necessarily $v(s) = -2$ and $w(s) = -2(N-2)$ for all s.

We will see that the other possibilities are incompatible with the fact that

$$\lim_{s\to-\infty} u(s) = \infty.$$

(B) We have $\lim_{s\to-\infty} v(s) = -\infty$. As in case (A)(2), $\lim_{s\to-\infty} w(s) = -\infty$, which is a contradiction with the nonexistence result for λ large. In fact if $\lambda > \lambda^*$ and we have a trajectory which is a radial solution such that

$$\lim_{s\to-\infty} (v(s), w(s)) = (-\infty, -\infty),$$

we can pick up s_1 such that $w(s_1) = -\lambda$. Since the autonomous system is invariant by translation in time we can assume that $s_1 = 0$. Then $w(0) = -\lambda e^{u(0)}$ and in the r variable $u(1) = 0$. Therefore the elliptic problem has a solution for $\lambda > \lambda^*$, which is a contradiction.

(C) We have $\liminf_{s\to-\infty} v(s) < \limsup_{s\to-\infty} v(s)$.
Under this hypothesis we have moreover

$$\liminf_{s\to-\infty} v(s) \le -2 \le \limsup_{s\to-\infty} v(s)$$

(with strict inequality at least on one side, as a consequence of the arguments in (A)).
As an application of the mean value theorem, there exists a decreasing sequence $\{s_n\}_{n\in\mathbb{N}}$ such that:

(i) $\lim_{n\to\infty} s_n = -\infty$;
(ii) $\frac{dv(s_n)}{ds} = 0$;
(iii) $v(s_{2n}) \geq -2 \geq v(s_{2n+1})$ for all n, with strict inequality at least on one side.
Then from (ii), $w(s_n) = (N-2)v(s_n)$.
After a finite number of indexes n, the trajectory must cross the manifold joining P_1 and P_2 which is the unstable manifold for P_1; then we must have

$$\lim_{s\to-\infty} (v(s), w(s)) = (0,0)$$

and this is a contradiction with the hypothesis that u is a singular solution.

As a conclusion, if u is a radial singular solution of (1.2.7) the only possibility is (A)(3). □

Then we have the following result.

Corollary 1.2.3. *The unique unbounded radial singular solution of* (1.2.7) *is given by*

$$\tilde{\lambda} = 2(N-2), \quad u(x) = \log\left(\frac{1}{|x|^2}\right).$$

By scaling $(x, u) \to (\alpha x, u(\alpha x) + 2\log\alpha)$ we can obtain the result in every ball and by the same arguments the unique radial solution for λ fixed of

$$\begin{cases} -\Delta u = \lambda e^u & \text{in } \mathbb{R}^N \setminus \{0\}, \\ u \in W^{1,2}_{\text{loc}}(\mathbb{R}^N) & e^u \in L^1_{\text{loc}}(\mathbb{R}^N) \end{cases}$$

is

$$u(x) = \log\left(\frac{1}{|x|^2}\right) - \log\left(\frac{\lambda}{2(N-2)}\right).$$

The following result describes the structure of the set of solutions in the radial setting.

Theorem 1.2.4.
1. *If $N \geq 10$, then $\lambda^* = \tilde{\lambda}$; for each $\lambda \in (0, \lambda^*)$ we have a unique radial regular solution, and $\lim_{\lambda\to\lambda^*} \underline{u}(\lambda) = u_*$ is the singular solution.*
2. *If $N \leq 9$, then $\tilde{\lambda} < \lambda^*$, and for $\lambda = \tilde{\lambda}$,* (1.2.7) *has infinitely many regular radial solutions, the values at the origin going to infinity.*

Moreover, in the second case:
(a) *There exists a positive constant $\epsilon_0 > 0$ such that if $0 < |\lambda - \tilde{\lambda}| < \epsilon_0$, then the corresponding problem* (1.2.7) *has a finite family of radial solutions.*
(b) *We have $\lim_{\lambda\to\lambda_*} \underline{u}(\lambda) = u_* \in L^\infty$.*

Proof. 1. The trajectory ϕ joining P_1 and P_2 is a stable manifold for the node P_2.

If we show that ϕ is a monotone curve contained in the region

$$-2 < v < 0, \quad -2(N-2) < w < 0,$$

then for each line $w = -\lambda$ we have a unique point of intersection, and therefore a unique regular radial solution for each $\lambda \in (0, 2(N-2))$.

First, by the direction field, it is easy to see that ϕ is below the line $w = (N-2)v$. Call R the straight line

$$w = \frac{N-2}{2}(v-2).$$

We will show that if $-2 < v < 0$, then $\frac{dw}{dv} < \frac{N-2}{2}$ along R; therefore, the trajectories (v, w) of the phase plane cross R from below, and this implies that the trajectory ϕ cannot cut the line R, since it starts from above. Since in the region

$$-2 < v < 0, \quad -2(N-2) < w < (N-2)v,$$

we have $\frac{dv}{ds} < 0$, it is sufficient to check that

$$\frac{dw}{ds} - \frac{N-2}{2}\frac{dv}{ds} > 0,$$

when $(v, w) \in R$, $-2 < v < 0$. But

$$\begin{aligned}
&\frac{dw}{ds} - \frac{N-2}{2}\frac{dv}{ds} \\
&\quad = (2+v)w - \left(\frac{N-2}{2}\right)(w - (N-2)v) \\
&\quad = (2+v)\left(\left(\frac{N-2}{2}\right)(v-2)\right) - \frac{N-2}{2}\left(\left(\frac{N-2}{2}\right)(v-2) - (N-2)v\right) \\
&\quad = (2+v)\left\{\left(\frac{N-2}{2}\right)(v-2) + \left(\frac{N-2}{2}\right)^2\right\}.
\end{aligned}$$

Then, since $N > 10$ and $-2 < v < 0$,

$$(v-2) + \left(\frac{N-2}{2}\right) > -4 + \left(\frac{N-2}{2}\right) > 0.$$

Therefore, if $N \geq 10$, the trajectory ϕ cannot cross the line R.

2. The straight line $w = -\tilde{\lambda}$ crosses the manifold ϕ infinitely many times. Each point of intersection s_j gives us a radial solution of (1.2.7) by scaling s in such a way that for $s = 0$ we have as initial value s_j. The rest is a consequence of the previous analysis. □

1.2.4 Stability of the singular solution

Finally, this is where the Hardy–Leray potential appears. If we look for the stability of the singular solution corresponding to P_2,

$$-2(N-2) = w = -\lambda e^{-2s} e^{u(s)}; \quad \text{therefore, } u(s) = \log(2(N-2)) - \log\lambda - 2s.$$

Since from the second equation we have $u(s) = -2s$, we conclude

$$\lambda = 2(N-2) \quad \text{and} \quad u_{2(N-2)}(r) = \log\left(\frac{1}{r^2}\right).$$

The stability depends directly on the sign of the linearized equation. Linearizing equation (1.2.5) in $u_{2(N-2)}(r)$ we find

$$-\Delta v = 2(N-2)e^{u_{2(N-2)}(r)} v \equiv \frac{2(N-2)}{|x|^2} v.$$

This fact is in some sense generic for the unbounded radial solutions to supercritical elliptic problems.

The associated energy functional is

$$E(v) = \int_{B_1} \left(|\nabla v|^2 - \frac{2(N-2)}{|x|^2} v^2\right) dx,$$

and, according to the Hardy inequality (2.2.1), $E(v)$ is positive defined if and only if

$$2(N-2) < \left(\frac{N-2}{2}\right)^2,$$

that is, if and only if $N > 10$. In fact, according to the improved Hardy inequalities (see below), in dimension $N = 10$ the singular solution is also stable.

A more general situation in which the Hardy inequality appears in this kind of analysis can be seen in the nice survey [99]. Recently, a general result has been obtained in [102] about the regularity of stable solutions to semilinear elliptic equations up to dimension 9.

The dynamical behavior of the unbounded solution can be seen in [257].

1.3 A Picone type inequality

We consider for $1 < p < \infty$ the space $W^{1,p}(\Omega)$, which is defined as the completion of $\mathcal{C}^\infty(\Omega)$, with respect to the norm

$$\|\phi\|_p = \left(\int_\Omega (|\phi|^p + |\nabla\phi|^p) dx\right)^{1/p},$$

and we denote by $W_0^{1,p}(\Omega)$ the completion of $\mathcal{C}_0^\infty(\Omega)$ with respect to the norm $\|.\|_p$.

By the Poincaré inequality, this norm is equivalent to the L^p-norm of the gradient.

We will present in this section an extension of a well-known Picone identity (see [259]) that will become a handy tool to be used in many situations. See also [14], [34] and [300].

Lemma 1.3.1. *Let $v > 0, u \geq 0$ be differentiable functions. Then if $1 < p < \infty$*

$$|\nabla u|^p \geq \nabla\Big(\frac{u^p}{v^{p-1}}\Big)|\nabla v|^{p-2}\nabla v. \tag{1.3.1}$$

Proof. Let $u, v \in \mathcal{C}^1(\Omega)$ be such that $u \geq 0$ and $v > 0$. Then to prove (1.3.1) we just have to prove that

$$R(u,v) = |\nabla u|^p - \nabla\Big(\frac{u^p}{v^{p-1}}\Big)|\nabla v|^{p-2}\nabla v \geq 0.$$

By a simple computation we obtain

$$R(u,v) = |\nabla u|^p + (p-1)\Big(\frac{u^p}{v^p}\Big)|\nabla v|^p - p\Big(\frac{u^{p-1}}{v^{p-1}}\Big)|\nabla v|^{p-2}\nabla v\nabla u.$$

We set $a = |\nabla u|$ and $b = (\frac{u}{v}|\nabla v|)^{p-1}$. Then using the Minkowski inequality we get

$$\Big(\frac{u^{p-1}}{v^{p-1}}\Big)|\nabla v|^{p-2}\nabla v\nabla u \leq \Big(\frac{u^{p-1}}{v^{p-1}}\Big)|\nabla v|^{p-1}|\nabla u| \leq \frac{1}{p}|\nabla u|^p + \frac{p-1}{p}\frac{u^p}{v^p}|\nabla v|^p.$$

Hence $R(u,v) \geq 0$. Note that if $R(u,v) = 0$, then we obtain $|\nabla u||\nabla v| = \langle\nabla u, \nabla v\rangle$ and therefore $\nabla u = \frac{u}{v}\nabla v$. Hence $\nabla(\frac{u}{v}) = 0$ and then $u = cv$; therefore equality is achieved on straight lines in the function space. □

We need the following integral form of Lemma 1.3.1.

Theorem 1.3.2. *Fix p so that $1 < p < \infty$. Assume $v \in W_0^{1,p}(\Omega)$ verifying $-\operatorname{div}(|\nabla v|^{p-2}\nabla v) = \nu$, a positive bounded Radon measure, with $v_{|\partial\Omega} = 0$ and $v \gneq 0$. Then for all $u \in W_0^{1,p}(\Omega)$*

$$\int_\Omega |\nabla u|^p\,dx \geq \int_\Omega \frac{|u|^p}{v^{p-1}}(-\operatorname{div}(|\nabla v|^{p-2}\nabla v))\,dx.$$

Proof. First step. Assume that $v \in W^{1,p}(\Omega)$ such that $v \geq \delta > 0$ in Ω and $u \in \mathcal{C}_0^\infty(\Omega)$. Without loss of generality we can assume that $u \geq 0$. Then there exists a family of regular functions $\{v_n\}$ such that

$$\begin{cases} v_n \to v & \text{in } W^{1,p}(\Omega),\ v_n \in \mathcal{C}^1(\Omega), \\ v_n \to v & \text{a. e., and } v_n > \frac{\delta}{2} \text{ in } \Omega. \end{cases} \tag{1.3.2}$$

As a consequence of the continuity of $-\operatorname{div}(|\nabla(\cdot)|^{p-2}\nabla(\cdot))$ as an operator from $W^{1,p}(\Omega)$ to $W^{-1,p'}(\Omega)$, $p' = \frac{p}{p-1}$, and by (1.3.1) applied to v_n we get

$$|\nabla u|^p \geq \nabla\left(\frac{u^p}{v_n^{p-1}}\right)|\nabla v_n|^{p-2}\nabla v_n.$$

Integrating by parts and by the Lebesgue *dominated convergence theorem* we obtain

$$\int_\Omega |\nabla u|^p\,dx \geq \int_\Omega \left(\frac{-\operatorname{div}(|\nabla v|^{p-2}\nabla v)}{v^{p-1}}\right)u^p, \qquad u \in C_0^\infty(\Omega),\ u \geq 0.$$

Second step. We consider the general case, i. e., $u \in W_0^{1,p}(\Omega), v \in W_0^{1,p}(\Omega)$, $v \geq 0$, and

$$-\operatorname{div}(|\nabla v|^{p-2}\nabla v) \geq 0$$

is a bounded Radon measure, $v|_{\partial\Omega} = 0, v \geq 0$ and not identically zero. By the strong maximum principle (see [310]) $v > 0$ in Ω. We call $v_m(x) = v(x) + \frac{1}{m}$, $m \in \mathbb{N}$. In this way we have

$$-\operatorname{div}(|\nabla v_m|^{p-2}\nabla v_m) = -\operatorname{div}(|\nabla v|^{p-2}\nabla v),$$

and $\{v_m\}$ converges in $W^{1,p}(\Omega)$ and almost everywhere. Therefore, by the first step, we obtain the result for $\phi \in C_0^\infty(\Omega)$, $\phi \geq 0$. Now the general case follows by a density argument, namely, if $u_n \to u$ in $W_0^{1,p}(\Omega)$, $u_n \in C_0^\infty(\Omega)$ and $u_n \geq 0$, then we have

$$\int_\Omega |\nabla u_n|^p\,dx \geq \int_\Omega \left(\frac{-\operatorname{div}(|\nabla v_n|^{p-2}\nabla v_n)}{v_n^{p-1}}\right)u_n^p\,dx = \int_\Omega \left(\frac{-\operatorname{div}(|\nabla v|^{p-2}\nabla v)}{v_n^{p-1}}\right)u_n^p\,dx.$$

By the hypothesis and Fatou's lemma we conclude the proof. □

Remark 1.3.3. Picone's inequality has a deep relationship with the spectrum of elliptic operators in divergence form. We think in the simplest case of the principal eigenvalue of the Laplacian; that is, let us assume that for $\lambda > 0$ and $v \in W_0^{1,2}(\Omega)$, $v > 0$, we have a positive Radon measure ν and a positive weight ρ such that

$$-\Delta v = \nu \geq \lambda\rho(x)v, \quad x \in \Omega.$$

By the Picone inequality we find in this case that

$$\int_\Omega |\nabla u|^2 dx \geq \lambda \int_\Omega \rho(x)u^2\,dx, \quad \text{for all } u \in W^{1,2}(\Omega),$$

which means that $\lambda \geq \lambda_1(\rho)$, the principal eigenvalue of the Laplacian with weight ρ.

Note that we can write

$$\lambda_1(\rho) = \sup\{\lambda > 0 : \exists v \in W^{1,2}(\Omega),\ \text{s. t.}\ -\Delta v \geq \lambda\rho(x)v\}.$$

This observation is used to define the first eigenvalue for elliptic operators in nondivergence form (see [58]) and for fully nonlinear elliptic operators homogeneous of degree one (see [121] and the references therein).

2 Looking at the Hardy–Leray potential

This chapter presents the behavior of the Hardy–Leray potential with respect to certain functional inequalities and their consequences.

2.1 Properties of the Hardy–Leray potential

Consider the Hardy–Leray potential

$$V(x) = \frac{1}{|x|^2}, \quad x \in \mathbb{R}^N,\ N \geq 3. \tag{2.1.1}$$

Lemma 2.1.1. *Let V be defined by* (2.1.1). *Then:*
1. $V \in \mathcal{M}^{\frac{N}{2},\infty}(\mathbb{R}^N)$, *the Marcinkiewicz space;*
2. $V \in L^p_{\mathrm{loc}}(\mathbb{R}^N)$, *if* $p < \frac{N}{2}$.

Proof. We check directly that $V \in \mathcal{M}^{\frac{N}{2},\infty}(\mathbb{R}^N)$. Indeed,

$$\left|\{x \in \mathbb{R}^N : |x|^{-2} > \mu\}\right| = \left|\{x \in \mathbb{R}^N : |x| < \mu^{-\frac{1}{2}}\}\right| = \frac{1}{N}\omega_{N-1}\mu^{-\frac{N}{2}},$$

where ω_{N-1} denotes the measure of the unit sphere. The result in (2) is an elementary consequence. □

Remark 2.1.2. The lack of integrability of the Hardy potential is the reason for which it does not satisfy the conditions of the above cited Cwikel–Lieb–Rosenblum theorem; see in particular inequality (1.1) in [166].

2.2 The Hardy–Leray inequality

The elementary proof of the Hardy–Leray inequality that we present here is due to L. Tartar (see [303]).

Theorem 2.2.1 (Hardy–Leray inequality). *Assume $N \geq 3$. For all $\phi \in C_0^\infty(\mathbb{R}^N)$ the following inequality holds:*

$$\left(\frac{N-2}{2}\right)^2 \int_{\mathbb{R}^N} \frac{\phi^2(x)}{|x|^2}\,dx \leq \int_{\mathbb{R}^N} |\nabla\phi(x)|^2\,dx. \tag{2.2.1}$$

Proof. The summability of both terms in the inequality is evident. Consider for $\lambda > 0$

$$\begin{aligned} 0 &\leq \int_{\mathbb{R}^N} \left\langle \nabla\phi(x) + \lambda\frac{x}{|x|^2}\phi(x), \nabla\phi(x) + \lambda\frac{x}{|x|^2}\phi(x)\right\rangle \\ &= \int_{\mathbb{R}^N} |\nabla\phi(x)|^2\,dx + 2\lambda \int_{\mathbb{R}^N} \left\langle \nabla\phi(x), \frac{x}{|x|^2}\phi(x)\right\rangle + \lambda^2 \int_{\mathbb{R}^N} \frac{\phi^2(x)}{|x|^2}\,dx. \end{aligned} \tag{2.2.2}$$

https://doi.org/10.1515/9783110606270-002

Now,

$$2\lambda\int_{\mathbb{R}^N}\left\langle\nabla\phi(x),\frac{x}{|x|^2}\phi(x)\right\rangle dx=-2\lambda\int_{\mathbb{R}^N}\phi(x)^2\operatorname{div}\left(\frac{x}{|x|^2}\right)dx=-\lambda(N-2)\int_{\mathbb{R}^N}\frac{\phi^2(x)}{|x|^2}\,dx.$$

Therefore, (2.2.2) becomes

$$0\le\int_{\mathbb{R}^N}|\nabla\phi(x)|^2\,dx+(\lambda^2-\lambda(N-2))\frac{\phi^2(x)}{|x|^2}\,dx,\quad\forall\lambda>0.$$

Since the quadratic function $f(\lambda)=\lambda^2-\lambda(N-2)$ has a minimum at $\lambda_0=\frac{N-2}{2}$ and

$$f(\lambda_0)=-\left(\frac{N-2}{2}\right)^2,$$

we conclude with inequality (2.2.1). □

We will denote

$$\Lambda_{N,2}=\left(\frac{N-2}{2}\right)^2. \tag{2.2.3}$$

Definition 2.2.2. We call $\mathcal{D}^{1,2}(\mathbb{R}^N)$ the completion of $C_0^\infty(\mathbb{R}^N)$ with respect to the seminorm

$$\|u\|_{1,2}=\left(\int_{\mathbb{R}^N}|\nabla u(x)|^2\,dx\right)^{\frac{1}{2}}.$$

Note that by the Sobolev inequality we have

$$S\left(\int_{\mathbb{R}^N}|u|^{2^*}\,dx\right)^{\frac{1}{2^*}}\le\left(\int_{\mathbb{R}^N}|\nabla u(x)|^2\,dx\right)^{\frac{1}{2}},\quad\forall u\in\mathcal{D}^{1,2}(\mathbb{R}^N), \tag{2.2.4}$$

where $2^*=\frac{2N}{N-2}$, $N\ge3$.

As a consequence of Theorem 2.2.1 we have the following results.

Corollary 2.2.3. *Inequality* (2.2.1) *holds for all* $u\in\mathcal{D}^{1,2}(\mathbb{R}^N)$.

Proof. The density of $C_0^\infty(\mathbb{R}^N)$ in $\mathcal{D}^{1,2}(\mathbb{R}^N)$ is sufficient to conclude this. □

2.3 Optimality and nonattainability of the constant $\Lambda_{N,2}$

We will try to analyze the Hardy–Leray in a closer way. The aim of this section is to prove that $\Lambda_{N,2}$ defined in (2.2.3) is the best constant in inequality (2.2.1) and, moreover, that it is not attained in the energy space $\mathcal{D}^{1,2}(\mathbb{R}^N)$.

2.3.1 Some preliminaries on symmetrization

We closely follow the classical work by G. Talenti [301] and the references therein (see also [302]).

Consider $E \subset \mathbb{R}^N$, a Lebesgue measurable set, and

$$u : E \to \mathbb{R},$$

a measurable function. We denote by $m(E)$ the Lebesgue measure of E.

We assume that:

- either $m(E) < \infty$ or
- u decays at infinity, that is, $m(\{x \in E : |u(x)| > t\})$ is finite for all $t > 0$.

Definition 2.3.1. Given a measurable function u defined in E, its distribution function is defined as

$$\begin{aligned} [0,\infty] : &\quad \to \quad [0,\infty], \\ t : &\quad \to \quad \mu(t) = m(\{x \in E : |u(x)| > t\}). \end{aligned}$$

The following properties are easy to check:

1. *μ is a nonincreasing function;*
2. *μ is right continuous;*
3. *$\mu(0)$ is the measure of the support of u and $\mu(\infty) = 0$;*
4. $\{t \geq 0\ \mu(t) = 0\} = [\operatorname{ess\,sup}|u|, \infty)$, *and as a consequence*, $\operatorname{support}(\mu) = [0, \operatorname{ess\,sup}|u|]$;
5. *for all $t > 0$,*

$$\lim_{s<t; s\to t} \mu(s) = \mu(t^-) = m(\{x \in E : |u(x)| \geq t\}),$$

and hence $\mu(t^-) - \mu(t) = m(\{x \in E : |u(x)| = t\})$ is the jump of μ at t.

We are able now to define the decreasing rearrangement of a function.

Definition 2.3.2. The decreasing rearrangement of a measurable function u as the distribution function of μ, given by Definition 2.3.1, is

$$u^*(s) = m(\{t \geq 0 : \mu(t) > s\}). \tag{2.3.1}$$

Proposition 2.3.3. *The following properties hold:*

1. *u^* is nonincreasing;*
2. *u^* is right continuous;*
3. $u^*(0) = \operatorname{ess\,sup}|u|$ *and* $u^*(+\infty) = 0$;
4. $\{s \geq 0 \mid u^*(s) = 0\} = [m(\operatorname{suppt}(u)), +\infty)$;

5. $\{t \geq 0 \mid \mu(t) \leq s\} = [u^*(s), +\infty)$ *for all* $s \geq 0$, *and thus*
$$u^*(s) = \min\{t \geq 0 : \mu(t) \leq s\};$$
6. $\{s \geq 0 : u^*(s) > t\} = [0, \mu(t))$ *for all* $t \geq 0$, *and as a consequence we find that* u *and* u^* *are equidistributed;*
7. $u^*(\mu(t)) \leq t$;
8. $\mu(u^*(s)) \leq s$ *for all* $s \geq 0$ *and* $\mu(u^*(s)^-) \geq s$ *for all* $0 \leq s \leq m(\text{suppt}(u))$;
9. $[\mu(t), \mu(t^-)) \subsetneqq \{s \geq 0 \mid u^*(s) = t\} \subsetneqq [\mu(t), \mu(t^-)]$.

For the proof we refer the reader to [301]. Note that (8) has the following consequence.

Let us take ϕ *with*
$$\phi : [0, \infty] \to [0, \infty],$$
a continuous, nondecreasing function such that $\phi(0) = 0$. *Then,*
$$\int_E \phi(|u(x)|)\, dx = \int_0^\infty \phi(u^*(s))\, ds. \tag{2.3.2}$$

Next we define the radial rearrangement.

Definition 2.3.4. The symmetric rearrangement of u, $u^\star$, is defined by the expression
$$u^\star(x) = u^*(\overline{\omega_N}|x|^N), \quad \forall x \in \mathbb{R}^N, \tag{2.3.3}$$
where $\overline{\omega_N}$ denotes the measure of the unit ball in $\mathbb{R}^N$.

We summarize the relevant properties of $u^\star$ in the following statement.

Proposition 2.3.5. *The following properties hold:*
1. $u^\star$ *is nonnegative, radial and radially decreasing with respect to distance to the origin;*
2. u *and* $u^\star$ *are equidistributed.*

It is easy to see that we have the following equivalent representations for u^* and $u^\star$, respectively:
$$u^* = \int_0^\infty \chi_{[0,\mu(t))}\, dt, \tag{2.3.4}$$
$$u^\star(x) = \int_0^\infty \chi_{\{x \in \mathbb{R}^N : \overline{\omega_N}|x|^N < \mu(t)\}}\, dt. \tag{2.3.5}$$

For the proof of all the properties above, see [301].

The results that are relevant to us come from rearrangements, which at the end provide certain inequalities in the radial setting. In this sense, let us state some results by Hardy–Littlewood [196], by F. Riesz [271] and by Polya–Szego [263] that easily follow from the above arguments.

Theorem 2.3.6. *Suppose u, v and w are measurable nonnegative functions.*

Then:

1. *(Hardy–Littlewood)*

$$\int_{\mathbb{R}^N} u(x)v(x)\,dx \le \int_0^\infty u^*(s)v^*(s)\,ds, \tag{2.3.6}$$

$$\int_{\mathbb{R}^N} u(x)v(x)\,dx \le \int_{\mathbb{R}^N} u^\star(x)v^\star(x)\,dx \tag{2.3.7}$$

and

2. *(F. Riesz)*

$$\int_{\mathbb{R}^N} u(x) \int_{\mathbb{R}^N} v(y)w(x-y)\,dy\,dx \le \int_{\mathbb{R}^N} u^\star(x) \int_{\mathbb{R}^N} v^\star(y)w^\star(x-y)\,dy\,dx. \tag{2.3.8}$$

Theorem 2.3.7 (Polya–Szego). *Assuming $u \in W^{1,p}(\mathbb{R}^N)$, $p \ge 1$, we have*

$$\int_{\mathbb{R}^N} \Phi(|\nabla u(x)|)\,dx \ge \int_{\mathbb{R}^N} \Phi(|\nabla u^\star(x)|)\,dx, \tag{2.3.9}$$

with $\Phi(t) = t^p$.

2.3.2 Constant $\Lambda_{N,2}$ is the best constant and is not attained

Step 1. *Optimality of the constant.* Following the idea of Hardy for the one dimensional case, we show that the best constant is $\Lambda_{N,2}$. Given $\epsilon > 0$, take the radial function

$$U_\epsilon(r) = \begin{cases} A_{N,\epsilon} & \text{if } r \in [0,1], \\ A_{N,\epsilon} r^{\frac{2-N}{2}-\epsilon} & \text{if } r > 1, \end{cases} \tag{2.3.10}$$

where $A_{N,\epsilon} = 2/(N-2+2\epsilon)$, whose derivative is

$$U'_\epsilon(r) = \begin{cases} 0 & \text{if } r \in [0,1], \\ -r^{-\frac{N}{2}-\epsilon} & \text{if } r > 1. \end{cases} \tag{2.3.11}$$

By a direct computation we get

$$\int_{\mathbb{R}^N} \frac{U_\epsilon^2(x)}{|x|^2}\,dx = \int_B \frac{U_\epsilon^2(x)}{|x|^2}\,dx + \int_{\mathbb{R}^N\setminus B} \frac{U_\epsilon^2(x)}{|x|^2}\,dx$$
$$= A_{N,\epsilon}^2 \omega_{N-1}\left(\int_0^1 r^{N-3}\,dr + \int_1^\infty r^{-(1+2\epsilon)}\,dr\right)$$
$$= A_{N,\epsilon}^2 \omega_{N-1}\int_0^1 r^{N-3}\,dr + A_{N,\epsilon}^2 \int_{\mathbb{R}^N} |\nabla U_\epsilon(x)|^2\,dx.$$

Then, given $A_{N,\epsilon} < \Lambda_{N,2}$, $\epsilon > 0$, there exists $U_\epsilon \in \mathcal{D}^{1,2}(\mathbb{R}^N)$ such that

$$A_{N,\epsilon}^2 \int_{\mathbb{R}^N} |\nabla U_\epsilon(x)|^2\,dx \le \int_{\mathbb{R}^N} \frac{U_\epsilon^2(x)}{|x|^2}\,dx.$$

STEP 2. *The same result is true in the unit ball $B_1(0) \subset \mathbb{R}^N$.* The argument of optimality in the case of the unit ball proceeds by approximation as follows. First, we remark that by the invariance under dilation the optimal constant will be the same for any ball. Second, let B_R be a ball with large radius. We take as test function $v(x) = \psi(x)U(x)$, where U is one of the approximate optimizers explicitly given above and $\psi \in \mathcal{C}_0^\infty(B_R)$ is a cutoff function which is identically 1 on B_{R-1} with $|\nabla\psi| \le m$. It is easily seen that for $R \gg 1$ the influence of ψ in the calculation of step 1 is negligible.

It is clear that the same argument works for any bounded domain Ω such that $0 \in \Omega$ and for the space $W_0^{1,2}(\Omega)$.

STEP 3. *Nonattainability of $\Lambda_{N,2}$ in $\mathbb{R}^N$.*

By a symmetrization argument, for all $u \in \mathcal{D}^{1,2}(\mathbb{R}^N)$, we have

$$\int_{\mathbb{R}^N} \frac{u(x)^2}{|x|^2}\,dx \le \int_{\mathbb{R}^N} \frac{(u^\star(x))^2}{|x|^2}\,dx \tag{2.3.12}$$

and

$$\int_{\mathbb{R}^N} |\nabla u(x)|^2\,dx \ge \int_{\mathbb{R}^N} |\nabla u^\star(x)|^2\,dx. \tag{2.3.13}$$

Then

$$\frac{\int_{\mathbb{R}^N} |\nabla u(x)|^2}{\int_{\mathbb{R}^N} \frac{u(x)^2}{|x|^2}\,dx} \ge \frac{\int_{\mathbb{R}^N} |\nabla u^\star(x)|^2\,dx}{\int_{\mathbb{R}^N} \frac{(u^\star(x))^2}{|x|^2}\,dx} \ge \Lambda_{N,2}. \tag{2.3.14}$$

As a consequence of the arguments above, if the minimizers exist, they must be radial. The corresponding Euler equation in radial coordinates is

$$u''(r) + (N-1)\frac{u'(r)}{r} + \Lambda_{N,2}\frac{u}{r^2} = 0, \tag{2.3.15}$$

which by elementary integration gives

$$u(r) = r^{-(\frac{N-2}{2})}(a_1 + a_2 \log(r)), \quad a_1, a_2 \in \mathbb{R}.$$

Note that the minimizers are not in $W^{1,2}_{\text{loc}}(\mathbb{R}^N)$. In others words, $\Lambda_{N,2}$ is not an eigenvalue in bounded domains containing the pole of the potential and with Dirichlet boundary data. □

We will call

$$u_2(x) = |x|^{-\frac{N-2}{2}} \tag{2.3.16}$$

the *pseudo-minimizer*.

As a consequence of the previous arguments, we find the following result.

Proposition 2.3.8. *Let $\Omega \subset \mathbb{R}^N$ be a bounded domain such that $0 \in \Omega$. Consider the linear operator*

$$\mathcal{L}_\lambda u \equiv -\Delta u - \frac{\lambda}{|x|^2} u \tag{2.3.17}$$

in $W^{1,2}(\Omega)$. Then:
1. *If $\lambda \le \Lambda_{N,2}$, $\mathcal{L}_\lambda$ is a positive operator.*
2. *If $\lambda > \Lambda_{N,2}$, $\mathcal{L}_\lambda$ is unbounded from below.*

Proof. (1) This is obvious from Hardy's inequality.

(2) As an easy consequence of the optimality of the constant and a density argument one proves the existence of $\phi \in C_0^\infty(\Omega)$ such that $\langle \mathcal{L}_\lambda \phi, \phi \rangle < 0$. We can assume, without any loss of generality, that $\|\phi\|_2 = 1$. Defining $u_\mu(x) = \mu^{N/2}\phi(\mu x)$ we have $\|u_\mu\|_2 = 1$ and the homogeneity of the operator allows us to conclude that $\langle \mathcal{L}_\lambda u_\mu, u_\mu \rangle = \mu^2 \langle \mathcal{L}_\lambda \phi, \phi \rangle < 0$. □

Remark 2.3.9. It is worth noting that the constant could be attained with other boundary conditions. This result is trivial with Neumann conditions for which the constant of the Rayleigh quotient is zero and the constant becomes the minimizer itself. For mixed boundary conditions, see [7]

2.4 Hardy's inequality in $W^{1,p}(\mathbb{R}^N)$, $1 < p < N$

The main point of this section is to discuss the following Hardy inequality for $1 < p < N$ in $\mathbb{R}^N$, which we include by completeness. For more details and applications, see [179].

Lemma 2.4.1. *Assume $1 < p < N$. Then if $u \in W^{1,p}(\mathbb{R}^N)$,*

(1) $\frac{u}{|x|} \in L^p(\mathbb{R}^N)$;

(2) *(Hardy inequality)*

$$\int_{\mathbb{R}^N} \frac{|u|^p}{|x|^p}\,dx \le \Lambda_{N,p}^{-1} \int_{\mathbb{R}^N} |\nabla u|^p\,dx$$

with $\Lambda_{N,p}^{-1} = (\frac{p}{N-p})^p$;

(3) *the constant* $\Lambda_{N,p}^{-1}$ *is optimal.*

Proof. Step 1. A density argument allows us to consider only smooth functions $u \in C_0^\infty(\mathbb{R}^N)$. Under this hypothesis we have the following identity:

$$|u(x)|^p = -\int_1^\infty \frac{d}{d\lambda}|u(\lambda x)|^p\,d\lambda = -p\int_1^\infty u^{p-1}(\lambda x)\langle x, \nabla u(\lambda x)\rangle\,d\lambda.$$

By using the Hölder inequality, it follows that

$$\begin{aligned}
\int_{\mathbb{R}^N} \frac{|u(x)|^p}{|x|^p}\,dx &= -p\int_1^\infty \int_{\mathbb{R}^N} \frac{u^{p-1}(\lambda x)}{|x|^{p-1}}\left\langle \frac{x}{|x|}, \nabla u(\lambda x)\right\rangle dx\,d\lambda \\
&= -p\int_1^\infty \frac{d\lambda}{\lambda^{N-1-p}} \int_{\mathbb{R}^N} \frac{u(y)^{p-1}}{|y|^{p-1}}\frac{\partial u(y)}{\partial r}\,dy = -\frac{p}{N-p}\int_{\mathbb{R}^N} \frac{u(y)^{p-1}}{|y|^{p-1}}\frac{\partial u(y)}{\partial r}\,dy \\
&\le \frac{p}{N-p}\left(\int_{\mathbb{R}^N} \frac{|u(y)|^p}{|y|^p}\,dy\right)^{(p-1)/p}\left(\int_{\mathbb{R}^N} \left|\frac{\partial u(y)}{\partial r}\right|^p dy\right)^{1/p},
\end{aligned}$$

and then we conclude that

$$\int_{\mathbb{R}^N} \frac{u^p(x)}{|x|^p}\,dx \le \left(\frac{p}{N-p}\right)^p \int_{\mathbb{R}^N} |\nabla u(x)|^p\,dx.$$

Step 2. Optimality of the constant. We proceed as in the case $p = 2$. Given $\varepsilon > 0$, consider the radial function

$$U(r) = \begin{cases} A_{N,p,\varepsilon} & \text{if } r \in [0,1], \\ A_{N,p,\varepsilon} r^{\frac{p-N}{p}-\varepsilon} & \text{if } r > 1, \end{cases} \tag{2.4.1}$$

where $A_{N,p,\varepsilon} = p/(N-p+p\varepsilon)$. Observe that its derivative is

$$U'(r) = \begin{cases} 0 & \text{if } r \in [0,1], \\ -r^{-\frac{N}{p}-\varepsilon} & \text{if } r > 1. \end{cases} \tag{2.4.2}$$

By direct computation we get

$$\int_{\mathbb{R}^N} \frac{U^p(x)}{|x|^p}\,dx = \int_B \frac{U^p(x)}{|x|^p}\,dx + \int_{\mathbb{R}^N - B} \frac{U^p(x)}{|x|^p}\,dx$$

$$= A^p_{N,p,\varepsilon}\omega_{N-1}\left(\int_0^1 r^{N-1-p}\,dr + \int_1^\infty r^{-(1+p\varepsilon)}\,dr\right)$$

$$= A^p_{N,p,\varepsilon}\omega_{N-1}\int_0^1 r^{N-1-p}\,dr + A^p_{N,p,\varepsilon}\int_{\mathbb{R}^N} |\nabla U(x)|^p\,dx.$$

We conclude as in the case $p = 2$. □

Corollary 2.4.2. *The same result is true in the unit ball $B \subset \mathbb{R}^N$.*

Proof. The proof of the first step is the same. The argument of optimality in the case of the unit ball proceeds by approximation as follows. First, we remark that by the invariance under dilation the optimal constant will be the same for any ball. Second, let B_R be a ball with large radius. We take as test function $v(x) = \psi(x)U(x)$, where U is one of the approximate optimizers explicitly given above and $\psi \in C_0^\infty(B_R)$ is a cutoff function which is identically 1 on B_{R-1} with $|\nabla\psi| \le m$. It is easily seen that for $R \gg 1$ the influence of ψ in the calculation of step 2 is negligible. □

Remark 2.4.3. The nonattainability of the optimal constant can be seen also as a consequence of the Hardy inequality with remainder term obtained in Theorem 2.5.1 below. See [6] for more details.

Therefore we have $\Lambda_{N,p} = (\frac{N-p}{p})^p$.

We remark that the *Euler equation* corresponding to the Hardy inequality is

$$-\Delta_p u \equiv -\operatorname{div}(|\nabla u|^{p-2}\nabla u) = \Lambda_{N,p}\frac{u^{p-1}}{|x|^p}.$$

See [179] for more details and some applications.

2.5 Hardy–Leray inequality with remainder term

The results in the previous section revealed that $\Lambda_{N,2}$ is not an eigenvalue, but more precisely, it is the infimum of the essential spectrum of the Laplace operator with respect to the weight given by the Hardy–Leray potential. This fact motivates the presence of a corrector term in inequality (2.2.1).

A short history of this question is as follows.

(a) Brezis–Vázquez in [97] obtained a remainder term for the classical Hardy–Sobolev inequality; precisely, if $1 \leq q < \frac{2N}{N-2}$, there exists a constant $C_q(|\Omega|) > 0$ such that

$$\int_\Omega |\nabla u|^2\,dx - \left(\frac{N-2}{2}\right)^2 \int_\Omega \frac{u^2}{|x|^2}\,dx \geq C_q(|\Omega|)\left(\int_\Omega |u|^q\,dx\right)^{\frac{2}{q}}, \tag{2.5.1}$$

for all $u \in W_0^{1,2}(\Omega)$.

(b) Vázquez–Zuazua in [311], among other results, improved the previous inequality by showing that if $1 < q < 2$, there exists $C_q(|\Omega|) > 0$ such that, for all $u \in W_0^{1,2}(\Omega)$,

$$\int_\Omega |\nabla u|^2\,dx - \left(\frac{N-2}{2}\right)^2 \int_\Omega \frac{u^2}{|x|^2}\,dx \geq C_q(|\Omega|)\left(\int_\Omega |\nabla u|^q\,dx\right)^{\frac{2}{q}}. \tag{2.5.2}$$

The improved inequality (2.5.2) proves that H, the completion of $C_0^\infty(\Omega)$ with respect to the norm

$$\|u\|_H^2 = \int_\Omega |\nabla u|^2\,dx - \left(\frac{N-2}{2}\right)^2 \int_\Omega \frac{u^2}{|x|^2}\,dx,$$

is a Hilbert space and, as a consequence, the improved inequality holds for $u \in H$.

This kind of inequalities have been extended in several directions, mainly the following.

(i) In [29], Adimurthi–Chaudhuri–Ramaswamy obtained by symmetrization arguments an extension of inequality (2.5.1) to the case of the Hardy–Sobolev inequality in the space $W_0^{1,p}(\Omega)$: if $1 < q < \frac{pN}{N-p}$, there exists $C_{p,q}(|\Omega|) > 0$ such that

$$\int_\Omega |\nabla u|^p\,dx - \left(\frac{N-p}{p}\right)^p \int_\Omega \frac{|u|^p}{|x|^p}\,dx \geq C_{p,q}(|\Omega|)\left(\int_\Omega |u|^q\,dx\right)^{\frac{p}{q}}, \quad \text{for all } u \in W_0^{1,p}(\Omega).$$

(ii) By using a change of variables that appeared in [116], Wang–Willem in [314] obtained the optimal remainder term related to (2.5.2) for the Caffarelli–Kohn–Nirenberg inequalities,

$$\int_{\mathbb{R}^N} |u|^2|x|^{-2(\gamma+1)}\,dx \leq \lambda_{N,2,\gamma}^{-1} \int_{\mathbb{R}^N} |\nabla u|^2|x|^{-2\gamma}\,dx,$$

where

$$\lambda_{N,2,\gamma} = \left(\frac{N-2(\gamma+1)}{2}\right)^2, \quad -\infty < \gamma < \frac{N-2}{2}.$$

Note that $\lambda_{N,2,0} = \Lambda_{N,2} = (\frac{N-2}{2})^2$. The precise result by Wang–Willem for $\gamma = 0$ reads as follows:

$$\int_\Omega |\nabla u|^2\,dx - \left(\frac{N-2}{2}\right)^2 \int_\Omega \frac{u^2}{|x|^2}\,dx \ \geq C(\Omega)\int_\Omega |\nabla u|^2\left(\log\left(\frac{R}{|x|}\right)\right)^{-2} dx, \qquad (2.5.3)$$

where R is such that $\overline{\Omega} \subset B_R(0)$ and $C(\Omega) > 0$, for all $u \in W_0^{1,2}(\Omega)$.

Finally in [6], the following general result is obtained with an elementary proof.

Theorem 2.5.1. *Let p, γ be such that $1 < p < N$ and $-\infty < \gamma < \frac{N-p}{p}$. Then for all q, $1 < q < p$, there exists a positive constant $C \equiv C(N,p,q,\gamma,\Omega)$ such that for all $\phi \in \mathcal{D}^{1,p}_{0,\gamma}(\Omega)$ we have*

$$\int_\Omega |\nabla\phi|^p|x|^{-p\gamma}\,dx - \left(\frac{N-p(\gamma+1)}{p}\right)^p \int_\Omega \frac{|\phi|^p}{|x|^{p(\gamma+1)}}\,dx \geq C\left(\int_\Omega |\nabla\phi|^q|x|^{-r\gamma}\,dx\right)^{\frac{p}{q}}, \qquad (2.5.4)$$

where $1 < q < p$ and $q \leq r < \infty$ if $\gamma \leq 0$, $1 \leq r < p + \rho(N,p,q,\gamma)$ if $\gamma > 0$, for some positive constant $\rho(N,p,q,\gamma)$, and the weighted Sobolev space $\mathcal{D}^{1,p}_{0,\gamma}(\Omega)$ is defined as the completion of $C_0^\infty(\Omega)$ with respect to the norm $\|\phi\|_{p,\gamma} = \left(\int_\Omega |\nabla\phi|^p|x|^{-p\gamma}dx\right)^{1/p}$.

The details about the functional setting and this general result can be seen in [6]. Here we only point out that for its proof one needs to overcome several difficulties. For instance, in the nonlinear setting, $p \neq 2$, Fourier analysis is not allowable and Schwarz symmetrization results are unknown due to the presence of weights. We recall that Schwarz symmetrization is based on an isoperimetric inequality, which in general is unknown in the weighted case.

In this section we limit ourselves to prove the result in the case $p = 2$ and $\gamma = 0$. This corresponds to the Hardy–Leray potential and includes an alternative proof of the optimal result in [314]. More precisely, we will prove the following result.

Theorem 2.5.2. *Assume $R > 0$ is such that $0 \in \Omega \subset \overline{\Omega} \subset B_R(0)$. Then there exists a positive constant $C \equiv C(N,\Omega)$ such that for all $\phi \in C_0^\infty(\Omega)$ we have*

$$\int_\Omega |\nabla\phi|^2\,dx - \left(\frac{N-2}{2}\right)^2 \int_\Omega \frac{\phi^2}{|x|^2}\,dx \geq C\int_\Omega |\nabla\phi|^2\left(\log\left(\frac{R}{|x|}\right)\right)^{-2} dx. \qquad (2.5.5)$$

We prove Theorem 2.5.2 as in [6] but using a different approach, perhaps more elementary than the previous one. This approach uses a Picone type inequality (see Theorem 1.3.2) and convenient test functions that, roughly speaking, *approximate the pseudo-minimizer*, the *ground state*.

Proof of Theorem 2.5.2. Consider u_2, the *pseudo-minimizer* defined in (2.3.16). Without loss of generality we can assume that $R = 1$. By a direct computation and using an approximation argument we obtain

$$\int_{\Omega} |\nabla\phi|^2\,dx - \left(\frac{N-2}{2}\right)^2 \int_{\Omega} \frac{\phi^2}{|x|^2}\,dx = \int_{\Omega} \left|\nabla\phi - \frac{\phi}{u_2}\nabla u_2\right|^2 dx.$$

Since $\overline{\Omega} \subset B_1(0)$, there exists a positive constant $C \equiv C(N,\Omega)$ such that

$$\left|\nabla\phi - \frac{\phi}{u_2}\nabla u_2\right|^2 \geq C\left(\log\left(\frac{1}{|x|}\right)\right)^{-2} \left|\nabla\phi - \frac{\phi}{u_2}\nabla u_2\right|^2.$$

Hence

$$\left|\nabla\phi - \frac{\phi}{u_2}\nabla u_2\right|^2 \geq C\left(\log\left(\frac{1}{|x|}\right)\right)^{-2} \left(|\nabla\phi|^2 + \left|\phi\frac{\nabla u_2}{u_2}\right|^2 - 2\frac{\phi}{u_2}\nabla u_2\nabla\phi\right).$$

Integrating and using Young's inequality we obtain

$$\begin{aligned}&\int_{\Omega} \left|\nabla\phi - \frac{\phi}{u_2}\nabla u_2\right|^2 dx \\ &\quad \geq C(1-\epsilon)\left\{\int_{\Omega} |\nabla\phi|^2\left(\log\left(\frac{1}{|x|}\right)\right)^{-2} dx - \frac{1}{\epsilon}\Lambda_{N,2}\int_{\Omega} \frac{\phi^2}{|x|^2}\left(\log\left(\frac{1}{|x|}\right)\right)^{-2} dx\right\}. \qquad (2.5.6)\end{aligned}$$

Hence, we just need to prove that

$$\begin{aligned} D_1 &= \inf_{\phi\in C_0^\infty(\Omega),\phi\neq 0} \frac{\int_\Omega |\nabla\phi - \frac{\phi}{u_2}\nabla u_2|^2\,dx}{\int_\Omega \frac{\phi^2}{|x|^2}(\log(\frac{1}{|x|}))^{-2}\,dx} \\ &= \inf_{\phi\in C_0^\infty(\Omega),\phi\neq 0} \frac{\int_\Omega |\nabla\phi|^2\,dx - (\frac{N-2}{2})^2\int_\Omega \frac{\phi^2}{|x|^2}\,dx}{\int_\Omega \frac{\phi^2}{|x|^2}(\log(\frac{1}{|x|}))^{-2}\,dx} > 0. \qquad (2.5.7)\end{aligned}$$

This follows by using the fact that $\bar{u}_2(x) = |x|^{-\frac{N-2}{2}}(\log(\frac{1}{|x|}))^{\frac{1}{2}}$ is a distributional solution to the equation

$$-\Delta w - \Lambda_{N,2}\frac{w}{|x|^2} = \frac{1}{4}\frac{w}{|x|^2}\left(\log\left(\frac{1}{|x|}\right)\right)^{-2}, \qquad (2.5.8)$$

and then using Picone's inequality with $\bar{u}_2$. Note that $\bar{u}_2 \in W^{1,q}(\Omega)$ for all $q < 2$. By an approximation argument as in [179], we can prove that the optimal constant in (2.5.7) is exactly $D_1 = \frac{1}{4}$ and, thus, it is independent of Ω.

Therefore by (2.5.6) and (2.5.7) we obtain

$$\begin{aligned}\int_{\Omega} |\nabla\phi|^2\,dx - \left(\frac{N-2}{2}\right)^2 \int_{\Omega} \frac{\phi^2}{|x|^2}\,dx &= \int_{\Omega} \left|\nabla\phi - \frac{\phi}{u_2}\nabla u_2\right|^2 dx \\ &\geq C\int_{\Omega} |\nabla\phi|^2\left(\log\left(\frac{1}{|x|}\right)\right)^{-2} dx. \qquad \square\end{aligned}$$

Next we prove that the result in Theorem 2.5.2 is sharp.

Theorem 2.5.3. *Inequality* (2.5.5) *in Theorem* 2.5.2 *is optimal in the sense that* $(\log(\frac{1}{|x|}))^{-2}$ *cannot be replaced by any weight of the form* $(\log(\frac{1}{|x|}))^{-2}g(x)$, *where* g *is a positive function such that* $g(x) \to \infty$ *as* $|x| \to 0$.

Proof. First we need the following result in the particular case $\alpha = 2$.

Lemma 2.5.4. *Assume that* $\alpha > 0$ *and let* Ω *be a bounded domain such that* $\overline{\Omega} \subset B_R(0)$, *where* $R < 1$. *We set*

$$A = \inf_{\phi \in C_0^\infty(\Omega), \phi \neq 0} \frac{\int_\Omega |\nabla \phi|^2 (\log(\frac{1}{|x|}))^{-\alpha}\, dx}{\int_\Omega \frac{\phi^2}{|x|^2} (\log(\frac{1}{|x|}))^{-\alpha}\, dx}.$$

Then $A = \Lambda_{N,2}$.

Proof. Fix $\phi \in C_0^\infty(\Omega)$. Then by using Picone's inequality and a suitable approximation argument we get the following pointwise inequality:

$$|\nabla \phi|^2 \left(\log\left(\frac{1}{|x|}\right)\right)^{-\alpha} \geq \nabla\left(\frac{\phi^2}{u_2}\right) \nabla u_2 \left(\log\left(\frac{1}{|x|}\right)\right)^{-\alpha}.$$

Integrating both sides of the previous inequality, we obtain

$$\int_\Omega |\nabla \phi|^2 \left(\log\left(\frac{1}{|x|}\right)\right)^{-\alpha} dx \geq \int_\Omega \frac{\phi^2}{u_2}\left(-\operatorname{div}\left(\left(\log\left(\frac{1}{|x|}\right)\right)^{-\alpha} \nabla u_2\right)\right) dx. \tag{2.5.9}$$

Now, by a direct computation we get

$$\begin{aligned} &-\operatorname{div}\left(\left(\log\left(\frac{1}{|x|}\right)\right)^{-\alpha} \nabla u_2\right) \\ &= \left(\log\left(\frac{1}{|x|}\right)\right)^{-\alpha} (-\Delta u_2) - \alpha\left(\log\left(\frac{1}{|x|}\right)\right)^{-(\alpha+1)} \frac{x}{|x|^2} \nabla u_2. \end{aligned}$$

Therefore,

$$\begin{aligned} &-\operatorname{div}\left(\left(\log\left(\frac{1}{|x|}\right)\right)^{-\alpha} \nabla u_2\right) \\ &= \Lambda_{N,2}\left(\log\left(\frac{1}{|x|}\right)\right)^{-\alpha} \frac{u_2}{|x|^2} + \alpha\left(\frac{N-2}{2}\right)\left(\log\left(\frac{1}{|x|}\right)\right)^{-\alpha-1} \frac{u_2}{|x|^2}. \end{aligned}$$

Hence, using (2.5.9), we conclude that $A \geq \Lambda_{N,2}$.

Note that

$$\int_\Omega |\nabla u_2|^2 \left(\log\left(\frac{1}{|x|}\right)\right)^{-\alpha} dx \leq \Lambda_{N,2} \int_0^{R_1} \frac{1}{r}\left(\log\left(\frac{1}{r}\right)\right)^{-\alpha} dr < \infty,$$

where $\Omega \subset B_{R_1}(0)$ and $R_1 < 1$. Moreover, a direct computation shows that

$$\frac{\int_{B_{R_1}(0)} |\nabla u_2|^2 (\log(\frac{1}{|x|}))^{-\alpha}\, dx}{\int_{B_{R_1}(0)} \frac{u_2^2}{|x|^2} (\log(\frac{1}{|x|}))^{-\alpha}\, dx} = \Lambda_{N,2}.$$

Hence, by an approximation argument as in [179], we obtain finally that $A = \Lambda_{N,2}$. □

We prove the optimality result in Theorem 2.5.3 by the following contradiction argument.

Assume by contradiction that there exist a constant $C > 0$ and a positive function $g(x)$ such that $g(x) \to \infty$ as $x \to 0$, satisfying

$$\int_\Omega |\nabla \phi|^2\, dx - \left(\frac{N-2}{2}\right)^2 \int_\Omega \frac{\phi^2}{|x|^2}\, dx \geq C \int_\Omega |\nabla \phi|^2 \left(\log\left(\frac{R}{|x|}\right)\right)^{-2} g(x)\, dx.$$

Taking $\Omega = B_\epsilon(0)$ with $0 < \epsilon$ small enough, there exists a positive constant C_ϵ such that $\Lambda_{N,2} C_\epsilon \gg D_1 = \frac{1}{4}$ and

$$\int_{B_\epsilon(0)} |\nabla \phi|^2\, dx - \left(\frac{N-2}{2}\right)^2 \int_{B_\epsilon(0)} \frac{\phi^2}{|x|^2}\, dx \geq C_\epsilon \int_{B_\epsilon(0)} |\nabla \phi|^2 \left(\log\left(\frac{R}{|x|}\right)\right)^{-2} dx.$$

Using Lemma 2.5.4 with $\alpha = 2$, we conclude that

$$\int_{B_\epsilon(0)} |\nabla \phi|^2\, dx - \left(\frac{N-2}{2}\right)^2 \int_{B_\epsilon(0)} \frac{\phi^2}{|x|^2}\, dx \geq C_\epsilon \Lambda_{N,2} \int_{B_\epsilon(0)} \frac{\phi^2}{|x|^2} \left(\log\left(\frac{1}{|x|}\right)\right)^{-2} dx.$$

Hence we get a contradiction with the optimality result in (2.5.7). □

2.5.1 A functional consequence

An important consequence of the Hardy inequality with remainder term is that

$$\|u\|_{\mathcal{H}(\Omega)} = \left(\int_\Omega \left(|\nabla u|^2 - \Lambda_{N,2} \frac{u^2}{|x|^2}\right) dx\right)^{1/2}$$

is a Euclidean norm on $\mathcal{C}_0^\infty(\Omega)$. Consider the Hilbert space $\mathcal{H}(\Omega)$ defined as the completion of $\mathcal{C}_0^\infty(\Omega)$ with respect to the norm $\|u\|_{\mathcal{H}(\Omega)}$. Then we have, with strict inclusions, that

$$W_0^{1,2}(\Omega) \subset \mathcal{H}(\Omega) \subset \bigcap_{1 \leq p < 2} W_0^{1,p}(\Omega), \tag{2.5.10}$$

and therefore the dual space of $\mathcal{H}(\Omega)$, denoted by $\mathcal{H}'$, satisfies the inclusions $\mathcal{H}' \subset W^{-1,2}$ and $L^m(\Omega) \subset \mathcal{H}'$ for all $m > \frac{2N}{N+2}$. This observation will be useful in the next sections.

3 Calderón–Zygmund theory and the Hardy–Leray potential

3.1 Introduction

Regularity results for solutions of elliptic equations in terms of the summability of a source term are nowadays classic. In the linear case we have two types of results: those by Stampacchia for variational operators with discontinuous coefficients and those by Calderón and Zygmund for general operators with regular coefficients. The results by Stampacchia (see [291]) can be summarized as follows in the paradigmatic case of the Laplacian.

Consider the solution to $-\Delta u = f$ with homogeneous Dirichlet boundary conditions. Then we know that:

- if $f \in L^m(\Omega)$, $m > \frac{N}{2}$, then $u \in W_0^{1,2}(\Omega) \cap L^\infty(\Omega)$;
- if $f \in L^m(\Omega)$, $\frac{2N}{N+2} \le m \le \frac{N}{2}$, then $u \in W_0^{1,2}(\Omega) \cap L^{m^{**}}(\Omega)$, $m^{**} = \frac{Nm}{N-2m}$;
- if $f \in L^m(\Omega)$, $1 < m < \frac{2N}{N+2}$, then $u \in W_0^{1,m^*}(\Omega)$, $m^* = \frac{Nm}{N-m}$.

Since the Laplacian can also be seen as an operator in *nondivergence form*, the results above are also a consequence of the $W^{2,m}$-regularity given by Calderón and Zygmund theory (see [110], [111] and [112]).

The aim of this chapter is to analyze the influence of a zero order perturbation term like $\lambda \frac{u}{|x|^2}$ in the regularity behavior of the solutions of elliptic problems.

If $0 \in \mathbb{R}^N \setminus \overline{\Omega}$, then the potential is a bounded weight and nothing new happens. Therefore, we will assume in this chapter that

$$0 \in \Omega.$$

We mainly follow in this chapter the reference [80]. We anticipate that the regularity depends in a deep way on the value of the spectral parameter λ. To be precise, we formulate the problem as follows. Consider the boundary problem

$$\begin{cases} -\Delta u = \lambda \frac{u}{|x|^2} + f & \text{in } \Omega,\ 0 \in \Omega, \\ u = 0 & \text{on } \partial\Omega, \end{cases} \tag{3.1.1}$$

where $f \in L^m(\Omega)$, $1 \le m \le \infty$, and

$$0 < \lambda < \left(\frac{N-2}{2}\right)^2 = \Lambda_{N,2}. \tag{3.1.2}$$

Note that

$$\alpha_-(\lambda) = \frac{N-2}{2} - \sqrt{\left(\frac{N-2}{2}\right)^2 - \lambda}, \quad \alpha_+(\lambda) = \frac{N-2}{2} + \sqrt{\left(\frac{N-2}{2}\right)^2 - \lambda} \tag{3.1.3}$$

https://doi.org/10.1515/9783110606270-003

are the roots of $\alpha^2 - (N-2)\alpha + \lambda = 0$. Such roots give rise to the functions $|x|^{-\alpha_-(\lambda)}$ and $|x|^{-\alpha_+(\lambda)}$, which turn out to be radial solutions to the homogenous problem

$$-\Delta w - \lambda \frac{w}{|x|^2} = 0. \tag{3.1.4}$$

The main goal of this chapter is to find under which conditions the Calderón–Zygmund–Stampacchia regularity is preserved by adding the Hardy potential term. To start we analyze two extreme cases.

3.1.1 Unbounded solution for $m > \frac{N}{2}$

Note that if $f \in L^m(\Omega)$, $m \geq \frac{N}{2}$, even if $f \in L^\infty$, then any solution $u \in L^1(\Omega)$ of (3.1.1) is unbounded. Indeed, assuming by simplicity that $f \geq 0$, by the maximum principle we have $u > 0$ and then

$$-\Delta u > 0.$$

Thus, $u \geq \eta$ in $B_r(0) \subset \Omega$ for a suitable $\eta > 0$. By weak comparison then $u \geq v$ in $B_r(0)$, where v is the solution to problem

$$\begin{cases} -\Delta v = \lambda \frac{\eta}{|x|^2} & \text{in } \Omega, \\ v = 0 & \text{on } \partial\Omega, \end{cases}$$

which is easily seen to be unbounded.

In the following example we can see a precise behavior in terms of λ. Consider the radial equation

$$-u''(r) - \frac{(N-1)}{r} u'(r) - \frac{\lambda}{r^2} u(r) = 1.$$

We look for a solution in the Sobolev space $W_0^{1,2}(\Omega)$ such that $u(1) = 0$. An elementary argument provides that such solution is

$$u(r) = A(r^{-\alpha_-(\lambda)} - r^2),$$

where $A = \frac{1}{2N+\lambda}$.

3.1.2 Nonexistence result for $m = 1$

The fact that the problem is linear, together with the previous section, shows by duality that, in general, there is no solution for data in $L^1(\Omega)$. The meaning of solution here is the following: given $f \in L^1(\Omega)$ we say that a function $u \in L^1(\Omega)$ such that $\frac{u}{|x|^2} \in L^1(\Omega)$ is a weak solution to equation

$$-\Delta u - \lambda \frac{u}{|x|^2} = f \tag{3.1.5}$$

if and only if

$$\int_\Omega u\left(-\Delta\phi - \lambda\frac{\phi}{|x|^2} - f\phi\right)dx = 0, \quad \text{for all } \phi \in C_0^\infty(\Omega).$$

We will say that the problem is well posed in $L^1(\Omega)$ if there exists a constant $C > 0$ such that

$$\|u\|_1 \le C\|f\|_1, \quad \text{for all } f \in L^1(\Omega).$$

Let us prove that problem (3.1.5) is not well posed in $L^1(\Omega)$. Assume that for each $f \in L^1(\Omega)$ we can find $u \in L^1(\Omega)$, a weak solution to problem (3.1.5). For $g \in L^\infty(\Omega)$ let $\psi \in W_0^{1,2}(\Omega)$ be the solution of

$$-\Delta\psi - \lambda\frac{\psi}{|x|^2} = g.$$

Then

$$\left|\int_\Omega f\psi\,dx\right| = \left|\int_\Omega gu\,dx\right| \le \|g\|_\infty\|u\|_1 \le C\|g\|_\infty\|f\|_1.$$

But then, taking the infimum on $f \in L^1(\Omega)$ with $\|f\|_1 \le 1$, we find that

$$\|\psi\|_\infty \le C\|g\|_\infty,$$

which is false in general as previously shown.

These previous results motivate why we will consider hereafter $1 < m < \frac{N}{2}$. We recall the classical Sobolev inequality. Let $1 \le p \le \infty$ and let $W^{1,p}(\mathbb{R}^N)$ be the Sobolev space of functions in $L^p(\mathbb{R}^N)$ with first derivatives in $L^p(\mathbb{R}^N)$.

Proposition 3.1.1. *Assume $1 \le p < N$. Then there exists a constant $S_{N,p}$ such that for every $u \in W^{1,p}(\mathbb{R}^N)$*

$$S_{N,p}\left(\int_{\mathbb{R}^N} |u|^{p^*}\,dx\right)^{\frac{1}{p^*}} \le \left(\int_{\mathbb{R}^N} |\nabla u|^p\,dx\right)^{\frac{1}{p}}, \quad \textit{where } p^* = \frac{Np}{N-p}. \tag{3.1.6}$$

See for instance [86] and [293] for a couple of different proofs.

Considering now

$$\mathcal{D}^{1,p}(\mathbb{R}^N) = \{f \in L^{p^*}(\mathbb{R}^N) \mid \nabla f \in [L^p(\mathbb{R}^N)]^N\},$$

the Sobolev inequality gives the continuous inclusion

$$\mathcal{D}^{1,p}(\mathbb{R}^N) \subset L^{p^*}(\mathbb{R}^N), \quad \text{with } S_{N,p}\|f\|_{L^{p^*}(\mathbb{R}^N)} \le \|\nabla f\|_{L^p(\mathbb{R}^N)}.$$

It is not too hard to prove that $W^{1,p}(\mathbb{R}^N)$ is dense in $\mathcal{D}^{1,p}(\mathbb{R}^N)$ with the metric of $L^{p^*}(\mathbb{R}^N)$. By an interpolation argument and the Sobolev inequality we obtain the following result.

Corollary 3.1.2. *Let* $1 \le p < N$. *Then*

$$W^{1,p}(\mathbb{R}^N) \subset L^q(\mathbb{R}^N), \quad \textit{for all } q \in [p, p^*],$$

with continuous injection.

Also, an iteration procedure allows to obtain the following result.

Corollary 3.1.3. *If* $p = N$, *then*

$$W^{1,N}(\mathbb{R}^N) \subset L^q(\mathbb{R}^N), \quad \textit{for all } q \in [p, \infty),$$

with continuous injection.

Finally we recall the classical result by Morrey.

Proposition 3.1.4. *If* $p > N$, *then*

$$W^{1,p}(\mathbb{R}^N) \subset L^\infty(\mathbb{R}^N),$$

with continuous injection. Moreover, for all $u \in W^{1,p}(\mathbb{R}^N)$ *we have*

$$|u(x) - u(y)| \le C|x - y|^{1-\frac{N}{p}} \|\nabla u\|_p, \quad \textit{for all } x, y \in \mathbb{R}^N,$$

with $C = C(N, p)$ *a universal constant.*

The reader can find the main properties of the Sobolev spaces for instance in the books [28], [86] and [239]. The constant of the Sobolev inequality (3.1.6), that is, when $1 \le p < N$, is the same for all bounded domains Ω. The difference is that in the whole $\mathbb{R}^N$ this constant is attained, while in every bounded domain there is no minimizer. The minimizers in $W^{1,p}(\mathbb{R}^N)$ are given, up to dilations and translation, by the function

$$U(x) = \frac{1}{(1 + |x|^{\frac{p}{p-1}})^{\frac{N-p}{p}}}.$$

The existence and uniqueness up to dilations and translation can be seen for instance in [302] and [230]. The exponent p^* is called the critical Sobolev exponent. Since for $p \ge N$ the Sobolev exponent is not finite, we can summarize the above by setting

$$p^* = \begin{cases} \frac{Np}{N-p} & \text{if } 1 \le p < N, \\ +\infty & \text{if } p > N. \end{cases}$$

In a domain $\Omega \subsetneq \mathbb{R}^N$ we define the Sobolev space with null trace, $W_0^{1,p}(\Omega)$, as the completion of $C_0^\infty(\Omega)$ with respect to the norm of $W^{1,p}(\mathbb{R}^N)$.

Finally, when $1 < p < \frac{N}{2}$ we can consider $W^{1,p^*}(\mathbb{R}^N)$, for which the *critical Sobolev exponent* is then

$$(p^*)^* := p^{**} = \frac{Np}{N - 2p}.$$

This exponent will appear in the next results. We also recall the classical truncation functions.

Definition 3.1.5. Let $T_k : \mathbb{R} \to \mathbb{R}$ be the truncation defined by $T_k(s) = \max\{\min\{k, s\}, -k\}$, $k > 0$. We also consider $G_k(s) = s - T_k(s)$.

Note that

$$T_k(s) = s \quad \text{if } |s| \le k, \quad T_k(s) = k\frac{s}{|s|} \quad \text{if } |s| \ge k.$$

We will use these truncations in order to obtain some estimates. To that end, we will consider the approximate Dirichlet problems

$$u_n \in W_0^{1,2}(\Omega) : -\Delta u_n = \lambda \frac{u_n}{|x|^2 + \frac{1}{n}} + f_n, \tag{3.1.7}$$

where $f_n(x) = T_n(f(x))$.

Observe that for every $n \in \mathbb{N}$, there is always a unique solution $u_n \in W_0^{1,2}(\Omega) \cap L^\infty(\Omega)$ of (3.1.7) (see for instance [291]).

3.2 Summability of finite energy solutions

In this section we will assume that $\frac{2N}{N+2} < m < \frac{N}{2}$. In this case, even though the right hand side of (3.1.1) belongs to the dual space, $W^{-1,2}(\Omega)$, the following surprising phenomenon shows up. *A restriction on λ must be assumed in order to have the same results as for the Laplacian. More precisely, for $m \ge 1$, let us define*

$$\lambda(m) = \frac{N(m-1)(N-2m)}{m^2}.$$

We will prove that this is the critical value of λ depending on the summability of f.

Theorem 3.2.1. *If the right hand side f of equation (3.1.1) belongs to $L^m(\Omega)$, $\frac{2N}{N+2} \le m < \frac{N}{2}$, and*

$$\lambda < \lambda(m) = \frac{N(m-1)(N-2m)}{m^2}, \tag{3.2.1}$$

then the corresponding weak solution u belongs to $L^{m^{**}}(\Omega)$, where we recall that $m^{**} = \frac{Nm}{N-2m}$.

Proof. Since for any $n \in \mathbb{N}$, the solution to (3.1.7), u_n, belongs to $L^\infty(\Omega)$, we can use $|u_n|^{\gamma-2}u_n$, with $\gamma = \frac{m(N-2)}{N-2m}$, as a test function in (3.1.7).

Since $m \ge \frac{2N}{N+2}$ we find that $\gamma \ge 2$. Therefore,

$$\frac{4(\gamma-1)}{\gamma^2} \int_\Omega |\nabla |u_n|^{\frac{\gamma}{2}}|^2 \le \lambda \int_\Omega \left(\frac{|u_n|^{\frac{\gamma}{2}}}{|x|} \right)^2 + \|f_n\|_{L^m} \left(\int_\Omega |u_n|^{(\gamma-1)m'} \right)^{\frac{1}{m'}},$$

and so

$$\left[\frac{4(\gamma-1)}{\gamma^2} - \lambda \left(\frac{2}{N-2} \right)^2 \right] \int_\Omega |\nabla |u_n|^{\frac{\gamma}{2}}|^2 \le \|f_n\|_{L^m} \left(\int_\Omega |u_n|^{(\gamma-1)m'} \right)^{\frac{1}{m'}}.$$

Note that in order to have an *a priori* estimate we need

$$\frac{4(\gamma-1)}{\gamma^2} > \lambda\left(\frac{2}{N-2}\right)^2, \tag{3.2.2}$$

that is,

$$(\gamma-1)(N-2)^2 > \lambda\gamma^2.$$

Given the definition of γ, this is true if

$$\lambda < \frac{N(m-1)(N-2m)}{m^2},$$

which justifies the restriction (3.2.1). Note that the definition of γ gives $\frac{\gamma 2^*}{2} = (\gamma-1)m' = m^{**}$.

Using the Sobolev inequality and denoting by $\mathcal{S}$ the optimal constant of the Sobolev embedding, we get

$$\mathcal{S}^2\left[\frac{4(\gamma-1)}{\gamma^2} - \lambda\left(\frac{2}{N-2}\right)^2\right]\left(\int_\Omega |u_n|^{m^{**}}\right)^{\frac{2}{2^*}-\frac{1}{m'}} \leq \|f_n\|_{L^m},$$

and thus

$$\left(\frac{\mathcal{S}}{\Lambda_{N,2}}\right)^2[\lambda(m)-\lambda]\|u_n\|_{L^{m^{**}}} \leq \|f_n\|_{L^m}. \tag{3.2.3}$$

Passing to the limit as n tends to infinity yields the result since u_n converges to the solution u of (3.1.1). □

3.2.1 Optimality of the condition on λ

We will see with the following example that the condition $\lambda < \lambda(m)$ is optimal.

Example 3.2.1. Consider problem (3.1.1) with $\Omega = B_1(0)$ and $f(x) = \frac{1}{|x|^\nu}$, with $\nu = \frac{N}{m}$. Passing to radial coordinates,

$$\left\{ -u''(r) - \frac{(N-1)}{r}u'(r) - \lambda\frac{u(r)}{r^2} = \frac{1}{r^\nu}, \quad u(1) = 0, \quad 0 < \lambda \leq \left(\frac{N-2}{2}\right)^2. \right. \tag{3.2.4}$$

We set

$$\alpha_\pm = -\left(\frac{N-2}{2}\right) \pm \sqrt{\left(\frac{N-2}{2}\right)^2 - \lambda},$$

that is, $\alpha_\pm \equiv -\alpha_\pm(\lambda)$ defined in (3.1.3). By elementary integration we find a family of solutions to (7.5.5):

$$u(r) = Ar^{\alpha_+} + Br^{\alpha_-} + Cr^{2-\nu}, \quad u(1) = 0, \quad A + B + C = 0,$$
$$C = -\frac{1}{(2-\nu)(1-\nu) + (N-1)(2-\nu) + \lambda}. \tag{3.2.5}$$

Note that if $B \neq 0$ the solutions are not in $W_0^{1,2}(\Omega)$. Therefore we consider $B = 0$, i. e.,

$$u(r) = \frac{1}{(2-\nu)(1-\nu) + (N-1)(2-\nu) + \lambda}[r^{\alpha_+} - r^{2-\nu}].$$

As $\nu = \frac{N}{m}$, the coefficient becomes

$$\frac{1}{(2-\nu)(1-\nu) + (N-1)(2-\nu) + \lambda} = \frac{1}{(2-\frac{N}{m})(1-\frac{N}{m}) + (N-1)(2-\frac{N}{m}) + \lambda}.$$

Observe that since $\lambda > N\frac{(m-1)(N-2m)}{m^2}$, we have $2-\nu > \alpha_+$, so that the summability of u coincides with the summability of r^{α_+} and r^{α_+} does not belong to $L^{m^{**}}$.

3.3 Existence of infinite energy solutions

As in the previous section we start with a result of existence under the restriction $\lambda < \lambda(m)$ and with $L^m(\Omega) \not\subset W^{-1,2}(\Omega)$, that is $1 < m < \frac{2N}{N+2}$. More precisely we have the following result.

Theorem 3.3.1. *If the right hand side f of equation* (3.1.1) *belongs to L^m, $1 < m < \frac{2N}{N+2}$, and*

$$\lambda < \lambda(m) = \frac{N(m-1)(N-2m)}{m^2},$$

then the weak solution u belongs to $W_0^{1,m^*}(\Omega)$.

Proof. Define for $\epsilon > 0$ the test function

$$v_\epsilon = ((\epsilon + |u_n|)^{\gamma-1} - \epsilon^{\gamma-1})\mathrm{sgn}(u_n), \quad \text{where } \gamma = \frac{(N-2)m}{N-2m}.$$

Use v_ϵ as test function in (3.1.7) to obtain

$$\begin{aligned}(\gamma-1)\int_\Omega \frac{|\nabla u_n|^2}{(\epsilon+|u_n|)^{2-\gamma}} &= \frac{4(\gamma-1)}{\gamma^2}\int_\Omega |\nabla[(\epsilon+|u_n|)^{\frac{\gamma}{2}} - \epsilon^{\frac{\gamma}{2}}]|^2 \\ &\le \lambda\int_\Omega \left(\frac{(\epsilon+|u_n|)^{\frac{\gamma}{2}} - \epsilon^{\frac{\gamma}{2}}}{|x|}\right)^2 dx \\ &\quad + \lambda\int_\Omega \frac{1}{|x|^2}(u_n v_\epsilon - [(\epsilon+|u_n|)^{\frac{\gamma}{2}} - \epsilon^{\frac{\gamma}{2}}]^2)\,dx + \int_\Omega f_n v_\epsilon\,dx. \end{aligned} \tag{3.3.1}$$

Using the Hardy and Sobolev inequalities we get

$$\mathcal{S}^2\Big[\frac{4(\gamma-1)}{\gamma^2}-\frac{\lambda}{\Lambda_{N,2}}\Big]\Big(\int_\Omega(\epsilon+|u_n|)^{\gamma\frac{2^*}{2}}\,dx\Big)^{\frac{2}{2^*}}$$
$$\leq \lambda\int_\Omega \frac{1}{|x|^2}(|u_n||(\epsilon+|u_n|)^{\gamma-1}-\epsilon^{\gamma-1}|-[(\epsilon+|u_n|)^{\frac{\gamma}{2}}-\epsilon^{\frac{\gamma}{2}}]^2)\,dx+\int_\Omega f_n v_\epsilon\,dx.$$

Now for fixed n and ϵ going to zero we have

$$\mathcal{S}^2\Big[\frac{4(\gamma-1)}{\gamma^2}-\frac{\lambda}{\Lambda_{N,2}}\Big]\Big(\int_\Omega |u_n|^{\gamma\frac{2^*}{2}}\,dx\Big)^{\frac{2}{2^*}}\leq \|f_n\|_m\Big(\int_\Omega |u_n|^{(\gamma-1)m'}\Big)^{\frac{1}{m'}}.$$

That is,

$$\Big(\frac{2\mathcal{S}}{N-2}\Big)^2\Big[\frac{N(m-1)(N-2m)}{m^2}-\lambda\Big]\|u_n\|_{L^{m^{**}}}\leq \|f_n\|_{L^m}.$$

Thus we have proved the following *a priori* estimate:

$$\Big(\frac{\mathcal{S}^2}{\Lambda_{N,2}}\Big)[\lambda(m)-\lambda]\|u_n\|_{L^{m^{**}}}\leq \|f_n\|_{L^m}. \tag{3.3.2}$$

Now we follow [69] in order to show that the sequence $\{u_n\}$ is bounded in $W_0^{1,m^*}(\Omega)$. Fix $\epsilon = R$ in (3.3.1) and use (3.3.2). Then we have

$$\int_\Omega \frac{|\nabla u_n|^2}{(R+|u_n|)^{(2-\gamma)}}\leq C(a,m,N,\mathcal{S},R)=M,$$

so that (using the Hölder inequality with exponents $\frac{2}{m^*}$ and $\frac{2}{2-m^*}$)

$$\int_\Omega |\nabla u_n|^{m^*}\leq \int_\Omega \frac{|\nabla u_n|^{m^*}}{(R+|u_n|)^{\frac{(2-\gamma)m^*}{2}}}(R+|u_n|)^{\frac{(2-\gamma)m^*}{2}}\leq M^{\frac{m^*}{2}}\Big(\int_\Omega (R+|u_n|)^{\frac{(2-\gamma)m^*}{2-m^*}}\Big)^{\frac{2-m^*}{2}}.$$

Thus we can deduce

$$\int_\Omega |\nabla u_n|^{m^*}\leq M^{\frac{m^*}{2}}\Big(\int_\Omega (R+|u_n|)^{m^{**}}\Big)^{\frac{2-m^*}{2}},$$

which implies the boundedness of the sequence $\{u_n\}$ in $W_0^{1,m^*}(\Omega)$. Therefore, since $m > 1$, there exists u such that, up to a subsequence,

$$u_n \rightharpoonup u,\quad \text{weakly in } W_0^{1,m^*}(\Omega) \text{ and almost everywhere,}$$
$$\text{and}\quad \frac{u_n}{|x|^2}\to\frac{u}{|x|^2}\ \text{strongly in } L^1(\Omega).$$

Hence we can pass to the limit in (3.1.7) to prove that u is a distributional solution to (3.1.1). □

The above result is optimal as the following result shows.

Proposition 3.3.2. *Let*

$$\lambda \geq \frac{N(m-1)(N-2m)}{m^2}, \quad 1 < m < \frac{2N}{N+2},$$

and let u_n be the solution of

$$u_n \in W_0^{1,2}(\Omega) : -\Delta u_n = \lambda \frac{u_n}{|x|^2 + \frac{1}{n}} + f_n.$$

Then an estimate of the type

$$\|u_n\|_{L^{m^{**}}} \leq c_0 \|f_n\|_{L^m} \tag{3.3.3}$$

for some nonnegative c_0 (independent of n) cannot hold.

Proof. Let $p > 1$ be such that $\frac{1}{p} + \frac{1}{m^{**}} = 1$, let g be in L^p and let $\{g_n\}$ be a sequence of smooth functions converging to g in L^p. Consider

$$z_n \in W_0^{1,2}(\Omega) : -\Delta z_n = \lambda \frac{z_n}{|x|^2 + \frac{1}{n}} + g_n.$$

Now let $f_n = |z_n|^{m'-2} z_n$; hence, choosing u_n as a test function in the equation for z_n and choosing z_n as a test function in (3.1.7), we obtain from (3.3.3)

$$\left| \int_\Omega f_n z_n \right| = \left| \int_\Omega g_n u_n \right| \leq \|g_n\|_p \|u_n\|_{m^{**}} \leq c_0 \|g\|_p \|f_n\|_m.$$

Hence,

$$\int_\Omega |z_n|^{m'} \leq c_0 \|g\|_p \left(\int_\Omega |z_n|^{(m'-1)m} \right)^{\frac{1}{m}},$$

which implies

$$\|z_n\|_{m'} \leq c_0 \|g\|_p.$$

But then, up to subsequences, z_n converges weakly in $L^{m'}$ to z (also in $W_0^{1,2}(\Omega)$), the unique weak solution of

$$z \in W_0^{1,2}(\Omega) : -\Delta z = \lambda \frac{z}{|x|^2} + g, \ g \in L^p.$$

Due to the choice of p, it turns out that z belongs to $L^{p^{**}}$, and this contradicts Example 3.2.1. Indeed, choosing $\lambda \geq \Lambda(p)$, we have $\lambda > \Lambda(p + \epsilon)$ for every $\epsilon > 0$. Hence, the solution in $W_0^{1,2}(\Omega)$ given by Example 3.2.1 corresponding to the L^p function $g(x) = \frac{1}{|x|^\nu}$, with $\nu = \frac{N}{p+\epsilon}$, belongs exactly to the Marcinkiewicz space M^{N/α_+}, and not to L^{N/α_+}. Since $\frac{N}{\alpha_+} \leq p^{**}$ and $\lambda \geq \Lambda(p)$, we thus have a contradiction. □

3.3.1 Complete blow-up for data in L^1

If $f \in L^1$, the solution of the approximated problems can go to infinity almost everywhere. We show this phenomenon with the following example.

Example 3.3.1. Let $\Omega = B_{\frac{1}{2}}(0)$, and let u_n in $W_0^{1,2}(\Omega)$ be the solution of

$$\begin{cases} -\Delta u_n = \lambda \frac{u_n}{1/n+|x|^2} + f_n & \text{in } \Omega, \\ u_n = 0 & \text{on } \partial\Omega, \end{cases}$$

with $f_n \in L^\infty$ being a sequence of nonnegative functions bounded in L^1. Since the solutions u_n are nonnegative, we have $u_n \geq z_n$, where z_n is the unique solution in $W_0^{1,2}(\Omega)$ of

$$\begin{cases} -\Delta z_n = f_n & \text{in } \Omega, \\ z_n = 0 & \text{on } \partial\Omega. \end{cases}$$

Therefore, by the maximum principle, $u_n \geq w_n$, where w_n is the unique solution in $W_0^{1,2}(\Omega)$ of

$$\begin{cases} -\Delta w_n = \lambda \frac{z_n}{|x|^2} + f_n & \text{in } \Omega, \\ w_n = 0 & \text{on } \partial\Omega. \end{cases}$$

Supposing that f_n is radially symmetric, so will be u_n, z_n and w_n. Passing to radial coordinates, we have the following straightforward calculations:

$$z_n = \frac{1}{N-2} \int_0^{1/2} \frac{s^{N-1}}{\max(\rho, s)^{N-2}} f_n(s)\, ds - \frac{2^{N-2}}{(N-2)\omega_{N-1}} \|f_n\|_1,$$

and therefore

$$\begin{aligned} w_n(\rho) &= \frac{\lambda}{N-2} \int_0^{1/2} \frac{s^{N-3}}{\max(\rho, s)^{N-2}} \int_0^{1/2} \left(\frac{t^{N-1} f_{n(t)}}{\max(s,t)^{N-2}}\, dt \right) ds + C \\ &\geq \int_0^{\rho} \frac{s^{N-3}}{\max(\rho, s)^{N-2}} \int_0^{1/2} \left(\frac{t^{N-1} f_{n(t)}}{\max(s,t)^{N-2}}\, dt \right) ds + C \\ &= C + \frac{\lambda}{N-2} \frac{1}{\rho^{N-2}} \int_0^{\rho} \frac{1}{s} \left[\int_0^{s} t^{N-1} f_n(t)\, dt \right] ds. \end{aligned}$$

Choose now $f(x) = \frac{1}{|x|^N(-\ln(|x|))^\alpha}$, with $1 < \alpha < 2$, so that f belongs to L^1, and let $f_n = T_n(f)$ as usual. Since f_n increases to f,

$$\lim_{n \to +\infty} \int_0^{s} t^{N-1} f_n(t)\, dt = \int_0^{s} t^{N-1} f(t)\, dt = \frac{1}{(\alpha-1)(-\ln(s))^{\alpha-1}}.$$

Hence, for every $|x| = \rho$ in Ω,

$$\lim_{n\to+\infty} w_n(\rho) \geq C + \frac{a}{(N-2)(\alpha-1)}\frac{1}{\rho^{N-2}}\int_0^\rho \frac{1}{s(-\ln(s))^{\alpha-1}} = +\infty.$$

Thus the limit of u_n is everywhere $+\infty$, and so no estimate can be obtained if the data are assumed to be just in L^1.

3.4 Necessary and sufficient conditions for solvability in weighted L^1-spaces

Let us present here a result of existence when the second member belongs to a suitable weighted L^1-space.

Theorem 3.4.1. *Let $\lambda \leq \Lambda_{N,2}$ and suppose that $f \in L^1(\Omega), f \geq 0$. Call $\gamma = \alpha_-(\lambda)$ defined in* (3.1.3). *Then problem* (3.1.1) *has a positive weak solution u if and only if f satisfies*

$$\int_{B_r(0)} |x|^{-\gamma} f\,dx < +\infty,$$

for some $B_r(0) \subset\subset \Omega$.

Moreover, if u is the unique weak solution to (3.1.1), *then*

$$\forall k \geq 0m \quad T_k(u) \in W^{1,2}(\Omega), \quad u \in L^q(\Omega), \quad \forall q \in \left(1, \frac{N}{N-2}\right), \tag{3.4.1}$$

$$u \in W_0^{1,q}(\Omega), \quad \textit{with } q < \frac{N}{N-1}, \tag{3.4.2}$$

Proof. Necessary condition: Consider u a positive weak solution to problem (3.1.1) and consider $\varphi_n \in L^\infty(\Omega) \cap W^{1,2}(\Omega)$ the positive solution to

$$\begin{cases} -\Delta\varphi_n = \lambda\frac{\varphi_{n-1}}{|x|^2+\frac{1}{n}} + 1 & \text{in } \Omega, \\ \varphi_n = 0 & \text{on } \partial\Omega, \end{cases}$$

where

$$\begin{cases} -\Delta\varphi_0 = 1 & \text{in } \Omega, \\ \varphi_0 = 0 & \text{on } \partial\Omega. \end{cases}$$

Then it is easy to check that $\varphi_0 \leq \varphi_1 \leq \varphi_{n-1} \leq \varphi_n \leq \varphi$, where φ is the pointwise limit, and then

$$\begin{cases} -\Delta\varphi = \lambda\frac{\varphi}{|x|^2} + 1 & \text{in } \Omega, \\ \varphi > 0 & \text{in } \Omega, \\ \varphi = 0 & \text{on } \partial\Omega. \end{cases} \tag{3.4.3}$$

Taking φ_n as a test function in (3.1.1), we get

$$\int_\Omega f\varphi_n\,dx \le \int_\Omega u\,dx = C < \infty.$$

Since $\{f\varphi_n\}_{n\in\mathbb{N}}$ is an increasing sequence uniformly bounded in $L^1(\Omega)$, applying the monotone convergence theorem and the local behavior of φ close to the origin, we obtain

$$\tilde{C}\int_{B_r(0)} |x|^{-\gamma} f\,dx \le \int_\Omega f\varphi\,dx \le C.$$

Sufficient condition: Assume that

$$\int_{B_r(0)} |x|^{-\gamma} f\,dx < +\infty,$$

for some $B_r(0) \subset\subset \Omega$; let us consider the sequence of energy solutions $u_n \in L^\infty(\Omega) \cap W^{1,2}(\Omega)$ to the following approximating problems:

$$\begin{cases} -\Delta u_n = \lambda \frac{u_{n-1}}{|x|^2+\frac{1}{n}} + f_n & \text{in } \Omega, \\ u_n(x) = 0 & \text{on } \partial\Omega, \end{cases} \tag{3.4.4}$$

where

$$\begin{cases} -\Delta u_0 = f_1 & \text{in } \Omega, \\ u_0(x) = 0 & \text{on } \partial\Omega, \end{cases}$$

$f_n = T_n(f)$ and $u_0 \le u_1 \le u_{n-1} \le u_n$ in $\mathbb{R}^N$. Since $f_n \ge 0$, $u_n(x) \ge 0$ in Ω. Take $\varphi \in W_0^{1,2}(\Omega)$, the positive energy solution to problem (3.4.3), as a test function in (3.4.4). It follows that

$$\int_\Omega u_n\,dx \le \int_\Omega f\varphi\,dx \le \tilde{C}\int_\Omega f|x|^{-\gamma} \le C.$$

Hence, since the sequence $\{u_n\}_{n\in\mathbb{N}}$ is increasing, we can define $u := \lim_{n\to\infty} u_n$, and conclude that $u \in L^1(\Omega)$. We claim that $\frac{u}{|x|^2} \in L^1(\Omega)$. Indeed, let ψ be the unique bounded positive solution to the problem

$$\begin{cases} -\Delta\psi = 1 & \text{in } \Omega, \\ \psi = 0 & \text{in } \partial\Omega. \end{cases}$$

Then $\psi \ge C$ in $B_r(0)$. By using ψ as a test function in (3.4.4),

$$\int_\Omega \frac{u_{n-1}}{|x|^2+\frac{1}{n}}\,dx \le \frac{1}{C}\int_{B_r(0)} \frac{\psi u_{n-1}}{|x|^2+\frac{1}{n}}\,dx + C(r)\int_{\Omega\setminus B_r(0)} u_n\,dx \le C,$$

and thus

$$\lambda \frac{u_{n-1}}{|x|^2 + \frac{1}{n}} + f_n \nearrow \lambda \frac{u}{|x|^2} + f \quad \text{strongly in } L^1(\Omega).$$

Testing with $T_k(u_n)$ in (3.4.4) and considering the previous estimates, we easily get that $T_k(u_n) \rightharpoonup T_k(u)$ weakly in $W_0^{1,2}(\Omega)$. Since the sequence $\{\lambda \frac{u_{n-1}}{|x|^2+\frac{1}{n}} + f_n\}_{n\in\mathbb{N}}$ converges strongly in $L^1(\Omega)$, by the classical results of [291], we reach $u \in L^\sigma(\Omega)$ for all $\sigma < \frac{N}{N-2}$ and $u \in W_0^{1,q}(\Omega)$ for all $q < \frac{N}{N-1}$. □

3.4.1 Uniqueness

In this section, we discuss the uniqueness of the solutions obtained as limit of solutions of the regular problems (3.1.7) (see [72]).

Theorem 3.4.2. *Under the assumptions of Theorem* 3.2.1 *or Theorem* 3.3.1, *the solution obtained as the limit of solutions of the regular problems* (3.1.7) *is unique.*

Proof. Consider the boundary value problems (3.1.7) and

$$z_n \in W_0^{1,2}(\Omega) : -\Delta z_n = \lambda \frac{z_n}{|x|^2 + \frac{1}{n}} + g_n, \tag{3.4.5}$$

where $\{g_n\}$ is any sequence converging to f in L^m. Therefore, since $m > 1$, there exists z such that, up to a subsequence, z_n converges weakly to z and z is a distributional solution to (3.1.1). Then

$$(u_n - z_n) \in W_0^{1,2}(\Omega) : -\Delta(u_n - z_n) = \lambda \frac{(u_n - z_n)}{|x|^2 + \frac{1}{n}} + f_n - g_n.$$

Here, we can repeat the calculation in previous theorems in order to obtain

$$\left(\frac{\mathcal{S}^2}{\Lambda_{N,2}}\right)[\Lambda(m) - \lambda]\,\|u_n - z_n\|_{L^{m^{**}}} \le \|f_n - g_n\|_{L^m}, \tag{3.4.6}$$

which implies that

$$\|u - z\|_{L^{m^{**}}} = 0,$$

and so the uniqueness follows. □

3.5 Further results

3.5.1 A different way to obtain the critical value for λ

Consider our problem:

$$\begin{cases} -\Delta u = \lambda \frac{u}{|x|^2} + g & \text{in } \Omega, \\ u = 0 & \text{on } \partial\Omega, \end{cases} \tag{3.5.1}$$

with $\lambda < \Lambda_{N,2}$. Define v as the *ground state transformation* of u, that is,

$$v(x) = |x|^{\gamma} u(x), \quad \text{where } \gamma = \frac{N-2}{2} - \sqrt{\Lambda_{N,2} - \lambda} > 0.$$

Then v solves problem

$$\begin{cases} -\operatorname{div}(|x|^{-2\gamma} \nabla v) = |x|^{-\gamma} g & \text{in } \Omega, \\ v = 0 & \text{on } \partial\Omega. \end{cases}$$

We set $f \equiv |x|^{-\gamma} g$.

Note that by definition,

$$0 < \gamma \leq \frac{N-2}{2} \quad \text{for all } 0 < \lambda \leq \Lambda_{N,2}.$$

We will show the existence of solutions of the original problem by studying the existence of *entropy solutions* of the transformed problem. To do that, we define the *weighted Sobolev space* $\mathcal{D}^{1,2}_{0,\gamma}(\Omega)$ as the completion of $\mathcal{C}^{\infty}_{0}(\Omega)$ with respect to the norm

$$\|\phi\|_{2,\gamma} = \left(\int_{\Omega} |\nabla \phi|^2 |x|^{-2\gamma} \, dx \right)^{1/2}.$$

By using the Caffarelli–Kohn–Nirenberg inequalities of [108] we have in particular the Sobolev inequality stated in the next theorem.

Theorem 3.5.1 (Sobolev inequality). *Let $u \in \mathcal{D}^{1,2}_{0,\gamma}(\Omega)$. Then there exists a positive constant $S = S(N, \gamma)$ such that*

$$S \left(\int_{\Omega} |u|^{2^*} |x|^{-2^* \gamma} \, dx \right)^{1/2^*} \leq \left(\int_{\Omega} |\nabla u|^2 |x|^{-2\gamma} \, dx \right)^{1/2}, \tag{3.5.2}$$

where $2^* = \frac{2N}{N-2}$.

Definition 3.5.2. Define $\mathcal{T}^{1,2}_{0,\gamma}(\Omega)$ to be the space of functions u such that $T_k(u) \in \mathcal{D}^{1,2}_{0,\gamma}(\Omega)$ for all $k > 0$.

Definition 3.5.3. Consider $f \in L^1(\Omega)$. We say that $u \in \mathcal{T}^{1,2}_{0,\gamma}(\Omega)$ is an entropy solution of

$$\begin{cases} -\operatorname{div}(|x|^{-2\gamma} \nabla v) = f & \text{in } \Omega, \\ v = 0 & \text{on } \partial\Omega, \end{cases} \tag{3.5.3}$$

if for every $k > 0$ and $\phi \in \mathcal{D}^{1,2}_{0,\gamma}(\Omega) \cap L^{\infty}(\Omega)$ we have

$$\int_{\Omega} |x|^{-2\gamma} \langle \nabla v, \nabla (T_k(v - \phi)) \rangle = \int_{\Omega} f T_k(v - \phi). \tag{3.5.4}$$

In the linear case, by uniqueness, the concept of entropy solution is equivalent to that of distributional solution and to the solution obtained by approximations. We refer to [68], [56], [132] and the references therein to justify the use of this concept of solution in a large class of problems.

Theorem 3.5.4. *Assume $f \in L^1(\Omega)$; then the problem*

$$\begin{cases} -\operatorname{div}(|x|^{-2\gamma}\nabla v) = f & \text{in } \Omega, \\ v = 0 & \text{on } \partial\Omega \end{cases} \tag{3.5.5}$$

has a unique entropy solution. If $f \geq 0$, then $v \geq 0$.

This is a particular case of Theorem 3.4.1, formulated for the *ground state transformed problem.*

It is worth pointing out that in the original problem the result above can be translated as follows. First at all, using the Hölder inequality, we check when $f = g|x|^{-\gamma} \in L^1(\Omega)$:

$$\int_\Omega |f|\,dx \leq \Big(\int_\Omega g^m\,dx\Big)^{\frac{1}{m}} \Big(\int_\Omega |x|^{-\gamma\frac{m}{m-1}}\,dx\Big)^{\frac{m-1}{m}}.$$

Thus, we need $\gamma\frac{m}{m-1} < N$. Since $\gamma = \frac{N-2}{2} - \sqrt{\Lambda_{N,2} - \lambda}$, after some elementary calculations we obtain again the condition $\lambda < \lambda(m)$.

3.5.2 A remark in the case $\lambda = \Lambda_{N,2}$

In this subsection we consider the problem in the critical value case $\lambda = \Lambda_{N,2}$, i. e.,

$$\begin{cases} -\Delta u - \Lambda_{N,2}\frac{u}{|x|^2} = g & \text{in } \Omega, \\ u = 0 & \text{on } \partial\Omega, \end{cases} \tag{3.5.6}$$

where g is a nonnegative function in $L^1(\Omega)$. According to Theorem 3.4.1, problem (3.5.6) is solvable in the weak sense if and only if $|x|^{-\frac{N-2}{2}} g \in L^1(\Omega)$.

There is a different approach to the critical problem (3.5.6) under consideration. As a consequence of the improved Hardy inequality proved in Section 2.5 we have defined in Section 2.5.1 a space larger than $W_0^{1,2}(\Omega)$, denoted $\mathcal{H}(\Omega)$. In general, if $f \in \mathcal{H}'$, the dual space, by the Lax–Milgram theorem the problem

$$-\Delta u - \frac{\Lambda_{N,2}}{|x|^2} = f, \quad u = 0 \text{ on } \partial\Omega,$$

has a unique solution $u \in \mathcal{H}(\Omega)$. Therefore, in particular, for $g \in L^m(\Omega)$ with $m > \frac{2N}{N+2}$ we can find a solution $u \in \mathcal{H}(\Omega)$.

4 Effect of the Hardy–Leray potential in the solvability of semilinear elliptic equations

4.1 Introduction

In this chapter we consider equations of the type

$$-\Delta u = \frac{\lambda u}{|x|^2} + u^p, \quad u \geq 0,$$

and we try to understand the conditions relating p and λ for the local solvability of such equations. As in Chapter 3 about the optimal summability, the optimal power for the solvability of the equation depends explicitly on the spectral parameter λ.

We will follow the original papers by Brezis, Dupaigne and Tesei, [89] and Dupaigne [145]. The organization of the chapter is as follows. In Section 4.2 we obtain the candidate to be the optimal power. We use a radial argumentation and look for local solutions in the whole $\mathbb{R}^N$. Section 4.3 will be devoted to collect some known results about elliptic equations and to introduce the *Newtonian capacity*. In Section 4.4 we discuss the existence problem. Finally in Section 4.5 we prove some nonexistence results that, as a byproduct, imply the optimality of the *critical power* obtained in Section 4.2.

4.2 The optimal power

Consider the equation

$$-\Delta u = \lambda \frac{u}{|x|^2} + u^p, \quad u \geq 0. \tag{4.2.1}$$

We try to calculate radial solutions, i. e., positive solutions to the ordinary differential equation

$$-u'' - \frac{(N-1)}{r} u' - \lambda \frac{u}{r^2} = u^p, \tag{4.2.2}$$

where $0 < \lambda \leq \Lambda_{N,2}$. For homogeneity reasons, we look for a solution of the form $\phi(r) = Ar^{-\beta}$, where $A, \beta > 0$. We obtain the algebraic identity

$$Ar^{-(\beta+2)}(-\beta(\beta+1) + (N-1)\beta - \lambda) = A^p r^{-p\beta},$$

which we solve by taking

$$(\beta + 2) = p\beta \quad \Leftrightarrow \quad \beta = \frac{2}{(p-1)}. \tag{4.2.3}$$

https://doi.org/10.1515/9783110606270-004

Since $A > 0$, we need

$$\beta^2 - (N-2)\beta + \lambda < 0.$$

This is equivalent to

$$\alpha_-(\lambda) < \beta < \alpha_+(\lambda), \tag{4.2.4}$$

where $\alpha_\pm(\lambda)$ were defined in (3.1.3). Therefore, since (4.2.3) holds,

$$p_-(\lambda) = 1 + \frac{2}{\alpha_-(\lambda)} \le p < p_+(\lambda) = 1 + \frac{2}{\alpha_+(\lambda)}. \tag{4.2.5}$$

Hence, we find a radial solution to equation (4.2.1) for fixed $\lambda \in (0, \Lambda_{N,2})$ if p verifies (4.2.5).

Some remarks about these critical exponents are in order:

$$p_+(\lambda) \longrightarrow \begin{cases} 2^* - 1 = \frac{N+2}{N-2} & \text{when } \lambda \to \lambda_{N,2}, \\ \infty & \text{when } \lambda \to 0, \end{cases}$$

$$p_-(\lambda) \longrightarrow \begin{cases} 2^* - 1 = \frac{N+2}{N-2} & \text{when } \lambda \to \lambda_{N,2}, \\ \frac{N}{N-2} & \text{when } \lambda \to 0. \end{cases}$$

Moreover p_+ is decreasing and p_- is increasing. In particular,

$$p_- < 2 * -1 = \frac{N+2}{N-2} < p_+ \quad \text{if } 0 < \lambda < \Lambda_{N,2}, \tag{4.2.6}$$

$$1 + \frac{2}{\alpha_+(\lambda)} = p_-(\lambda) < p < p_+(\lambda) = 1 + \frac{2}{\alpha_-(\lambda)}. \tag{4.2.7}$$

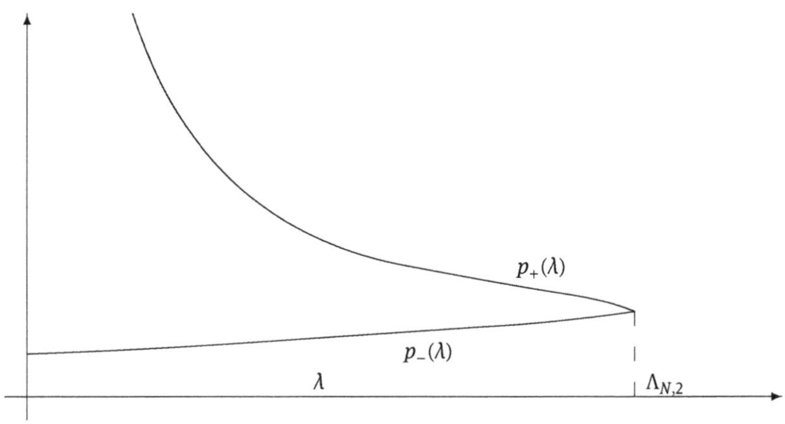

Note the continuity with respect to the semilinear case with $\lambda = 0$ and the new feature by the effect of the Hardy–Leray potential, that is, for $\lambda \in (0, \Lambda_{N,2})$ we have a bounded interval of p for which we find radial solutions.

We analyze the local summability of $\phi(r)$. Since $\beta + 2 < \alpha_+ + 2 < N$, we have:

- $\phi(|x|) \in L^1_{\text{loc}}(\mathbb{R}^N)$;
- $\frac{\phi(|x|)}{|x|^2} \in L^1_{\text{loc}}(\mathbb{R}^N)$
  and
- $\phi^p(|x|) \in L^1_{\text{loc}}(\mathbb{R}^N)$.

In this chapter we will prove the following important result.

The exponent $p_+(\lambda)$ represents the threshold for weak solvability of the Dirichlet problem for equation (4.2.1) *in the case of bounded domains $\Omega \subset \mathbb{R}^N$ containing the origin.*

4.3 Some previous results for linear elliptic equations

Before we begin to study the above statement, let us recall some basic result of potential theory and linear partial differential equations. For details we refer to [149, Sections 4.7 and 4.8] and [291], respectively. We will also use the truncations given in Definition 3.1.5. Recall that if

$$\mathcal{D}^{1,2}(\mathbb{R}^N) = \{f \in L^{2^*}(\mathbb{R}^N) \mid \nabla f \in [L^2(\mathbb{R}^N)]^N\},$$

the Sobolev inequality gives the continuous inclusion

$$\mathcal{D}^{1,2}(\mathbb{R}^N) \subset L^{2^*}(\mathbb{R}^N).$$

More specifically we have the functional inequality

$$S\|f\|_{L^{2^*}(\mathbb{R}^N)} \le \|\nabla f\|_{L^2(\mathbb{R}^N)}.$$

It is not too hard to prove that $W^{1,2}(\mathbb{R}^N)$ is dense in $\mathcal{D}^{1,2}(\mathbb{R}^N)$ with the metric of $L^{2^*}(\mathbb{R}^N)$.

4.3.1 Elliptic equations with measure data

To complete the landscape of existence results for linear elliptic equations, we also recall the classical result by G. Stampacchia in Section 9 of his fundamental paper [291]. We will denote by $\mathcal{M}(\Omega)$ the Radon measures in Ω with finite total variation. For $\mu \in \mathcal{M}(\Omega)$, $|\mu|_\omega$ denotes the total variation of μ. A detailed presentation of the main properties of the Radon measures can be found in [149]. In particular we will use the fact that for all $\mu \in \mathcal{M}(\Omega)$ there exists a sequence $f_n \in \mathcal{C}_0^\infty(\Omega)$ such that $f_n \rightharpoonup \mu$ weak-$*$ in $\mathcal{M}(\Omega)$. We next proceed to estate the classical result by Stampacchia.

Theorem 4.3.1. *Let $A(x)$ be an $N \times N$ symmetric matrix with measurable bounded coefficients in a bounded domain $\Omega \subset \mathbb{R}^N$ and such that there exists $\alpha > 0$ verifying*

$$\langle A(x)\xi, \xi\rangle \ge \alpha|\xi|^2, \quad \textit{for all } x \in \Omega \textit{ and for all } \xi \in \mathbb{R}^N.$$

If $\mu \in \mathcal{M}(\Omega)$, then there exists a unique solution $u \in W^{1,q}(\Omega)$, $q \in (1, \frac{N}{N-1})$, to the equation

$$-\operatorname{div}(A(x)\nabla u(x)) = f(x), \quad x \in \Omega.$$

Moreover, there exist $C = C(N, q, \alpha)$ such that

$$\|u\|_{W_0^{1,q}(\Omega)} \leq C|\mu|_\Omega,$$

and for all $k > 0$, $T_k(u) \in W_0^{1,2}(\Omega)$.

The original proof by Stampacchia is a consequence of a duality argument taking advantage of the linearity of the problem. For an alternative proof the reader can see Theorem 11.6 in [64]. The results by Stampacchia are related to the well-known results by De Giorgi solving the Hilbert problem (see [137]). As a consequence of Theorem 4.3.1 we also obtain in particular the following result.

Corollary 4.3.2. *If $v_n \in W^{1,q}(\Omega)$ is a solution to $-\operatorname{div}(A(x)\nabla v_n(x)) = f_n(x)$, $x \in \Omega$, and $f_n \to f$ in $L^1(\Omega)$, then $v_n \to v$ in $L^1(\Omega)$ and $v \in W_0^{1,q}(\Omega)$ is the solution to*

$$-\operatorname{div}(A(x)\nabla v(x)) = f(x), \quad x \in \Omega.$$

A final remark is that the existence is also true for a larger class of nonlinear elliptic operators; however, the uniqueness for measure data is a different question. See for instance [132] and the references therein.

Another ingredient to prove nonexistence results comes from the following Kato inequality. See [209] for the proof in a general setting.

Lemma 4.3.3. *Let Ω be a bounded domain in $\mathbb{R}^N$. Assume that $u \in L^1_{\mathrm{loc}}(\Omega)$ and $\Delta u \in L^1_{\mathrm{loc}}(\Omega)$. Then Δu^+ is a Radon measure and*

$$\Delta u^+ \geq \chi_{\{u \geq 0\}} \Delta u \quad \text{in } \mathcal{D}'(\Omega).$$

This inequality has been extended to the case in which Δu is a Radon measure. See Theorem 1.1 in [96]

4.3.2 Newtonian capacity

The capacity that we introduce next is related to the *Newtonian potential*, $|x|^{2-N}$, if $N \geq 3$. The significance of the concept of capacity from this point of view is contained in the fact that in many circumstances sets of zero capacity and only these can appear as sets of singularities for certain classes of functions or as null-sets that can be ignored.

Let us motivate, at least in an informal way, the concept of *Newtonian capacity* in dimension $N = 3$. We refer here to the survey by I. Markina [237].

Let $D \subset \mathbb{R}^3$ be a bounded domain representing a conducting body. Assume that $D \subset B_{r_0}(0)$ and consider for $R > r_0$ $S_R(0)$ the spherical surface of radius R surrounding D and connected with earth. If the body D is charged with an electrical charge, q, this charge will be distributed over D and will induce a charge on $S_R(0)$ just to reach the equilibrium. The charged body defines an electric field, $\vec{F}(x)$, in $\mathbb{R}^3$, that is a potential vector field with potential $\rho(x)$, that is, $\vec{F}(x) = -\nabla\rho(x)$. The vector field $\vec{F}(x)$ vanishes in D and there exists a distribution of charge in between the surfaces ∂D and $S_R(0)$ that we suppose have continuous densities, ϕ_1 on ∂D and ϕ_2 on $S_R(0)$. Then the potential

$$\rho_R(x) = \int_{\partial D} \frac{\phi_1(y)}{|x-y|} d\sigma(y) + \int_{S_R(0)} \frac{\phi_2(y)}{|x-y|} d\sigma(y),$$

is a continuous function in $\mathbb{R}^3$.

Since $S_R(0)$ is connected to earth we have $\rho_R = 0$ on $S_R(0)$. Moreover as $0 = \vec{F}(x) = -\nabla\rho(x)$ for $x \in D$, ρ_R is constant in D and therefore by continuity it is constant in ∂D. Call such constant V. The value of $\rho_R(x)$ is called the difference of potential between ∂D and $S_R(0)$.

It is known that the value of the constant V is proportional to the charge q and the proportionality factor is called *the capacity C of the condenser formed by ∂D and $S_R(0)$*, i. e.,

$$C = \frac{q}{V}.$$

Fixing the domain D and $R > r_0$ we can summarize the properties of ρ_R as follows. Call O_R the open set limited by ∂D and the spherical surface $S_R(0)$. Then ρ_R verifies

$$\begin{cases} \Delta\rho_R(x) = 0 & x \in O_R, \\ \rho_R(x) = V & x \in \partial D, \\ \rho_R(x) = 0 & x \in S_R(0). \end{cases} \tag{4.3.1}$$

By classical elliptic theory, there exists ρ_∞ such that $\lim_{R\to\infty} \rho_R(x) = \rho_\infty(x)$ and that

$$\begin{cases} \Delta\rho_\infty(x) = 0 & x \in \mathbb{R}^3 \setminus \overline{D}, \\ \rho_\infty(x) = V & x \in \partial D, \\ \lim_{|x|\to\infty} \rho_\infty(x) = 0. \end{cases} \tag{4.3.2}$$

It is natural to expect that for a suitable measure ϕ_∞,

$$\rho_\infty(x) = \int_{\partial D} \frac{d\phi_\infty(y)}{|x-y|}, \quad \text{with} \int_{\partial D} d\phi_\infty(y) = q.$$

The measure $\phi_\infty(y)$ is called the *equilibrium distribution* of the charge q and the function ρ_∞ the equilibrium potential. Therefore for every charge q distributed on D there is a corresponding equilibrium potential $\rho_\infty(x)$ which is $\rho_\infty(x) = V_\infty$ for all $x \in D$.

The constant $C_\infty = \frac{q}{V_\infty}$ *is called the capacity of the body* D. Then we can say that the capacity of a body D is the value of the charge q on D that produces the unit difference of potential between ∂D and ∞ under the equilibrium distribution of charge.

All the previous considerations can be extended to every $N > 2$ by considering the Riesz potential $|x-y|^{2-N}$, which for $N = 3$ is $|x-y|^{-1}$. Hereafter we will consider a general dimension $N > 2$.

We will look for a variational definition of the capacity that will give a more useful way to handle the concept.

Let $\mathcal{M}^+(\mathbb{R}^N)$ be the set of positive finite Radon measures in $\mathbb{R}^N$. For $\mu \in \mathcal{M}^+(\mathbb{R}^N)$ we recall that the Newtonian of μ is given by

$$u_\mu(x) = \int \frac{d\mu(y)}{|x-y|^{N-2}} \quad \text{and} \quad -\Delta u_\mu(x) = c_N \mu(x). \tag{4.3.3}$$

The energy of μ is given by

$$\mathcal{E}(\mu) = \iint \frac{d\mu(x)\,d\mu(y)}{|x-y|^{N-2}}.$$

The distribution of equilibrium of a charge q distributed on a body D is the measure ν which minimizes the energy, that is,

$$\mathcal{E}(\nu) \le \mathcal{E}(\mu), \quad \forall \mu \in \mathcal{M}^+(\mathbb{R}^N) \text{ such that } \int d\mu(y) = q.$$

Note that we can write the energy in terms of the Newtonian potential,

$$\mathcal{E}(\mu) = \int u_\mu(x)\,d\mu(x), \quad \text{where } u_\mu(x) = \int \frac{d\mu(y)}{|x-y|^{N-2}}.$$

Normalizing by putting $C(D) = \nu(D)$ we have

$$u_\mu(x) = 1 \quad \text{if } x \in D$$

and

$$\mathcal{E}(\nu) = \int_D d\nu(x) = \nu(D) = C(D).$$

Therefore, we define the capacity of a compact set D as follows:

$$C(D) = \inf\{c_N \mathcal{E}(\mu) \mid \mu \in \mathcal{M}^+(\mathbb{R}^N), u_\mu(x) > 1 \text{ in } D\}.$$

Observe that since $-\Delta u_\mu = c_N \mu$ in D and $-\Delta u_\mu = 0$ in $\mathbb{R}^N \setminus D$, integrating by parts we get

$$\int_{\mathbb{R}^N} |\nabla u_\mu(x)|^2\,dx = \int_{\mathbb{R}^N} u_\mu(x)(-\Delta u_\mu(x))\,dx = c_N \int_{\mathbb{R}^N} u_\mu(x)\,d\mu(x) = c_N \mathcal{E}(\mu) = C(D).$$

Let us give a precise definition of capacity.

Definition 4.3.4. If $A \subset \mathbb{R}^N$ set

$$\operatorname{cap}_{1,2}(A) = \inf\left\{ \int_{\mathbb{R}^N} |\nabla f|^2 \, dx \mid f \in \mathcal{D}^{1,2}(\mathbb{R}^N),\ A \subset \operatorname{Int}\{f \geq 1\} \right\}. \tag{4.3.4}$$

We call $\operatorname{cap}_{1,2}(A)$ the Newtonian capacity of A.

Using regularization we find that

$$\operatorname{cap}_{1,2}(K) = \inf\left\{ \int_{\mathbb{R}^N} |\nabla f|^2 \, dx \mid f \in C_0^\infty(\mathbb{R}^N),\ f \geq \chi_K \right\}$$

for each compact $K \subset \mathbb{R}^N$.

It is quite standard to prove the following statement.

Lemma 4.3.5. $\operatorname{cap}_{1,2}$ *is an outer measure on* $\mathbb{R}^N$.

We have, moreover, the following result.

Lemma 4.3.6. *Call* $\mathcal{H}^r(A)$ *the r-Hausdorff measure of* $A \subset \mathbb{R}^N$.
1. *If* $\mathcal{H}^{N-2}(A) < \infty$, *with* $N > 2$, *then* $\operatorname{cap}_{1,2}(A) = 0$.
2. *If* $\operatorname{cap}_{1,2}(A) = 0$, *then* $\mathcal{H}^r(A) = 0$ *for all* $r > N-2$.
3. *If* $f \in W^{1,2}(\mathbb{R}^N)$, *then* f *admits a* $\operatorname{cap}_{1,2}$*-quasicontinuous representative.*

The proof of this results and other properties of the capacity can be seen in the beautiful handbooks [149] and [236].

Application: eliminable singularities

We will also use the following results about eliminable singularities and comparison (see [89] and the references therein).

Lemma 4.3.7. *Let* $E \subset\subset \Omega$ *be a closed set such that* $\operatorname{cap}_{1,2}(E) = 0$ *and assume that* $u, f \in L^1(\Omega \setminus E)$ *are two nonnegative functions such that*

$$-\Delta u \geq f \quad \text{in } \mathcal{D}'(\Omega \setminus E).$$

Then $u, f \in L^1(\Omega)$ *and*

$$-\Delta u \geq f \quad \text{in } \mathcal{D}'(\Omega).$$

Furthermore if $\Omega_1 \subset\subset \Omega$ *is a smooth domain and* $v \in L^1(\Omega_1)$ *is the solution of*

$$\begin{cases} -\Delta v = f & \text{in } \Omega_1 \\ v = 0 & \text{on } \partial\Omega_1, \end{cases}$$

then

$$u \geq v \quad \text{a.e. in } \Omega_1 \tag{4.3.5}$$

Proof. Consider $u_k = T_k(u)$. By using Kato's inequality in Lemma 4.3.3 we obtain

$$-\Delta u_k \geq f\chi_{\{u<k\}} \quad \text{in } \mathcal{D}'(\Omega \setminus E). \tag{4.3.6}$$

We call $f_k := f\chi_{\{u<k\}}$. Since $-\Delta u_k$ is a nonnegative distribution on $\Omega \setminus E$ it extends to a nonnegative measure, μ_k, on $\Omega \setminus E$. Therefore by using Theorem 4.3.1 for measure data we have $u_k \in W_0^{1,p}(\Omega)$ and, since u_k is bounded, then $u_k \in W_{\text{loc}}^{1,2}(\Omega \setminus E)$. Next we prove that $u_k \in W_{\text{loc}}^{1,2}(\Omega)$. Since $\text{cap}_{1,2}(E) = 0$, there exist $\eta_n \in \mathcal{C}_0^\infty(\Omega)$ such that:
(i) $\eta_n \equiv 1$ in a neighborhood of E;
(ii) $\|\eta_n\|_{W_0^{1,2}(\Omega)} \to 0 = \text{cap}_{1,2}(E)$.

Take $\phi \geq 0$, $\phi \in \mathcal{C}_0^\infty(\Omega)$ and define

$$\phi_n = \phi(1-\eta_n) \in \mathcal{C}_0^\infty(\Omega \setminus E). \quad \text{Note that therefore, } \phi_n \to \phi \text{ in } W_0^{1,2}(\Omega).$$

Since $\Delta u_k \leq 0$, using Young's inequality we obtain the following estimate:

$$\begin{aligned}\int_\Omega |\nabla u_k|^2 \phi_n^2 &\leq \int_\Omega e^{(k-u_k)} |\nabla u_k|^2 \phi_n^2 = -e^k \int_\Omega \langle \nabla(e^{-u_k}), \nabla u_k \rangle \phi_n^2 \\ &= e^k \Big(2 \int_\Omega e^{-u_k} \phi_n \langle \nabla \phi_n, \nabla u_k \rangle + \int_\Omega e^{-u_k} \Delta u_k \phi_n^2 \Big) \\ &\leq 2e^k \int_\Omega e^{-u_k} \phi_n \langle \nabla \phi_n, \nabla u_k \rangle \\ &\leq C_k \int_\Omega e^{-u_k} |\nabla \phi_n|^2 + \frac{1}{2} \int_\Omega e^{-u_k} |\nabla u_k|^2 \phi_n^2.\end{aligned}$$

Hence,

$$\int_\Omega \big|\nabla(u_k \phi_n)\big|^2 \leq C \int_\Omega |\nabla u_k|^2 \phi_n^2 + \int_\Omega u_k^2 |\nabla \phi_n|^2 \leq C_k' \int_\Omega |\nabla \phi_n|^2,$$

which, passing to the limit as $n \to \infty$, implies $u_k \in W_{\text{loc}}^{1,2}(\Omega)$.

We next prove that

$$-\Delta u_k \geq f_k \quad \text{in } \mathcal{D}'(\Omega). \tag{4.3.7}$$

For ϕ and ϕ_n as above, since (4.3.6) holds,

$$\int_{\Omega \setminus E} u_k(-\Delta \phi_n) \geq \int_{\Omega \setminus E} f_k \phi_n.$$

For the left hand side we have

$$\lim_{n\to\infty} \int_\Omega u_k(-\Delta \phi_n) = \lim_{n\to\infty} \int_\Omega \langle \nabla u_k \nabla \phi_n \rangle = \int_\Omega \langle \nabla u_k \nabla \phi \rangle = \int_\Omega u_k(-\Delta \phi),$$

while the right hand side verifies

$$\lim_{n\to\infty}\int_\Omega f_k\phi_n = \int_\Omega f_k\phi.$$

This gives (4.3.7). In particular u_k is superharmonic in Ω and, therefore, for almost every $x \in \Omega$ and all balls $B \subset \Omega$ centered at x,

$$u_k(x) \geq \frac{1}{|B|}\int_B u_k(y)dy.$$

Since $u_k \to u$ a. e. $x \in \Omega$, and by Fatou's lemma we find that

$$\int_B u(y)dy < +\infty,$$

that is, $u \in L^1_{\text{loc}}(\Omega)$. Then passing to the limit in (4.3.7) we find that $f \in L^1_{\text{loc}}(\Omega)$ and

$$-\Delta u \geq f, \quad \text{in } \mathcal{D}'(\Omega).$$

We now prove the comparison result (4.3.5). Consider a mollifier η_n and call $u_n = u * \eta_n$, $f_n = f * \eta_n$. For n large enough we have

$$-\Delta u_n \geq f_n \quad \text{and} \quad u_n \geq 0 \text{ in } \Omega'.$$

Now, by the classical comparison principle, $u_n \geq v_n$ in Ω', where v_n is the solution to

$$\begin{cases} -\Delta v_n = f_n & \text{in } \Omega', \\ v_n = 0 & \text{on } \partial\Omega'. \end{cases}$$

Note that since $u_n \to u$, $f_n \to f$ in $L^1(\Omega')$ as $n \to \infty$, using Corollary 4.3.2 we have $v_n \to v$ in $L^1(\Omega')$, from which we easily obtain (4.3.5). □

The following result is part of the folklore of the subject and can be found for instance in [41].

Lemma 4.3.8. *Assume that $u \geq 0$ in Ω, $u \not\equiv 0$, $u \in L^1_{\text{loc}}(\Omega)$ and $\frac{u}{|x|^2} \in L^1_{\text{loc}}(\Omega)$. If u satisfies $-\Delta u - \lambda\frac{u}{|x|^2} \geq 0$ in the sense of distributions, then there exist a positive constant C and a small ball $B_R(0) \subset \Omega$ such that $u(x) \geq C|x|^{-\alpha_-(\lambda)}$ in $B_R(0)$, where $\alpha_-(\lambda)$ is defined in (3.1.3).*

Proof. By using the strong maximum principle and the comparison result it is not difficult to obtain $u \geq \eta$ in a small ball $B_r(0)$. Fixing $R > 0$, let $w \in W^{1,2}(B_R(0))$ be the unique positive solution to problem

$$\begin{cases} -\Delta w - \lambda\frac{w}{|x|^2} = 0 & \text{in } B_R(0), \\ w = \eta & \text{on } \partial B_R(0). \end{cases} \tag{4.3.8}$$

By a direct computation we obtain $w(r) = Cr^{-\alpha_-(\lambda)}$ and $C = \eta R^{\alpha_-(\lambda)}$. Since u is a supersolution to problem (4.3.8), using the weak comparison principle we conclude that $u \geq w$ in $B_R(0)$, thus $u \geq C|x|^{-\alpha_-(\lambda)}$ in $B_R(0)$ and the result follows. □

Note that our previous Theorem 3.4.1, which will also be used at some point, goes in the same direction as this result.

4.3.3 The Ambrosetti–Rabinowitz mountain pass theorem

Seeking some completeness in our exposition, we will give a proof of the classical *mountain pass theorem* by Ambrosetti and Rabinowitz; see [37]. We present a proof based on the Ekeland variational principle that requires a minimal regularity in the functional (this is perhaps the reason of why it is so useful).

The variational principle of Ekeland

The idea of the *ε-variational principle* of Ekeland can be described in the following way.

Assume that f is a lower semicontinuous real valued function defined on the metric space $(\mathcal{M}, d)$ such that $f(x) \geq \beta$ for all $x \in \mathcal{M}$.

The principle deals with the construction of minimizing sequences with some given control. More precisely, we construct sequences verifying

$$\inf_{x \in \mathcal{M}} \{f(x)\} + \varepsilon > f(x_\varepsilon)$$

and

$$f(y) \geq f(x_\varepsilon) - \varepsilon d(x_\varepsilon, y).$$

We can read geometrically these conditions by saying that for all $\varepsilon > 0$ the graph of f is above the *inverted cone* of amplitude ϵ with vertex at $(x_\epsilon, f(x_\epsilon))$. See the original paper [148]; we follow here the proof in [238].

Theorem 4.3.9. *Let $\mathcal{M}$ be a complete metric space and let*

$$\phi : \mathcal{M} \to (-\infty, \infty]$$

be a proper function such that:
(i) $\phi(y) \geq \beta$;
(ii) ϕ *is lower semicontinuous.*

Given $\varepsilon > 0$ and $u \in \mathcal{M}$ such that

$$\phi(u) \leq \inf_{\mathcal{M}} \phi + \varepsilon,$$

there exists $v \in \mathcal{M}$ such that:
1. $\phi(u) \geq \phi(v)$;
2. $d(u, v) \leq 1$;
3. *if $v \neq w \in \mathcal{M}$, then $\phi(w) \geq \phi(v) - \varepsilon d(v, w)$.*

Proof. Fixing $\varepsilon > 0$, we define the following order relation on $\mathcal{M}$: we say

$$w \preceq v, \quad \text{if and only if} \quad \phi(w) + \varepsilon d(w, v) \leq \phi(v).$$

Consider $u_0 = u$ and by recurrence define the sequence $\{u_n\}$ as follows: for $n \in \mathbb{N}$ we take

$$S_n = \{w \in \mathcal{M} : w \preceq u_n\}.$$

Choosing $u_{n+1} \in S_n$ such that

$$\phi(u_{n+1}) \leq \inf_{S_n}\{\phi\} + \frac{1}{n+1},$$

we get $u_{n+1} \preceq u_n$ and $S_{n+1} \subset S_n$. The lower semicontinuity of ϕ implies that S_n is a closed set. Now, if $w \in S_{n+1}$, we have $w \preceq u_{n+1} \preceq u_n$ and then,

$$\varepsilon d(w, u_{n+1}) \leq \phi(u_{n+1}) - \phi(w) \leq \inf_{S_n}\{\phi\} + \frac{1}{n+1} - \inf_{S_n}\{\phi\} = \frac{1}{n+1}.$$

If we set diameter$(S_{n+1}) = \delta_{n+1}$, then $\delta_{n+1} \leq \frac{2}{\varepsilon(n+1)}$ and so, $\lim_{n\to\infty} \delta_{n+1} = 0$. The completeness of $\mathcal{M}$ implies $\bigcap_{n=1}^{\infty} S_n = \{v\}$ for some $v \in \mathcal{M}$. But, in particular, $v \in S_0$; hence $v \preceq u_0 = u$, i. e., $\phi(v) \leq \phi(u) + \varepsilon d(u, v) \leq \phi(u)$ and

$$d(u, v) \leq \frac{\phi(u) - \phi(v)}{\varepsilon} \leq \varepsilon^{-1}\Big(\inf_{\mathcal{M}}\{\phi\} + \varepsilon - \inf_{\mathcal{M}}\{\phi\}\Big) = 1.$$

Then, $d(u, v) \leq 1$.

To get (3), suppose that $w \preceq v$. Then for all $n \in \mathbb{N}$, $w \preceq u_n$, i. e.,

$$w \in \bigcap_{n=1}^{\infty} S_n,$$

and then $w = v$.

So we conclude that if $w \neq v$, then $\phi(w) \geq \phi(v) - \varepsilon d(v, w)$. □

The results that we explain below show how to use the variational principle to find *critical points* of functionals. This will be used to prove the mountain pass theorem.

Corollary 4.3.10. *Let $\mathcal{X}$ be a Banach space, and let $\varphi : \mathcal{X} \to \mathbb{R}$ be a differentiable lower bounded function in $\mathcal{X}$. Then for all $\varepsilon > 0$ and for all $u \in \mathcal{X}$ such that*

$$\varphi(u) \leq \inf_{\mathcal{X}} \varphi + \varepsilon,$$

there exists $v \in \mathcal{X}$ verifying

$$\varphi(v) \leq \varphi(u),$$
$$\|u - v\|_{\mathcal{X}} \leq \varepsilon^{1/2},$$
$$\|\varphi'(v)\|_{\mathcal{X}'} \leq \varepsilon^{1/2}.$$

Proof. In Theorem 4.3.9 take $\mathcal{M} = \mathcal{X}$, $\phi = \varphi$, $\varepsilon > 0$, $\lambda = \frac{1}{\varepsilon^{1/2}}$, $d(\cdot,\cdot) = \|\cdot - \cdot\|_{\mathcal{X}}$. For each u we obtain $v \in \mathcal{X}$ such that

$$\varphi(v) \le \varphi(u),$$
$$\|u - v\|_{\mathcal{X}} \le \varepsilon^{1/2},$$

and for all $w \neq v$

$$\varphi(w) > \varphi(v) - \varepsilon^{1/2}\|(v - w)\|.$$

Taking in particular $w = v + th$ with $t > 0$ and $h \in \mathcal{X}$, $\|h\| = 1$. Then

$$\varphi(v + th) - \varphi(v) > -\varepsilon^{1/2}t.$$

This implies $-\varepsilon^{1/2} \le \langle \varphi'(v), h\rangle$ for all $h \in \mathcal{X}$, $\|h\| = 1$, and hence

$$\|\varphi'(v)\|_{\mathcal{X}'} \le \varepsilon^{1/2}.$$

□

Corollary 4.3.11. *If $\mathcal{X}$ and φ are as in Corollary 4.3.10, then for all minimizing sequences of φ, $\{u_k\} \subset \mathcal{X}$, there exists a minimizing sequence $\{v_k\} \subset \mathcal{X}$ such that*

$$\varphi(v_k) \le \varphi(u_k),$$
$$\|u_k - v_k\|_{\mathcal{X}} \to 0, \quad k \to \infty,$$
$$\|\varphi'(v_k)\|_{\mathcal{X}'} \to 0, \quad k \to \infty.$$

Proof. If $\varphi(u_k) \to c = \inf_{\mathcal{X}} \varphi$, define $\varepsilon_k = \varphi(u_k) - c$ if this quantity is positive and $\varepsilon_k = \frac{1}{k}$ if $\varphi(u_k) = c$. Then for ε_k we take the v_k that comes from Corollary 4.3.10. □

In the spirit of the previous corollary, we find a notion of *almost minimum* for φ. The problem now is to get *almost critical points* of different types, for instance *mountain pass type* critical points in the sense of [37].

Here we will describe how the variational principle of Ekeland separates the geometric aspect of the problem of finding critical points of a functional from the analytical one. Clearly, what we will need is some type of compactness property to complete the search of critical points.

The mountain pass theorem

We start with some hypotheses and notation:

H1 Consider $\mathcal{X}$, a Banach space, $\mathcal{K} \subset \mathbb{R}^N$ a compact metric space and $\mathcal{K}_0 \subset \mathcal{K}$ a closed set.

H2 Let $\chi : \mathcal{K}_0 \to \mathcal{X}$ be a continuous function.

H3 Define

$$\mathcal{M} = \{g \in \mathcal{C}(\mathcal{K}, \mathcal{X}) : g(s) = \chi(s), s \in \mathcal{K}_0\}.$$

We recall the definition of *subdifferential* in the case of the norm in $\mathcal{X}$,

$$\partial\|x_0\| = \{p \in \mathcal{X}' : \|x_0\| - \|x\| \le p(x_0 - x), \forall x \in \mathcal{X}\},$$

where $\mathcal{X}'$ denotes the dual space of $\mathcal{X}$.

We will use the following abstract result.

Lemma 4.3.12. *Let $\mathcal{X}$ be a Banach space. Then for the norm*

$$\|\cdot\| : \mathcal{X} \to \mathbb{R}$$

we have

$$\partial\|x_0\| = \{p \in \mathcal{X}' : p(x_0) = \|x_0\|, \|p\|_{\mathcal{X}'} = 1\}.$$

Proof. Given $x_0 \in \mathcal{X}$ by the Hahn–Banach theorem there exists $p \in \mathcal{X}'$ such that $p(x_0) = \|x_0\|$ and $\|p\|_{\mathcal{X}'} = 1$. For such p we have

$$p(x - x_0) = p(x) - p(x_0) \le \|p\|_{\mathcal{X}'}\|x\| - \|x_0\| \le \|x\| - \|x_0\|,$$

and hence $\|x_0\| - \|x\| \le p(x_0 - x)$ and, therefore, $p \in \partial\|x_0\|$. Reciprocally, if $p \in \partial\|x_0\|$, we get

$$\|x_0\| - \|x\| \le p(x_0 - x) \quad \forall x \in \mathcal{X}.$$

In particular for $x = \lambda x_0$, $\lambda > 0$, we find $\|x_0\|(1 - \lambda) \le (1 - \lambda)p(x_0)$, so $(p(x_0) - \|x_0\|)(1 - \lambda) \ge 0$ for all $\lambda \in \mathbb{R}^+$. Then $p(x_0) = \|x_0\|$. Moreover $\|x_0\| - \|x\| \le p(x_0) - p(x) \le \|x_0\| - p(x)$ implies $p(x) < \|x\|$ and then $\|p\|_{\mathcal{X}'} \le 1$ since $p(x_0) = \|x_0\|$ and $\|p\|_{\mathcal{X}'} = 1$. □

Corollary 4.3.13. *Let $\mathcal{C}(\mathcal{K}, \mathbb{R})$ with the supremum norm. Denote by $\mathcal{M}(\mathcal{K})$ the Radon measures on $\mathcal{K}$. Then*

$$\partial\|f\|_\infty = \left\{\mu \in \mathcal{M}(\mathcal{K}) : \mu \ge 0, \int_{\mathcal{K}} d\mu = 1,\ \operatorname{supp}(\mu) \subset \{t : f(t) = \|f\|_\infty\}\right\}.$$

Proof. It is clear that the Radon measures verifying the conditions above are the subdifferential of the norm. More precisely, by Lemma 4.3.12 we have:

(i) If the conditions are satisfied, then $\|\mu\| = 1$ and

$$\int_{\mathcal{K}} f(x)\, d\mu = \|f\|_\infty \int_{\operatorname{supp}\mu} d\mu = \|f\|_\infty.$$

(ii) If $\|\mu\| = 1$ and $\int_{\mathcal{K}} f(x)\, d\mu = \|f\|_\infty$, then $\mu > 0$ and

$$\operatorname{supp}(\mu) \subset \{s \mid f(s) = \|f\|_\infty\}. \qquad \square$$

Theorem 4.3.14. *Consider $\mathcal{X}$, $\mathcal{K}$, $\mathcal{K}_0$ and $\mathcal{M}$ verifying* ***H1-H2-H3****. Let $u : \mathcal{X} \to \mathbb{R}$ be a functional such that:*

(1) *u is continuous;*

(2) *u is Gateaux differentiable and, moreover,*

$$u' : \mathcal{X} \longrightarrow \mathcal{X}'$$

is continuous from the strong to weak-$*$ topologies.

Consider

$$\begin{cases} \alpha = \inf_{\varphi \in \mathcal{M}} \max_{s \in \mathcal{K}} u(\varphi(s)), \\ \alpha_1 = \max_{\chi(\mathcal{K}_0)} u, \end{cases}$$

and assume that $\alpha > \alpha_1$. Then for all $\varepsilon > 0$ and for all $\varphi \in \mathcal{M}$ such that $\max_{s \in \mathcal{K}} u(\varphi(s)) \leq \alpha + \varepsilon$ there exists $v_\varepsilon \in \mathcal{X}$ such that

$$\begin{aligned} \alpha - \varepsilon \leq u(v_\varepsilon) &\leq \max_{s \in \mathcal{K}} u(\varphi(s)), \\ d(v_\varepsilon, \varphi(\mathcal{K})) &\leq \varepsilon^{1/2}, \\ \|u'(v_\varepsilon)\| &\leq \varepsilon^{1/2}. \end{aligned}$$

Proof. We can assume that $0 < \varepsilon < \alpha - \alpha_1$. Let $\varphi \in \mathcal{M}$ verifying

$$\max_{s \in \mathcal{K}} u(\varphi(s)) \leq \alpha + \varepsilon.$$

Define

$$I : \mathcal{M} \to \mathbb{R} \quad \text{by } I(c) = \max_{s \in \mathcal{K}} u(c(s))$$

for $c \in \mathcal{M}$. By hypothesis $\alpha = \inf_{c \in \mathcal{M}} I(c) > \alpha_1$. Moreover I is lower semicontinuous, because if $\|c_k - c\|_\infty \to 0$ we have

$$I(c) = u(c(s_0)) = \lim_{k \to \infty} u(c_k(s_0)) \leq \lim_{k \to \infty} I(c_k).$$

By the Ekeland variational principle applied to $I : \mathcal{M} \to \mathbb{R}$, given $\varepsilon > 0$ and $\varphi \in \mathcal{M}$ such that $I(\varphi) \leq \alpha + \varepsilon$ there exists $c_\varepsilon \in \mathcal{M}$ so that:

(i) $I(c_\varepsilon) \leq I(\varphi) \leq \alpha + \varepsilon$;

(ii) $I(c) \geq I(c_\varepsilon) - \varepsilon^{1/2}\|c - c_\varepsilon\|_\infty$ if $c \in \mathcal{M}$;

(iii) $\|c_\varepsilon - \varphi\|_\infty \leq \varepsilon^{1/2}$.

Consider $y : \mathcal{K} \to \mathcal{X}$ continuous such that $y(\mathcal{K}_0) = 0$. For $h \neq 0$, $h \in \mathbb{R}$ from (ii) we conclude that $I(c_\varepsilon + yh) - I(c_\varepsilon) \geq -\varepsilon^{1/2}|h|\|y\|_\infty$. Thus,

$$\frac{1}{|h|} I(c_\varepsilon + yh) - I(c_\varepsilon) \geq -\varepsilon^{1/2}\|y\|_\infty.$$

By the definition of I,

$$\begin{aligned} I(c_\varepsilon + yh) - I(c_\varepsilon) &= \max u(c_\varepsilon + yh) - \max u(c_\varepsilon) \\ &= \max_{s\in\mathcal{K}}\{u(c_\varepsilon(s)) + h\langle u'(c_\varepsilon(s)), y(s)\rangle + o(h)\} - \max_{s\in\mathcal{K}} u(c_\varepsilon(s)). \end{aligned}$$

To simplify the notation we write

$$f(s) = u(c_\varepsilon(s)) \quad \text{and} \quad g(s) = \langle u'(c_\varepsilon(s)), y(s)\rangle.$$

We have by hypothesis $f \in C(\mathcal{K}, \mathbb{R})$; also $g \in C(\mathcal{K}, \mathbb{R})$ because:

(a) by the Banach–Alaoglu theorem $\partial\|f\|_\infty$ is weak-$*$ compact;

(b) by hypothesis if $v_n \to v$ in $\mathcal{X}$, then

$$\langle u'(v_n), y\rangle \to \langle u'(v), y\rangle$$

for all $y \in \mathcal{X}$;

(c) if $u'(v_n) \rightharpoonup u'(v)$ is weak-$*$ in $\mathcal{X}'$ and $y_n \to y$ in $\mathcal{X}$, then

$$\langle u'(v_n), y_n\rangle \to \langle u'(v), y\rangle, \quad \text{as } n \to \infty.$$

Then the inequalities above can be written as

$$\begin{aligned} -\varepsilon^{1/2}\|y\|_\infty &\leq \lim_{h\to 0} \frac{1}{h}\{I(c_\varepsilon + hy) - I(c_\varepsilon)\} \\ &= \lim_{h\to 0} \frac{1}{h}\{\|(f + hg)\|_\infty - \|f\|_\infty\} \\ &\leq \sup\left\{\int_0^1 g d\mu : \mu \in \partial\|f\|_\infty\right\}. \end{aligned}$$

Define $F : \partial\|f\|_\infty \to \mathbb{R}$ by $F(\mu) = \int_K g\, d\mu$ and consider a weak-$*$ convergent sequence $\mu_n \rightharpoonup \mu$ in $\partial\|f\|_\infty$. Then

$$F(\mu_n) = \int_{\mathcal{K}} g\, d\mu_n \to \int_{\mathcal{K}} g\, d\mu = F(\mu),$$

so the supremum is attained. As a consequence we have

$$-e^{1/2}\|y\|_\infty \leq \max\left\{\int_{\mathcal{K}} \{u'(c_\varepsilon(s)), y(s)\}\, d\mu : \mu \in \partial\|f\|_\infty\right\}.$$

Consider $\Gamma = \{\gamma \in \mathcal{C}(\mathcal{K}, \mathcal{X}) : \gamma(s) = 0 \text{ if } s \in \mathcal{K}_0, \|\gamma\|_\infty \leq 1\}$. Then,

$$-\varepsilon^{1/2} \leq -\varepsilon^{1/2}\|\gamma\|_\infty \leq \inf_{\gamma\in\Gamma}\left\{\max\left\{\int_{\mathcal{K}} \{u'(c_\varepsilon(s)), \gamma(s)\}\, d\mu : \mu \in \partial\|f\|_\infty\right\}\right\}.$$

Take $\{\mu_n(\gamma)\} \subset \partial\|f\|_\infty$ such that for fixed γ,

$$\int_{\mathcal{K}} \{u'(c_\varepsilon(s)), \gamma(s)\}\, d\mu_n(\gamma) \longrightarrow M(\gamma),\ n \to \infty,$$

where

$$M(\gamma) = \max_{u\in\partial\|f\|_\infty} \left\{\int_{\mathcal{K}} \{u'(c_\varepsilon(s)), \gamma(s)\}\, d\mu\right\}.$$

Then by choosing a convenient weak-$*$ convergent subsequence $\mu_n(\gamma) \rightharpoonup \mu(\gamma)$ we have

$$\int_{\mathcal{K}} \{u'(c_\varepsilon(s)), \gamma(s)\}\, d\mu(\gamma) = M(\gamma).$$

Then by Lemma 4.3.13,

$$\begin{aligned}
-\varepsilon^{1/2} &\leq \inf_{\gamma\in\Gamma} M(\gamma) = \inf_{\gamma\in\Gamma} \int_{\mathcal{K}} \{u'(c_\varepsilon(s)), \gamma(s)\}\, d\mu(\gamma) \\
&\leq \max_{\mu\in\partial\|f\|_\infty} \inf_{\gamma\in\Gamma} \int_{\mathcal{K}} \{u'(c_\varepsilon(s)), \gamma(s)\}\, d\mu = \max_{\mu\in\partial\|f\|_\infty} \left(-\int_{\mathcal{K}} \|u'(c_\varepsilon(s))\|_{\mathcal{X}'}\, d\mu\right) \\
&= -\min\{\|u'(c_\varepsilon(s))\|_{\mathcal{X}'} : s \in \{t \in \mathcal{K} : u(c_\varepsilon(t)) = \|f\|_\infty\}\}.
\end{aligned}$$

Hence,

$$\min_{s\in\{t\in\mathcal{K}:u(c_\varepsilon(t))=\|f\|_\infty\}} \|u'(c_\varepsilon(s))\|_{\mathcal{X}'} \leq \varepsilon^{1/2}.$$

In other words, there exists $s_\varepsilon \in \mathcal{K}$ such that if $v_\varepsilon = c_\varepsilon(s_\varepsilon)$, then:
(i) $u(v_\varepsilon) = \max_{s\in\mathcal{K}} u(c_\varepsilon(s)) = I(c_\varepsilon) \leq \inf I + \varepsilon^{1/2}$;
(ii) $\|u'(v_\varepsilon)\| \leq \varepsilon^{1/2}$;
(iii) $u(v_\varepsilon) \geq \alpha_1$;
(iv) $d(v_\varepsilon, \varphi(\mathcal{K})) \leq \varepsilon^{1/2}$. □

The classical condition of compactness is the condition of Palais–Smale that we formalize in the following definition.

Definition 4.3.15. Let $\mathcal{X}$ be a Banach space and $U : \mathcal{X} \longrightarrow \mathbb{R}$ a Gateaux differentiable functional; U verifies the Palais–Smale condition to the level $c \in \mathbb{R}$ if and only if for all sequences $\{x_k\} \subset \mathcal{X}$ verifying

(i) $U(x_k) \to c$ as $k \to \infty$;
(ii) $U'(x_k) \to 0$ in $\mathcal{X}'$ for $k \to \infty$,

there exists a subsequence $x_{k_j} \to x$ in $\mathcal{X}$.

A sequence verifying (i) and (ii) is called a Palais–Smale sequence for U.

With these notions we can formulate the *mountain pass theorem.*

Theorem 4.3.16. *Let $\mathcal{X}$ be a Banach space and $U : \mathcal{X} \to \mathbb{R}$ a continuous functional verifying:*
(i) *U is Gateaux differentiable;*
(ii) *U' is continuous from $\mathcal{X}$ to $\mathcal{X}'$ with the weak-$*$ topology.*

Assume that there exist u_0, $u_1 \in \mathcal{X}$ and a ball B_r, centered in u_0 and with radius r such that if $u_1 \in \mathcal{X} - B_r$ and $|x - u_0| = r$, then

$$\inf_{\partial B_r} U(x) > \max\{U(u_0), U(u_1)\}.$$

Consider

$$\Gamma = \{g \in \mathcal{C}([0,1], \mathcal{X}) : g(0) = u_0, g(1) = u_1\}$$

and

$$c = \inf_{f\in\Gamma} \sup_{s\in[0,1]} U(f(s)).$$

If U satisfies the Palais–Smale condition at level c, then c is a critical value for U and $c > \max\{U(u_0), U(u_1)\}$.

Proof. Take $\mathcal{K} = [0,1]$, $\mathcal{K}_0 = \{0,1\}$, $\chi(0) = u_0$, $\chi(1) = u_1$ and $\mathcal{M} = \Gamma$; since $c_0 = \inf_{\partial B_r} U(x) > \max\{U(u_0), U(u_1)\} = c_1$ and because all paths $g \in \Gamma$ have nonempty intersections with ∂B_r, Theorem 4.3.14 implies that for a given sequence $\{\varepsilon_k\}$ with $\varepsilon_k \to 0$ we can find a sequence $\{x_k\} \subset \mathcal{X}$ such that
(a) $U(x_k) \to c$;
(b) $U'(x_k) \to 0$.

By the Palais–Smale condition for level c, there exists a subsequence such that

$$c = \lim_{k\to\infty} U(x_k) = U(x),$$
$$0 = \lim_{k\to\infty} U'(x_k) = U'(x).$$

This says that x is a critical point at level c. □

4.4 Results on existence

Consider the equation

$$-\Delta u = \frac{\lambda u}{|x|^2} + u^p + f(x), \quad u \geq 0 \text{ in } \Omega \subset \mathbb{R}^N. \tag{4.4.1}$$

We will assume $f \in L^1_{\mathrm{loc}}(\Omega)$ and $f \geq 0$ in Ω.

Let us consider the notion of solution that will be used here.

Definition 4.4.1. We say that $u \in L^1_{\mathrm{loc}}(\Omega)$, $u \geq 0$, is a very weak supersolution (subsolution) to equation (4.4.1) if $\frac{u}{|x|^2} \in L^1_{\mathrm{loc}}(\Omega)$, $u \in L^p_{\mathrm{loc}}(\Omega)$ and $\forall \phi \in C_0^\infty(\Omega)$ such that $\phi \geq 0$, we have

$$\int_\Omega (-\Delta\phi) u \, dx \geq (\leq) \int_\Omega \Big(u^p + \lambda \frac{u}{|x|^2} + f\Big)\phi \, dx.$$

If u is a very weak super- and subsolution, then we say that u is a very weak solution.

The idea in the definition is to give sense to the equation in a distributional sense. According to (4.2.6) the results on existence will be analyzed in separated cases.

CASE 1. $0 \leq \lambda < \Lambda_{N,2}$, $p < \frac{N+2}{N-2}$.

In this case we can use the variational framework. Consider the energy functional

$$J_\lambda(u) = \frac{1}{2}\int_\Omega |\nabla u|^2 - \frac{\lambda}{2}\int_\Omega \frac{u^2}{|x|^2} - \frac{1}{p+1}\int_\Omega u^{p+1},$$

which is well defined in the Sobolev space $W_0^{1,2}(\Omega)$. Set, for $u \in W_0^{1,2}(\Omega)$, $g(t) = J_\lambda(tu)$.

Note that by using the Hardy, Hölder and Sobolev inequalities

$$g(t) \geq t^2 \frac{1}{2}\Big(1 - \frac{\lambda}{\Lambda_{N,2}}\Big)\int_\Omega |\nabla u|^2 - t^{p+1}\frac{1}{(p+1)S}|\Omega|^{1-\frac{p+1}{2_*}}\Big(\int_\Omega |\nabla u|^2\Big)^{\frac{p+1}{2}}.$$

If $0 < p < 1$, for t close to 0 we have $g(t) = J_\lambda(tu) < 0$ and for $t \to \infty$, $g(t) = J_\lambda(tu) \to \infty$. Moreover, since $p + 1 < 2$, using the Young inequality we find

$$J_\lambda(u) \geq \frac{1}{4}\Big(1 - \frac{\lambda}{\Lambda_{N,2}}\Big)\int_\Omega |\nabla u|^2 - C(\Omega, p, N),$$

and therefore

$$0 > \alpha = \inf_{u \in H_0^1(\Omega), u \neq 0} J_\lambda(u).$$

By, for instance, Corollary 4.3.11 we find a bounded Palais–Smale sequence $\{u_k\}_{k\in\mathbb{N}}$, i. e.,

$$J_\lambda(u_k) \to \alpha \quad \text{as } k \to \infty \text{ and } J_\lambda'(u_k) \to 0 \text{ in } W^{-1,2}(\Omega).$$

Therefore there exists a relabeled subsequence $\{u_k\}_{k\in\mathbb{N}}$ such that:

1. $u_k \rightharpoonup u$ weakly in $W_0^{1,2}(\Omega)$;
2. by the Rellich theorem, $u_k \to u$ strongly in $L^p(\Omega)$ for all $1 \le q < \frac{2N}{N-2}$;
3. $u_k \to u$ almost everywhere.

Next we check the Palais–Smale condition. By hypothesis we have

$$J'_\lambda(u_k) = -\Delta u_k - \lambda \frac{u_k}{|x|^2} - u_k^p - y_k,$$

where $y_k \to 0$ in $W^{-1,2}(\Omega)$. Then we test with $(u - u_k)$ and obtain

$$\int_\Omega \langle \nabla u_k, \nabla(u_k - u)\rangle \, dx = \int_\Omega \lambda \frac{u_k}{|x|^2}(u_k - u) \int_\Omega u_k(u_k - u)\, dx + \langle y_k, (u_k - u)\rangle.$$

Adding to both members of the identity the term

$$-\int_\Omega \langle \nabla u, \nabla(u_k - u)\rangle$$

and observing that this tends to 0 as $k \to \infty$, we get

$$\int_\Omega |\nabla(u_k - u)|^2 \, dx = \int_\Omega \lambda \frac{u_k}{|x|^2}(u_k - u) \int_\Omega u_k(u_k - u)\, dx + \langle y_k, (u_k - u)\rangle - \int_\Omega \langle \nabla u, \nabla(u_k - u)\rangle.$$

Hence

$$\int_\Omega |\nabla(u_k - u)|^2 \, dx = o(1), k \to \infty,$$

and u is a solution to the semilinear problem and we realize a minimum.

In this case there are arguments of minimization that arrive at the same result. Note that any minimizing sequence, $\{u_n\}$, is bounded in $W_0^{1,2}(\Omega)$. Hence, by the Rellich theorem, there exists a subsequence $\{u_{n_k}\}$ weakly convergent to some $u \in W_0^{1,2}(\Omega)$. Moreover $J_\lambda(u)$ is weakly lower semicontinuous (up to a subsequence) and then $\alpha = J_\lambda(u)$ and $J'_\lambda(u) = 0$. In particular u is a solution to the semilinear problem.

If $1 < p < \frac{N+2}{N-2}$, J_λ has the geometry of *mountain pass*. Indeed

$$J_\lambda(u) \ge \frac{1}{2}\left(1 - \frac{\lambda}{\Lambda_{N,2}}\right)\int_\Omega |\nabla u|^2 - \frac{1}{(p+1)S}|\Omega|^{1-\frac{p+1}{2*}}\left(\int_\Omega |\nabla u|^2\right)^{\frac{p+1}{2}}.$$

Hence $J_\lambda(u) \ge \beta > 0$. Moreover taking $v \in W_0^{1,2}(\Omega)$ with $\|\nabla v\|_2$ large enough, $J_\lambda(v) < 0$. From Theorem 4.3.16, there exists a Palais–Smale sequence for the mini-max level

$$c = \inf_{f \in \Gamma} \sup_{s \in [0,1]} J_\lambda(f(s)),$$

where

$$\Gamma = \{g \in \mathcal{C}([0,1], W_0^{1,2}(\Omega)) : g(0) = 0, g(1) = v\}.$$

Since $1 < p < \frac{N+2}{N-2}$ such Palais–Smale sequence admits a convergent subsequence and we conclude with the Ambrosetti–Rabinowitz theorem. As we can see, it is sufficient to reproduce the case $0 < p < 1$ to get the Palais–Smale condition.

Remark 4.4.2. In general if $\frac{N+2}{N-2} \leq p$ by the corresponding Pohozaev identity we have in general no solution different from zero on star-shaped domains. Indeed

$$\frac{N}{p+1} \int_{\Omega} u^{p+1}\, dx - \frac{(N-2)}{2} \int_{\Omega} u^{p+1}\, dx = \frac{1}{2} \int_{\partial\Omega} \frac{\partial u}{\partial \nu}^2 \langle x, \nu \rangle d\sigma.$$

However, we will be able to prove existence for suitable small data f.

Case 2. $0 \leq \lambda = \Lambda_{N,2}, p < p^+(\Lambda_{N,2}) = \frac{N+2}{N-2}$. By using the improved Hardy inequalities (see Theorem 2.5.2) we observed in Section 2.5.1 that

$$\|u\|_H^2 = \int \left(|\nabla u|^2 - \lambda \frac{u^2}{|x|^2} \right) dx$$

is a Euclidean norm on $\mathcal{C}_0^\infty(\Omega)$. We have defined in Section 2.5.1 the Hilbert space $\mathcal{H}(\Omega)$ by completion of $\mathcal{C}_0^\infty(\Omega)$ with respect to such norm. Also by the improved Hardy inequality we have the compact embedding

$$\mathcal{H}(\Omega) \hookrightarrow L^q(\Omega), \quad \text{for all } q < 2^*.$$

Therefore, as in the first case above, we could find a nontrivial solution by the mountain pass theorem.

Case 3. $0 \leq \lambda < \Lambda_{N,2}, p < p^+(\lambda)$.

We will study precisely the following problem:

$$\begin{cases} -\Delta u = \frac{\lambda u}{|x|^2} + u^p + tf & \text{in } \Omega, \\ u \geq 0 & \text{in } \Omega, \\ u = 0 & \text{on } \partial\Omega, \end{cases} \tag{4.4.2}$$

where $\Omega \subset \mathbb{R}^N$, $N \geq 3$, is such that $0 \in \Omega$, $0 < \lambda < \Lambda_{N,2}$, $t > 0$ and $f \gneqq 0$.

Theorem 4.4.3. *Assume the hypotheses above and $0 \leq p < p_+(\lambda)$. Then if $t > 0$ is small enough, problem (4.4.2) has a distributional solution.*

Proof. Consider $B_R(0) \supset \Omega$. Note that since $p < p_+(\lambda)$, we have $\alpha_-(\lambda)p < \alpha_-(\lambda) + 2$. Therefore we can choose $\beta \in (\alpha_(\lambda)-, \alpha_+(\lambda))$ close to $\alpha_-(\lambda)$ such that $\beta p < \beta + 2$. Fixing β as above, we prove that there exists $A > 0$ and $t > 0$ such that

$$w = A|x|^{-\beta} \in W^{1,2}(\Omega) \quad \text{is a supersolution to (4.4.2).}$$

Indeed,

$$-\Delta w - \lambda w|x|^2 = -AP(-\beta)|x|^{-(\beta+2)}, \quad \text{where } P(s) = s(s-1) + (N-1)s + \lambda.$$

Since $\beta \in (\alpha_-, \alpha_+)$, $P(-\beta) < 0$. We need A and t such that

$$-\frac{1}{2}AP(-\beta)|x|^{-(\beta+2)} \geq A^p|x|^{-p\beta} \quad \text{and} \quad -\frac{1}{2}AP(-\beta)|x|^{-(\beta+2)} \geq tf.$$

Since $p\beta - \beta - 2 < 0$ the first inequality is verified if

$$A \leq \left(-\frac{1}{2}P(-\beta)R^{p\beta-\beta-2}\right)^{\frac{1}{p-1}}.$$

Therefore it is sufficient to pick t such that

$$-\frac{1}{2}AP(-\beta)R^{-(\beta+2)} \geq t\|f\|_\infty.$$

Then we have

$$\begin{cases} -\Delta w - \frac{\lambda w}{|x|^2} \geq w^p + tf & \text{in } \Omega, \\ w \geq 0 & \text{on } \partial\Omega. \end{cases}$$

Now we use a monotone iteration to construct the solution. Consider u_0, the solution to

$$\begin{cases} -\Delta u_0 = tf & \text{in } \Omega, \\ u_0 = 0 & \text{on } \partial\Omega \end{cases}$$

and define for $k \geq 1$

$$\begin{cases} -\Delta u_k = \frac{\lambda}{|x|^2}u_{k-1} + u_{k-1}^p + tf & \text{in } \Omega, \\ u_k = 0 & \text{on } \partial\Omega. \end{cases}$$

Since $w \geq u_0 > 0$, by recurrence we find

$$0 \leq u_{k-1} \leq u_k \leq w, \quad \text{for all } k \in \mathbb{N}.$$

Then there exists $u(x) = \lim_{k\to\infty} u_k(x)$ and the convergence is in $L^p(\Omega)$ by dominated convergence. Therefore u is a distributional solution. □

Remark 4.4.4. In fact the minimal solution found above is in the sense of the energy and by the elliptic regularity classical solution in $\Omega \setminus \{0\}$. See for details the paper by L. Dupaigne [145].

Theorem 4.4.5. *Under the same hypotheses as Theorem* 4.4.3 *the set*

$$I = \{t > 0 \mid \textit{problem } (4.4.2) \textit{ has a solution}\}$$

is a bounded interval.

Proof. Notice that I is an interval just because if $\lambda_1 < \lambda_2 \in I$ and u_{λ_2} is a solution to (4.4.2), then u_{λ_2} is a supersolution to problem (4.4.2) for λ_1, and hence it is sufficient to proceed as in Theorem 4.4.3.

Assume that $\lambda \in I$ and u_λ is a solution to (4.4.2). Consider $\phi_1 > 0$ as the eigenfunction corresponding to λ_1, the principal eigenvalue of the Laplacian in Ω. Then testing the equation for u_λ with ϕ_1 we find, using the Young inequality,

$$\int_\Omega \frac{\lambda}{|x|^2} u_\lambda \phi_1 + \int_\Omega u_\lambda^p \phi_1 + \lambda \int_\Omega f\phi_1 = \lambda_1 \int_\Omega u_\lambda \phi_1 \le \frac{1}{2} \int_\Omega u_\lambda^p \phi_1 + C \int_\Omega \phi_1.$$

Therefore,

$$\int_\Omega \frac{\lambda}{|x|^2} u_\lambda \phi_1 + \frac{1}{2} \int_\Omega u_\lambda^p \phi_1 + \lambda \int_\Omega f\phi_1 \le C \int_\Omega \phi_1,$$

and then necessarily I is bounded. □

4.5 Results on nonexistence

We will prove the following nonexistence result.

Theorem 4.5.1. *Let $0 < \lambda \le \Lambda_{N,2}$ and $p \ge p^+(\lambda)$. Assume $u \in L^p(B_R(0) \setminus \{0\})$, $u \ge 0$, satisfies*

$$-\Delta u - \frac{\lambda}{|x|^2} u \ge u^p, \quad \text{in } \mathcal{D}'(B_R(0) \setminus \{0\}).$$

Then $u \equiv 0$.

Proof. Assume by contradiction that there exists a nontrivial solution $u \ge 0$. Then by Theorem 3.4.1 in particular we have

$$\int_\Omega \frac{u^p}{|x|^{\alpha_-(\lambda)}} < \infty.$$

Also by Lemma 4.3.8 in a ball $B_r(0)$ and a suitable constant, m, we have,

$$u \ge m|x|^{-\alpha_-(\lambda)}.$$

By using the Hölder inequality we obtain

$$\int_\Omega \frac{u}{|x|^{\alpha_-(\lambda)+2}} \le \left(\int_\Omega u^p |x|^{-\alpha_-(\lambda)}\right)^{\frac{1}{p}} \left(\int_\Omega |x|^{-\alpha_-(\lambda)-2\frac{p}{p-1}}\right)^{\frac{p-1}{p}} < \infty,$$

because as $p \geq p^+(\lambda)$ and $\lambda < \Lambda_{N,2}$, $-\alpha_-(\lambda) - 2\frac{p}{p-1} > -N$. Therefore,

$$\int_\Omega \frac{u}{|x|^{\alpha_-(\lambda)-2}} < \infty.$$

If we assume that $p > p^+(\lambda)$, by using the Picone inequality in $B_r(0)$, we find that for all $\phi \in C_0^\infty(B_r(0))$

$$\int_{B_r(0)} |\nabla\phi|^2\,dx \geq \int_{B_r(0)} \left(\frac{\Delta u}{u}\right)\phi^2\,dx \geq \int_{B_r(0)} \frac{\phi^2}{|x|^{\alpha_-(\lambda)(p-1)}}\,dx,$$

and since

$$\alpha_-(\lambda)(p-1) > \alpha_-(\lambda)(p^+(\lambda)-1) = 2,$$

we reach a contradiction with the Hardy inequality.

Note that the previous argument does not work in general when $p = p^+(\lambda)$. To overcome this situation, we consider

$$w(x) = A|x|^{-\alpha_-(\lambda)}\log\left(\frac{1}{|x|}\right)$$

with $A > 0$ and we try to find A in such a way that w becomes a subsolution to the equation.

By a direct computation we find that

$$-\Delta w - \lambda\frac{w}{|x|^2} = A(2\sqrt{\Lambda_{N,2}-\lambda})|x|^{-\alpha_-(\lambda)-2}.$$

Moreover,

$$-\Delta u - \frac{\lambda}{|x|^2}u \geq m|x|^{-\alpha_-(\lambda)-2}.$$

Taking

$$A = m(2\sqrt{\Lambda_{N,2}-\lambda}) + \lambda\log\left(\frac{1}{r}\right)^{-1}, \quad C = Ar^{-\alpha_-(\lambda)}\log\left(\frac{1}{r}\right)$$

and defining

$$z = u + C - w,$$

we have $z \in L^1(B_r(0), |x|^{-\alpha_-(\lambda)-2})$. Moreover, by a direct calculation again,

$$-\Delta z - \lambda\frac{z}{|x|^2} \geq 0 \text{ in } B_r(0), \quad z = C - w \geq 0 \text{ on } \partial B_r,$$

and, by the strong maximum principle, $z > 0$ in $B_r(0)$. Choosing $0 < A_1 < A$ and $r_1 < r$ small enough we prove that

$$u(x) \geq A_1 |x|^{-\alpha_-(\lambda)} \log\left(\frac{1}{|x|}\right) \quad \text{in } B_{r_1}(0).$$

Again by the Picone inequality in $B_{r_1}(0)$, we find that

$$\int_{B_{r_1}(0)} |\nabla \phi|^2 \, dx \geq \int_{B_{r_1}(0)} \left(\frac{\Delta u}{u}\right) \phi^2 \, dx \geq \int_{B_{r_1}(0)} \frac{1}{|x|^2} \log\left(\frac{1}{|x|}\right)^{p^+(\lambda)-1} \phi^2 \, dx$$

for all $\phi \in C_0^\infty(B_{r_1}(0))$, which is a contradiction with the Hardy–Leray inequality. □

4.5.1 Complete blow-up

As a byproduct of the previous existence proof and the nonexistence result we are able to prove the complete blow-up; that is, the solutions u_n of any truncated problem verify that $u_n(x) \to \infty$ as $n \to \infty$ for all $x \in \Omega$.

Theorem 4.5.2. *Assume that $0 < \lambda \leq \Lambda_{N,2}$ holds, $p > 1$ and $t > 0$. If problem (4.4.2) has no weak solution, then there exists complete blow-up.*

Proof. We argue by contradiction. Consider u_n as a solution to the truncated problem,

$$\begin{cases} -\Delta u_n = g_n(u_n) + \lambda a_n(x) u_n + cf & \text{in } \Omega, \\ u_n > 0 & \text{in } \Omega, \\ u_n = 0 & \text{on } \partial\Omega, \end{cases} \tag{4.5.1}$$

where $a_n(x) = T_n(\frac{1}{|x|^2})$ and $g_n(s) = T_n(u^p)$. Assume by contradiction that

$$\int_\Omega g_n(u_n)\delta(x) \, dx + \int_\Omega a_n(x) u_n \delta(x) \, dx \leq C < \infty, \quad \text{for all } n \in \mathbb{N}.$$

Consider ψ_1 as a solution to the problem

$$\begin{cases} -\Delta \psi_1 = 1 & \text{in } \Omega, \\ \psi_1 = 0 & \text{on } \partial\Omega, \end{cases}$$

and use it as a test function in problem (4.5.1). We get

$$\int_\Omega u_n(-\Delta \psi_1) - \int_\Omega a_n u_n \psi_1 = \int_\Omega g_n(u_n)\psi_1 + t \int_\Omega f\psi_1.$$

Hence

$$\int\limits_{\Omega} u_n \le C \quad \text{and, since } \{u_n\}_{n\in\mathbb{N}} \text{ is nondecreasing, } u_n \nearrow u \text{ in } L^1(\Omega) \text{ for some } u$$

by monotone convergence. Since a_n and g_n are nondecreasing, again by monotony, we could pass to the limit in the truncated problems, getting a solution to (4.4.2), which is a contradiction. □

4.6 Further results and comments

- **The critical case in $\mathbb{R}^N$, $N \ge 3$.** It is well known that the minimizers to the variational problem

$$S = \inf_{u\in\mathcal{D}^{1,2}(\mathbb{R}^N)} \frac{\int_{\mathbb{R}^N} |\nabla u|^2\, dx}{(\int_{\mathbb{R}^N} |u|^{\frac{2N}{N-2}}\, dx)^{\frac{N-2}{N}}}, \tag{4.6.1}$$

where $\mathcal{D}^{1,2}(\mathbb{R}^N)$ is the completion of $C_0^\infty(\mathbb{R}^N)$ with respect to the norm

$$\|u\|_{\mathcal{D}^{1,2}(\mathbb{R}^N)} = \Big(\int\limits_{\mathbb{R}^N} |\nabla u|^2\, dx\Big)^{\frac{1}{2}},$$

are given by the function

$$\varphi(x) = \frac{N(N-2)^{\frac{N-2}{4}}}{(1+|x|^2)^{\frac{N-2}{2}}}$$

and all its dilations and translations, i. e.,

$$\varphi_{\lambda,x_0}(x) = \lambda^{\frac{2-N}{2}} \varphi\Big(\frac{(x-x_0)}{\lambda}\Big).$$

The Euler–Lagrange equation corresponding to the minimization problem (4.6.1) is

$$-\Delta u = S u^{2^*-1}, \quad x \in \mathbb{R}^N \quad 2^* = \frac{2N}{N-2}.$$

Susanna Terracini in [304] studied, among others questions, the problem

$$-\Delta u - \lambda \frac{u}{|x|^2} = S_\lambda u^{2*-1}, \quad 0 < \lambda < \Lambda_{N,2}, \quad x \in \mathbb{R}^N \setminus \{0\}, \tag{4.6.2}$$

for which the positive solutions are

$$\eta_\lambda(x) = \frac{(N(N-2)\rho^2(\lambda))^{\frac{N-2}{4}}}{|x|^{(1-\rho(\lambda))}(1+|x|^{2\rho(\lambda)})^{\frac{N-2}{2}}} \quad \text{where } \rho(\lambda) = \Big(1 - \frac{4\lambda}{(N-2)^2}\Big)^{\frac{1}{2}}$$

and

$$S_\lambda = S\left(1 - \frac{4\lambda}{(N-2)^2}\right)^{\frac{N-1}{N}},$$

with S defined in (4.6.1).

Note that in a bounded star-shaped domain $\Omega \subset \mathbb{R}^N$, equation (4.6.2) has no positive solution. To see this, it is sufficient to use a Pohozaev type argument. See [261] and the extensions in [268].

In [10] the reader can find results for the problem

$$\begin{cases} -\Delta_p u = \frac{\lambda}{|x|^p}|u|^{p-1}u + |u|^{p^*-1}u, & \text{in } \mathbb{R}^N, \\ u(x) > 0, & u \in \mathcal{D}^{1,p}(\mathbb{R}^N), \end{cases} \tag{4.6.3}$$

where $N \geq 3, \lambda > 0, 1 < p < N$ and $p^* = Np/(N-p)$ is the critical Sobolev exponent. Here $\mathcal{D}^{1,p}(\mathbb{R}^N)$ denotes the space obtained as the completion of the class of smooth functions with compact support with respect to the norm

$$\|u\|_{\mathcal{D}^{1,p}(\mathbb{R}^N)} = \left(\int_{\mathbb{R}^N} |\nabla u|^p \, dx\right)^{1/p}.$$

The potential $1/|x|^p$ is related to the Hardy–Sobolev inequality in Lemma 2.4.1. Equation (4.6.3) in bounded domains is studied with related nonlinearities, among others, in the articles [14], [87], [179], [184] and in the references quoted in these papers. We may add that there are still many open questions in this framework.

- **Other related problems.** One can find similar problems involving the Caffarelli–Kohn–Nirenberg inequalities in [108]; see also Section 6.2.3 of this monograph. More results along these directions can be found in the papers [15], [16], [7] and [5]. The papers [116] and [154] develop the theory in the case of $\mathbb{R}^N$. The results here are focalized in the attainability of the optimal constant and the symmetry of the minimizers of the Caffarelli–Kohn–Nirenberg inequalities.

5 The Hardy–Leray potential in semilinear heat equations

5.1 Introduction

In this chapter we will study the following problem:

$$\begin{cases} u_t - \Delta u = \lambda \frac{u}{|x|^2} + u^p + f & \text{in } \Omega_T \equiv \Omega \times (0,T), \\ u(x,t) > 0 & \text{in } \Omega_T, \\ u(x,t) = 0 & \text{on } \partial\Omega \times (0,T), \\ u(x,0) = u_0(x) & \text{if } x \in \Omega, \end{cases} \tag{5.1.1}$$

where either Ω is an open bounded domain in $\mathbb{R}^N$ such that $0 \in \Omega$, or $\Omega = \mathbb{R}^N$, $N \geq 3$, $p > 1$ and $\lambda > 0$. We assume that f and u_0 are nonnegative measurable functions under some summability hypotheses that will be presented.

There exists a great amount of literature dealing with the case $\lambda = 0$. See for instance the papers [171], [227], [316], the book [269] and the references therein. In that case, existence and uniqueness of local solutions in time are well known for all $p > 1$, at least for regular initial data. Moreover, this solution is bounded for small time. On the other hand, Fujita found in [171], in some way, a surprising result, namely, that for $1 < p \leq 1 + \frac{2}{N}$, any positive solution to the semilinear heat equation ((5.1.1) with $\lambda = 0$) blows up in a finite time in the $L^\infty(\mathbb{R}^N)$-norm.

As in the elliptic setting studied in Chapter 4, the case $\lambda > 0$ is quite different.

1. In the pioneering work by P. Baras and J. Goldstein, [41], the linear problem is studied. Since in Chapter 13 we study the theory for the fractional heat equation and we introduce a slightly different proof for the linear case that it is applicable to the heat equation, we do not emphasize this point in this chapter.
2. It is not difficult to show that any positive solution of (5.1.1) is unbounded (see Section 5.2.1).
3. Some restrictions on p are needed to insure the existence of solution even in the weakest sense. More precisely, if $p_+(\lambda) = 1 + \frac{2}{\alpha_-(\lambda)}$, the critical exponent of the associated elliptic problem, we will prove that for $p \geq p_+(\lambda)$ there is no distributional solution. Hereafter, we will say that problem (5.1.1) *blows up completely and instantaneously* if the solutions to the truncated problems (with the weight $\frac{\lambda}{|x|^2+\frac{1}{n}}$ instead of $\frac{\lambda}{|x|^2}$) converge to infinity for every $(x,t) \in \Omega_T$ as $n \to \infty$. This is the case if $p \geq p_+(\lambda)$.
4. There exists a Fujita type exponent, $p_+(\lambda) > F(\lambda) = 1 + \frac{2}{N-\alpha_-(\lambda)} > F(0)$. That is, for all $1 < p < F(\lambda)$ and for all nonnegative initial data, the solution blows up in a finite time, and $F(\lambda)$ is optimal with this property. That is, the Hardy potential produces a shift in the Fujita exponent with respect to that corresponding to $\lambda = 0$. Note that $\lim_{\lambda \to 0^+} F(\lambda) = F(0)$.

https://doi.org/10.1515/9783110606270-005

The objective of this chapter is to explain the influence of the Hardy term on the existence or nonexistence of solutions by proving the results explained above. The results in this chapter were obtained in the paper [21].

5.2 Preliminaries and tools

This section contains some preliminaries and certain tools that will be used in the next sections. We start with the different concepts of solution that will be used in the next sections, namely, *very weak solution* and *solution obtained as limit of approximations.*

Definition 5.2.1. We say that $u \in \mathcal{C}([0,T];L^1_{\mathrm{loc}}(\Omega))$ is a very weak supersolution (subsolution) to the semilinear equation in problem (5.1.1) if $\frac{u}{|x|^2} \in L^1_{\mathrm{loc}}(\Omega_T)$, $u^p \in L^1_{\mathrm{loc}}(\Omega_T)$, $f \in L^1_{\mathrm{loc}}(\Omega_T)$ and, for all $\phi \in \mathcal{C}_0^\infty(\Omega_T)$ such that $\phi \geq 0$, we have

$$\int_0^T\int_\Omega (-\phi_t - \Delta\phi)u\,dx\,dt \geq \text{(respectively } \leq\text{)} \int_0^T\int_\Omega \left(\lambda\frac{u}{|x|^2} + u^p + f\right)\phi\,dx\,dt. \tag{5.2.1}$$

If u is a very weak super- and subsolution, then we say that u is a very weak solution. In particular if u is a very weak solution to (5.1.1), then $u \in \mathcal{C}([0,T];L^1_{\mathrm{loc}}(\Omega)) \cap L^p((0,T);L^p_{\mathrm{loc}}(\Omega))$.

The previous definition is *local in nature,* among other things, because there is no reference to the boundary data and because only the regularity is needed to give distributional sense to the equation. We will use the general framework given in Definition 5.2.1 to prove nonexistence with *local arguments.*

To prove existence with L^1-data, we will consider *solutions obtained as limit of approximations* (see for instance [129] and [265]). For the heat operator the *solutions obtained as limit of approximations* coincide with the so-called *renormalized solutions.* See [61] for the details. For the reader's convenience we reformulate the existence and regularity theory with positive L^1-data for the heat equation. Recall the usual truncation operator,

$$T_n(s) = \begin{cases} s & |s| \leq n, \\ n\,\mathrm{sign}(s) & |s| \geq n. \end{cases}$$

Consider $f \in L^1((0,T)\times\Omega)$, $f \geq 0$, and let $\{f_n\}$ be a sequence of bounded functions such that:

(i) $f_n(x,t) \leq f(x,t)$, $(x,t) \in (0,T)\times\Omega$, and
(ii) $f_n \to f$ in $L^1((0,T)\times\Omega)$.

In a similar way for $u_0 \in L^1(\Omega)$, $u_0 \geq 0$, let $u_{0,n}$ be a sequence of bounded functions such that:

(i) $u_{0,n}(x) \le u_0(x)$, $x \in \Omega$, and
(ii) $u_{0,n} \to u_0$ in $L^1(\Omega)$.

Let u_n be the classical solution to the heat problem

$$\begin{cases} u_{nt} - \Delta u_n = f_n & \text{in } \Omega_T \equiv \Omega \times (0,T), \\ u_n(x,t) > 0 & \text{in } \Omega_T, \\ u_n(x,t) = 0 & \text{on } \partial\Omega \times (0,T), \\ u_n(x,0) = u_{0,n}(x) & \text{if } x \in \Omega. \end{cases} \tag{5.2.2}$$

Then
(a) $u_n \in L^2((0,T); W_0^2(\Omega)) \cap C([0,T]; L^2(\Omega))$ and $u_{nt} \in L^2((0,T); W^{-1,2}(\Omega))$;
(b) u_n verifies

$$\int_0^T \langle u_{nt}, \phi\rangle\, dt + \int_0^T\int_\Omega \langle \nabla u_n, \nabla\phi\rangle\, dx\, dt = \int_0^T\int_\Omega f_n\phi\, dx\, dt, \quad \forall \phi \in L^2((0,T); W_0^2(\Omega));$$

(c) $\{u_n\}$ is bounded in $L^q((0,T); W_0^{1,q}(\Omega))$ for all $1 \le q < \frac{N+2}{N+1}$ and for all $k > 0$, $\{T_k(u_n)\}$ is bounded in $L^2((0,T); W_0^{1,2}(\Omega))$.

As a consequence, up to a subsequence, there exists $u \in L^q((0,T); W_0^{1,q}(\Omega))$ for all $1 \le q < \frac{N+2}{N+1}$ such that

$$u_n \rightharpoonup u \quad \text{weakly in } L^q((0,T); W_0^{1,q}(\Omega)),\ \forall q \in [1, \frac{N+2}{N+1}).$$

The reader can find the details in [42] and in [66] in a more general case. With these previous properties one can prove that

$$\nabla u_n \to \nabla u, \quad \text{a.e. in } \Omega.$$

Therefore by Fatou's and Vitali's theorems, it follows that

$$u_n \to u \quad \text{strongly in } L^q((0,T); W_0^{1,q}(\Omega)),\ \forall q \in [1, \frac{N+2}{N+1})$$

and

$$T_k(u_n) \to T_k(u) \quad \text{strongly in } L^2((0,T); W_0^{1,2}(\Omega)),\ \forall k > 0.$$

Hence u is a solution to the problem

$$\begin{cases} u_t - \Delta u = f & \text{in } \Omega_T \equiv \Omega \times (0,T), \\ u(x,t) > 0 & \text{in } \Omega_T, \\ u(x,t) = 0 & \text{on } \partial\Omega \times (0,T), \\ u(x,0) = u_0(x) & \text{if } x \in \Omega. \end{cases} \tag{5.2.3}$$

More details can be found, for instance, in [265] and [4], in a more general setting.

By the strong maximum principle for the truncated problems (5.2.2), we find that for $f \geq 0$, $u_0 \geq 0$, then $u > 0$ in $\Omega \times (0, T)$. Also, it is easy to check a strong comparison principle for solutions obtained as limit of approximations. In [4] it is proved in particular that the solutions obtained as limit of approximations for the heat equations coincide with the distributional solutions and that problem (5.2.3) has a unique distributional solution. Therefore, in the process of constructing a solution, u, as limit of approximations, we see that all subsequences of $\{u_n\}$ converge to u.

5.2.1 Local behavior of supersolutions of the linear equation

The behavior of any positive supersolution in a neighborhood of the origin is obtained looking for a self-invariant solution to the homogenous linear equation

$$v_t - \Delta v - \lambda \frac{v}{|x|^2} = 0 \quad \text{in } \mathbb{R}^N. \tag{5.2.4}$$

More precisely, we look for a solution to (5.2.4) of the form $v(r,t) = t^{-\mu}\phi(\frac{r}{t^\nu})$. Therefore, by setting $s = \frac{r}{t^\nu}$ and taking $\nu = \frac{1}{2}$, we find the equation

$$\begin{aligned} 0 &= v_t - v'' - \left(\frac{N-1}{r}\right)v' - \lambda \frac{v}{r^2} \\ &\equiv (-t^{-(\mu+1)})\left(\phi''(s) + \left(\frac{(N-1)}{s} + \frac{s}{2}\right)\phi'(s) + \left(\mu + \frac{\lambda}{s^2}\right)\phi(s)\right). \end{aligned} \tag{5.2.5}$$

If we set $\phi(s) = s^{-\alpha}e^{-\beta s^\gamma}$, then

$$\begin{aligned} \phi'(s) &= \left(-\frac{\alpha}{s} - \gamma\beta s^{\gamma-1}\right)\phi(s), \\ \phi''(s) &= \left(\frac{\alpha}{s^2} - \beta\gamma(\gamma-1)s^{\gamma-2} + \left(\frac{\alpha}{s} + \beta\gamma s^{\gamma-1}\right)^2\right)\phi(s). \end{aligned} \tag{5.2.6}$$

From (5.2.5) and (5.2.6), it is sufficient to choose $\alpha = \alpha_-(\lambda)$ given in (3.1.3), $\beta = \frac{1}{4}$, $\gamma = 2$ and $\mu = \frac{N-\alpha_-(\lambda)}{2}$ to verify

$$v_t - v'' - \left(\frac{N-1}{r}\right)v' - \lambda \frac{v}{r^2} = 0.$$

We find in this way that

$$v(x,t) = t^{-\frac{N}{2}+\alpha_-(\lambda)}r^{-\alpha_-(\lambda)}e^{-\frac{|x|^2}{4t}}, \tag{5.2.7}$$

and then it is easy to check that

$$\int_{\mathbb{R}^N} |x|^{-\alpha_-(\lambda)}v(x,t)\,dx = C.$$

Note that, for $\lambda = 0$, we recover the fundamental kernel of the heat equation.

The unboundedness of any very weak supersolution is obtained in the following result, which also gives a quantitative information; see for instance [6].

A perturbation of the linear problem

As an example of the borderline behavior of the Hardy potential, we prove the existence of a solution $\psi \in L^2((0,T);\mathcal{D}^{1,2}(\mathbb{R}^N))$ to the problem

$$\begin{cases} \psi_t - \Delta\psi - \lambda\frac{\psi}{|x|^2} = aB(x,t)\psi & \text{in } \mathbb{R}^N,\ t \in (0,T), \\ \psi(x,0) = \psi_0(x), \end{cases} \tag{5.2.8}$$

with $\sup_{t\in[0,T]}\int_{\mathbb{R}^N} B^{\frac{N}{2}}(x,t)\,dx \le C(T)$, $\psi_0 \in L^1(\mathbb{R}^N)\cap L^\infty(\mathbb{R}^N)$ and a small.

The function B is in $L^\infty([0,T], L^{\frac{N}{2}}(\mathbb{R}^N))$ with respect to the spatial variables, while the Hardy potential is only in the Marcinkiewicz space $\mathcal{M}^{\frac{N}{2}}(\mathbb{R}^N)$. The following result shows how B can be interpreted as a small perturbation to the linear problem.

Theorem 5.2.2. *There exists $a_0 > 0$ such that for all $a \le a_0$, problem (5.2.8) has a unique solution $\psi \in L^2(0,T,\mathcal{D}^{1,2}(\mathbb{R}^N))$.*

Proof. We define the following mapping from $L^2((0,T);L^{2^*}(\mathbb{R}^N))$ into $L^2((0,T);L^{2^*}(\mathbb{R}^N))$:

$$\begin{aligned} H: L^2((0,T);L^{2^*}(\mathbb{R}^N)) &\longrightarrow L^2((0,T);L^{2^*}(\mathbb{R}^N)), \\ v &\longrightarrow H(v) = u, \end{aligned}$$

where $u \in L^2((0,T);L^{2^*}(\mathbb{R}^N))$ is the solution to

$$\begin{cases} u_t - \Delta u - \lambda\frac{u}{|x|^2} = aB(x,t)v & \text{in } \mathbb{R}^N,\ t \in (0,T), \\ u(x,0) = \psi_0(x). \end{cases}$$

It is clear that $u \in L^2((0,T);\mathcal{D}^{1,2}(\mathbb{R}^N))$ and then H is well defined.

We claim that for $a \le a_0$ small enough the operator H is a contraction. Denote $w = u_1 - u_2$. It follows that $w_t - \Delta w - \lambda\frac{w}{|x|^2} = aB(x,t)(v_1 - v_2)$. Hence using w as a test function and by the Hardy inequality it follows that

$$\begin{aligned} &\frac{1}{2}\int_{\mathbb{R}^N} w^2(x,T)\,dx - \frac{1}{2}\int_{\mathbb{R}^N} w^2(x,0)\,dx + \left(\frac{\Lambda_{N,2}-\lambda}{\Lambda_{N,2}}\right)\int_0^T\int_{\mathbb{R}^N} |\nabla w|^2\,dx\,dt \\ &\le a\int_0^T\int_{\mathbb{R}^N} B(x,t)(v_1 - v_2)w\,dx\,dt. \end{aligned}$$

Applying Hölder's, Sobolev's and Young's inequalities,

$$
\begin{aligned}
&CS\int_0^T\Big(\int_{\mathbb{R}^N} w^{2^*}\,dx\Big)^{\frac{2}{2^*}}dt\\
&\quad\le a\int_0^T\Big(\int_{\mathbb{R}^N} w^{2^*}\,dx\Big)^{\frac{1}{2^*}}\Big(\int_{\mathbb{R}^N}(v_1-v_2)^{2^*}\,dx\Big)^{\frac{1}{2^*}}\Big(\int_{\mathbb{R}^N}B(x,t)^{\frac{N}{2}}\,dx\Big)^{\frac{2}{N}}dt\\
&\quad\le aC\varepsilon\int_0^T\Big(\int_{\mathbb{R}^N} w^{2^*}\,dx\Big)^{\frac{2}{2^*}}dt+aC(T)C'(\varepsilon)\int_0^T\Big(\int_{\mathbb{R}^N}(v_1-v_2)^{2^*}\,dx\Big)^{\frac{2}{2^*}}dt.
\end{aligned}
$$

Hence, choosing a small, we have

$$\|H(v_1-v_2)\|_{L^2((0,T);L^{2^*}(\mathbb{R}^N))}\le\overline{C}\|v_1-v_2\|_{L^2((0,T);L^{2^*}(\mathbb{R}^N))},$$

with $\overline{C}<1$. Thus the claim follows and then by the classical Banach fixed point theorem we reach the conclusion. □

The next result will be the key for many arguments in the proofs of nonexistence.

Lemma 5.2.3. *Assume that u is a nonnegative function defined in Ω such that $u\not\equiv 0$, $u\in L^1_{\mathrm{loc}}(\Omega_T)$ and $\frac{u}{|x|^2}\in L^1_{\mathrm{loc}}(\Omega_T)$. If u satisfies $u_t-\Delta u-\lambda\frac{u}{|x|^2}\ge 0$ in $\mathcal{D}'(\Omega_T)$ with $\lambda\le\Lambda_{N,2}$ and $B_{r_1}(0)\subset\subset\Omega$, then there exists a constant $C=C(N,r_1,t_1,t_2)$ such that for each cylinder $B_r(0)\times(t_1,t_2)\subset\subset\Omega_T$, $0<r<r_1$,*

$$u\ge C|x|^{-\alpha_-(\lambda)}\quad\text{in } B_r(0)\times(t_1,t_2),$$

where $-\alpha_-(\lambda)$ is given in (3.1.3). In particular, for r conveniently small we can assume $u>1$ in $B_r(0)\times(t_1,t_2)$.

Proof. Since $u\not\equiv 0$, using the strong maximum principle for the heat equation, it follows that for any cylinder $B_{r_1}(0)\times(T_1,T_2)$, there exists $\eta>0$ such that $u\ge\eta>0$ in $B_{r_1}(0)\times(T_1,T_2)$.

Let $w\in L^2((T_1,T_2);W^{1,2}(B_{r_1}(0)))$ be the unique positive solution to the problem

$$
\begin{cases}
w_t-\Delta w-\lambda\frac{w}{|x|^2}=0 & \text{in } B_{r_1}(0)\times(T_1,T_2),\\
w=\eta & \text{on } \partial B_{r_1}(0)\times(T_1,T_2),\\
w(x,T_1)=0 & \text{in } B_{r_1}(0).
\end{cases}
\tag{5.2.9}
$$

It is clear that $w>0$ in $B_{r_1}(0)\times(T_1,T_2)$. We will prove that

$$w(x,t)\ge C|x|^{-\alpha_-(\lambda)}\quad\text{in } B_r(0)\times(t_1,t_2)\subset\subset B_{r_1}(0)\times(T_1,T_2).$$

Indeed, define $v(x,t) = |x|^{\alpha_-(\lambda)} w(x,t)$, that is, the so-called *ground state transformation* of w. Since $w \in L^2((T_1,T_2); W^{1,2}(B_{r_1}(0)))$, we have

$$v \in L^2((T_1,T_2); W^{1,2}_{\alpha_-(\lambda)}(B_{r_1}(0))) \cap \mathcal{C}((T_1,T_2); L^2_{\alpha_-(\lambda)}(B_{r_1}(0))),$$

where $L^2_{\alpha_-(\lambda)}(B_{r_1}(0))$ and $W^{1,2}_{\alpha_-(\lambda)}(B_{r_1}(0))$ are the weighted Lebesgue and Sobolev spaces defined as the complexion of $\mathcal{C}_0^\infty(B_{r_1}(0))$ with respect to the norms

$$\|\phi\|^2_{L^2_{\alpha_-(\lambda)}} = \int_{B_{r_1}(0)} |\phi|^2 |x|^{-2\alpha_-(\lambda)}\, dx,$$

$$\|\phi\|^2_{W^{1,2}_{\alpha_-(\lambda)}} = \int_{B_{r_1}(0)} |\phi|^2 |x|^{-2\alpha_-(\lambda)}\, dx + \int_{B_{r_1}(0)} |\nabla\phi|^2 |x|^{-2\alpha_-(\lambda)}\, dx,$$

respectively. Moreover v solves the problem

$$\begin{cases} |x|^{-2\alpha_-(\lambda)} v_t - \operatorname{div}(|x|^{-2\alpha_-(\lambda)} \nabla v) = 0 & \text{in } B_{r_1}(0) \times (T_1,T_2), \\ v = \eta r_1^{\alpha_-(\lambda)} & \text{on } \partial B_{r_1}(0) \times (T_1,T_2), \\ v(x,T_1) = 0 & \text{in } B_{r_1}(0). \end{cases} \tag{5.2.10}$$

Since v is in the corresponding energy space and the weight $|x|^{-2\alpha_-(\lambda)}$ is in the Muckenhoupt class, we can apply the Harnack inequality obtained in [190] (see also [6] and [123]) and we conclude that $v \geq C$ in $B_r(0) \times (t_1,t_2) \subset\subset B_{r_1}(0) \times (T_1,T_2)$. Hence

$$w(x,t) \geq C|x|^{-\alpha_-(\lambda)} \quad \text{in } B_r(0) \times (t_1,t_2).$$

Since u is a supersolution to problem (5.2.9), by using the weak comparison principle we conclude that $u \geq w$ in $B_r(0) \times (T_1,T_2)$, thus $u \geq C|x|^{-\alpha_-(\lambda)}$ in $B_r(0) \times (t_1,t_2)$ and the result follows. □

5.2.2 A technical remark on existence

Note that the direct use of the *very weak supersolutions* has serious difficulties. For this reason, the following result is an important tool in order to prove the nonexistence result. Roughly speaking, given a very weak supersolution, we will show that there exists a minimal solution to (5.1.1), at least defined in a subcylinder, which is obtained as a limit of solutions to some approximated problems in the sense stated at the beginning of this section. For such a class of solutions, the flexibility of calculus is bigger and allows us to compare and to obtain *a priori* estimates.

Lemma 5.2.4. *If $\bar{u} \in \mathcal{C}([0,T); L^1_{loc}(\tilde{\Omega}))$ is a very weak supersolution to the equation in* (5.1.1) *with $\lambda \leq \Lambda_{N,2}$, $f \in L^1(\Omega_T)$ and $\tilde{\Omega} \supset\supset \Omega$, then there exists a minimal solution to problem* (5.1.1) *obtained as limit of approximations.*

Proof. If $\bar{u}$ is a supersolution to (5.1.1) with $\lambda \le \Lambda_{N,2}$ we construct a sequence

$$\{v_n\} \in \mathcal{C}([0,T); L^1(\Omega)) \cap L^p([0,T); L^p(\Omega)),$$

starting with

$$\begin{cases} v_{0t} - \Delta v_0 = f & \text{in } \Omega \times (0,T), \\ v_0(x,0) = T_1(\bar{u}(x,0)) & \text{if } x \in \Omega, \\ v_0(x,t) = 0 & \text{on } \partial\Omega \times (0,T). \end{cases} \tag{5.2.11}$$

By the comparison principle for the heat equation, it follows that $v_0 \le \bar{u}$ in $\Omega \times (0,T)$. By iteration we define

$$\begin{cases} v_{nt} - \Delta v_n = \lambda \frac{v_{n-1}}{|x|^2 + \frac{1}{n}} + v_{n-1}^p + f & \text{in } \Omega \times (0,T), \\ v_n(x,0) = T_n(\bar{u}(x,0)) & \text{if } x \in \Omega, \\ v_n(x,t) = 0 & \text{on } \partial\Omega \times (0,T). \end{cases} \tag{5.2.12}$$

Using again, in a recurrent way, weak comparison for the linear heat equation, it follows that $v_0 \le \cdots \le v_{n-1} \le v_n \le \bar{u}$ in $\Omega \times (0,T)$, so we obtain the pointwise limit $v = \lim v_n$ that verifies $v \le \bar{u}$ and

$$\begin{cases} v_t - \Delta v = \lambda \frac{v}{|x|^2} + v^p + f & \text{in } \Omega \times (0,T), \\ v(x,t) = 0 & \text{on } \partial\Omega \times (0,T), \\ v(x,0) = \bar{u}(x,0) & \text{if } x \in \Omega, \end{cases} \tag{5.2.13}$$

in the weak sense. Moreover, according to the estimates for the sequence $\{v_n\}$ and using the techniques in [129] and [60], it is easy to check that v has the regularity of a solution obtained as limit of approximations to (5.1.1) in $\Omega \times (0,T)$. That is, $v \in L^q((0,T); W_0^{1,q}(\Omega))$, $\forall q \in [1, \frac{N+2}{N+1})$ and $T_k(u_n) \to T_k(u)$ strongly in $L^2((0,T); W_0^{1,2}(\Omega))$, for all $k > 0$. □

Remark 5.2.5. Note that if w is a very weak positive supersolution to problem

$$w_t - \Delta w - \lambda \frac{w}{|x|^2} = g, \tag{5.2.14}$$

then g must satisfy $\int_0^T \int_{B_r(0)} |x|^{-\alpha_-(\lambda)} g \, dx < \infty$. To see this, it suffices to consider as a test function in (5.2.14) a truncation of φ, the solution to the equation $-\Delta\varphi - \lambda \frac{\varphi}{|x|^2} = 1$, and the result follows. Since we are considering positive solutions to problem (5.1.1), by setting $g = u^p + f$, we obtain

$$\int_{t_1}^{t_2} \int_{B_r(0)} |x|^{-\alpha_-(\lambda)} (u^p + f) \, dx \, dt < \infty, \quad \text{for all } B_r(0) \times (t_1, t_2) \subset\subset \Omega_T.$$

This necessary condition is inspired in the elliptic Theorem 3.4.1 and will be useful in the forthcoming arguments.

5.3 Nonexistence results: $p \geq p_+(\lambda)$

We will see here that for the heat equation perturbed with the Hardy potential, the critical power to obtain existence is also $p_+(\lambda)$, the same value as in the elliptic case and obtained in (4.2.5). Recall that the main properties of $p_-(\lambda)$ and $p_+(\lambda)$ are

$$p_+(\lambda) \to 2^* - 1 = \frac{N+2}{N-2} \text{ as } \lambda \to \Lambda_{N,2}, \quad p_+(\lambda) \to \infty \text{ as } \lambda \to 0,$$
$$p_-(\lambda) \to 2^* - 1 = \frac{N+2}{N-2} \text{ as } \lambda \to \Lambda_{N,2}, \quad p_-(\lambda) \to \frac{N}{N-2} \text{ as } \lambda \to 0.$$

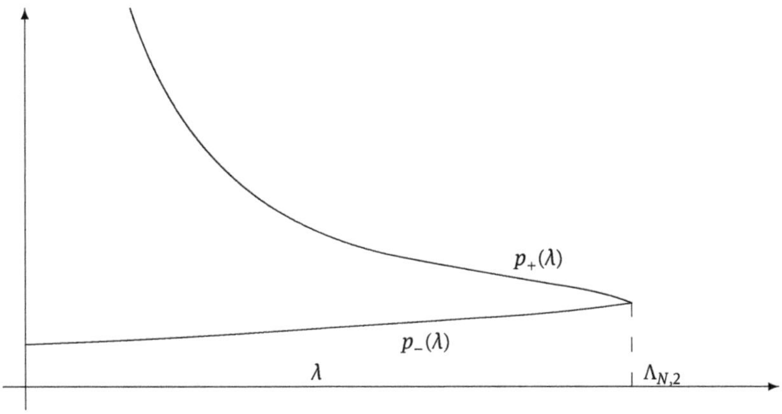

It is clear that $p_+(\lambda)$ and $p_-(\lambda)$ are respectively decreasing and increasing functions on λ and, thus,

$$1 < p_-(\lambda) \leq 2^* - 1 \leq p_+(\lambda).$$

Theorem 5.3.1. *If $p \geq p_+(\lambda)$, then problem* (5.1.1) *has no positive very weak supersolution. In the case where $f \equiv 0$, the unique nonnegative very weak supersolution is $u \equiv 0$.*

Proof. Without loss of generality, we can assume $f \in L^\infty(\Omega_T)$. We argue by contradiction. Assume that $\tilde{u}$ is a very weak supersolution.

If $\lambda > \Lambda_{N,2} = (\frac{N-2}{2})^2$, then it is sufficient to consider $\tilde{u}$ as a very weak supersolution to the problem

$$\begin{cases} v_t - \Delta v = \lambda \frac{v}{|x|^2} + f_1 & \text{in } \Omega \times (0,T), \\ v > 0 & \text{in } \Omega \times (0,T), \\ v = 0 & \text{on } \partial\Omega \times (0,T), \end{cases} \tag{5.3.1}$$

where $f_1(x) = v^p + f$. Hence the nonexistence result follows using [68] and [6].

For the case $\lambda \leq \Lambda_{N,2}$ we argue again by contradiction. If $\tilde{u}$ is a very weak supersolution to (5.1.1), then $\tilde{u}_t - \Delta\tilde{u} - \lambda\frac{\tilde{u}}{|x|^2} \geq 0$ in $\mathcal{D}'(\Omega_T)$.

Since $\tilde{u}$ is also a very weak supersolution in any $B_R(0) \times (T_1, T_2) \subset\subset \Omega_T$, then by Lemma 5.2.4, the problem

$$\begin{cases} u_t - \Delta u = \lambda \frac{u}{|x|^2} + u^p + f & \text{in } B_R(0) \times (T_1, T_2), \\ u(x,t) > 0 & \text{in } B_R(0) \times (T_1, T_2), \\ u(x,t) = 0 & \text{on } \partial B_R(0) \times (T_1, T_2), \\ u(x,t_1) = \tilde{u}(x, T_1) & \text{if } x \in B_R(0), \end{cases} \tag{5.3.2}$$

has a minimal solution u obtained by approximation of truncated problems in $B_R(0) \times (T_1, T_2)$. In particular $u = \lim v_n$, with $v_n \in L^\infty(B_R(0) \times (T_1, T_2))$ and v_n being a solution to (5.2.12) in $B_R(0) \times (T_1, T_2)$. Note that since $u_t - \Delta u - \lambda \frac{u}{|x|^2} \geq 0$ in $\mathcal{D}'(B_{r_1}(0) \times (T_1, T_2))$ and using Lemma 5.2.3, there exists a cylinder $B_r(0) \times (t_1, t_2)$, with $0 < r < r_1 < R$, $0 < T_1 < t_1 < t_2 < T_2 \leq T$ and there exists a constant $C = C(N, r_1, t_1, t_2)$ such that $u \geq C|x|^{-\alpha_-(\lambda)}$ and $u > 1$ in $B_r(0) \times (T_1, T_2)$ for suitable r. In particular, since $u \in L^1_{\text{loc}}(\Omega_T)$, we have $\log(u) \in L^p(B_r(0) \times (t_1, t_2))$, for all $p \in [1, \infty)$. By a scaling, we can assume that the cylinder is $B_r(0) \times (0, \tau)$.

We divide the proof in three cases.

STEP 1. *The case $p > p_+(\lambda)$*. Consider $\phi \in C_0^\infty(B_r(0))$ and take $\frac{|\phi|^2}{v_n}$ as a test function in problems (5.2.12). Applying the Picone inequality in Section 1.3, we obtain

$$\begin{aligned} \int_0^\tau \int_{B_r(0)} v_n^{p-1} \phi^2 \, dx \, dt &\leq \int_0^\tau \int_{B_r(0)} \frac{|\phi|^2}{v_n} v_{nt} \, dx \, dt + \int_0^\tau \int_{B_r(0)} (-\Delta v_n) \frac{|\phi|^2}{v_n} \, dx \, dt \\ &\leq \int_{B_r(0)} |\log v_n(x,\tau)| \phi^2 \, dx + \int_0^\tau \int_{B_r(0)} |\nabla \phi|^2 \, dx \, dt. \end{aligned}$$

Therefore, passing to the limit as n tends to infinity and thanks to Lemma 5.2.3, we have

$$\begin{aligned} \int_{B_r(0)} |\log u(x,\tau)| \phi^2 \, dx + \int_0^\tau \int_{B_r(0)} |\nabla \phi|^2 \, dx \, dt &\geq \int_0^\tau \int_{B_r(0)} u^{p-1} \phi^2 \, dx \, dt \\ &\geq C \int_0^\tau \int_{B_r(0)} \frac{\phi^2}{|x|^{(p-1)\alpha_-(\lambda)}} \, dx \, dt. \end{aligned}$$

Using the Hölder and Sobolev inequalities, it follows that

$$\begin{aligned} \int_{B_r(0)} |\log(u(x,\tau))| |\phi|^2 \, dx &\leq \Big(\int_{B_r(0)} |\phi|^{2^*} \, dx \Big)^{\frac{2}{2^*}} \Big(\int_{B_r(0)} |\log u(x,\tau)|^{\frac{N}{2}} \, dx \Big)^{\frac{2}{N}} \\ &\leq \Big(\int_{B_r(0)} |\log u(x,\tau)|^{\frac{N}{2}} \, dx \Big)^{\frac{2}{N}} \mathcal{S}^{-1} \int_{B_r(0)} |\nabla \phi|^2 \, dx, \end{aligned}$$

where $\mathcal{S}$ is the optimal constant in the Sobolev embedding. Thus we have

$$\left[1+\left(\int_{B_r(0)} |\log u(x,\tau)|^{\frac{N}{2}}\,dx\right)^{\frac{2}{N}} \mathcal{S}^{-1}\right] \int_{B_r(0)} |\nabla\phi|^2\,dx\,dt \geq C \int_{B_r(0)} \frac{\phi^2}{|x|^{(p-1)\alpha_-(\lambda)}}\,dx\,dt.$$

Since $p > p_+(\lambda)$, we have $(p-1)\alpha_-(\lambda) > 2$ and we obtain a contradiction with the Hardy inequality.

Step 2. *The case $p = p_+(\lambda)$ and $\lambda < \Lambda_{N,2}$.*

Fix the cylinder $B_\eta(0)\times(0,\tau)$ as above, and consider $w(x,t) = |x|^{-\alpha_-(\lambda)}(t^2(\log(\frac{1}{|x|}))^\beta + 1)$ defined for $(x,t) \in B_\eta(0)\times(0,\tau)$, for some $\beta > 0$ that will be chosen below. Since $\lambda < \Lambda_{N,2}$, we have $w \in \mathcal{C}([0,\tau], L^2(B_\eta(0))) \cap L^2((0,\tau), W^{1,2}(B_\eta(0)))$. By a direct computation we get

$$\begin{aligned} w_t - \Delta w - \lambda\frac{w}{|x|^2} &= \frac{t}{|x|^{2+\alpha_-(\lambda)}}\left\{2|x|^2\left(\log\left(\frac{1}{|x|}\right)\right)^\beta + \beta(t)\left(\log\left(\frac{1}{|x|}\right)\right)^{\beta-1}\right.\\ &\quad\left.\times\left[(N-2-2\alpha_-(\lambda)) + (1-\beta)\left(\left(\log\left(\frac{1}{|x|}\right)\right)^{-1}\right]\right\}. \end{aligned}$$

Note that

$$w^{p_+(\lambda)} = |x|^{-(2+\alpha_-(\lambda))}\left[t^2\left(\log\left(\frac{1}{|x|}\right)\right)^\beta + 1\right]^{p_+(\lambda)}.$$

If we set

$$h(x,t) = \left[t^2\left(\log\left(\frac{1}{|x|}\right)\right)^\beta + 1\right]^{1-p_+(\lambda)},$$

then

$$w^{p_+(\lambda)}h(x,t) = |x|^{-(2+\alpha_-(\lambda))}\left[t^2\left(\log\left(\frac{1}{|x|}\right)\right)^\beta + 1\right].$$

Hence, there exists $\bar{t}_2 \in (0,\tau)$ such that

$$w_t - \Delta w - \lambda\frac{w}{|x|^2} \leq \beta h(x,t) w^{p_+(\lambda)} \quad \text{in } B_\eta(0)\times(0,\bar{t}_2).$$

Denote $u_1 = c_1 u$, for a certain constant $c_1 > 0$. Then we have

$$u_{1t} - \Delta u_1 - \lambda\frac{u_1}{|x|^2} \geq c_1^{1-p_+(\lambda)} u_1^{p_+(\lambda)}.$$

Consider $c_0 > 0$ in the same way as C in Lemma 5.2.3 after a scaling. For fixed $c_1 > 0$ such that $c_1 c_0 \geq 1$ and for a suitable small β we have

$$c_1^{1-p_+(\lambda)} \geq \beta\|h\|_{L^\infty}.$$

Since $c_1c_0 \geq 1$, we have $u_1 \geq w(x,t)$ on $\partial B_\eta(0) \times (0,\tau)$, $u_1(x,0) \geq w(x,0)$ for $x \in B_\eta(0)$ and

$$u_{1t} - \Delta u_1 - \lambda \frac{u_1}{|x|^2} \geq \beta h(x,t) u_1^{p_+(\lambda)}.$$

CLAIM: We have $u_1 \geq w$ in $B_\eta(0) \times (0, \bar{t}_2)$.

Indeed, consider $v = w - u_1$. Then

$$v_t - \Delta v - \lambda \frac{v}{|x|^2} \leq \beta h(x,t)(w^{p_+(\lambda)} - u_1^{p_+(\lambda)}).$$

Hence using Kato's inequality (see [255]) we obtain the estimate

$$(v_+)_t - \Delta v_+ - \lambda \frac{v_+}{|x|^2} \leq p_+(\lambda)\beta h(x,t) w^{p_+-1} v_+ \leq p_+(\lambda)\beta |x|^{-2} v_+. \tag{5.3.3}$$

Since $\frac{v_+^2}{|x|^2} \in L^1(B_\eta(0))$, an approximation argument and integration by parts gives

$$v_+ \in L^2((0,\bar{t}_2), W_0^{1,2}(B_\eta(0))).$$

Therefore choosing β small enough such that $\lambda + p_+(\lambda)\beta < \Lambda_{N,2}$ and using v_+ as a test function in (5.3.3) we deduce that $v_+ \equiv 0$ and the claim is proved.

To finish the proof in this step, we use the same argument as in STEP 1, namely, for all $\phi \in C_0^\infty(B_\eta(0))$ we arrive at the following inequality:

$$c_2 \int_0^\tau \int_{B_\eta(0)} u^{p-1}|\phi|^2\,dx \leq \int_0^\tau \int_{B_\eta(0)} |\nabla\phi|^2\,dx = \tau \int_{B_\eta(0)} |\nabla\phi|^2\,dx, \tag{5.3.4}$$

where $c_2 > 0$ is independent of ϕ. Using the result of the claim we obtain for $r << \eta$,

$$c_3 \int_{B_r(0)} \frac{|\phi|^2}{|x|^2}\left(\log\left(\frac{1}{|x|}\right)\right)^{\beta(p-1)} dx \leq \int_{B_r(0)} |\nabla\phi|^2\,dx,$$

a contradiction with the Hardy inequality. Hence the result follows. □

STEP 3. *The case $p = p_+(\lambda)$ and $\lambda = \Lambda_{N,2}$.*

Here we have $\alpha_-(\lambda) = \frac{N-2}{2}$ and $p_+(\Lambda_{N,2}) = 2^* - 1$. If $\tilde{u}$ is a supersolution to (5.1.1), then thanks to Lemma 5.2.3 and Remark 5.2.5 it follows that

$$\tau C^p \int_{B_r(0)} |x|^{-\alpha_-(\lambda)(p+1)}\,dx \leq \int_0^\tau \int_{B_r(0)} |x|^{\alpha_-(\lambda)} u^p\,dx\,dt < \infty.$$

Since $\alpha_-(\lambda)(p_+(\Lambda_{N,2}) + 1) = N$, we reach a contradiction and the result follows.

Remark 5.3.2. It is worth pointing out that for $\lambda = 0$, $p_+(\lambda) = \infty$ and we get the existence of a local solution for the heat equation for all $p > 1$, at least for a suitable initial datum.

5.4 Instantaneous and complete blow-up results

The nonexistence result obtained above for $p \geq p_+(\lambda)$ is very strong in the sense that a complete and instantaneous blow-up phenomenon occurs in two different senses.

(a) If u_n is the solution to problem (5.1.1), where the Hardy potential is replaced by the bounded weight $1/(|x|^2 + \frac{1}{n})$, then $u_n(x,t) \to \infty$ as $n \to \infty$.

(b) If u_n is the solution to problem (5.1.1), with $p = p_n < p_+(\lambda)$ and $p_n \to p_+(\lambda)$ as $n \to \infty$, then $u_n(x,t) \to \infty$ as $n \to \infty$.

In both cases, (x,t) is an arbitrary point in $\Omega \times (0,T)$.

5.4.1 Blow-up for the approximated problems when $p \geq p_+(\lambda)$

Let us prove now the blow-up behavior for some approximated problems.

Theorem 5.4.1. *Let $u_n \in C((0,T); L^1(\Omega)) \cap L^p((0,T); L^p(\Omega))$ be a solution to the problem*

$$\begin{cases} u_{nt} - \Delta u_n = \dfrac{u_n^p}{1+\frac{1}{n}u_n^p} + \lambda a_n(x) u_n + cf & \text{in } \Omega_T \equiv \Omega \times (0,T), \\ u_n(x,t) = 0 & \text{on } \partial\Omega \times (0,T), \\ u_n(x,0) = 0 & \text{if } x \in \Omega, \end{cases} \tag{5.4.1}$$

with $f \neq 0$, $a_n(x) = \frac{1}{|x|^2+\frac{1}{n}}$ and $p \geq p_+(\lambda)$. Then $u_n(x_0,t_0) \to \infty$, $\forall (x_0,t_0) \in \Omega \times (0,T)$.

Proof. Without loss of generality, we can assume that $f \in L^\infty(\Omega_T)$ and $\lambda \leq \Lambda_{N,2}$. The existence of a positive solution to problem (5.4.1) follows using classical subsupersolution arguments. By using the monotonicity of the nonlinear term and the coefficient a_n, we can assume the existence of a minimal solution u_n such that $u_n \leq u_{n+1}$, for all $n \geq 1$. Therefore to get the blow-up result we just have to show the complete blow-up for the family of minimal solutions denoted by u_n.

Suppose by contradiction that there exists $(x_0,t_0) \in \Omega_T$ such that

$$u_n(x_0,t_0) \to C_0 < \infty \quad \text{as } n \to \infty.$$

Using the classical Harnack inequality proved by J. Moser in [244] and J. Nash [250], there exist $s > 0$ and a positive constant $C = C(N,s,t_0,\beta)$ such that

$$\iint_{R^-} u_n(x,t)\,dx\,dt \leq C \operatorname*{ess\,inf}_{R^+} u_n,$$

where $R^- = B_s(x_0)\times(t_0 - \frac{3}{4}\beta, t_0 - \frac{1}{4}\beta)$ and $R^+ = B_s(x_0)\times(t_0 - \frac{1}{2}\beta, t_0 + \frac{1}{2}\beta)$. We can suppose that $0 \in B_s(x_0)$, since otherwise, we may consider $B_\delta(y) \subset B_s(x_0)$ such that

$$\iint_{R_y^-} u_n(x,t)\,dx\,dt \leq \iint_{R^-} u_n(x,t)\,dx\,dt \leq C \operatorname*{ess\,inf}_{R^+} u_n \leq C \operatorname*{ess\,inf}_{R_y^+} u_n,$$

with $R_y^- = B_\delta(y) \times (t_0 - \frac{3}{4}\beta, t_0 - \frac{1}{4}\beta)$ and $R_y^+ = B_\delta(y) \times (t_0 - \frac{1}{2}\beta, t_0 + \frac{1}{2}\beta)$. In a recurrent way, we get

$$\iint_{R^-} u_n(x,t)\,dx\,dt \le C \operatorname*{ess\,inf}_{R^+} u_n \le C u_n(x_0, t_0) \le C',$$

with $R^- = B_r(0) \times (t_1, t_2)$ and $R^+ = B_r(0) \times (t_3, t_4)$, $t_0 \in (t_3, t_4)$.

Therefore, by the monotone convergence theorem, there exists $u \ge 0$ such that $u_n \uparrow u$ strongly in $L^1(B_r(0) \times (t_1, t_2))$.

Let $\overline{\varphi}$ be the solution to the problem

$$\begin{cases} \overline{\varphi}_t - \Delta\overline{\varphi} = \chi_{B_r(0)\times[T-t_2,T-t_1]} & \text{in } \Omega_T, \\ \overline{\varphi}(x,t) = 0 & \text{on } \partial\Omega \times (0,T), \\ \overline{\varphi}(x,0) = 0 & \text{in } \Omega, \end{cases}$$

and consider the translation in time $\varphi(x,t) = \overline{\varphi}(T-t,x)$ in such a way that

$$\begin{cases} -\varphi_t - \Delta\varphi = \chi_{B_r(0)\times[t_1,t_2]} & \text{in } \Omega_T, \\ \varphi(x,t) = 0 & \text{on } \partial\Omega \times (0,T), \\ \varphi(x,T) = 0 & \text{in } \Omega. \end{cases}$$

Take φ as a test function in (5.4.1). Then we have

$$\begin{aligned} C' &\ge \int_{t_1}^{t_2} \int_{B_r(0)} u_n(x,t)\,dx\,dt \\ &= \int_0^T \int_\Omega \frac{u_n^p}{1 + \frac{1}{n}u_n^p}\varphi\,dx\,dt + \lambda \int_0^T \int_\Omega a_n(x) u_n \varphi\,dx\,dt + c \int_0^T \int_\Omega f\varphi\,dx\,dt. \end{aligned}$$

By the monotone convergence theorem, as $n \to \infty$,

$$\begin{aligned} \frac{u_n^p}{1 + \frac{1}{n}u_n^p} &\to u^p \quad \text{in } L^1_{\mathrm{loc}}(B_r(0) \times (t_1, t_2)), \\ a_n(x) u_n &\nearrow \frac{u}{|x|^2} \quad \text{in } L^1_{\mathrm{loc}}(B_r(0) \times (t_1, t_2)). \end{aligned}$$

Thus it follows that u is a very weak supersolution to (5.4.1) in $B_{r_1}(0) \times (\bar{t}_1, \bar{t}_2) \subset\subset B_r(0) \times (t_1, t_2)$, a contradiction with Theorem 5.3.1. □

5.4.2 Blow-up when $p_n \to p_+(\lambda)$

The following blow-up result, for a sequence $p_n \uparrow p_+(\lambda)$, turns out to be very interesting.

Theorem 5.4.2. *Assume that p_n satisfies $p_n < p_+(\lambda)$, $p_n \to p_+(\lambda)$ as $n \to \infty$ and $f \gneqq 0$. Let $u_n \in C([0,T); L^1_{loc}(\Omega))$ be a very weak supersolution to the problem*

$$\begin{cases} u_{nt} - \Delta u_n \geq \lambda \frac{u_n}{|x|^2} + u_n^{p_n} + f & \text{in } \Omega_T, \\ u_n(x,t) = 0 & \text{on } \partial\Omega \times (0,T), \\ u_n(x,0) = 0 & \text{in } \Omega. \end{cases} \tag{5.4.2}$$

Then $u_n(x_0,t_0) \to \infty$, $\forall (x_0,t_0) \in \Omega \times (0,T)$.

Proof. Without loss of generality we can again assume that $f \in L^\infty(\Omega_T)$. Suppose by contradiction that there exists a subsequence denoted p_n and a supersolution u_n such that for some point $(x_0,t_0) \in \Omega_T$ we have $u_n(x_0,t_0) \to C_0 < \infty$, $\forall n \in \mathbb{N}$. Without loss of generality we can assume that $p_n(\lambda) = 1 + \frac{2}{\alpha_-(\lambda)+\frac{1}{n}}$. Thanks to the classical Harnack inequality, there exists $s > 0$ and a positive constant $C = C(N,s,t_0,\beta)$ such that

$$\iint\limits_{R^-} u_n(x,t)\,dx\,dt \leq C \operatorname*{ess\,inf}_{R^+} u_n \leq CC_0,$$

where $R^- = B_s(x_0) \times (t_0 - \frac{3}{4}\beta, t_0 - \frac{1}{4}\beta)$ and $R^+ = B_s(x_0) \times (t_0 - \frac{1}{2}\beta, t_0 + \frac{1}{2}\beta)$. As in the discussion of the proof of Theorem 5.4.1, we can suppose that $0 \in B_s(x_0)$.

If $u_n \in C([0,T); L^1_{loc}(\Omega))$ is a very weak supersolution to problem (5.4.2), then there exists a minimal solution to (5.4.2) in $\Omega_1 \times (t_1,t_2) \subset\subset \Omega_T$ with $0 \in \Omega_1 \subset\subset \Omega$ obtained by approximation. Let us denote by $v_n \leq u_n$ this minimal solution. Then v_n solves

$$\begin{cases} v_{nt} - \Delta v_n = \lambda \frac{v_n}{|x|^2} + v_n^{p_n} + f & \text{in } \Omega_1 \times (t_1,t_2), \\ v_n(x,t) = 0 & \text{on } \partial\Omega_1 \times (t_1,t_2), \\ v_n(x,t_1) = 0 & \text{in } \Omega_1. \end{cases} \tag{5.4.3}$$

Let ϕ be the solution to the problem

$$\begin{cases} -\phi_t - \Delta\phi = 1 & \text{in } \Omega_1 \times (t_1,t_2), \\ \phi = 0 & \text{on } \partial\Omega_1 \times (t_1,t_2), \\ \phi(x,t_1) = 0 & \text{in } \Omega_1. \end{cases}$$

Using ϕ as a test function in (5.4.3) and since $v_n \leq u_n$, we get

$$C \geq \int\limits_{t_1}^{t_2}\int\limits_{\Omega_1} v_n(x,s)\,dx\,ds = \int\limits_{t_1}^{t_2}\int\limits_{\Omega_1} g_n(x,s)\phi\,dx\,ds,$$

where $g_n(x,t) = \lambda \frac{v_n}{|x|^2} + v_n^{p_n} + f$. Thus

$$\int\limits_{t_1}^{t_2}\int\limits_{\Omega_1} g_n\phi\,dx\,ds \leq C \quad \text{for all } n.$$

Hence, it follows that g_n is uniformly bounded in $L^1_{loc}(\Omega_1 \times (t_1, t_2))$ and then $g_n \rightharpoonup \mu$ in the sense of measures.

Let $\varphi \in C_0^\infty(\Omega_1)$. Taking $T_k(v_n)\cdot\varphi$ as a test function in (5.4.3) and using the previous boundedness, we get

$$\int_{\Omega_1} \Theta_k(v_n(x,t_2))\varphi\,dx + \int_{t_1}^{t_2}\int_{\Omega_1} |\nabla T_k(v_n)|^2 \varphi\,dx\,ds + \int_{t_1}^{t_2}\int_{\Omega_1} \Theta_k(v_n)(-\Delta\varphi)\,dx\,ds = \int_{t_1}^{t_2}\int_{\Omega_1} g_n(x,s)\varphi T_k(v_n)\,dx\,ds,$$

where $\Theta_k(s) = \int_0^s T_k(\sigma)d\sigma$. Thus,

$$\int_{\Omega_1} \Theta_k(v_n(x,t_2))\varphi\,dx + \int_{t_1}^{t_2}\int_{\Omega_1} |\nabla T_k(v_n)|^2 \varphi\,dx\,ds \le c\int_{t_1}^{t_2}\int_{\Omega_1} \Theta_k(v_n)\,dx\,ds + k\int_{t_1}^{t_2}\int_{\Omega_1} g_n(x,s)\varphi\,dx\,ds \le C.$$

Hence, there exists a nonnegative function v such that

$$T_k(v_n) \rightharpoonup T_k(v), \quad \text{weakly in } L^2_{loc}((t_1,t_2), W^{1,2}_{loc}(\Omega_1)).$$

Moreover, by the results in [66], we obtain $v_n \to v$ strongly in $L^q_{loc}((t_1,t_2), L^q_{loc}(\Omega_1)) \equiv L^q_{loc}(\Omega_1 \times (t_1,t_2))$ with $q < \frac{N}{N-2}$.

Let $\psi \in C_0^\infty(B_r(0)\times(t_1,t_2))$ be a nonnegative function. By the Fatou lemma it follows that

$$\lim_{n\to\infty}\int_{t_1}^{t_2}\int_{B_r(0)} g_n(x,s)\psi\,dx\,ds \ge \int_{t_1}^{t_2}\int_{B_r(0)} v^{p_+(\lambda)}\psi\,dx\,ds + \lambda\int_{t_1}^{t_2}\int_{B_r(0)} \frac{v\psi}{|x|^2}\,dx\,ds + \int_{t_1}^{t_2}\int_{B_r(0)} f\psi\,dx\,ds.$$

Therefore, using ψ as a test function in (5.4.3) and passing to the limit as $n \to \infty$, it follows that

$$-\int_{t_1}^{t_2}\int_{B_r(0)} v\psi_t\,dx\,ds + \int_{t_1}^{t_2}\int_{B_r(0)} v(-\Delta\psi)\,dx\,ds \ge \int_{t_1}^{t_2}\int_{B_r(0)} v^{p_+}(\lambda)\psi\,dx\,ds + \lambda\int_{t_1}^{t_2}\int_{B_r(0)} \frac{v\psi}{|x|^2}\,dx\,ds + \int_{t_1}^{t_2}\int_{B_r(0)} f\psi\,dx\,ds.$$

Hence v is a very weak supersolution to (5.1.1) obtained by approximation and then we reach a contradiction. □

Remark 5.4.3. We can also obtain nonexistence results by using the *local existence result* in Lemma 5.2.4 and some contradiction with the Hardy inequality. Note also that a nonexistence result and Lemma 5.2.4 give us complete and instantaneous blow-up.

5.5 Existence of solutions: $p < p_+(\lambda)$

The goal of this section is to study the complementary interval of powers $1 < p < p_+(\lambda)$ and prove that under some suitable hypotheses on f and u_0, problem (5.1.1) has a positive solution. For the existence result we will consider the case $f \equiv 0$. For the case $f \not\equiv 0$, see Remark 5.5.4 below.

First of all, recall that if $0 < \lambda \le \Lambda_{N,2}$ and $1 < p < p_+(\lambda)$, the stationary problem

$$- \Delta u = \lambda \frac{u}{|x|^2} + u^p \text{ in } \Omega, \quad u = 0 \text{ on } \partial\Omega, \tag{5.5.1}$$

has a positive very weak supersolution w in the following cases:

(A) For $0 < \lambda < \Lambda_{N,2}$, we have the following.
  - **(I)** If $1 < p < \frac{N+2}{N-2}$ and Ω is a bounded domain, there exists a positive solution to problem (5.5.1) using the classical mountain pass theorem in the Sobolev space $W_0^{1,2}(\Omega)$. See [89].
  - **(II)** If $\frac{N+2}{N-2} < p < p_+(\lambda)$, there exists a positive solution in $\mathbb{R}^N$ as was found in Chapter 4.

(B) If $\lambda = \Lambda_{N,2}$, then $p_+(\Lambda_{N,2}) = \frac{N+2}{N-2} = p_-(\Lambda_{N,2})$. Let $H(\Omega)$ be the Hilbert space completion of $C_0^\infty(\Omega)$ with respect to the norm

$$\|\phi\|^2 = \int_\Omega \left(|\nabla\phi|^2 - \Lambda_{N,2} \frac{\phi^2}{|x|^2} \right) dx,$$

introduced in Chapter 2, Section 2.5.1. Since $p < \frac{N+2}{N-2} = 2^* - 1$, classical variational methods in the space $H(\Omega)$ allow us to prove the existence of a positive solution w to the stationary problem (5.5.1). See details in [181].

Theorem 5.5.1. *Assume that* $0 < \lambda \le \Lambda_{N,2}$ *and* $1 < p < p_+(\lambda)$. *Suppose that* $u_0(x) \le \overline{w}$, *where* $\overline{w}$ *is a supersolution to the stationary problem*

$$-\Delta u = \lambda \frac{u}{|x|^2} + u^p \textit{ in } \Omega, \quad u(x) = 0 \textit{ on } \partial\Omega.$$

Then for all $T > 0$, *the problem*

$$\begin{cases} u_t - \Delta u = \lambda \frac{u}{|x|^2} + u^p & \textit{in } \Omega_T \equiv \Omega \times (0,T), \\ u(x,t) = 0 & \textit{on } \partial\Omega \times (0,T), \\ u(x,0) = u_0(x) & \textit{if } x \in \Omega \end{cases} \tag{5.5.2}$$

has a global solution.

Proof. The proof uses the classical subsupersolution argument (see Lemma 5.2.4). □

Remark 5.5.2. With the results above we find the optimality of the power $p_+(\lambda)$, which is one of the main objectives. Nevertheless, it will be interesting to know the optimal class of initial data for which there exists a solution and the regularity of such solutions according to the regularity of the data. It is also an interesting question to know the asymptotic behavior as $t \to \infty$ of such solutions.

In this direction we have the following necessary condition.

Proposition 5.5.3. *Assume that $\lambda \le \Lambda_{N,2}$. If problem* (5.1.1) *has a very weak supersolution, then there exists $r > 0$ such that the initial value function, u_0, satisfies*

$$\int_{B_r(0)} |x|^{-\alpha_-(\lambda)} u_0(x)\,dx < \infty, \quad \textit{where } \alpha_-(\lambda) \textit{ is defined in } (3.1.3).$$

Proof. We argue by contradiction. Suppose that u is a very weak supersolution to (5.1.1) with $u(x,0) = u_0(x)$ in Ω such that $\forall r > 0$, $\int_{B_r(0)} |x|^{-\alpha_-(\lambda)} u_0(x)\,dx = \infty$. We consider the sequence $\{u_n\}$ of minimal solutions to the problems

$$\begin{cases} u_{nt} - \Delta u_n = \lambda \frac{u_n}{|x|^2+\frac{1}{n}} + u_n^p & \text{in } \Omega \times (0,T), \\ u_n(x,t) = 0 & \text{on } \partial\Omega \times (0,T), \\ u_n(x,0) = T_n(u_0(x)) & \text{in } \Omega, \end{cases} \tag{5.5.3}$$

with $1 < p < p_+(\lambda)$ and T depending just on the supersolution. For every n, let us consider the positive eigenfunction φ_n to the associated problem

$$-\Delta\varphi_n - \frac{\lambda\varphi_n}{|x|^2 + \frac{1}{n}} = c_n\varphi_n, \quad \text{with} \int_{B_r(0)} \varphi_n^2\,dx = 1.$$

Up to a subsequence we find that $\varphi_n \to \varphi$ in $L^s(\Omega)$, $s < \frac{2N}{N-2}$ and $c_n \to c$ with

$$-\Delta\varphi - \frac{\lambda\varphi}{|x|^2} = c\varphi, \quad \int_{B_r(0)} \varphi^2\,dx = 1 \text{ and } \varphi \ge C|x|^{-\alpha_-(\lambda)},$$

in a neighborhood of the origin. Define $\psi_n = \frac{\varphi_n}{\|\varphi_n\|_1}$; then $\psi_n \to \psi = \frac{\varphi}{\|\varphi\|_1}$ in $L^1(\Omega)$. Taking ψ_n as a test function in (5.5.3), we get

$$\frac{d}{dt}\int_{B_r(0)} \psi_n u_n(x,t)\,dx + c_n \int_{B_r(0)} u_n(x,t)\psi_n\,dx \ge \int_{B_r(0)} u_n^p \psi_n\,dx.$$

Since $p > 1$, we can apply Jensen's inequality and conclude that

$$\frac{d}{dt}\int_{B_r(0)} \psi_n u_n(x,t)\,dx + c_n \int_{B_r(0)} \psi_n u_n(x,t)\,dx \ge \Big(\int_{B_r(0)} \psi_n u_n(x,t)\,dx\,dt\Big)^p.$$

Note that if $Y_n(t) = \int_{B_r(0)} \psi_n u_n(x,t)\,dx$, then $Y_n'(t) + cY_n(t) \ge [Y_n(t)]^p$.

Using the hypothesis on u_0 and the definition of ψ_n it follows that $\lim_{n\to\infty} Y_n(0) = \infty$. We set $Z_n(t) = e^{ct}Y_n(t)$. Then $Z_n'(t) \geq e^{-c(p-1)t}Z_n^p(t)$. Thus $Z_n(t)$ is an increasing function and therefore

$$Z_n(t) \geq Z_n(0) = Y_n(0) \to \infty \quad \text{as } n \to \infty \text{ uniformly in } t \in (0,T).$$

Integrating the differential equation of Z_n, it follows that

$$\frac{1}{p-1}\left(\frac{1}{Z_n^{p-1}(0)} - \frac{1}{Z_n^{p-1}(t)}\right) \geq \frac{1}{c(p-1)}(1 - e^{-c(p-1)t}), \quad t > 0.$$

Hence we conclude that

$$Y_n^{p-1}(0) = Z_n^{p-1}(0) \leq \frac{c}{1 - e^{-c(p-1)t}} < \infty, \quad \forall t > 0,$$

a contradiction with the hypothesis on the initial datum. Hence the result follows. □

Remark 5.5.4.

1. An open question that is left here is whether or not the previous necessary condition is also sufficient for *local existence*.
2. In the presence of a source term $f \gneqq 0$, if $f(x,t) \leq \frac{c_0(t)}{|x|^2}$ with $c_0(t)$ *sufficiently small*, then the above computation allows us to prove the existence of a supersolution. Hence, the existence of a minimal solution to problem (5.1.1) follows for all $p < p_+(\lambda)$.

5.6 Cauchy problem

In [171], H. Fujita considered the initial value problem

$$\begin{cases} u_t = \Delta u + u^p, & x \in \mathbb{R}^N,\ t > 0, \\ u(x,0) = u_0(x) \geq 0, & x \in \mathbb{R}^N, \end{cases} \tag{5.6.1}$$

where $1 < p < \infty$.

Fujita showed that if $1 < p < 1 + \frac{2}{N}$, then for all positive initial datum u_0 there exists $T > 0$ such that the solution to problem (5.6.1) satisfies $\|u(\cdot,t_n)\|_\infty \to \infty$ as $t_n \to T$. Moreover, if $p > 1 + \frac{2}{N}$, then there are both global solutions for small data and nonglobal solutions for large data. The number $F(0) = 1 + \frac{2}{N}$ is often called the critical Fujita blow-up exponent for the heat equation. Moreover it is proved that for $p = 1 + \frac{2}{N}$, a suitable norm of the solution goes to infinity in a finite time. We refer to [316] for a simple proof of this last fact.

In this section we study the Fujita type behavior for the Cauchy problem with the Hardy potential. More precisely, we consider

$$u_t - \Delta u = \lambda \frac{u}{|x|^2} + u^p \text{ in } \mathbb{R}^N, \quad u(x,0) = u_0(x) \gneqq 0 \text{ in } \mathbb{R}^N, \tag{5.6.2}$$

with $0 < \lambda \le \Lambda_{N,2}$ and $1 < p < p_+(\lambda)$. Since, clearly, $u \notin L^\infty$, in our case we have the blow-up in L^∞ for free. The problem to study is how to obtain the blow-up in a suitable Lebesgue space with a weight.

We have found in Section 5.2.1 that $v(r,t) = t^{-\frac{N}{2}+\alpha_-(\lambda)} r^{-\alpha_-(\lambda)} e^{-\frac{r^2}{4t}}$ is a self-invariant solution to the homogenous linear equation (5.2.4), that is,

$$v_t - \Delta v - \lambda \frac{v}{|x|^2} = 0 \quad \text{in } \mathbb{R}^N.$$

Moreover, $\int_{\mathbb{R}^N} |x|^{-\alpha_-(\lambda)} v(x,t)\,dx = C$. In particular, for $\lambda = 0$ (equivalently $\alpha_-(\lambda) = 0$), we get the fundamental solution of the heat equation. Incidentally, it is worthy to point out that, in particular for $\lambda = 0$, our arguments below give a proof for the Fujita result on the semilinear heat equation.

5.6.1 Subsolution blow-up in a finite time for small p

According to the behavior of the self-similar solution v to (5.2.4) and since any positive solution to (5.6.2) is a supersolution to (5.2.4), it is natural to formulate the following definition.

Definition 5.6.1. Considering $u(x,t)$, a positive solution to (5.6.2), we say that u blows up in a finite time if there exists $T^* < \infty$ such that

$$\lim_{t\to T^*} \int_{B_r(0)} |x|^{-\alpha_-(\lambda)} u(x,t)\,dx = \infty,$$

for any ball $B_r(0)$.

In order to find a Fujita type exponent for (5.6.2) where the blow-up phenomenon occurs in the sense of Definition 5.6.1, we follow closely the arguments used in [175] which are in some way constructive. More precisely, for $r = |x|$, we look for a family of subsolutions to problem (5.6.2) in the form $w(r,t,T) = (T-t)^{-\theta}\zeta(\frac{r}{(T-t)^\beta})$, with $\theta, \beta > 0$ to be chosen and $\zeta > 0$ a smooth function. Denoting $s = \frac{r}{(T-t)^\beta}$, it follows that

$$\begin{aligned} w_t(r,t) &= (T-t)^{-\theta-1}\Big(\theta\zeta + \frac{\beta r}{(T-t)^\beta}\zeta'(s)\Big), \\ w_r(r,t) &= (T-t)^{-\theta-\beta}\zeta'(s), \\ w_{rr}(r,t) &= (T-t)^{-\theta-2\beta}\zeta''(s). \end{aligned} \tag{5.6.3}$$

We need to have

$$w_t - w_{rr} - \frac{(N-1)}{r} w_r - \lambda \frac{w}{r^2} \le w^p. \tag{5.6.4}$$

By setting $\theta = \frac{1}{p-1}$, $\beta = \frac{1}{2}$ and replacing the value in (5.6.3), (5.6.4) becomes

$$-\zeta''(s) - \left(\frac{N-1}{s} - \frac{1}{2}s\right)\zeta'(s) - \left(\frac{\lambda}{s^2} - \theta\right)\zeta(s) \le \zeta^p(s).$$

We will try to find $\zeta(s)$ of the form $\zeta(s) = A\phi(cs)$ with $\phi(s) = s^{-\alpha_-(\lambda)}e^{-\frac{s^2}{4}}$, $A > 0$ and $c > 0$. Since $\alpha_-(\lambda)^2 - (N-2)\alpha_-(\lambda) + \lambda = 0$, we need the inequality

$$-(cs)^2\left(\frac{c^2}{4} + \frac{1}{4}\right) + \frac{c^2}{2} - c^2\alpha_-(\lambda) - \frac{\alpha_-(\lambda)}{2} + \frac{c^2(N-1)}{2} + \frac{1}{p-1} \le A^{p-1}(cs)^{-\alpha_-(\lambda)(p-1)}e^{-\frac{1}{4}(cs)^2(p-1)}$$

to be satisfied. Fixing $c = 1$, there exists s_0 such that for $s \ge s_0$,

$$\gamma_1(s) \equiv -s^2\left(\frac{1}{4} + \frac{1}{4}\right) + \frac{1}{2} - c^2\alpha_-(\lambda) - \frac{\alpha_-(\lambda)}{2} + \frac{(N-1)}{2} + \frac{1}{p-1} \le 0.$$

Set $M = \max\{\gamma_1(s) \mid 0 \le s \le s_0\}$; for A large enough we conclude that

$$M < A^{p-1}s^{-\alpha_-(\lambda)(p-1)}e^{-\frac{1}{4}s^2(p-1)} \quad \text{if } 0 \le s \le s_0.$$

Then for such an A

$$\gamma_1(s) \le A^{p-1}s^{-\alpha_-(\lambda)(p-1)}e^{-\frac{1}{4}s^2(p-1)}.$$

That is, we find a family of subsolutions to (5.6.2) given by

$$w(r,t,T) = A(T-t)^{-\frac{1}{p-1}}\left(\frac{r}{(T-t)^{\frac{1}{2}}}\right)^{-\alpha_-(\lambda)} e^{-\frac{1}{4}\frac{r^2}{T-t}}. \tag{5.6.5}$$

Let $B_r(0)$ be a ball in $\mathbb{R}^N$. Integrating we obtain

$$\int_{B_r(0)} |x|^{-\alpha_-(\lambda)}w(|x|,t,T)\,dx = C(T-t)^{-\frac{1}{p-1}+\frac{N}{2}-\frac{\alpha_-(\lambda)}{2}} \int_0^{\frac{r}{(T-t)^{1/2}}} \phi(s)s^{N-\alpha_-(\lambda)-1}\,ds.$$

Note that $-\frac{1}{p-1} + \frac{N}{2} - \frac{\alpha_-(\lambda)}{2} < 0$ implies $p < 1 + \frac{2}{N-\alpha_-(\lambda)}$.

Therefore if $p < 1 + \frac{2}{N-\alpha_-(\lambda)}$, we find

$$\lim_{t\to T} \int_{B_r(0)} |x|^{-\alpha_-(\lambda)}w(|x|,t,T)\,dx = \infty.$$

Hence a candidate to the Fujita exponent for problem (5.6.2) is $F(\lambda) = 1 + \frac{2}{N-\alpha_-(\lambda)}$. It is clear that if $0 < \lambda \le \Lambda_{N,2}$, then

$$1 < 1 + \frac{2}{N} < F(\lambda) < p_-(\lambda) \le \frac{N+2}{N-2} \le p_+(\lambda) < \infty.$$

5.6.2 Global supersolutions for $F(\lambda) < p < p_+(\lambda)$

The main goal of this subsection is to get a family of global supersolutions to problem (5.6.2) for $F(\lambda) < p < p_+(\lambda)$, where $F(\lambda) = 1 + \frac{2}{N-\alpha_-(\lambda)}$ as above. According to the previous results, the interval $F(\lambda) < p < p_+(\lambda)$ is the natural range in order to have some chances to obtaining global supersolutions.

We look for a family of radial supersolutions. We start with a w such that

$$w_t - w_{rr} - \frac{(N-1)}{r} w_r - \lambda \frac{w}{r^2} \geq w^p. \tag{5.6.6}$$

Assume that $w(r,t,T) = (T+t)^{-\theta} g(\frac{r}{(T+t)^\beta})$ (see [171] and [174]), with $\theta, \beta > 0$ to be chosen, and where g is a smooth bounded positive function. Denoting $s = \frac{r}{(T+t)^\beta}$, it follows that

$$\begin{aligned} w_t(r,t) &= -(T+t)^{-\theta-1}\left(\theta g + \frac{\beta r}{(T+t)^\beta} g'(s)\right), \\ w_r(r,t) &= (T+t)^{-\theta-\beta} g'(s), \\ w_{rr}(r,t) &= (T+t)^{-\theta-2\beta} g''(s). \end{aligned}$$

In order to get homogeneity in equation (5.6.6), it is sufficient to choose $\theta = \frac{1}{p-1}$ and $\beta = \frac{1}{2}$. Therefore (5.6.6) gives

$$g''(s) + \left(\frac{N-1}{s} + \frac{1}{2}s\right) g'(s) + \left(\frac{\lambda}{s^2} + \theta\right) g(s) + g^p(s) \leq 0.$$

Let us consider $\alpha_-(\lambda) < \gamma < \frac{2}{p-1}$. If we take $g(s) = A\phi(cs)$, with $\phi(s) = s^{-\gamma} e^{-\frac{s^2}{4}}$, $A > 0$, $c > 0$, then we need

$$\begin{aligned} &\frac{c^2[\gamma^2 - (N-2)\gamma + \lambda]}{(cs)^2} + (cs)^2\left[\frac{c^2}{4} - \frac{1}{4}\right] \\ &+ c^2\gamma - \frac{c^2}{2} - \frac{c^2(N-1)}{2} - \frac{\gamma}{2} + \frac{1}{p-1} \\ &+ A^{p-1}(cs)^{-(p-1)\gamma} e^{-\frac{(cs)^2}{4}(p-1)} \leq 0. \end{aligned}$$

Let $G(c,s)$ be defined as $G(c,s) = c^2(s - \frac{N}{2}) - \frac{s}{2} + \frac{1}{p-1}$. Since $p > F(\lambda)$, we have

$$G(1, \alpha_-(\lambda)) = -\frac{N\alpha_-(\lambda)}{2} + \frac{1}{p-1} < 0,$$

and hence by continuity there exist $c < 1$ and $\alpha_-(\lambda) < \gamma < \frac{2}{p-1}$ such that $G(c,\gamma) < 0$. As $c < 1$, we obtain $(cs)^2[\frac{c^2}{4} - \frac{1}{4}] \leq 0$. Moreover, since $\gamma < \frac{2}{p-1}$ and $c^2[\gamma^2 - (N-2)\gamma + \lambda] \leq 0$, choosing A small enough it follows that

$$\frac{c^2[\gamma^2 - (N-2)\gamma + \lambda]}{(cs)^2} + A^{p-1}(cs)^{-(p-1)\gamma} e^{-\frac{(cs)^2}{4}} \leq 0.$$

Therefore, we have found a family of supersolutions w defined by

$$w(r,t,T) = Ac^{-\gamma}(T+t)^{\frac{\gamma}{2}-\frac{1}{p-1}} r^{-\gamma} e^{-c^2 \frac{r^2}{4(T+t)}}. \tag{5.6.7}$$

Note that $w(x,0,T) = AT^{-\frac{1}{p-1}} a^{-\gamma}|x|^{-\gamma} e^{-a^2\frac{|x|^2}{4}}$, where $a = \frac{c}{T^{\frac{1}{2}}} < 1$.

Remark 5.6.2. Now we can conjecture that the Fujita exponent is $F(\lambda) = 1 + \frac{2}{N-\alpha_-(\lambda)}$. The next sections will show that this conjecture is true.

5.6.3 Local existence results for $1 < p < p_+(\lambda)$

We prove that problem (5.6.2) has a local positive solution for a class of suitable initial data. The main result is the following theorem.

Theorem 5.6.3. *Assume that* $1 < p < p_+(\lambda)$ *and*

$$u_0(x) \le T^{-\theta}|x|^{-\beta}, \text{ with } T > 0, \quad \alpha_-(\lambda) < \beta < \alpha_+(\lambda) \text{ such that } \beta + \alpha_-(\lambda) < N.$$

Then problem (5.6.2) *has a local positive solution* $u \in L^2(0,T; W^{1,2}_{\text{loc}}(\mathbb{R}^N))$.

Proof. Let $w(x,t) = \frac{1}{(T-t)^{\theta}}|x|^{-\beta}$ with $\alpha_-(\lambda) < \beta < \alpha_+(\lambda)$ and $\beta + \alpha_-(\lambda) < N$. It follows that $\beta(N-1) - \beta(\beta+1) - \lambda > 0$. Since $p < p_+(\lambda)$, we have $\beta < p\beta < \beta + 2$. Hence for $T > 1$, choosing θ large, we get easily that w is a supersolution to (5.6.2) in $\mathbb{R}^N \times (0,T)$. It is clear by the hypothesis on u_0 that $u_0(x) \le w(x,0)$.

To prove the existence of a solution u we follow by approximation. Let B_n be the ball in $\mathbb{R}^N$ with radius n and centered at the origin. We consider

$$v_n \in L^2((0,T), W^{1,2}_0(B_{n+1})), \quad \forall T > 0,$$

the weak solutions to the following approximated problems:

$$\begin{cases} v_{nt} - \Delta v_n = \lambda \frac{1}{|x|^2 + \frac{1}{n}} \tilde{v}_{n-1} + \tilde{v}^p_{n-1} & \text{in } B_{n+1},\ t > 0, \\ v_n(x,0) = u_0(x) & \text{in } B_{n+1},\ t > 0, \\ v_n(x,t) = 0 & \text{on } \partial B_{n+1},\ t > 0, \end{cases} \tag{5.6.8}$$

with

$$\begin{cases} v_{0t} - \Delta v_0 = 0 & \text{in } B_1,\ t > 0, \\ v_0(x,0) = u_0(x) & \text{in } B_1,\ t > 0, \\ v_0(x,t) = 0 & \text{on } \partial B_1,\ t > 0, \end{cases}$$

and $\tilde{v}_{n-1} = v_{n-1}$ in B_n, $\tilde{v}_{n-1} = 0$ in $\mathbb{R}^N \setminus B_n$. Applying the classical comparison principle, we conclude that $0 < v_0 \le v_1 \le \cdots \le v_{n-1} \le v_n \le w$ in $B_{n+1} \times (0, T_1)$ with $T_1 < T$. Hence there exists $u \in L^2(0,T_1,L^2_{\mathrm{loc}}(\mathbb{R}^N))$ such that $v_n \uparrow u$ strongly in $L^2((0,T_1),L^2_{\mathrm{loc}}(\mathbb{R}^N))$ and $u \le w$. Using the monotonicity of v_n and the dominated convergence theorem it follows that $v_n \to u$ strongly $L^p(K\times(0,T_1))$ for all compact sets $K \subset \mathbb{R}^N$. Take $\phi \in C_0^\infty(\mathbb{R}^N \times (0,T_1))$. Then using ϕ as a test function in (5.6.8) and by letting $n \to \infty$, we easily get that u solves problem (5.6.2) with $u(x,0) = u_0(x)$. It is clear that also $u \in L^2(0,T_1;W^{1,2}_{\mathrm{loc}}(\mathbb{R}^N))$. □

5.6.4 Global existence for $F(\lambda) < p < p_+(\lambda)$ and small data

In this subsection we consider the class of initial data

$$\mathcal{F}_D = \{u_0 : \mathbb{R}^N \to \mathbb{R} \mid 0 \le u_0(x) \le A|x|^{\alpha_-(\lambda)} e^{-D^2\frac{|x|^2}{4}}\}.$$

Let $w(r,t,T)$ be the family of supersolutions founded in (5.6.7). Choosing D such that

$$\left(\frac{2(\gamma-\alpha)}{D^2-a^2}\right)^{\frac{\gamma\alpha_-(\lambda)}{2}} e^{-\frac{\gamma\alpha_-(\lambda)}{2}} \le AT^{-\frac{1}{p-1}} a^{-\gamma},$$

we get $u_0(x) \le w(|x|,0,T)$. Thus w is a supersolution to (5.6.2). It is clear that $w \in L^2(0,T;W^{1,2}(\mathbb{R}^N))$ for all $T > 0$. Hence using an iteration argument as in Section 5.6.3 we get the existence of a global solution u to problem (5.6.2) with $u \le w$. We also have $u \in L^2(0,T;W^{1,2}(\mathbb{R}^N))$ for all $T > 0$ and $u(x,t) \to 0$ as $t \to \infty$ at least in the pointwise sense. Therefore we obtain the following result.

Theorem 5.6.4. *Assume that $u_0 \in \mathcal{F}_D$. Then for $F(\lambda) < p < p_+(\lambda)$, problem* (5.6.2) *has a minimal positive global energy solution $u \in L^2(0,T;W^{1,2}(\mathbb{R}^N))$ for all $T > 0$ such that $u(x,t) \to 0$ for $t \to \infty$.*

Remark 5.6.5. It would be very interesting to look for a general class of initial data such that problem (5.6.2) has a global solution. The imposed condition in Theorem 5.6.4 is far from optimal.

5.6.5 Blow-up result for $p < F(\lambda)$

In this subsection we prove that $F(\lambda)$ behaves like a Fujita type exponent, namely, we will prove that, for $p < F(\lambda)$, any solution to (5.6.2) blows up in a finite time in an appropriate sense and for any initial datum.

Let us begin by analyzing the properties of the subsolutions $w(r,t,T)$ defined in (5.6.5).

(i) By construction, $w(r,t,T)$ blows up in a finite time in the sense of the local weighted L^1-norm.

(ii) Assume that $p < F(\lambda) = 1 + \frac{2}{N-\alpha_-(\lambda)}$. Let us denote by $\overline{u}(x,t)$ a time translation of a solution to (5.6.2), namely, if $u(x,t)$ is a solution to (5.6.2), then $\overline{u}(x,t) = u(x,t+T)$. Since $\overline{u}(x,t)$ is a supersolution to the homogenous equation (5.2.4) with the same initial values, it is sufficient to check that $v(x,T) \geq w(r,0,T)$, in order to get that $\overline{u}(x,0) \geq w(r,0,T)$. This immediately follows because $p < F(\lambda) = 1 + \frac{2}{N-\alpha_-(\lambda)}$.

(iii) Consider $\overline{u}(x,t) \geq w(x,t)$, $\forall t < T$. Let us write $h(x,t) = w(x,t) - \overline{u}(x,t)$. Then, $h_t - \Delta h \leq \lambda \frac{h}{|x|^2} + w^p - \overline{u}^p$. Applying Kato's inequality (see [255]), it follows that

$$\begin{aligned} & h_t^+ - \Delta h_+ \leq \lambda \frac{h_+}{|x|^2} + p w^{p-1} h_+ \quad \text{in } \mathbb{R}^N,\ t \in (0,T_1),\ T_1 < T, \\ & h_+(x,0) = 0. \end{aligned} \tag{5.6.9}$$

Note that $h_+ \leq w$. Since $\lambda \frac{w^2}{|x|^2} + p w^{p+1} \in L^1(\mathbb{R}^N \times (0,T_1))$, we have $\lambda \frac{h_+^2}{|x|^2} + p w^{p-1} h_+^2 \in L^1(\mathbb{R}^N \times (0,T_1))$. Therefore using $T_k(h_+)$ as a test function in (5.6.9), it follows that

$$\int_{\mathbb{R}^N} \Theta_k(h_+)(x,T_1) + \int_0^{T_1} \int_{\mathbb{R}^N} |\nabla T_k(h_+)|^2 \, dx \, dt \leq \int_0^{T_1} \int_{\mathbb{R}^N} \left(\lambda \frac{h_+^2}{|x|^2} + p w^{p-1} h_+^2 \right) dx \, dt \leq C$$

for all $k > 0$. Hence, letting $k \to \infty$ we obtain $h^+ \in L^2(0,T_1,W^{1,2}(\mathbb{R}^N))$. Since $p < 1 + \frac{2}{N-\alpha_1}$, we have $\alpha_1(p-1) < 2$ and then there exists $C(T,T_1)$ such that $\forall \epsilon > 0$, $w^{p-1} \leq \epsilon \frac{1}{|x|^2} + C(T,T_1)$. Taking h_+ as a test function in (5.6.9), we get

$$\frac{\partial}{\partial t} \int_{\mathbb{R}^N} h_+^2(x,t) \, dx \leq C(T) \int_{\mathbb{R}^N} h_+^2(x,t) \, dx.$$

Applying Gronwall's inequality, we immediately deduce that $h_+ \equiv 0$ and then $\overline{u}(x,t) \geq w(x,t)$, $\forall t < T$.

Since we have found a family of subsolutions that has finite blow-up, we can deduce that for $p < F(\lambda) = 1 + \frac{2}{N-\alpha_-(\lambda)}$, the solutions to (5.6.2) blow up in a finite time.

Hence we have shown the following result.

Theorem 5.6.6. *Assume that $p \leq F(\lambda) = 1 + \frac{2}{N-\alpha_-(\lambda)}$. Then u blows up in a finite time in the sense of Definition* 5.6.1.

The critical case $p = F(\lambda)$ is considered in the last section.

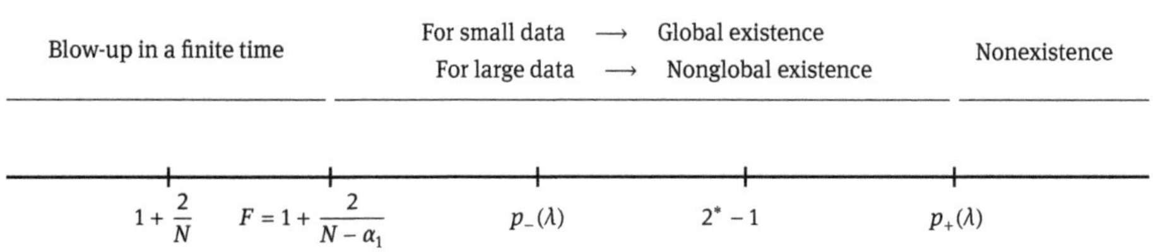

Fujita exponent for the heat equation with Hardy potential.

5.7 Further results and remarks

5.7.1 Problems involving the p-Laplacian heat equation

In the papers [30] and [179] the following types of problems were considered:

$$\begin{cases} u_t - \Delta_p u = \frac{\lambda}{|x|^p}|u|^{p-2}u + \mu|u|^{q-1}u & x \in \Omega,\ t > 0,\ \lambda \in \mathbb{R}, \\ u(x,0) = f(x) & x \in \Omega, \\ u(x,t) = 0 & x \in \partial\Omega,\ t > 0, \end{cases} \tag{5.7.1}$$

where $-\Delta_p u \equiv -\operatorname{div}(|\nabla u|^{p-2}\nabla u), f(x) \geq 0$ satisfying convenient regularity assumptions, Ω is a bounded domain in $\mathbb{R}^N$ such that $0 \in \Omega$, $q > 0$ and $1 < p < N$. The elliptic part of each problem is the Euler–Lagrange equation of the Hardy inequality obtained in Section 2.4. The analysis reveals that the behavior depends on p, more precisely on the Harnack inequality for the parabolic operator. This means that there is a different behavior according to whether $p \geq 2$ or $1 \leq p < 2$. See for instance [18] and [130] for a more general class of problems involving the Caffarelli–Kohn–Nirenberg inequalities. We recall that for $1 \leq p < 2$ there is the finite time extinction of solutions. For a general study of the p-Laplacian heat equation the reader is referred to the classical book by E. Di Benedetto [134]. The existence and regularity results depend in general on the relation between λ and the best constant in Hardy's inequality defined in Lemma 2.4.1.

We omit more details about these kinds of problems and we refer to the quoted articles.

5.7.2 Problems associated to the Caffarelli–Kohn–Nirenberg inequalities

The paper by L. Caffarelli, R. Kohn and L. Nirenberg [107] appeared in the year 1982. It contains the closest result up to now to solve the conjecture by J. Leray in [225]. More precisely, what they proved was that the set of singularities, S, of a class of solutions of the Navier–Stokes equations verifies

$$\mathcal{H}^1(S) = 0,$$

where $\mathcal{H}^1$ denotes the Hausdorff 1-measure. As a byproduct of the arguments in the proof of this result, the authors also proved in [108] a family of weighted inequalities, often known as *Caffarelli–Kohn–Nirenberg inequalities*. See also Section 6.2.3.

In the article [130] one of the authors analyzed the family of problems

$$\begin{cases} u_t - \operatorname{div}(|x|^{-p\gamma}|\nabla u|^{p-2}\nabla u) = \lambda \frac{u^{p-2}u}{|x|^{p(\gamma+1)}}, & \text{in } \Omega \times (0,\infty),\ 0 \in \Omega, \\ u(x,t) = 0, & \text{on } \partial\Omega \times (0,\infty) \\ u(x,0) = \psi(x) \geq 0, \end{cases} \tag{5.7.2}$$

when $1 < p < N$, $-\infty < (\gamma + 1) < \frac{N}{p}$, under certain hypotheses for the initial data. This family includes linear and quasilinear parabolic evolution problems. Note that if $p = 2$ and $\gamma = 0$ we find again the heat equation perturbed by the Hardy–Leray potential. One can also find results for nonlinear reaction terms in the literature; see for instance [16] and [18].

5.7.3 The borderline case $p = F(\lambda)$

We consider in this section the borderline case $p = F(\lambda)$. The argument will be different and technically more complicated than in the subcritical case. We follow some technical ideas used in [313]. We argue by contradiction. Assume that u is a global solution to problem (5.6.2) in such a way that for some ball $B_r(0)$,

$$\int_{B_r(0)} |x|^{-\alpha_-(\lambda)} u(x,t)\,dx < \infty \quad \text{for all } t > 0.$$

We set $v(x,t) = |x|^{\alpha_-(\lambda)} u(x,t)$. Therefore v satisfies

$$|x|^{-2\alpha_-(\lambda)} v_t - \operatorname{div}(|x|^{-2\alpha_-(\lambda)} \nabla v) = |x|^{\alpha_-(\lambda)(p+1)} v^p, \tag{5.7.3}$$

with

$$\int_{B_r(0)} |x|^{-2\alpha_-(\lambda)} v(x,t)\,dx < \infty \quad \text{for all } t > 0. \tag{5.7.4}$$

Choose θ such that $\frac{1}{p} < \theta < 1$. Let $\psi \in \mathcal{C}^2(\mathbb{R}^N)$ be a cutoff function such that:

1. $\psi = 1$ in $B_1(0)$, $\psi = 0$ in $\mathbb{R}^N \backslash B_2(0)$ and $0 \le \psi \le 1$;
2. $|\Delta\psi| \le C(\theta)\psi^\theta$.

For each $l \in \mathbb{N}$, consider the scaled cutoff function $\psi_l(x) = \psi(\frac{x}{l})$. Using ψ_l as a test function in (5.7.3), it follows that

$$\frac{d}{dt}\int_{\mathbb{R}^N} |x|^{-2\alpha_-(\lambda)} v\psi_l\,dx = -\int_{\mathbb{R}^N} v \operatorname{div}(|x|^{-2\alpha_-(\lambda)} \nabla\psi_l)\,dx + \int_{\mathbb{R}^N} |x|^{\alpha_-(\lambda)(p+1)} v^p \psi_l\,dx. \tag{5.7.5}$$

In the sequel we denote by C any positive constant that is independent of v and $l \in \mathbb{N}$. Using the Hölder inequality, we easily find

$$\begin{aligned}
&\int_{\mathbb{R}^N} v|\operatorname{div}(|x|^{-2\alpha_-(\lambda)} \nabla\psi_l)|\,dx \\
&\quad \le \Big(\int_{\mathbb{R}^N} |x|^{\alpha_-(\lambda)(p+1)} v^p \psi_l\,dx\Big)^{\frac{1}{p}} \Big(\int_{l \le |x| \le 2l} |x|^{\frac{p+1}{p-1}\alpha_-(\lambda)} \frac{|\operatorname{div}(|x|^{-2\alpha_-(\lambda)} \nabla\psi_l)|^{p'}}{\psi_l^{\frac{1}{p-1}}}\,dx\Big)^{\frac{1}{p'}} \\
&\quad \le C l^{\frac{N\alpha_-(\lambda)}{p'} - 2} \Big(\int_{\mathbb{R}^N} |x|^{\alpha_-(\lambda)(p+1)} v^p \psi_l\,dx\Big)^{\frac{1}{p}},
\end{aligned}$$

where in the last estimate we have used the fact that for $1 \le l \le |x| \le 2l$,

$$\frac{|\operatorname{div}(|x|^{-2\alpha_-(\lambda)}\nabla\psi_l)|^{p'}}{\psi_l^{\frac{1}{p-1}}} = \left(\frac{|\operatorname{div}(|x|^{-2\alpha_-(\lambda)}\nabla\psi_l)|}{\psi_l^{\theta}}\right)^{p'} \psi_l^{\theta p' - \frac{1}{p-1}}$$
$$\le C(\theta)|x|^{-2p'\alpha_-(\lambda)} l^{-2p'} \le C(\theta) l^{-2p'(\alpha_-(\lambda)+1)}.$$

Note that since $\frac{1}{p} < \theta < 1$, we have $\theta p' - \frac{1}{p-1} > 0$. We call $w_l(t) = \int_{\mathbb{R}^N} |x|^{-2\alpha_-(\lambda)} v\psi_l\, dx$. Then

$$\frac{d}{dt} w_l(t) \ge (\mathrm{int}_{\mathbb{R}^N} |x|^{\alpha_-(\lambda)(p+1)} v^p \psi_l\, dx)^{\frac{1}{p}} [-Cl^{\frac{N\alpha_-(\lambda)}{p'}-2} + (|x|^{\alpha_-(\lambda)(p+1)} v^p \psi_l\, dx)^{\frac{p-1}{p}}]. \tag{5.7.6}$$

Using again the Hölder inequality, it follows that

$$\int_{\mathbb{R}^N} |x|^{\alpha_-(\lambda)(p+1)} v^p \psi_l\, dx \ge C_0 w_l^p(t) l^{-(p-1)(N\alpha_-(\lambda))}. \tag{5.7.7}$$

We finally obtain

$$\frac{d}{dt} w_l(t) \ge C w_l(t) l^{-\frac{N\alpha_-(\lambda)}{p'}} [-Cl^{\frac{N\alpha_-(\lambda)}{p'}-2} + w_l^{p-1}(t) l^{-\frac{p-1}{p'}(N\alpha_-(\lambda))}].$$

Since $p = 1 + \frac{2}{N\alpha_-(\lambda)}$, we have $\frac{N\alpha_-(\lambda)}{p'} - 2 = -\frac{p-1}{p'}(N\alpha_-(\lambda))$, and hence,

$$\frac{d}{dt} w_l(t) \ge C_1 w_l(t) l^{-2}(-C_2 + w_l^{p-1}(t)). \tag{5.7.8}$$

Note that w_l is an increasing function in l. Under the hypothesis that u is a global solution, from (5.7.8) necessarily

$$w_l^{p-1}(t) \le C_2 \quad \text{for all } t > 0 \text{ and } l > 0.$$

In fact, by contradiction, if for some $\epsilon > 0$, there exist $t_0 > 0$, $l_0 > 0$ such that $w_{l_0}^{p-1}(t_0) \ge C_2 + \epsilon$, then for any $l \ge l_0$ we have $\frac{dw_l}{dt} > 0$ in the time in which w_l is defined. Consider now the solution to the initial value problem

$$y'(t) = C_1 y(t) l^{-2}(-C_2 + y^{p-1}(t)), \quad y(t_0) = (C_2 + \epsilon)^{\frac{1}{p-1}}.$$

Since the solution blows up in an explicit finite time, T, obtained by elementary integration and by classical comparison arguments, we conclude that there exists $T^* \le T$, such that

$$w_l(t) \uparrow \infty \quad \text{as } t \uparrow T^*.$$

But this is a contradiction with (5.7.4), which proves the uniform estimate for w_l. Moreover, since C_2 is independent of l, by letting $l \to \infty$ we obtain

$$\int_{\mathbb{R}^N} |x|^{-2\alpha_-(\lambda)} v(x,t)\,dx \le C_2 \quad \text{for all } t > 0. \tag{5.7.9}$$

Also we have

$$\begin{aligned}\int_{\mathbb{R}^N} v|\operatorname{div}(|x|^{-2\alpha_-(\lambda)}\nabla\psi_l)|\,dx &\le \int_{l\le|x|\le 2l} v\frac{|\operatorname{div}(|x|^{-2\alpha_-(\lambda)}\nabla\psi_l)|}{\psi_l^\theta}\psi_l^\theta\,dx \\ &\le \frac{C(\theta)}{l^2}\int_{\mathbb{R}^N} |x|^{-2\alpha_-(\lambda)} v(x,t)\,dx \le Cl^{-2}.\end{aligned}$$

Hence from (5.7.5), by letting $l \to \infty$ we conclude that

$$\frac{d}{dt}\int_{\mathbb{R}^N} |x|^{-2\alpha_-(\lambda)} v(x,t)\,dx \ge \frac{1}{2}\int_{\mathbb{R}^N} |x|^{\alpha_-(\lambda)(p+1)} v^p(x,t)\,dx, \tag{5.7.10}$$

in the distributional sense.

For $\tau \in (0,1)$, we write (5.7.8) as

$$\frac{dw_l}{dt} \ge C_1 w^{1-\tau} l^{-2}(-C_2 w_l^\tau + w_l^{p-(1-\tau)}), \tag{5.7.11}$$

in the distributional sense.

Next, consider $\phi_l(x) = 1 - \psi_l(x)$ and a cutoff function, $\xi \in C_0^\infty(\mathbb{R}^N)$, such that $0 \le \xi \le 1$, $\xi(x) = 1$ for $|x| \le 2$ and $\xi(x) = 0$ if $|x| \ge 3$. Define $\xi_k(x) = \xi(\frac{x}{k})$.

Using $\phi_l\xi_k$ as a test function in (5.7.3) it follows that

$$\begin{aligned}\frac{d}{dt}\int_{\mathbb{R}^N} |x|^{-2\alpha_-(\lambda)} v\phi_l\xi_k\,dx &\le \int_{\mathbb{R}^N} + \int_{\mathbb{R}^N} |x|^{\alpha_-(\lambda)(p+1)} v^p\phi_l\xi_k\,dx \\ &\le \int_{\mathbb{R}^N} v|\operatorname{div}(|x|^{-2\alpha_-(\lambda)}\nabla\phi_l)|\,dx + \int_{\mathbb{R}^N} v|\operatorname{div}(|x|^{-2\alpha_-(\lambda)}\nabla\xi_k)|\,dx \\ &\quad + \int_{\mathbb{R}^N} |x|^{\alpha_-(\lambda)(p+1)} v^p\phi_l\xi_k\,dx.\end{aligned}$$

Since $\nabla\phi_l = -\nabla\psi_l$, by similar methods as above, we obtain

$$\int_{\mathbb{R}^N} v|\operatorname{div}(|x|^{-2\alpha_-(\lambda)}\nabla\phi_l)|\,dx \le Cl^{-2+\frac{N\alpha_-(\lambda)}{p'}}\Big(\int_{\mathbb{R}^N} x|^{\alpha_-(\lambda)(p+1)} v^p\,dx\Big)^{\frac{1}{p}}.$$

Now for $k \gg l$ we have

$$\int_{\mathbb{R}^N} v|\operatorname{div}(|x|^{-2\alpha_-(\lambda)}\nabla\xi_k)|\,dx \le Ck^{-2}\int_{\mathbb{R}^N} |x|^{-2\alpha_-(\lambda)} v\phi_l\,dx.$$

Thus, we have

$$\frac{d}{dt}\int_{\mathbb{R}^N} |x|^{-2\alpha_-(\lambda)} v\phi_l\xi_k\,dx$$
$$\le Cl^{-2+\frac{N\alpha_-(\lambda)}{p'}}\Big(\int_{\mathbb{R}^N}\frac{v^p}{|x|^{\alpha_-(\lambda)(p+1)}}\,dx\Big)^{\frac{1}{p}} + Ck^{-2}\int_{\mathbb{R}^N}\frac{v\phi_l}{|x|^{2\alpha_-(\lambda)}}\,dx + \int_{\mathbb{R}^N}|x|^{\alpha_-(\lambda)(p+1)}v^p\phi_l\xi_k\,dx.$$

Therefore by letting $k\to\infty$ and by Young's inequality we have

$$\frac{d}{dt}\int_{\mathbb{R}^N}|x|^{-2\alpha_-(\lambda)}v\phi_l\,dx \le 2\int_{\mathbb{R}^N}|x|^{\alpha_-(\lambda)(p+1)}v^p\,dx + C_3 l^{-2p'+N\alpha_-(\lambda)}. \tag{5.7.12}$$

Consider

$$A = \sup_{t>0,l>0} w_l(t) = \sup_{t>0}\int_{\mathbb{R}^N}|x|^{-2\alpha_-(\lambda)}v(x,t)\,dx.$$

Then, using (5.7.9) and the fact that u is a global solution, we obtain $0 < A < \infty$. Since w_l is increasing in l, then for all $\epsilon > 0$ there exist $t_0 \ge 0$ and $l_0 \ge 2$ such that $w_{\frac{l_0}{2}}(t_0) \ge A - \epsilon$.

From (5.7.10) and by integrating in time, we obtain

$$\frac{1}{2}\int_{t_0}^{\infty}\int_{\mathbb{R}^N}|x|^{\alpha_-(\lambda)(p+1)}v^p\,dx\,dt \le \sup_{t>0}\int_{\mathbb{R}^N}|x|^{-2\alpha_-(\lambda)}v(x,t)\,dx - w_{\frac{l_0}{2}}(t_0) \le \epsilon.$$

Let $s \ge t_0$. By (5.7.12), a direct computation provides

$$\int_{\mathbb{R}^N}|x|^{-2\alpha_-(\lambda)}v(x,s)\phi_{\frac{l_0}{2}}(x)\,dx$$
$$\le 2\int_{t_0}^{s}\int_{\mathbb{R}^N}|x|^{\alpha_-(\lambda)(p+1)}v^p\,dx + C_4\Big(\frac{l_0}{2}\Big)^{-2p'+N\alpha_-(\lambda)}(s-t_0) + \int_{\mathbb{R}^N}|x|^{-2\alpha_-(\lambda)}v(x,s)\phi_{\frac{l_0}{2}}(x)\,dx$$
$$\le 3\epsilon + C_4\Big(\frac{l_0}{2}\Big)^{-2p'+N\alpha_-(\lambda)}(s-t_0).$$

Now setting $l = l_0$ in (5.7.11) it follows that

$$\frac{dw_{l_0}}{dt} \ge C_1 w^{1-\tau} l_0^{-2}\Big(-C_2\Big(\int_{l_0\le|x|\le 2l_0}|x|^{-2\alpha_-(\lambda)}v\,dx\Big)^{\tau} + w_{l_0}^{p-(1-\tau)}\Big)$$
$$\ge C_1 w^{1-\tau} l_0^{-2}\Big(-C_2\Big(3\epsilon + C_4\Big(\frac{l_0}{2}\Big)^{-2p'+N\alpha_-(\lambda)}(s-t_0)\Big)^{\tau} + w_{l_0}^{p-(1-\tau)}\Big)$$

in (t_0, ∞) in the distributional sense. Fix now $\epsilon_0 \in (0, A)$ and $M > 0$ such that $C_2(3\epsilon + M)^\tau \le \frac{1}{2}(A - \epsilon_0)^{p-(1-\tau)}$. Then, by letting

$$t_1 = t_0 + \frac{M}{C_4}\left(\frac{l_0}{2}\right)^{2p'-(N\alpha_-(\lambda))},$$

it follows that

$$\frac{dw_{l_0}(t)}{dt} \ge \frac{C}{2} l_0^{-2} w_{l_0}^p(t) \quad \text{for all } t \in (t_0, t_1),$$

in the distributional sense. After integration in time and using the fact that $w_{l_0}(t)$ is increasing for $t \in (t_0, t_1)$, we obtain

$$\begin{aligned} w_{l_0}(t_1) &\ge w_{l_0}(t_0) + \frac{C}{2} l_0^{-2}(A - \epsilon_0)^p (t_1 - t_0) \\ &\ge w_{l_0}(t_0) + \frac{C}{2} l_0^{-2}(A - \epsilon_0)^p \frac{M}{C_3}\left(\frac{l_0}{2}\right)^{2p'-(N\alpha_-(\lambda))}. \end{aligned}$$

Since $p = 1 + \frac{2}{N\alpha_-(\lambda)}$, we have $2p' - (N\alpha_-(\lambda)) = 2$, and therefore

$$w_{l_0}(t_1) \ge w_{l_0}(t_0) + \frac{C}{2}(A - \epsilon_0)^p \frac{M}{C_3} 2^{-2p'+(N\alpha_-(\lambda))}.$$

We set $\varrho = \frac{C}{2}(A - \epsilon_0)^p \frac{M}{C_3} 2^{-2p'+(N\alpha_-(\lambda))}$. Then

$$w_{l_0}(t_1) \ge w_{l_0}(t_0) + \varrho \ge A - \epsilon_0 + \varrho.$$

As w_l is an increasing function in l, using the same argument as above, we find

$$w_{2l_0}(t_2) \ge w_{2l_0}(t_1) + \varrho \ge A - \epsilon_0 + 2\varrho,$$

where

$$t_2 = t_1 + \frac{M}{C_4} l_0^{2p'-(N\alpha_-(\lambda))}.$$

Hence by an iteration argument it follows that

$$w_{2^{i-1}l_0}(t_i) \ge w_{2^{i-1}l_0}(t_1) + i\varrho \ge A - \epsilon_0 + i\varrho,$$

where

$$t_i = t_{i-1} + \frac{M}{C_4}(2^{i-2} l_0)^{2p'-(N\alpha_-(\lambda))}.$$

Finally we conclude that

$$\sup_{t>0} \int_{\mathbb{R}^N} |x|^{-2\alpha_-(\lambda)} v(x,t)\, dx \geq w_{2^{i-1} l_0}(t_i) \geq i\varrho \to \infty \quad \text{as } i \to \infty,$$

a contradiction with (5.7.9).

Therefore, for all initial data $u_0(x) \geq 0$, $u_0(x) \not\equiv 0$, there exists $T < \infty$ such that

$$\lim_{t \uparrow T} \int_{B_r(0)} |x|^{-\alpha_1} u(x,t)\, dx = +\infty. \qquad \square$$

6 Elliptic equations with a nonlinearity on the gradient and the Hardy–Leray potential

6.1 Introduction

In this chapter we study the solvability of the quasilinear elliptic equation

$$-\Delta u = |\nabla u|^q + \lambda \frac{u}{|x|^2} + cf, \quad x \in \Omega \subset \mathbb{R}^N, \; N \geq 3, \; 1 < q \leq 2, \tag{6.1.1}$$

where Ω is a domain such that $0 \in \Omega$ and the existence of solutions in particular for the Dirichlet problem. We assume that λ, c are positive real numbers and f is a nonnegative function under some extra hypotheses that will be presented later.

As in the previous chapters, we will see that the behavior of equation (6.1.1) is deeply related to the values $\alpha_{\pm}(\lambda)$ defined in (3.1.3), the exponents of the radial weak solutions to

$$-\Delta u - \lambda \frac{u}{|x|^2} = 0,$$

which we know can be written as $u(r) = c_1|x|^{-\alpha_+(\lambda)} + c_2|x|^{-\alpha_-(\lambda)}$, $c_1, c_2 \in \mathbb{R}$.

As in the semilinear case, the problem under consideration will be to describe, for each $\lambda \in (0, \Lambda_{N,2}]$ fixed, the optimal exponent $q_+(\lambda)$ for which there is a solution of (6.1.1), at least in a weak sense, when $1 < q < q_+(\lambda) < 2$. It is clear that the variational techniques are not useful in the quasilinear setting and that the difficulties are considerably bigger than in the semilinear case studied in Chapter 4.

It is worth pointing out that this type of quasilinear problems appears in several contexts. For instance, for the quadratic case $q = 2$, in the simplest situation of $\lambda = 0$, problem (6.1.1) is the stationary counterpart of the Kardar–Parisi–Zhang model (see [205]) and of some flame propagation models (see [57]). The existence of solutions in this case can be seen for instance in [212], [75] and the references therein. For stronger results on classification and multiplicity of solutions, we refer to [8]. Moreover, equation (6.1.1) can be read as the Hamilton–Jacobi equation

$$|\nabla u|^q + \lambda \frac{u}{|x|^2} + cf = 0$$

with the viscosity term given by the Laplacian. As a consequence of the nonexistence results in this paper, the reader could check without difficulty that the *vanishing viscosity method* by P. Lax does not produce a solution for the first order equation. See for instance [232] for details and applications of this. The physical motivation for $q < 2$ can be seen in [213]. There is a huge amount of references for the case $q \leq 2$; see for instance [74], [75], [76], [78], [156], [157], [214], [249] and [266].

When $\lambda > 0$, that is, when the Hardy potential is present, the behavior is different. The study of this behavior is the main goal of this chapter. Section 6.2 starts with some auxiliary results about quasilinear nonlinearities that have some interest

https://doi.org/10.1515/9783110606270-006

in itself and try to be self-contained. In Section 6.3 we discuss the candidate to be the critical exponent $q_+(\lambda)$ and prove that if $q > q_+(\lambda)$ problem (6.1.1) has not a *very weak supersolution* in the sense presented below (see Definition 6.3.1). In Section 6.4 we analyze the complete blow-up of some approximated problem. This is crucial for a better understanding of the nonexistence results. In order to prove that in fact $q_+(\lambda)$ is the threshold for existence, we must show existence of solutions for $q < q_+(\lambda)$. These results are obtained in Sections 6.5 and 6.6. Most of them come from paper [17].

6.2 Some auxiliary results

In this section we will obtain some results that will be used later.

6.2.1 Comparison results

We describe here two comparison results. The first one is related to finite energy solutions and will use the arguments in [266]. The second one relates to data in $L^1(\Omega)$ and was obtained in [31].

We start by proving the comparison for finite energy solutions. More precisely, we prove the following result.

Proposition 6.2.1. *Let $w \in W^{1,2}(\Omega)$ be such that*

$$\begin{cases} -\Delta w \le \langle h(x), \nabla w\rangle & \Omega, \\ w \le 0 & \partial\Omega, \end{cases} \tag{6.2.1}$$

where $h \in L^q(\Omega)$, $q > N$. Then $w \le 0$ in Ω.

Proof. Taking $w_k := (w-k)^+$ as a test function in (6.2.1) and applying the Hölder and Sobolev inequalities, we obtain

$$\begin{aligned} \int_{\Omega_k} |\nabla w_k|^2\,dx &\le |\Omega|^\alpha \Big(\int_{\Omega_k} |\nabla w_k|^2\,dx\Big)^{\frac12}\Big(\int_{\Omega_k} |h|^q\,dx\Big)^{\frac1q}\Big(\int_{\Omega_k} |w_k|^{\frac{2N}{N-2}}\,dx\Big)^{\frac{N-2}{2N}} \\ &\le |\Omega|^\alpha \mathcal{S}\Big(\int_{\Omega_k} |\nabla w_k|^2\,dx\Big)\Big(\int_{\Omega_k} |h|^q\,dx\Big)^{\frac1q}, \end{aligned} \tag{6.2.2}$$

where $\mathcal{S}$ is the optimal constant in the Sobolev inequality, $\alpha = 1 - (\frac12 + \frac1N + \frac{N-2}{2N})$ and

$$\Omega_k = \{x \in \Omega : w(x) > k, |\nabla w| > 0\}.$$

Let us argue by contradiction and assume that $\sup w > 0$.

Denote $M = \sup w$. Then either $M = +\infty$ and, since $w \in W^{1,2}(\Omega)$, we have

$$\lim_{k\to M} |\Omega_k| = 0, \tag{6.2.3}$$

or $M < \infty$ and again by the fact that $w \in W^{1,2}(\Omega)$, we have $\nabla w = 0$ a. e. in $\{x \in \Omega \mid w = M\}$. By Stampacchia's theorem, therefore, we find again (6.2.3). As a consequence,

$$\|h\|_{L^q(\Omega_k)} \to 0 \quad \text{as } k \to M,$$

and then there exists $k_0 < M$ such that

$$|\Omega|^\alpha S\|h\|_{L^q(\Omega_k)} \le \frac{1}{2}.$$

Hence from (6.2.2) we find that

$$\int_{\Omega_{k_0}} |\nabla w_k|^2\,dx = 0 \quad \text{and then } w \le k_0 < M,$$

which gives a contradiction. □

Lemma 6.2.2. *There exists a unique solution $u \in W_0^{1,2}(\Omega)$, with $u \ge 0$ in Ω, to the problem*

$$\begin{cases} -\Delta u = \langle h(x), \nabla u\rangle + f & x \in \Omega, \\ u = 0 & x \in \partial\Omega, \end{cases}$$

where $h : \Omega \to \mathbb{R}^N$ is such that $|h| \in L^q(\Omega)$, $q > N$, and $f \in W^{-1,2}(\Omega)$. In particular we have

$$\|v\|_{W_0^{1,2}(\Omega)} \le C\|f\|_{W^{-1,2}(\Omega)}. \tag{6.2.4}$$

Proof. Let us define the operator

$$\begin{aligned} T : W_0^{1,2}(\Omega) &\longrightarrow W_0^{1,2}(\Omega), \\ u &\mapsto T(u) = v, \end{aligned}$$

such that v verifies

$$-\Delta v = \langle h(x), \nabla u\rangle + f.$$

The continuity of T follows easily. We show that T is compact, i. e., that for any sequence (u_n) bounded in $W_0^{1,2}(\Omega)$ there exists a subsequence of $v_n = T(u_n)$ which converges strongly to some $v \in W_0^{1,2}(\Omega)$. We note that $\langle h(x), \nabla u\rangle \in W^{-1,2}(\Omega)$, since for any $\varphi \in W_0^{1,2}(\Omega)$ one has

$$\begin{aligned} &\left|\int_\Omega \langle h(x), \nabla u\rangle \varphi\,dx\right| \\ &\quad \le C(\Omega)\Big(\int_\Omega |h(x)|^q\,dx\Big)^{\frac{1}{q}}\Big(\int_\Omega |\nabla u|^2\,dx\Big)^{\frac{1}{2}}\Big(\int_\Omega \varphi^{2^*}\,dx\Big)^{\frac{1}{2^*}} \\ &\quad \le C(\Omega)\Big(\int_\Omega |\nabla\varphi|^2\Big)^{\frac{1}{2}}, \end{aligned}$$

by the Sobolev inequality. The fact that also $f \in W^{-1,2}(\Omega)$ yields that the sequence $v_n = T(u_n)$ is uniformly bounded in $W_0^{1,2}(\Omega)$. By Rellich's theorem we get that for a subsequence, still labeled with n, it holds that

$$\begin{aligned} v_n \to v &\quad \text{strongly in } L^2(\Omega), \\ \nabla v_n \rightharpoonup \nabla v &\quad \text{weakly in } (L^2(\Omega))^N, \\ v_n \to v &\quad \text{a. e. in } \Omega, \end{aligned} \tag{6.2.5}$$

with $v \in W_0^{1,2}(\Omega)$. We take as test function $v_n - v$ to arrive at

$$\begin{aligned} \int_\Omega |\nabla(v_n - v)|^2\,dx &= \int_\Omega \langle h(x), \nabla u_n\rangle (v_n - v)\,dx + \int_\Omega f(v_n - v)\,dx \\ &\quad - \int_\Omega \nabla v \nabla(v_n - v)\,dx \\ &= \int_\Omega \langle h(x), \nabla v\rangle (v_n - v)\,dx + o(1), \end{aligned}$$

due to the weak convergence $\nabla v_n \rightharpoonup \nabla v$ in $(L^2(\Omega))^N$, the strong convergence $v_n \to v$ in $L^2(\Omega)$ and the fact that $f \in W^{-1,2}(\Omega)$. With the remaining term we argue as follows:

$$\begin{aligned} \left|\int_\Omega \langle h(x), \nabla v\rangle (v_n - v)\,dx\right| &\le \int_\Omega |h(x)||\nabla v||T_m(v_n - v)|\,dx \\ &\quad + \int_{\Omega_m} |h(x)||\nabla v||v_n - v|\,dx := I_1 + I_2, \end{aligned}$$

where $T_m(s) = \max\{-m, \min(s, m)\}$ is the usual truncation and $\Omega_m = \Omega_m^n =: \Omega \cap \{|v_n - v| > m\}$. Observe that

$$\begin{aligned} I_1 \le \left(\int_\Omega |h(x)|^q\,dx\right)^{\frac{1}{q}} \left(\int_\Omega |\nabla v|^2\,dx\right)^{\frac{1}{2}} \\ \times \left(\int_\Omega |T_m(v_n - v)|^{2^*}\,dx\right)^{\frac{1}{2^*}}. \end{aligned}$$

The strong convergence $v_n \to v$ in $L^2(\Omega)$ implies the strong convergence of $T_m(v_n - v)$ to zero in $L^r(\Omega)$ for any $r \ge 1$, in particular for $r = 2^*$, thus $I_1 = o(1)$ as $n \to \infty$.

Concerning the second term, we have

$$\begin{aligned} I_2 \le \left(\int_{\Omega_m} |h(x)|^q\,dx\right)^{\frac{1}{q}} \left(\int_{\Omega_m} |\nabla v|^2\,dx\right)^{\frac{1}{2}} \\ \times \left(\int_{\Omega_m} |v_n - v|^{2^*}\,dx\right)^{\frac{1}{2^*}} \end{aligned}$$

$$\leq C\Big(\int_{\Omega_m} |h(x)|^q\,dx\Big)^{\frac{1}{q}}\Big(\int_{\Omega_m} |\nabla(v_n - v)|^2\,dx\Big)^{\frac{1}{2}}$$

$$\leq C\Big(\int_{\Omega_m} |h(x)|^q\,dx\Big)^{\frac{1}{q}}.$$

Summing up, we have shown that

$$\int_{\Omega} |\nabla(v_n - v)|^2\,dx \leq C\sup_n\Big(\int_{\Omega_m} |h(x)|^q\,dx\Big)^{\frac{1}{q}}.$$

We proved that $v_n - v$ is uniformly bounded in $L^2(\Omega)$ and thus $|\Omega_m^n| \to 0$ as $n \to \infty$. Therefore,

$$\lim_{n\to\infty} \int_{\Omega} |\nabla(v_n - v)|^2\,dx = 0,$$

and hence v_n is strongly convergent in $W_0^{1,2}(\Omega)$, which concludes the compactness of T.

In order to apply the fixed point theorem, we show that there exists $M > 0$, such that $\|u\|_{W_0^{1,2}(\Omega)} \leq M$, for every $u \in W_0^{1,2}(\Omega)$ and every $\sigma \in [0,1]$ verifying $u = \sigma T(u)$.

Arguing by contradiction, we assume that for every n there exist $u_n \in W_0^{1,2}(\Omega)$ and $\sigma_n \in [0,1]$ such that $u_n = \sigma_n T(u_n)$ and $\|u_n\|_{W_0^{1,2}(\Omega)} \geq n$. We consider $w_n = \frac{u_n}{\|u_n\|_{W_0^{1,2}(\Omega)}}$ that satisfies

$$-\Delta w_n = \sigma_n \langle h(x), \nabla w_n\rangle + \sigma_n \frac{f}{\|u_n\|_{W_0^{1,2}(\Omega)}}.$$

Since $\|u_n\|_{W_0^{1,2}(\Omega)} \to \infty$, applying the Kato inequality (see [209, 96]), this identity implies that for n large

$$-\Delta w_n^+ \leq C|h(x)||\nabla w_n^+|.$$

Then, by Lemma 6.2.1 it follows that $w^+ \equiv 0$, which contradicts the fact that

$$\|w_n\|_{W_0^{1,2}(\Omega)} = 1.$$

We have proved that T has a unique fixed point, which is the desired solution.

Finally we show the last estimate of the proposition. Set $v_1 = \frac{v}{\|f\|_{W^{-1,2}(\Omega)}}$, then repeating the same computation as in the previous step, we reach

$$\|v_1\|_{W_0^{1,2}(\Omega)} \leq C(h),$$

and hence (6.2.4) follows. □

We now formulate and prove the comparison in the weak setting. See also [31], where the result was obtained by using the isoperimetric inequality.

Proposition 6.2.3. *Let $w \in L^1(\Omega)$ be such that $\Delta w \in L^1(\Omega)$, satisfying*

$$\begin{cases} -\Delta w \le \langle h(x), \nabla w \rangle & \Omega, \\ w \le 0 & \partial\Omega, \end{cases} \tag{6.2.6}$$

where $h \in (L^{N+\epsilon}(\Omega))^N$. Then $w \le 0$ in Ω.

To prove Proposition 6.2.3 we will use the following result for the adjoint operator.

Lemma 6.2.4. *Assume $h \in L^q(\Omega, \mathbb{R}^N)$, $q > N$. Then there exists a unique solution $\phi \in W_0^{1,2}(\Omega) \cap L^\infty(\Omega)$ to the problem*

$$-\Delta\phi = -\operatorname{div}(\phi \cdot h) + 1, \text{ in } \Omega, \quad \phi = 0 \text{ on } \partial\Omega. \tag{6.2.7}$$

Moreover $\phi > 0$ in Ω.

Proof. If the norm of h is small enough, this is a direct consequence of the Lax–Milgram result. For the general case we use the Schauder fixed point theorem.

Consider the operator

$$\begin{aligned} T : W_0^{1,2}(\Omega) &\to W_0^{1,2}(\Omega), \\ \phi &\mapsto T(\phi) = v \end{aligned}$$

such that

$$-\Delta v = -\operatorname{div}\big(h(x)\phi\big) + 1, \quad v_{|\partial\Omega} \equiv 0,$$

which is clearly continuous. We need to check that T is compact. Note that

$$\operatorname{div}\big(h(x)\phi\big) \in W^{-1,2}(\Omega).$$

Indeed, for $\psi \in W_0^{1,2}(\Omega)$, by the Sobolev inequality, we have

$$\begin{aligned} &\int_\Omega \operatorname{div}\big(h(x)\phi\big)\psi \, dx \\ &\quad \le \Big(\int_\Omega |h(x)|^q \, dx\Big)^{\frac{1}{q}} \Big(\int_\Omega |\phi|^{2^*} \, dx\Big)^{\frac{1}{2^*}} \Big(\int_\Omega |\nabla\psi|^2 \, dx\Big)^{\frac{1}{2}} \\ &\quad \le C\Big(\int_\Omega |\nabla\psi|^2 \, dx\Big)^{\frac{1}{2}}. \end{aligned}$$

Consider a sequence (ϕ_n) bounded in $W_0^{1,2}(\Omega)$. Set $\phi \in W_0^{1,2}(\Omega)$ such that for a subsequence

$$\begin{aligned} \phi_n &\to \phi \quad \text{strongly in } L^2(\Omega), \\ \nabla\phi_n &\rightharpoonup \nabla\phi \quad \text{weakly in } \big(L^2(\Omega)\big)^N, \\ \phi_n &\to \phi \quad \text{a. e. in } \Omega. \end{aligned} \tag{6.2.8}$$

Then $v_n = T(\phi_n)$ is also a bounded sequence in $W_0^{1,2}(\Omega)$. Thus, there exists $v \in W_0^{1,2}(\Omega)$ such that, up to a subsequence,

$$\begin{aligned} v_n &\to v \quad \text{strongly in } L^2(\Omega),\\ \nabla v_n &\rightharpoonup \nabla v \quad \text{weakly in } (L^2(\Omega))^N,\\ v_n &\to v \quad \text{a. e. in } \Omega. \end{aligned} \tag{6.2.9}$$

Therefore taking $v_n - v$ as a test function we obtain

$$\begin{aligned} &\int_\Omega |\nabla(v_n - v)|^2\,dx = \int_\Omega \operatorname{div}(h(x)(\phi_n - \phi))(v_n - v)\,dx\\ &\int_\Omega \operatorname{div}(h(x)T_m(\phi_n - \phi))(v_n - v)\,dx + \int_{\Omega_m} \operatorname{div}(h(x)G_m(\phi_n - \phi))(v_n - v)\,dx\\ &\quad\le \int_\Omega T_m(\phi_n - \phi)\langle h(x), \nabla(v_n - v)\rangle\,dx + \int_{\Omega_m} G_m(\phi_n - \phi)\langle h(x), \nabla(v_n - v)\rangle\,dx\\ &\quad\equiv I_1 + I_2, \quad \text{where } \Omega_m = \Omega_m^n =: \Omega \cap \{|\phi_n - \phi| > m\}. \end{aligned}$$

Note that

$$|I_1| \le \|h\|_q \|T_m(\phi_n - \phi)\|_{2^*} \|\nabla(v_n - v)\|_2 |\Omega|^\alpha = o(1), \quad \text{as } n \to \infty.$$

Moreover, if $\alpha = 1 - (\frac{1}{2^*} + \frac{1}{2} + \frac{1}{q}) > 0$,

$$|I_2| = \left|\int_{\Omega_m} G_m(\phi_n - \phi)\langle h(x), \nabla(v_n - v)\rangle\,dx\right| \le |\Omega_m|^\alpha \|h\|_q \|\phi_n - \phi\|_{2^*} \|\nabla(v_n - v)\|2 = o(1),$$

because $|\Omega_m^n| \to 0$.

Finally, let ϕ and $\sigma \in [0,1]$ satisfy that $\sigma T(\phi) = \phi$, that is

$$-\Delta\phi = -\sigma \operatorname{div}(h(x)\phi) + 1. \tag{6.2.10}$$

Let us show that there exists a constant $M > 0$ such that $\|\phi\|_{W_0^{1,2}(\Omega)} \le M$. We consider $v \in W_0^{1,2}(\Omega)$ a nonnegative function such that

$$-\Delta v = \sigma\langle h(x)\nabla v\rangle + |\phi|^{2^*-2}\phi, \quad v_{|\partial\Omega} \equiv 0. \tag{6.2.11}$$

The existence of such a solution is granted by taking $f = |\phi|^{2^*-2}\phi$ in Lemma 6.2.4. Multiplying by v and ϕ, respectively, equations (6.2.10) and (6.2.11) and subtracting we get

$$\begin{aligned}\int_\Omega |\phi|^{2^*}\,dx = \sigma\int_\Omega v\,dx &\le C\Big(\int_\Omega v^{2^*}\,dx\Big)^{\frac{1}{2^*}}\\ &\le C\Big(\int_\Omega |\nabla v|^2\,dx\Big)^{\frac12} \le C\||\phi|^{2^*-2}\phi\|_{W^{-1,2}(\Omega)}\\ &\le C\Big(\int_\Omega \phi^{2^*}\,dx\Big)^{\frac{2^*-1}{2^*}},\end{aligned}$$

where we have used also (6.2.4). Hence,

$$\Big(\int_\Omega \phi^{2^*}\,dx\Big)^{\frac{1}{2^*}} \le C. \tag{6.2.12}$$

To get a bound for the gradient, we multiply (6.2.10) by ϕ. Integrating by parts, we obtain

$$\begin{aligned}\int_\Omega |\nabla\phi|^2\,dx = \sigma\int_\Omega h(x)\phi\nabla\phi\,dx + \sigma\int_\Omega \phi\,dx &\le C\Big(\int_\Omega \phi^2\,dx\Big)^{\frac12}\\ &+ \Big(\int_\Omega \phi^{2^*}\,dx\Big)^{\frac{1}{2^*}}\Big(\int_\Omega |\nabla\phi|^2\,dx\Big)^{\frac12}\Big(\int_\Omega |h(x)|^q\,dx\Big)^{\frac1q}.\end{aligned}$$

Therefore,

$$\Big(\int_\Omega |\nabla\phi|^2\,dx\Big)^{\frac12} \le \Big(\int_\Omega \phi^{2^*}\,dx\Big)^{\frac{1}{2^*}}\Big(\int_\Omega |h(x)|^q\,dx\Big)^{\frac1q}.$$

Then $\|\phi\|_{W_0^{1,2}(\Omega)} \le C$.

By the Schauder fixed point theorem (see Theorem 11.3 in [186]), we have shown that there exists a unique solution to (6.2.7), $\phi \in W_0^{1,2}(\Omega)$.

The positivity of ϕ follows easily.

To conclude the proof, we show that $\phi \in L^\infty(\Omega)$. To this end, we argue as in [17] and consider the following test function $v = \operatorname{sign}(\phi)(|\phi| - k)^+$ in (6.2.7). We find that

$$\int_{A(k)} |\nabla\phi|^2\,dx - \int_{A(k)} \langle h(x), \nabla\phi\rangle\phi\,dx = \int_{A(k)} v\,dx, \tag{6.2.13}$$

where $A(k) = \{x \in \Omega : |\phi(x)| > k\}$. Note that $\nabla\phi = \nabla v$ in $A(k)$ and also that $v = 0$ in $\Omega \setminus A(k)$. Since $\phi \in W_0^{1,2}(\Omega)$, $|A(k)| \to 0$ as $k \to \infty$. Our purpose is to show that there exists k_0 such that $|A(k)| = 0$ for any $k \geq k_0$. We treat first the integral on the right hand side,

$$\int_{A(k)} v\,dx \leq \Big(\int_{A(k)} v^{2^*}\,dx\Big)^{\frac{1}{2^*}} |A(k)|^{1-\frac{1}{2^*}}. \tag{6.2.14}$$

Using the Sobolev inequality and the fact that $\nabla v = \nabla \phi$ in $A(k)$ we get

$$\int_{A(k)} v\,dx \leq CS\Big(\int_{A(k)} |\nabla \phi|^2\,dx\Big)^{\frac{1}{2}} |A(k)|^{1-\frac{1}{2^*}}.$$

By Hölder's inequality with exponents $\frac{1}{2} + \frac{1}{2^*} + \frac{1}{q} + \frac{1}{\delta} = 1$, we find

$$\begin{aligned}
&\int_{A(k)} \langle h(x), \nabla \phi\rangle \phi\,dx \\
&\quad\leq \Big(\int_{A(k)} |h(x)|^q\,dx\Big)^{\frac{1}{q}} \Big(\int_{A(k)} \phi^{2^*}\,dx\Big)^{\frac{1}{2^*}} \\
&\qquad\times \Big(\int_{A(k)} |\nabla \phi|^2\,dx\Big)^{\frac{1}{2}} |A(k)|^{\frac{1}{\delta}} \\
&\quad\leq C\Big(\int_{A(k)} |\nabla \phi|^2\,dx\Big)^{\frac{1}{2}} |A(k)|^{\frac{1}{\delta}} \leq \eta\Big(\int_{A(k)} |\nabla \phi|^2\,dx\Big)^{\frac{1}{2}},
\end{aligned}$$

with $\eta \ll 1$ if k is sufficiently large. Thus we can absorb this term into the first integral in (6.2.13). Then, using (6.2.14) and Sobolev's inequality we arrive at

$$\Big(\int_{A(k)} v^{2^*}\,dx\Big)^{\frac{1}{2^*}} \leq C|A(k)|^{1-\frac{1}{2^*}}. \tag{6.2.15}$$

Since, if $0 < k < m$, then $A(m) \subset A(k)$, it is easy to deduce that

$$\begin{aligned}
|A(m)|^{\frac{1}{2^*}}(m-k) &= \Big(\int_{A(m)} (m-k)^{2^*_\gamma}\,dx\Big)^{\frac{1}{2^*}} \leq \Big(\int_{A(m)} v^{2^*}\,dx\Big)^{\frac{1}{2^*}} \\
&\leq \Big(\int_{A(k)} v^{2^*}\,dx\Big)^{\frac{1}{2^*}}.
\end{aligned}$$

Plugging this inequality into (6.2.15) we have shown that

$$\mu(A(m)) \leq \frac{C}{(m-k)^{2^*}} \mu(A(k))^{2^*-1},$$

with $2^* - 1 > 1$. Then, the classical Stampacchia theorem (see [291]) implies that there exists k_0 such that $\mu(A(k)) = 0$ for any $k \geq k_0$, as desired. □

To finish this section we proceed to prove the comparison result.

Proof of Proposition 6.2.3. Since $u, \Delta u \in L^1(\Omega)$, using the representation in terms of the Green function and deriving the integral, we find that $\nabla u \in (L^1(\Omega))^N$ (see Lemma 1 in [209]). Using now the Kato inequality [209], we have

$$-\Delta w_+ \le \langle h(x), \nabla w_+\rangle \text{ in } \Omega, \quad w_+ = 0 \text{ on } \partial\Omega,$$

and hence we can assume without loss of generality that $w \ge 0$ in Proposition 6.2.3.

Taking as a test function $\phi \in L^\infty(\Omega)$, $\phi \ge 0$, verifying (6.2.7), we find that

$$\begin{aligned}\int_\Omega (-\Delta w)\phi\, dx &\le \int_\Omega \langle h(x), \nabla w\rangle \phi\, dx = -\int_\Omega \operatorname{div}(h(x)\phi) w\, dx\\ &= \int_\Omega (-\Delta\phi) w\, dx - \int_\Omega w\, dx.\end{aligned}$$

Therefore,

$$\int_\Omega w\, dx \le 0, \quad \text{thus } w \le 0 \text{ in } \Omega. \qquad \square$$

6.2.2 A quantitative version of the maximum principle

We present here the version of the maximum principle given by H. Brezis and X. Cabre in [87]. Denote by $\delta(x)$ the distance function to the boundary of Ω, that is, $\delta(x) = d(x, \partial\Omega)$.

Proposition 6.2.5. *Consider $h, f(x) \ge 0$ and $f \in L^\infty(\Omega)$. Let v be the solution of*

$$\begin{cases} -\Delta v(x) = f(x) & \text{in } \Omega,\\ v(x) = 0 & \text{on } \partial\Omega,\end{cases}$$

where $\Omega \subset \mathbb{R}^N$ is a smooth bounded domain.

Then, there exists a positive constant $c = c(\Omega)$ such that

$$\frac{v(x)}{\delta(x)} \ge c\int_\Omega f(y)\delta(y)\, dy, \quad \text{for all } x \in \Omega.$$

Proof. Consider a compact set $K \subset \Omega$ and $0 < r < \frac{d(K,\partial\Omega)}{2}$. Let $\{B_r(x_i)\}_1^m$ be a finite covering of K by balls of radius r, each contained in Ω.

Let $\phi_1, \phi_2 \dots \phi_m$ be the solutions to the problems

$$\begin{cases} -\Delta\phi_i(x) = \chi_{B_r(x_i)} & \text{in } \Omega,\\ \phi_i(x) = 0 & \text{on } \partial\Omega,\end{cases} \qquad i = 1, 2, \dots, m,$$

where $\chi_{B_r(x_i)}$ is the characteristic function of the ball $B_r(x_i)$. By the Hopf boundary lemma, there exists a positive constant c such that

$$\phi_i(x) \geq c\delta(x), \quad \text{for all } x \in \Omega \text{ and all } i = 1, 2, \dots, m,$$

with $\delta(x) = d(x, \partial\Omega)$. Consider $x \in K$ and a ball of the covering, $B_r(x_i)$, containing x. Note that $B_r(x_i) \subset B_{2r}(x) \subset \Omega$. Since $-\Delta v > 0$ in Ω, that is, v is superharmonic in Ω, we find that

$$\begin{aligned} v(x) &\geq \frac{1}{|B_{2r}|} \int_{B_{2r}(x)} v(y)\,dy = c(r) \int_{B_{2r}(x)} v(y)\,dy \geq c(r) \int_{B_r(x_i)} v(y)\,dy \\ &= c(r) \int_{\Omega} v(y)(-\Delta\phi_i(y))\,dy = c(r) \int_{\Omega} f(y)\phi_i(y)\,dy \geq c(r)c \int_{\Omega} f(y)\delta(y)\,dy. \end{aligned}$$

Assume, moreover, that K is such that $\Omega \setminus K$ has a smooth boundary. We have the estimate above and to conclude the global result we have to get the lower estimate of v in $\Omega \setminus K$. To do that, consider the auxiliary problem

$$\begin{cases} -\Delta\psi = 0 & \text{in } \Omega, \\ \psi(x) = 0 & \text{on } \partial\Omega, \\ \psi(x) = 1 & \text{on } \partial K. \end{cases}$$

By the Hopf lemma we have

$$\psi(x) \geq c\delta(x), \quad \text{for all } x \in \Omega \setminus K.$$

Since v is superharmonic and

$$v(x) \geq c \int_{\Omega} f(y)\delta(y)\,dy \quad \text{on } \partial K,$$

the maximum principle implies that

$$v(x) \geq c\Big(\int_{\Omega} f(y)\delta(y)\,dy\Big)\phi(x) \geq c_1\Big(\int_{\Omega} f(y)\delta(y)\,dy\Big)\delta(x) \quad \text{for all } x \in \Omega \setminus K,$$

which proves the result. □

6.2.3 Caffarelli–Kohn–Nirenberg inequalities

In the proofs of nonexistence that we will perform next, we need some of the inequalities that appear in the paper by Caffarelli, Kohn and Nirenberg [108]. More precisely, for $\gamma < \frac{N-p}{p}$, we consider the weighted Sobolev space $D_{\gamma}^{1,p}(\Omega)$ as the closure of $C_0^{\infty}(\Omega)$

with respect to the norm $\|.\|_{D^{1,p}_\gamma(\Omega)}$ defined by

$$\|\phi\|^p_{D^{1,p}_\gamma} = \int_\Omega (|\phi|^p + |\nabla\phi|^p)|x|^{-p\gamma}\,dx.$$

Using a Poincaré type inequality, we know that the space $D^{1,p}_{0,\gamma}(\Omega)$ may be also defined as the completion of $C_0^\infty(\Omega)$ with respect to the $L^p(|x|^{-p\gamma}\,dx)$-norm of the gradient. We recall now the Hardy–Sobolev inequality that we will use systematically in the next chapters.

Lemma 6.2.6. *Suppose $1 < p < N$ and $-\infty < \gamma < \frac{N-p}{p}$. Then, for all $u \in C_0^\infty(\Omega)$, we have*

$$\int_\Omega |u|^p|x|^{-p(\gamma+1)}\,dx \le \Lambda^{-1}_{N,p,\gamma}\int_\Omega |\nabla u|^p|x|^{-p\gamma}\,dx, \quad \Lambda_{N,p,\gamma} = \left(\frac{N-p(\gamma+1)}{p}\right)^p. \tag{6.2.16}$$

Moreover, $\Lambda^{-1}_{N,p,\gamma}$ is optimal and it is not achieved.

As we said, the proof can be seen in [108]. Alternatively, one can try to make a direct proof following the arguments in Chapter 2 for $p = 2$ and $p \neq 2$. The elliptic equations involving coefficients related to Caffarelli–Kohn–Nirenberg inequalities have been studied in the article [15].

6.3 Nonexistence results: exponent $q \geq q_+(\lambda)$

The main goal in this section is to show that $q_+(\lambda)$ represents the threshold for the exponent q, in a such way that problem (6.1.1) has no positive supersolution in the very weak sense if $q > q_+(\lambda)$. In the whole section, we use the concept of *very weak (sub/super)solution*, which, roughly speaking, is the more general form for which the equation has a meaning in the distributional sense. We provide a precise definition of this concept that is the translation to the quasilinear case of the same concept for semilinear problems.

Definition 6.3.1. We say that $u \in L^1_{\text{loc}}(\Omega)$ is a very weak supersolution (subsolution) to equation (6.1.1) when, if $\frac{u}{|x|^2} \in L^1_{\text{loc}}(\Omega)$, $|\nabla u|^q \in L^1_{\text{loc}}(\Omega)$ and $\forall\phi \in C_0^\infty(\Omega)$ such that $\phi \geq 0$, we have

$$\int_\Omega (-\Delta\phi)u\,dx \geq (\leq) \int_\Omega \left(|\nabla u|^q + \lambda\frac{u}{|x|^2} + f\right)\phi\,dx.$$

If u is a very weak super- and subsolution, then we say that u is a very weak solution.

If $\lambda > \Lambda_{N,2} \equiv (\frac{N-2}{2})^2$, the nonexistence result of positive very weak solution to problem (6.1.1) is a consequence of the optimality of $\Lambda_{N,2}$ as bound in the Hardy–Leray inequality. See for instance [13]. Then, hereafter we will assume $0 < \lambda \leq \Lambda_{N,2}$.

We recall the lower estimate of u near the origin obtained in Lemma 4.3.8 and the necessary condition on the source term obtained in Theorem 3.4.1 to have a solution.

To find the optimal exponent we search for a solution in the form $u(x) = A|x|^{-\beta}$ of the equation. Hence by a direct computation as in the semilinear case, we obtain

$$\beta = \frac{2-q}{q-1} \quad \text{and} \quad \beta^q A^{q-1} = \beta(N-\beta-2) - \lambda. \tag{6.3.1}$$

Since the left hand side is positive, the right hand side must be positive, but the second member is positive if and only if

$$\alpha_-(\lambda) < \beta < \alpha_+(\lambda), \quad \text{where } \alpha_\pm(\lambda) \text{ are defined by (3.1.3).}$$

Since (6.3.1) holds, $\alpha_-(\lambda) < \beta < \alpha_+(\lambda)$ is equivalent to

$$q_-(\lambda) \equiv \frac{2+\alpha_+(\lambda)}{\alpha_+(\lambda)+1} < q < \frac{2+\alpha_-(\lambda)}{\alpha_-(\lambda)+1} \equiv q_+(\lambda). \tag{6.3.2}$$

Hence the heuristic guess for optimal exponent seems to be $q_+(\lambda)$. The main result on nonexistence in this direction is the following.

Theorem 6.3.2. *Assume that $f \geq 0$. Let $q_+(\lambda) = \frac{2+\alpha_-(\lambda)}{1+\alpha_-(\lambda)}$, where $\alpha_-(\lambda)$ is defined in* (3.1.3). *If $q \geq q_+(\lambda)$, then equation* (6.1.1) *has no positive very weak supersolution. In the case where $f \equiv 0$, the unique nonnegative very weak supersolution is $u \equiv 0$.*

Proof. We divide the proof into three steps.

First step: $q > q_+(\lambda)$. Assume by contradiction that equation (6.1.1) has a very weak supersolution u. Then

$$-\Delta u - \lambda \frac{u}{|x|^2} \gneq 0.$$

Therefore, by Lemma 4.3.8, there exists a positive constant C and a small ball $B_r(0) \subset \mathbb{R}^N$ such that $u(x) \geq C|x|^{-\alpha_-(\lambda)}$ in $B_r(0)$. Take $\phi \in C_0^\infty(B_r(0))$. Using $|\phi|^{q'}$, $q' = \frac{q}{q-1}$, as a test function in (6.1.1) and by the Hölder and Young inequalities we obtain

$$c_1\lambda \int\limits_{B_r(0)} \frac{u|\phi|^{q'}}{|x|^2}\, dx \leq \int\limits_{B_r(0)} |\nabla\phi|^{q'}\, dx, \tag{6.3.3}$$

where c_1 is a positive constant that is independent of u and ϕ. Using the lower estimate for u in $B_r(0)$ we obtain

$$c_2\lambda \int\limits_{B_r(0)} \frac{|\phi|^{q'}}{|x|^{2+\alpha_-(\lambda)}}\, dx \leq \int\limits_{B_r(0)} |\nabla\phi|^{q'}\, dx.$$

We recall that $q > q_+(\lambda)$, and hence $2+\alpha_-(\lambda) > q'$ and we reach a contradiction with the classical Hardy inequality for $W_0^{1,q'}(B_r(0))$ proved in Section 2.4. Then, the result follows.

SECOND STEP: $q = q_+(\lambda)$ and $\lambda < \Lambda_{N,2}$. Again we argue by contradiction. Assume that equation (6.1.1) has a very weak supersolution u. As above, by Lemma 4.3.8 there is a positive constant c_0 such that

$$u(x) \geq \frac{c_0}{|x|^{\alpha_-(\lambda)}} \quad \text{in some ball } B_\eta(0) \subset\subset \Omega. \tag{6.3.4}$$

Without loss of generality we assume that $\eta = e^{-1}$. Using Lemma 4.3.8 we obtain

$$\int_{B_\eta(0)} |\nabla u|^{q_+(\lambda)}|x|^{-\alpha_-(\lambda)}\,dx < \infty \quad \text{and} \quad \int_{B_\eta(0)} \frac{u}{|x|^{2+\alpha_-(\lambda)}}\,dx < \infty. \tag{6.3.5}$$

Let $w(x) = |x|^{-\alpha_-(\lambda)}(\log(\frac{1}{|x|}))^\beta$, where β is a positive small constant that we will choose below. Since $\lambda < \Lambda_{N,2}$, we have $w \in W^{1,2}(B_\eta(0))$ and, in particular, $w \in W^{1,q_+(\lambda)}(B_\eta(0))$. By a direct computation we obtain

$$-\Delta w - \lambda\frac{w}{|x|^2} = \frac{\beta}{|x|^{2+\alpha_-(\lambda)}}\left(\log\left(\frac{1}{|x|}\right)\right)^{\beta-1}\left[(N-2-2\alpha_-(\lambda)) + (1-\beta)(\left(\log\left(\frac{1}{|x|}\right)\right)^{-1}\right].$$

Note that $|\nabla w| = |x|^{-\alpha_-(\lambda)-1}(\alpha_-(\lambda)\log(\frac{1}{|x|}) + \beta(\log(\frac{1}{|x|}))^{\beta-1}))$, thus

$$\begin{aligned}|\nabla w|^{q_+(\lambda)}&\left(\alpha_-(\lambda)\log\left(\frac{1}{|x|}\right) + \beta\left(\left(\log\left(\frac{1}{|x|}\right)\right)^{-1}\right)\right)^{1-q_+(\lambda)}\\ &= |x|^{-\alpha_-(\lambda)-2}\left(\alpha_-(\lambda)\log\left(\frac{\eta}{|x|}\right) + \beta\left(\left(\log\left(\frac{1}{|x|}\right)\right)^{-1}\right)\right).\end{aligned}$$

Since $|x| \leq e^{-1}$, choosing β small enough, we conclude that

$$-\Delta w - \lambda\frac{w}{|x|^2} \leq \beta^{\frac{1}{2}}|\nabla w|^{q_+(\lambda)}h(x),$$

where $h(x) = (\alpha_-(\lambda)\log(\frac{1}{|x|}) + \beta((\log(\frac{1}{|x|}))^{-1})^{1-q_+(\lambda)}$. Observe that this is bounded in the ball $B_\eta(0)$. Set $u_1 \equiv c_1 u$. Then

$$-\Delta u_1 - \lambda\frac{u_1}{|x|^2} \geq c_1^{1-q_+(\lambda)}|\nabla u_1|^{q_+(\lambda)}.$$

Let c_0 be a fixed constant satisfying (6.3.4) when $\eta = e^{-1}$ and take $c_1 > 0$ such that $c_1 c_0 \geq 1$. Then for β suitable small we have

$$c_1^{1-q_+(\lambda)} \geq \|h\|_\infty \beta^{\frac{1}{2}}.$$

Since $c_1 c_0 \geq 1$ we obtain $u_1(x) \geq w(x)$ for $|x| = e^{-1}$ and, moreover,

$$-\Delta u_1 - \lambda\frac{u_1}{|x|^2} \geq \beta^{\frac{1}{2}}h(x)|\nabla u_1|^{q_+(\lambda)}.$$

Claim. $u_1 \geq w$. We write $v = w - u_1$. Using the regularity of w and (6.3.5) we obtain $v \in W^{1,q_+(\lambda)}(B_\eta(0))$, $v \leq 0$ on $\partial B_\eta(0)$ and

$$\int_{B_\eta(0)} \frac{|v|}{|x|^{2+\alpha_-(\lambda)}}\,dx < \infty, \qquad \int_{B_\eta(0)} |\nabla v|^{q_+(\lambda)}|x|^{-\alpha_-(\lambda)}\,dx < \infty. \tag{6.3.6}$$

By a direct computation it follows that

$$-\Delta v - \lambda\frac{v}{|x|^2} \leq q_+(\lambda)h(x)\beta^{\frac{1}{2}}|\nabla w|^{q_+(\lambda)-2}\nabla w\nabla v \equiv a(x)\nabla v,$$

where the vector field $a(x) = -\beta^{\frac{1}{2}}q_+(\lambda)\frac{x}{|x|^2}$ belongs to $L^q(B_\eta(0))$ for all $q < N$. Observe that the vector field $a(x)$ does not have enough regularity to apply the comparison argument used in [31]. To overcome this lack of regularity we proceed as follows. Using the Kato inequality (see [209] and the extension in [96]) we get

$$-\Delta v_+ - \lambda\frac{v_+}{|x|^2} + q_+(\lambda)\beta^{\frac{1}{2}}\left\langle\frac{x}{|x|^2}, \nabla v_+\right\rangle \leq 0 \quad\text{and}\quad \int_{B_\eta(0)} |\nabla v_+|^{q_+(\lambda)}|x|^{-\alpha_-(\lambda)}\,dx < \infty,$$

and, since $\frac{\alpha_-(\lambda)}{q_+(\lambda)} < \frac{N-2}{2}$, the Caffarelli–Kohn–Nirenberg inequality applied to v_+ gives

$$\int_{B_\eta(0)} \frac{v_+^{q_+(\lambda)}}{|x|^{q_+(\lambda)+\alpha_-(\lambda)}}\,dx < \infty. \tag{6.3.7}$$

Define $\gamma = \frac{\beta^{\frac{1}{2}}q_+(\lambda)}{2}$ and consider the weight $|x|^{-2\gamma}$. Then for suitable β, one has $2\gamma < N-2$ and hence $|x|^{-2\gamma}$ is an admissible weight to use the Caffarelli–Kohn–Nirenberg inequalities (see [108]). Thus, we have

$$-\operatorname{div}(|x|^{-2\gamma}\nabla v_+) - \lambda\frac{v_+}{|x|^{2(\gamma+1)}} = |x|^{-2\gamma}\left(-\Delta v_+ + q_+(\lambda)\left\langle\frac{x}{|x|^2}, \nabla v_+\right\rangle - \lambda\frac{v_+}{|x|^2}\right) \leq 0 \tag{6.3.8}$$

(a detailed study of these equations related to the Caffarelli–Kohn–Nirenberg inequalities can be found in [17]). Moreover, there exists $\sigma_1 > 2 + \alpha_-(\lambda)$, depending only on N and λ, such that

$$\int_{B_\eta(0)} \frac{v_+}{|x|^{\sigma_1}}\,dx < \infty. \tag{6.3.9}$$

Indeed,

$$\begin{aligned}\int_{B_\eta(0)} \frac{v_+}{|x|^{\sigma_1}}\,dx &= \int_{B_\eta(0)} \frac{v_+}{|x|^{\frac{q_+(\lambda)+\alpha_-(\lambda)}{q_+(\lambda)}}}\frac{1}{|x|^{\sigma_1-\frac{q_+(\lambda)+\alpha_-(\lambda)}{q_+(\lambda)}}}\,dx\\ &\leq \left(\int_{B_\eta(0)} \frac{v_+^{q_+(\lambda)}}{|x|^{q_+(\lambda)+\alpha_-(\lambda)}}\,dx\right)^{\frac{1}{q_+(\lambda)}}\left(\int_{B_\eta(0)} \frac{1}{|x|^{q_+'(\lambda)(\sigma_1-\frac{q_+(\lambda)+\alpha_-(\lambda)}{q_+(\lambda)})}}\,dx\right)^{\frac{1}{q_+'(\lambda)}}.\end{aligned}$$

Denote $\theta(\sigma_1) = q'_+(\sigma_1 - \frac{q_+(\lambda)+\alpha_-(\lambda)}{q_+(\lambda)})$. Since $q_+(\lambda) = \frac{2+\alpha_-(\lambda)}{1+\alpha_-(\lambda)}$, the conjugate of $q_+(\lambda)$ is $q'_+ = 2 + \alpha_-(\lambda)$. Hence,

$$\theta(\sigma_1) = (2 + \alpha_-(\lambda))(\sigma_1 - 1) - \alpha_-(\lambda)(1 + \alpha_-(\lambda)).$$

By a direct computation we get $\theta(2 + \alpha_-(\lambda)) = 2(1 + \alpha_-(\lambda)) = N - 2\sqrt{\Lambda_{N,2} - \lambda} < N$. Thus, there exists $\sigma_1 > 2 + \alpha_-(\lambda)$ such that $\theta(\sigma_1) < N$ and

$$\int_{B_\eta(0)} |x|^{-(q'_+(\lambda)(\sigma_1 - \frac{q_+(\lambda)+\alpha_-(\lambda)}{q_+(\lambda)}))}\, dx < \infty.$$

Thus (6.3.9) holds.

To finish, the idea should be to use φ, the solution to problem

$$\begin{cases} -\operatorname{div}(|x|^{-2\gamma}\nabla\varphi) - \lambda\frac{\varphi}{|x|^{2(\gamma+1)}} = \frac{1}{|x|^{2(\gamma+1)}} & \text{in } B_\eta(0), \\ \varphi = 0 & \text{on } \partial B_\eta(0), \end{cases}$$

as a test function in (6.3.8). However, a direct calculation shows that

$$\varphi(x) = \frac{1}{|x|^a} - \frac{1}{\eta^a}, \quad \text{where } a = \frac{N - 2(\gamma+1)}{2} - \sqrt{\left(\frac{N - 2(\gamma+1)}{2}\right)^2 - \lambda},$$

which does not have the required regularity to be used directly as a test function in (6.3.8). Therefore, we need to consider the approximating sequence

$$\varphi_n(x) = \frac{1}{(|x| + \frac{1}{n})^a} - \frac{1}{(\eta + \frac{1}{n})^a}.$$

We see that $\varphi_n \in C^1(B_\eta(0))$, $\varphi_n = 0$ on $\partial B_\eta(0)$,

$$\nabla\varphi_n(x) = -\frac{a}{(|x| + \frac{1}{n})^{a+1}}\frac{x}{|x|}$$

and

$$-\operatorname{div}(|x|^{-2\gamma}\nabla\varphi_n) = |x|^{-2\gamma}\left(\frac{a(N-1-2\gamma)}{|x|(|x| + \frac{1}{n})^{a+1}} - \frac{a(a+1)}{(|x| + \frac{1}{n})^{a+2}}\right).$$

Note that

$$\int_{B_\eta(0)} |x|^{-2\gamma}|\nabla v_+||\nabla\varphi_n|\, dx < \infty \quad \text{and} \quad \int_{B_\eta(0)} \frac{v_+\varphi_n}{|x|^{2(\gamma+1)}}\, dx < \infty.$$

Choosing φ_n as a test function in (6.3.8) we obtain

$$\int_{B_\eta(0)} v(-\operatorname{div}(|x|^{-2\gamma}\nabla\varphi_n))\, dx - \lambda \int_{B_\eta(0)} \frac{v_+\varphi_n}{|x|^{2(\gamma+1)}}\, dx \le 0. \tag{6.3.10}$$

By the definition of w_n we have

$$\frac{v_+\varphi_n}{|x|^{2(\gamma+1)}} \leq \frac{v_+}{|x|^{a+2(\gamma+1)}} + \frac{2}{\eta^a}\frac{v_+}{|x|^{2(\gamma+1)}}.$$

Since $a+2(\gamma+1) \to 2+\alpha_-(\lambda)$ as $\gamma \to 0$, choosing β small we can find γ small such that $a+2(\gamma+1) < \sigma_1$. Hence

$$\frac{v_+\varphi_n}{|x|^{2(\gamma+1)}} \leq \frac{Cv_+}{|x|^{\sigma_1}}.$$

Then by the definition of σ_1 and using the dominated convergence theorem, we easily prove that

$$\int\limits_{B_\eta(0)} \frac{v_+\varphi_n}{|x|^{2(\gamma+1)}}\,dx \to \int\limits_{B_\eta(0)} \frac{v_+}{|x|^{a+2(\gamma+1)}}\,dx - \frac{1}{\eta^a}\int\limits_{B_\eta(0)} \frac{v_+}{|x|^{2(\gamma+1)}}\,dx \quad \text{as } n \to \infty.$$

We deal now with the first term in (6.3.10),

$$\begin{aligned} |v_+ \operatorname{div}(|x|^{-2\gamma}\nabla\varphi_n)| &= \left|\left(\frac{a(N-1-2\gamma)v_+}{|x|^{1+2\gamma}(|x|+\frac{1}{n})^{a+1}} - \frac{a(a+1)v_+}{|x|^{2\gamma}(|x|+\frac{1}{n})^{a+2}}\right)\right| \\ &\leq \frac{a(N-1-2\gamma)v_+}{|x|^{1+2\gamma}(|x|+\frac{1}{n})^{a+1}} + \frac{a(a+1)v_+}{|x|^{2\gamma}(|x|+\frac{1}{n})^{a+2}}. \end{aligned}$$

As above it is not difficult to see that

$$\frac{a(N-1-2\gamma)v_+}{|x|^{1+2\gamma}(|x|+\frac{1}{n})^{a+1}} + \frac{a(a+1)v_+}{|x|^{2\gamma}(|x|+\frac{1}{n})^{a+2}} \leq \frac{a(N+a-2\gamma)v_+}{|x|^{\sigma_1}}$$

and then, by the dominated convergence theorem, we obtain

$$\int\limits_{B_\eta(0)} v_+ \operatorname{div}(|x|^{-2\gamma}\nabla\varphi_n)\,dx \to \int\limits_{B_\eta(0)} \frac{a(N-a-2(\gamma+1))v_+}{|x|^{2(1+\gamma)+a}}\,dx \quad \text{as } n \to \infty.$$

Hence, passing to the limit in (6.3.10) and since

$$a(N-a-2(\gamma+1)) - \lambda = 0,$$

we have

$$\int\limits_{B_\eta(0)} v(-\operatorname{div}(|x|^{-2\gamma}\nabla\varphi_n))\,dx - \lambda\int\limits_{B_\eta(0)} \frac{v_+\varphi_n}{|x|^{2(\gamma+1)}}\,dx \to \frac{1}{\eta^a}\int\limits_{B_\eta(0)} \frac{v_+}{|x|^{2(1+\gamma)}}\,dx,$$

as $n \to \infty$. Thus, according to (6.3.10), $\int_{B_\eta(0)} \frac{v_+}{|x|^{2(1+\gamma)}}\,dx \leq 0$, and hence $v_+ \equiv 0$ and then $u_1 \geq w$.

To finish the proof in this case we use the same argument as in the first step. More precisely, for every $\phi \in C_0^\infty(B_r(0))$, $0 < r << \eta$, we have

$$c_1 \int_{B_r(0)} \frac{u_1|\phi|^{q'_+(\lambda)}}{|x|^2}\,dx \le \int_{B_r(0)} |\nabla\phi|^{q'_+(\lambda)}\,dx, \tag{6.3.11}$$

where $c_1 > 0$ is independent of ϕ. Using the result of the claim and the fact that $q'_+(\lambda) = \alpha_-(\lambda) + 2$, we obtain

$$c_2 \int_{B_r(0)} \frac{|\phi|^{q'_+(\lambda)}}{|x|^{q'_+(\lambda)}}\left(\log\left(\frac{1}{|x|}\right)\right)^\beta dx \le \int_{B_r(0)} |\nabla\phi|^{q'_+(\lambda)}\,dx,$$

a contradiction with Hardy's inequality in $W_0^{1,q'_+(\lambda)}(B_r(0))$. Hence the result follows.

THIRD STEP: $q = q_+(\lambda)$ and $\lambda = \Lambda_{N,2}$. Assume by contradiction that problem (6.1.1) has a positive very weak supersolution u. In this case $\alpha_-(\lambda) = \frac{N-2}{2}$ and $q_+(\lambda) = \frac{N+2}{N}$. Hence by Lemma 4.3.8 we obtain $u(x) \ge c|x|^{-\alpha_-(\lambda)}$ and by Lemma 4.3.8

$$\int_{B_\eta(0)} |\nabla u|^{q_+(\lambda)}|x|^{-\alpha_-(\lambda)}\,dx < \infty.$$

We consider $\phi \in C_0^\infty(B_\eta(0))$ such that $\phi \ge 0$ and $\phi = 1$ in $B_{\eta_1}(0)$. Then by the regularity of u we have $\int_{B_\eta(0)} |\nabla(\phi u)|^{q_+(\lambda)}|x|^{-\alpha_-(\lambda)}\,dx < \infty$. Since $\frac{\alpha_-(\lambda)}{q_+(\lambda)} = \frac{N(N-2)}{2(N+2)} < N$, we can apply the Caffarelli–Kohn–Nirenberg inequalities to obtain

$$C_1 \int_{B_\eta(0)} (\phi u)^{q_+(\lambda)}|x|^{-\alpha_-(\lambda)}\,dx \le \int_{B_\eta(0)} |\nabla(\phi u)|^{q_+(\lambda)}|x|^{-\alpha_-(\lambda)}\,dx < \infty,$$

$$\int_{B_{\eta_1}(0)} u^{q_+(\lambda)}|x|^{-\alpha_-(\lambda)}\,dx < \infty \quad \text{for } \eta_1 < \eta.$$

Therefore we conclude that $u \in \mathcal{D}^{1,q'_+(\lambda)}_{\alpha_-(\lambda)}(B_{\eta_1}(0))$, that we recall is the weighted Sobolev space defined as the completion of $C^\infty(\overline{B_\eta(0)})$ with respect to the norm $\|\cdot\|_{\mathcal{D}^{1,q_+(\lambda)}_{\alpha_-(\lambda)}}$, defined by

$$\|\phi\|^{q_+(\lambda)}_{\mathcal{D}^{1,q_+(\lambda)}_{\alpha_-(\lambda)}} = \int_{B_{\eta_1}(0)} |\phi|^{q_+(\lambda)}|x|^{-\alpha_-(\lambda)}\,dx + \int_{B_{\eta_1}(0)} |\nabla\phi|^{q_+(\lambda)}|x|^{-\alpha_-(\lambda)}\,dx.$$

It is not difficult to see that for every $\phi \in \mathcal{D}^{1,q_+(\lambda)}_{\alpha_-(\lambda)}(B_\eta(0))$ we have

$$C_2 \int_{B_{\eta_1}(0)} \frac{|\phi|^{q_+(\lambda)}}{|x|^{\alpha_-(\lambda)+q_+(\lambda)}}\,dx \le \int_{B_{\eta_1}(0)} |\phi|^{q_+(\lambda)}|x|^{-\alpha_-(\lambda)}\,dx + \int_{B_{\eta_1}(0)} |\nabla\phi|^{q_+(\lambda)}|x|^{-\alpha_-(\lambda)}\,dx,$$

where $C_2 > 0$ is independent of ϕ; in particular,

$$\int_{B_{\eta_1}(0)} \frac{u^{q_+(\lambda)}}{|x|^{\alpha_-(\lambda)+q_+(\lambda)}}\,dx < \infty. \tag{6.3.12}$$

Using again the fact that $u(x) \geq c|x|^{-\alpha_-(\lambda)}$ and since $\alpha_-(\lambda) + q_+(\lambda) + \alpha_-(\lambda)q_+(\lambda) = N$, we reach a contradiction with (6.3.12).

Hence the nonexistence result follows. □

Remark 6.3.3.

1. Note that $q_+(\lambda) < 2$, for all $\lambda \in (0, \Lambda_{N,2}]$. Hence for $q = 2$ we easily obtain the nonexistence result from the first step in Theorem 6.3.2. See [19]. Moreover, $q_+(\lambda) \to \frac{N+2}{N}$ if $\lambda \to \Lambda_{N,2}$ and $q_+(\lambda) \to 2$ if $\lambda \to 0$. As a consequence, we have encountered a discontinuity with the known results for $\lambda = 0$. See, for instance, [194].
2. If $1 < q \leq \frac{N}{N-1}$, then problem (6.1.1) has no very weak positive solution in $\mathbb{R}^N$. This follows using the results in [31] and [194]. For the reader's convenience we include a proof.
   We argue by contradiction. Assume that (6.1.1) has a positive solution u in $\mathbb{R}^N$ with $1 < q \leq \frac{N}{N-1}$. It is not difficult to see, using the strong maximum principle, that for any compact set $K \subset \Omega$ there exists a positive constant $c(K)$ such that $u \geq c(K)$. Let $\phi \in C_0^\infty(\Omega)$. Then using $|\phi|^{q'}$ as a test function in (6.1.1) we obtain

   $$q' \int_{\mathbb{R}^N} |\nabla u||\nabla \phi||\phi|^{(q'-1)}\,dx \geq \int_{\mathbb{R}^N} |\nabla u|^q |\phi|^{q'}\,dx + \lambda \int_{\mathbb{R}^N} \frac{u}{|x|^2}|\phi|^{q'}\,dx.$$

   Using Young inequalities we conclude that

   $$\int_{\mathbb{R}^N} |\nabla \phi|^{q'}\,dx \geq c_1\lambda \int_{\mathbb{R}^N} \frac{u}{|x|^2}|\phi|^{q'}\,dx. \tag{6.3.13}$$

   Since $q' > N$, we have $\mathrm{Cap}_{1,q'}(K) = 0$ for any compact set of $\mathbb{R}^N$. Thus, there exists a sequence $\{\phi_n\} \subset C_0^\infty(\mathbb{R}^N)$ such that $\phi_n \geq \chi_K$ and $\|\nabla \phi_n\|_{L^{q'}(\mathbb{R}^N)} \to 0$ as $n \to \infty$. Inserting this in (6.3.13) we reach a contradiction.
3. Despite the previous remark, in bounded domains there are no restrictions on p from below. This follows by the fact that the relative q-capacity, $q > N$, of a ball with respect to a concentric bigger ball is not zero. See [239, page 106].

6.4 Blow-up result

As a consequence of the nonexistence result, we obtain the next blow-up behavior for approximated problems.

Theorem 6.4.1. *Assume that $p \geq q_+(\lambda)$. If $u_n \in W_0^{1,p}(\Omega)$ is a solution to problem*

$$\begin{cases} -\Delta u_n = |\nabla u_n|^p + \lambda a_n(x) u_n + \alpha f & \text{in } \Omega, \\ u_n > 0 & \text{in } \Omega, \\ u_n = 0 & \text{on } \partial\Omega, \end{cases} \tag{6.4.1}$$

with $f \geq 0, f \not\equiv 0$ and $a_n(x) = \frac{1}{|x|^2 + \frac{1}{n}}$, then $u_n(x_0) \to \infty$, $\forall x_0 \in \Omega$.

To prove Theorem 6.4.1, we need the following lemma by F. Murat [248], which is a bit more general than Lemma 5.2 in [31].

Lemma 6.4.2. *Assume that $\{u_n\}$ is a sequence of positive functions such that $\{u_n\}$ is uniformly bounded in $W_{\mathrm{loc}}^{1,p}(\Omega)$ for some $1 < p \leq 2$ with $u_n \rightharpoonup u$ weakly in $W_{\mathrm{loc}}^{1,p}(\Omega)$ and that $u_n \leq u$ for all $n \in \mathbb{N}$. Assume that $-\Delta u_n \geq 0$ in $\mathcal{D}'(\Omega)$ and if $p < 2$, that the sequence $\{T_k(u_n)\}$ is uniformly bounded in $W_{\mathrm{loc}}^{1,2}(\Omega)$ for k fixed. Then $\nabla T_k(u_n) \to \nabla T_k(u)$ strongly in $(L_{\mathrm{loc}}^2(\Omega))^N$.*

Proof. Note that the hypothesis on the boundedness of $\{T_k(u_n)\}$ in $W_{\mathrm{loc}}^{1,2}(\Omega)$ is needed just when $p < 2$. Therefore by hypothesis we conclude that

$$\|\nabla T_k(u)\|_{L^2(K)} \leq \|\nabla T_k(u_n)\|_{L^2(K)} \quad \text{for all bounded regular domain } K \subset\subset \Omega.$$

Let $\phi \in C_0^\infty(\Omega)$ be a positive function. Since $u_n \leq u$ we get

$$\int_\Omega -\Delta u_n (T_k(u_n)\phi)\,dx = \int_\Omega \phi |\nabla T_k(u_n)|^2\,dx + \int_\Omega T_k(u_n) \nabla\phi \nabla u_n\,dx. \tag{6.4.2}$$

On the other hand, using again that $u_n \leq u$,

$$\begin{aligned} \int_\Omega -\Delta u_n (T_k(u)\phi)\,dx &= \int_\Omega \phi \nabla u_n \nabla T_k(u)\,dx + \int_\Omega T_k(u) \nabla\phi \nabla u_n\,dx \\ &= \int_\Omega \phi \nabla T_k(u_n) \nabla T_k(u)\,dx + \int_\Omega T_k(u) \nabla\phi \nabla u_n\,dx \\ &\leq \frac{1}{2}\int_\Omega \phi |\nabla T_k(u_n)|^2\,dx + \frac{1}{2}\int_\Omega \phi |\nabla T_k(u)|^2\,dx + \int_\Omega T_k(u) \nabla\phi \nabla u_n\,dx. \end{aligned}$$

Thus by the above computation and (6.4.2), we have

$$\frac{1}{2}\int_\Omega \phi |\nabla T_k(u_n)|^2\,dx \leq \frac{1}{2}\int_\Omega \phi |\nabla T_k(u)|^2\,dx + \int_\Omega (T_k(u) - T_k(u_n)) \nabla\phi \nabla u_n\,dx.$$

Hence we conclude that

$$\limsup_{n\to\infty} \int_\Omega \phi(|\nabla T_k(u_n)|^2 - |\nabla T_k(u)|^2)\,dx \leq 0 \quad \text{for all positive test functions } \phi.$$

We set $w_n = \phi T_k(u_n)$ and $w = \phi T_k(u)$, where ϕ is a positive test function. Then $w_n \rightharpoonup w$ weakly in $W^{1,2}_{\rm loc}(\Omega)$. Note that $w_n \to w$ strongly in $L^2_{\rm loc}(\Omega)$. Therefore, using the above computation we easily get

$$0 \le \limsup_{n\to\infty}(\|\nabla w_n\|_{L^2_{\rm loc}(\Omega)} - \|\nabla w\|_{L^2_{\rm loc}(\Omega)}) \le 0.$$

Thus, by the definition of the weak limit and using the strong convergence of w_n to w in $L^2_{\rm loc}(\Omega)$ we conclude the proof. □

Lemma 6.4.3. *Let g be a positive function such that $g \in L^p(\Omega)$ with $p > \frac{N}{2}$ and $s > 0$. Assume that w_1, w_1 are positive functions such that $w_1, w_1 \in W^{1,2}_0(\Omega) \cap L^\infty(\Omega)$ verifying*

$$\begin{cases} -\Delta w_1 \le \frac{|\nabla w_1|^p}{1+s|\nabla w_1|^p} + g & \text{in } \Omega, \\ w_1 = 0 & \text{on } \partial\Omega \end{cases} \tag{6.4.3}$$

and

$$\begin{cases} -\Delta w_2 \ge \frac{|\nabla w_2|^p}{1+s|\nabla w_2|^p} + g & \text{in } \Omega, \\ w_2 = 0 & \text{on } \partial\Omega. \end{cases} \tag{6.4.4}$$

Then $w_2 \ge w_1$ in Ω.

Proof. Consider $w = w_1 - w_2$. Then $w \in W^{1,2}_0(\Omega) \cap L^\infty(\Omega)$. We will prove that $w^+ = 0$. By (6.4.3) and (6.4.4) it follows that

$$-\Delta w \le \frac{|\nabla w_1|^p}{1+s|\nabla w_1|^p} - \frac{|\nabla w_2|^p}{1+s|\nabla w_2|^p}.$$

Now for $x, y \in \mathbb{R}^N$ we define the function γ by setting

$$\gamma(t) = T(t|x| + (1-t)|y|), \quad \text{where } T(t) = \frac{|t|^p}{1+s|t|^p}.$$

For $x = \nabla w_1$ and $y = \nabla w_2$ we have

$$\frac{|\nabla w_1|^p}{1+s|\nabla w_1|^p} - \frac{|\nabla w_2|^p}{1+s|\nabla w_2|^p} = \gamma(1) - \gamma(0) = \gamma'(\theta).$$

Since

$$\big|\gamma(1) - \gamma(0)\big| \le \big||\nabla w_1| - |\nabla w_2|\big|\big|T'(\theta)\big| \le |\nabla w_1 - \nabla w_2|\big|T'(\theta)\big|$$

and $|T'(t)| = p\frac{|t|^{p-1}}{(1+s|t|^p)^2} \le C$, we conclude that

$$\left|\frac{|\nabla w_1|^p}{1+s|\nabla w_1|^p} - \frac{|\nabla w_2|^p}{1+s|\nabla w_2|^p}\right| \le C|\nabla w|.$$

Hence, it follows that

$$-\Delta w \le C|\nabla w|, \quad w \in W_0^{1,2}(\Omega) \cap L^\infty(\Omega).$$

Using the Kato inequality we get

$$-\Delta w_+ \le C|\nabla w_+|, \quad 0 \le w_+ \in W_0^{1,2}(\Omega) \cap L^\infty(\Omega).$$

Therefore, from the maximum principle in Proposition 6.2.1 (see also Lemma 4.6 of [31]) it follows that $w_+ \equiv 0$ and then we obtain the comparison result. □

Now we are able to prove the blow-up result.

Proof of Theorem 6.4.1. Without loss of generality, we can assume that $f \in L^\infty(\Omega)$ and that λ is small. Assume the existence of $x_0 \in \Omega$ such that $u_n(x_0) \le C$ for all n. Using the maximum principle in Proposition 6.2.5, there exists a structural positive constant C' (independent of u_n), such that

$$C \ge u_n(x_0) \ge C'(\Omega)\delta(x_0) \int_\Omega (\lambda a_n(x)u_n + |\nabla u_n|^p + f)\delta(x)\,dx, \tag{6.4.5}$$

where $\delta(x) = \operatorname{dist}(x, \partial\Omega)$. Let $\phi \in C_0^\infty(\Omega)$ be a positive function. Using $T_k(u_n)\phi$ as a test function in (6.4.1), we can prove that $T_k(u_n)$ is uniformly bounded in $W_{\mathrm{loc}}^{1,2}(\Omega)$.

For n fixed, we consider $v_j \in W_0^{1,2}(\Omega) \cap L^\infty(\Omega)$, the minimal positive solution to problem

$$\begin{cases} -\Delta v_j = \lambda a_n(x)v_j + \frac{|\nabla v_j|^p}{1+\frac{1}{j}|\nabla v_j|^p} + \alpha f & \text{in } \Omega, \\ v_j = 0 & \text{on } \partial\Omega. \end{cases} \tag{6.4.6}$$

Using an iteration argument as in [31], it follows that $v_j \le v_{j+1}$ and $v_j \le u_n$ for every n. Define

$$w_n = \lim_{j\to\infty} v_j \le u_n.$$

Claim. The following statements hold:

(a) $\{T_k(w_n)\}$ is bounded in $W_{\mathrm{loc}}^{1,2}(\Omega)$;
(b) $w_n \in W_{\mathrm{loc}}^{1,p}(\Omega)$;
(c) w_n is a supersolution to problem (6.4.1);
(d) $w_n \le w_{n+1}$.

Assume the claim holds. As before, there exists a positive structural constant C' such that

$$C \ge w_n(x_0) \ge C'(\Omega)\delta(x_0) \int_\Omega (\lambda a_n(x)w_n + |\nabla w_n|^p + f)\delta(x)\,dx.$$

Since $\{w_n\}$ is a monotone sequence we conclude that

$$a_n(x)w_n \nearrow \frac{w}{|x|^2} \quad \text{in } L^1_{\text{loc}}(\Omega) \text{ and } \int_\Omega |\nabla w_n|^p \delta(x) \le C'.$$

Thus $\{w_n\}$ is bounded in $W^{1,p}_{\text{loc}}(\Omega)$. Hence using (a) in the claim and Lemma 6.4.2, we conclude that

$$T_k(w_n) \to T_k(w) \quad \text{strongly in } W^{1,2}_{\text{loc}}(\Omega).$$

Since w_n is a supersolution to (6.4.1), letting $n \to \infty$ we obtain that w satisfies also

$$-\Delta w \ge |\nabla w|^p + \lambda \frac{w}{|x|^2} + cf,$$

a contradiction with Theorem 6.3.2. □

Proof of the claim. Statements (a) and (d) follow directly from equation (6.4.6) by an application of the corresponding inequality of type (6.4.5) and the fact that a_n is a nondecreasing sequence. To prove (b) we consider separately two cases: (i) $p \le 2$ and (ii) $p > 2$.

For $p \le 2$ we have

$$T_k(v_j) \to T_k(w_n) \quad \text{as } j \to \infty \text{ strongly in } W^{1,2}_{\text{loc}}(\Omega). \tag{6.4.7}$$

To obtain the convergence, we use a nonlinear test function as in [70] (see also [74] and [71]). Consider $\phi(s) = se^{\frac{1}{4}s^2}$, in such a way that $\phi'(s) - |\phi(s)| \ge \frac{1}{2}$. For $\psi \in C_0^\infty(\Omega)$, $\psi \ge 0$, take $\phi(T_k(v_j) - T_k(w_n))\psi(x)$ as a test function in equation (6.4.6). We obtain from the left hand side

$$\int_\Omega \nabla v_j \phi'(T_k(v_j) - T_k(w_n))\nabla(T_k(v_j) - T_k(w_n))\psi \, dx$$
$$= \int_\Omega |\nabla(T_k(v_j) - T_k(w_n))|^2 \phi'(T_k(v_j) - T_k(w_n))\psi \, dx + o(1).$$

We set $H(\nabla v_j) = \frac{|\nabla v_j|^p}{1+\frac{1}{j}|\nabla v_j|^p}$. Then, the right hand side could be estimated by

$$\int_\Omega H(\nabla v_j)\phi(T_k(v_j) - T_k(w_n))\psi \, dx \le \delta \int_\Omega |\nabla T_k(v_j) - \nabla T_k(w_n)|^2 |\phi(T_k(v_j) - T_k(w_n))|\psi \, dx + o(1),$$

where $\delta \le 1$. Since

$$\int_\Omega (\lambda a_n(x)v_j + cf)\phi(T_k(v_j) - T_k(w_n))\psi(x) \, dx \to 0 \quad \text{as } m \to \infty,$$

we conclude the required convergence and in particular the almost everywhere convergence up to a subsequence.

In the case $p \geq 2$ the result is directly obtained as follows: since

$$\int_\Omega |\nabla v_j|^2\,dx = \int_\Omega (-\Delta v_j)v_j\,dx \leq \int_\Omega (-\Delta v_j)u_n\,dx \leq \Big(\int_\Omega |\nabla v_j|^2\,dx\Big)^{\frac{1}{2}}\Big(\int_\Omega |\nabla u_n|^2\,dx\Big)^{\frac{1}{2}},$$

we have $v_j \rightharpoonup w_n$ weakly in $W_0^{1,2}(\Omega)$ as $j \to \infty$. By using the last inequality and the weak lower semicontinuity of the norm it turns out that

$$\int_\Omega |\nabla w_n|^2\,dx \leq \liminf_{j\to\infty}\int_\Omega |\nabla v_j|^2\,dx \leq \limsup_{j\to\infty}\int_\Omega |\nabla v_j|^2\,dx.$$

Moreover, taking into account that $-\Delta v_j \geq 0$,

$$\int_\Omega |\nabla v_j|^2\,dx = \int_\Omega (-\Delta v_j)v_j\,dx \leq \int_\Omega (-\Delta v_j)w_n\,dx \leq \Big(\int_\Omega |\nabla v_j|^2\,dx\Big)^{\frac{1}{2}}\Big(\int_\Omega |\nabla w_n|^2\,dx\Big)^{\frac{1}{2}}.$$

Hence

$$\limsup_{j\to\infty}\int_\Omega |\nabla v_j|^2\,dx \leq \int_\Omega |\nabla w_n|^2\,dx.$$

Then we conclude the strong convergence in $W_0^{1,2}(\Omega)$. In particular we have the almost everywhere convergence of the gradients. To conclude the proof of (b) it is sufficient to observe that

$$\begin{aligned} C \geq u_n(x_0) &\geq C'(\Omega)\delta(x_0)\int_\Omega \Big(\lambda a_n(x)v_j + \frac{|\nabla v_j|^p}{1+\frac{1}{j}|\nabla v_j|^p} + f\Big)\delta(x)\,dx \\ &\geq C'(\Omega)\delta(x_0)\int_\Omega (\lambda a_n(x)w_n + |\nabla w_n|^p + f)\delta(x)\,dx, \end{aligned} \tag{6.4.8}$$

by Fatou's lemma. To prove (c), we use a nonnegative test function in problem (6.4.6) and we pass to the limit by Fatou's lemma. □

6.5 Existence result: $1 < p < q_+(\lambda)$ and $\lambda < \Lambda_{N,2}$

We consider $\alpha_+(\lambda)$ and $\alpha_-(\lambda)$ defined in (3.1.3). Together with the critical exponent $q_+(\lambda) \equiv \frac{2+\alpha_-(\lambda)}{1+\alpha_-(\lambda)}$ we define $q_-(\lambda) \equiv \frac{2+\alpha_+(\lambda)}{1+\alpha_+(\lambda)}$. Clearly $q_-(\lambda) \leq q_+(\lambda)$.

We can now state the following result of existence of solutions in the whole $\mathbb{R}^N$.

Theorem 6.5.1. *Assume that $q_-(\lambda) < q < q_+(\lambda)$, where $q_-(\lambda), q_+(\lambda)$ are given above. Then problem (6.1.1) with $f \equiv 0$ has a very weak solution $u > 0$ in $\mathbb{R}^N$.*

Proof. We search a solution in the form $u(x) = A|x|^{-\beta}$. Hence by a direct computation we obtain $\beta = \frac{2-q}{q-1}$ and

$$\beta^q A^{q-1} = \beta(N - \beta - 2) - \lambda.$$

To have $A > 0$ we need $\beta \in (\alpha_-(\lambda), \alpha_+(\lambda))$, which is equivalent to $q_-(\lambda) < q < q_+(\lambda)$. Note that $u \in L^1_{loc}(\Omega)$, $\frac{u}{|x|^2} \in L^1_{loc}$ and, since $\frac{N}{N-1} < q_-(\lambda) < q$, we get $|\nabla u|^q \in L^1_{loc}$ (compare with Remark 6.3.3). Hence the result follows. □

Remark 6.5.2. The solution w in Theorem 6.5.1 is in the space $W^{1,2}_{loc}(\mathbb{R}^N)$ if and only if $q > \frac{N+2}{N}$. Note that for all $\lambda \in [0, \Lambda_{N,2})$, $\frac{N+2}{N} \in (q_-(\lambda), q_+(\lambda))$ and if $\lambda = \Lambda_{N,2}$, then $\frac{N+2}{N} = q_-(\lambda) = q_+(\lambda)$.

We deal now with the existence of solutions to Dirichlet problem in bounded domains. We will use the following result (Lemma 5.3 in [31]).

Lemma 6.5.3. *If $u_n \rightharpoonup u$ weakly in $W^{1,2}_0(\Omega)$ and*
1. $0 \le u_n \le u$;
2. $-\Delta u_n \ge 0$ *in* $\mathcal{D}'(\omega)$;

then

$$\|\nabla u_n - \nabla u\|_2 \to 0 \quad \text{as } n \to \infty.$$

Proof. Note that

$$\int_\Omega |\nabla u_n|^2 = \int_\Omega u_n(-\Delta u_n) \le \int_\Omega u(-\Delta u_n) = \int_\Omega \langle \nabla u \nabla u_n \rangle \le \Big(\int |\nabla u|^2\Big)^{\frac{1}{2}} \Big(\int |\nabla u_n|^2\Big)^{\frac{1}{2}}.$$

Therefore, by the weak semicontinuity of the norm,

$$\Big(\int |\nabla u|^2\Big)^{\frac{1}{2}} \le \liminf_{n\to\infty} \Big(\int |\nabla u_n|^2\Big)^{\frac{1}{2}} \le \limsup_{n\to\infty} \Big(\int |\nabla u_n|^2\Big)^{\frac{1}{2}} \le \Big(\int |\nabla u|^2\Big)^{\frac{1}{2}},$$

which allows us to conclude the proof. □

Theorem 6.5.4. *Assume that $1 < p < q_+(\lambda)$, where $q_+(\lambda) = \frac{2+\alpha_-(\lambda)}{1+\alpha_-(\lambda)}$. Then there exists c_0 such that if $c < c_0$ and $f(x) \le \frac{1}{|x|^2}$, then problem*

$$\begin{cases} -\Delta u = |\nabla u|^p + \lambda \frac{u}{|x|^2} + cf & \text{in } \Omega, \\ u = 0 & \text{on } \partial\Omega \end{cases} \tag{6.5.1}$$

has a very weak positive solution u.

Proof. We start by assuming that for $c > 0$ and $f(x) \le \frac{1}{|x|^2}$, we are able to find a radial positive supersolution $\overline{w} \in W^{1,p}(\Omega)$ to problem (6.5.1) such that

$$\exists s > 0, \quad \text{for which } \frac{\overline{w}^{1+s}}{|x|^2}, \overline{w}^{\frac{(2-p)s+p}{2-p}} \in L^1(\Omega). \tag{6.5.2}$$

Consider $a_n(x) = \frac{1}{|x|^2+\frac{1}{n}} \uparrow |x|^{-2}$, $f_n = \min\{f, n\} \uparrow f$. Then problem

$$\begin{cases} -\Delta u_n = \lambda a_n(x) u_n + \frac{|\nabla u_n|^p}{1+\frac{1}{n}|\nabla u_n|^p} + c f_n & \text{in } \Omega, \\ u_n = 0 & \text{on } \partial\Omega \end{cases} \tag{6.5.3}$$

has a minimal positive solution $u_n \in W_0^{1,2}(\Omega) \cap L^\infty(\Omega)$. By Lemma 6.4.3 and using the comparison principle given in Proposition 6.2.3, we get $u_n \le u_{n+1}$ and $u_n \le \overline{w}$ for every n. Hence $\overline{u} = \lim_{n\to\infty} u_n \le w$. Define $\phi_n = (1+u_n)^s - 1$, where s is as in (6.5.2). Then using ϕ_n as a test function in (6.5.3), we have

$$\int_\Omega \frac{|\nabla u_n|^2}{(1+u_n)^{1-s}}\,dx \le C_1, \qquad \int_\Omega |\nabla u_n|^p (1+u_n)^s\,dx \le C_2.$$

In particular

$$\frac{1}{k}\int_\Omega |\nabla T_k u_n|^2 \le C_3, \qquad \int_\Omega |\nabla u_n|^p \le C_4.$$

We claim now that

$$\nabla T_k v_n \to \nabla T_k v \quad \text{as } n \to \infty \text{ strongly in } W_0^{1,p}(\Omega).$$

To prove the claim we consider nonlinear test functions structured by $\phi(s) = s e^{\frac{1}{4}s^2}$, which is appropriate because $\phi'(s) - |\phi(s)| \ge \frac{1}{2}$ (see for instance [71]). That is, we consider $\phi(T_k v_n - T_k v)$ as a test function in (6.5.3) and then

$$\begin{aligned} &\int_\Omega \nabla v_n \nabla(T_k v_n - T_k v)\phi'(T_k v_n - T_k v)\,dx \\ &\quad = \int_\Omega \frac{|\nabla v_n|^p}{1+\frac{1}{n}|\nabla v_n|^p}\phi(T_k u_n - T_k u)\,dx + \int_\Omega (a_n v_n + f_n)\phi(T_k u_n - T_k u)\,dx. \end{aligned}$$

It is clear that

$$\lambda \int_\Omega (a_n v_n + f_n)\phi(T_k u_n - T_k u)\,dx \le \lambda \int_\Omega \left(\frac{\overline{w}}{|x|^2} + f\right)\phi(T_k u_n - T_k u)\,dx \to 0 \quad \text{as } n \to \infty.$$

We deal now with the other terms. We have

$$\begin{aligned}\int_\Omega \nabla v_n \nabla(T_k v_n - T_k v)\phi'(T_k v_n - T_k v)\,dx &= \int_\Omega \nabla T_k v_n \nabla(T_k v_n - T_k v)\phi'(T_k v_n - T_k v)\,dx \\ &= \int_\Omega \big|\nabla T_k(v_n) - \nabla T_k(v)\big|^2 \phi'(T_k v_n - T_k v)\,dx \\ &\quad + \int_\Omega \nabla T_k(v)\nabla(T_k(v_n) - T_k v)\phi'(T_k v_n - T_k v)\,dx \\ &= \int_\Omega \big|\nabla T_k(v_n) - \nabla T_k(v)\big|^2 \phi'(T_k v_n - T_k v)\,dx + o(1),\end{aligned}$$

where we have used the monotonicity of $\{v_n\}$. On the other hand, since $p \le 2$, there exists $\varepsilon \le 1$ and a nonnegative constant $C(\varepsilon)$ such that

$$\begin{aligned}&\int_\Omega \frac{|\nabla v_n|^p}{1 + \frac{1}{n}|\nabla v_n|^p}\phi(T_k v_n - T_k v) \\ &\le \int_\Omega |\nabla T_k v_n|^p |\phi(T_k v_n - T_k v)|\,dx \\ &\le \varepsilon \int_\Omega \big|\nabla T_k(v_n)\big|^2 |\phi(T_k v_n - T_k v)|\,dx + C_\varepsilon \int_\Omega |\phi(T_k v_n - T_k v)|\,dx \\ &\le \varepsilon \int_\Omega |\nabla T_k v_n - \nabla T_k v|^2 |\phi(T_k v_n - T_k v)|\,dx - \varepsilon \int_\Omega |\nabla T_k v|^2 |\phi(T_k v_n - T_k v)|\,dx \\ &\quad + 2\varepsilon \int_\Omega \nabla T_k v_n \nabla T_k v |\phi(T_k v_n - T_k v)|\,dx + C_\varepsilon \int_\Omega |\phi(T_k v_n - T_k v)|\,dx.\end{aligned}$$

The fact that $T_k v_n \rightharpoonup T_k v$ weakly in $W_0^{1,2}(\Omega)$ gives

$$\int_\Omega \frac{|\nabla v_n|^p}{1 + \frac{1}{n}|\nabla v_n|^p}\phi(T_k u_n - T_k u) \le \varepsilon \int_\Omega |\nabla T_k v_n - \nabla T_k v|^2 |\phi(T_k v_n - T_k v)|\,dx + o(1).$$

Hence it follows that

$$\int_\Omega |\nabla T_k v_n - \nabla T_k v|^2 \big(\phi'(T_k v_n - T_k v) - \varepsilon|\phi(T_k v_n - T_k v)|\big)\,dx \le o(1).$$

Since $\phi'(s) - \varepsilon|\psi(s)| \ge \frac{1}{2}$, we obtain $\int_\Omega |\nabla T_k v_n - \nabla T_k v|^2 \le o(1)$ and then

$$\nabla T_k v_n \to \nabla T_k v \quad \text{as } n \to \infty \text{ strongly in } W_0^{1,p} \text{ for all } k > 0,$$

as claimed.

Note that, as a consequence, we can find a subsequence such that $\nabla v_{n_j} \to \nabla v$ almost everywhere in Ω. Define now $\psi_n = (1 + G_k(v_n))^s - 1$, where as usual $G_k(t) = t - T_k(t)$, the truncation function. Then using ψ_n as a test function in (6.5.3), by the Hölder inequality we obtain

$$\int_{\Omega} \frac{|\nabla G_k(v_n)|^2}{(1 + G_k(v_n))^{1-s}}\, dx \le C_0 \left\{ \int_{v_n \ge k} (1 + \overline{w})^{\frac{(2-p)s+p}{2-p}}\, dx + \lambda \int_{v_n \ge k} \frac{\overline{w}^{1+s}}{|x|^2}\, dx + \int_{v_n \ge k} f(x)(\overline{w} + 1)^s\, dx \right\}.$$

It is clear by the hypotheses on $\overline{w}$ that

$$\int_{v_n \ge k} (1+\overline{w})^{\frac{(2-p)s+p}{2-p}}\, dx + \lambda \int_{v_n \ge k} \frac{\overline{w}^{1+s}}{|x|^2}\, dx + \int_{v_n \ge k} f(x)(\overline{w}+1)^s\, dx \to 0 \quad \text{as } k \to \infty \text{ uniformly in } n.$$

Hence we conclude that

$$\limsup_{k\to\infty} \int_{v_n \ge k} |\nabla v_n|^p\, dx \le \limsup_{k\to\infty} \int_{\Omega} |\nabla G_k(v_n)|^p (1 + G_k(v_n))^s\, dx = 0 \tag{6.5.4}$$

uniformly in n. To prove that

$$|\nabla v_n|^p \to |\nabla v|^p \quad \text{strongly in } L^1(\Omega),$$

we will use Vitali's theorem (see for instance [64] pag. 16). So, we only have to show the equi-integrability of $\{|\nabla v_n|^p|\}$. Let $E \subset \Omega$ be a measurable set. Then,

$$\int_E |\nabla v_n|^p\, dx \le \int_E |\nabla T_k v_n|^p\, dx + \int_{\{v_n \ge k\}\cap E} |\nabla v_n|^p\, dx.$$

By the fact that $T_k(v_n) \to T_k(v)$ strongly in $W_0^{1,p}(\Omega)$, the integral $\int_E |\nabla T_k(v_n)|^p\, dx$ is uniformly small if $|E|$ is small enough. On the other hand, by (6.5.4) we obtain

$$\int_{\{v_n \ge k\}\cap E} |\nabla v_n|^p\, dx \le \int_{\{v_n \ge k\}} |\nabla v_n|^p\, dx \to 0 \quad \text{as } k \to \infty \text{ uniformly in } n.$$

The equi-integrability of $|\nabla v_n|^q$ follows immediately, and the strong convergence holds. Hence, since

$$\nabla u_n \to \nabla \overline{u} \quad \text{as } n \to \infty \text{ strongly in } L^p(\Omega),$$

we conclude that $\overline{u}$ is a very weak solution to problem (6.5.1).

It is worthy to point out that for the values of p for which a supersolution in $W_0^{1,2}(\Omega)$ exists (in particular if $1 < p \le q_-(\lambda)$), the proof is easier and, moreover, the solution $\overline{u} \in W_0^{1,2}(\Omega)$. In this last case it suffices to take u_n as test function and to use Lemma 6.5.3.

To finish the proof we have to find the required supersolution. We will consider two cases:

(i) $q_-(\lambda) < p < q_+(\lambda)$;
(ii) $1 < p \le p_-(\lambda)$.

Case (i). $q_-(\lambda) < p < q_+(\lambda)$. Consider u, the radial solution obtained in Theorem 6.5.1. Then $\frac{u^{1+s}}{|x|^2}, u^{\frac{(2-p)s+p}{2-p}} \in L^1(\Omega)$ for all $0 < s < \frac{p(N-1)-N}{2-p} < 1$. Define $v(x)$ to be the unique solution to problem

$$\begin{cases} -\Delta v = 0 & \text{in } \Omega, \\ v = u & \text{on } \partial\Omega. \end{cases}$$

Note that $v \in \mathcal{C}^\infty(\Omega)$ and $0 < c_1 \le v \le c_2$ for some positive constants c_1 and c_2. We set $\overline{w} = t(u - v)$, $t > 0$. Then $\overline{w} \in W_0^{1,p}(\Omega)$, $w \ge 0$ in Ω and

$$\begin{aligned} -\Delta\overline{w} - \lambda\frac{\overline{w}}{|x|^2} &= t\left(-\Delta u - \lambda\frac{u}{|x|^2}\right) + t\lambda\frac{v}{|x|^2} = t|\nabla u|^p + t\lambda\frac{v}{|x|^2} \\ &\ge t\left(\frac{1}{(1+\epsilon)^{p-1}}|\nabla\overline{w}|^p t^{-p} - \left(\frac{1+\frac{1}{\epsilon}}{1+\epsilon}\right)^{p-1}|\nabla v|^p\right) + t\lambda\frac{v}{|x|^2}, \end{aligned}$$

where in the last estimate we have used the following elemental inequality:

$$|a+b|^p \le (1+\epsilon)^{p-1}|a|^p + \left(1+\frac{1}{\epsilon}\right)^{p-1}|b|^p.$$

Taking $t = \frac{1}{1+\epsilon}$, we conclude that

$$-\Delta\overline{w} - \lambda\frac{\overline{w}}{|x|^2} \ge |\nabla\overline{w}|^p + \left(\frac{\lambda}{1+\epsilon}\right)\frac{v}{|x|^2} - \left(\frac{1}{\epsilon^{p-1}(1+\epsilon)}\right)|\nabla v|^p.$$

Hence choosing ϵ large enough there exists a positive constant c_0 such that

$$\frac{\lambda}{1+\epsilon}\frac{v}{|x|^2} - \frac{1}{\epsilon^{p-1}(1+\epsilon)}|\nabla v|^p \ge \frac{c_0}{|x|^2}.$$

Since $|x|^2 f(x) < 1$, $\overline{w} \in W_0^{1,p}(\Omega)$ is a supersolution to problem (6.5.1) if $c < c_0$.

Case (ii). $1 < p \le q_-(\lambda)$. We start by getting a supersolution in a ball, i. e., $\Omega = B_R(0)$. Without loss of generality we will assume $R = 1$.

Since $p \le q_-(\lambda)$, there exists $\beta \in (\alpha_-(\lambda), \alpha_+(\lambda))$, close to $\alpha_-(\lambda)$, such that $p(\beta+1) < \beta + 2$.

Define $\overline{w}(x) \equiv A(|x|^{-\beta} - 1)$. Then $\overline{w} \in W_0^{1,p}(B_R(0))$ and

$$-\Delta\overline{w} - \lambda\frac{\overline{w}}{|x|^2} = A(\beta(N-\beta-2) - \lambda)|x|^{-\beta-2} + \frac{A}{|x|^2}.$$

Since $\beta \in (\alpha_-(\lambda), \alpha_+(\lambda))$, we have $\beta(N-\beta-2) - \lambda > 0$.

Hence choosing $A^{p-1} = \frac{\beta(N-\beta-2)-\lambda}{\beta^p}$ we obtain

$$-\Delta\overline{w} - \lambda\frac{\overline{w}}{|x|^2} \geq |\nabla\overline{w}|^p + \frac{A}{|x|^2},$$

that is, if $c_0 = A$, $\overline{w}$ is a supersolution to (6.5.1) in $B_1(0)$ for all $c < c_0$.

In the case of a general domain Ω that contains the origin we consider a ball $B_R(0)$ such that $\Omega \subset B_{\frac{R}{2}}(0)$. We have the corresponding supersolution in $B_R(0)$ found above, for which we perform the same arguments as in the first case. □

6.6 The critical case, $\lambda \equiv \Lambda_{N,2}$ and $p < \frac{N+2}{N}$

Assume that $\lambda = \Lambda_{N,2}$ and $p < p_+ \equiv \frac{N+2}{N}$ and consider the function

$$w(x) = \left|\frac{x}{R}\right|^{-\frac{N-2}{2}} \left(\log\left(\frac{R}{|x|}\right)\right)^{1/2} - A,$$

where $A = (\frac{r}{R})^{-\frac{N-2}{2}}(\log(\frac{R}{|r|}))^{1/2}$. Then $w(x) = 0$ if $|x| = r$. It is easy to check that $w \in W_0^{1,q}(B_r(0))$ for all $q < 2$ and

$$-\Delta w - \Lambda_{N,2}\frac{w}{|x|^2} = \frac{1}{4}\frac{w}{|x|^2}\left(\log\left(\frac{R}{|x|}\right)\right)^{-2} + A\frac{\Lambda_{N,2}}{|x|^2}.$$

Since $|\nabla w(x)| = R^{\frac{N-2}{2}}(\log(\frac{R}{|x|}))^{\frac{1}{2}}|x|^{-\frac{N}{2}}(\frac{N-2}{2} + \frac{1}{2}(\log(\frac{R}{|x|}))^{-1})$ and $p < \frac{N+2}{N}$, we get for a suitable positive constant c

$$|\nabla w|^p \leq c\frac{w}{|x|^2}\left(\log\left(\frac{R}{|x|}\right)\right)^{-2}.$$

Hence, up to a positive constant c_1, $c_1 w$ is a supersolution to problem

$$\begin{cases} -\Delta w = |\nabla w|^p + \Lambda_{N,2}\frac{w}{|x|^2} + c_0 f & \text{in } B_r(0), \\ w = 0 & \text{on } \partial B_r(0), \end{cases} \tag{6.6.1}$$

where $|x|^2 f$ is bounded and c_0 is small.

To prove the existence of solution, we consider the approximated problems

$$\begin{cases} -\Delta u_n = \Lambda_{N,2}a_n(x)u_n + \frac{|\nabla v_n|^p}{1+\frac{1}{n}|\nabla v_n|^p} + cf_n & \text{in } B_r(0), \\ v_k = 0 & \text{on } \partial B_r(0), \end{cases} \tag{6.6.2}$$

where $a_n(x) = \min\{n, \frac{1}{|x|^2}\}$ and $f_n(x) = T_n(f(x))$. It is easy to check that $\Lambda_{N,2} < \lambda_1(a_n)$, the principal eigenvalue of the Laplacian with weight a_n. Then by similar arguments to those used previously, we can prove that there exists a minimal solution u_n of (6.6.2). Since, in particular, $w \in W_0^{1,1}(B_r(0))$, by Theorem 4.3 in [31] we conclude that $\{u_n\}$ is increasing in n and that $u_n \le w$ in $B_r(0)$. Hence $u_n \uparrow u$ pointwise and $u \le w$. It is easy to see that $u \in L^q(B_r(0))$ for all $q < 2^*$.

Consider, as in the sublinear case, $H(B_r(0))$, the completion of $C_0^\infty(B_r(0))$ with respect to the norm

$$\|\phi\|_{H(B_r(0))} = \Big(\int_{B_r(0)} |\nabla\phi|^2\,dx - \Lambda_{N,2}\int_{B_r(0)} \frac{\phi^2}{|x|^2}\,dx\Big)^{1/2}.$$

It is well known that $W_0^{1,2}(B_r(0)) \subset H(B_r(0)) \subset W_0^{1,q}(B_r(0))$ for all $q < 2$. We could check that $w \notin H(B_r(0))$; however, $\{u_n\}$ is bounded in $H(B_r(0))$. Indeed, taking u_n as a test function in (6.6.2), we have

$$\begin{aligned}\|u_n\|^2_{H(B_r(0))} &= \int_{B_r(0)} |\nabla u_n|^2\,dx - \Lambda_{N,2}\int_{B_r(0)} \frac{u_n^2}{|x|^2}\,dx \le \int_{B_r(0)} |\nabla u_n|^p u_n\,dx + c_0\int_{B_r(0)} \frac{u_n}{|x|^2}\,dx\\ &\le \int_{B_r(0)} |\nabla u_n|^p w\,dx + c_0\int_{B_r(0)} \frac{w}{|x|^2}\,dx.\end{aligned}$$

Using the Hölder and Young inequalities and the improved Hardy–Sobolev inequalities (see [6] and [314]) we obtain

$$\begin{aligned}\int_{B_r(0)} |\nabla u_n|^p w\,dx &= \int_{B_r(0)} |\nabla u_n|^p\Big(\log\Big(\frac{R}{|x|}\Big)\Big)^{-p}\Big(\log\Big(\frac{R}{|x|}\Big)\Big)^{p} w\,dx\\ &\le \delta\int_{B_r(0)} |\nabla u_n|^2\Big(\log\Big(\frac{R}{|x|}\Big)\Big)^{-2}dx + C(\delta)\int_{B_r(0)} w^{\frac{2}{2-p}}\Big(\log\Big(\frac{R}{|x|}\Big)\Big)^{\frac{2p}{2-p}}dx\\ &\le \delta\|u_n\|^2_{H(B_r(0))} + C(\delta)\int_{B_r(0)} w^{\frac{2}{2-p}}\Big(\log\Big(\frac{R}{|x|}\Big)\Big)^{\frac{2p}{2-p}}dx.\end{aligned}$$

Since $p < \frac{N+2}{N}$, we have

$$\int_{B_r(0)} w^{\frac{2}{2-p}}(\log(\frac{R}{|x|}))^{\frac{2p}{2-p}}\,dx < \infty.$$

Hence choosing δ small we conclude that $\|u_n\|^2_{H(B_r(0))} \le C$. Hence, $u_n \rightharpoonup u$ weakly in $H(B_r(0))$ and then $\|u\|^2_{H(B_r(0))} \le \|u_n\|^2_{H(B_r(0))}$. We will prove that $u_n \to u$ strongly in $H(B_r(0))$. Since we are in the Hilbertian framework, it suffices to prove

$$\lim_{n\to\infty}\|u_n\|^2_{H(B_r(0))} = \|u\|^2_{H(B_r(0))}.$$

Consider the linear form $F_n : H(B_r(0)) \to \mathbb{R}$, $F_n \equiv -\Delta u_n - \Lambda_{N,2} a_n(x) u_n$. By the regularity of u_n we find that $F_n \in H'(B_r(0))$, the dual space of $H(B_r(0))$.

Since $u_n \le u$, by the fact that $-\Delta u_n - \Lambda_{N,2} a_n(x) u_n \ge 0$, we get

$$\begin{aligned}\|u_n\|^2_{B_r(0)} &\le \int\limits_{H(B_r(0))} (-\Delta u_n - \Lambda_{N,2} a_n(x) u_n) u_n \, dx = \langle F_n, u_n \rangle \\ &\le \int\limits_{B_r(0)} (-\Delta u_n - \Lambda_{N,2} a_n(x) u_n) u \, dx = \langle F_n, u \rangle.\end{aligned}$$

If $\{F_n\}$ is uniformly bounded in $H'(B_r(0))$ we are done because then $F_n \rightharpoonup F$ in the weak-$*$ topology of $H'(B_r(0))$ and in particular if $\phi \in \mathcal{C}_0^\infty(B_r(0))$, we obtain

$$\langle F_n, \phi \rangle \to \int\limits_{B_r(0)} \left(\nabla u \nabla \phi - \Lambda_{N,2} \frac{u\phi}{|x|^2} \right) dx.$$

Hence $F = -\Delta u - \Lambda_{N,2} \frac{u}{|x|^2} \in H'(B_r(0))$. Thus, by the density,

$$\langle F_n, u \rangle \to \langle F, u \rangle = \|u\|^2_{H(B_r(0))}.$$

As a byproduct we find the strong convergence of $\{u_n\}_{n\in\mathbb{N}}$ to u in $H(B_r(0))$. Therefore, u is a weak solution to problem (6.6.1).

To finish we just need to prove that $\{F_n\}$ is uniformly bounded in $H'(B_r(0))$.

Consider $\phi \in \mathcal{C}_0^\infty(B_r(0))$. Then

$$|\langle F_n, \phi \rangle| = \left| \int\limits_{B_r(0)} \phi(-\Delta u_n - \Lambda_{N,2} a_n(x) u_n) \, dx \right| \le \int\limits_{B_r(0)} |\nabla u_n|^p |\phi| \, dx + c_0 \int\limits_{B_r(0)} |f||\phi| \, dx.$$

Using the hypothesis on f we obtain

$$\int\limits_{B_r(0)} |f||\phi| \, dx \le C(f) \|\phi\|_{H(B_r(0))}.$$

For the other term, using the fact that $\{u_n\}$ is bounded in $H(B_r(0))$, we have

$$\begin{aligned}\int\limits_{B_r(0)} |\nabla u_n|^p |\phi| \, dx &= \int\limits_{B_r(0)} |\nabla u_n|^p \left(\log\left(\frac{R}{|x|} \right) \right)^{-p} \left(\log\left(\frac{R}{|x|} \right) \right)^{p} |\phi| \, dx \\ &\le \left(\int\limits_{B_r(0)} |\nabla u_n|^2 \left(\log\left(\frac{R}{|x|} \right) \right)^{-2} dx \right)^{\frac{p}{2}} \left(\int\limits_{B_r(0)} |\phi|^{\frac{2}{2-p}} \left(\log\left(\frac{R}{|x|} \right) \right)^{\frac{2p}{2-p}} dx \right)^{\frac{2-p}{2}} \\ &\le C \left(\int\limits_{B_r(0)} |\phi|^{\frac{2}{2-p}} \left(\log\left(\frac{R}{|x|} \right) \right)^{\frac{2p}{2-p}} dx \right)^{\frac{2-p}{2}}.\end{aligned}$$

Since $p < \frac{N+2}{2}$, we have $\frac{2}{2-p} < 2^*$. Therefore using the properties of $H(B_r(0))$, there exists a constant $C_1 > 0$ such that

$$\int_{B_r(0)} |\phi|^{\frac{2}{2-p}} \left(\log\left(\frac{R}{|x|}\right)\right)^{\frac{2p}{2-p}} dx \le C_1 \|\phi\|_{H(B_r(0))}.$$

As a conclusion we obtain

$$|\langle F_n, \phi\rangle| \le C\|\phi\|_{H(B_r(0))}.$$

Remark 6.6.1.

1. As before, we can consider the case of a general domain Ω that contains the origin and prove that $t(w-v)$ is a supersolution, where w is defined above, v is a harmonic function such that $v = w$ on the boundary of Ω and $t > 0$. Then the existence result follows using the same computation as in the proof of Theorem 6.5.4. Hence the problem

$$\begin{cases} -\Delta u = \Lambda_{N,2}\frac{u}{|x|^2} + |\nabla u|^p + cf & \text{in } \Omega, \\ u = 0 & \text{on } \partial\Omega \end{cases}$$

   has a positive solution $u \in H(\Omega)$ if $|x|^2 f$ is bounded and c is small.
2. A proper definition of the *gradient* associated to the operator $-\Delta - \frac{\lambda}{|x|^2}I$ provides existence of solutions. Indeed, in [15] the following example is studied:

$$-\Delta u - \Lambda_{N,2}\frac{u}{|x|^2} = \left|\nabla u + \left(\frac{N-2}{2}\right)\frac{u}{|x|^2}x\right|^2 |x|^{\frac{N-2}{2}} + \lambda f(x),$$

   for x in Ω, $u = 0$ on $\partial\Omega$ and $\Lambda_{N,2} = (\frac{N-2}{2})^2$ and with f under some hypotheses of summability.

6.7 Further remarks

The following interesting questions are still open:

1. Fixing $1 < p < q_+(\lambda)$, obtain the optimal class of functions – according to their summability – in order to have existence of a *very weak solution* to the Dirichlet problem with data in such a class.
2. Assume that, for λ fixed and for f in a given class, we are able to find a very weak solution, u. What is the regularity of u in terms of the regularity of f?
3. Complete the study of results on uniqueness and nonuniqueness. We recall that for $\lambda = 0$ there are some results on multiplicity of unbounded solutions. See for instance [157] (for a ball) and [8], where all the solutions are characterized in a bounded domain.

As in the previous chapters, another line of study is to replace the Laplacian by other elliptic operators, such as the p-Laplacian. This problem, with a power of the gradient on the left hand side and with the Hardy–Leray potential, has been studied in [19] and [20].

7 The heat equation with nonlinearity on the gradient and the Hardy–Leray potential

7.1 Introduction

This chapter deals with the existence and nonexistence of positive solutions to the following problem:

$$\begin{cases} u_t - \Delta u = |\nabla u|^p + \lambda \frac{u}{|x|^2} + f(x,t) & \text{in } \Omega_T \equiv \Omega \times (0,T), \\ u(x,t) > 0 & \text{in } \Omega_T, \\ u(x,t) = 0 & \text{on } \partial\Omega \times (0,T), \\ u(x,0) = u_0(x) & \text{if } x \in \Omega, \end{cases} \tag{7.1.1}$$

where Ω is either an open bounded domain in $\mathbb{R}^N$ such that $0 \in \Omega$, or $\Omega = \mathbb{R}^N$, $N \geq 3$, $p > 1$ and $\lambda > 0$. We suppose that f and u_0 are positive measurable functions with some hypotheses that will be specified later. For the theory of quasilinear parabolic problem with general data we refer for instance to [61], [62], [67], [70], [77], [265], [267] and the references therein.

The general contents of the chapter may be described by saying that we want to understand the effect of the *uncertainty principle* expressed by the Hardy inequality in this kind of evolution models.

Problem (7.1.1) can be seen as a Hamilton–Jacobi equation with a viscosity term; see [127], [232], [55] and the references therein. As we explained in Chapter 6 for the case $\lambda = 0$ and $p = 2$, this model, proposed by Kardar–Parisi–Zhang, appears in the physical theory of growth and roughening of surfaces; see [205]. If $p = 2$ and $\lambda = 0$, a general multiplicity result for positive solutions was proved in [9], where the authors showed a direct connection between problem (7.1.1) and semilinear problems with measure data.

For general $p > 1$ and $\lambda = 0$, the model is known as the generalized Kardar–Parisi–Zhang equation. This was introduced by Krug–Spohn in [213]. Existence results are obtained in [187] and [55] for the Cauchy problem ($\Omega = \mathbb{R}^N$) and for regular initial data u_0 and $f \equiv 0$. Note that the results by B. Gilding, M. Guedda and R. Kersner in [187] show that there is no Fujita type exponent for the Cauchy problem with $\lambda = 0$.

One of the main results in this chapter is to prove that the optimal power for existence of solutions in the case of the stationary problem studied in Chapter 6, that is,

$$q_+(\lambda) \equiv \frac{2 + \alpha_-(\lambda)}{1 + \alpha_-(\lambda)}, \quad 0 < \lambda \leq \Lambda_{N,2}, \tag{7.1.2}$$

with $\alpha_-(\lambda)$ defined in (3.1.3), is also the critical power for the parabolic equation (7.1.1). We recall that the critical power only depends on λ and N. We will also use the related

https://doi.org/10.1515/9783110606270-007

parameter

$$q_-(\lambda) = \frac{2 + \alpha_+(\lambda)}{1 + \alpha_+(\lambda)}, \tag{7.1.3}$$

with $\alpha_-(\lambda)$ defined again in (3.1.3). Note that $q_-(\lambda) \le q_+(\lambda)$.

In order to show that $q_+(\lambda)$ is the critical power, we will prove that if $p \ge q_+(\lambda)$ there is no nontrivial *local solution*, even in the distributional sense. In the opposite case, if $p < q_+(\lambda)$, there exist positive solutions under some suitable additional hypotheses on the data. It is due to this behavior depending on the spectral parameter λ that the problem becomes so interesting.

Note that the nonlinear perturbations of the heat equation that we are considering do not satisfy the so-called *sign condition* (see [221]). To prove the existence of solution we use the classical arguments of truncation and compactness that allow to pass to the limit. We need to obtain uniform estimates under the presence of the Hardy potential and the lack of monotonicity of the nonlinear terms. These difficulties make us prove some additional comparison lemmas and new estimates. We will also use the results on almost everywhere convergence of the gradients obtained in [73] (see also [67]).

In the second part of this chapter we analyze the behavior of the Cauchy problem ($\Omega = \mathbb{R}^N$), which brings us some surprises with respect to the case $\lambda = 0$. Indeed, if $\Lambda_{N,2} \ge \lambda > 0$, there exists a Fujita type exponent. This is a consequence of the singularities introduced by the Hardy potential.

The present chapter is organized as follows. In Section 7.2 we present some preliminaries and prove results that will be used in the next sections. Section 7.2.1 is devoted to prove a new maximum principle for the heat equation, which has a precedent in the maximum principles obtained in Chapter 6 for the Laplace equation and has some interest in itself. The techniques for the proof in the parabolic case are very different to those in the stationary case.

As a consequence, we obtain a comparison principle and then a uniqueness result for some approximated problems. In Section 7.2.2 we find a lower estimate for the local behavior of a general supersolution and some *a priori* estimates for *very weak solutions* in the sense of Definition 7.2.1.

Section 7.3 deals with the nonexistence of nontrivial local solutions to problem (7.1.1) for supercritical powers. It is in this place where the Caffarelli–Kohn–Nirenberg inequalities play an important role.

In Section 7.4 we prove the existence result in bounded domains for $p < q_+(\lambda)$ under some additional hypotheses on u_0 and f. The idea is to use the fact that the associated elliptic problem has a positive solution. Since the supersolution of the elliptic problem is a supersolution to the parabolic problem under some hypotheses on the data, using an iteration argument and the comparison principle obtained in Section 7.2.1, we get the existence of a positive solution to the parabolic problem in an appropriate sense. Whence, $q_+(\lambda)$ is the *critical exponent* for existence of positive solutions to problem (7.1.1).

In Section 7.5 we analyze the Cauchy problem (7.5.2), with $p < p_+(\lambda)$. The most relevant fact in this case is the existence of a *Fujita type exponent*. This result comes in some way as a surprise, because the quasilinear problem with $\lambda = 0$ does not present a similar behavior. See [187]. The study is performed as follows. In Section 5.6.1 we find a family of subsolutions to (7.5.2) blowing up in finite time, for $1 < p < F(\lambda) = 1 + \frac{1}{N-\alpha_-(\lambda)+1} < p_+(\lambda)$, in a natural weighted norm. The interval $F(\lambda) < p < p_+(\lambda)$ is studied in Section 7.5.2, where we find a family of supersolutions to problem (7.5.2) defined for all time $t > 0$.

With all these previous results in hand, we conjecture that $F(\lambda)$ is really the Fujita type exponent for problem (7.5.2). To prove that this is true, we proceed in several steps. In Section 7.5.3 we prove a local in time existence theorem, for a suitable class of initial data and $1 < p < p_+(\lambda)$. Moreover in Section 7.5.4, using the global supersolutions found in Section 7.5.2, we obtain that for small initial data and $F(\lambda) < p < p_+(\lambda)$, there exist global solutions. Section 7.5.5 is devoted to prove Theorem 7.5.8, the blow-up with respect to a natural norm, which in particular shows that $F(\lambda)$ is the Fujita exponent for (7.5.2). Note that this is a major difference with respect to the case of the heat equation ($\lambda = 0$) with a gradient term, for which given any $p > 1$ and for small initial datum, there exists a global solution; see [187].

Almost all new results on the topic presented in this chapter were obtained in [23].

7.2 Preliminaries and tools

As in the elliptic case, despite the fact that weights do not appear in the principal part of the operator in problem (7.1.1), we will need some properties of weighted Sobolev spaces related to the Caffarelli–Kohn–Nirenberg inequalities. See Section 6.2.3.

We recall the definition of the k-truncation function $T_k(\sigma)$,

$$T_k(\sigma) = \begin{cases} \sigma & |\sigma| \le k, \\ k\,\mathrm{sign}(\sigma) & |\sigma| > k \end{cases} \tag{7.2.1}$$

and

$$G_k(\sigma) = \sigma - T_k(\sigma). \tag{7.2.2}$$

As in previous situations, we define the concept of solution in the more general distributional sense.

Definition 7.2.1. Consider the equation

$$u_t - \Delta u = |\nabla u|^p + \lambda \frac{u}{|x|^2} + f(x,t) \quad \text{in } \Omega \times (0,T). \tag{7.2.3}$$

We say that $u \in \mathcal{C}((0,T); L^1_{\mathrm{loc}}(\Omega))$ is a very weak supersolution (subsolution) to problem (7.2.3) if $\frac{|u|}{|x|^2} \in L^1_{\mathrm{loc}}(\Omega_T)$, $|\nabla u|^p \in L^1_{\mathrm{loc}}(\Omega_T)$, $f \in L^1_{\mathrm{loc}}(\Omega_T)$ and for all $\phi \in \mathcal{C}_0^\infty(\Omega \times (0,T))$

such that $\phi \geq 0$, we have

$$\int_0^T \int_\Omega (-\phi_t - \Delta\phi) u \, dx \, dt \geq (\leq) \int_0^T \int_\Omega \left(|\nabla u|^p + \lambda \frac{u}{|x|^2} + cf \right) \phi \, dx \, dt. \tag{7.2.4}$$

If u is a very weak super- and subsolution, then we say that u is a very weak solution.

Note that all these notions of solution are *local in nature*. The nonexistence result in this framework is the strongest possible. It is clear that if u is a very weak supersolution (subsolution) to (7.1.1), then

$$u \in \mathcal{C}((0,T); L^1_{\text{loc}}(\Omega)) \cap L^p((0,T); W^{1,p}_{\text{loc}}(\Omega)).$$

Assume u is a solution to problem (7.1.1) such that $\frac{u}{|x|^2} \in L^1(\Omega_T)$, $u \in L^p((0,T); W_0^{1,p}(\Omega))$, $f \in L^1(\Omega_T)$ and $u_0(x) \in L^1(\Omega)$. Then u can be recovered as a limit of truncated problems. Indeed, let us define $g(x,t) \equiv |\nabla u|^p + \lambda \frac{u}{|x|^2} + f$; then $g \in L^1(\Omega_T)$. Consider now $\{g_n\}_{n\in\mathbb{N}} \subset L^1(\Omega_T) \cap L^\infty(\Omega_T)$, $g_n \to g$ in $L^1(\Omega_T)$ and $\{\phi_n\}_{n\in\mathbb{N}} \subset L^1(\Omega) \cap L^\infty(\Omega)$, $\phi_n \to u_0$ in $L^1(\Omega)$. Solving the problems

$$u_{nt} - \Delta u_n = g_n \text{ in } \Omega_T, \quad u_n(x,0) = \phi_n(x) \text{ in } \Omega, \quad u(x,t) = 0 \text{ on } (x,t) \in \partial\Omega \times (0,T),$$

the theory of *solutions obtained as limit of approximations* says that

$$u_n \to u \text{ in } L^q(0,T; W_0^{1,q}(\Omega)), \quad q < \max\left\{p, \frac{N+2}{N+1}\right\}$$

and $u \in \mathcal{C}([0,T], L^1(\Omega))$. See Section 5.2.

Since $u_n \in L^2((0,T); W_0^2(\Omega)) \cap \mathcal{C}([0,T]; L^2(\Omega))$ and $u_{nt} \in L^2((0,T); W^{-1,2}(\Omega))$ the relevant fact is that we can take test functions $\phi \in L^2((0,T); W_0^{1,2}(\Omega)) \cap L^\infty(\Omega \times (0,T))$. See for instance [129], [60], [61] and [267] for more general operators.

The concept of *solutions obtained by approximations* is equivalent in this setting to *entropy solution* or *renormalized solution*. See, for instance, [258] for a proof of the equivalence.

The previous remarks motivate the strategy to follow in order to solve problem (7.1.1). We will consider some truncated problems (see (7.2.10) below) and then we will obtain uniform estimates that permit to pass to the limit and find a solution at least in the renormalized, or entropy, setting.

7.2.1 Maximum principle and comparison results

We begin by proving a maximum principle that we will use systematically here. This result is the parabolic counterpart of a result for elliptic equations obtained in Section 6.2; see also [31]. The techniques used in the proof of the parabolic case are completely different from those used in the elliptic case.

Lemma 7.2.2 (Maximum principle). *Let $h(x,t)$ be a measurable function such that $|h| \in L^{2r_0}([0,T]; L^{2p_0}(\Omega))$, where $p_0, r_0 > 1$ and $\frac{N}{2p_0} + \frac{1}{r_0} < 1$. Assume that $w(x,t) \geq 0$ verifies:*

(i) $w \in \mathcal{C}((0,T); L^1(\Omega)) \cap L^{r_1}([0,T]; W_0^{1,p_1}(\Omega))$, *where $r_1, p_1 \geq 1$ such that*

$$\frac{N}{2p_1} + \frac{1}{r_1} > \frac{N+1}{2};$$

(ii) *w is a subsolution to problem*

$$\begin{cases} w_t - \Delta w \leq |h||\nabla w| & \text{in } \Omega_T, \\ w(x,t) = 0 & \text{on } \partial\Omega \times (0,T), \\ w(x,0) = 0 & \text{in } \Omega. \end{cases} \tag{7.2.5}$$

Then $w \equiv 0$.

Proof. Since $w \in L^{r_1}([0,T]; W_0^{1,p_1}(\Omega))$ with $r_1, p_1 \geq 1$ such that $\frac{N}{2p_1} + \frac{1}{r_1} > \frac{N+1}{2}$, using the Gagliardo–Nirenberg interpolation inequality (see [173] and [251]) it follows that $w \in L^{r_2}([0,T]; L^{p_2}(\Omega))$ with $r_2, q_2 \geq 1$ and $\frac{N}{2p_2} + \frac{1}{r_2} > \frac{N}{2}$.

As $\frac{N}{2p_0} + \frac{1}{r_0} < 1$, we have $\frac{N}{2p_0'} + \frac{1}{r_0'} > \frac{N}{2}$. We choose $\beta > 0$ such that

$$\frac{1}{\beta+1}\left(\frac{N}{2p_0'} + \frac{1}{r_0'}\right) > \frac{N}{2}.$$

We claim that

$$\int_0^T\int_\Omega w^{\beta-1}|\nabla w|^2\,dx\,dt < \infty. \tag{7.2.6}$$

For $\varepsilon > 0$, consider $m_\varepsilon(s) \equiv (\frac{s}{1+\varepsilon s})^\beta$ and define $D_\varepsilon(s) = \int_0^s m_\varepsilon(\tau)d\tau$. Using $m_\varepsilon(w)$ as a test function in (7.2.5), it follows that

$$\int_\Omega D_\varepsilon(w(x,T))\,dx + \beta\int_0^T\int_\Omega \left(\frac{w}{1+\varepsilon w}\right)^{\beta-1}\frac{|\nabla w|^2}{(1+\varepsilon w)^2}\,dx\,dt \leq \int_0^T\int_\Omega \left(\frac{w}{1+\varepsilon w}\right)^{\beta}|h||\nabla w|\,dx\,dt.$$

From the Young inequality we get

$$\int_0^T\int_\Omega \left(\frac{w}{1+\varepsilon w}\right)^{\beta}|h||\nabla w|\,dx\,dt \leq \eta_1\int_0^T\int_\Omega \left(\frac{w}{1+\varepsilon w}\right)^{\beta-1}|\nabla w|^2\,dx\,dt + \eta_2\int_0^T\int_\Omega |h|^2(x,t)w^{\beta+1}\,dx\,dt.$$

On the other hand, by the Hölder inequality,

$$\begin{aligned}\int_0^T\int_\Omega |h|^2 w^{\beta+1}\,dx\,dt &\leq \int_0^T\left(\int_\Omega |h|^{2p_0}\,dx\right)^{\frac{1}{p_0}}\left(\int_\Omega w^{p_0'(\beta+1)}\,dx\right)^{\frac{1}{p_0'}}dt \\ &\leq \left(\int_0^T\left(\int_\Omega |h|^{2p_0}\,dx\right)^{\frac{r_0}{p_0}}dt\right)^{\frac{1}{r_0}}\left(\int_0^T\left(\int_\Omega w^{p_0'(\beta+1)}\,dx\right)^{\frac{r_0'}{p_0'}}dt\right)^{\frac{1}{r_0'}}.\end{aligned}$$

Using the hypothesis on h and the choice of β, it follows that

$$\left(\int_0^T\left(\int_\Omega |h|^{2p_0}\,dx\right)^{\frac{r_0}{p_0}}dt\right)^{\frac{1}{r_0}}\left(\int_0^T\left(\int_\Omega w^{p_0'(\beta+1)}\,dx\right)^{\frac{r_0'}{p_0'}}dt\right)^{\frac{1}{r_0'}}<\infty.$$

Letting $\varepsilon\to 0$ and using Fatou's lemma, we find that

$$\int_\Omega w^{\beta+1}(x,T)\,dx+\beta\int_0^T\int_\Omega w^{\beta-1}|\nabla w|^2\,dx\,dt<\infty,$$

and then the claim follows.

By the assumption on h, there exists $q>2$ such that $|h|^q\in L^{r_2}([0,T],L^{p_2}(\Omega))$ and

$$\frac{N}{2p_2}+\frac{1}{r_2}<1.$$

Let ψ be the solution to the problem

$$\begin{cases}\psi_t-\Delta\psi=|h(x,T-t)|^q\psi & \text{in } \Omega_T,\\ \psi(x,t)=0 & \text{on } \partial\Omega\times(0,T),\\ \psi(x,0)=1 & \text{in } \Omega.\end{cases}\tag{7.2.7}$$

Using the classical parabolic regularity theory (see for instance [220], [39]), we obtain $\psi\in L^\infty(\Omega_T)$. Considering $\psi_1(x,t)=\psi(x,T-t)$, ψ_1 solves

$$\begin{cases}-\frac{d\psi_1}{dt}-\Delta\psi_1=|h(x,t)|^q\psi_1 & \text{in } \Omega_T,\\ \psi_1(x,t)=0 & \text{on } \partial\Omega\times(0,T),\\ \psi_1(x,T)=1 & \text{in } \Omega.\end{cases}\tag{7.2.8}$$

Taking $(\frac{w}{1+\varepsilon w})^\beta\psi_1$ as a test function in (7.2.5), passing to the limit as $\varepsilon\to 0$ and using the estimate (7.2.6), it follows that

$$\begin{aligned}&\frac{1}{\beta+1}\int_\Omega w^{\beta+1}(x,t)\psi_1(x,t)\,dx+\beta\int_0^t\int_\Omega w^{\beta-1}|\nabla w|^2\psi_1\,dx\,ds\\&\quad+\frac{1}{\beta+1}\int_0^t\int_\Omega w^{\beta+1}((-\psi_1)_t-\Delta\psi_1)\,dx\,ds\le\int_0^t\int_\Omega |h||\nabla w|w^\beta\psi_1\,dx\,ds\\&=\int_0^t\int_\Omega |h||\nabla w|w^{\theta_1+\theta_2+\theta_3}\psi_1\,dx\,ds,\end{aligned}$$

with $\theta_1=\frac{\beta-1}{2}$, $\theta_2=\frac{\beta+1}{q}$ and $\theta_3=\frac{(q-2)(\beta+1)}{2q}$. Note that $\theta_1+\theta_2+\theta_3=\beta$. The Young inequality with $\theta_1,\theta_2,\theta_3$ and the definition of ψ_1 gives

$$\frac{1}{\beta+1}\int_\Omega w^{\beta+1}(x,t)\psi_1(x,t)\,dx+\beta\int_0^t\int_\Omega w^{\beta-1}|\nabla w|^2\psi_1\,dx\,ds$$
$$+\frac{1}{\beta+1}\int_0^t\int_\Omega w^{\beta+1}|h|^q\psi_1\,dx\,ds\le \eta_1\int_0^t\int_\Omega w^{\beta-1}|\nabla w|^2\psi_1\,dx\,ds$$
$$+\eta_2\int_0^t\int_\Omega w^{\beta+1}|h|^q\psi_1\,dx\,ds+\eta_3\int_0^t\int_\Omega w^{\beta+1}(x,s)\psi_1(x,s)\,dx\,ds.$$

Choosing η_1 and η_2 small, we get

$$\frac{1}{\beta+1}\int_\Omega w^{\beta+1}(x,t)\psi_1(x,t)\,dx\le\eta_3\int_0^t\int_\Omega w^{\beta+1}(x,s)\psi_1(x,s)\,dx\,ds.$$

Since $w\ge 0$ and $\psi_1>0$ in $\Omega\times(0,t)$ for $t<T$, by Gronwall's inequality we obtain $w\equiv 0$. □

For the reader convenience we recall that a function $F:\Omega\times\mathbb{R}\times\mathbb{R}^N\to\mathbb{R}$ is called a Caratheodory function if is measurable in $(x,t)\in\Omega\times\mathbb{R}$ for ξ fixed and continuous in $\xi\in\mathbb{R}^N$ for a.e $(x,t)\in\Omega\times\mathbb{R}$.

As a consequence of Lemma 7.2.2 we have the following comparison result.

Lemma 7.2.3 (Comparison principle). *Let $u,v\in\mathcal{C}((0,T);L^1(\Omega))\cap L^p((0,T);W_0^{1,p}(\Omega))$, for some $p>1$, with $|u_t-\Delta u|\in L^1(\Omega_T)$ and $|v_t-\Delta v|\in L^1(\Omega_T)$. Consider a Caratheodory function $H(x,t,s)$ such that $H(x,t,\cdot)\in\mathcal{C}^1(\mathbb{R}^N)$ for all $(x,t)\in\Omega_T$ with*

$$\sup_{s\in\mathbb{R}^N}|\nabla_s H(x,t,s)|=h(x,t)\in L^{2r}([0,T];L^{2p}(\Omega)),\quad \text{for } p,r>1 \text{ and } \frac{N}{2p}+\frac{1}{r}<1.$$

Assume that u and v verify

$$\begin{cases} u_t-\Delta u\ge H(x,t,\nabla u)+f & \text{in }\Omega,\\ u(x,0)=u_0(x) & \text{in }\Omega,\end{cases}\qquad \begin{cases} v_t-\Delta v\le H(x,t,\nabla v)+f & \text{in }\Omega,\\ v(x,0)=v_0(x) & \text{in }\Omega,\end{cases}\tag{7.2.9}$$

where $f\in L^1(\Omega_T)$, $u_0,v_0\in L^1(\Omega)$ and $v_0(x)\le u_0(x)$ in Ω. Then $v\le u$ in Ω_T.

Proof. Denoting $w=v-u$, it is sufficient to prove that $w_+=0$. For $(x,t)\in\Omega_T$, consider $J(\sigma)=H(x,t,\sigma\nabla v+(1-\sigma)\nabla u)$, $\sigma\in[0,1]$. By hypothesis, there exists $\theta\in(0,1)$ such that

$$H(x,t,\nabla v)-H(x,t,\nabla u)=\langle\nabla_s H(x,t,\theta\nabla v+(1-\theta)\nabla u),\nabla w\rangle.$$

From (7.2.9), we conclude that w satisfies $w_t-\Delta w\le|h(x,t)||\nabla w|$ in Ω_T. Therefore, applying Kato inequality (see Lemma 3.2 in [255]) we have

$$\begin{cases} w_t^+-\Delta w^+\le|h(x,t)||\nabla w^+| & \text{in }\Omega_T,\\ w^+(x,t)=0 & \text{on }\partial\Omega\times(0,T),\\ w^+(x,0)=0 & \text{in }\Omega.\end{cases}$$

By the hypothesis on h and using Lemma 7.2.2, we conclude that $w_+\equiv 0$. □

The following consequences of this will be very useful.

Corollary 7.2.4. *Assume $u_0 \in L^1(\Omega)$, $f \in L^1(\Omega_T)$ and $H(s) = \frac{|s|^p}{1+\frac{1}{n}|s|^p}$, $p > 1$. For $n \in \mathbb{N}$ consider the problem*

$$\begin{cases} u_{nt} - \Delta u_n - \lambda a_n(x) u_n = \frac{|\nabla u_n|^p}{1+\frac{1}{n}|\nabla u_n|^p} + f_n & \text{in } \Omega_T, \\ u_n(x,t) = 0 & \text{on } \partial\Omega \times (0,T), \\ u_n(x,0) = u_{n0}(x) & \text{if } x \in \Omega, \end{cases} \tag{7.2.10}$$

where $a_n(x) = \min\{\frac{1}{|x|^2}, n\}$, $u_{n0} = T_n(u_0)$ and $f_n = T_n(f)$. Then (7.2.10) has a unique positive solution u_n. Moreover, $u_n \le u_{n+1}$ for every $n \in \mathbb{N}$.

Corollary 7.2.5. *Assume that $1 \le p < \frac{N+2}{N+1}$ and $f \in L^1(\Omega_T)$. Let u, v be two functions such that $u, v \in L^p((0,T); W^{1,p}(\Omega))$, $u_t - \Delta u, v_t - \Delta v \in L^1(\Omega_T)$ and*

$$\begin{cases} u_t - \Delta u \ge |\nabla u|^p + f & \text{in } \Omega_T, \\ u(x,0) = u_0(x) & \text{in } \Omega, \end{cases} \qquad \begin{cases} v_t - \Delta v \le |\nabla v|^p + f & \text{in } \Omega_T, \\ v(x,0) = v_0(x) & \text{in } \Omega, \end{cases} \tag{7.2.11}$$

where $v \le u$ on $\partial\Omega \times (0,T)$, $u_0, v_0 \in L^1(\Omega)$ and $v_0(x) \le u_0(x)$ in Ω. Then $v \le u$ in Ω_T.

Proof. Consider $w = v - u$. It is clear that $w \in L^p((0,T); W^{1,p}(\Omega))$, $w \le 0$ on $\partial\Omega \times (0,T)$, $w_0(x) \le 0$ in Ω and $w_t - \Delta w \in L^1(\Omega_T)$. In order to conclude, it is sufficient to prove that $w_+ = 0$. We have

$$w_t - \Delta w \le |\nabla v|^p - |\nabla u|^p \text{ in } \Omega_T.$$

Since $1 \le p < \frac{N+2}{N+1} < 2$, it follows that

$$\begin{cases} w_t - \Delta w \le a(x,t)|\nabla w| & \text{in } \Omega_T, \\ w \le 0 & \text{on } \partial\Omega \times (0,T),\ w_0(x) \le 0 \text{ in } \Omega, \\ w \in L^p(0,T,W^{1,p}(\Omega)), & w_t - \Delta w \in L^1(\Omega_T), \end{cases}$$

with $a(x,t) \le p|\nabla u|^{p-1}$ if $p > 1$ and $a(x,t) = 1$ if $p = 1$. Using Kato's inequality we have

$$\begin{cases} (w_+)_t - \Delta w_+ \le a(x,t)|\nabla w_+|, \\ w_+(x,0) = 0 \quad \text{in } \Omega, \\ w_+ \in L^p(0,T,W^{1,p}(\Omega)). \end{cases}$$

Since $p < \frac{N+2}{N+1}$, we have $a(x,t) \in L^{2r}([0,T]; L^{2q}(\Omega))$ for $q, r > 1$ and $\frac{N}{2q} + \frac{1}{r} < 1$. Using the previous comparison lemma, we conclude that $w_+ = 0$. □

Remark 7.2.6. As a consequence of Corollary 7.2.5, if $1 \le p < \frac{N+2}{N+1}$, then the problem

$$\begin{cases} u_t - \Delta u = |\nabla u|^p & \text{in } \Omega_T, \\ u = 0 & \text{on } \partial\Omega \times (0,T), \\ u_0(x) = u_0 & \text{in } \Omega \end{cases}$$

has a unique positive solution $u \in \mathcal{C}([0,T]; L^1(\Omega)) \cap L^p((0,T), W_0^{1,p}(\Omega))$. It is worthy to point out that since $\alpha_-(\lambda) < N$, we have $\frac{N+2}{N+1} < q_+(\lambda)$, where $q_+(\lambda)$ is the critical power given in (7.1.2).

7.2.2 Local behavior of a very weak supersolution to problem (7.1.1)

We begin by analyzing the behavior of any positive solution in a neighborhood of the origin. According to Lemma 5.2.3, if u is a nonnegative supersolution defined in Ω, that is, $u \gneqq 0$, $u \in L^1_{\text{loc}}(\Omega_T)$, $\frac{u}{|x|^2} \in L^1_{\text{loc}}(\Omega_T)$ and u satisfies $u_t - \Delta u - \lambda \frac{u}{|x|^2} \geq 0$ in $\mathcal{D}'(\Omega_T)$ with $\lambda \leq \Lambda_{N,2}$ and $B_{r_1}(0) \subset\subset \Omega$, then there exists a constant $C = C(N, r_1, t_1, t_2)$ such that for each cylinder $B_r(0) \times (t_1, t_2) \subset\subset \Omega_T$, $0 < r < r_1$,

$$u \geq C|x|^{-\alpha_-(\lambda)} \quad \text{in } B_r(0) \times (t_1, t_2),$$

where $\alpha_-(\lambda)$ is given in (3.1.3). By choosing r small enough, we may assume that $u > 1$ in $B_r(0) \times (t_1, t_2)$.

We obtain some *a priori* local integral estimates for a very weak solution to problem (7.2.3).

Lemma 7.2.7. *Assume that u is a nonnegative very weak supersolution to* (7.2.3) *with $\lambda \leq \Lambda_{N,2}$ and α_- is as in* (3.1.3). *Then in each cylinder $B_r(0) \times (t_1, t_2)$ with $0 < t_1 < t_2 \leq T$ we have*

1. $\int_{t_1}^{t_2} \int_{B_r(0)} |\nabla u|^p |x|^{-\alpha_-(\lambda)}\, dx\, dt < \infty$;
2. $\int_{t_1}^{t_2} \int_{B_r(0)} \frac{u}{|x|^{2+\alpha_-(\lambda)}}\, dx\, dt < \infty$;
3. $\int_{t_1}^{t_2} \int_{B_r(0)} f|x|^{-\alpha_-(\lambda)}\, dx\, dt < \infty$.

Moreover, if $|\nabla u|^p + \lambda \frac{|u|}{|x|^2} + f \in L^1(\Omega_T)$ and u is a solution to problem (7.1.1) *obtained as a limit of approximations, then*

$$\sup_{t \in (t_1, t_2)} \int_{\Omega} u(x,t)|x|^{-\alpha_-(\lambda)}\, dx < \infty. \tag{7.2.12}$$

Proof. Since $u \in \mathcal{C}((0,T); L^1_{\text{loc}}(\Omega)) \cap L^p((0,T); W^{1,p}_{\text{loc}}(\Omega))$, fixing $0 < T_1 < t_1 < t_2 < T_2 < T$, in particular, $u \in L^p((T_1, T_2); W^{1,p}(B_r(0)))$ for all $B_r(0) \subset \Omega$. Moreover, there exists a positive constant c_0 such that $u(x,t) \geq c_0$ for all $(x,t) \in B_r(0) \times (T_1, T_2)$. Consider $\bar{w}$, the solution to

$$\begin{cases} \bar{w}_t - \Delta \bar{w} = \lambda \frac{\bar{w}}{|x|^2} + 1 & \text{in } B_r(0) \times (T_1, T_2), \\ \bar{w}(x,t) = 0 & \text{on } \partial B_r(0) \times (T_1, T_2), \\ \bar{w}(x, T_1) = 0 & \text{in } B_r(0), \end{cases} \tag{7.2.13}$$

with $\lambda \leq \Lambda_{N,2}$. It is clear that $w \in L^2(0,T;W_0^{1,2}(\Omega))$ if $\lambda < \Lambda_{N,2}$ and $w \in L^2(0,T;H(\Omega))$ if $\lambda = \Lambda_{N,2}$, where $H(\Omega)$ is the Hilbert space defined in Section 2.5.1. According to Lemma 5.2.3, there exists a constant $C = C(N, r_1, T_1, T_2)$ such that $\bar{w} \geq C|x|^{-\alpha_-}$ in $B_r(0) \times (T_1, T_2)$. Then $w(x,t) = \bar{w}(x, T_2 + T_1 - t)$ satisfies

$$\begin{cases} -w_t - \Delta w = \lambda \frac{w}{|x|^2} + 1 & \text{in } B_r(0) \times (T_1, T_2), \\ w(x,t) = 0 & \text{on } \partial B_r(0) \times (T_1, T_2), \\ w(x, T_2) = 0 & \text{in } B_r(0). \end{cases}$$

Observe that $w(x,t) \geq C|x|^{-\alpha_-}$ in $B_r(0) \times (t_1, t_2)$.

Let us consider the sequence $\{u_n\}$ of solutions to the approximated problems

$$\begin{cases} u_{nt} - \Delta u_n = \frac{|\nabla u|^p}{1+\frac{1}{n}|\nabla u|^p} + \lambda \frac{u_n}{|x|^2} + T_n(f) & \text{in } B_r(0) \times (T_1, T_2), \\ u_n(x,t) = 0 & \text{on } \partial B_r(0) \times (T_1, T_2), \\ u_n(x, T_1) = c_0 & \text{in } B_r(0). \end{cases} \tag{7.2.14}$$

Since $\lambda \leq \Lambda_{N,2}$, using the weak comparison principle for the operator $v_t - \Delta v - \lambda \frac{v}{|x|^2}$, it follows that $u_n \leq u_{n+1}$ and $u_n \leq u$, $\forall n \in \mathbb{N}$.

Taking w as a test function in (7.2.14), we find

$$\int_{T_1}^{T_2} \int_{B_r(0)} u_n \, dx\, dt = \int_{T_1}^{T_2} \int_{B_r(0)} \frac{|\nabla u|^p}{1 + \frac{1}{n}|\nabla u|^p} w \, dx\, dt + \int_{T_1}^{T_2} \int_{B_r(0)} T_n(f) w \, dx\, dt.$$

Hence passing to the limit as $n \to \infty$ and using Fatou's lemma we obtain

$$\int_{T_1}^{T_2} \int_{B_r(0)} |\nabla u|^p w \, dx\, dt < \infty \quad \text{and} \quad \int_{T_1}^{T_2} \int_{B_r(0)} f w \, dx\, dt < \infty.$$

Taking into account the lower estimate of w near the origin given by Lemma 5.2.3, we have

$$\int_{t_1}^{t_2} \int_{B_r(0)} |\nabla u|^p |x|^{-\alpha_-(\lambda)} \, dx\, dt < \infty \quad \text{and} \quad \int_{t_1}^{t_2} \int_{B_r(0)} f |x|^{-\alpha_-(\lambda)} \, dx\, dt < \infty.$$

To get the estimate on the term $\frac{u}{|x|^{2+\alpha_-(\lambda)}}$, consider v_n, the solution to

$$\begin{cases} v_{nt} - \Delta v_n = \frac{|\nabla u|^p}{1+\frac{1}{n}|\nabla u|^p} + \lambda \frac{T_n(u)}{|x|^2} + f & \text{in } B_r(0) \times (T_1, T_2), \\ v_n(x,t) = 0 & \text{on } \partial B_r(0) \times (T_1, T_2), \\ v_n(x, T_1) = c_0 & \text{if } x \in B_r(0). \end{cases} \tag{7.2.15}$$

It is clear that $\{v_n\}$ is an increasing sequence on n such that $v_n \le u$. Hence taking again w as a test function in (7.2.15) and using the same computation as above, it follows that

$$\int_{t_1}^{t_2}\int_{B_r(0)} \frac{u}{|x|^{2+\alpha_-(\lambda)}}\,dx\,dt < \infty.$$

We prove now estimate (7.2.12). Fixing t_1, t_2 such that $(t_1, 2t_2) \subset (0, T)$, let φ be the solution to the following elliptic problem:

$$\begin{cases} -\Delta\varphi = \lambda\frac{\varphi}{|x|^2} + 1 & \text{in } \Omega,\\ \varphi(x) = 0 & \text{on } \partial\Omega. \end{cases} \tag{7.2.16}$$

Note that $\varphi \simeq C|x|^{-\alpha_-(\lambda)}$ in $B_r(0)$. Taking $(2t_2 - t)\varphi$ as a test function in (7.2.15) and integrating in $\Omega \times (t, 2t_2)$ with $t_1 < t < t_2$, it follows that

$$\begin{aligned} &-(2t_2 - t)\int_\Omega v_n(x,t)\varphi\,dx + \int_t^{2t_2}\int_\Omega v_n(x,\sigma)\varphi\,dx\,d\sigma + \int_t^{2t_2}\int_\Omega (2t_2-\sigma)u(x,\sigma)\Big(\frac{\lambda\varphi}{|x|^2}+1\Big)\,dx\,d\sigma \\ &= \int_t^{2t_2}\int_\Omega \Big(\frac{|\nabla u|^p}{1+\frac{1}{n}|\nabla u|^p} + \lambda\frac{T_n(u)}{|x|^2}\Big)(2t_2-\sigma)\varphi\,dx\,d\sigma + c\int_t^{2t_2}\int_\Omega f(2t_2-\sigma)\varphi\,dx\,d\sigma. \end{aligned}$$

Hence

$$2t_2\int_{t_1}^{2t_2}\int_\Omega u_n(x,\sigma)\Big(\frac{\lambda\varphi}{|x|^2}+1\Big)\,dxd\sigma \ge (2t_2-t)\int_\Omega u(x,t)\varphi\,dx \ge c(2t_2-t)\int_{B_r(0)} u_n(x,t)|x|^{-\alpha_-(\lambda)}\,dx.$$

Thus by letting $n \to \infty$ we conclude that

$$\int_{B_r(0)} u(x,t)|x|^{-\alpha_-(\lambda)}\,dx \le \frac{t_2}{(2t_2-t)}\int_{t_1}^{2t_2}\int_\Omega u(x,\sigma)(c|x|^{-\alpha_-(\lambda)-2}+1)\,dx\,d\sigma \le C \quad \text{for all } t \in (t_1, t_2).$$

Then (7.2.12) follows. □

Remark 7.2.8. Since the blow-up in the L^∞-norm is instantaneous, estimate (7.2.12) suggests the *natural norm* to study nontrivial blow-up results.

7.2.3 Passing to the limits in locally truncated problems

Consider for fixed $n \in \mathbb{N}$ and $k \ge 1$ the truncated problems

$$\begin{cases} v_{kt} - \Delta v_k = \lambda T_n(\frac{1}{|x|^2})v_k + \frac{|\nabla v_k|^p}{1+\frac{1}{k}|\nabla v_k|^p} + f & \text{in } \Omega_1 \times (0,T),\\ v_k(x,t) = 0 & \text{on } \partial\Omega_1 \times (0,T),\\ v_k(x,0) = T_n(\tilde{u}_0) & \text{if } x \in \Omega_1, \end{cases} \tag{7.2.17}$$

where Ω_1 is any bounded regular domain with $\Omega_1 \subset\subset \Omega$, $0 < \lambda \leq \Lambda_{N,2}$, $1 < p < q_+(\lambda)$, $u_0 \in L^1(\Omega)$ and $f \in L^\infty(\Omega_T)$. We suppose that $\tilde{u} \in C([0,T); L^1_{\text{loc}}(\Omega))$ is a very weak supersolution to (7.2.3) with $1 < p \leq 2$. Then, clearly, $\tilde{u}$ is a supersolution to (7.2.17). Hence, using Lemma 7.2.3 we find that $v_k \leq v_{k+1}$ and $v_k \leq \tilde{u}$ for every k. Therefore, there exists u_n satisfying $u_n = \lim_{k\to\infty} v_k \leq \tilde{u}$ a. e. and in $L^1(\Omega_T)$.

For simplicity of notation we set

$$g_k(x,t) = \frac{|\nabla v_k|^p}{1 + \frac{1}{k}|\nabla v_k|^p} + \lambda a_n(x) v_k + f, \quad \text{where } a_n = T_n\left(\frac{1}{|x|^2}\right).$$

We prove that g_k is uniformly bounded in $L^1(\Omega_1 \times (0,T), \phi\, dx\, dt)$, where ϕ is the solution to the problem

$$\begin{cases} -\Delta\phi = 1 & \text{in } \Omega_1, \\ \phi = 0 & \text{on } \partial\Omega_1. \end{cases} \tag{7.2.18}$$

Note that $\phi(x) \backsimeq c\,\text{dist}(x, \partial\Omega_1)$. Indeed, using ϕ as a test function in (7.2.17) and since $v_k \leq \tilde{u}$, we have

$$\begin{aligned} &\int_{\Omega_1} \tilde{u}(x,T)\phi\, dx + \int_0^T\int_{\Omega_1} \tilde{u}(x,s)\, dx\, ds \\ &\quad \geq \int_{\Omega_1} v_k(x,T)\phi\, dx + \int_0^T\int_{\Omega_1} v_k(x,s)\, dx\, ds \geq \int_0^T\int_{\Omega_1} g_k(x,s)\phi\, dx\, ds, \end{aligned}$$

thus

$$\int_0^T\int_{\Omega_1} g_k\phi\, dx\, ds \leq C \quad \text{for all } k.$$

It is not difficult to show that u solves

$$u_t - \Delta u = \mu \text{ in } \mathcal{D}'(\Omega_1 \times (0,T)), \quad \text{where } g_k \rightharpoonup \mu \text{ weakly in the measures sense.}$$

The almost everywhere pointwise convergence of the gradient is well known, but for the self-contained arguments, we include a proof. To that end, we follow the ideas in [67]. We refer also to the papers [60] and [258], where strong results about the convergence of truncating approximation are obtained.

Lemma 7.2.9. *Assume that* $\{v_k\}$ *is defined by* (7.2.17). *Then*

$$\nabla v_k \to \nabla u_n a.e. (x,t) \in \Omega_1 \times (0,T).$$

Proof. Using $T_m(v_k) \cdot \phi$ as a test function in (7.2.17) we get

$$\int_{\Omega_1} \Theta_m(v_k(x,T))\phi\,dx + \int_0^T \int_{\Omega_1} |\nabla T_m(v_k)|^2 \phi\,dx\,ds + \int_0^T \int_{\Omega_1} \Theta_m(v_k)(-\Delta\phi)\,dx\,ds = \int_0^T \int_{\Omega_1} g_k(x,s)\phi T_m(v_k)\,dx\,ds,$$

where $\Theta_m(s) = \int_0^s T_m(\sigma)\,d\sigma$. Thus, by the previous boundedness, we find

$$\int_{\Omega_1} \Theta_m(v_k(x,T))\phi\,dx + \int_0^T \int_{\Omega_1} |\nabla T_m(v_k)|^2 \phi\,dx\,ds + \int_0^T \int_{\Omega_1} \Theta_m(v_k)\,dx\,ds \le m \int_0^T \int_{\Omega_1} g_k(x,s)\phi\,dx\,ds \le Cm.$$

Then

$$\int_0^T \int_{\Omega_1} |\nabla T_m(v_k)|^2 \phi\,dx\,ds \le Cm,$$

and, therefore, up to a subsequence,

$$\nabla T_m(v_k) \rightharpoonup \nabla T_m(u_n) \quad \text{weakly in } \left(L^2(\Omega_1 \times (0,T), \phi\,dx\,dt)\right)^N.$$

Since g_k is bounded in $L^1(\Omega_1 \times (0,T), \phi\,dx\,dt)$, we obtain

$$\int_0^T \int_{\Omega_1} |\nabla v_k|^q \phi\,dx\,dt < C, \quad q < \frac{N+2}{N+1}\ \forall k \in \mathbb{N}.$$

See Lemma 2.2 in [67]. To prove the a. e. convergence of the gradients it is sufficient to show that for some $\theta > 0$, we have

$$\int_0^T \int_{\Omega_1} |\nabla(v_k - u_n)|^\theta \phi\,dx\,dt \to 0 \quad \text{as } k \to \infty.$$

As in [67], we denote by $\omega(k,\nu,m,\varepsilon)$ any quantity that satisfies

$$\lim_{\varepsilon \to 0^+} \limsup_{m\to\infty} \limsup_{\nu\to\infty} \limsup_{k\to\infty} w(k,\nu,m,\varepsilon) = 0,$$

and by $w^{\nu,m,\varepsilon}(k)$ any quantity that goes to zero as $k \to \infty$ for a fixed ν, m and ε.

We perform the following Landes–Mustonen time regularization; see [221].

For $v \in L^2(0,T;W_0^{1,2}(\Omega_1))$, we define

$$v_\nu(x,t) = \int_{-\infty}^{t} \overline{v}(x,s)e^{\nu(s-t)}\,ds,$$

where

$$\overline{v}(x,t) = \begin{cases} v(x,t) & \text{if } t \in [0,T], \\ 0 & \text{if } t \notin [0,T]. \end{cases}$$

It is easy to check that $v_\nu \to v$ strongly in $L^2((0,T);W_0^{1,2}(\Omega_1))$ and $(v_\nu)' = \nu(v - v_\nu)$ in the weak sense, i. e.,

$$\langle (v_\nu)', w\rangle = \nu \int_{\Omega_T} (v - v_\nu)w\,dx\,dt \quad \text{for all } w \in L^2((0,T);W_0^{1,2}(\Omega_1)).$$

See [222] for details.

Since $\phi \in W_0^{1,2}(\Omega_1)$, we have $T_m(v_k)\phi \in L^2((0,T);W_0^{1,2}(\Omega_1))$ and $(T_m(v_k)\phi)_\nu = \phi(T_m(u_n))_\nu$.

Consider $0 < 2\theta < 1$ and the splitting,

$$\int_0^T\int_{\Omega_1} |\nabla(v_k - u_n)|^{2\theta}\phi\,dx\,dt = \int_{u_n \ge m} |\nabla(v_k - u_n)|^{2\theta}\phi\,dx\,dt + \int_{u_n < m} |\nabla(v_k - u_n)|^{2\theta}\phi\,dx\,dt$$
$$\equiv I_{m,k} + J_{m,k}.$$

We estimate the first integral:

$$\int_{u_n \ge m} |\nabla(v_k - u_n)|^{2\theta}\phi\,dx\,dt \le c\left(\int_0^T\int_{\Omega_1} [|\nabla v_k| + |\nabla u_n|]^q\phi\,dx\,dt\right)^{\frac{2\theta}{q}} (\text{meas}\{|u_n(x,t)| \ge m\})^{1-\frac{2\theta}{q}}.$$

Since $\text{meas}\{|u_n(x,t)| \ge m\} \to 0$ as $m \to \infty$, it follows that

$$I_{m,k} \to 0 \quad \text{as } m \to \infty. \tag{7.2.19}$$

We deal now with $J_{m,k}$. We have

$$J_{m,k} = \int_0^T\int_{\Omega_1} |\nabla(T_m(v_k) - T_m(u_n))|^{2\theta}\phi\,dx\,dt.$$

Using the definition of $(T_m(u_n))_\nu$ and the monotonicity of v_k it follows that

$$\lim_{k\to\infty} \chi_{\{|v_k - (T_m(u_n))_\nu| > \varepsilon\}} = \chi_{\{|u_n - (T_m(u_n))_\nu| > \varepsilon\}} \quad \text{for all } \nu \text{ and } m \in \mathbb{N},$$

$$\lim_{\nu\to\infty}\chi_{\{|u_n-(T_m(u_n))_\nu|>\varepsilon\}}=\chi_{\{|u_n-T_m(u_n)|>\varepsilon\}}\quad\text{for all } m\in\mathbb{N}.$$

For simplicity, we set $\Omega_{1T}\equiv\Omega_1\times(0,T)$. Hence we have

$$\begin{aligned}J_{m,k}\le&\int_0^T\int_\Omega|\nabla(T_m(v_k)-T_m(u_n))|^{2\theta}\phi\chi_{\{|v_k-(T_m(u_n))_\nu|>\varepsilon\}}\,dx\,dt\\&+\int_0^T\int_\Omega|\nabla(T_m(v_k)-T_m(u_n))|^{2\theta}\phi\chi_{\{|v_k-(T_m(u_n))_\nu|\le\varepsilon\}}\,dx\,dt.\end{aligned}\tag{7.2.20}$$

Since $|\nabla(T_m(v_k)-T_m(u_n))|^{2\theta}\phi$ is bounded in $L^{\frac{q}{2\theta}}(\Omega_1\times(0,T))$ and since $\chi_{\{|u_n-(T_m(u_n))_\nu|>\varepsilon\}}$ goes to zero almost everywhere, we conclude that

$$\int_0^T\int_{\Omega_1}|\nabla(T_m(v_k)-T_m(u_n))|^{2\theta}\phi\chi_{\{|v_k-(T_m(u_n))_\nu|>\varepsilon\}}\,dx\,dt=\omega(k,\nu,m,\varepsilon).\tag{7.2.21}$$

On the other hand we have

$$\begin{aligned}&\int_0^T\int_{\Omega_1}|\nabla(T_m(v_k)-T_m(u_n))|^{2\theta}\phi\chi_{\{|v_k-(T_m(u_n))_\nu|\le\varepsilon\}}\,dx\,dt\\&\quad\le C(\text{meas}(\Omega_1\times(0,T)))^{1-\theta}\left(\int_0^T\int_{\Omega_1}|\nabla(T_m(v_k)-T_m(u_n))|^2\phi\chi_{\{|v_k-(T_m(u_n))_\nu|\le\varepsilon\}}\,dx\,dt\right)^\theta.\end{aligned}$$

Note that

$$\begin{aligned}&\int_0^T\int_{\Omega_1}|\nabla(T_m(v_k)-T_m(u_n))|^2\phi\chi_{\{|v_k-(T_m(u_n))_\nu|\le\varepsilon\}}\,dx\,dt\\&\quad\le\int_0^T\int_{\Omega_1}|(\nabla v_k-T_m(u_n))|^2\phi\chi_{\{|v_k-(T_m(u_n))_\nu|\le\varepsilon\}}\,dx\,dt\\&\quad\le\int_0^T\int_{\Omega_1}\nabla v_k\nabla(v_k-T_m(u_n))\phi\chi_{\{|v_k-(T_m(u_n))_\nu|\le\varepsilon\}}\,dx\,dt\\&\qquad-\int_0^T\int_{\Omega_1}\nabla T_m(u_n)\nabla(v_k-T_m(u_n))\phi\chi_{\{|v_k-(T_m(u_n))_\nu|\le\varepsilon\}}\,dx\,dt.\end{aligned}$$

Using the weak convergence of $T_m(v_k)$ and the monotonicity of $\{v_k\}$ we obtain

$$\int_0^T\int_\Omega\nabla T_m(u_n)\nabla(v_k-T_m(u_n))\phi\chi_{\{|v_k-(T_m(u_n))_\nu|\le\varepsilon\}}\,dx\,dt=\omega^{\nu,m,\varepsilon}(k).\tag{7.2.22}$$

We deal now with the first integral in (7.2.20). Since $(T_m(u_n))_\nu\phi$ converges strongly to $T_m(u_n)\phi$ in $L^2((0,T);W_0^{1,2}(\Omega_1))$ as $\nu\to\infty$, it follows that

$$\int_0^T\int_\Omega |\nabla v_k - T_m(u_n))|^2 \phi\chi_{\{|v_k-(T_m(u_n))_\nu|\le\varepsilon\}}\,dx\,dt$$
$$=\int_0^T\int_\Omega \nabla v_k\nabla(v_k-(T_m(u_n))_\nu)\phi\chi_{\{|v_k-(T_m(u_n))_\nu|\le\varepsilon\}}\,dx\,dt+\omega^{m,\varepsilon}(k,\nu).$$

Using $T_\varepsilon(v_k-(T_m(u_n))_\nu)\phi$ as a test function in (7.2.17), it follows that

$$\langle (v_k)_t, T_\varepsilon(v_k-(T_m(u_n)_\nu)\phi\rangle+\int_{\Omega_T}\phi\nabla v_k\nabla T_\varepsilon(v_k-(T_m(u_n))_\nu)\,dx\,dt$$
$$+\int_{\Omega_T}T_\varepsilon(v_k-(T_m(u_n))_\nu)\nabla v_k\nabla\phi\,dx\,dt=\int_0^T\int_\Omega \phi T_\varepsilon(v_k-(T_m(u_n))_\nu)g_k\,dx\,dt$$
$$\le\varepsilon\int_0^T\int_\Omega g_k\phi\,dx\,dt\le C\varepsilon.$$

A simple variation of Lemma 3.1 in [67] allows us to prove that

$$\langle (v_k)_t, T_\varepsilon(v_k-(T_m(u_n))_\nu)\phi\rangle\ge w^{\nu,m,\varepsilon}(k).$$

On the other hand, we can check that

$$\int_0^T\int_\Omega |T_\varepsilon(v_k-(T_m(u_n))_\nu)||\nabla v_k||\nabla\phi|\,dx\,dt\le C\varepsilon.$$

Thus we conclude that

$$\int_0^T\int_\Omega \phi\nabla v_k\nabla T_\varepsilon(v_k-(T_m(u_n))_\nu)\,dx\,dt\le w(k,\nu,m,\varepsilon). \tag{7.2.23}$$

Putting together (7.2.21), (7.2.22) and (7.2.23) it follows that

$$J_{m,k}=w(k,\nu,m,\varepsilon). \tag{7.2.24}$$

Hence, from (7.2.19) and (7.2.24) the result follows. □

We will apply now Lemma 7.2.9 to obtain the following results.

Lemma 7.2.10. *Assume that $\lambda \leq \Lambda_{N,2}$ and $f \in L^\infty(\Omega_T)$. Let Ω_1 be a bounded domain such that $0 \in \Omega_1 \subset\subset \Omega$. Then there exists u_n, a minimal solution to the problem*

$$\begin{cases} u_{nt} - \Delta u_n = \lambda T_n(\frac{1}{|x|^2})u_n + |\nabla u_n|^p + f & \text{in } \Omega_1 \times (0,T), \\ u_n(x,t) = 0 & \text{on } \partial\Omega_1 \times (0,T), \\ u_n(x,0) = T_n(u_0) & \text{if } x \in \Omega_1. \end{cases} \tag{7.2.25}$$

Moreover if $\tilde{u} \in \mathcal{C}([0,T]; L^1_{loc}(\Omega))$ is a very weak supersolution to (7.2.3) in $\Omega \times (0,T)$ with $u_0(x) \leq \tilde{u}(x,0)$, then $u_n \leq u_{n+1} \leq \tilde{u}$ almost everywhere.

Proof. Consider $\{v_k\}$, the sequence of solutions to problems (7.2.17) in $\Omega_1 \times (0,T)$, $k \geq 1$. Using Lemma 7.2.3 we find that $v_k \leq v_{k+1}$ and $v_k \leq \tilde{u}$ for every k. Since $a_n(x) = T_n(\frac{1}{|x|^2}) \in L^\infty(\Omega_1 \times (0,T))$, standard arguments (see for instance [131] and the references therein) show that there exists u_n such that $v_k \rightharpoonup u_n$ weakly in $L^2(0,T; W^{1,2}_0(\Omega_1))$. Therefore, we have $u_n \in L^2(0,T; W^{1,2}_0(\Omega_1)) \cap \mathcal{C}([0,T], L^2(\Omega_1))$, $u_n \leq \tilde{u}$ a. e. and u_n solves (7.2.25). Moreover, by construction and by Lemma 7.2.3, we conclude that u_n is minimal and the same argument shows that $u_n \leq u_{n+1} \leq \tilde{u}$. □

Corollary 7.2.11. *Let $\tilde{u} \in \mathcal{C}([0,T]; L^1_{loc}(\Omega))$ be a very weak supersolution to (7.2.3). Then, under the hypotheses of Lemma 7.2.10, there exists a positive function u defined in $\Omega_1 \times (0,T)$ such that $u \in L^\infty(0,T; L^1(\Omega_1))$, $\frac{|u|}{|x|^2} \in L^1(\Omega_1 \times (0,T))$, $|\nabla u|^p \in L^1_{loc}(\Omega_1 \times (0,T))$ and*

$$\int_0^T \int_{\Omega_1} u(-\varphi_t - \Delta\varphi)\, dx\, dt - \int_0^T \int_{\Omega_1} \lambda \frac{u\varphi}{|x|^2}\, dx\, dt \geq \int_0^T \int_{\Omega_1} |\nabla u|^p \varphi\, dx\, dt + \int_0^T \int_{\Omega_1} f\varphi$$

holds for all $\varphi \in \mathcal{C}^\infty_0(\Omega_T)$, $\varphi \geq 0$.

Moreover, if u_n is the minimal solution to problem (7.2.25) in $\Omega_1 \times (0,T)$, then:

1. *$u_n \uparrow u \leq \tilde{u}$, a. e. in $\Omega_1 \times (0,T)$, and in $L^1(\Omega_1 \times (0,T), dx\, dt)$;*
2. *$a_n(x)u_n \uparrow \frac{u}{|x|^2}$ a. e. and in $L^1(\Omega_1 \times (0,T), dx\, dt)$;*
3. *for all $t \in [0,T]$:*
   (a) *$\lim_{n\to\infty} \int_{\Omega_1} u_n(x,t)\, dx = \int_{\Omega_1} u(x,t)\, dx \leq \int_{\Omega_1} \tilde{u}(x,t)\, dx < +\infty$;*
   (b) *$\lim_{n\to\infty} \int_{\Omega_1} a_n(x)u_n(x,t)\, dx = \int_{\Omega_1} \frac{u(x,t)}{|x|^2}\, dx \leq \int_{\Omega_1} \frac{\tilde{u}(x,t)}{|x|^2}\, dx < +\infty$;*
4. *$\nabla u_n \to \nabla u$ a. e. and $|\nabla T_m u_n| \to |\nabla T_m u|$ in $L^1(\Omega_1 \times (0,T), \phi\, dx\, dt)$, $\forall m > 0$;*
5. *$\int_0^T \int_{\Omega_1} |\nabla u|^p \phi\, dx\, dt \leq \liminf_{n\to\infty} \int_0^T \int_{\Omega_1} |\nabla u_n|^p \phi\, dx\, dt \leq C$;*

where ϕ is the solution to the retrograde problem

$$\begin{cases} -\phi_t - \Delta\phi = 1 & \text{in } \Omega_1 \times (0,T), \\ \phi = 0 & \text{on } \partial\Omega_1 \times (0,T), \\ \phi(x,T) = 0. \end{cases} \tag{7.2.26}$$

Proof. Items (1), (2) and (3) are a direct consequence of the monotonicity of the sequence $\{u_n\}$ and the fact that $u_n \le \tilde{u}$. Using $T_m(u_n)\phi$ as a test function in (7.2.25) we get

$$\int_0^T \int_{\Omega_1} |\nabla T_m(u_n)|^2 \phi \, dx \, ds \le Cm,$$

and therefore, up to a subsequence, we have

$$\nabla T_m(u_n) \rightharpoonup \nabla T_m(u) \quad \text{weakly in } L^2(\Omega_1 \times (0,T), \phi \, dx \, dt).$$

By Lemma 7.2.9, $\nabla u_n \to \nabla u$ a. e. Using the Vitali theorem we get the convergence of the gradients in item (4). Finally item (5) is a direct consequence of Fatou's lemma. □

7.3 Nonexistence and blow-up results

In this section we will assume that $\lambda > 0$. To prove nonexistence and blow-up results, without loss of generality, we suppose that $\tilde{u}(x,0) \equiv \tilde{u}_0 = 0$ and f is a nonnegative function such that $f \in L^\infty(\Omega_T)$.

7.3.1 Nonexistence

The main goal in this subsection is to find a necessary and sufficient condition on p in such a way that problem (7.1.1) has no positive solution. We will use sistematically the *Hardy-Sobolev inequality with weights* of Caffarelli-Kohn-Nirenberg. See for example [15].

The threshold exponent $q_+(\lambda)$ coincides with the critical one in the elliptic case, as obtained in (6.3.2). More precisely we will prove the following nonexistence result.

Theorem 7.3.1. *Let $q_+(\lambda)$ be defined by* (7.1.2). *If $p \ge q_+(\lambda)$, then problem* (7.1.1) *has no positive very weak supersolution $\tilde{u}$ such that $\tilde{u}_t - \Delta \tilde{u} \in L^1_{\mathrm{loc}}(\Omega_T)$. In the case $f \equiv 0$, the unique nonnegative very weak supersolution is $\tilde{u} \equiv 0$.*

Proof. We argue by contradiction. Assume that $\tilde{u}$ is a very weak supersolution to (7.1.1) such that $\tilde{u}_t - \Delta \tilde{u} \in L^1_{\mathrm{loc}}(\Omega_T)$ with $\tilde{u}_0 \equiv 0$ and $f \in L^\infty(\Omega_T)$.

If $\lambda > \Lambda_{N,2} = (\frac{N-2}{2})^2$, then it is sufficient to consider $\tilde{u}$ as a very weak supersolution to the problem

$$\begin{cases} v_t - \Delta v = \lambda \dfrac{v}{|x|^2} + f_1 & \text{in } \Omega \times (0,T), \\ v > 0 & \text{in } \Omega \times (0,T), \\ v = 0 & \text{on } \partial\Omega \times (0,T), \end{cases} \tag{7.3.1}$$

where $f_1(x) = |\nabla v|^p + cf$. The nonexistence result follows from the result in [68]; see also [6] for an alternative proof.

Let us consider the case $\lambda \le \Lambda_{N,2}$. If $\tilde{u}$ is a very weak supersolution to (7.1.1), then by Corollary 7.2.11 there exists a supersolution u which is the limit of the sequence $\{u_n\}$ of the minimal solution to the truncated problems (7.2.25). Since $u_t - \Delta u - \lambda \frac{u}{|x|^2} \ge 0$ in $\mathcal{D}'(\Omega_T)$, by Lemma 5.2.3, there exists a cylinder $B_r(0) \times (T_1, T_2)$ with $0 < r < r_1$ $0 < T_1 < T_2 \le T$ and a constant $C = C(N, r_1, T_1, T_2)$ such that $u \ge C|x|^{-\alpha_-}$ and $u > 1$ in $B_r(0) \times (T_1, T_2)$.

Fixing $0 < T_1 < t_1 < t_2 < T_2$ and letting $\phi \in C_0^\infty(B_r(0))$, using $\frac{|\phi|^{p'}}{u_n^m}$ as a test function in (7.2.25) with $p' = \frac{p}{p-1}$ and $0 < m \le 1$, passing to the limit as $n \to \infty$, it follows that

$$\frac{1}{1-m} \int_{B_r(0)} u^{1-m}(x, t_2)|\phi|^{p'}\, dx\, dt + p' \int_{t_1}^{t_2} \int_{B_r(0)} \frac{|\phi|^{p'-2}\phi \nabla\phi \nabla u}{u^m}\, dx\, dt$$
$$\ge \int_{t_1}^{t_2} \int_{B_r(0)} |\nabla u|^p \frac{|\phi|^{p'}}{u^m}\, dx\, dt + \lambda \int_{t_1}^{t_2} \int_{B_r(0)} u^{1-m} \frac{|\phi|^{p'}}{|x|^2}\, dx\, dt \quad \text{if } 0 < m < 1 \tag{7.3.2}$$

and

$$\int_{B_r(0)} \log(u(x, t_2))|\phi|^{p'}\, dx\, dt + p' \int_{t_1}^{t_2} \int_{B_r(0)} \frac{|\phi|^{p'-2}\phi \nabla\phi \nabla u}{u}\, dx\, dt$$
$$\ge \int_{t_1}^{t_2} \int_{B_r(0)} |\nabla u|^p \frac{|\phi|^{p'}}{u}\, dx\, dt + \lambda \int_{t_1}^{t_2} \int_{B_r(0)} \frac{|\phi|^{p'}}{|x|^2}\, dx\, dt \quad \text{if } m = 1. \tag{7.3.3}$$

We start by analyzing inequality (7.3.2). Using the Young inequality, for $\varepsilon > 0$ there exists $C(\varepsilon)$ such that

$$p' \int_{t_1}^{t_2} \int_{B_r(0)} \frac{|\phi|^{p'-1}|\nabla\phi||\nabla u|}{u^m}\, dx\, dt \le \varepsilon \int_{t_1}^{t_2} \int_{B_r(0)} \frac{|\phi|^{p'}|\nabla u|^p}{u^m}\, dx\, dt + C(\varepsilon) \int_{t_1}^{t_2} \int_{B_r(0)} \frac{|\nabla\phi|^{p'}}{u^m}\, dx\, dt. \tag{7.3.4}$$

Since $u(x,t) \ge C|x|^{-\alpha_-(\lambda)}$ in $B_r(0) \times (t_1, t_2)$, there exists a positive constant $C \equiv C(\varepsilon, N, r, t_1, t_2)$ such that

$$\int_{t_1}^{t_2} \int_{B_r(0)} \frac{|\nabla\phi|^{p'}}{u^m}\, dx\, dt \le C \int_{B_r(0)} |\nabla\phi|^{p'} |x|^{m\alpha_-(\lambda)}\, dx,$$
$$\lambda \int_{t_1}^{t_2} \int_{B_r(0)} u^{1-m} \frac{|\phi|^{p'}}{|x|^2}\, dx\, dt \ge C \int_{B_r(0)} \frac{|\phi|^{p'}}{|x|^{2+(1-m)\alpha_-(\lambda)}}\, dx. \tag{7.3.5}$$

Hence, for ε small enough, from (7.3.2), (7.3.4) and (7.3.5),

$$\frac{1}{1-m}\int_{B_r(0)} u^{1-m}(x,t_2)|\phi|^{p'}\,dx + C\int_{B_r(0)} |\nabla\phi|^{p'}|x|^{m\alpha_-(\lambda)}\,dx \ge C\int_{B_r(0)} \frac{|\phi|^{p'}}{|x|^{2+(1-m)\alpha_-(\lambda)}}\,dx. \tag{7.3.6}$$

By a similar computation for the case $m = 1$, using inequality (7.3.3), we obtain

$$\int_{B_r(0)} \log(u(x,t_2))|\phi|^{p'}\,dx + C\int_{B_r(0)} |\nabla\phi|^{p'}|x|^{\alpha_-(\lambda)}\,dx \ge C\int_{B_r(0)} \frac{|\phi|^{p'}}{|x|^2}\,dx. \tag{7.3.7}$$

We divide the proof in four steps.

Step 1. $p_+(\lambda) < p \le 2$.

Since $p_+(\lambda) < p$, we have $p' < \alpha_-(\lambda) + 2$. There are two possibilities.

– If $\alpha_-(\lambda) < p' < \alpha_- + 2$, it follows that $|x|^{-\frac{N}{p'}\alpha_-(\lambda)} \in L^{1+\varepsilon}(\Omega)$. Then the Hölder and Sobolev inequalities give

$$\begin{aligned}
&\int_{B_r(0)} \log(u(x,t_2))|\phi|^{p'}\,dx \\
&\le \Big(\int_{B_r(0)} |x|^{\frac{(p')^*\alpha_-(\lambda)}{p'}}|\phi|^{(p')^*}\,dx\Big)^{\frac{p'}{(p')^*}}\Big(\int_{B_r(0)} (\log u(x,t_2))^{\frac{N}{p'}}|x|^{-\frac{N}{p'}\alpha_-(\lambda)}\,dx\Big)^{\frac{p'}{N}} \\
&\le \mathcal{S}^{-1}\Big(\int_{B_r(0)} |\nabla\phi|^{p'}|x|^{\alpha_-(\lambda)}\,dx\Big)\Big(\int_{B_r(0)} (\log u(x,t_2))^{\frac{N}{p'}}|x|^{-\frac{N}{p'}\alpha_-(\lambda)}\,dx\Big)^{\frac{p'}{N}},
\end{aligned}$$

where $(p')^* = \frac{Np'}{N-p'}$ and $\mathcal{S}$ is the optimal constant in the Sobolev inequality. Moreover, from (7.3.7) we obtain

$$\begin{aligned}
&\Big(\int_{B_r(0)} |\nabla\phi|^{p'}|x|^{\alpha_-(\lambda)}\,dx\Big)\Big[\mathcal{S}^{-1}\Big(\int_{B_r(0)} (\log u)^{\frac{N}{p'}}|x|^{-\frac{N}{p'}\alpha_-(\lambda)}\,dx\Big)^{\frac{p'}{N}} + C(\varepsilon,t_1,t_2)\Big] \\
&\ge C(N,r_0,t_1,t_2)\int_{B_r} \frac{|\phi|^{p'}}{|x|^2}\,dx.
\end{aligned}$$

Hence,

$$\int_{B_r(0)} |\nabla\phi|^{p'}|x|^{\alpha_-(\lambda)}\,dx \ge C\int_{B_r(0)} \frac{|\phi|^{p'}}{|x|^2}\,dx \quad \text{for all } \phi \in C_0^\infty(B_r(0)),$$

where the constant $C > 0$ is independent of ϕ. Since $p' - \alpha_-(\lambda) < 2$, we reach a contradiction with the Hardy–Sobolev inequality, which in this case says

$$\int_{B_r(0)} |\nabla\phi|^{p'}|x|^{\alpha_-(\lambda)}\,dx \ge \Lambda_{N,p',\alpha_-}\int_{B_r(0)} \frac{\phi^{p'}}{|x|^{p'-\alpha_-(\lambda)}}\,dx. \tag{7.3.8}$$

– The second possibility is the complementary case, $p' \le \alpha_-(\lambda)$. With this hypothesis we use the family of inequalities (7.3.2). Using the Hölder inequality, since u is the limit of approximations, by the regularity results in Lemma 7.2.7, it follows that

$$\begin{aligned}
&\frac{1}{1-m}\int_{B_r(0)} u^{1-m}(x,t_2)|\phi|^{p'}\,dx \\
&\quad\le \Big(\int_{B_r(0)} u(x,t_2)|x|^{-\alpha_-(\lambda)}\,dx\Big)^{1-m}\Big(\int_{B_r(0)} |\phi|^{\frac{p'}{m}}|x|^{\frac{1-m}{m}\alpha_-(\lambda)}\,dx\Big)^{m} \\
&\quad\le C\Big(\int_{B_r(0)} |\phi|^{\frac{p'}{m}}|x|^{\frac{1-m}{m}\alpha_-(\lambda)}\,dx\Big)^{m}.
\end{aligned}$$

Consider $a = (\frac{1}{m} - 1)N$, $b = N - \frac{1}{m}(N - p')$, $\theta = \frac{am\alpha_-(\lambda)}{p'}$ and $\gamma = \alpha_-(\lambda)[\frac{1-m}{m} - \frac{am}{p'}]$. For these values, it is clear that $b < p'$ for all $0 < m < 1$. Choosing m such that $\frac{p'}{(p')^*} < m < 1$, we find that $a < (p')^*$. Thus $a + b = \frac{p'}{m}$ and $\theta + \gamma = \frac{1-m}{m}\alpha_-(\lambda)$. Then by the Hölder and Sobolev inequalities we conclude that

$$\begin{aligned}
&\frac{1}{1-m}\int_{B_r(0)} u^{1-m}(x,t_2)|\phi|^{p'}\,dx \\
&\quad\le C\Big(\int_{B_r(0)} |x|^{\theta(\frac{1}{1-m})}\,dx\Big)^{1-m}\Big(\int_{B_r(0)} |\phi|^{(p')^*}|x|^{(p')^*\frac{m\alpha_-(\lambda)}{p'}}\,dx\Big)^{\frac{ma}{(p')^*}} \\
&\qquad\times\Big(\int_{B_r(0)} |\phi|^{p'}|x|^{\gamma\frac{p'}{b}}\,dx\Big)^{\frac{m((p')^*-a)}{(p')^*}} \\
&\quad\le CS^{-1}\Big(\int_{B_r(0)} |\nabla\phi|^{p'}|x|^{m\alpha_-(\lambda)}\,dx\Big)^{\frac{ma}{p'}}\Big(\int_{B_r(0)} |\phi|^{p'}|x|^{\gamma\frac{q}{b}}\,dx\Big)^{\frac{m((p')^*-a)}{(p')^*}} \\
&\quad\le \overline{C}(\varepsilon)S^{-1}\frac{CS^{-1}ma}{q}\int_{B_r(0)} |\nabla\phi|^{p'}|x|^{m\alpha_-(\lambda)}\,dx + \varepsilon\frac{CS^{-1}m((p')^*-a)}{(p')^*}\int_{B_r(0)} |\phi|^{p'}|x|^{\gamma\frac{p'}{b}}\,dx.
\end{aligned}$$

Moreover, from the above computation and by (7.3.6), it follows that

$$(\overline{C}(\varepsilon)S^{-1} + C(\varepsilon, t_1, t_2))\Big(\int_{B_r(0)} |\nabla\phi|^{p'} |x|^{m\alpha_-(\lambda)}\, dx\Big)$$
$$\geq C(N, r_0, t_1, t_2) \int_{B_r(0)} \frac{|\phi|^{p'}}{|x|^{2+(1-m)\alpha_-(\lambda)}}\, dx - \varepsilon \frac{CS^{-1}m((p')^* - a)}{(p')^*} \int_{B_r(0)} |\phi|^{p'} |x|^{\gamma\frac{p'}{b}}\, dx.$$

Let us recall the Hardy–Sobolev inequality with weight that we need in this case,

$$\int_{B_r(0)} |\nabla\phi|^{p'} |x|^{m\alpha_-(\lambda)}\, dx \geq \Lambda_{N,p',m} \int_{B_r(0)} \frac{|\phi|^{p'}}{|x|^{p'-\alpha_-(\lambda)m}}\, dx.$$

Since $2 + (1-m)\alpha_-(\lambda) > p' - m\alpha_-(\lambda)$ if and only if $q_+(\lambda) < p$, in order to get a contradiction with Hardy's inequality, it is sufficient to choose the above parameters such that $-\frac{\gamma p'}{b} \leq 2 + (1-m)\alpha_-(\lambda)$.

(a) If $\gamma\frac{p'}{b} \geq 0$, then the result follows directly.

(b) If $-\gamma\frac{p'}{b} > 0$, by the definition of the parameters we have

$$-\gamma\frac{p'}{b} = \frac{p'\alpha_-(\lambda)}{b}\Big[\frac{am}{p'} - \frac{1-m}{m}\Big].$$

Hence we need to find the value of m and the decomposition in a, b, θ, γ, such that

$$\frac{p'\alpha_-(\lambda)}{b}\Big[\frac{am}{p'} - \frac{1-m}{m}\Big] \leq 2 + (1-m)\alpha_-(\lambda).$$

Since $\frac{p'}{(p')^*} < m < 1$, we have $b > 0$.

(i) If $\gamma \geq 0$, then the result follows directly from case (a).

(ii) If $\gamma < 0$, choosing m close to 1, it follows that

$$\Big|\gamma\frac{q}{b}\Big| = \frac{q\alpha_-(\lambda)}{b}\Big[\frac{am}{q} - \frac{1-m}{m}\Big] \leq 2 + (1-m)\alpha_-(\lambda).$$

This finishes case (b).

Therefore the nonexistence result follows in this case.

STEP 2. $p = q_+(\lambda)$ and $\lambda < \Lambda_{N,2}$. We will proceed in a similar way as in the elliptic case. Note that, by definition, $q_+(\lambda) \leq 2$. As above from Lemma 5.2.3 and the comments at the beginning of the Section 7.2.2, we know that for a suitable positive constant c_0,

$$u(x,t) \geq \frac{c_0}{|x|^{\alpha_-(\lambda)}} \quad \text{in } B_\eta(0) \times (t_1, t_2). \tag{7.3.9}$$

Without loss of generality we can assume that $\eta \le e^{-1}$. From the construction of u and by using the regularity results in Lemma 7.2.7, we obtain

$$\int_{t_1}^{t_2}\int_{B_\eta(0)} |\nabla u|^{q_+(\lambda)}|x|^{-\alpha_-(\lambda)}\,dx\,dt < \infty \quad \text{and} \quad \int_{t_1}^{t_2}\int_{B_\eta(0)} \frac{u}{|x|^{2+\alpha_-(\lambda)}}\,dx\,dt < \infty. \tag{7.3.10}$$

Consider the function $w(x,t) = |x|^{-\alpha_-(\lambda)}((t-t_1)^2(\log(\frac{1}{|x|}))^\beta + 1)$ defined in $(x,t) \in B_\eta(0) \times (t_1,t_2)$, with $\beta > 0$ that will be chosen below. Since $\lambda < \Lambda_{N,2}$, we have

$$w \in \mathcal{C}([t_1,t_2], L^2(B_\eta(0))) \cap L^2((t_1,t_2), W^{1,2}(B_\eta(0)))$$

and by a direct computation we obtain

$$\begin{aligned} &w_t - \Delta w - \lambda\frac{w}{|x|^2}\\ &= \frac{(t-t_1)}{|x|^{2+\alpha_-}}\left\{2|x|^2\left(\log\left(\frac{1}{|x|}\right)\right)^\beta + \beta(t-t_1)\left(\log\left(\frac{1}{|x|}\right)\right)^{\beta-1}\right.\\ &\quad\left.\times\left[(N-2-2\alpha_-(\lambda)) + (1-\beta)\left(\left(\log\left(\frac{1}{|x|}\right)\right)^{-1}\right]\right\}. \end{aligned}$$

Note that

$$|\nabla w| = |x|^{-\alpha_-(\lambda)-1}\left\{(t-t_1)^2\left(\alpha_-(\lambda)\left(\log\left(\frac{1}{|x|}\right)\right)^\beta + \beta\left(\log\left(\frac{1}{|x|}\right)\right)^{\beta-1}\right) + \alpha_-(\lambda)\right\},$$

thus

$$\begin{aligned} &|\nabla w|^{q_+(\lambda)}\left\{(t-t_1)^2\left(\alpha_-(\lambda)\left(\log\left(\frac{1}{|x|}\right)\right)^\beta + \beta\left(\log\left(\frac{1}{|x|}\right)\right)^{\beta-1}\right) + \alpha_-(\lambda)\right\}^{1-q_+(\lambda)}\\ &= |x|^{-\alpha_-(\lambda)-2}\left\{(t-t_1)^2\left(\alpha_-(\lambda)\left(\log\left(\frac{1}{|x|}\right)\right)^\beta + \beta\left(\log\left(\frac{1}{|x|}\right)\right)^{\beta-1}\right) + \alpha_-(\lambda)\right\}. \end{aligned}$$

If we set

$$h(x,t) = \left\{(t-t_1)^2\left(\alpha_-(\lambda)\left(\log\left(\frac{1}{|x|}\right)\right)^\beta + \beta\left(\log\left(\frac{1}{|x|}\right)\right)^{\beta-1}\right) + \alpha_-(\lambda)\right\}^{1-q_+(\lambda)},$$

then $0 \le h(x,t) \le \alpha_-(\lambda)^{1-q_+(\lambda)}$ in the same cylinder. Since $|x| \le e^{-1}$, choosing β small enough, there exists $\bar t_1 \in (t_1,t_2)$ such that

$$w_t - \Delta w - \lambda\frac{w}{|x|^2} \le \beta|\nabla w|^{q_+(\lambda)}h(x,t) \quad \text{in } B_\eta(0)\times(t_1,\bar t_1)$$

and $w(x,t_1) = |x|^{-\alpha_-(\lambda)}$ for $x \in B_\eta(0)$. Consider $u_1 \equiv c_1 u$. Then

$$(u_1)_t - \Delta u_1 - \lambda\frac{u_1}{|x|^2} \ge c_1^{1-p}|\nabla u_1|^{q_+(\lambda)} \quad \text{in } B_\eta(0)\times(t_1,\bar t_1).$$

Let c_0 be a constant satisfying (7.3.9) and take $c_1 > 0$ such that $c_1 c_0 \geq 2$. Then for β conveniently small we have

$$c_1^{1-q_+(\lambda)} \geq \|h\|_\infty \beta.$$

Since $c_1 c_0 \geq 2$ we obtain $u_1(x,t) \geq w(x,t)$ for $|x| = \eta$, $w(x,t_1) \leq u_1(x,t_1)$ and

$$(u_1)_t - \Delta u_1 - \lambda \frac{u_1}{|x|^2} \geq \beta h(x,t)|\nabla u_1|^{q_+(\lambda)}.$$

CLAIM. We have $u_1 \geq w$ in $B_\eta(0) \times (t_1, \bar{t}_1)$.

Set $v = w - u_1$. Using the regularity of w, by (7.3.10) it follows that

$$v \in \mathcal{C}([t_1,\bar{t}_1], L^1(B_\eta(0))) \cap L^{q_+(\lambda)}((t_1,\bar{t}_1), W^{1,q_+(\lambda)}(B_\eta(0))),$$

$v(x,t) \leq 0$ on $\partial B_\eta(0)$, $v(x,t_1) \leq 0$ for $x \in B_\eta(0)$ and

$$\int_{t_1}^{\bar{t}_1} \int_{B_\eta(0)} \frac{|v|}{|x|^{2+\alpha_-(\lambda)}}\, dx\, dt < \infty, \qquad \int_{t_1}^{\bar{t}_1} \int_{B_\eta(0)} |\nabla v|^{q_+(\lambda)} |x|^{-\alpha_-(\lambda)}\, dx\, dt < \infty. \tag{7.3.11}$$

By a direct computation we obtain

$$v_t - \Delta v - \lambda \frac{v}{|x|^2} \leq q_+(\lambda) h(x,t) \beta |\nabla w|^{q_+(\lambda)-2} \nabla w \nabla v \equiv a(x,t) \nabla v,$$

where

$$a(x,t) \equiv \beta q_+(\lambda) h(x,t) |\nabla w|^{q_+(\lambda)-2} \nabla w = -\beta q_+(\lambda) h(x,t) |\nabla w|^{q_+(\lambda)-1} \frac{x}{|x|}.$$

Since $h(x,t) = (|\nabla w||x|^{\alpha_-(\lambda)+1})^{1-q_+(\lambda)}$, we have $a(x,t) = -\beta q_+(\lambda) \frac{x}{|x|^2} \in L^\infty([t_1,\bar{t}_1]; L^q(B_\eta(0)))$ for all $q < N$. Note that due to the regularity of the vector field a, we cannot apply the comparison principle, Lemma 7.2.3. To overcome this lack of regularity, we proceed as follows.

Using the Kato inequality we have

$$(v_+)_t - \Delta v_+ - \lambda \frac{v_+}{|x|^2} + q_+(\lambda)\beta \left\langle \frac{x}{|x|^2}, \nabla v_+ \right\rangle \leq 0 \quad \text{and} \quad \int_{t_1}^{\bar{t}_1} \int_{B_\eta(0)} |\nabla v_+|^{p_+} |x|^{-\alpha_-(\lambda)}\, dx\, dt < \infty.$$

Since $\frac{\alpha_-(\lambda)}{q_+(\lambda)} < \frac{N-2}{2}$, by the Hardy–Sobolev inequality applied to v_+ we obtain

$$\int_{t_1}^{\bar{t}_1} \int_{B_\eta(0)} \frac{v_+^{q_+(\lambda)}}{|x|^{q_+(\lambda)+\alpha_-(\lambda)}}\, dx\, dt < \infty. \tag{7.3.12}$$

Define $\gamma = \frac{\beta q_+(\lambda)}{2}$ and consider the weight $|x|^{-2\gamma}$. For suitable β, it follows that $2\gamma < N-2$. Thus, $|x|^{-2\gamma}$ is an admissible weight in order to have Caffarelli–Kohn–Nirenberg inequalities. Hence,

$$|x|^{-2\gamma}(v_+)_t - \operatorname{div}(|x|^{-2\gamma}\nabla v_+) - \lambda\frac{v_+}{|x|^{2(\gamma+1)}}$$
$$= |x|^{-2\gamma}\left((v_+)_t - \Delta v_+ + q_+(\lambda)\beta\left\langle \frac{x}{|x|^2}, \nabla v_+\right\rangle - \lambda\frac{v_+}{|x|^2}\right) \le 0. \tag{7.3.13}$$

We claim that there exists $\sigma_1 > 2 + \alpha_-(\lambda)$, depending only on N and λ, such that

$$\int_{t_1}^{\bar{t}_1}\int_{B_\eta(0)} \frac{v_+}{|x|^{\sigma_1}}\,dx\,dt < \infty. \tag{7.3.14}$$

Indeed,

$$\int_{t_1}^{\bar{t}_1}\int_{B_\eta(0)} \frac{v_+}{|x|^{\sigma_1}}\,dx\,dt = \int_{t_1}^{\bar{t}_1}\int_{B_\eta(0)} \frac{v_+}{|x|^{\frac{q_+(\lambda)+\alpha_-(\lambda)}{q_+(\lambda)}}}\frac{1}{|x|^{\sigma_1 - \frac{q_+(\lambda)+\alpha_-(\lambda)}{q_+(\lambda)}}}\,dx\,dt$$
$$\le \left(\int_{t_1}^{\bar{t}_1}\int_{B_\eta(0)} \frac{v_+^{q_+(\lambda)}}{|x|^{q_+(\lambda)+\alpha_-(\lambda)}}\,dx\,dt\right)^{\frac{1}{q_+(\lambda)}}\left(\int_{t_1}^{\bar{t}_1}\int_{B_\eta(0)} \frac{1}{|x|^{p'_+(\sigma_1 - \frac{q_+(\lambda)+\alpha_-(\lambda)}{q_+(\lambda)})}}\,dx\,dt\right)^{\frac{1}{q'_+(\lambda)}}.$$

Denote $\theta(\sigma_1) = q'_+(\lambda)(\sigma_1 - \frac{q_+(\lambda)+\alpha_-(\lambda)}{q_+(\lambda)})$.

Observe again that the conjugate of $q_+(\lambda) = \frac{2+\alpha_-(\lambda)}{1+\alpha_-(\lambda)}$ is $q'_+(\lambda) = 2 + \alpha_-(\lambda)$. Hence we have $\theta(\sigma_1) = (2+\alpha_-(\lambda))(\sigma_1 - 1) - \alpha_-(\lambda)(1+\alpha_-(\lambda))$ and $\theta(2+\alpha_-(\lambda)) = 2(1+\alpha_-(\lambda)) = N - 2\sqrt{\Lambda_{N,2} - \lambda} < N$. Thus, there exists $\sigma_1 > 2 + \alpha_-(\lambda)$ such that $\theta(\sigma_1) < N$ and then

$$\int_{B_\eta(0)} |x|^{-(q'_+(\lambda)(\sigma_1 - \frac{q_+(\lambda)+\alpha_-(\lambda)}{q_+(\lambda)}))}\,dx < \infty.$$

Hence (7.3.14) holds.

We would like to use φ, the solution to the problem

$$\begin{cases} -\operatorname{div}(|x|^{-2\gamma}\nabla\varphi) - \lambda\frac{\varphi}{|x|^{2(\gamma+1)}} = \frac{1}{|x|^{2(\gamma+1)}} & \text{in } B_\eta(0), \\ \varphi = 0 & \text{on } \partial B_\eta(0), \end{cases}$$

as a test function in (7.3.13). However, a direct calculation shows that

$$\varphi(x) = \frac{\eta^a}{\lambda}\left(\frac{1}{|x|^a} - \frac{1}{\eta^a}\right), \quad \text{where } a = \frac{N-2(\gamma+1)}{2} - \sqrt{\left(\frac{N-2(\gamma+1)}{2}\right)^2 - \lambda}. \tag{7.3.15}$$

Thus, φ does not have the required regularity. Instead, we consider the approximated sequence,

$$\varphi_n(x) = \frac{\eta^a}{\lambda}\left(\frac{1}{(|x| + \frac{1}{n})^a} - \frac{1}{(\eta + \frac{1}{n})^a}\right),$$

with $\varphi_n \in \mathcal{C}^1(B_\eta(0))$, $\varphi_n = 0$ on $\partial B_\eta(0)$, $\nabla\varphi_n(x) = -\frac{\eta^a}{\lambda}\frac{a}{(|x|+\frac{1}{n})^{a+1}}\frac{x}{|x|}$ and

$$-\operatorname{div}(|x|^{-2\gamma}\nabla\varphi_n)) = \frac{\eta^a}{\lambda}|x|^{-2\gamma}\left(\frac{a(N-1-2\gamma)}{|x|(|x| + \frac{1}{n})^{a+1}} - \frac{a(a+1)}{(|x| + \frac{1}{n})^{a+2}}\right).$$

Note that

$$\int_{t_1}^{\bar{t}_1}\int_{B_\eta(0)} |x|^{-2\gamma}|\nabla v_+||\nabla\varphi_n|\,dx\,dt < \infty \quad \text{and} \quad \int_{t_1}^{\bar{t}_1}\int_{B_\eta(0)} \frac{v_+\varphi_n}{|x|^{2(\gamma+1)}}\,dx\,dt < \infty.$$

Choosing φ_n as a test function in (7.3.13) and since by construction $v_+(x, t_1) = 0$, we get

$$\int_{t_1}^{\bar{t}_1}\int_{B_\eta(0)} v_+(-\operatorname{div}(|x|^{-2\gamma}\nabla\varphi_n))\,dx\,dt - \lambda\int_{t_1}^{\bar{t}_1}\int_{B_\eta(0)} \frac{v_+\varphi_n}{|x|^{2(\gamma+1)}}\,dx\,dt \le 0. \tag{7.3.16}$$

By the definition of φ_n we have

$$\frac{v_+\varphi_n}{|x|^{2(\gamma+1)}} \le \frac{\eta^a}{\lambda}\left(\frac{v_+}{|x|^{a+2(\gamma+1)}} + \frac{2}{\eta^a}\frac{v_+}{|x|^{2(\gamma+1)}}\right).$$

Since a is defined by (7.3.15), we have $a + 2(\gamma+1) \to 2 + \alpha_-(\lambda)$ as $\gamma \to 0$. By choosing β small enough, we find γ small in order to have $a + 2(\gamma+1) < \sigma_1$. Hence $\frac{v_+\varphi_n}{|x|^{2(\gamma+1)}} \le \frac{Cv_+}{|x|^{\sigma_1}}$. Therefore by definition of σ_1 and the dominated convergence theorem, we prove that

$$\int_{t_1}^{\bar{t}_1}\int_{B_\eta(0)} \frac{v_+\varphi_n}{|x|^{2(\gamma+1)}}\,dx\,dt \to \frac{\eta^a}{\lambda}\int_{t_1}^{\bar{t}_1}\int_{B_\eta(0)} \frac{v_+}{|x|^{a+2(\gamma+1)}}\,dx\,dt - \frac{1}{\lambda}\int_{t_1}^{\bar{t}_1}\int_{B_\eta(0)} \frac{v_+}{|x|^{2(\gamma+1)}}\,dx\,dt \quad \text{as } n \to \infty.$$

We now deal with the first term in (7.3.16),

$$\begin{aligned}|v_+\operatorname{div}(|x|^{-2\gamma}\nabla\varphi_n)| &= \frac{\eta^a}{\lambda}\left|\left(\frac{a(N-1-2\gamma)v_+}{|x|^{1+2\gamma}(|x| + \frac{1}{n})^{a+1}} - \frac{a(a+1)v_+}{|x|^{2\gamma}(|x| + \frac{1}{n})^{a+2}}\right)\right| \\ &\le \frac{\eta^a}{\lambda}\frac{a(N-1-2\gamma)v_+}{|x|^{1+2\gamma}(|x| + \frac{1}{n})^{a+1}} + \frac{\eta^a}{\lambda}\frac{a(a+1)v_+}{|x|^{2\gamma}(|x| + \frac{1}{n})^{a+2}}.\end{aligned}$$

As above it is not difficult to see that

$$\frac{\eta^a}{\lambda}\left(\frac{a(N-1-2\gamma)v_+}{|x|^{1+2\gamma}(|x|+\frac{1}{n})^{a+1}}+\frac{a(a+1)v_+}{|x|^{2\gamma}(|x|+\frac{1}{n})^{a+2}}\right)\le\frac{\eta^a}{\lambda}\frac{a(N+a-2\gamma)v_+}{|x|^{\sigma_1}},$$

and then by the dominated convergence theorem again we obtain

$$\int_{t_1}^{\bar t_1}\int_{B_\eta(0)} v_+\,\mathrm{div}(|x|^{-2\gamma}\nabla\varphi_n)\,dx\,dt\to\frac{\eta^a}{\lambda}\int_{t_1}^{\bar t_1}\int_{B_\eta(0)}\frac{a(N-a-2(\gamma+1))v_+}{|x|^{2(1+\gamma)+a}}\,dx\,dt\quad\text{as } n\to\infty.$$

Hence passing to the limit in (7.3.16) and since $a(N-a-2(\gamma+1))-\lambda=0$, we have

$$\int_{t_1}^{\bar t_1}\int_{B_\eta(0)} v_+(-\mathrm{div}(|x|^{-2\gamma}\nabla\varphi_n))\,dx\,dt-\lambda\int_{t_1}^{\bar t_1}\int_{B_\eta(0)}\frac{v_+\varphi_n}{|x|^{2(\gamma+1)}}\,dx\,dt\to\int_{t_1}^{\bar t_1}\int_{B_\eta(0)}\frac{v_+}{|x|^{2(1+\gamma)}}\,dx\,dt,$$

as $n\to\infty$. Thus, according to (7.3.16), $\int_{t_1}^{\bar t_1}\int_{B_\eta(0)}\frac{v_+}{|x|^{2(1+\gamma)}}\,dx\,dt\le 0$, and hence $v_+\equiv 0$.

Therefore $u_1\ge w$ in $(t_1,\bar t_1)\times B_\eta(0)$. To finish the proof we use the same argument as in the first case. More precisely, fixing $\hat t_1\in(t_1,\bar t_1)$, consider $\phi\in C_0^\infty(B_r(0))$ with $0<r<<\eta$. Then there exist positive constants C_1, C_2 depending on $\varepsilon,\hat t_1,\bar t_1$ and N such that

$$\begin{aligned}\frac{1}{1-m}&\int_{B_r(0)} u^{1-m}(x,\hat t_1)|\phi|^{p'}\,dx+C_1\int_{B_r(0)}|\nabla\phi|^{p'}|x|^{m\alpha_-(\lambda)}\left(\log\frac{1}{|x|}\right)^{-\beta m}dx\\ &\ge C_2\int_{B_r(0)}\frac{|\phi|^{p'}}{|x|^{2+(1-m)\alpha_-(\lambda)}}\left(\log\frac{1}{|x|}\right)^{\beta(1-m)}dx.\end{aligned}\tag{7.3.17}$$

By similar arguments to those in (7.3.9), it follows that

$$\begin{aligned}\frac{1}{1-m}&\int_{B_r(0)} u^{1-m}(x,\hat t_1)|\phi|^{p'}\,dx\\ &\le\left(\int_{B_r(0)} u(x,\hat t_1)|x|^{-\alpha_-(\lambda)}\,dx\right)^{1-m}\left(\int_{B_r(0)}|\phi|^{\frac{p'}{m}}|x|^{\frac{1-m}{m}\alpha_-(\lambda)}\,dx\right)^{m}\\ &\le\bar C(\varepsilon)S^{-1}\frac{CS^{-1}ma}{p'}\int_{B_r(0)}|\nabla\phi|^{p'}|x|^{m\alpha_-(\lambda)}\,dx+\varepsilon\frac{CS^{-1}m((p')^*-a)}{(p')^*}\int_{B_r(0)}|\phi|^{p'}|x|^{\gamma\frac{p'}{b}}\,dx,\end{aligned}$$

with $a=(\frac{1}{m}-1)N$, $b=N-\frac{1}{m}(N-p')$ and $\gamma=\alpha_-(\lambda)[\frac{1-m}{m}-\frac{am}{p'}]$. Moreover, $b<p'$ for all $0<m<1$. Choosing m such that $\frac{p'}{(p')^*}<m<1$, we get $a<(p')^*$. Therefore, from the above computation and by (7.3.17) we have

$$[\overline{C}(\varepsilon)S^{-1} + C(\varepsilon,\hat{t}_1,\bar{t}_1)]\Big(\int\limits_{B_r(0)} |\nabla\phi|^{p'}|x|^{m\alpha_-(\lambda)}\Big(\log\frac{1}{|x|}\Big)^{-\beta m}\Big)$$
$$\geq C(N,r,\hat{t}_1,\bar{t}_1)\int\limits_{B_r(0)} \frac{|\phi|^{p'}}{|x|^{2+(1-m)\alpha_-(\lambda)}}\Big(\log\frac{1}{|x|}\Big)^{\beta(1-m)}dx$$
$$-\varepsilon\frac{CS^{-1}m((p')^*-a)}{(p')^*}\int\limits_{B_r(0)} |\phi|^q|x|^{\gamma\frac{p'}{b}}\,dx.$$

With the same argument of the first step, there exists a constant $C > 0$ such that

$$\int\limits_{B_r(0)} |\nabla\phi|^{p'}|x|^{m\alpha_-(\lambda)}\,dx \geq C\int\limits_{B_r(0)} \frac{|\phi|^{p'}}{|x|^{2+(1-m)\alpha_-(\lambda)}}\Big(\log\frac{1}{|x|}\Big)^{\beta m}dx \quad \text{for all } \phi. \tag{7.3.18}$$

Since $p' = 2+\alpha_-$, from the Hardy inequality we know that

$$\int\limits_{B_r(0)} |\nabla\phi|^{p'}|x|^{m\alpha_-(\lambda)}\,dx \geq \Lambda_{N,m,p}\int\limits_{B_r(0)} \frac{|\phi|^{p'}}{|x|^{p'-\alpha_-(\lambda)m}}\,dx = \Lambda_{N,m,p}\int\limits_{B_r(0)} \frac{|\phi|^{p'}}{|x|^{2+(1-m)\alpha_-(\lambda)}}\,dx. \tag{7.3.19}$$

Putting together (7.3.18) and (7.3.19) we reach a contradiction with the optimality of the Hardy constant. Hence the result also follows in this critical case.

STEP 3. $p = q_+(\lambda)$ and $\lambda = \Lambda_{N,2}$.

In this case $\alpha_-(\lambda) = \frac{N-2}{2}$ and $q_+(\lambda) = \frac{N+2}{N}$. We know that $u(x,t) \geq c|x|^{-\alpha_-(\lambda)}$ in $B_\eta(0)\times(t_1,t_2)$ and

$$\int\limits_{t_1}^{t_2}\int\limits_{B_\eta(0)} |\nabla u|^{q_+(\lambda)}|x|^{-\alpha_-(\lambda)}\,dx\,dt < \infty.$$

We consider $\phi \in C_0^\infty(B_{\eta_1}(0))$ such that $\phi \geq 0$ and $\phi = 1$ in $B_{\eta_1}(0)$, with $\eta_1 < \eta$. Thanks to the regularity of u, we get $\int_{t_1}^{t_2}\int_{B_\eta(0)} |\nabla(\phi u)|^{q_+(\lambda)}|x|^{-\alpha_-(\lambda)}\,dx\,dt < \infty$. Since $\frac{\alpha_-(\lambda)}{q_+(\lambda)} = \frac{N(N-2)}{2(N+2)} < N$, we can apply the Caffarelli–Kohn–Nirenberg inequalities to obtain

$$C_1\int\limits_{t_1}^{t_2}\int\limits_{B_\eta(0)} (\phi u)^{q_+(\lambda)}|x|^{-\alpha_-(\lambda)}\,dx\,dt \leq \int\limits_{t_1}^{t_2}\int\limits_{B_\eta(0)} |\nabla(\phi u)|^{q_+(\lambda)}|x|^{-\alpha_-(\lambda)}\,dx\,dt < \infty,$$

and hence

$$\int\limits_{t_1}^{t_2}\int\limits_{B_{\eta_1}(0)} u^{q_+(\lambda)}|x|^{-\alpha_-(\lambda)}\,dx\,dt < \infty \quad \text{for some } \eta_1 < \eta.$$

Therefore we conclude that $u \in L^{q_+(\lambda)}((t_1,t_2),\mathcal{D}^{1,p_+}_{\frac{\alpha_-(\lambda)}{q_+(\lambda)}}(B_{\eta_1}(0)))$.

For all $\phi \in L^{p_+}((t_1,t_2), \mathcal{D}^{1,p_+}_{\frac{\alpha_-(\lambda)}{q_+(\lambda)}}(B_{\eta_1}(0)))$, we have

$$C_2 \int_{t_1}^{t_2} \int_{B_{\eta_1}(0)} \frac{|\phi|^{q_+(\lambda)}}{|x|^{\alpha_-(\lambda)+q_+(\lambda)}}\,dx\,dt \le \int_{t_1}^{t_2} \int_{B_{\eta_1}(0)} |\phi|^{q_+(\lambda)}|x|^{-\alpha_-(\lambda)}\,dx\,dt$$
$$+ \int_{t_1}^{t_2} \int_{B_{\eta_1}(0)} |\nabla\phi|^{q_+(\lambda)}|x|^{-\alpha_-(\lambda)}\,dx\,dt,$$

where $C_2 > 0$ is independent of ϕ. In particular,

$$\int_{t_1}^{t_2} \int_{B_{\eta_1}(0)} \frac{u^{q_+(\lambda)}}{|x|^{\alpha_-(\lambda)+q_+(\lambda)}}\,dx\,dt < \infty. \tag{7.3.20}$$

Using the fact that $u(x,t) \ge c|x|^{-\alpha_-(\lambda)}$ in $B_{\eta_1}(0) \times (t_1,t_2)$ and since

$$\alpha_-(\Lambda_{N,2})q_+(\Lambda_{N,2}) + \alpha_-(\Lambda_{N,2}) + q_+(\Lambda_{N,2}) = N,$$

we reach a contradiction with (7.3.20). Hence the nonexistence result follows in this case.

STEP 4. $p > 2$. This case is now very simple. Indeed, for some constant $C(p)$, we have $|\nabla\tilde{u}|^p \ge |\nabla\tilde{u}|^2 - C(p)$ and therefore

$$\tilde{u}_t - \Delta\tilde{u} \ge \lambda\frac{\tilde{u}}{|x|^2} + |\nabla\tilde{u}|^2 - C(p) + f \quad \text{in } \Omega\times(0,T).$$

By Lemma 5.2.3, there exists a positive constant $c_0(N,r,t_1,t_2)$ such that

$$u(x,t) \ge \frac{c_0}{|x|^{\alpha_-(\lambda)}} \quad \text{in } B_r(0)\times(t_1,t_2) \subset\subset \Omega_T.$$

Thus,

$$\tilde{u}_t - \Delta\tilde{u} \ge (\lambda-\varepsilon)\frac{\tilde{u}}{|x|^2} + |\nabla\tilde{u}|^2 + \frac{\varepsilon c_0}{|x|^{2+\alpha_-}} - C(p) + f \quad \text{in } B_r(0)\times(t_1,t_2).$$

Choosing r small enough, it follows that

$$\tilde{u}_t - \Delta\tilde{u} \ge (\lambda-\varepsilon)\frac{\tilde{u}}{|x|^2} + |\nabla\tilde{u}|^2 + f \quad \text{in } B_r(0)\times(t_1,t_2).$$

Since $q_+(\lambda-\varepsilon) < 2$, we conclude as in the first step. □

Remark 7.3.2. Note that there is a lack of continuity with respect to the case $\lambda = 0$. The same thing happens in the elliptic problem (see Chapter 6). The results in [187], show that under some regularity and boundedness conditions on f and u_0, there is a positive solution to problem $u_t - \Delta u = |\nabla u|^p + f$ for any $p \ge 1$. However, for $\lambda > 0$, there is no solution if $p \ge q_+(\lambda)$. Note that $q_+(\lambda) \to 2$ as $\lambda \to 0$, so the Hardy potential is a singular perturbation of the above problem.

7.3.2 Complete and instantaneous blow-up

As a consequence of the nonexistence result for $p \geq q_+(\lambda)$, we obtain instantaneous and complete blow-up for the following approximated problems.

Theorem 7.3.3. *Let $\widetilde{u}_n \in C((0,T); L^1(\Omega)) \cap L^p((0,T); W_0^{1,p}(\Omega))$ be a solution to the problem*

$$\begin{cases} \widetilde{u}_{nt} - \Delta \widetilde{u}_n = |\nabla \widetilde{u}_n|^p + \lambda a_n(x)\widetilde{u}_n + f & \text{in } \Omega_T, \\ \widetilde{u}_n(x,t) = 0 & \text{on } \partial\Omega \times (0,T), \\ \widetilde{u}_n(x,0) = 0 & \text{if } x \in \Omega, \end{cases} \tag{7.3.21}$$

with $f \neq 0$, $a_n(x) = T_n(\frac{1}{|x|^2})$ and $p \geq q_+(\lambda)$. Then $\widetilde{u}_n(x_0,t_0) \to \infty$, $\forall (x_0,t_0) \in \Omega \times (0,T)$.

Proof. Without loss of generality, we can assume that $f \in L^\infty(\Omega_T)$ and $\lambda \leq \Lambda_{N,2}$.

Let us suppose by contradiction that there exists $(x_0,t_0) \in \Omega_T$ such that

$$\widetilde{u}_n(x_0,t_0) \to C_0 < \infty, \quad n \to \infty.$$

Assume first that $q_+(\lambda) \leq p \leq 2$. By the same arguments as in Lemma 7.2.10, for all $n \in \mathbb{N}$ we construct a minimal solution u_n to (7.3.21) such that $u_n \leq u_{n+1}$ $\forall n \in \mathbb{N}$ and $u_n(x_0,t_0) \leq \widetilde{u}_n(x_0,t_0) \to C_0 < \infty$.

The classical Harnack inequality for the heat equation implies that there exists $s > 0$ and a positive constant $C = C(N,s,t_0,\beta)$ such that

$$\iint\limits_{R^-} u_n(x,t)\,dx\,dt \leq C \operatorname*{ess\,inf}_{R^+} u_n,$$

where $R^- = B_s(x_0) \times (t_0 - \frac{3}{4}\beta, t_0 - \frac{1}{4}\beta)$ and $R^+ = B_s(x_0) \times (t_0 + \frac{1}{4}\beta, t_0 + \beta)$. We can suppose that $0 \in B_s(x_0)$, since otherwise, we may consider $B_\delta(y) \subset B_s(x_0)$ such that

$$\iint\limits_{R_y^-} u_n(x,t)\,dx\,dt \leq \iint\limits_{R^-} u_n(x,t)\,dx\,dt \leq C \operatorname*{ess\,inf}_{R^+} u_n \leq C \operatorname*{ess\,inf}_{R_y^+} u_n,$$

with $R_y^- = B_\delta(y) \times (t_0 - \frac{3}{4}\beta, t_0 - \frac{1}{4}\beta)$ and $R_y^+ = B_\delta(y) \times (t_0 + \frac{1}{4}\beta, t_0 + \beta)$. In a recurrent way, we get

$$\iint\limits_{R^-} u_n(x,t)\,dx\,dt \leq C \operatorname*{ess\,inf}_{R^+} u_n \leq C u_n(x_0,t_0) \leq C',$$

with $R^- = B_r(0) \times (t_1,t_2)$ and $R^+ = B_r(0) \times (t_3,t_4)$.

Denote $g_n(x,t) = |\nabla u_n|^p + \lambda a_n(x)u_n + f$. Then u_n solves

$$u_{nt} - \Delta u_n = g_n \quad \text{in } B_r(0) \times (t_1,t_2). \tag{7.3.22}$$

Let ϕ be the solution to the problem

$$\begin{cases} -\phi_t - \Delta\phi = 1 & \text{in } B_r(0) \times (t_1, t_2), \\ \phi = 0 & \text{on } \partial B_r(0) \times (t_1, t_2), \\ \phi(x, t_2) = 0. \end{cases}$$

Using ϕ as a test function in (7.3.22), we have

$$\int_{t_1}^{t_2} \int_{B_r(0)} u_n(x,s)\, dx\, ds = \int_{t_1}^{t_2} \int_{B_r(0)} g_n(x,s)\phi\, dx\, ds,$$

thus

$$\int_0^T \int_\Omega g_n \phi\, dx\, ds \le C' \quad \text{for all } n.$$

Therefore, $\nabla u_n \rightharpoonup \nabla u$ weakly in $L^p(B_r(0) \times (t_1, t_2); \phi\, dx\, dt)$ and thanks to the monotone convergence theorem, $a_n(x)u_n \nearrow \frac{u}{|x|^2}$ in $L^1(B_r(0) \times (t_1, t_2); \phi\, dx\, dt)$.

In order to reach a contradiction with Theorem 7.3.1, we prove that u is a supersolution to (7.3.21) in $\mathcal{D}'(B_r(0) \times (t_1, t_2))$, i. e., in the sense of Corollary 7.2.11.

Since $u_n \uparrow u$ in $L^1(B_r(0) \times (t_1, t_2); \phi\, dx\, dt)$, we have $-\Delta u_n \to -\Delta u$ in the sense of distributions. Using Lemma 7.2.9, we know that $\nabla u_n \to \nabla u$ almost everywhere. We can apply Fatou's lemma and we have

$$\int_0^T \int_\Omega |\nabla u|^p \phi\, dx \le \int_0^T \int_\Omega |\nabla u_n|^p \phi\, dx$$

for all $\phi \in C_0^\infty(\Omega \times (0, T))$, $\phi \ge 0$. Thus we conclude that u is a supersolution to (7.2.3) in the sense of Corollary 7.2.11 and therefore we reach a contradiction with the nonexistence result of Theorem 7.3.1.

Suppose now that $p > 2$. Since $\tilde{u}_n$ is a solution to (7.3.21), it follows that $\tilde{u}_n$ is a supersolution to the problem

$$\begin{cases} w_t - \Delta w = f & \text{in } \Omega_T \equiv \Omega \times (0, T), \\ w(x,t) = 0 & \text{on } \partial\Omega \times (0, T), \\ w(x, 0) = 0 & \text{if } x \in \Omega. \end{cases}$$

Moreover using the fact that $f \gneqq 0$, and by the strong maximum principle for the heat equation it follows that

$$\tilde{u}_n(x,t) \ge w(x,t) > \eta_0 \quad \text{for all } (x,t) \in B_r(0) \times (T_1, T_2).$$

Fixing $\eta_0 < 1$, since $p > 2$, there exists $\eta_1 > 0$ such that $|\nabla u_n|^p \geq \eta_1 |\nabla u_n|^2 - \eta_0$. Therefore we conclude that

$$\widetilde{u}_{nt} - \Delta \widetilde{u}_n \geq \eta_1 |\nabla \widetilde{u}_n|^2 - \eta_0 + \lambda \eta_0 a_n(x) \widetilde{u}_n + f \quad \text{in } B_r(0) \times (T_1, T_2).$$

Since $\lim_{n\to\infty} \lim_{|x|\to 0} a_n(x) = \infty$, for fixed $\lambda > 0$, there exists $n_0 \in \mathbb{N}$ and $r_1 > 0$ such that for $n \geq n_0$, we have $-\eta_0 + \lambda \eta_0 a_n(x) \widetilde{u}_n \geq \frac{\lambda \eta_0 a_n(x) \widetilde{u}_n}{2}$ in $B_{r_1}(0)$. Thus, from the last computation, it follows that

$$\widetilde{u}_{nt} - \Delta \widetilde{u}_n \geq \eta_1 |\nabla \widetilde{u}_n|^2 + \lambda \eta_0 \frac{a_n(x) \widetilde{u}_n}{2} + f \quad \text{in } B_{r_1}(0) \times (T_1, T_2).$$

Therefore for some $\eta_2 > 0$, $\eta_2 \widetilde{u}_n$ is a supersolution to the problem

$$\begin{cases} w_t - \Delta w = |\nabla w|^2 + \lambda a_n(x) w + f & \text{in } B_{r_1}(0) \times (T_1, T_2), \\ w(x,t) = 0 & \text{on } \partial B_{r_1}(0) \times (T_1, T_2), \\ w(x,0) = 0 & \text{if } x \in B_{r_1}(0). \end{cases} \tag{7.3.23}$$

Proceeding as in the case $p \leq 2$, we reach the same contradiction. □

Remark 7.3.4. Note that the nonexistence results of this section are local in nature and then imply also nonexistence for the Cauchy problem, i. e., when $\Omega = \mathbb{R}^N$.

7.4 Existence results

In this section we consider $0 < \lambda \leq \Lambda_{N,2}$ and $1 < p < q_+(\lambda)$, the complementary interval to the nonexistence result studied in the previous section. The goal here is to show that, under the hypothesis of existence of a suitable supersolution w, there exists a positive solution to problem (7.1.1). This result, together with the nonexistence obtained in the previous section, shows that $q_+(\lambda)$ is the optimal power to have existence.

We begin by proving the following general existence result.

Theorem 7.4.1. *Consider $r > 0$ such that $\Omega \subset\subset B_r(0)$. Let w be a positive function defined in $B_r(0) \times [0, T_1)$ verifying:*

(i) $w \in L^p((0,T_1); W^{1,p}(B_r(0))) \cap C([0,T_1]; L^2(B_r(0)))$;

(ii) *there exists $s > 0$ such that* $\frac{w^{1+s}}{|x|^2}, w^{\frac{(2-p)s+p}{2-p}} \in L^1(B_r(0) \times (0,T))$;

(iii) $0 \leq u_0(x) \leq w(x,0)$;

(iv) *w is a very weak supersolution to equation (7.1.1) with $f = 0$.*

Then there exists u, with at least the same regularity of w, which is a solution to the problem

$$\begin{cases} u_t - \Delta u = |\nabla u|^p + \lambda \frac{u}{|x|^2} & \text{in } \Omega_T \equiv \Omega \times (0,T), \\ u(x,t) = 0 & \text{on } \partial\Omega \times (0,T), \\ u(x,0) = u_0(x) & \text{if } x \in \Omega. \end{cases} \tag{7.4.1}$$

Proof. Assume that the initial datum satisfies $0 \lneqq u_0(x) \le w(x,0)$. Since $\Omega \subset\subset B_r(0)$, w is a supersolution to (7.4.1). Moreover, it is clear that $v(x,t) = 0$ is a strict subsolution to (7.4.1). We construct a solution to (7.4.1) as a limit of approximated problems. Consider $v_n \in L^2((0,T), W_0^{1,2}(\Omega))$, the weak solution to the following approximated problem:

$$\begin{cases} v_{nt} - \Delta v_n = \lambda \frac{1}{|x|^2+\frac{1}{n}} v_n + \frac{|\nabla v_n|^p}{1+\frac{1}{n}|\nabla v_n|^p} & \text{in } \Omega,\ t>0, \\ v_n(x,t) = 0 & \text{on } \partial\Omega,\ t>0, \\ v_n(x,0) = u_0(x) & \text{in } \Omega,\ t>0. \end{cases} \tag{7.4.2}$$

By Lemma 7.2.3, we conclude that $0 < v_1 \le \cdots \le v_{n-1} \le v_n \le w$ in $\Omega\times(0,T_1)$. Hence there exists $u \in L^p((0,T_1);L^p(\Omega))$ such that $v_n \uparrow u$ strongly in $L^2(\Omega\times(0,T_1))$ and $u \le w$. Since $\frac{w^{1+s}}{|x|^2} \in L^1(\Omega\times(0,T_1))$, using the dominated convergence theorem we get $\frac{1}{|x|^2+\frac{1}{n}} v_n \to \frac{u}{|x|^2}$ strongly in $L^1(\Omega\times(0,T_1))$. To show that u is a solution to problem (7.4.1), we need to prove that $\frac{|\nabla v_n|^p}{1+\frac{1}{n}|\nabla v_n|^p} \to |\nabla u|^p$ strongly in $L^1(\Omega\times(0,T_1))$. Although the arguments are well known we include here the details for completeness.

Claim. *The sequence $\{v_n\}$ defined in (7.4.2) is uniformly bounded in $L^p((0,T_1); W_0^{1,p}(\Omega))$.*

To prove the claim we consider $h(\sigma) = (\sigma+1)^s - 1$ and its primitive $H(\sigma) = \frac{1}{s+1}(\sigma + 1)^{s+1} - \sigma$. Using $h(v_n)$ as a test function in (7.4.2), it follows that

$$\begin{aligned} &\int_\Omega H(v_n(x,T))\,dx + s\int_0^{T_1}\int_\Omega (v_n+1)^{s-1}|\nabla v_n|^2\,dx\,dt \\ &\quad\le \int_0^{T_1}\int_\Omega (v_n+1)^s|\nabla v_n|^p\,dx\,dt + \lambda\int_0^{T_1}\int_\Omega \frac{(1+w)^s w}{|x|^2}\,dx\,dt + \int_\Omega H(u_0(x))\,dx \\ &\quad\le \varepsilon\int_0^{T_1}\int_\Omega (v_n+1)^{s-1}|\nabla v_n|^2\,dx\,dt \\ &\qquad + C(\varepsilon)\int_0^{T_1}\int_\Omega (v_n+1)^{\frac{(2-p)s+p}{2-p}}\,dx\,dt + \lambda\int_0^{T_1}\int_\Omega \frac{(1+w)^s w}{|x|^2}\,dx\,dt + C(s,u_0) \\ &\quad\le \varepsilon\int_0^{T_1}\int_\Omega (v_n+1)^{s-1}|\nabla v_n|^2\,dx\,dt + C(\varepsilon,w,u_0). \end{aligned}$$

Hence we conclude that

$$\frac{1}{s+1}\int_\Omega (v_n(x,T)+1)^{s+1}\,dx + (s-\varepsilon)\int_0^{T_1}\int_\Omega (v_n+1)^{s-1}|\nabla v_n|^2\,dx\,dt \le C(\varepsilon,w,T_1).$$

Thus

$$\int_0^{T_1}\int_\Omega (v_n+1)^{s-1}|\nabla v_n|^2\,dx\,dt \le C_1 \quad\text{and}\quad \int_0^{T_1}\int_\Omega (v_n+1)^{s}|\nabla v_n|^p\,dx\,dt \le C_2.$$

As a consequence, we have

$$\frac{1}{k^{1-s}}\int_0^{T_1}\int_\Omega \big|\nabla T_k(v_n)\big|^2\,dx\,dt \le C_3 \quad\text{and}\quad \int_0^{T_1}\int_\Omega |\nabla v_n|^p\,dx\,dt \le C_4.$$

Hence $v_n \rightharpoonup u$ weakly in $L^p(0,T_1,W_0^{1,p}(\Omega))$ and $T_k(v_n) \rightharpoonup T_k(u)$ weakly in $L^2(0,T_1,$ $W_0^{1,2}(\Omega))$.

Define $\theta_k(\sigma) = (1+G_k(\sigma))^s - 1$, whose primitive is

$$\Theta_k(\sigma) = \begin{cases} 0 & \text{if } 0 \le \sigma < k, \\ \frac{1}{s+1}(1+\sigma-k)^{s+1} - (\sigma-k) & \text{if } \sigma \ge k. \end{cases} \tag{7.4.3}$$

Consider $\theta_k(v_n)$ as a test function in (7.4.2). Then,

$$\begin{aligned} &\int_\Omega \Theta_k(v_n(x,T_1))\,dx + s\int_0^{T_1}\int_\Omega \big(1+G_k(v_n)\big)^{s-1}\big|\nabla G_n(v_n)\big|^2\,dx\,dt \\ &\quad\le \int_0^{T_1}\int_\Omega \big(1+G_k(v_n)\big)^s - 1)|\nabla v_n|^p\,dx\,dt + \lambda\int_0^{T_1}\int_\Omega \frac{((1+G_k(v_n))^s-1))v_n}{|x|^2}\,dx\,dt \\ &\qquad + \int_\Omega \Theta_k(u_0(x))\,dx \\ &\quad\le \int_0^{T_1}\int_\Omega \big((1+G_k(v_n)\big)^s - 1)|\nabla v_n|^p\,dx\,dt + \lambda\int_0^{T_1}\int_\Omega \frac{((1+G_k(w))^s-1))w}{|x|^2}\,dx\,dt \\ &\qquad + \int_\Omega \Theta_k(u_0(x))\,dx. \end{aligned}$$

Using the regularity of w it follows that

$$\int_0^{T_1}\int_\Omega \frac{(1+G_k(w))^s-1)w}{|x|^2}\,dx\,dt \to 0 \quad \text{as } k\to\infty.$$

In the same way, we have $\int_\Omega \Theta_k(u_0(x))\,dx \to 0$ as $k\to\infty$. Hence

$$\limsup_{k\to\infty}\int\int_{\{v_n\ge k\}} |\nabla v_n|^p\,dx\,dt \le \limsup_{k\to\infty}\int_0^{T_1}\int_\Omega \big|\nabla G_k(v_n)\big|^p\big(1+G_k(v_n)\big)^s\,dx\,dt = 0. \tag{7.4.4}$$

By Lemma 7.2.9 we obtain $\nabla v_n \to \nabla u$ a. e. in $\Omega \times (0, T_1)$.

STRONG CONVERGENCE OF THE NONLINEAR TERMS. To obtain the L^1-strong convergence of $\frac{|\nabla v_n|^p}{1+\frac{1}{n}|\nabla v_n|^p} \to |\nabla u|^p$, we will use Vitali's lemma (see for instance [64] page 16).

Let $E \subset \Omega \times (0, T)$ be a measurable set such that $|E| < \varepsilon$. We have

$$\iint_E \frac{|\nabla v_n|^p}{1+\frac{1}{n}|\nabla v_n|^p}\,dx\,dt = \iint_{E\cap\{v_n\geq k\}} \frac{|\nabla v_n|^p}{1+\frac{1}{n}|\nabla v_n|^p}\,dx\,dt + \iint_{E\cap\{v_n<k\}} \frac{|\nabla v_n|^p}{1+\frac{1}{n}|\nabla v_n|^p}\,dx\,dt$$
$$\leq \iint_{E\cap\{v_n\geq k\}} |\nabla v_n|^p\,dx\,dt + \iint_{E\cap\{v_n<k\}} |\nabla T_k(v_n)|^p\,dx\,dt.$$

From (7.4.4) we obtain

$$\limsup_{k\to\infty} \iint_{E\cap\{v_n\geq k\}} |\nabla v_n|^p\,dx\,dt = 0.$$

On the other hand, since $T_k(v_n)$ is bounded in $L^2((0,T_1); W_0^{1,2}(\Omega))$ and $p < 2$, using the Hölder inequality it follows that

$$\iint_{E\cap\{v_n<k\}} |\nabla T_k(v_n)|^p\,dx\,dt \leq \left(\iint_{E\cap\{v_n<k\}} |\nabla T_k(v_n)|^2\,dx\,dt\right)^{\frac{p}{2}} |E|^{\frac{2-p}{2}} \leq C|E|^{\frac{2-p}{2}} \leq C\varepsilon^{\frac{2-p}{2}}.$$

Therefore, by Vitali's lemma

$$\frac{|\nabla v_n|^p}{1+\frac{1}{n}|\nabla v_n|^p} \to |\nabla u|^p \quad \text{in } L^1(\Omega \times (0,T_1)) \text{ as } n \to \infty,$$

and thus u is a weak solution to problem (7.1.1). □

To complete the previous existence result we need to find a supersolution to (7.4.1). One way to proceed is using the elliptic results in Chapter 6.

Consider $\alpha_-(\lambda)$ and $\alpha_+(\lambda)$ as in (3.1.3), $q_+(\lambda)$ as in (6.3.2) and $q_-(\lambda)$ as defined in (7.1.3). It follows that

$$\frac{N}{N-1} < q_-(\lambda) \leq \frac{N+2}{N} \leq q_+(\lambda) < 2 \quad \text{if } 0 < \lambda \leq \Lambda_{N,2}.$$

Moreover, if $q_-(\lambda) < p < q_+(\lambda)$, then by setting $w(x) = A|x|^{-\beta}$, where $\beta = \frac{2-p}{p-1}$, $A > 0$, we obtain that $w > 0$ satisfies

$$-\Delta w = |\nabla w|^p + \lambda \frac{w}{|x|^2} \quad \text{in } \mathbb{R}^N.$$

Note that $w \in L^1_{\mathrm{loc}}(\mathbb{R}^N)$, $\frac{w}{|x|^2} \in L^1_{\mathrm{loc}}(\mathbb{R}^N)$ and since $\frac{N}{N-1} < q_-(\lambda) < p$, we have $|\nabla w|^p \in L^1_{\mathrm{loc}}(\mathbb{R}^N)$. It is clear that $w \in W^{1,2}_{\mathrm{loc}}(\mathbb{R}^N)$ if and only if $p > \frac{N+2}{N}$. As a consequence we prove the next existence result.

Theorem 7.4.2. *Assume that $\lambda < \Lambda_{N,2}$ and $1 < p < q_+(\lambda)$. Then problem (7.4.1) has a supersolution $\overline{w}$ satisfying the regularity assumptions of Theorem 7.4.1. Moreover, if $u_0 \le \overline{w}$, problem (7.4.1) has a solution.*

Proof. We divide the proof in two steps:
(i) $1 < p \le q_-(\lambda)$;
(ii) $q_-(\lambda) < p < q_+(\lambda)$.

(i) Assume $1 < p \le q_-(\lambda)$. Since $p \le q_-(\lambda)$, there exists $\beta \in (\alpha_-(\lambda), \alpha_+(\lambda))$, close to $\alpha_-(\lambda)$, such that $p(\beta+1) < \beta+2$. Define $\overline{w}(x) \equiv |x|^{-\beta}$, where $A^{p-1} = \frac{\beta(N-\beta-2)-\lambda}{\beta^p}$. Then $\overline{w} \in W^{1,2}(\Omega)$ and

$$-\Delta\overline{w} - \lambda\frac{\overline{w}}{|x|^2} \ge |\nabla\overline{w}|^p.$$

It is clear that $\overline{w}$ is a supersolution to problem (7.4.1) if $u_0(x) \le \overline{w}$. Since $\overline{w}$ satisfies the regularity hypotheses required in Theorem 7.4.1, the existence result follows if $u_0 \le w$.

(ii) Assume now that $p_-(\lambda) < p < q_+(\lambda)$. Since $\frac{N+2}{N} \in (p_-(\lambda), q_+(\lambda))$, we begin by assuming that $p > \frac{N+2}{N}$. Let $w(x) = A|x|^{-\beta}$ with $\beta = \frac{2-p}{p-1}$ and $\beta^p A^{p-1} = \beta(N-\beta-2) - \lambda$. Then $\beta \in (\alpha_-, \alpha_+)$ and $A > 0$. Now $w \in W^{1,2}_{\mathrm{loc}}(\mathbb{R}^N)$ satisfies the regularity hypotheses of Theorem 7.4.1 and

$$-\Delta w = \lambda\frac{w}{|x|^2} + |\nabla w|^p.$$

Hence the existence result follows.

Assume that $p_-(\lambda) < p \le \frac{N+2}{N}$. Since $\lambda < \Lambda_{N,2}$, we fix λ_1 such that $\lambda < \lambda_1 < \Lambda_{N,2}$ and $q_+(\lambda_1) > \frac{N+2}{N}$. Consider $p_1 > \frac{N+2}{N}$ in such a way that $p_1 < p_+(\lambda_1)$. As in the previous case, we define $w_1(x) \equiv A_1|x|^{\beta_1}$ with $\beta_1 = \frac{2-p_1}{p_1-1}$ and $\beta_1^p A_1^{p-1} = \beta_1(N-\beta-2) - \lambda_1$. Moreover $w_1 \in W^{1,2}_{\mathrm{loc}}(\mathbb{R}^N)$ satisfies

$$-\Delta w_1 = \lambda_1\frac{w_1}{|x|^2} + |\nabla w_1|^{p_1}.$$

Since $p < p_1$ and $\lambda < \lambda_1$, a direct computation shows that

$$-\Delta w_1 \ge \lambda\frac{w_1}{|x|^2} + \theta|\nabla w_1|^p,$$

where $\theta > 0$. Therefore, setting $w_2(x) = \theta^{\frac{1}{\theta-1}} w_1(x) = \theta^{\frac{1}{\theta-1}} A_1|x|^{\beta_1}$, we obtain

$$-\Delta w_2 \ge \lambda\frac{w_2}{|x|^2} + |\nabla w_2|^p.$$

As above if $u_0(x) \le w_2(x)$, then w_2 is a supersolution to problem (7.4.1); using Theorem 7.4.1 we conclude the proof. □

Note that if $\lambda = \Lambda_{N,2}$, then $p_+(\Lambda_{N,2}) = \frac{N+2}{N}$. Therefore, using [14] we get the existence of a positive constant $c_1 > 0$ such that, if

$$w(x) = \left|\frac{x}{R}\right|^{-\frac{N-2}{2}} \left(\log\left(\frac{R}{|x|}\right)\right)^{1/2} - A, \quad \text{with } A = \left(\frac{r}{R}\right)^{-\frac{N-2}{2}} \left(\log\left(\frac{R}{|r|}\right)\right)^{1/2}, \; r < R, \tag{7.4.5}$$

then $c_1 w(x)$ is a supersolution to the problem

$$-\Delta w = \Lambda_{N,2}\frac{1}{|x|^2} w + |\nabla w|^p + cg \text{ in } B_r(0) \quad \text{and} \quad w_n = 0 \text{ on } \partial B_r(0),$$

where $0 \le g(x) \le \frac{1}{|x|^2}$ and c is small enough.

Choosing $r > 0$ such that $\Omega \subset B_{\frac{r}{2}}(0)$, it follows that $w \in W^{1,q}(\Omega)$ for all $q < 2$. In fact, one has $w \in H(\Omega)$, as defined in Section 2.5.1. It is not difficult to verify that w satisfies the regularity assumptions in Theorem 7.4.1. In this way, we have proved the following result.

Theorem 7.4.3. *Let $\lambda = \Lambda_{N,2}$ and assume that $u_0(x) \le w$ as given in (7.4.5). Then problem (7.4.1) with $\lambda = \Lambda_{N,2}$ has a minimal solution $v \in \mathcal{C}([0,T]; L^1(\Omega)) \cap L^2(0,T; H(\Omega))$.*

Remark 7.4.4. We can also prove with similar arguments that the problem

$$\begin{cases} u_t - \Delta u = |\nabla u|^p + \lambda \frac{u}{|x|^2} + cf(x,t) & \text{in } \Omega_T, \\ u(x,t) = 0 & \text{on } \partial\Omega \times (0,T), \\ u(x,0) = u_0(x) & \text{if } x \in \Omega \end{cases}$$

has a positive solution if $f(x,t) \le \frac{1}{|x|^2}$ and $c \le c(T)$ is a suitable small constant.

It would be interesting to find necessary and sufficient conditions on f and u_0 to get the existence of a positive solution to the above problem.

7.5 Cauchy problem

In [187], the authors consider the initial value problem

$$\begin{cases} u_t = \Delta u + |\nabla u|^p, & x \in \mathbb{R}^N, \; t > 0, \\ u(x,0) = u_0(x) \ge 0, & x \in \mathbb{R}^N, \end{cases} \tag{7.5.1}$$

with $p > 1$. They prove that problem (7.5.1) has a global positive solution under some condition on the initial datum u_0. That means that there is a great difference with the semilinear heat equation because here there is no Fujita type exponent.

We will see, however, that the behavior is quite different under the influence of a Hardy term, i. e., if we consider instead the problem

$$u_t - \Delta u = \lambda \frac{u}{|x|^2} + |\nabla u|^p \text{ in } \mathbb{R}^N, \; t > 0, \; \Lambda_{N,2} \ge \lambda > 0, \quad u(x,0) = u_0(x) \text{ in } \mathbb{R}^N. \tag{7.5.2}$$

First of all, from the previous sections, we know that the above equation has no local supersolution for $p \geq q_+(\lambda)$.

In this section we will prove that a new different feature holds for the Cauchy problem (7.5.2): the existence of a Fujita type exponent $F(\lambda) < q_+(\lambda)$ if $\lambda \in (0, \Lambda_{N,2}]$.

In Section 5.2.1 we have found a self-invariant solution to the homogenous linear equation

$$v_t - \Delta v - \lambda \frac{v}{|x|^2} = 0 \quad \text{in } \mathbb{R}^N \times (0, \infty). \tag{7.5.3}$$

This gives us a lower bound for the behavior at zero and at infinity of every supersolution to (7.5.2) and, in particular, for $\lambda = 0$ we get the fundamental solution of the heat equation.

To find the Fujita type exponent we proceed with the same strategy as in the semilinear case.

7.5.1 A class of subsolutions to (7.5.2) for small p, blow-up in a finite time

To guess the possible Fujita exponent we start by studying the values of p for which there exists a class of subsolutions to equation (7.5.3) blowing up in a finite time.

More precisely, we look for a family of subsolutions to problem (7.5.2) in the form

$$w(r, t, T) = (T - t)^{-\theta} \zeta\left(\frac{r}{(T - t)^\beta}\right),$$

with $\theta, \beta > 0$ to be chosen and $\zeta > 0$ a smooth function that will be related to the self-similar solution of the linear problem. Denoting $s = \frac{r}{(T-t)^\beta}$, it follows that

$$\begin{aligned} w_t &= (T - t)^{-\theta-1}\left(\theta\zeta + \frac{\beta r}{(T - t)^\beta}\zeta'(s)\right), \\ w_r(r, t) &= (T - t)^{-\theta-\beta}\zeta'(s), \\ w_{rr}(r, t) &= (T - t)^{-\theta-2\beta}\zeta''(s). \end{aligned} \tag{7.5.4}$$

We need to have

$$w_t - w_{rr} - \frac{(N - 1)}{r} w_r - \lambda \frac{w}{r^2} \leq |w_r|^p. \tag{7.5.5}$$

By setting $\theta = \frac{2-p}{2(p-1)}$, $\beta = \frac{1}{2}$ and replacing the value in (7.5.4), inequality (7.5.5) becomes

$$-\zeta''(s) - \left(\frac{N - 1}{s} - \frac{1}{2}s\right)\zeta'(s) - \left(\frac{\lambda}{s^2} - \theta\right)\zeta(s) \leq |\zeta'|^p(s). \tag{7.5.6}$$

Assume that $\zeta(s) = A\phi(cs)$ with $\phi(s) = s^{-\alpha_-(\lambda)}e^{-\frac{s^2}{4}}$, $A > 0$, $c > 0$. Then (7.5.6) is equivalent to

$$\begin{aligned}
&- cs^2\left(\frac{c}{4} + \frac{1}{4}\right) + \frac{1}{s^2}(c\alpha_-(\lambda)(N-1) - c^2[\alpha_-(\lambda) + \alpha_-(\lambda)^2] - \lambda) \\
&\quad + \frac{c(N-\alpha_-(\lambda))}{2} - c^2\left(\alpha_-(\lambda) - \frac{1}{2}\right) + \frac{(2-p)}{2(p-1)} \\
&\le A^{p-1}c^{(p-1)(1-\alpha_-)}\left(\frac{\alpha_-(\lambda)}{cs} + \frac{sc}{2}\right)^p s^{-\alpha_-(\lambda)(p-1)}e^{-\frac{1}{4}(cs)^2(p-1)}.
\end{aligned}$$

If s is large enough, then the above inequality holds for all $A > 0$ and $c > 0$. Now, if s is close to 0, since $\alpha_-(\lambda)^2 - (N-2)\alpha_-(\lambda) + \lambda = 0$, choosing $c \in (0,1]$ such that $c\alpha_-(\lambda)(N-1) - c^2[\alpha_-(\lambda) + \alpha_-(\lambda)^2] - \lambda \le 0$, the above inequality holds for all $A > 0$. It remains to obtain a control in a compact interval, for which it is sufficient to choose $A > 0$ large enough. Hence we can find a family of subsolutions to (7.5.2) given by

$$w(r,t,T) = A(T-t)^{-\frac{2-p}{2(p-1)}}\left(\frac{cr}{(T-t)^{\frac{1}{2}}}\right)^{-\alpha_-(\lambda)} e^{-\frac{1}{4}\frac{cr^2}{T-t}}. \tag{7.5.7}$$

Remark 7.5.1. It is worth to point out that in the case $\lambda = 0$, namely, $\alpha_-(0) = 0$, the previous inequality is reduced to

$$-cs^2\left(\frac{c}{4} + \frac{1}{4}\right) + \frac{cN}{2} + \frac{c^2}{2} + \theta \le A^{p-1}c^{2p-1}\left(\frac{s}{2}\right)^p e^{-\frac{1}{4}(cs)^2(p-1)},$$

which is not true near $s = 0$ since $\frac{c^2}{2} + \frac{cN}{2} + \frac{(2-p)}{2(p-1)} > 0$. This gives a quantitative justification to the influence of $\lambda > 0$, in order to have a Fujita type exponent.

Fix $r > 0$. Since

$$\int_{B_r(0)} |x|^{-\alpha_-(\lambda)} w(x,t,T)\,dx = C(T-t)^{-\frac{(2-p)}{2(p-1)} + \frac{N}{2} - \frac{\alpha_-(\lambda)}{2}} \int_0^{\frac{cr}{\sqrt{(T-t)}}} \phi(s)s^{N-\alpha_-(\lambda)-1}\,ds,$$

for $p < 1 + \frac{1}{N-\alpha_-(\lambda)+1}$, that is, $-\frac{(2-p)}{2(p-1)} + \frac{N}{2} - \frac{\alpha_-(\lambda)}{2} < 0$, it follows that

$$\lim_{t\to T} \int_{B_r(0)} |x|^{-\alpha_-(\lambda)} w(x,t,T)\,dx = \infty.$$

We now analyze some properties of the subsolution $w(r,t,T)$ found in (7.5.7).

(i) The function $w(r,t,T)$ blows up in a finite time in the sense of the local weighted L^1-norm in Definition 7.5.7.

(ii) Assume that $p < F(\lambda) = 1 + \frac{1}{N-\alpha_-(\lambda)+1}$.
We denote by $\overline{u}(x,t) = u(x,t+T)$ a time translation of a positive solution $u(x,t)$ to (7.5.2). Since $\overline{u}(x,t)$ is a supersolution to the homogenous equation (7.5.3) with

the same initial datum, in order to get $\overline{u}(x,0) \geq w(r,0,T)$, it is sufficient to check that for v defined in (5.2.7) and some $T > 0$, one has $v(x,T) \geq w(r,0,T)$. This immediately follows because $p < F(\lambda) = 1 + \frac{1}{N-\alpha_-(\lambda)+1}$.

(iii) Call $G(x,t) = |\nabla w|^{p-1}$. Since $p < F(\lambda)$, there exists $q > 2$ such that

$$\sup_{t\in[0,T_1]} \int_{\mathbb{R}^N} |G(x,t)|^{\frac{qN}{2}}\, dx = C(T_1) < \infty.$$

Remark 7.5.2. It is natural to conjecture that $F(\lambda) = 1 + \frac{1}{N-\alpha_-(\lambda)+1}$ is optimal with this blowing-up property, i. e., it could be seen as the Fujita exponent for problem (7.5.2).

Observe that if $0 < \lambda \leq \Lambda_{N,2}$, then

$$F(0^+) = \lim_{\lambda\downarrow 0} F(\lambda) = \frac{N+2}{N+1} < F(\lambda) < q_-(\lambda) \leq \frac{N+2}{N} \leq q_+(\lambda) \leq 2.$$

The first inequality shows a discontinuity of the Fujita exponent at $\lambda = 0$.

7.5.2 A class of global supersolutions to (7.5.2) for $F(\lambda) < p < q_+(\lambda)$

Consider $F(\lambda) = 1 + \frac{1}{N-\alpha_-(\lambda)+1}$ as in the previous section. We look for a family of supersolutions to equation (7.5.2), i. e., we look for w such that

$$w_t - w_{rr} - \frac{(N-1)}{r} w_r - \lambda \frac{w}{r^2} \geq |w'|^p, \tag{7.5.8}$$

for $F(\lambda) < p < q_+(\lambda)$.

Assume that $w(r,t,T) = (T+t)^{-\theta} g(\frac{r}{(T+t)^\beta})$ (see for instance [175] and [174]), with g being a smooth bounded positive function and $\theta, \beta > 0$ to be conveniently chosen. Calling $s = \frac{r}{(T+t)^\beta}$, it follows that:

- $w_t = -(T+t)^{-\theta-1}(\theta g + \frac{\beta r}{(T+t)^\beta} g'(s))$;
- $w_r(r,t) = (T+t)^{-\theta-\beta} g'(s)$;
- $w_{rr}(r,t) = (T+t)^{-\theta-2\beta} g''(s)$.

In order to get homogeneity in the equation, it is sufficient to choose $\theta = \frac{(2-p)}{2(p-1)}$ and $\beta = \frac{1}{2}$. Therefore (7.5.8) gives

$$g''(s) + \left(\frac{N-1}{s} + \frac{1}{2}s\right) g'(s) + \left(\frac{\lambda}{s^2} + \theta\right) g(s) + |g'(s)|^p \leq 0. \tag{7.5.9}$$

Consider γ such that $\alpha_-(\lambda) < \gamma < \frac{2}{p-1}$ and $g(s) = A\phi(cs)$ with, as above,

$$\phi(s) = s^{-\gamma} e^{-\frac{s^2}{4}}, \quad A > 0,\ c > 0.$$

Hence substituting in (7.5.9), we have

$$\frac{c^2[\gamma^2-(N-2)\gamma+\lambda]}{(cs)^2}+(cs)^2\left[\frac{c^2}{4}-\frac{1}{4}\right]+c^2\gamma$$
$$-\frac{c^2}{2}-\frac{c^2(N-1)}{2}-\frac{\gamma}{2}+\frac{(2-p)}{2(p-1)}+A^{p-1}c^{p-1}\left[\frac{\gamma}{cs}+\frac{sc}{2}\right]^{p-1}(cs)^{-\gamma(p-1)}e^{-\frac{1}{4}(cs)^2(p-1)}\le 0.$$

Since $\alpha_-(\lambda)<\gamma$ and $p>1+\frac{1}{N-\alpha_-(\lambda)+1}$, it is sufficient to choose $c<1$. Then

$$c^2\gamma-\frac{c^2}{2}-\frac{c^2(N-1)}{2}-\frac{\gamma}{2}+\frac{(2-p)}{2(p-1)}\le 0\quad\text{and}\quad(cs)^2\left[\frac{c^2}{4}-\frac{1}{4}\right]\le 0.$$

Moreover, from the fact that $\gamma<\frac{2}{p-1}$ and $c^2[\gamma^2-(N-2)\gamma+\lambda]\le 0$, choosing A small enough we get

$$\frac{c^2[\gamma^2-(N-2)\gamma+\lambda]}{(cs)^2}+A^{p-1}c^{p-1}\left[\frac{\gamma}{cs}+\frac{sc}{2}\right]^{p-1}(cs)^{-\gamma(p-1)}e^{-\frac{1}{4}(cs)^2(p-1)}\le 0.$$

So, our family of supersolutions is of the form

$$w(r,t,T)=Ac^{-\gamma}(T+t)^{\frac{\gamma}{2}-\frac{2-p}{2(p-1)}}r^{-\gamma}e^{-c^2\frac{r^2}{4(T+t)}}. \tag{7.5.10}$$

Note that $w(r,0,T)=AT^{-\frac{2-p}{2(p-1)}}a^{-\gamma}|x|^{-\gamma}e^{-\frac{-a^2|x|^2}{4}}$, where $a=\frac{c}{T^{\frac{1}{2}}}<1$.

7.5.3 Local existence for $1<p<q_+(\lambda)$

As in Theorem 7.4.1, we begin by proving a local existence result under the hypothesis of existence of a supersolution.

Theorem 7.5.3. *Let w be a nonnegative function defined in $\mathbb{R}^N\times[0,T_1)$, verifying:*
(i) $w\in L^p((0,T_1);W^{1,p}_{loc}(\mathbb{R}^N))\cap C([0,T_1];L^2_{loc}(\mathbb{R}^N))$;
(ii) $\exists s>0$ *such that* $\frac{w^{1+s}}{|x|^2},w^{\frac{(2-p)s+p}{2-p}}\in L^1(K\times(0,T))$, *for all compact sets* $K\subset\mathbb{R}^N$;
(iii) $0\le u_0(x)\le w(x,0)$ *and* $u_0^{s+1}\in L^1_{loc}(\mathbb{R}^N)$;
(iv) w *is a very weak supersolution to equation* (7.5.2).

Then there exists a solution u to (7.5.2) satisfying $u(x,0)=u_0$ with, at least, the same regularity as w.

Proof. The proof follows closely the arguments used in Theorem 7.4.1. The only difference is that in this case the estimates are local in space and we need to use a suitable cutoff function. For the reader's convenience we give some of the details. Consider an initial datum $0\lneqq u_0(x)\le w(x,0)$, where $w(x,t)$ satisfies the above hypotheses. We construct a global solution as a limit of approximated problems in bounded domains. Let B_n be the ball in $\mathbb{R}^N$ with radius n and centered at the origin. We consider

$$v_n \in L^2((0,T);W_0^{1,2}(B_{n+1})) \cap L^\infty((0,T);W_0^{1,p}(B_{n+1})), \quad \forall T > 0,$$

the weak solutions to the following approximated problems:

$$\begin{cases} v_{nt} - \Delta v_n = \lambda \frac{1}{|x|^2+\frac{1}{n}} \tilde{v}_{n-1} + \frac{|\nabla v_n|^p}{1+\frac{1}{n}|\nabla v_n|^p} & \text{in } B_{n+1},\ t > 0, \\ v_n(x,0) = u_0(x) & \text{in } B_{n+1},\ t > 0, \\ v_n(x,t) = 0 & \text{on } \partial B_{n+1},\ t > 0, \end{cases} \tag{7.5.11}$$

with

$$\begin{cases} v_{0t} - \Delta v_0 = 0 & \text{in } B_1,\ t > 0, \\ v_0(x,0) = u_0(x) & \text{in } B_1,\ t > 0, \\ v_0(x,t) = 0 & \text{on } \partial B_1,\ t > 0 \end{cases}$$

and $\tilde{v}_{n-1} = v_{n-1}$ in B_n, $\tilde{v}_{n-1} = 0$ in $\mathbb{R}^N \setminus B_n$. Applying the comparison Lemma 7.2.3, we conclude that $0 < \tilde{v}_0 \le \tilde{v}_1 \le \cdots \le \tilde{v}_{n-1} \le \tilde{v}_n \le w$ in B_{n+1}. Hence there exists $u \in L^p((0,T_1);L^p_{\mathrm{loc}}(\mathbb{R}^N))$ such that $\tilde{v}_n \uparrow u$ strongly in $L^2((0,T_1);L^2_{\mathrm{loc}}(\mathbb{R}^N))$ and $u \le w$. Since $\frac{w^{1+s}}{|x|^2} \in L^1(K \times (0,T_1))$ for all compact sets $K \subset \mathbb{R}^N$, using the dominated convergence theorem we easily get $\frac{1}{|x|^2+\frac{1}{n}}\tilde{v}_{n-1} \to \frac{u}{|x|^2}$ strongly in $L^1(K \times (0,T_1))$.

Note that to conclude we have to prove that

$$\frac{|\nabla \tilde{v}_n|^p}{1+\frac{1}{n}|\nabla \tilde{v}_n|^p} \to |\nabla u|^p \quad \text{strongly in } L^1_{\mathrm{loc}}(\mathbb{R}^N \times (0,T_1)).$$

We claim that $\{\tilde{v}_n\}$ is uniformly bounded in $L^p((0,T_1);W^{1,p}_{\mathrm{loc}}(\mathbb{R}^N))$. To prove the claim we use $((v_n+1)^s - 1)\phi^2$ as a test function in (7.5.11), with $\phi \in C_0^\infty(\mathbb{R}^N)$ being a fixed nonnegative function and $n \ge n_0$ is chosen such that $\operatorname{supp}\phi \subset\subset B_{n_0}(0)$.

Let us consider $H(\sigma) = \frac{1}{s+1}(\sigma+1)^{s+1} - \sigma$. It follows that

$$\begin{aligned} &\int_{\mathbb{R}^N} H(v_n(x,T))\phi^2\,dx + s\int_0^{T_1}\int_{\mathbb{R}^N}(v_n+1)^{s-1}|\nabla v_n|^2\phi^2\,dx\,dt \\ &\quad + 2\int_0^{T_1}\int_{\mathbb{R}^N}((v_n+1)^s - 1)\phi\nabla v_n\nabla\phi\,dx\,dt \\ &\le \int_0^{T_1}\int_{\mathbb{R}^N}(v_n+1)^s|\nabla v_n|^p\phi^2\,dx\,dt + \lambda\int_0^{T_1}\int_{\mathbb{R}^N}\frac{(1+w)^s w}{|x|^2}\phi^2\,dx\,dt + \int_{\mathbb{R}^N} H(u_0(x))\phi^2\,dx \\ &\le \varepsilon\int_0^{T_1}\int_{\mathbb{R}^N}(v_n+1)^{s-1}|\nabla v_n|^2\phi^2\,dx\,dt + C(\varepsilon)\int_0^{T_1}\int_{\mathbb{R}^N}(v_n+1)^{\frac{(2-p)s+p}{2-p}}\phi^2\,dx\,dt \end{aligned}$$

$$+\lambda\int_0^{T_1}\int_{\mathbb{R}^N}\frac{(1+w)^s w}{|x|^2}\phi^2\,dx\,dt + C(s,u_0) \le \varepsilon\int_0^{T_1}\int_{\mathbb{R}^N}(v_n+1)^{s-1}|\nabla v_n|^2\phi^2\,dx\,dt$$
$$+ C(\varepsilon, w, u_0).$$

Hence we conclude that

$$\frac{1}{s+1}\int_{\mathbb{R}^N}(v_n(x,T)+1)^{s+1}\phi^2\,dx + (s-\varepsilon)\int_0^{T_1}\int_{\mathbb{R}^N}(v_n+1)^{s-1}|\nabla v_n|^2\phi^2\,dx\,dt \le C(\varepsilon, w, T_1, \phi).$$

Thus

$$\int_0^{T_1}\int_{\mathbb{R}^N}(v_n+1)^{s-1}|\nabla v_n|^2\phi^2\,dx\,dt \le C_1 \quad\text{and}\quad \int_0^{T_1}\int_{\mathbb{R}^N}(v_n+1)^{s}|\nabla v_n|^p\phi^2\,dx\,dt \le C_2.$$

As a consequence, we have

$$\frac{1}{k^{1-s}}\int_0^{T_1}\int_{\mathbb{R}^N}|\nabla T_k(v_n)|^2\phi^2\,dx\,dt \le C_3 \quad\text{and}\quad \int_0^{T_1}\int_{\mathbb{R}^N}|\nabla v_n|^p\phi^2\,dx\,dt \le C_4.$$

Hence, using the definition of $\tilde{v}_n$ we obtain:
(i) $\tilde{v}_n \rightharpoonup u$ weakly in $L^p((0,T_1); W^{1,p}_{\text{loc}}(\mathbb{R}^N))$ and
(ii) $T_k(\tilde{v}_n) \rightharpoonup T_k(u)$ weakly in $L^2((0,T_1); W^{1,2}_{\text{loc}}(\mathbb{R}^N))$.

Let be $\theta_k(\sigma)$ defined in (7.4.3) and consider $\theta_k(v_n)\phi^2$ as a test function in (7.5.11). Then

$$\begin{aligned}
&\int_{\mathbb{R}^N}\theta_k(v_n(x,T_1))\phi^2\,dx + s\int_0^{T_1}\int_{\mathbb{R}^N}\phi^2(1+G_k(v_n))^{s-1}|\nabla G_n(v_n)|^2\,dx\,dt\\
&\quad + 2\int_0^{T_1}\int_{\mathbb{R}^N}((1+G_k(v_n))^s-1)\phi\nabla v_n\nabla\phi\,dx\,dt\\
&\le \int_0^{T_1}\int_{\mathbb{R}^N}((1+G_k(v_n))^s-1)\phi^2|\nabla v_n|^p\,dx\,dt + \int_{\mathbb{R}^N}\theta_k(u_0(x))\phi^2\,dx\\
&\quad + \lambda\int_0^{T_1}\int_{\mathbb{R}^N}\frac{((1+G_k(v_n))^s-1))v_n}{|x|^2}\phi^2\,dx\,dt\\
&\le \int_0^{T_1}\int_{\mathbb{R}^N}((1+G_k(v_n))^s-1)\phi^2|\nabla v_n|^p\,dx\,dt + \int_{\mathbb{R}^N}\theta_k(u_0(x))\phi^2\,dx\\
&\quad + \lambda\int_0^{T_1}\int_{\mathbb{R}^N}\frac{(1+G_k(w))^s-1)w}{|x|^2}\phi^2\,dx\,dt.
\end{aligned}$$

Thus

$$\int_0^{T_1}\int_{\mathbb{R}^N} ((1+G_k(v_n))^s-1)|\nabla v_n||\nabla\phi||\phi|\,dx\,dt$$

$$\le \varepsilon\int_0^{T_1}\int_{\{v_n\ge k\}} ((1+G_k(v_n))^s-1)|\nabla v_n||\nabla\phi||\phi|\,dx\,dt$$

$$\le \varepsilon\int_0^{T_1}\int_{\mathbb{R}^N} ((1+G_k(v_n))^s|\nabla G_k(v_n)|^2\phi^2\,dx\,dt + C(\varepsilon)\int_0^{T_1}\int_{\{v_n\ge k\}} ((1+G_k(v_n))^s-1)|\nabla\phi|^2\,dx\,dt.$$

Since $\phi\in C_0^\infty(\mathbb{R}^N)$ is fixed, the regularity of w gives

$$\int_0^{T_1}\int_{\mathbb{R}^N} \frac{((1+G_k(w))^s-1))w}{|x|^2}\phi\,dx\,dt + \int_0^{T_1}\int_{\{v_n\ge k\}} ((1+G_k(v_n))^s-1)|\nabla\phi|^2\,dx\,dt \to 0 \quad \text{as } k\to\infty.$$

In the same way, we have $\int_{\mathbb{R}^N}\theta_k(u_0(x))\phi^2\,dx\to 0$ as $k\to\infty$. Hence we conclude that

$$\limsup_{k\to\infty}\int\int_{\{v_n\ge k\}} |\nabla v_n|^p\,dx\,dt \le \limsup_{k\to\infty}\int_0^{T_1}\int_{\mathbb{R}^N} |\nabla G_k(v_n)|^p(1+G_k(v_n))^s\,dx\,dt = 0.$$

From Lemma 7.2.9 we obtain that $\nabla v_n\to\nabla u$ a. e. To conclude we have to prove the strong convergence of $\frac{|\nabla v_n|^p}{1+\frac{1}{n}|\nabla v_n|^p}$ to $|\nabla u|^p$ in $L^1(K\times(0,T_1))$. This last statement follows using the same arguments as in the proof of Theorem 7.4.1. □

Note that using a separation of variables argument we can find a local supersolution for initial datum u_0 such that $u_0(x)\le A|x|^{-\beta}$ with β close to $\alpha_-(\lambda)$. More precisely we have the following result.

Proposition 7.5.4. *Assume that $p<q_+(\lambda)$ and*

$$u(x,0)\le T^{-\theta}|x|^{-\beta}, \quad \text{with } T>0, \alpha_-(\lambda)<\beta<\alpha_+(\lambda) \text{ such that } \beta+\alpha_-(\lambda)<N.$$

Then (7.5.2) has a local solution u in $\mathbb{R}^N\times(0,T)$ for some $T>0$, satisfying that for any ball $B_r(0), r>0$,

$$\int_{B_r(0)} |x|^{-\alpha_-(\lambda)}u(x,t)\,dx<\infty, \quad \text{for each } 0<t<T. \tag{7.5.12}$$

Proof. Consider $w(x,t)=\frac{1}{(T-t)^\theta}|x|^{-\beta}$ with $\alpha_-(\lambda)<\beta<\alpha_+(\lambda)$ and $(\beta+\alpha_-(\lambda))<N$; it follows that $\beta(N-1)-\beta(\beta+1)-\lambda>0$. Moreover, since $p<q_+(\lambda)$, we have $\beta<(\beta+1)p<\beta+2$. A direct computation shows that for $T>1$ and θ large, w is a supersolution to (7.5.2). Since $u_0(x)\le w(x,0)$, using Theorem 7.5.3, there exists a minimal solution u to (7.5.2) such that (7.5.12) holds. □

7.5.4 Global existence for small data and $F(\lambda) < p < q_+(\lambda)$

We consider the class of initial data

$$\mathcal{F}_D = \{u_0 : \mathbb{R}^N \to \mathbb{R} | 0 \le u_0(x) \le A|x|^{-\alpha_-(\lambda)} e^{-D^2 \frac{|x|^2}{4}}\}$$

and the class of supersolutions found in Section 7.5.2 given by

$$w(r,t,T) = Ac^{-\gamma}(T+t)^{\frac{\gamma}{2} - \frac{2-p}{2(p-1)}} r^{-\gamma} e^{-\frac{r^2}{4(T+t)}}.$$

In order to have $u_0(x) \le w(x,0)$ for some supersolution, we must choose D verifying

$$\left(\frac{2(\gamma - \alpha_-(\lambda))}{D^2 - a^2}\right)^{\frac{\gamma - \alpha_-(\lambda)}{2}} e^{-\frac{\gamma - \alpha_-(\lambda)}{2}} \le AT^{-\frac{p-2}{2(p-1)}} a^{-\gamma}.$$

Furthermore, since $2(\alpha_-(\lambda) + 1) < N$ and

$$p\,(\alpha_-(\lambda) + 1) < q_+(\lambda)\,(\alpha_-(\lambda) + 1) = \alpha_-(\lambda) + 2 < N,$$

for $\gamma > \alpha_-(\lambda)$, with γ near $\alpha_-(\lambda)$, we have $2(\gamma + 1) < N$ and $(\gamma + 1)p < N$, so $w \in L^2((0,T), W^{1,2}(\mathbb{R}^N))$ and $|\nabla w|^p$ belongs to $L^1((0,T) \times \mathbb{R}^N)$.

Note that γ can be chosen as close to $\alpha_-(\lambda)$ as we want. This fact will be used to reach an extra regularity property on w. Therefore using Theorem 7.5.3 there exists a minimal solution $u \le w$ defined in $\mathbb{R}^N \times (0,T)$.

Since $w \in L^2((0,T); W^{1,2}(\mathbb{R}^N))$ with $w^{\frac{2}{2-p}} \in L^\infty((0,T); L^1(\mathbb{R}^N))$, we claim that the solution u, obtained in Theorem 7.5.3, is an energy solution to (7.5.2), i. e., $u \in L^2((0,T); W^{1,2}(\mathbb{R}^N))$ for all $T > 0$. To prove the claim we consider v_n, the solution of (7.5.11). Following the computations of Theorem 7.5.3 and setting $s = 1$, it follows that

$$\begin{aligned}
\frac{1}{2}\int_{B_n} v_n^2(x,T)\,dx + 2\int_0^T\int_{B_n} |\nabla v_n|^2\,dx\,dt &\le \int_0^T\int_{B_n} v_n|\nabla v_n|^p\,dx\,dt + \frac{1}{2}\int_{B_n} u_0^2(x)\,dx \\
&\le \varepsilon \int_0^T\int_{B_n} |\nabla v_n|^2\,dx\,dt + C(\varepsilon)\int_0^T\int_{B_n} v_n^{\frac{2}{2-p}}\,dx\,dt + C(u_0) \\
&\le \varepsilon \int_0^T\int_{B_n} |\nabla v_n|^2\,dx\,dt + C(\varepsilon)\int_0^T\int_{B_n} w^{\frac{2}{2-p}}\,dx\,dt + C(u_0).
\end{aligned}$$

Hence we conclude that

$$\frac{1}{2}\int_{B_n} v_n^2(x,T)\,dx + (s-\varepsilon)\int_0^T\int_{B_n} |\nabla v_n|^2\,dx\,dt \le C(\varepsilon)\int_0^T\int_{B_n} w^{\frac{2}{2-p}}\,dx\,dt + C(s,u_0).$$

Since w has an exponential decay at infinity, we have just to consider the behavior of w near 0. But as $\frac{2\gamma}{2-p} < N$, we have

$$\frac{1}{2}\int_{B_n} v_n^2(x,T)\,dx + (1-\varepsilon)\int_0^T\int_{B_n} |\nabla v_n|^2\,dx\,dt \le C_1 + C_2\int_{B_1} |x|^{\frac{-2\gamma}{2-p}}\,dx\,dt < C.$$

Therefore $\{v_n\}$ is bounded in $L^2((0,T);W^{1,2}(\mathbb{R}^N))$ and then

$$v_n \rightharpoonup u \quad \text{weakly in } L^2((0,T);W^{1,2}(\mathbb{R}^N)).$$

A similar computation allows us to prove that $\{v_{nt}\}$ is bounded in $L^2((0,T);W^{-1,2}(\mathbb{R}^N))$.

We show now that $v_n \to u$ strongly in $L^2((0,T);W^{1,2}(\mathbb{R}^N))$. Let ϕ be a positive test function such that $0 \le \phi \le 1$, $\phi = 1$ in $B_R(0)$ and $\phi = 0$ in $\mathbb{R}^N \backslash B_{2R}(0)$, with R very large. For $n >> R$, using $(v_n - u)\phi$ as a test function in (7.5.11), it follows that

$$\begin{aligned}\int_0^T\int_{\mathbb{R}^N} v_{nt}(v_n-u)\phi\,dx\,dt &+ \int_0^T\int_{\mathbb{R}^N} \phi\nabla v_n\nabla(v_n-u)\,dx\,dt + \int_0^T\int_{\mathbb{R}^N}(v_n-u)\nabla\phi\nabla v_n\,dx\,dt\\ &= \lambda\int_0^T\int_{\mathbb{R}^N}\frac{1}{|x|^2+\frac{1}{n}}\tilde{v}_{n-1}(v_n-u)\phi\,dx\,dt + \int_0^T\int_{\mathbb{R}^N}\frac{|\nabla v_n|^p}{1+\frac{1}{n}|\nabla v_n|^p}(v_n-u)\phi\,dx\,dt. \qquad (7.5.13)\end{aligned}$$

We claim that

$$\lambda\int_0^T\int_{\mathbb{R}^N}\left|\frac{1}{|x|^2+\frac{1}{n}}\tilde{v}_{n-1}(v_n-u)\phi\right|dx\,dt + \int_0^T\int_{\mathbb{R}^N}\left|\frac{|\nabla v_n|^p}{1+\frac{1}{n}|\nabla v_n|^p}(v_n-u)\phi\right|dx\,dt \to 0 \quad \text{as } n\to\infty,$$

independently of the choice of ϕ. To prove the claim we use the fact that v_n and u are dominated by w.

The first term is dominated by $\lambda\frac{w^2}{|x|^2}$, and hence it is sufficient to apply the dominated convergence theorem.

To estimate the second term, we follow the general arguments used above. Indeed, using Hölder's inequality, we have

$$\begin{aligned}\int_0^T\int_{\mathbb{R}^N}\left|\frac{|\nabla v_n|^p}{1+\frac{1}{n}|\nabla v_n|^p}(v_n-u)\phi\right|dx\,dt &\le \int_0^T\int_{\mathbb{R}^N}|\nabla v_n|^p|v_n-u|\,dx\,dt\\ &\le \left(\int_0^T\int_{\mathbb{R}^N}|\nabla v_n|^2\,dx\,dt\right)^{\frac{p}{2}}\left(\int_0^T\int_{\mathbb{R}^N}|v_n-u|^{\frac{2}{2-p}}\,dx\,dt\right)^{\frac{2-p}{2}}.\end{aligned}$$

Thanks to the choice of γ and the condition on p, from the dominated convergence theorem we get $\int_0^T\int_{\mathbb{R}^N}|v_n-u|^{\frac{2}{2-p}}\,dx\,dt \to 0$ as $n\to\infty$. Since $\{v_n\}$ is bounded in $L^2((0,T);W^{1,2}(\mathbb{R}^N))$, the claim follows.

We deal now with the fist term in (7.5.13). It is clear that

$$\int_0^T\int_{\mathbb{R}^N}(v_n-u)\nabla\phi\nabla v_n\,dx\,dt\to 0\quad\text{as } n\to\infty.$$

Hence,

$$\begin{aligned}
&\int_0^T\int_{\mathbb{R}^N} v_{nt}(v_n-u)\phi\,dx\,dt+\int_0^T\int_{\mathbb{R}^N}\phi\nabla v_n\nabla(v_n-u)\,dx\,dt\\
&\quad=\int_0^T\int_{\mathbb{R}^N}(v_n-u)_t(v_n-u)\phi\,dx\,dt+\int_0^T\int_{\mathbb{R}^N}u_t(v_n-u)\phi\,dx\,dt\\
&\qquad+\int_0^T\int_{\mathbb{R}^N}\phi|\nabla(v_n-u)|^2\,dx\,dt+\int_0^T\int_{\mathbb{R}^N}\phi\nabla u\nabla(v_n-u)\,dx\,dt\\
&\quad=\frac{1}{2}\int_{\mathbb{R}^N}(v_n-u)^2\phi\,dx+\int_0^T\int_{\mathbb{R}^N}\phi|\nabla(v_n-u)|^2\,dx\,dt\\
&\qquad+\int_0^T\int_{\mathbb{R}^N}u_t(v_n-u)\phi\,dx\,dt+\int_0^T\int_{\mathbb{R}^N}\phi\nabla u\nabla(v_n-u)\,dx\,dt.
\end{aligned}$$

Using the weak convergence of $\{v_n\}$ and a duality argument, it is not difficult to see that

$$\int_0^T\int_{\mathbb{R}^N}u_t(v_n-u)\phi\,dx\,dt+\int_0^T\int_{\mathbb{R}^N}\phi\nabla u\nabla(v_n-u)\,dx\,dt=o(1),$$

independently of the choice of ϕ. Hence by letting $R\to\infty$ we have

$$\frac{1}{2}\int_{\mathbb{R}^N}(v_n-u)^2\,dx+\int_0^T\int_{\mathbb{R}^N}|\nabla(v_n-u)|^2\,dx\,dt\to 0\quad\text{as } n\to\infty.$$

Thus the strong convergence follows. As a consequence, we get $u\in L^2((0,T);W^{1,2}(\mathbb{R}^N))$ and $u\le w$. Moreover, $u(x,t)\to 0$ as $t\to\infty$ for all $x\in\mathbb{R}^N\setminus\{0\}$.

Hence we have proved the following theorem.

Theorem 7.5.5. *Assume that $F(\lambda)<p<q_+(\lambda)$. Then for all $u_0\in\mathcal{F}_D$, the Cauchy problem (7.5.2) has a positive global energy solution $u\in L^2(0,T;W^{1,2}(\mathbb{R}^N))$.*

Remark 7.5.6.

1. The local existence result proved in Proposition 7.5.4 holds for all $p<q_+(\lambda)$ at least for a particular class of initial data.

2. The required conditions on the initial data u_0 in Proposition 7.5.4 and in Theorem 7.5.5 are far from being optimal.

7.5.5 Blow-up in a finite time for $p < F(\lambda)$ and any positive initial datum

The L^∞-blow-up is instantaneous in problem (7.5.2) because the solutions are unbounded. This means that the L^∞-blow-up is trivial in this case. Due to the behavior of self-similar solutions to (7.5.3) and the fact that any positive solution to (7.5.2) is a supersolution to (7.5.3), it is natural to make the following definition.

Definition 7.5.7. Consider $u(x,t)$, a positive solution to (7.5.2). Then we say that u blows up in a finite time if there exists $T^* < \infty$ such that

$$\lim_{t\to T^*} \int_{B_r(0)} |x|^{-\alpha_-(\lambda)} u(x,t)\,dx = \infty,$$

for any ball $B_r(0)$.

Theorem 7.5.8. *Assume that $p < F(\lambda) = 1 + \frac{1}{N-\alpha_-(\lambda)+1}$. Then any positive solution u to problem (7.5.2) blows up in a finite time in the sense of Definition 7.5.7.*

Proof. Consider $w(x,t,T)$, a subsolution of the family defined in (7.5.7), and denote by $\overline{u}(x,t) = u(x,t+T)$ the time translation of u, which is a supersolution of the linear equation (5.2.4) with the same initial datum. Therefore to see that $\overline{u}(x,0) \geq w(|x|,0,T)$ it is sufficient to prove that v defined by (5.2.7), for some $T > 0$, satisfies $v(x,T) \geq w(|x|,0,T)$. This last inequality follows by the hypothesis $p < F(\lambda) = 1 + \frac{1}{N-\alpha_-(\lambda)+1}$.

With these premises, we will prove that $\overline{u}(x,t) \geq w(x,t,T)$, $\forall t < T$.

Denote $k(x,t) = w(x,t,T) - \overline{u}(x,t)$. It follows immediately that

$$k_t - \Delta k \leq \lambda \frac{k}{|x|^2} + |\nabla w|^p - |\nabla \overline{u}|^p \quad \text{in } \mathbb{R}^N,\ t \in (0,T_1),\ \text{for every } T_1 < T.$$

Since $p < F(\lambda)$, the right hand side belongs to $L^1_{\mathrm{loc}}(\mathbb{R}^N \times (0,T))$. Applying the parabolic Kato inequality, we have

$$\begin{aligned} &(k_+)_t - \Delta k_+ \leq \lambda \frac{k_+}{|x|^2} + pG|\nabla k_+| \quad \text{in } \mathbb{R}^N,\ t \in (0,T_1),\ T_1 < T, \\ &k_+(x,0) = 0, \end{aligned} \tag{7.5.14}$$

where $G = |\nabla w|^{p-1}$. Using $\varphi_\epsilon = \frac{k_+}{1+\epsilon k_+}$ as a test function in (7.5.14), one can check that $k_+ \in L^2((0,T_1); W^{1,2}(\mathbb{R}^N))$.

Fix $\gamma > 0$ such that $(\gamma+1)\lambda < \Lambda_{N,2}$. Then we will prove that

$$\int_0^{T_1} \int_{\mathbb{R}^N} k_+^{\gamma-1} |\nabla k_+|^2\,dx\,dt < \infty. \tag{7.5.15}$$

To see this, consider $\varepsilon > 0$ and $m_\varepsilon(s) \equiv (\frac{s}{1+\varepsilon s})^\gamma$. Define $M_\varepsilon(s) = \int_0^s m_\varepsilon(y)dy$. Using $m_\varepsilon(k_+)$ as a test function in (7.5.14), we have

$$\int_{\mathbb{R}^N} M_\varepsilon(k_+(x,T_1))\,dx + \gamma \int_0^{T_1}\int_{\mathbb{R}^N} \left(\frac{k_+}{1+\varepsilon k_+}\right)^{\gamma-1} \frac{|\nabla k_+|^2}{(1+\varepsilon k_+)^2}\,dx\,dt$$
$$\leq \int_0^{T_1}\int_{\mathbb{R}^N} \left(\frac{k_+}{1+\varepsilon k_+}\right)^{\gamma} G|\nabla k_+|\,dx\,dt.$$

Using now Young's inequality

$$\int_0^{T_1}\int_{\mathbb{R}^N} \left(\frac{k_+}{1+\varepsilon k_+}\right)^{\gamma} G|\nabla k_+|\,dx\,dt$$
$$\leq \epsilon \int_0^{T_1}\int_{\mathbb{R}^N} \left(\frac{k_+}{1+\varepsilon k_+}\right)^{\gamma-1} |\nabla k_+|^2\,dx\,dt + C(\epsilon)\int_0^{T_1}\int_{\mathbb{R}^N} |G|^2(x,t)k_+^{\gamma+1}\,dx\,dt.$$

Since $k_+ \leq w$ and $2(p-1)(\alpha_-(\lambda)+1) + \alpha_-(\lambda)(\gamma+1) < N$, we get

$$\int_0^{T_1}\int_{\mathbb{R}^N} |G|^2(x,t)k_+^{\gamma+1}\,dx\,dt < C.$$

Hence, by letting $\varepsilon \to 0$ and using Fatou's lemma, we conclude that

$$\int_{\mathbb{R}^N} k_+^{\gamma+1}\,dx + \gamma\int_0^{T_1}\int_{\mathbb{R}^N} k_+{}^{\gamma-1}|\nabla k_+|^2\,dx\,dt < \infty,$$

and then (7.5.15) follows.

Note that since $p < F(\lambda)$, there exists $q > 2$ such that

$$\sup_{t\in[0,T_1]} \int_{\mathbb{R}^N} |G(x,t)|^{\frac{qN}{2}}\,dx = C(T_1) < \infty.$$

For $a > 0$ sufficiently small and γ as above, i. e., $\lambda(\gamma+1) < \Lambda_{N,2}$, we consider ψ, the solution to the problem

$$\begin{cases} -\psi_t - \Delta\psi - \lambda(\gamma+1)\frac{\psi}{|x|^2} = a|G(x,t)|^q\psi & \text{in } \mathbb{R}^N,\ t\in(0,T_1),\\ \psi(x,T_1) = \phi(x) & \text{in } \mathbb{R}^N, \end{cases} \tag{7.5.16}$$

where $0 < \phi \in L^1(\mathbb{R}^N)\cap L^\infty(\mathbb{R}^N)$. We refer to Lemma 5.2.2 for the proof of the existence of such a ψ. Using $\psi(\frac{k_+}{1+\varepsilon k_+})^\gamma$ as a test function in (7.5.14), passing to the limit as $\varepsilon \to 0$

and by the summability result in the claim, it follows that

$$\begin{aligned}\frac{1}{\gamma+1}\int_{\mathbb{R}^N} k_+^{\gamma+1}(x,T_1)\psi(x,T_1)\,dx + \gamma\int_0^{T_1}\int_{\mathbb{R}^N} k_+^{\gamma-1}|\nabla k_+|^2\psi\,dx\,ds \\ + \frac{1}{\gamma+1}\int_0^{T_1} k_+^{\gamma+1}(-\psi_t - \Delta\psi)\,dx\,ds \le \int_0^{T_1}\int_{\mathbb{R}^N} |G||\nabla k_+|k_+^{\gamma}\psi\,dx\,ds \\ = \int_0^{T_1}\int_{\mathbb{R}^N} |G||\nabla k_+|v_+^{\theta_1+\theta_2+\theta_3}\psi\,dx\,ds,\end{aligned}$$

with $\theta_1 = \frac{\gamma-1}{2}$, $\theta_2 = \frac{\gamma+1}{q}$ and $\theta_3 = \frac{(q-2)(\gamma+1)}{2q}$ (observe that $\theta_1 + \theta_2 + \theta_3 = \gamma$).

Hence by the definition of ψ and the Young inequality, we have

$$\begin{aligned}\frac{1}{\gamma+1}\int_{\mathbb{R}^N} k_+^{\gamma+1}(x,T_1)\psi(x,T_1)\,dx + \gamma\int_0^{T_1}\int_{\mathbb{R}^N} k_+^{\gamma-1}|\nabla k_+|^2\psi\,dx\,ds \\ + \frac{a}{\gamma+1}\int_0^{T_1}\int_{\mathbb{R}^N} k_+^{\gamma+1}|G|^q\psi\,dx\,ds \le \eta_1\int_0^{T_1}\int_{\mathbb{R}^N} k_+^{\gamma-1}|\nabla k_+|^2\psi\,dx\,ds \\ + \eta_2\int_0^{T_1}\int_{\mathbb{R}^N} k_+^{\gamma+1}|G|^q\psi\,dx\,ds + \eta_3\int_0^{T_1}\int_{\mathbb{R}^N} k_+^{\gamma+1}(x,s)\psi(x,s)\,dx\,ds.\end{aligned}$$

Choosing η_1 and η_2 small, we get

$$\frac{1}{\gamma+1}\int_{\Omega} k_+^{\gamma+1}(x,T_1)\psi(x,T_1)\,dx \le \eta_3\int_0^{T_1}\int_{\Omega} k_+^{\gamma+1}(x,s)\psi(x,s)\,dx\,ds.$$

Since $k_+ \ge 0$ and $\psi > 0$ in $\mathbb{R}^N \times (0,T_1)$ for $t < T_1$, by Gronwall's inequality we obtain $k_+ \equiv 0$ and therefore $w(x,t,T) \le u(x,T+t)$.

Since $T_1 < T$ is arbitrary and $w(x,t,T)$ blows up as $t \to T$, we conclude that for $p < F(\lambda) = 1 + \frac{1}{N-\alpha_1+1}$, any positive solution to (7.5.2) blows up in a finite time. □

7.6 Further remarks

- The arguments used above cannot be applied to deal with the critical exponent $p = F(\lambda) = 1 + \frac{1}{N-\alpha_-(\lambda)+1}$. To the best of our knowledge, the analysis of the problem in the case $p = F(\lambda)$ remains open.

- A good example for training could be the problems related to Caffarelli–Kohn–Nirenberg inequalities

$$\begin{cases} u_t - \operatorname{div}(|x|^{-2\gamma}\nabla u) = \lambda \frac{u^\alpha}{|x|^{2(\gamma+1)}} + \beta|x|^{-\mu}|\nabla u|^q & \text{in } \Omega \times (0,T), \\ u \geq 0 \quad \text{in } \Omega \times (0,T) \text{ and } u = 0 \text{ on } \partial\Omega \times (0,T), \\ u(x,0) = u_0(x) \quad \text{in } \Omega, \end{cases} \tag{7.6.1}$$

where $\Omega \subset \mathbb{R}^N$ $(N \geq 2)$ is a bounded regular domain such that $0 \in \Omega$, $\lambda > 0$, $\alpha \geq 0$, $-\infty < \gamma < \frac{N-2}{2}$, u_0 is a positive function such that $u_0 \in L^m(\Omega)$ for some $m \geq 1$, $\mu \in \mathbb{R}$, $q > 1$ and $\beta = 0$ or $\beta = 1$. Some particular results for the problem when $\beta = 0$ can be seen in [4] and [130]. Note that in [130] problems more general than (7.6.1) are considered. In general, problems with a term in the gradient seem to be interesting and more difficult to tackle than the case $p = 2$.

8 Fractional Laplacian type operators

8.1 Introduction

Let us start this chapter with the observation that if we consider the Laplace operator $-\Delta$ and $u, v \in C_0^\infty(\mathbb{R}^N)$, then an integration by parts (Green's formula!) gives

$$\langle -\Delta u, v\rangle = \int_{\mathbb{R}^N} (-\Delta u) v\, dx = \int_{\mathbb{R}^N} \nabla u \cdot \nabla v\, dx.$$

In particular, for $v = u$,

$$\langle -\Delta u, u\rangle = \int_{\mathbb{R}^N} |\nabla u|^2\, dx > 0, \quad \text{if } u \neq 0,$$

that is, $-\Delta$ is a positive operator. Thus, using the spectral theory, we can define fractional powers of such operator. The fractional powers of the Laplacian have been considered from many points of view, including potential theory, Levy processes, semigroup theory and pseudo-differential operators. A reference to find equivalent definitions of the fractional Laplacian is the paper [219]. We will study the potential theory approach and its connection with the Fourier transform. A first appearance of the fractional Laplacian from this point of view is the seminal paper by M. Riesz [272].

To describe correctly this precise framework, let us recall the following basic results of *distribution theory*. We will denote the nonnegative integers by $\mathbb{Z}_+$. Consider $\mathcal{S}(\mathbb{R}^N)$, the L. Schwartz class, that is,

$$\mathcal{S}(\mathbb{R}^N) = \Big\{f \in C^\infty(\mathbb{R}^N) : \forall \alpha, \beta \in \mathbb{Z}_+{}^N,\ \exists C_{\alpha,\beta} > 0,\ \sup_{x\in\mathbb{R}^N} |x^\beta D^\alpha f(x)| \le C_{\alpha,\beta}\Big\}. \tag{8.1.1}$$

Definition 8.1.1 (Fourier transform). For $f \in \mathcal{S}(\mathbb{R}^N)$, its Fourier transform is defined as

$$\mathcal{F}(f)(\xi) = \int_{\mathbb{R}^N} f(x) e^{-2\pi i\langle x,\xi\rangle}\, dx. \tag{8.1.2}$$

The inverse Fourier transform is given by

$$\mathcal{F}^{-1}(f)(\xi) = \int_{\mathbb{R}^N} f(x) e^{2\pi i\langle x,\xi\rangle}\, dx,$$

so that we have the inversion formula

$$\mathcal{F}^{-1}(\mathcal{F}(f)) = f,$$

and the Parseval–Plancherel identity holds:

$$\int_{\mathbb{R}^N} f(x)\overline{h(x)}\, dx = \int_{\mathbb{R}^N} \mathcal{F}(f)(\xi)\overline{\mathcal{F}(h)(\xi)}\, d\xi, \quad \int_{\mathbb{R}^N} |f(x)|^2\, dx = \int_{\mathbb{R}^N} |\mathcal{F}(f)(\xi)|^2\, d\xi. \tag{8.1.3}$$

https://doi.org/10.1515/9783110606270-008

The reader can find the classical theory of Fourier transform in any textbook on real or harmonic analysis; see for instance [189], [144], [231], [270] and [296].

Let $\mathcal{S}'(\mathbb{R}^N)$ be the dual space of $\mathcal{S}(\mathbb{R}^N)$. Given $\lambda > 0$, let us denote by δ^λ the dilation operator that for a function f is defined as $\delta^\lambda f(x) = f(\lambda x)$. The operator δ^λ has a natural extension to $u \in \mathcal{S}'(\mathbb{R}^N)$ by means of

$$\langle \delta^\lambda u, f\rangle = \langle u, \lambda^{-N}\delta^{\lambda^{-1}} f\rangle, \quad \text{for all } f \in \mathcal{S}(\mathbb{R}^N).$$

Also, let ρ be an element of the rotation group in $\mathbb{R}^N$ that we identify with the corresponding operator $\rho f(x) = f(\rho x)$. This defines an operator acting on each $u \in \mathcal{S}'(\mathbb{R}^N)$ by the pairing

$$\langle \rho u, f\rangle = \langle u, \rho^{-1} f\rangle, \quad \text{for all } f \in \mathcal{S}(\mathbb{R}^N).$$

We say that $u \in \mathcal{S}'(\mathbb{R}^N)$ is a *homogeneous distribution* of degree α if

$$\langle u, \delta^\lambda f\rangle = \lambda^{-N-\alpha}\langle u, f\rangle, \quad \text{for all } f \in \mathcal{S}(\mathbb{R}^N).$$

We say that $u \in \mathcal{S}'(\mathbb{R}^N)$ is rotation invariant if $\rho u = u$ for all rotation ρ in $\mathbb{R}^N$. The following result is easy to prove (see [293]).

Lemma 8.1.2. *Let $u \in \mathcal{S}'(\mathbb{R}^N)$ be a homogeneous distribution of degree α. Then its Fourier transform is a tempered distribution of degree $-N-\alpha$.*

If $u \in \mathcal{S}'(\mathbb{R}^N)$ is a rotation-invariant, tempered distribution, its Fourier transform is also rotation-invariant.

In the case of radial functions we have explicitly the following formula: let $f \in L^1(\mathbb{R}^N)$ be so that $f(x) = f_0(|x|)$ for some 1-variable function f_0 defined on $(0,\infty)$. Then, $\mathcal{F}(f)$ is radial and

$$\mathcal{F}(f)(\xi) = \int_0^\infty f_0(r)\widehat{d\sigma}(r\xi)r^{N-1}\,dr, \tag{8.1.4}$$

where $\widehat{d\sigma}$ denotes the Fourier transform of the surface measure on the unit sphere, that is,

$$\widehat{d\sigma}(\xi) = \int_{S^{N-1}} e^{-2\pi i\langle\omega,\xi\rangle}\,d\sigma(\omega) = \int_{S^{N-1}} \cos(2\pi\langle\omega,\xi\rangle)\,d\sigma(\omega). \tag{8.1.5}$$

As is well known, we have $\widehat{d\sigma}(\xi) = \frac{2\pi}{|\xi|^{\frac{N-2}{2}}} J_{\frac{N-2}{2}}(2\pi|\xi|)$, where J_ν denotes the Bessel function of the first kind and index ν (see [294]). Hence, we have the expression

$$\mathcal{F}(f)(\xi) = 2\pi\int_0^\infty f_0(r)\left(\frac{r}{|\xi|}\right)^{\frac{N-2}{2}} J_{\frac{N-2}{2}}(2\pi r|\xi|)r\,dr. \tag{8.1.6}$$

The details can be seen in Chapter 2 of the book [189] or Chapter 4 of [296].

A spectral definition of the *s*-fractional Laplacian

Note that if $u \in \mathcal{S}(\mathbb{R}^N)$, then $-\Delta u \in \mathcal{S}(\mathbb{R}^N)$, and hence

$$\mathcal{F}(-\Delta u)(\xi) = 4\pi^2|\xi|^2\mathcal{F}(u)(\xi).$$

Therefore, it is natural to define for $0 < s < 1$ the s-fractional Laplacian by

$$\mathcal{F}[(-\Delta)^s u](\xi) := 2^{2s}\pi^{2s}|\xi|^{2s}\mathcal{F}(u)(\xi). \tag{8.1.7}$$

The fractional Laplacian represents one of the simplest examples of pseudo-differential operators, despite its symbol being not regular. From the original contributions by A. P. Calderón and A. Zygmund about singular integral operators we recall the following formula, which gives the relationship between the singular operators and the *square root* of $-\Delta$. See [110], [111] and [112].

Let R_j, $j = 1, 2, \ldots, N$, be the Riesz transforms, i. e., the singular integral operator associated to the kernel $K_j(x) = \frac{1}{\gamma}\frac{x_j}{|x|^{N+1}}$, where $\gamma = \frac{\pi^{N+1}}{\Gamma(\frac{N+1}{2})}$. The corresponding Fourier multiplier is then

$$m_k(\xi) = -i\frac{\xi_k}{|\xi|}.$$

By using the Fourier transform we find the following Calderón–Zygmund formula:

$$(-\Delta)^{\frac{1}{2}} = \sum_{k=1}^{N} R_k\partial_k.$$

See [112].

A first consequence of the formula above is that $(-\Delta)^{\frac{1}{2}}$ is a *nonlocal operator*.

In the last years an increasing interest to study the *fractional Laplacian* has appeared. This has been basically motivated by the applications to different areas of mathematics, physics and economy. The fractional Laplacian is a paradigmatic example of Levy processes. Regarding this approach, there is a very extensive literature. We refer for instance to [38], [43], [207] and to the nice survey by E. Valdinoci [309] for more details on this topic.

Nonlocal operators related to the fractional Laplacian naturally appear in elasticity problems [285], water waves [125], [126], [297], crystal dislocation [305], thin obstacle problems [103], phase transition [33], [101], [290], flame propagation [104], stratified materials [278], water waves [125], [126], [297] and quasigeostrophic flows [109], [235], among other phenomena.

Remark 8.1.3. The general theory of pseudo-differential operators was developed in the 1970s by Calderón, Kohn, Nirenberg, Treves, Hörmander, Fefferman, Beals and Kumano-Go, among a large list of researchers. We refer for instance to the books [201], [217], [307] and the articles [252], [253], [254], [52] [50], [51], [113] and [160]. The theory was later extended to include symbols with less regular coefficients by Bony, Meyer, Sjöstrand and others.

Looking at the identity (8.1.7) and in order to study the behavior of $(-\Delta)^s$, $0 < s < 1$, let us describe the integral expression of the operator, that is, the *kernel* corresponding to the multiplier $(2\pi|\xi|)^{2s}$. Since $|\xi|^{2s}$ is a homogeneous tempered distribution of degree $2s$, the s-fractional Laplacian should be represented by the convolution (in the principal value sense) with a kernel $-(N+2s)$-homogeneous of the form

$$K_{N,s}(x) = \frac{C_{N,s}}{|x|^{N+2s}},$$

which unfortunately is *not a locally integrable function* in $\mathbb{R}^N$. Therefore we must justify the meaning of the operator $(-\Delta)^s$ acting on functions in $\mathcal{S}(\mathbb{R}^N)$.

We can find an answer to the above problem in the paper by M. Riesz [272]. We will follow Chapter V in the book by E. M. Stein, [293] and Chapter I of the book by N. Landkof [223], restricting our attention just to the case in which $0 < s < 1$.

In the next subsection we will explain some classical results about the Riesz potential. They will be the key to formally describe the integral representation of the fractional Laplacian. In the rest of this chapter we will study some *integro-differential operators, with measurable, bounded coefficients,* that are nonlocal and have a behavior similar to $(-\Delta)^s$. The reader can see the basic results that will be used of Potential Theory, for instance, in [197] and [236].

The study of problems in the framework of integro-differential equations is quite recent and has raised a great interest particularly in connection with problems involving nonlocal effects. We will analyze the basic theory of these operators related to existence, uniqueness and regularity for the solution to the Dirichlet problem. All this will be formulated in the next section. In this part of the chapter we will follow the exposition in [224].

8.1.1 Riesz potentials

Recall the definition of the Gamma function

$$\Gamma(z) = \int_0^\infty e^{-s} s^{z-1}\, ds, \quad \Re(z) > 0.$$

Using an iterative argument, the measure of the $(N-1)$-Hausdorff measure of the unit sphere is given by

$$\omega_{N-1} = \frac{2\pi^{\frac{N}{2}}}{\Gamma(\frac{N}{2})}.$$

See the details, for instance, on page 74 of [306]. Note that $\Gamma(z+1) = z\Gamma(z)$ if z is neither 0 nor a negative integer.

Consider the Riesz potentials in $\mathbb{R}^N$,

$$k_{N,\alpha}(x) = \frac{c_{N,\alpha}}{|x|^{N-\alpha}}, \quad 0 < \alpha < N.$$

We have $k_{N,\alpha} \in L^1_{\text{loc}}(\mathbb{R}^N)$ and, therefore, it defines an absolutely continuous measure for which, as we will see next,

$$\mathcal{F}^{-1}(k_{N,\alpha})(\xi) = c_{N,\alpha}\pi^{\frac{N}{2}-\alpha}\frac{\Gamma(\frac{\alpha}{2})}{\Gamma(\frac{N-\alpha}{2})}|\xi|^{-\alpha}.$$

The constant $c_{N,\alpha}$ will be picked up in order to normalize the previous inverse Fourier transform. To do that, we compute the inverse Fourier transform of the tempered distribution $f(x) = \frac{1}{|x|^{N-\alpha}}$; by using polar coordinates, the inverse Fourier transform of f is given by

$$\begin{aligned}\mathcal{F}^{-1}(f)(\xi) &= \int_{\mathbb{R}^N} f(y)e^{2\pi i\langle\xi,y\rangle}\,dy = \frac{2\pi}{|\xi|^{\frac{N}{2}-1}}\int_0^\infty r^{\alpha-\frac{N}{2}}J_{\frac{N}{2}-1}(2\pi|\xi|r)\,dr \\ &= (2\pi)^{\frac{N}{2}-\alpha}\frac{1}{|\xi|^\alpha}\int_0^\infty t^{\alpha-\frac{N}{2}}J_{\frac{N}{2}-1}(t)dt = \pi^{\frac{N}{2}-\alpha}\frac{\Gamma(\frac{\alpha}{2})}{\Gamma(\frac{N-\alpha}{2})}\frac{1}{|\xi|^\alpha}\end{aligned}$$

(see [293]). Hence by taking

$$c_{N,\alpha} = \pi^{\alpha-\frac{N}{2}}\frac{\Gamma(\frac{N-\alpha}{2})}{\Gamma(\frac{\alpha}{2})} \tag{8.1.8}$$

we obtain

$$\mathcal{F}^{-1}(k_{N,\alpha})(\xi) = \frac{1}{|\xi|^\alpha}.$$

Summarizing, calling

$$g_\alpha(x) = \frac{1}{|x|^\alpha}, \quad \text{we have } \mathcal{F}(g_\alpha)(x) = k_{N,\alpha}(x) = \frac{c_{N,\alpha}}{|x|^{N-\alpha}}.$$

Observe also that $c_{N,\alpha} \cdot c_{N,N-\alpha} = 1$. (8.1.9)

Definition 8.1.4. For $0 < \alpha < N$ we define the Riesz operator of order α, I_α, by means of the Fourier multiplier formula

$$\mathcal{F}(I_\alpha(f))(\xi) = \frac{1}{|\xi|^\alpha}\mathcal{F}(f)(\xi) \equiv g_\alpha(\xi)\mathcal{F}(f)(\xi).$$

By the previous calculations we have the convolution integral defining I_α,

$$I_\alpha f(x) = \pi^{\alpha-\frac{N}{2}}\frac{\Gamma(\frac{N-\alpha}{2})}{\Gamma(\frac{\alpha}{2})}\int_{\mathbb{R}^N}\frac{f(y)}{|x-y|^{N-\alpha}}\,dy.$$

A property of the Riesz potentials that we must have in mind is *the semigroup property*, that is,

$$I_\alpha(I_\beta(f)) = I_{\alpha+\beta}(f), \quad \text{for all } f \in \mathcal{S}(\mathbb{R}^N),\ 0 < \alpha, \beta,\ \alpha+\beta < N.$$

Remark 8.1.5. Note that the previous computations are done in the *distribution sense*, that is, the integrability at infinity is provided by the test function and a duality argument. In particular, a solution $S \in \mathcal{S}'(\mathbb{R}^N)$ to the problem

$$(-\Delta)^s S(x) = \delta_0, \quad x \in \mathbb{R}^N, \quad \delta_0 \text{ being the Dirac measure},$$

is given by

$$|2\pi\xi|^{2s}\mathcal{F}(S)(\xi) = 1.$$

According to the previous calculation

$$S(x) = (2\pi)^{-2s} c_{N,2s} \frac{1}{|x|^{N-2s}}, \quad \text{where } c_{N,2s} \text{ is given by (8.1.8).} \tag{8.1.10}$$

In this context, S represents the *fundamental solution* of $(-\Delta)^s$ and coincides with the kernel of the Riesz potential I_{2s}.

In particular, if $f \in \mathcal{S}(\mathbb{R}^N)$ we find that the solution to $(-\Delta)^s u = f$, modulo s-harmonic functions, is given by

$$u(x) = (-\Delta)^{-s} f(x) = f * S(x) = (2\pi)^{-2s} c_{N,2s} \int_{\mathbb{R}^N} \frac{f(y)}{|x-y|^{N-2s}}\, dy.$$

8.1.2 Analytic continuation and the fractional Laplacian formula

We focus now our attention on finding the expression of the fractional Laplacian in terms of a kernel. It is clear that the calculus developed for the Riesz potential does not work *mutatis mutandis* for the multiplier $|\xi|^{2s}$ because the candidate to be the kernel is not locally integrable. The way to solve this problems follows the suggestions in the M. Riesz paper [272], as done in [223].

Since $\Gamma(\frac{N-\alpha}{2})$ has poles in $\alpha = N + 2m$, $m = 0, 1, 2 \ldots$, $c_{N,\alpha}$, as a function in $\alpha \in \mathbb{C}$, is regular unless $\alpha = N + 2m$, $m = 0, 1, 2 \ldots$. This remark will be one of the key ingredients to define the s-fractional Laplacian. Set $\tilde{c}_{N,\alpha} = (2\pi)^{-\alpha} c_{N,\alpha}$. Given a function $u \in \mathcal{S}(\mathbb{R}^N)$, define for $x \in \mathbb{R}^N$ fixed

$$v(x, \alpha) = \tilde{c}_{N,\alpha} \int_{\mathbb{R}^N} \frac{u(x+y)}{|y|^{N-\alpha}}\, dy.$$

We see that $v(x, \alpha)$ is an analytic function in $\Re\alpha > 0$, with singular points at $\alpha = N + 2m$, $m = 0, 1, 2 \ldots$. Following the idea in the paper by M. Riesz, [272], the way to define the kernel expression of $(-\Delta)^s$, $0 < s < 1$, is to make an analytic extension of $v(x, \alpha)$ to $-2 < \Re\alpha < 0$ and take such analytic extension as the definition of the fractional operator. To justify this, note that if $\Re\alpha > 0$ and $\alpha \neq N + 2m$, $m = 0, 1, 2 \dots$, we can split the integral defining $v(x, \alpha)$ as follows. Consider B the unit ball in $\mathbb{R}^N$. Then

$$v(x,\alpha) = \widetilde{c}_{N,\alpha} \int_{\mathbb{R}^N \setminus B} \frac{u(x+y)}{|y|^{N-\alpha}}\,dy + \widetilde{c}_{N,\alpha} \int_B \frac{u(x+y)-u(x)}{|y|^{N-\alpha}}\,dy + \widetilde{c}_{N,\alpha} u(x) \int_B \frac{1}{|y|^{N-\alpha}}\,dy. \tag{8.1.11}$$

Since

$$\widetilde{c}_{N,\alpha} u(x) \int_B \frac{1}{|y|^{N-\alpha}}\,dy = \frac{\widetilde{c}_{N,\alpha}}{\alpha} u(x) \omega_{N-1},$$

this term is clearly analytic in $\Re\alpha \leq 0$ and, obviously, the first term in the splitting also is analytic in $\Re\alpha \leq 0$. It remains to study the second term. To this point we will use P. Pizzetti's formula in [260].

Proposition 8.1.6. *Let $u \in \mathcal{S}(\mathbb{R}^N)$ and define*

$$S_u(r) = \frac{1}{\omega_{N-1}} \int_{S^{N-1}} u(r\sigma)\,d\sigma.$$

Then the following expansion holds:

$$S_u(r) = \sum_{k=0}^{\infty} \frac{r^{2k} \Delta^k u(0)}{2^k k! n(n+2) \cdots (n+2k-2)},$$

where Δ is the classical Laplacian.

A detailed proof using the Fourier transform can be found in [228].

Consider the mean value of $u(x + \cdot)$ in $|y| = r$, that is,

$$\bar{u}(x,r) = \frac{1}{\omega_{N-1}} \int_{|y|=r} u(x+y)\,d\omega(y).$$

Then

$$\int_B \frac{u(x+y)-u(x)}{|y|^{N-\alpha}}\,dy = \omega_{N-1} \int_0^1 (\bar{u}(x,r) - u(x)) r^{\alpha-1}\,dr, \tag{8.1.12}$$

and by the Pizzetti formula, $\bar{u}(x,r) - u(x) = O(r^2)$. Therefore the integral in (8.1.12) is uniformly bounded.

As a consequence, $v(x,\alpha)$ can be analytically continued to $-2 < \Re\alpha \leq 0$. Note that if $-2 < \Re\alpha \leq 0$,

$$\frac{\widetilde{c}_{N,\alpha}}{\alpha} u(x) \omega_{N-1} = -\widetilde{c}_{N,\alpha} u(x) \int_{\mathbb{R}^N \setminus B} \frac{dy}{|y|^{N-\alpha}},$$

then the splitting (8.1.11) can be written as

$$v(x,\alpha) = \widetilde{c}_{N,\alpha} \int_{\mathbb{R}^N} \frac{u(x+y)-u(x)}{|y|^{N-\alpha}}\,dy. \tag{8.1.13}$$

With these calculations we reach finally some meaning to the operator defined from the kernel corresponding to the multiplier $(2\pi|\xi|)^{2s}$, $0 < s < 1$: if $-\alpha = 2s$, then we have the following definition.

Definition 8.1.7. For $0 < s < 1$ and $u \in \mathcal{S}(\mathbb{R}^N)$ we define the s-fractional Laplacian by

$$(-\Delta)^s u(x) =: C_{N,s} P.V. \int_{\mathbb{R}^N} \frac{u(x) - u(y)}{|x - y|^{N+2s}} dy := C_{N,s} \lim_{\epsilon \to 0} \int_{|x-y|>\epsilon} \frac{u(x) - u(y)}{|x - y|^{N+2s}} dy, \quad (8.1.14)$$

where we have set

$$C_{N,s} = -\widetilde{c}_{N,-2s} = 4^s \pi^{-\frac{N}{2}} \frac{\Gamma(\frac{(N+2s)}{2})}{|\Gamma(-s)|}. \quad (8.1.15)$$

In [223] the case $s > 1$ is also considered.

The first remark is that the operator $(-\Delta)^s$, the fractional power of the Laplacian, is nonlocal. For that, remember that *a linear operator is local if and only if it is a differential operator.*

The following result, a Plancherel type identity, justifies that the operator defined in Definition 8.1.7 is really the fractional power of the Laplacian, $0 < s < 1$.

Lemma 8.1.8. *Let $N \geq 1$ and $0 < s < 1$. Then for all $u \in \mathcal{S}(\mathbb{R}^N)$ we have*

$$\int_{\mathbb{R}^N} |\xi|^{2s} |\hat{u}(\xi)|^2 \, d\xi = \frac{1}{2} \widetilde{C}_{N,s} \int_{\mathbb{R}^N} \int_{\mathbb{R}^N} \frac{|u(x) - u(y)|^2}{|x - y|^{N+2s}} \, dx \, dy, \quad (8.1.16)$$

where $\widetilde{C}_{N,s} = (2\pi)^{-2s} C_{N,s} = \pi^{2s-\frac{N}{2}} \frac{\Gamma(\frac{(N+2s)}{2})}{|\Gamma(-s)|}$.

Proof. Fixing y, we call $\overline{y} = x - y$. By the Plancherel identity and taking into account that

$$\widehat{u(\cdot + \overline{y})}(\xi) = e^{2\pi i \langle \xi, \overline{y} \rangle} \hat{u}(\xi),$$

we obtain

$$\int_{\mathbb{R}^N} \int_{\mathbb{R}^N} \frac{|u(x) - u(x + \overline{y})|^2}{|\overline{y}|^{N+2s}} \, dx \, d\overline{y} = \int_{\mathbb{R}^N} \left(\int_{\mathbb{R}^N} |\overline{y}|^{-N-2s} |e^{2\pi i \langle \xi, \overline{y} \rangle} - 1|^2 \, d\overline{y} \right) |\hat{u}(\xi)|^2 \, d\xi.$$

Now, if $\theta = \xi/|\xi|$, using polar coordinates

$$\int_{\mathbb{R}^N} |\overline{y}|^{-N-2s} |e^{2\pi i \langle \xi, \overline{y} \rangle} - 1|^2 \, d\overline{y} = |\xi|^{2s} \int_0^\infty \int_{S_{N-1}} |e^{2\pi i r \langle \omega, \theta \rangle} - 1|^2 \, d\theta r^{-2s-1} \, dr.$$

Observe that, by rotation invariance, the integral defines an expression independent of θ. To compute this constant we proceed as follows:

$$c = \int_0^\infty \int_{S_{N-1}} |e^{2\pi i r\langle\omega,\theta\rangle} - 1|^2\, d\theta r^{-2s-1}\, dr = \int_0^\infty \int_{S_{N-1}} (2 - 2\cos(2\pi r\langle\omega,\theta\rangle))\, d\theta r^{-2s-1}\, dr$$
$$= 2\int_0^\infty (\omega_{N-1} - 2\pi^{\frac{N}{2}}(2\pi r)^{-\frac{(N-2)}{2}} J_{\frac{(N-2)}{2}}(2\pi r)) r^{-2s-1}\, dr,$$

where $J_{\frac{(N-2)}{2}}$ denotes, as before, the Bessel function of the first kind of degree $\frac{(N-2)}{2}$ (see [167] and [27]). Since $\omega_{N-1} = 2\frac{\pi^{\frac{N}{2}}}{\Gamma(\frac{N}{2})}$ the final expression of c follows from the identity

$$\int_0^\infty r^{-z}\left(J_{\frac{(N-2)}{2}}(r) - 2^{-\frac{(N-2)}{2}}\Gamma\left(\frac{N}{2}\right)^{-1} r^{\frac{(N-2)}{2}}\right) dr = 2^{-z}\frac{\Gamma(\frac{(N-2z)}{4})}{\Gamma(\frac{(N+2z)}{4})},$$

for $\frac{N}{2} < \operatorname{Re} z < \frac{(N+4)}{2}$; see [319]. □

8.1.3 Some elementary properties of the fractional Laplacian

Assume $u \in \mathcal{S}(\mathbb{R}^N)$. Since $(-\Delta)^s$ is a Fourier multiplier operator, we find that the fractional Laplacian commutes with derivatives,

$$\partial_i(-\Delta)^s u(x) = \mathcal{F}^{-1}(2\pi i\xi_i|2\pi i\xi|^{2s}\mathcal{F}(u)(\xi))(x) = (-\Delta)^s(\partial_i u)(x).$$

It also satisfies the *semigroup property*:
1. $(-\Delta)^1 = -\Delta$, $(-\Delta)^0 = \mathrm{Id}$;
2. $(-\Delta)^{s_1} \circ (-\Delta)^{s_2} = (-\Delta)^{s_1+s_2}$.

We also have the following Leibniz formula for $s \in (0,1)$

$$(-\Delta)^s(uv) = (-\Delta)^s(u)v + u(-\Delta)^s(v) - B(u,v), \quad \text{where}$$
$$B_{2s}(u,v) = C_{N,s}\, P.V. \int_{\mathbb{R}^N} \frac{(u(x)-u(y))(v(x)-v(y))}{|x-y|^{N+2s}}\, dy, \tag{8.1.17}$$

which is easily obtained from the pointwise identity

$$u(x)v(x) - u(y)v(y) \equiv v(x)(u(x)-u(y)) + u(x)(v(x)-v(y)) - (u(x)-u(y))(v(x)-v(y)).$$

Integration by parts formulas

The next results extend to the fractional case for both Gauss and Green formulas.

Lemma 8.1.9. *Let $u, v \in C^2(\mathbb{R}^N) \cap L^\infty(\mathbb{R}^N)$ with $\Omega \subset \mathbb{R}^N$ being a domain. Then:*

1. *(Gauss formula)*

$$\int_\Omega (-\Delta)^s u(x)\,dx = \int_{\mathbb{R}^N\setminus\Omega} \mathcal{N}_s u(x)\,dx,$$

where

$$\mathcal{N}_s u(x) =: C_{N,s}\int_\Omega \frac{u(x)-u(y)}{|x-y|^{N+2s}}\,dy, \quad \textit{for all } x\in\mathbb{R}^N\setminus\Omega;$$

2. *(Green formula)*

$$\begin{aligned}&\frac{1}{2}C_{N,s}\int\int_{\mathcal{D}_\Omega}\frac{(u(x)-u(y))(v(x)-v(y))}{|x-y|^{N+2s}}\,dx\,dy\\&\quad=\int_\Omega v(y)(-\Delta)^s u(y)\,dy+\int_{\mathbb{R}^N\setminus\Omega} v(x)\mathcal{N}_s u(x)\,dx,\end{aligned}$$

where $\mathcal{D}_\Omega = \mathbb{R}^{2N}\setminus(\mathcal{C}\Omega\times\mathcal{C}\Omega)$.

Proof. Note that by symmetry in x, y we have

$$\int_\Omega\int_\Omega \frac{u(x)-u(y)}{|x-y|^{N+2s}}\,dx\,dy = 0.$$

Therefore:

1.

$$\begin{aligned}\int_\Omega(-\Delta)^s u(x)\,dx &= C_{N,s}\int_\Omega\int_\Omega\frac{u(x)-u(y)}{|x-y|^{N+2s}}\,dx\,dy+\int_\Omega C_{N,s}\int_{\mathbb{R}^N\setminus\Omega}\frac{u(x)-u(y)}{|x-y|^{N+2s}}\,dy\,dx\\&=\int_{\mathbb{R}^N\setminus\Omega}C_{N,s}\int_\Omega\frac{u(x)-u(y)}{|x-y|^{N+2s}}\,dx\,dy := \int_{\mathbb{R}^N\setminus\Omega}\mathcal{N}_s u(x)\,dx;\end{aligned}$$

2.

$$\begin{aligned}&\frac{1}{2}C_{N,s}\iint_{\mathcal{D}_\Omega}\frac{(u(x)-u(y))(v(x)-v(y))}{|x-y|^{N+2s}}\,dx\,dy\\&\quad=\frac{1}{2}C_{N,s}\iint_{\mathcal{D}_\Omega}v(x)\frac{(u(x)-u(y))}{|x-y|^{N+2s}}\,dx\,dy-\frac{1}{2}C_{N,s}\iint_{\mathcal{D}_\Omega}v(y)\frac{(u(x)-u(y))}{|x-y|^{N+2s}}\,dx\,dy\\&\quad=C_{N,s}\iint_{\mathcal{D}_\Omega}v(x)\frac{(u(x)-u(y))}{|x-y|^{N+2s}}\,dx\,dy\\&\quad=C_{N,s}\int_\Omega\int_{\mathbb{R}^N}v(x)\frac{(u(x)-u(y))}{|x-y|^{N+2s}}\,dx\,dy+C_{N,s}\int_\Omega\int_{\mathbb{R}^N\setminus\Omega}v(x)\frac{(u(x)-u(y))}{|x-y|^{N+2s}}\,dx\,dy\\&\quad=\int_\Omega v(x)(-\Delta)^s u(x)\,dx+\int_{\mathbb{R}^N\setminus\Omega}v(x)\mathcal{N}_s u(x)\,dx.\end{aligned}$$

□

Remark 8.1.10. These *integration by parts formulas* have been used in [141] to give a formulation to the Neumann problem for the fractional Laplacian. See also [47] for the Neumann problem in the case of higher order fractional operators.

Note that if $u = 0 = v$ in $\mathbb{R}^N \setminus \Omega$, the Green formula in (2) becomes

$$\frac{1}{2}C_{N,s} \iint_{\mathcal{D}_\Omega} \frac{(u(x) - u(y))(v(x) - v(y))}{|x - y|^{N+2s}}\, dx\, dy = \int_\Omega v(y)(-\Delta)^s u(y)\, dy.$$

This expression will be used to formulate the Dirichlet problem.

8.2 Analytical preliminaries related to $(-\Delta)^s$

If $u \in \mathcal{S}(\mathbb{R}^N)$, in general $(-\Delta)^s(u) \notin \mathcal{S}(\mathbb{R}^N)$. Indeed, the lack of regularity of the multiplier gives a lack of rapid decay. Nevertheless, by using the Fourier transform of its distributional derivatives, one can realize that $(-\Delta)^s(u) \in \mathcal{C}^\infty(\mathbb{R}^N)$. Also, from the integral representation of the fractional Laplacian, it is easy to see that if ϕ is a function in the Schwartz class, then

$$|(-\Delta)^s \phi(x)| \le C \frac{1}{1 + |x|^{N+2s}}, \tag{8.2.1}$$

where C depends on integrability bounds of ϕ and its first two derivatives. To see this, observe that $|(-\Delta)^s \phi(x)| \le C$, for all x. So, it suffices to prove

$$|(-\Delta)^s \phi(x)| \le C \frac{1}{|x|^{N+2s}}, \qquad \text{for } |x| > 1.$$

We write

$$\begin{aligned} P.V. \int_{\mathbb{R}^N} \frac{\phi(x) - \phi(y)}{|x - y|^{N+2s}}\, dy = & \int_{\{|x-y|<\frac{|x|}{2}\}} \frac{\phi(x) - \phi(y) - \nabla\phi(x) \cdot (x - y)}{|x - y|^{N+2s}}\, dy \\ & + \int_{\{|x-y|\ge\frac{|x|}{2}\}} \frac{\phi(x) - \phi(y)}{|x - y|^{N+2s}}\, dy = I + II, \end{aligned}$$

where in I we have used the fact that, by symmetry,

$$\int_{\{\epsilon<|x-y|<\frac{|x|}{2}\}} \frac{\nabla\phi(x) \cdot (x - y)}{|x - y|^{N+2s}}\, dy = 0.$$

Now, for $x, y \in \mathbb{R}^N$ with $|x-y| < \frac{|x|}{2}$, Taylor's theorem says that $\exists \xi \in \mathbb{R}^N$, with $|x-\xi| < \frac{|x|}{2}$, so that

$$|\phi(x) - \phi(y) - \nabla\phi(x) \cdot (x - y)| \le \frac{1}{2}|D^2\phi(\xi)||x - y|^2.$$

Since $|\xi| \sim |x|$, we have

$$|I| \le C \frac{1}{|x|^M} \int_{\{|x-y|<\frac{|x|}{2}\}} \frac{1}{|x - y|^{N+2s-2}}\, dy = \mathcal{O}_M\Big(\frac{1}{|x|^{M+2s-2}}\Big), \qquad \forall M \gg 1.$$

We split II in two pieces, II_1 and II_2, according to whether $|y| \le 2|x|$ or $|y| > 2|x|$, respectively. For the first one we have

$$|II_1| \le \left(\frac{2}{|x|}\right)^{N+2s} \int_{|y|\le 2|x|} (|\phi(x)| + |\phi(y)|)\,dy \le C\frac{1}{|x|^{N+2s}}\left(|x|^N|\phi(x)| + \int_{|y|\le 2|x|} |\phi(y)|\,dy\right).$$

Thus, $|II_1| \le C\frac{1}{|x|^{N+2s}}$. For the second piece, we use that $|\phi(x) - \phi(y)| \le |\nabla\phi(\eta)||x-y|$, for some $\eta \in \mathbb{R}^N$ with $|\eta| \sim |y|$, since in this case $|y| > 2|x|$. Hence,

$$|II_2| \le C\int_{|y|>2|x|} \frac{|x-y|}{|y|^M}\frac{1}{|x-y|^{N+2s}}\,dy = \mathcal{O}_M\left(\frac{1}{|x|^{M+2s-1}}\right), \quad \forall M \gg 1.$$

This finishes our estimate.

According to (8.2.1), it is natural to extend the definition of the operator to a larger class of functions. To that end, consider the space

$$\mathcal{L}^s(\mathbb{R}^N) := \left\{u : \mathbb{R}^N \to \mathbb{R} \text{ measurable: } \int_{\mathbb{R}^N} \frac{|u(x)|}{1+|x|^{N+2s}}\,dx < \infty\right\},$$

endowed with the norm

$$\|u\|_{\mathcal{L}^s(\mathbb{R}^N)} := \int_{\mathbb{R}^N} \frac{|u(x)|}{1+|x|^{N+2s}}\,dx.$$

As was pointed out in [287], we have the following result. This will allow us to better understand the pointwise meaning of the fractional Laplacian operator.

Lemma 8.2.1. *Let $s \in (0,1)$. If $u \in \mathcal{L}^s(\mathbb{R}^N) \cap C^{2s+\gamma}_{\rm loc}(\mathbb{R}^N)$ (or $C^{1,2s+\gamma-1}$ if $2s > 1$) for some $\gamma > 0$, then $(-\Delta)^s u$ is a continuous function and is well defined as the following principal value:*

$$(-\Delta)^s u(x) := C_{N,s}\, P.V. \int_{\mathbb{R}^N} \frac{u(x)-u(y)}{|x-y|^{N+2s}}\,dy.$$

Proof. Fix a bounded domain $\Omega \subset \mathbb{R}^N$. By mollification, there exists a sequence $\{u_k\} \subset \mathcal{S}(\mathbb{R}^N)$, uniformly bounded in $C^{2s+\gamma}(\Omega)$ (or $C^{1,2s+\gamma-1}(\Omega)$ if $2s > 1$) and such that

$$u_k \to u \quad \text{uniformly in } \Omega \text{ and in } \mathcal{L}^s(\mathbb{R}^N).$$

The uniform boundedness in $C^{2s+\gamma}(\Omega)$ (or $C^{1,2s+\gamma-1}(\Omega)$ if $2s > 1$) and the decay of u implies that

$$(-\Delta)^s(u_k)(x) \to C_{N,s}\int_{\mathbb{R}^N} \frac{u(x)-u(y)}{|x-y|^{N+2s}}\,dy, \quad \text{uniformly in } \Omega.$$

On the other hand, it is clear that

$$(-\Delta)^s(u_k)(x) \to (-\Delta)^s(u)(x) \quad \text{in } \mathcal{S}'(\mathbb{R}^N).$$

Therefore, by the uniqueness of the limit,

$$(-\Delta)^s(u)(x) = C_{N,s}\int_{\mathbb{R}^N}\frac{u(x)-u(y)}{|x-y|^{N+2s}}\,dy \quad \text{if } x\in\Omega.$$

Since Ω is arbitrary we conclude that $(-\Delta)^s u(x) \in \mathcal{C}(\mathbb{R}^N)$. □

If $u \in \mathcal{L}^s(\mathbb{R}^N)$, then $(-\Delta)^s u$ can be defined in the weak sense as a tempered distribution, that is, we can compute the duality product $\langle(-\Delta)^s u, \varphi\rangle$, for every φ in the Schwartz class. With more generality we have the following definition.

Definition 8.2.2. Given $f \in L^1(\mathbb{R}^N)$ we say that a measurable function $u : \mathbb{R}^N \to \mathbb{R}$ is a weak solution to $(-\Delta)^s u = f$ if one has

$$\int_{\mathbb{R}^N}(-\Delta)^s u\phi\,dx = \int_{\mathbb{R}^N} f\phi\,dx, \quad \text{for all } \phi\in\mathcal{S}(\mathbb{R}^N).$$

We observe the following properties, which can be obtained by an easy calculation.

Assume that $u, v \in \mathcal{S}(\mathbb{R}^N)$.

1. It holds that

$$\langle(\Delta)^s u, v\rangle := \int_{\mathbb{R}^N}(-\Delta)^s uv\,dx = \frac{1}{2}C_{N,s}\,P.V\int_{\mathbb{R}^N}\int_{\mathbb{R}^N}\frac{(u(x)-u(y))(v(x)-v(y))}{|x-y|^{N+2s}}\,dy\,dx.$$

2. As a consequence, $(-\Delta)^s$ is a symmetric operator,

$$\langle(-\Delta)^s u, v\rangle = \langle u, (-\Delta)^s v\rangle.$$

3. Also, $\langle(-\Delta)^s u, v\rangle$ is a definite positive bilinear form in $\mathcal{S}(\mathbb{R}^N)$. The associated Gagliardo seminorm is defined by

$$|||u|||_s^2 = \frac{1}{2}C_{N,s}\,P.V\int_{\mathbb{R}^N}\int_{\mathbb{R}^N}\frac{(u(x)-u(y))^2}{|x-y|^{N+2s}}\,dy\,dx.$$

We define $\dot{H}^s(\mathbb{R}^N)$ as the homogeneous Hilbert space that results by completion with respect to the seminorm $|||\cdot|||_s$.

We also consider $H^s(\mathbb{R}^N)$ as the completion of $\mathcal{S}(\mathbb{R}^N)$ with respect to the norm

$$\|u\|_{H^s} = \|u\|_s = \left(\|u\|_2^2 + \frac{C_{N,s}}{2}\,P.V\int_{\mathbb{R}^N}\int_{\mathbb{R}^N}\frac{(u(x)-u(y))^2}{|x-y|^{N+2s}}\,dy\,dx\right)^{1/2}.$$

The space $H^s(\mathbb{R}^N)$ is a Hilbert space.

Therefore, we can extend Lemma 8.1.8 to $H^s(\mathbb{R}^N)$. In this way, we have the following result.

Proposition 8.2.3. *Let $N \geq 1$ and $0 < s < 1$. Then for all $u \in H^s(\mathbb{R}^N)$*

$$\int_{\mathbb{R}^N} |\xi|^{2s} |\mathcal{F}(u)(\xi)|^2 \, d\xi = \frac{\widetilde{C}_{N,s}}{2} \int_{\mathbb{R}^N} \int_{\mathbb{R}^N} \frac{|u(x) - u(y)|^2}{|x - y|^{N+2s}} \, dx \, dy,$$

where $\widetilde{C}_{N,s}$ is the constant defined in Lemma 8.1.8.

The class $C_0^\infty(\mathbb{R}^N)$ is dense in $H^s(\mathbb{R}^N)$ (see for instance Theorem 7.38 in [28]). Moreover, the associated scalar product in $H^s(\mathbb{R}^N)$ can be reformulated, modulo constants, as follows:

$$\begin{aligned} \langle u, v \rangle_{H^s(\mathbb{R}^N)} &:= \langle (-\Delta)^s u, v \rangle + (u, v) \\ &:= \frac{C_{N,s}}{2} P.V. \int_{\mathbb{R}^N} \int_{\mathbb{R}^N} \frac{(u(x) - u(y))(v(x) - v(y))}{|x - y|^{N+2s}} \, dx \, dy + \int_{\mathbb{R}^N} uv \, dx. \end{aligned}$$

Summarizing, we obtain the following useful formulation, which includes the corresponding *integration by parts* (see also [138]).

Proposition 8.2.4. *Let $s \in (0, 1)$ and $u \in H^s(\mathbb{R}^N)$. Then,*

$$\langle (-\Delta)^s u, u \rangle = \|(-\Delta)^{\frac{s}{2}} u\|_{L^2(\mathbb{R}^N)}^2 = \|(2\pi|\xi|)^s \mathcal{F} u\|_{L^2(\mathbb{R}^N)}^2.$$

The dual space of $H^s(\mathbb{R}^N)$ is defined by

$$H^{-s}(\mathbb{R}^N) = \{f \in \mathcal{S}'(\mathbb{R}^N) / |\xi|^{-s} \mathcal{F}(f) \in L^2(\mathbb{R}^N)\}.$$

The following properties are now immediate:
1. $(-\Delta)^s : H^s(\mathbb{R}^N) \to H^{-s}(\mathbb{R}^N)$ is a continuous operator;
2. $(-\Delta)^s$ is a symmetric operator in $H^s(\mathbb{R}^N)$, that is,
$$\langle (-\Delta)^s u, v \rangle = \langle u, (-\Delta)^s v \rangle, \quad u, v \in H^s(\mathbb{R}^N);$$
3. denoting also by $\langle \cdot, \cdot \rangle$ the natural duality product between $H^s(\mathbb{R}^N)$ and $H^{-s}(\mathbb{R}^N)$, we have
$$|\langle (-\Delta)^s u, v \rangle| \leq \|u\|_{H^s(\mathbb{R}^N)} \|v\|_{H^s(\mathbb{R}^N)}.$$

The *regional Sobolev space in a domain* $\Omega \subset \mathbb{R}^N$ is defined by

$$H^s(\Omega) = \left\{ u \in L^2(\Omega) \mid \frac{|u(x) - u(y)|}{|x - y|^{\frac{N}{2}+s}} \in L^2(\Omega \times \Omega) \right\},$$

with the norm

$$\|u\|_{H^s(\Omega)} = \left(\|u\|_{L^2(\Omega)}^2 + \frac{C_{N,s}}{2} \int_\Omega \int_\Omega \frac{|u(x) - u(y)|^2}{|x - y|^{N+2s}} \, dy \, dx \right)^{\frac{1}{2}}.$$

One natural problem is to know when the restriction to Ω of functions of $H^s(\mathbb{R}^N)$ coincides with $H^s(\Omega)$. According to this, let us make the following definition.

Definition 8.2.5. Given $\Omega \subset \mathbb{R}^N$, an open set, we say that Ω has the extension property for $H^s(\Omega)$ if for any $u \in H^s(\Omega)$ there exists $\tilde{u} \in H^s(\mathbb{R}^N)$ such that $\tilde{u}|_\Omega = u$ and

$$\|\tilde{u}\|_s \le C(n, s, \Omega)\|u\|_{H^s(\Omega)}.$$

The following result gives a class of domains with the *extension property with respect to* $H^s(\Omega)$.

Theorem 8.2.6. *Assume $\Omega \subset \mathbb{R}^N$, an open set with Lipschitz bounded boundary. Then $H^s(\Omega)$ is continuously embedded in $H^s(\mathbb{R}^N)$, that is, Ω has the extension property with respect to $H^s(\Omega)$.*

A proof can be seen, for example, in [138] and more details can be found in the references therein.

The next step is to study the following *fractional Sobolev inequality*, which will be used in a systematic way hereafter. The elementary proof that we present here appears in Proposition 15.5 of the book [264], where a more general case is studied. This proof is due to H. Brezis.

Theorem 8.2.7. *Let $s \in (0, 1)$ and $N > 2s$. There exists a constant $S(N, s)$ such that for any $u \in H^s(\mathbb{R}^N)$ we have*

$$\|u\|^2_{L^{2^*_s}(\mathbb{R}^N)} \le S(N, s) \int_{\mathbb{R}^N} \int_{\mathbb{R}^N} \frac{|u(x) - u(y)|^2}{|x - y|^{N+2s}}\, dx\, dy,$$

where

$$2^*_s = \frac{2N}{N - 2s}$$

is the Sobolev critical exponent.

Proof. By density it is sufficient to do the proof for $u \in C_0^\infty(\mathbb{R}^N)$. Given $x, y \in \mathbb{R}^N$ we trivially have

$$|u(x)| \le |u(x) - u(y)| + |u(y)|.$$

Consider a ball $B_r(x)$ and take the average in $B_r(x)$ on both sides of the inequality. Then

$$|u(x)| \le \fint_{B_r(x)} |u(x) - u(y)|\, dy + \fint_{B_r(x)} |u(y)|\, dy.$$

By the Hölder inequality we have

$$\fint_{B_r(x)} |u(x) - u(y)|\, dy \le \left(\fint_{B_r(x)} |u(x) - u(y)|^2\, dy\right)^{\frac{1}{2}} \le \left(r^{N+2s} \fint_{B_r(x)} \frac{|u(x) - u(y)|^2}{|x - y|^{N+2s}}\, dy\right)^{\frac{1}{2}}$$

and for any $p > 1$,

$$\fint_{B_r(x)} |u(y)|\,dy \le \left(\fint_{B_r(x)} |u(y)|^p\,dy\right)^{\frac{1}{p}}.$$

Hence,

$$|u(x)| \le r^s\left(\frac{1}{\omega_N}\int_{\mathbb{R}^N} \frac{|u(x)-u(y)|^2}{|x-y|^{N+2s}}\,dy\right)^{\frac{1}{2}} + r^{-\frac{N}{p}}\left(\frac{1}{\omega_N}\int_{\mathbb{R}^N}|u(y)|^p\,dy\right)^{\frac{1}{p}}. \tag{8.2.2}$$

Since, in particular, u is bounded and with compact support, $u \in L^p(\mathbb{R}^N)$ for all $p \ge 1$. Therefore the right hand side of the previous inequality is finite for all $x \in \mathbb{R}^N$. Call

$$f(r) = Ar^s + Br^{-\frac{N}{p}},$$

where

$$A = \left(\frac{1}{\omega_N}\int_{\mathbb{R}^N} \frac{|u(x)-u(y)|^2}{|x-y|^{N+2s}}\,dy\right)^{\frac{1}{2}} \quad \text{and} \quad B = \left(\frac{1}{\omega_N}\int_{\mathbb{R}^N}|u(y)|^p\,dy\right)^{\frac{1}{p}}.$$

Minimizing in $r > 0$ we obtain that the minimum is attained at

$$r_{\min} = C(N,p,s)\left(\frac{B}{A}\right)^{\frac{p}{ps+N}}$$

and, substituting in (8.2.2), we get

$$|u(x)| \le C_1\left(\int_{\mathbb{R}^N} \frac{|u(x)-u(y)|^2}{|x-y|^{N+2s}}\,dy\right)^{\frac{\frac{1}{2p}}{\frac{s}{N}+\frac{1}{p}}}\left(\int_{\mathbb{R}^N}|u(y)|^p\,dy\right)^{\frac{\frac{s}{pN}}{\frac{s}{N}+\frac{1}{p}}}.$$

If we choose p such that $\frac{1}{p} = \frac{1}{2} - \frac{s}{N}$, then

$$\int_{\mathbb{R}^N}|u(x)|^p\,dx \le C_1^p\left(\int_{\mathbb{R}^N}\int_{\mathbb{R}^N} \frac{|u(x)-u(y)|^2}{|x-y|^{N+2s}}\,dy\,dx\right)\left(\int_{\mathbb{R}^N}|u(y)|^p\,dy\right)^{\frac{2s}{N}},$$

which implies

$$\|u\|_{\frac{2N}{N-2s}} \le S(N,s)\left(\int_{\mathbb{R}^N}\int_{\mathbb{R}^N} \frac{|u(x)-u(y)|^2}{|x-y|^{N+2s}}\,dy\,dx\right)^{\frac{1}{2}}.$$ □

As an immediate consequence of Theorem 8.2.7 we can formulate the following results.

Corollary 8.2.8. *If $0 < s < 1$ and $N > 2s$, then*

$$H^s(\mathbb{R}^N) \subset L^q(\mathbb{R}^N) \quad \text{if } q \in [2, 2_s^*],$$

with continuous embedding.

Proof. This result follows just from applying Sobolev embedding and interpolation between $L^2(\mathbb{R}^N)$ and $L^{2_s^*}(\mathbb{R}^N)$. □

Corollary 8.2.9. *Consider $s \in (0,1)$ and $N > 2s$. Let $\Omega \subset \mathbb{R}^N$ be a domain with the extension property respect to $H^s(\Omega)$. Then there exists $C = C(N,s,q,\Omega)$ such that for any $u \in H^s(\Omega)$*

$$\|u\|_q \le C\|u\|_{H^s(\Omega)}, \quad \text{for any } q \in [2, 2_s^*].$$

Moreover, if Ω is bounded, then $H^s(\Omega)$ is continuously embedded in $L^q(\Omega)$ for $q \in [1, 2_s^*]$.

Proof. To verify the result we apply the extension theorem, the Sobolev inequality and again interpolation. □

In Theorem 7.1 and Corollary 7.2 of [138] we find the compactness result extending to the fractional case Rellich results. We will use precisely the following result.

Theorem 8.2.10. *Let $0 < s < 1$ and $N > 2s$. Let $q \in [1, 2_s^*)$ and let $\Omega \subset \mathbb{R}^N$ be a bounded domain verifying the extension property for $H^s(\Omega)$. Assume that $\mathcal{A} \subset L^2(\Omega)$ satisfies:*
1. *There exists a constant $M > 0$ such that $\|u\|_2 \le M$ for any $u \in \mathcal{A}$.*
2. *It holds that*

$$\sup_{u\in\mathcal{A}} \int_\Omega \int_\Omega \frac{|u(x)-u(y)|^2}{|x-y|^{N+2s}}\, dy\, dx < +\infty.$$

Then $\mathcal{A}$ is precompact in $L^q(\Omega)$.

We refer to [138] for the proof. More functional properties related to fractional Sobolev spaces can be seen for instance in [28], [138], [239] and the references therein.

8.2.1 Some remarks on the regularity of solutions in the whole $\mathbb{R}^N$

We will try to summarize the behavior of the s-Laplacian operating in spaces of Hölder functions. We follow closely Section 2 in the paper by L. Silvestre [287]. The first result shows the behavior of $(-\Delta)^s$ with respect to Hölder continuous function spaces.

We call $C^{0,\alpha}(\mathbb{R}^N)$, $0 < \alpha < 1$, the class of bounded continuous functions such that the seminorm

$$[u]_{0,\alpha} = \sup_{x\neq y} \frac{|u(x)-u(y)|}{|x-y|^\alpha} < \infty.$$

The norm in $C^{0,\alpha}(\mathbb{R}^N)$ is defined by

$$\|u\|_{0,\alpha} = \|u\|_\infty + [u]_{0,\alpha}.$$

For instance, the class of continuous functions with compact support and with the α-modulus of continuity are a subspace of $C^{0,\alpha}(\mathbb{R}^N)$.

In a similar way we denote by $C^{k,\alpha}(\mathbb{R}^N)$, $k \in \mathbb{N}$, $0 < \alpha < 1$, the class of functions which have continuous bounded derivatives until the order k and the derivative of order k belongs to $C^{0,\alpha}(\mathbb{R}^N)$. The norm in $C^{k,\alpha}(\mathbb{R}^N)$ is defined, for instance, by

$$\|u\|_{k,\alpha} = \sum_{j=0}^{k} |D^j u|_0 + [D^k]_{0,\alpha},$$

where

$$|D^j u|_0 = \max_{|\beta|=k} \sup_{\mathbb{R}^N} |D^\beta u| \quad \text{and} \quad [D^k]_{0,\alpha} = \max_{|\beta|=k} [D^\beta u]_{0,\alpha}.$$

More properties of the spaces of Hölder continuous functions can be seen in the literature, for instance on page 51 and onwards of [186].

Proposition 8.2.11. *Consider* $1 \geq \alpha > 2s > 0$. *Then there exists a positive constant* $c = c(N, s, \alpha)$ *such that, for all* $u \in C^{0,\alpha}$, *one has* $(-\Delta)^s u \in C^{0,\alpha-2s}$ *and*

$$|(-\Delta)^s u|_{C^{0,\alpha-2s}} \leq c|u|_{C^{0,\alpha}}.$$

Proof. The difference $|(-\Delta)^s u(x_1) - (-\Delta)^s u(x_2)|$ is given by

$$|(-\Delta)^s u(x_1) - (-\Delta)^s u(x_2)| = C_{N,s} \left| \int_{\mathbb{R}^N} \frac{u(x_1) - u(x_1 + y) - u(x_2) + u(x_2 + y)}{|y|^{N+2s}} \, dy \right|.$$

For a ball B_r with $r > 0$ we split the integral in two pieces, the integral over B_r and the integral over its complementary set, that is, we consider

$$I_1 = C_{N,s} \left| \int_{B_r} \frac{u(x_1) - u(x_1 + y) - u(x_2) + u(x_2 + y)}{|y|^{N+2s}} \, dy \right|$$

and

$$I_2 = C_{N,s} \left| \int_{\mathbb{R}^N \setminus B_r} \frac{u(x_1) - u(x_1 + y) - u(x_2) + u(x_2 + y)}{|y|^{N+2s}} \, dy \right|.$$

To estimate I_1 the regularity of u gives $|u(x_i) - u(x_i + y)| \leq |u|_{C^{0,\alpha}} |y|^\alpha$. Then

$$I_1 \leq C_{N,s} |u|_{C^{0,\alpha}} \int_{B_r} \frac{|y|^\alpha}{|y|^{N+2s}} \, dy = c|u|_{C^{0,\alpha}} r^{\alpha-2s}.$$

For I_2 we use that $|u(x_1 + y) - u(x_2 + y)| \leq |u|_{C^{0,\alpha}} |x_1 - x_2|^\alpha$; it follows that

$$I_2 \leq C_{N,s} \int_{\mathbb{R}^N \setminus B_r} 2|u|_{C^{0,\alpha}} |x_1 - x_2|^\alpha \frac{|y|^\alpha}{|y|^{N+2s}} \, dy \leq C_{N,s} 2|u|_{C^{0,\alpha}} |x_1 - x_2|^\alpha r^{-2s}.$$

In particular, choosing $r = |x_1 - x_2|$ we conclude that

$$I_1 + I_2 \leq c(N, s, \alpha)|x_1 - x_2|^{\alpha-2s}.$$ □

As a direct consequence of Proposition 8.2.11 and the commutation of the fractional Laplacian with derivatives we have the following result.

Corollary 8.2.12. *Assume $1 \geq \alpha > 2s > 0$. There exists a positive constant $c_1 = c_1(\alpha, s, N)$ such that for every $u \in C^{1,\alpha}$,*

$$|(-\Delta)^s u|_{C^{1,\alpha-2s}} \leq c_1 |u|_{C^{1,\alpha}}.$$

Since we are considering fractional powers, it is possible to consider Hölder regularity $0 < \alpha < 2s < 2$. In this case we have the following result.

Proposition 8.2.13. *If $0 < \alpha < 2s$ and $0 < s < 1$, then there exists a constant $c_1 = c_1(\alpha, s, N)$ such that for every $u \in C^{1,\alpha}$,*

$$|(-\Delta)^s u|_{C^{0,\alpha-2s+1}} \leq c_1 |u|_{C^{1,\alpha}}.$$

Proof.

- If $s < \frac{1}{2}$ we have

$$|(-\Delta)^s u(x_1) - (-\Delta)^s u(x_2)| \leq I_1 + I_2,$$

where I_1 and I_2 are as in Proposition 8.2.11. To estimate I_1, since $u \in C^{1,\alpha}$,

$$\begin{aligned} &|u(x_1) - u(x_1 + y) - u(x_2) + u(x_2 + y)| \\ &\leq |\langle \nabla u(x_1) - \nabla u(x_2), y\rangle| + |u|_{C^{1,\alpha}} |y|^{1+\alpha} \\ &\leq |u|_{C^{1,\alpha}} (|y||x_1 - x_2|^\alpha + |y|^{1+\alpha}). \end{aligned}$$

Then

$$I_1 \leq |u|_{C^{1,\alpha}} (r^{1-2s}|x_1 - x_2|^\alpha + r^{1+\alpha-2s}).$$

And I_2 follows in a similar way.

- If $s \geq \frac{1}{2}$ we use the semigroup decomposition, i. e.,

$$(-\Delta)^s = (-\Delta)^{s-\frac{1}{2}} (-\Delta)^{\frac{1}{2}},$$

and the Calderón–Zygmund formula,

$$(-\Delta)^{\frac{1}{2}} = \sum_{k=1}^{N} R_k \partial_k.$$

Since $u \in C^{1,\alpha}$ and the Riesz transform maps continuously $C^{0,\alpha}(\mathbb{R}^N)$ into $C^{0,\alpha}(\mathbb{R}^N)$ (see the Lemma below and Theorem 2 in [112]) we conclude that

$$v(x) = (-\Delta)^{\frac{1}{2}} u(x) = \sum_{k=1}^{N} R_k \partial_k u(x) \in C^{0,\alpha}.$$

Therefore

$$(-\Delta)^s u(x) = (-\Delta)^{s-\frac{1}{2}} v(x) \in \mathcal{C}^{0,\alpha-2s+1}.$$

The estimates follow directly from the previous results. □

To do this section self-contained, we will present the proof of the result by Calderón–Zygmund used in the particular case of the Riesz transforms, which is specially transparent.

Lemma 8.2.14. *Let R_k be the Riesz transform in the direction k. Assume $0 < \alpha \leq 1$. Then*

$$R_k : \mathcal{C}^{0,\alpha}(\mathbb{R}^N) \to \mathcal{C}^{0,\alpha}(\mathbb{R}^N)$$

continuously, that is, there exists a constant $c = c(N,\alpha)$ such that

$$|R_k u|_{\mathcal{C}^{0,\alpha}(\mathbb{R}^N)} \leq c|u|_{\mathcal{C}^{0,\alpha}(\mathbb{R}^N)}.$$

Proof. We can assume by invariance under translations that we estimate the Hölder condition at $x = 0$, i. e., it is sufficient to prove that

$$\left|R_k u(x) - R_k u(0)\right| \leq c|u|_{\mathcal{C}^{0,\alpha}(\mathbb{R}^N)}|x|^\alpha.$$

Fix $x \in \mathbb{R}^N$.

Since the kernel of R_k is odd ($K_k(y) = C\frac{y_k}{|y|^{N+1}}$), we have

$$\int_{\epsilon<|y|<R} K_k(y)\,dy = 0 \quad \text{for every } 0 < \epsilon < R.$$

Then

$$\begin{aligned}
&R_k u(x) - R_k u(0)\\
&\quad = \lim_{\epsilon\to 0}\left(\int_{\epsilon<|y|<2|x|} (u(x-y) - u(x))K_k(y)\,dy - \int_{\epsilon<|y|<2|x|} (u(-y) - u(0))K_k(y)\,dy\right)\\
&\quad\quad + \lim_{R\to\infty}\left(\int_{2|x|<|y|<R} (u(x-y) - u(x))K_k(y)\,dy - \int_{2|x|<|y|<R} (u(-y) - u(0))K_k(y)\,dy\right)\\
&\quad =: \lim_{\epsilon\to 0} I_\epsilon + \lim_{R\to\infty} I_R.
\end{aligned}$$

The regularity of u allows us to write

$$\begin{aligned}
|I_\epsilon| &\leq |u|_{\mathcal{C}^{0,\alpha}(\mathbb{R}^N)} \int_{|y|<2|x|} |y|^{-N+\alpha}\,dy\\
&\leq 2|u|_{\mathcal{C}^{0,\alpha}(\mathbb{R}^N)}\omega_{N-1} \int_0^{2|x|} t^{\alpha-1}dt = 2^{1+\alpha}\omega_{N-1}|u|_{\mathcal{C}^{0,\alpha}(\mathbb{R}^N)}|x|^\alpha.
\end{aligned}$$

To estimate I_R we remark that $|\nabla K_k(y)| \le \frac{C_N}{|y|^{N+1}}$. Then

$$\big|K_k(x-y) - K_k(y)\big| \le C\frac{|x|}{|y|^{N+1}}, \quad \text{if } |y| \ge 2|x|,$$

and therefore

$$|I_R| \le \left|\int_{2|x|<|x-y|<R} (u(y)-u(0))K_k(x-y)\,dy - \int_{2|x|<|y|<R} (u(y)-u(0))K_k(-y)\,dy\right|.$$

We observe that $\{2|x| < |y| < R\} \subset \{|y| < 3|x|\} \cup \{R-|x| < |y| < R+|x|\} := C_1 \cup C_2$, and hence

$$\begin{aligned}
|I_R| &\le \int_{2|x|<|x-y|<R} \big|u(y)-u(0)\big|\big|K_k(x-y)-K_k(y)\big|\,dy + \int_{C_1\cup C_2} \big|u(y)-u(0)\big|\big|K_k(y)\big|\,dy \\
&\le c\int_{2|x|<|x-y|} |y|^\alpha \frac{|x|}{|y|^{N+1}}\,dy + c\int_{|y|<3|x|} \frac{|y|^\alpha}{|y|^N}\,dy \\
&\quad + 2\|u\|_\infty \int_{R-|x|<|y|<R+|x|} \frac{dy}{|y|^N} \\
&\le c\int_{|x|}^{\infty} |x|\frac{dt}{t^{2-\alpha}} + c\int_0^{3|x|} \frac{dt}{t^{1-\alpha}} + c\int_{R-|x|}^{R+|x|} \frac{dt}{t} \le c|x|^\alpha + \log\left(\frac{R+|x|}{R-|x|}\right).
\end{aligned}$$

Finally, we observe that for all fixed $x \in \mathbb{R}^N$, $\lim_{R\to\infty} \log(\frac{R+|x|}{R-|x|}) = 0$. This gives the result. □

The simple iteration of Corollary 8.2.12 and Proposition 8.2.13 implies the following consequence.

Corollary 8.2.15. *Assume that $k+\alpha-2s \notin \mathbb{Z}$. Then if $u \in C^{k,\alpha}$, $(-\Delta)^s u \in C^{m,\beta}$, where m is the integral part of $k+\alpha-2s$ and $\beta = k+\alpha-2s-m$.*

Let us recall now the solution of the Dirichlet problem in a ball. Let B_r be the ball of radius r. The solution to the Dirichlet problem

$$\begin{cases} (-\Delta)^s v(x) = 0 & \text{in } B_r, \\ v(x) = g(x) & \text{in } \mathbb{R}^N \setminus B_r \end{cases}$$

is given by

$$v(x) = \int_{\mathbb{R}^N\setminus B_r} P(x,y)g(y)\,dy, \tag{8.2.3}$$

where

$$P(x,y) = c_{N,s}\left(\frac{r^2-|x|^2}{|y|^2-r^2}\right)^s \frac{1}{|x-y|^N} \tag{8.2.4}$$

is the Poisson kernel. See [223] for more details and the proof of the Poisson formula.

To finish, we will explain a couple of results that show behavior similar to that of the fractional Laplacian and the local Laplacian.

Proposition 8.2.16. *Consider the equation $(-\Delta)^s u = f$, where $u \in L^\infty(\mathbb{R}^N)$ and $f \in \mathcal{C}^{0,\alpha}(\mathbb{R}^N)$, $s > 0$ and $\alpha \in (0,1]$.*

1. *If $\alpha + 2s \le 1$, then $u \in \mathcal{C}^{0,\alpha+2s}$ and there exists a constant $c = c(\alpha, s, N)$ such that*
$$|u|_{\mathcal{C}^{0,\alpha+2s}} \le c\{\|u\|_{L^\infty} + |f|_{\mathcal{C}^{0,\alpha}}\}.$$
2. *If $\alpha + 2s > 1$, then $u \in \mathcal{C}^{1,\alpha+2s-1}$ and there exists a constant $c_1 = c_1(\alpha, s, N)$ such that*
$$|u|_{\mathcal{C}^{1,\alpha+2s-1}} \le c_1\{\|u\|_{L^\infty} + |f|_{\mathcal{C}^{0,\alpha}}\}.$$

Proof. The regularity is a local property and then it suffices to prove it in a neighborhood of the origin by the translation invariance of the inverse operator. By the semigroup property we can moreover subordinate the result to the properties of the inverse of the Laplacian. To do this, consider a cutoff function $\phi \in \mathcal{C}_0^\infty(\mathbb{R}^N)$ such that $\phi(x) \ge 0$, $\phi(x) = 0$ if $|x| > 2$ and $\phi(x) = 1$ if $|x| \le 1$.

Define
$$u_0(x) = \int_{\mathbb{R}^N} \frac{\phi(y)f(y)}{|x-y|^{N-2s}}\,dy.$$

It is clear that in the ball B_1, $(-\Delta)^s u_0(x) = (-\Delta)^s u(x)$ and then $u_0 - u(x)$ coincides in $B_{\frac{1}{2}}$ with the solution of the problem
$$(-\Delta)^s v(x) = 0 \text{ in } B_{\frac{1}{2}}, \quad v(x) = u_0 - u(x) \text{ in } \mathbb{R}^N \setminus B_{\frac{1}{2}}.$$

Looking at the regularity of the Poisson kernel, it is easily seen that v is regular in $B_{\frac{1}{2}}$ and that its respective norms in $\mathcal{C}^{0,\alpha+2s}$ and $\mathcal{C}^{1,\alpha+2s-1}$ can be estimated by the L^∞-norms of $u_0 - u$ and f. Therefore, to obtain the result, everything is reduced to obtain the regularity of u_0 in $B_{\frac{1}{2}}$.

We can use now the semigroup property as follows:
$$u_0 = (-\Delta)^{-s}(f\phi) = (-\Delta)^{1-s} \circ \Delta^{-1}(f\phi).$$

The classical result about the regularity for the Laplacian, $\Delta^{-1}(f\phi) \in \mathcal{C}^{2,\alpha}$ and its norm only depends on the $\mathcal{C}^{0,\alpha}$ property of the datum f. Hence it is sufficient to apply Corollary 8.2.15 to conclude the proof. □

Note that by the same strategy of Proposition 8.2.16 but using $\mathcal{C}^{1,\alpha}$ instead of the $\mathcal{C}^{2,\alpha}$ estimates for the Laplacian, we find the following result.

Proposition 8.2.17. *Consider the equation $(-\Delta)^s u = f$, where $u \in L^\infty(\mathbb{R}^N$ and $f \in L^\infty(\mathbb{R}^N)$, $s > 0$ and $\alpha \in (0,1]$.*

1. *If $2s \le 1$, then $u \in \mathcal{C}^{0,\alpha}$ for any $0 < \alpha < s$ and there exists a constant $c = c(\alpha, s, N)$ such that*
$$|u|_{\mathcal{C}^{0,\alpha}} \le c\{\|u\|_{L^\infty} + \|f\|_{L^\infty}\}.$$

2. *If $2s > 1$, then $u \in \mathcal{C}^{1,\alpha}$ for any $0 < \alpha < 2s-1$ and there exists a constant $c_1 = c_1(\alpha, s, N)$ such that*

$$|u|_{\mathcal{C}^{1,\alpha}} \le c_1\{\|u\|_{L^\infty} + \|f\|_{L^\infty}\}.$$

8.3 The Dirichlet problem: the variational framework

To study the homogeneous Dirichlet problem, consider the space $H_0^s(\Omega)$ defined as

$$H_0^s(\Omega) = \{u \in H^s(\mathbb{R}^N) \text{ with } u = 0 \text{ a. e. in } \mathbb{R}^N \setminus \Omega\},$$

endowed with the norm

$$|\|u\||_{H_0^s(\Omega)} = \left(\int_\Omega |u(x)|^2\, dx + \iint_{\mathcal{D}_\Omega} \frac{|u(x) - u(y)|^2}{|x-y|^{N+2s}}\, dx\, dy\right)^{1/2},$$

where

$$\mathcal{D}_\Omega = \mathbb{R}^{2N} \setminus (\mathcal{C}\Omega \times \mathcal{C}\Omega). \tag{8.3.1}$$

Note that $\mathcal{C}_0^\infty(\Omega)$ is dense in $H_0^s(\Omega)$. Hence, if $u \in H_0^s(\Omega)$, we have $u = 0$ a. e. in $\mathbb{R}^N \setminus \Omega$ and we obtain the identity

$$\int_{\mathbb{R}^N}\int_{\mathbb{R}^N} \frac{|u(x) - u(y)|^2}{|x-y|^{N+2s}}\, dx\, dy = \iint_{\mathcal{D}_\Omega} \frac{|u(x) - u(y)|^2}{|x-y|^{N+2s}}\, dx\, dy.$$

Here, we will prove the following Poincaré type result.

Theorem 8.3.1. *Let $\Omega \subset \mathbb{R}^N$ be a bounded Lipschitz domain. Then there exists a constant $C > 0$ such that*

$$\|u\|_{L^2(\Omega)} \le C\left(\iint_{\mathcal{D}_\Omega} \frac{|u(x) - u(y)|^2}{|x-y|^{N+2s}}\, dx\, dy\right)^{1/2}, \quad \text{for any } u \in H_0^s(\Omega). \tag{8.3.2}$$

Proof. We argue by contradiction. Assume that there exists a sequence $\{u_n\}_{n\in\mathbb{N}}$ such that

$$\|u_n\|_{L^2(\Omega)} = 1 \quad \text{and} \quad \left(\iint_{\mathcal{D}_\Omega} \frac{|u_n(x) - u_n(y)|^2}{|x-y|^{N+2s}}\, dx\, dy\right)^{1/2} \le \frac{1}{n}.$$

Since H_0^s is a Hilbert space, any bounded subset is weakly precompact, that is, there exist a function $u \in H_0^s(\Omega)$ and a subsequence (that we relabel with the same notation) $\{u_n\}_{n\in\mathbb{N}}$ such that $u_n \rightharpoonup u$ weakly in H_0^s. By using the Fatou lemma we have

$$\left(\iint_{\mathcal{D}_\Omega} \frac{|u(x) - u(y)|^2}{|x-y|^{N+2s}}\, dx\, dy\right)^{1/2} \le \liminf_{n\to\infty}\left(\iint_{\mathcal{D}_\Omega} \frac{|u_n(x) - u_n(y)|^2}{|x-y|^{N+2s}}\, dx\, dy\right)^{1/2} = 0,$$

but then $u = 0$. Now by Theorem 8.2.10, we must have

$$0 = \|u\|_{L^2(\Omega)} = \lim_{n\to\infty} \|u_n\|_{L^2(\Omega)} = 1,$$

a contradiction. □

The result in Theorem 8.3.1 shows that $\|\|\cdot\|\|_{H_0^s}$ is equivalent to the norm

$$\|u\|_{H_0^s} = \left(\frac{1}{2}C_{N,s}\iint_{\mathcal{D}_\Omega}\frac{|u(x)-u(y)|^2}{|x-y|^{N+2s}}\,dx\,dy\right)^{1/2} \tag{8.3.3}$$

and this, in turn, is associated to the scalar product

$$\langle u, v\rangle_{H_0^s(\Omega)} = \frac{1}{2}C_{N,s}\iint_{\mathcal{D}_\Omega}\frac{(u(x)-u(y))(v(x)-v(y))}{|x-y|^{N+2s}}\,dx\,dy. \tag{8.3.4}$$

Therefore, the pair $(H_0^s(\Omega), \|\cdot\|_{H_0^s(\Omega)})$ is a Hilbert space with an equivalent norm.

The dual space of $H_0^s(\Omega)$ is

$$H^{-s} = \{f \in \mathcal{S}'(\mathbb{R}^N) \mid |\xi|^{-s}\mathcal{F}f \in L^2(\mathbb{R}^N)\}.$$

By the Riesz Theorem, H^{-s} is also a Hilbert space.

We have the following properties of the operator $(-\Delta)^s$ defined in $H_0^s(\Omega)$, which are easy to check:
1. $(-\Delta)^s : H_0^s(\Omega) \to H^{-s}$ is a continuous operator;
2. $(-\Delta)^s$ is symmetric, that is
$$\langle(-\Delta)^s u, v\rangle = \langle u, (-\Delta)^s v\rangle, \quad u, v \in H_0^s(\Omega);$$
3. $\langle(-\Delta)^s u, v\rangle = \langle u, v\rangle_{H_0^s(\Omega)}$.

We can formulate the Dirichlet problem in this variational framework as follows: given $f \in H^{-s}$ we look for $u \in H_0^s(\Omega)$ with

$$\begin{cases} (-\Delta)^s u = f & \text{in } \Omega, \\ u = 0 & \text{in } \mathbb{R}^N \setminus \Omega, \end{cases} \tag{8.3.5}$$

such that for all $v \in H_0^s(\Omega)$ the following identity holds:

$$\langle u, v\rangle_{H_0^s(\Omega)} = \langle f, v\rangle_{\text{duality}}.$$

The existence and uniqueness of solution to problem 8.3.5 is a direct consequence of the Lax–Milgram theorem (see for instance [86]).

We can define the Green operator as the inverse of the fractional Laplacian, i. e.,

$$\begin{aligned} G : H^{-s} &\to H_0^s(\Omega), \\ f &\to G(f) = u, \quad \text{the solution to (8.3.5).} \end{aligned} \tag{8.3.6}$$

With more generality, we want to extend the fractional operator to a class of operators with variable coefficients that give rise to nonlocal models parallel to the divergence-form local operator

$$-\operatorname{div}(A(x)\nabla u). \tag{8.3.7}$$

Here $A(x)$ is a measurable, symmetric, bounded matrix valued function satisfying

$$\lambda|\xi|^2 \le \langle A(x)\xi, \xi\rangle \le \lambda^{-1}|\xi|^2, \tag{8.3.8}$$

for some $0 < \lambda < 1$.

We are interested in some integral expressions that correspond to the nonlocal version of problem (8.3.7). We will consider the following class of kernels.

Definition 8.3.2. We say that the measurable function

$$\mathcal{K} : \mathbb{R}^N \times \mathbb{R}^N \setminus \{(x,x) : x \in \mathbb{R}^N\} \to [0,\infty)$$

is a (symmetric) $2s$-*kernel* if it satisfies:
(i) $\mathcal{K}(x,y) = \mathcal{K}(y,x)$;
(ii) there exist $0 < s < 1$ and $0 < \lambda \le 1$ such that for any $(x,y) \in \mathbb{R}^N \times \mathbb{R}^N$, $x \neq y$,

$$\lambda \le \mathcal{K}(x,y)|x-y|^{N+2s} \le \lambda^{-1}.$$

Associated to every $2s$-kernel $\mathcal{K}(x,y)$ we define a principal value integral operator of the following form:

$$\mathcal{L}(u)(x) = \mathcal{L}_{\mathcal{K}}(u)(x) = P.V. \int_{\mathbb{R}^N} (u(x) - u(y))\mathcal{K}(x,y)\, dy. \tag{8.3.9}$$

We will refer to $\mathcal{L}$ as a nonlocal $2s$-operator.

Clearly, $\mathcal{L}$ is well defined if, say, $u \in \mathcal{S}(\mathbb{R}^N)$. Associated to $\mathcal{K}(x,y)$, and consequently to $\mathcal{L}$ too, we define the bilinear form

$$\mathcal{E} : \mathcal{S}(\mathbb{R}^N) \times \mathcal{S}(\mathbb{R}^N) \to \mathbb{R}$$

as

$$\mathcal{E}(u,v) = \frac{1}{2}\int_{\mathbb{R}^N}\int_{\mathbb{R}^N} (u(x)-u(y))(v(x)-v(y))\mathcal{K}(x,y)\, dx\, dy, \quad u,v \in \mathcal{S}(\mathbb{R}^N). \tag{8.3.10}$$

Remark 8.3.3. Many results for the $2s$-kernels can be extended to a large class of kernels. For instance in [207], condition (*ii*) is assumed for $|x-y| \le 1$, while for $|x-y| > 1$ it suffices to have an estimate of the type

$$\mathcal{K}(x,y) \le \frac{M}{|x-y|^{N+\alpha}},$$

for some $M \geq 1$ and some $\alpha > 0$. However, to keep it short we restrict ourselves just to the previous case.

Extensions to a quasilinear setting are also considered in the literature. See [49], [218] and the references therein. We do not consider in this book this kind of models.

It is natural to consider the following Dirichlet problem:

$$\begin{cases} \mathcal{L}u = f & \text{in } \Omega, \\ u = 0 & \text{in } \mathbb{R}^N \setminus \Omega, \end{cases} \tag{8.3.11}$$

where $\Omega \subset \mathbb{R}^N$, $N \geq 2$, is a bounded domain with Lipschitz boundary, $\mathcal{L}$ is defined by (8.3.9) and $f \in H^{-s}(\Omega)$.

Here we recall the definition of *finite energy solution* for such a problem.

Definition 8.3.4. For $f \in H^{-s}(\Omega)$ we say that $u \in H_0^s(\Omega)$ is a *finite energy solution* to (8.3.11) if

$$\mathcal{E}(u, w) = \langle f, w \rangle, \quad \forall w \in H_0^s(\Omega),$$

where $\langle \cdot, \cdot \rangle$ represents the duality between $H_0^s(\Omega)$ and $H^{-s}(\Omega)$.

Observe that since $L^{(2_s^*)'}(\Omega) \subset H^{-s}(\Omega)$, $u \in H_0^s(\Omega)$ is a finite energy solution to (8.3.11) with $f \in L^{(2_s^*)'}(\Omega)$ if it satisfies

$$\frac{1}{2} \iint_{D_\Omega} (u(x) - u(x))(w(x) - w(y))\mathcal{K}(x, y)\, dx\, dy = \int_\Omega f(x)w(x)\, dx, \quad \forall w \in H_0^s(\Omega).$$

Due to the functional framework in which we set such a Dirichlet problem, we can state the first existence result for finite energy solutions as a straightforward consequence of the Lax–Milgram theorem.

Theorem 8.3.5. *Let $f \in H^{-s}(\Omega)$. Then there exists a unique $u \in H_0^s(\Omega)$ such that for all $v \in H_0^s(\Omega)$ the following identity holds:*

$$\mathcal{E}(u, v) = \langle f, v \rangle.$$

Moreover, if f belongs to $H^{-s}(\Omega) \cap L^1(\Omega)$ and is positive, then u is positive too.

8.3.1 Elementary estimates

In order to investigate summability properties of the solutions in terms of the summability of the data, in this section we will analyze the behavior of the Stampacchia truncation operators that give control of the level set of a function in the Sobolev space. We recall for any $k \geq 0$ the definition of the truncation functions $T_k(\sigma)$ and $G_k(\sigma)$, $\sigma \in \mathbb{R}$,

$$T_k(\sigma) = \max\{-k; \min\{k, \sigma\}\} \quad \text{and} \quad G_k(\sigma) = \sigma - T_k(\sigma). \tag{8.3.12}$$

Proposition 8.3.6. *Let v be a function in $H_0^s(\Omega)$.*
(i) *If $\psi \in \mathrm{Lip}(\mathbb{R})$, then $\psi(v) \in H_0^s(\Omega)$. In particular, for any $k \geq 0$, $T_k(v), G_k(v) \in H_0^s(\Omega)$.*
(ii) *For any $k \geq 0$*

$$\|G_k(v)\|^2_{H_0^s(\mathbb{R}^N)} \leq \int_{\mathbb{R}^N} G_k(v)(-\Delta)^s v\, dx. \tag{8.3.13}$$

(iii) *For any $k \geq 0$*

$$\|T_k(v)\|^2_{H_0^s(\mathbb{R}^N)} \leq \int_{\mathbb{R}^N} T_k(v)(-\Delta)^s v\, dx. \tag{8.3.14}$$

Remark 8.3.7. If $u \in H_0^s(\Omega)$, then $T_k u = 0$ and $G_k u = 0$ in $\mathbb{R}^N \setminus \Omega$. In particular,

$$\|G_k(v)\|^2_{H_0^s(\Omega)} \leq \int_\Omega G_k(v)(-\Delta)^s v\, dx,$$
$$\|T_k(v)\|^2_{H_0^s(\Omega)} \leq \int_\Omega T_k(v)(-\Delta)^s v\, dx.$$

Proof. Consider as above $D_\Omega = \mathbb{R}^{2N} \setminus (\mathcal{C}(\Omega) \times \mathcal{C}(\Omega))$.

(i) We have (skipping the constant $C_{N,s}$ for the moment)

$$\begin{aligned}\|\psi(v)\|^2_{H_0^s(\Omega)} &= \iint_{D_\Omega} \frac{(\psi(v(x)) - \psi(v(y)))(\psi(v(x)) - \psi(v(y)))}{|x-y|^{N+2s}}\, dx\, dy\\ &\leq L_\psi \iint_{D_\Omega} \frac{(v(x) - v(x))(v(x) - v(y))}{|x-y|^{N+2s}}\, dx\, dy = L_\psi^2 \|v\|^2_{H_0^s(\Omega)}.\end{aligned}$$

(ii) Due to (i) we are allowed to write the integral:

$$\int_\Omega G_k(v)(-\Delta)^s v\, dx.$$

Hence we have

$$\int_\Omega G_k(v)(-\Delta)^s v\, dx = \iint_{D_\Omega} \frac{(v(x) - v(x))(G_k(v(x)) - G_k(v(y)))}{|x-y|^{N+2s}}\, dx\, dy.$$

Let us split the set D_Ω in the following way:

$$D_\Omega = D_1 \cup D_2 \cup D_3 \cup D_4,$$

where

$$\begin{aligned}D_1 &= \{(x,y) \in D_\Omega : |v(x)| \leq k, |v(y)| \geq k\},\\ D_2 &= \{(x,y) \in D_\Omega : |v(x)| \geq k, |v(y)| \geq k\},\\ D_3 &= \{(x,y) \in D_\Omega : |v(x)| \leq k, |v(y)| \leq k\},\\ D_4 &= \{(x,y) \in D_\Omega : |v(x)| \geq k, |v(y)| \leq k\}.\end{aligned}$$

Note that since for any $k \geq 0$ and any $s \in \mathbb{R}$, $s = T_k(s) + G_k(s)$,

$$
\begin{aligned}
&(v(x) - v(x))(G_k(v(x)) - G_k(v(y))) \\
&\quad = (G_k(v(x)) - G_k(v(y)))^2 + (T_k(v(x)) - T_k(v(y)))(G_k(v(x)) - G_k(v(y))).
\end{aligned}
$$

Now we observe that the latter term turns out to be positive: indeed

$$(T_k(v(x)) - T_k(v(y)))(G_k(v(x)) - G_k(v(y))) \equiv 0 \quad \text{in } D_2 \text{ and in } D_3.$$

On the other hand

$$
\begin{aligned}
&(T_k(v(x)) - T_k(v(y)))(G_k(v(x)) - G_k(v(y)))\chi_{D_1} \\
&\quad = (v(x) - k\operatorname{sign}(v(y)))(-(|v(y)| - k)^+ \operatorname{sign}(v(y)))\chi_{D_1} \\
&\quad = (k - v(x)\operatorname{sign}(v(y)))(|v(y)| - k)^+ \geq 0,
\end{aligned}
$$

since $|v(x)| \leq k$ in D_1. In the same way, it follows that

$$
\begin{aligned}
&(T_k(v(x)) - T_k(v(y)))(G_k(v(x)) - G_k(v(y)))\chi_{D_4} \\
&\quad = (k\operatorname{sign}(v(x)) - v(y))((|v(x)| - k)^+ \operatorname{sign}(v(x)))\chi_{D_4} \geq 0.
\end{aligned}
$$

Consequently

$$
\begin{aligned}
\|G_k(v)\|^2_{H^s_0(\Omega)} &= \iint_{D_\Omega} \frac{|G_k(u(x)) - G_k(u(y))|^2}{|x - y|^{N+2s}}\,dx\,dy \\
&\leq \iint_{D_\Omega} \frac{(v(x) - v(x))(G_k(u(x)) - G_k(u(y)))}{|x - y|^{N+2s}}\,dx\,dy = \int_\Omega G_k(v)(-\Delta)^s v\,dx.
\end{aligned}
$$

(iii) The idea of the proof is basically the same as that of (ii). Let us consider for $k \geq 0$,

$$\int_\Omega T_k(v)(-\Delta)^s v\,dx = \iint_{D_\Omega} \frac{(v(x) - v(x))(T_k(v(x)) - T_k(v(y)))}{|x - y|^{N+2s}}\,dx\,dy.$$

Arguing as above we have

$$
\begin{aligned}
&(v(x) - v(y))(T_k(v(x)) - T_k(v(y))) \\
&\quad = (T_k(v(x)) - T_k(v(y)))^2 + (G_k(v(x)) - G_k(v(y)))(T_k(v(x)) - T_k(v(y)))
\end{aligned}
$$

and the last term is either positive (in $D_1 \cup D_4$) or zero (over $D_2 \cup D_3$). Thus

$$
\begin{aligned}
\|T_k(v)\|^2_{H^s_0(\Omega)} &\leq \iint_{D_\Omega} \frac{(T_k(u(x)) - T_k(u(y)))^2}{|x - y|^{N+2s}}\,dx\,dy \\
&\leq \iint_{D_\Omega} \frac{(v(x) - v(x))(T_k(u(x)) - T_k(u(y)))}{|x - y|^{N+2s}}\,dx\,dy = \int_\Omega T_k(v)(-\Delta)^s v\,dx. \quad \square
\end{aligned}
$$

We recall that when we deal with fractional derivatives, the chain rule is no longer true. By elementary calculus, if the function involved is convex or concave, an inequality holds. When applied to a fractional operator, we have the following classical elementary result.

Proposition 8.3.8. *Assume that $\Phi : \mathbb{R} \to \mathbb{R}$ is a Lipschitz convex function such that $\Phi(0) = 0$. Then if $u \in H_0^s(\Omega)$ we have*

$$\mathcal{L}\Phi(u) \le \Phi'(u)\mathcal{L}(u) \quad a.e. \text{ in } \Omega. \tag{8.3.15}$$

On the other hand, if $\Psi : \mathbb{R} \to \mathbb{R}$ is a Lipschitz concave function, then if $u \in H_0^s(\Omega)$ we have

$$\mathcal{L}\Psi(u) \ge \Psi'(u)\mathcal{L}(u) \quad a.e. \text{ in } \Omega. \tag{8.3.16}$$

Proof. The proof relies on elementary properties of convex and concave functions and the fact that the kernel defining $\mathcal{L}$ is nonnegative: to prove (8.3.15) we use that Φ satisfies $\Phi(a) - \Phi(b) \le \Phi'(a)(a - b), \forall a, b$, whereas (8.3.16) follows from the opposite inequality for the concave function Ψ. □

A consequence of the previous result is the following corollary, which will be used in the next sections.

Corollary 8.3.9. *Consider a nonnegative function $u \in H_0^s(\Omega)$; then, for any positive k, it follows that*

$$\mathcal{L}T_k(u) \ge \mathcal{L}u \cdot \chi_{\{x\in\Omega, u(x)\le k\}}, \quad x \in \Omega.$$

Proof. It suffices to apply (8.3.16) with the choice $\Phi(s) = T_k(s)$, which, restricted to positive values, turns out to be a concave function. □

8.3.2 Elliptic Kato inequality

In the folklore of the area, the extension of the Kato inequality to the fractional Laplacian is well known. For the sake of completeness, we include next a proof of the Kato inequality for weak solutions.

Theorem 8.3.10. *If $f \in L^1(\Omega)$ and $u \in H^s(\mathbb{R}^N)$ is a weak solution to*

$$\mathcal{L}u = f \quad \text{in } \mathbb{R}^N,$$

then

$$\mathcal{L}|u| \le f \operatorname{sign}(u) \quad \text{in the weak sense.} \tag{8.3.17}$$

That is, we have

$$\int_{\mathbb{R}^N} |u|\mathcal{L}\psi \, dx \le \int_{\mathbb{R}^N} \operatorname{sign}(u) f\psi \, dx, \quad \forall \psi \in \mathcal{C}_0^\infty, \ \psi \ge 0.$$

Proof. Consider for $\epsilon \in (0,1)$ the regularization of $|s|$ given by

$$\varphi_\epsilon(s) = \left(|s|^2 + \epsilon^2\right)^{\frac{1}{2}} - \epsilon.$$

It is easy to verify that
(i) $\varphi_\epsilon(s) \to |s|$ uniformly as $\epsilon \to 0$;
(ii) $\varphi_\epsilon'(s)$ is bounded uniformly with respect to ϵ;
(iii) $\varphi_\epsilon(s)$ is convex.

If we assume that u is smooth enough, since φ_ϵ is convex and Lipschitz and satisfies $\varphi_\epsilon(0) = 0$, then

$$\begin{aligned}\mathcal{L}\varphi_\epsilon(u(x)) &= \int_{\mathbb{R}^N} (\varphi_\epsilon(u(x)) - \varphi_\epsilon(u(y)))k(x,y)\,dy \\ &\le \int_{\mathbb{R}^N} \varphi_\epsilon'(u(x))(u(x) - u(y))k(x,y)\,dy = \varphi_\epsilon'(u(x))\mathcal{L}u(x).\end{aligned}$$

Now if $\psi \in C_0^\infty$, $\psi \ge 0$, we have

$$\begin{aligned}\int_{\mathbb{R}^N} \varphi_\epsilon(u(x))\mathcal{L}\psi(x)\,dx &= \int_{\mathbb{R}^N} \mathcal{L}\varphi_\epsilon(u(x))\psi(x)\,dx \\ &\le \int_{\mathbb{R}^N} \varphi_\epsilon'(u(x))\mathcal{L}u(x)\psi(x)\,dx = \int_{\mathbb{R}^N} \varphi_\epsilon'(u(x))f(x)\psi(x)\,dx.\end{aligned}$$

Passing to the limit with respect to ε in the previous inequality we obtain

$$\int_{\mathbb{R}^N} |u|\mathcal{L}\psi\,dx \le \int_{\mathbb{R}^N} \operatorname{sign}(u)f\psi\,dx,$$

as desired. □

8.3.3 Weak maximum principle and comparison results

The next results are basic to prove existence of solutions by monotony arguments. Their proofs are elementary.

Theorem 8.3.11. *Let $u \in H^s(\mathbb{R}^N)$ be the energy solution to the problem*

$$\begin{cases} \mathcal{L}(u) = f & \text{in } \Omega, \\ u = g & \text{in } \mathbb{R}^N \setminus \Omega, \end{cases}$$

where $f \ge 0$ in Ω and $g \ge 0$ in $\mathbb{R}^N \setminus \Omega$.

Then $u \ge 0$ in Ω.

Proof. We consider $u = u_+ - u_-$, the difference between the positive and negative parts of u. Since u is the energy solution,

$$\frac{1}{2}\iint_{D_\Omega} (u(x) - u(y))(w(x) - w(y))\mathcal{K}(x,y)\,dx\,dy = \int_\Omega f(x)w(x)\,dx, \quad \forall w \in H_0^s(\Omega),$$

and then

$$\int_\Omega f(x)\phi(x)\,dx \geq 0 \quad \text{for all } \phi \in H_0^s(\Omega),\ \phi \geq 0.$$

We observe that for $\phi = u_-$,

$$\begin{aligned}(u(x) - u(y))(\phi(x) - \phi(y)) &= (u(x) - u(y))(u_-(x) - u_-(y))\\ &= (u_+(x) - u_+(y))(u_-(x) - u_-(y)) - (u_-(x)\\ &\quad - u_-(y))(u_-(x) - u_-(y)).\end{aligned}$$

Moreover

$$\begin{aligned}&\iint_{D_\Omega} (u(x) - u(y))(u_-(x) - u_-(y))\mathcal{K}(x,y)\,dx\,dy\\ &= \int_\Omega\int_\Omega (u(x) - u(y))(u_-(x) - u_-(y))\mathcal{K}(x,y)\,dx\,dy\\ &\quad + 2\int_\Omega dx \int_{\mathbb{R}^N\setminus\Omega} (u(x) - g(y))u_-(x)\mathcal{K}(x,y)\,dy.\end{aligned}$$

Therefore, according to the algebraic identity above and since

$$(u_+(x) - u_+(y))(u_-(x) - u_-(y)) \leq 0,$$

we find that

$$\int_\Omega\int_\Omega (u(x)-u(y))(u_-(x)-u_-(y))\mathcal{K}(x,y)\,dx\,dy \leq -\int_\Omega\int_\Omega (u_-(x)-u_-(y))^2\mathcal{K}(x,y)\,dx\,dy \leq 0.$$

But also $g \geq 0$, and hence

$$\int_\Omega dx \int_{\mathbb{R}^N\setminus\Omega} (u(x) - g(y))u_-(x)\mathcal{K}(x,y)\,dy \leq 0.$$

Hence if we assume that $u_- \not\equiv 0$ we get a contradiction. □

As an immediate consequence we obtain the following comparison result.

Corollary 8.3.12. *Let $u \in H^s(\mathbb{R}^N)$ and $v \in H^s(\mathbb{R}^N)$ be solutions to the problems*

$$\begin{cases}(-\Delta)^s u = f_1 & \text{in } \Omega,\\ u = g_1 & \text{in } \mathbb{R}^N \setminus \Omega,\end{cases} \qquad \begin{cases}(-\Delta)^s v = f_2 & \text{in } \Omega,\\ v = g_2 & \text{in } \mathbb{R}^N \setminus \Omega,\end{cases}$$

respectively. If $f_1 \leq f_2$ and $g_1 \leq g_2$, then $u(x) \leq v(x)$, for all $x \in \mathbb{R}^N$.

8.3.4 Some interpolation results

We will need the following classical *interpolation theorem* as a tool in the next sections. First, we recall the definition of the Marcinkiewicz spaces, $\mathcal{M}^p(\Omega)$, as

$$\mathcal{M}^p(\Omega) = \Big\{f : \Omega \to \mathbb{R} \mid \text{ measurable such that } \sup_{\lambda>0} \lambda^p \big|\{x \in \Omega \mid |f(x)| \geq \lambda\}\big| < \infty\Big\}.$$

The Marcinkiewicz space, $\mathcal{M}^p(\Omega)$, coincides with the Lorenz space $L^{p,\infty}(\Omega)$, or weak-$L^p(\Omega)$. See for instance [114, Chapter 1] and [296, Chapter 5] for the definition and main properties of these spaces.

Theorem 8.3.13. *Let* $1 < p_0, p_1, q_0, q_1 < \infty$ *and* $\theta \in (0,1)$. *Define p and q in the following way:*

$$\frac{1}{p} = \frac{1-\theta}{p_0} + \frac{\theta}{p_1} \quad \textit{and} \quad \frac{1}{q} = \frac{1-\theta}{q_0} + \frac{\theta}{q_1}.$$

If T is a linear map with

$$T : L^{p_0}(\Omega) \to L^{q_0}(\Omega) \quad \textit{and} \quad T : \mathcal{M}^{p_1}(\Omega) \to L^{q_1}(\Omega),$$

then we have

$$\|Tf\|_{L^p} \leq c\|f\|_{L^q}.$$

Proof. See for instance [295, 296]. □

Next, we recall the Gagliardo–Nirenberg inequality.

Theorem 8.3.14 (Gagliardo–Nirenberg inequality). *Let v be a function in* $L^\rho(\mathbb{R}^N)$ *such that* $(-\Delta)^{\frac{s}{2}} v \in L^h(\mathbb{R}^N)$ *with* $h \geq 1$ *and* $\rho \geq 1$. *Then there exists a positive constant c, depending only on* N, s, h *and* ρ, *such that*

$$\|v\|_{L^\eta(\mathbb{R}^N)} \leq c\big\|(-\Delta)^{\frac{s}{2}} v\big\|^{\theta}_{L^h(\mathbb{R}^N)} \|v\|^{1-\theta}_{L^\rho(\mathbb{R}^N)}, \tag{8.3.18}$$

for every η *and* θ *satisfying*

$$0 \leq \theta \leq 1, \quad 1 \leq \eta < +\infty, \quad \frac{1}{\eta} = \theta\Big(\frac{1}{h} - \frac{s}{N}\Big) + \frac{1-\theta}{\rho}.$$

For a complete account on results regarding extensions of the Gagliardo–Nirenberg type inequalities to the fractional setting, see the paper by Brezis and Mironescu [92] and the more recent work [93].

In particular we are interested in the following result.

Theorem 8.3.15. *Assume that* v *belongs to* $W^{s_1,p_1}(\mathbb{R}^N) \cap W^{s_2,p_2}(\mathbb{R}^N)$, *for some* s_i *and* $p_i > 0$, $i = 1, 2$. *Then for* $0 \le s_1 < s_2 < \infty$, $1 < p_1 < \infty$, $1 < p_2 < \infty$,

$$s = \theta s_1 + (1-\theta)s_2, \quad \frac{1}{p} = \frac{\theta}{p_1} + \frac{1-\theta}{p_2},$$

we have $\|v\|_{W^{s,p}(\mathbb{R}^N)} \le c\|v\|_{W^{s_1,p_1}(\mathbb{R}^N)}\|v\|_{W^{s_2,p_2}(\mathbb{R}^N)}$.

Proof. See [92, Corollary 3.2]. □

8.4 Elliptic problem: finite energy setting

We will study the summability of the *variational solutions* in terms of the summability of the source term.

Consider $u \in H_0^s(\Omega)$, the solution to the Dirichlet problem

$$\begin{cases} \mathcal{L}u = f & \text{in } \Omega, \\ u = 0 & \text{in } \mathbb{R}^N \setminus \Omega, \end{cases} \tag{8.4.1}$$

where $\Omega \subset \mathbb{R}^N$, $N \ge 2$, is a bounded domain with Lipschitz boundary and $f \in H^{-s}(\Omega)$.

Observe that since $L^{(2_s^*)'}(\Omega) \subset H^{-s}(\Omega)$, $u \in H_0^s(\Omega)$ is a finite energy solution to (8.3.11) with $f \in L^{(2_s^*)'}(\Omega)$ if it satisfies

$$\frac{1}{2}\iint_{D_\Omega} (u(x)-u(y))(w(x)-w(y))\mathcal{K}(x,y)\,dx\,dy = \int_\Omega f(x)w(x)\,dx, \quad \forall w \in H_0^s(\Omega).$$

We can deal, from now on, with positive solutions of problems with positive data. Indeed, such an assumption is not restrictive, since the general case can be obtained by decomposing the datum into its positive and negative parts and then dealing with the two data separately, thanks to the linearity of the operator.

Note that the existence of solution can also be obtained exploiting the variational structure of the equation. Indeed, the equation in (8.4.1) turns out to be the Euler–Lagrange equation associated to the following quadratic functional in $H_0^s(\Omega)$:

$$J(u) = \frac{1}{2}\mathcal{E}(u,u) - \langle u, f\rangle. \tag{8.4.2}$$

Actually it is not hard to see that the above functional is lower bounded, coercive and weakly lower semicontinuous (with respect to the weak topology of $H_0^s(\Omega)$). Hence we can deduce the existence of solution of (8.4.1) as a minimum of such a functional.

As a consequence of the uniqueness given by the Lax–Milgram theorem we can define the following continuous map:

$$\begin{aligned} T : H^{-s}(\Omega) &\to H_0^s(\Omega), \\ f &\to T(f) = u, \quad \text{the solution to (8.4.1).} \end{aligned} \tag{8.4.3}$$

Note that in the case of $\mathcal{L} = (-\Delta)^s$, T happens to be the Green operator.

Once the existence and uniqueness in the energy framework are guaranteed, we turn to the study of the summability of the solution to the Dirichlet problem depending on the summability of the datum f. More precisely, by Theorem 8.2.7, $L^{\frac{2N}{N+2s}}(\Omega) \subset H^{-s}(\Omega)$ and then if $f \in L^{\frac{2N}{N+2s}}(\Omega)$, the unique solution u to problem (8.4.1) satisfies $u \in L^{\frac{2N}{N-2s}}(\Omega)$ and

$$\|u\|_{L^{\frac{2N}{N-2s}}(\Omega)} \leq c\|f\|_{L^{\frac{2N}{N+2s}}(\Omega)}.$$

The goal is to improve the inequality by assuming that $f \in L^m(\Omega)$ for $m > \frac{2N}{N+2s}$.

8.4.1 Bounded solutions: Moser and Stampacchia methods

For local problems (i. e., problems associated to operators of the type (8.3.7)–(8.3.8)), in the seminal paper [291] G. Stampacchia proved the boundedness of energy solutions through estimates on their level sets. See too [292] to find general results for elliptic operators in divergence form. Moreover, using a different type of arguments (namely, nonlinear test functions) J. Moser in [245] obtained the same type of results with L^∞-estimates for such solutions. We want to point out that in the fractional and probabilistic framework, the Stampacchia method has been used by Fukushima in [172]. For the reader's convenience, we provide the details of the results under a partial differential equation formulation and give the method by Moser in the fractional setting. The result that we prove is the following.

Theorem 8.4.1. *Let f belong to $L^m(\Omega)$, with $m > \frac{N}{2s}$. There exists a constant C, only depending on λ, N, Ω, $\|u\|_{H_0^s(\Omega)}$, $\|f\|_{L^m(\Omega)}$ and s, such that the unique energy solution of* (8.4.1) *satisfies*

$$\|u\|_{L^\infty(\Omega)} \leq C. \tag{8.4.4}$$

We present two alternative proofs, reminiscent of those by Moser and Stampacchia. They represent different approaches and contain useful techniques applicable in many other cases.

Boundedness of the solution via the Moser method

Proof. Assume with loss of generality that $|\Omega| = 1$. For $\beta > 1$ and $T > 0$ large, set the following convex function:

$$\Phi(\sigma) = \Phi_T(\sigma) = \begin{cases} \sigma^\beta & \text{if } 0 \leq \sigma < T, \\ \beta T^{\beta-1}\sigma - (\beta-1)T^\beta & \text{if } \sigma \geq T. \end{cases}$$

Since Φ is Lipschitz (with constant $L_\Phi = \beta T^{\beta-1}$) and $\Phi(0) = 0$, we have $\Phi(u) \in H_0^s(\Omega)$ and, using (8.3.15),

$$\mathcal{L}\Phi(u) \leq \Phi'(u)\mathcal{L}u. \tag{8.4.5}$$

Since Φ is positive, multiplying both sides of (8.4.5) by $\Phi(u)$ itself, we deduce

$$\lambda\mathcal{S}\|\Phi(u)\|^2_{L^{\frac{2N}{N-2s}}(\Omega)} \le \lambda\|\Phi(u)\|^2_{H_0^s(\Omega)} \le \int_\Omega \Phi(u)\mathcal{L}\Phi(u).$$

On the other hand, using one more time the convexity of Φ,

$$\begin{aligned}\int_\Omega \Phi(u)\mathcal{L}\Phi(u) &\le \int_\Omega \Phi(u)\Phi'(u)\mathcal{L}u = \int_\Omega \Phi(u)\Phi'(u)f \\ &\le \|f\|_{L^m(\Omega)}\|\Phi(u)\Phi'(u)\|_{L^{\frac{m}{m-1}}(\Omega)} \le \|f\|_{L^m(\Omega)}\|\beta u^{2\beta-1}\|_{L^{m'}(\Omega)},\end{aligned}$$

where we have used that $\Phi(r) \le r^\beta$ and that $\Phi'(r) \le \beta r^{\beta-1}$.

Combining both estimates, writing $p = \frac{2N}{N-2s}$ and $q = 2m'$ and letting $T \to \infty$ we arrive at

$$\begin{aligned}\|u\|_{L^{\beta p}(\Omega)} = \left(\int_\Omega u^{\beta p}\,dx\right)^{\frac{1}{\beta p}} &\le (c\beta\|f\|_{L^m(\Omega)})^{\frac{1}{2\beta}}\left(\int_\Omega u^{\frac{2\beta-1}{2}q}\,dx\right)^{\frac{1}{\beta q}} \\ &\le (c\beta\|f\|_{L^m(\Omega)})^{\frac{1}{2\beta}}\left(1 + \int_\Omega u^{\beta q}\,dx\right)^{\frac{1}{\beta q}}.\end{aligned}$$

The rest is a recurrence argument. Taking $\beta_0 = 1$, we define $\beta_j = (\frac{p}{q})^j$ so that the following identity holds: $\beta_{j+1}q = \beta_j p$. Let us denote

$$A_j = \left(\int_\Omega u^{\beta_j p}\right)^{\frac{1}{\beta_j p}}, \quad c_j = (c\|f\|_{L^m(\Omega)}\beta_j)^{\frac{1}{2\beta_j}}.$$

We have the recurrence formula $A_{j+1} \le c_{j+1}(1 + A_j^{\beta_j p})^{\frac{1}{\beta_j p}}$. By renormalizing if needed, we may assume that $A_0 = 1$ and then, since we are in a probability space, $A_j \ge 1, \forall j$. Taking logarithms,

$$\log A_{j+1} \le \log c_{j+1} + \frac{1}{\beta_j p}\log(1 + A_j^{\beta_j p}) \le \log c_{j+1} + \frac{1}{\beta_j p} + \log A_j,$$

where we have used that $\log(1 + x) \le 1 + \log x$ if $x \ge 1$. By induction,

$$\log A_{j+1} \le \sum_{k=1}^{j+1}\log c_k + \sum_{k=0}^{j}\frac{1}{\beta_k p} + \log A_0.$$

Observe also that $A_0 \le \|u\|_{H_0^s(\Omega)}$ and that the two series

$$\sum_{k=1}^\infty \log c_k = \sum_{k=1}^\infty \frac{1}{2\beta_k}\log(C\beta_k\|f\|_{L^m(\Omega)}), \quad \sum_{k=1}^\infty \frac{1}{\beta_k p},$$

are convergent. This ends the proof since $\lim_{j\to\infty} A_j = \|u\|_{L^\infty(\Omega)}$. □

Boundedness of solution via the Stampacchia method

In this subsection we give another proof of the boundedness of the solution of (8.4.1), for $f \in L^p(\Omega)$, $p > \frac{N}{2s}$. In this case the proof follows the idea of the Stampacchia theorem for second order elliptic equations with bounded coefficients.

The proof uses the following numerical iteration result, which is part of a more general result in [291].

Lemma 8.4.2. *Let $\psi : \mathbb{R}^+ \to \mathbb{R}^+$ be a nonincreasing function such that*

$$\psi(h) \le \frac{M\psi(k)^\delta}{(h-k)^\gamma}, \qquad \forall h > k > 0,$$

where $M > 0$, $\delta > 1$ and $\gamma > 0$. Then $\psi(d) = 0$, where $d^\gamma = M\psi(0)^{\delta-1}2^{\frac{\delta\gamma}{\delta-1}}$.

Proof. Consider the sequence $k_r = d - \frac{d}{2^r}$, $r = 0, 1, 2 \dots$. In this way $k_0 = 0$ and by induction we have

$$\psi(k_r) \le \frac{\psi(0)}{2^{\mu r}}, \quad \text{where } \mu = \frac{\gamma}{\delta - 1}. \tag{8.4.6}$$

Indeed, the inequality holds for $r = 0$ with an equality. Assume that the inequality holds for r. Since, by hypothesis,

$$\psi(k_{r+1}) \le \frac{M}{(k_r - k_{r+1})^\gamma}\psi(k_r)^\delta,$$

we have

$$\begin{aligned}\psi(k_{r+1}) &\le M\Big(\frac{2^{r+1}}{d}\Big)^\gamma\Big(\frac{\psi(0)}{2^{\mu r}}\Big)^\delta \\ &= \Big(\frac{M\psi(0)^{\delta-1}}{d^\gamma}2^{\frac{\gamma\delta}{\gamma-1}}\Big)\frac{\psi(0)}{2^{\mu(r+1)}} = \frac{\psi(0)}{2^{\mu(r+1)}}.\end{aligned}$$

Passing to the limit as $r \to \infty$ we obtain $\psi(d) = 0$. □

Proof of Theorem 8.4.1. Let $k > 0$ and choose $w = G_k(u)$ as a test function in the energy formulation of (8.4.1). Hence, thanks to (8.3.13), we deduce that

$$\iint_{D_\Omega} |G_k(u(x)) - G_k(u(y))|^2 \mathcal{K}(x,y)\,dx\,dy \le \int_{A_k} f(x)G_k(u(x))\,dx,$$

where $A_k = \{x \in \Omega : u \ge k\}$. Applying the Sobolev inequality on the left hand side,

$$\lambda S^{-2}\|G_k(u)\|^2_{L^{2^*_s}(\Omega)} \le \iint_{D_\Omega} |G_k(u(x)) - G_k(u(y))|^2 \mathcal{K}(x,y)\,dx\,dy$$

and the Hölder inequality on the right hand side,

$$\Big|\int_{A_k} f(x)G_k(u(x))\,dx\Big| \le \|f\|_{L^m(\Omega)}\|G_k(u)\|_{L^{2^*_s}(\Omega)}|A_k|^{1-\frac{1}{2^*_s}-\frac{1}{m}},$$

we have

$$\lambda \|G_k(u)\|_{L^{2^*_s}(\Omega)} \le \mathcal{S}^2 \|f\|_{L^m(\Omega)} |A_k|^{1-\frac{1}{2^*_s}-\frac{1}{m}}.$$

Consequently, for any $h > k$, we have $A_h \subset A_k$ and $G_k(u)\chi_{A_h} \ge (h-k)$ and thus

$$(h-k)|A_h|^{\frac{1}{2^*_s}} \le \frac{\mathcal{S}^2}{\lambda} \|f\|_{L^m(\Omega)} |A_k|^{1-\frac{1}{2^*_s}-\frac{1}{m}}.$$

Manipulating the above inequality we deduce that

$$|A_h| \le \frac{\mathcal{S}^{22^*_s} \|f\|^{2^*_s}_{L^m(\Omega)} |A_k|^{2^*_s(1-\frac{1}{2^*_s}-\frac{1}{m})}}{\lambda^{2^*_s}(h-k)^{2^*_s}}.$$

Since $m > \frac{N}{2s}$ we have

$$2^*_s\left(1 - \frac{1}{2^*_s} - \frac{1}{m}\right) > 1.$$

Now we apply Lemma 8.4.2 with the choice $\psi(\sigma) = |A_\sigma|$. Consequently there exists k_0 such that $\psi(k) \equiv 0$ for any $k \ge k_0$ and thus $\operatorname{ess\,sup}_\Omega u \le k_0$. □

8.4.2 The limit case $m = \frac{N}{2s}$: exponential summability

As in the local case, we stress that for the limit case $f \in L^{N/2s}(\Omega)$ the boundedness of the solutions cannot be expected, but solutions satisfy an exponential summability. Here we state the result for such a case.

Theorem 8.4.3. *Assume that $f \in L^{\frac{N}{2s}}(\Omega)$. Then the unique energy solution u to (8.4.1) satisfies the following estimate:*

$$\exists \alpha > 0 \quad s.t. \quad \int_\Omega e^{\alpha u}\, dx < \infty. \tag{8.4.7}$$

In particular, $u \in L^q(\Omega)$ for every $q < \infty$.

Proof. In order to prove the regularity of the energy solution (the existence follows by Theorem 8.3.5), let us consider for any $T > 0$ the following convex function:

$$\Phi(\sigma) = \Phi_T(\sigma) = \begin{cases} e^{\alpha\sigma} - 1 & \text{if } 0 \le \sigma < T, \\ \alpha e^{\alpha T}(\sigma - T) + e^{\alpha T} - 1 & \text{if } \sigma \ge T, \end{cases}$$

where $\alpha > 0$ will be fixed later. Let u be the unique $H^s_0(\Omega)$ solution to (8.4.1) and since $\Phi_T(\sigma)$ is a convex and globally Lipschitz function with $\Phi_T(0) = 0$, $\Phi_T(u)$ is an admissible test function. Thus, according to the Sobolev inequality and (8.3.15) we deduce that

$$\lambda\mathcal{S}\|\Phi(u)\|^2_{L^{2^*_s}(\Omega)} \le \int_\Omega \Phi'(u)\Phi(u)\mathcal{L}u = \int_\Omega \Phi'(u)\Phi(u)f. \tag{8.4.8}$$

Observe that for $u < T$ we have $\Phi'(u) = \alpha e^{\alpha u} = \alpha\Phi(u) + \alpha$, and so by (8.4.8) we split the last integral in two parts:

$$\lambda S\|\Phi(u)\|^2_{L^{2_s^*}(\Omega)} \le \alpha \int_{\{x\in\Omega:u<T\}} [\Phi(u)^2 f + \Phi(u) f]\, dx + \alpha e^{\alpha T} \int_{\{x\in\Omega:u\ge T\}} \Phi(u) f\, dx. \tag{8.4.9}$$

Using the Hölder inequality we have

$$\alpha \int_{\{x\in\Omega:u<T\}} [\Phi(u)^2 f + \Phi(u) f]\, dx \le \alpha\|\Phi(u)\|^2_{L^{2_s^*}(\Omega)}\|f\|_{L^{\frac{N}{2s}}(\Omega)}(1 + |\Omega|^{\frac{N-2s}{2N}}) + c(f),$$

and moreover, since Φ is increasing, we get

$$\alpha e^{\alpha T} \int_{\{x\in\Omega:u\ge T\}} \Phi(u) f\, dx \le \alpha \frac{e^{\alpha T}}{\Phi(T)} \int_{\{u\ge T\}} \Phi(u)^2 f \le 2\alpha\|\Phi(u)\|^2_{L^{2_s^*}(\Omega)}\|f\|_{L^{\frac{N}{2s}}(\Omega)},$$

provided that $\alpha T > 1$. Putting together the above estimates in (8.4.9) and fixing α small enough, we get a bound for the $L^{2_s^*}(\Omega)$-norm of $\Phi(u)$ that implies (8.4.7) once one takes T sufficiently large (depending on f, but not on u). □

8.4.3 A Calderón–Zygmund type result

In the finite energy setting we can obtain an optimal result on summability that extends to the fractional case of the Calderón–Zygmund theory. More precisely we have the following result.

Theorem 8.4.4. *Let f be a positive function that belongs to $L^m(\Omega)$, with $\frac{2N}{N+2s} \le m < \frac{N}{2s}$. Then, there exists a constant $c = c(N, m, s) > 0$ such that the unique energy solution to*

$$\begin{cases} \mathcal{L}u = f & \text{in } \Omega, \\ u = 0 & \text{in } \mathbb{R}^N \setminus \Omega \end{cases}$$

satisfies

$$\|u\|_{L^{m_s^{**}}(\Omega)} \le c\|f\|_{L^m(\Omega)}, \quad \text{where } m_s^{**} = \frac{mN}{N - 2ms}. \tag{8.4.10}$$

Proof. The existence of a unique energy solution was obtained in Theorem 8.3.5. Hence we just have to deal with the regularity of the solution.

Define the following convex, differentiable function:

$$\Phi(\sigma) = \begin{cases} \sigma^\beta & \text{if } 0 \le \sigma \le T, \\ \beta T^{\beta-1}(\sigma - T) + T^\beta & \text{if } \sigma > T, \end{cases} \tag{8.4.11}$$

with $\beta = \frac{m_s^{**}}{2_s^*} > 1$.

Note that since Φ is convex and $\Phi(0) = 0$, according to Proposition 8.3.6, $\Phi(u)$ is an admissible test function. Hence by (8.3.15) we have

$$\mathcal{L}\Phi(u) \le f(x)\Phi'(u) \quad \text{weakly in } \Omega.$$

Multiplying the above inequality one more time by $\Phi(u)$, we deduce that

$$\mathcal{E}(\Phi(u), \Phi(u)) \le \int_\Omega f\Phi(u)\Phi'(u)\,dx.$$

The Hölder and Sobolev inequalities yield

$$\lambda\mathcal{S}\|\Phi(u)\|^2_{L^{2_s^*}(\Omega)} \le \|f\|_{L^m(\Omega)}\Big(\int_\Omega |\Phi'(u)\Phi(u)|^{m'}\Big)^{\frac{1}{m'}}. \tag{8.4.12}$$

An elementary computation shows that there exists a constant c, independent of T, such that

$$\forall\sigma \ge 0, \quad |\Phi'(\sigma)\Phi(\sigma)|^{m'} \le c|\Phi(\sigma)|^{2_s^*},$$

so that

$$\lambda\mathcal{S}\Big(\int_\Omega |\Phi(u)|^{2_s^*}\Big)^{\frac{2}{2_s^*}-\frac{1}{m'}} \leqslant c\|f\|_{L^m(\Omega)}.$$

Finally, since $\beta 2_s^* = m'(2\beta - 1) = m_s^{**}$ and $\frac{2}{2_s^*} - \frac{1}{m'} = \frac{1}{m_s^{**}}$, we conclude, taking the limit as T diverges, that

$$\|u\|_{L^{m_s^{**}}(\Omega)} \leqslant c\|f\|_{L^m(\Omega)},$$

where c depends on $\lambda, \mathcal{S}, s, N, m$ and Ω. □

8.4.4 Further fractional regularity

This section is devoted to prove an interesting regularity property of the energy solutions of (8.4.1).

Assume that Ω has a bounded Lipschitz boundary, so that we ensure that Ω has the extension property. By Theorem 8.2.6 any function in $H_0^s(\Omega)$ extended as 0 outside Ω belongs to $W^{s,2}(\mathbb{R}^N)$. Thus, we can apply the interpolation result that we stated in Theorem 8.3.15 and we get the following regularity result.

Theorem 8.4.5. *Let u be an energy solution of (8.4.1) with f in $L^m(\Omega)$, $m \ge (2_s^*)'$. Then u belongs to $W_0^{\sigma,p}(\Omega)$, where*

$$\sigma = \theta s \quad \text{and} \quad \frac{1}{p} = \frac{1}{m} + \theta\Big(\frac{1}{2} - \frac{1}{m}\Big) - \frac{2s}{N}(1-\theta), \quad \theta \in (0,1). \tag{8.4.13}$$

The Sobolev space $W_0^{\sigma,p}(\Omega)$, $\sigma \in (0,1)$, $p > 1$, is defined as the completion of $\mathcal{C}_0^\infty(\Omega)$ with respect to the seminorm

$$||u||^p_{W_0^{\sigma,p}} := \int_\Omega \int_\Omega \frac{|u(x) - u(y)|^p}{|x-y|^{N+p\sigma}} dxdy.$$

The reader can see the functional properties of this space for example in [28] and in [138]. Below, in Chapter 10, we will return to these Sobolev spaces.

Proof. The proof easily follows applying Theorem 8.3.15 and interpolating between the Sobolev space $H_0^s(\Omega) \equiv W_0^{s,2}(\Omega)$ (i. e., $\theta = 1$) and $L^{m_s^{**}}(\Omega)$ (i. e., $\theta = 0$). □

Observe that the regularity given by the above theorem does not appear in the literature for the local case ($s = 1$), where one looks for estimates either on u or on Δu.

8.5 A fractional Picone inequality and applications to sublinear problems

We formulate an extension of an inequality obtained by Picone in [259]. To be more precise, Picone considers that if $u, v \in \mathcal{C}^2(a,b)$ and $u > \delta > 0$ in (a,b), then

$$(v')^2 \geq \left(\frac{v^2}{u}\right)' u'.$$

The corresponding extension of this pointwise inequality to higher dimensions is that if $\Omega \subset \mathbb{R}^N$, $u, v \in \mathcal{C}^2(\Omega)$ and $u > \delta > 0$ on Ω, then

$$|\nabla v|^2 \geq \left\langle \nabla\left(\frac{v^2}{u}\right), \nabla u \right\rangle.$$

Moreover, for any $1 < p < \infty$,

$$|\nabla v|^p \geq \left\langle \nabla\left(\frac{v^p}{u^{p-1}}\right), |\nabla u|^{p-2}\nabla u \right\rangle;$$

see Section 1.3.

The useful integral form was obtained in a quite general framework in [14], where also some interesting applications were obtained. It is precisely this integral form that we prove for the fractional operator $\mathcal{L}$. More precisely, we obtain the following result.

Theorem 8.5.1 (Picone inequality). *Consider $u, v \in H_0^s(\Omega)$ and assume that $\mathcal{L}u$ is a positive bounded Radon measure in Ω, with $u \gneq 0$. Then,*

$$\int_\Omega \frac{\mathcal{L}u}{u} v^2 \, dx \leq \mathcal{E}(v,v). \tag{8.5.1}$$

Proof. Let us recall first that for any $\phi, \psi \in H_0^s(\Omega)$, we have, thanks to the symmetry of $\mathcal{L}$,

$$\int_\Omega \phi \mathcal{L}\psi \, dx = \int_\Omega \mathcal{L}\phi\psi \, dx = \mathcal{E}(\phi, \psi) \le \lambda^{-1} \|\phi\|_{H_0^s(\Omega)} \|\psi\|_{H_0^s(\Omega)}. \tag{8.5.2}$$

We set, for any $k, \eta > 0$, $v_k = T_k(v)$, $\tilde{u} = u + \eta$ and we define $w = \frac{v_k^2}{\tilde{u}}$; actually it is not hard to see that $w \in H_0^s(\Omega)$. We want to prove that, $\forall u, v \in H_0^s(\Omega)$ and $\forall k, \eta > 0$ and thanks to (8.5.2),

$$\mathcal{E}(u, w) = \int \mathcal{L}u \frac{v_k^2}{\tilde{u}} \le \mathcal{E}(v_k, v_k) \le \lambda^{-1} \|T_k(v)\|^2_{H_0^s(\Omega)}. \tag{8.5.3}$$

Once we have proved such an inequality, (8.5.1) follows by letting k diverge and η vanish, using the monotone convergence theorem and Fatou's lemma (this is why we ask $\mathcal{L}u$ to be a positive Radon measure). Observe that

$$\begin{aligned}(u(x) - u(y))(w(x) - w(y)) &= (\tilde{u}(x) - \tilde{u}(y))\left(\frac{v_k^2(x)}{\tilde{u}(x)} - \frac{v_k^2(y)}{\tilde{u}(y)}\right) \\ &= v_k^2(x) + v_k^2(y) - v_k^2(x)\frac{\tilde{u}(y)}{\tilde{u}(x)} - v_k^2(y)\frac{\tilde{u}(x)}{\tilde{u}(y)}.\end{aligned} \tag{8.5.4}$$

Thus, by the Young inequality we deduce

$$v_k^2(x)\frac{\tilde{u}(y)}{\tilde{u}(x)} + v_k^2(y)\frac{\tilde{u}(x)}{\tilde{u}(y)} \ge 2v_k(y)v_k(x)$$

and so, gathering together the above inequality with (8.5.4), we get

$$(u(x) - u(y))(w(x) - w(y)) \le (v_k(x) - v_k(y))^2.$$

Hence, (8.5.3) follows by changing the above inequality into the corresponding integral against the kernel $\mathcal{K}(x, y)$. □

Some extensions of the Picone inequality

The general Picone inequality in Section 1.3 can be extended to the fractional framework by observing the next result.

Lemma 8.5.2. *Let $1 < p < \infty$, $1 < q \le p$ and let u_1, u_2 be two measurable functions with $u_2 \geqslant 0$ and $u_1 > 0$. Then the following inequality holds:*

$$\begin{aligned}&|u_1(x) - u_1(y)|^{p-2}(u_1(x) - u_1(y))\left[\frac{u_2(x)^q}{u_1(x)^{q-1}} - \frac{u_2(y)^q}{u_1(y)^{q-1}}\right] \\ &\quad \leqslant |u_2(x) - u_2(y)|^q |u_1(x) - u_1(y)|^{p-q}.\end{aligned} \tag{8.5.5}$$

Now we can state Picone's inequality related to the fractional p-Laplacian defined by

$$(-\Delta)_p^s u(x) := P.V. \int_{\mathbb{R}^N} \frac{|u(x)-u(y)|^{p-2}(u(x)-u(y))}{|x-y|^{N+ps}}\,dy. \tag{8.5.6}$$

Proposition 8.5.3. *Consider $u, v \in W_0^{s,p}(\Omega)$ with $u \geq 0$. Assume that $(-\Delta)_p^s u$ is a positive bounded Radon measure in Ω. Then*

$$2\int_\Omega (-\Delta)_p^s u \frac{v^p}{u^{p-1}} \leqslant \|v\|^p_{W_0^{s,p}(\Omega)}. \tag{8.5.7}$$

The detailed proofs can be found in [49].

An application

Next we present an application of the Picone inequality to a semilinear problem first studied by Brezis–Oswald in [95], and then by Brezis–Kamin in [90]. To be more precise, consider the semilinear problem

$$\begin{cases} \mathcal{L}u = f(x,u) & \text{in } \Omega, \\ u > 0 & \text{in } \Omega, \\ u \equiv 0 & \text{in } \mathbb{R}^N \setminus \Omega, \end{cases} \tag{8.5.8}$$

where $f : \Omega \times [0,\infty) \to \mathbb{R}$ is a Caratheodory function. The aim is to give some condition on f to have uniqueness of solutions to (8.5.8).

Observe that if $\mathcal{L}$ is a linear differential operator of second order, this problem is classical and was solved by H. Brezis and L. Oswald in 1986 (see [95] and [90]). Indeed, requiring that the nonlinearity satisfies

$$\sigma \to \frac{f(x,\sigma)}{\sigma} \quad \text{is decreasing for a. e. } x \in \Omega, \tag{8.5.9}$$

they proved uniqueness for solutions of

$$\begin{cases} -\Delta u = f(x,u) & \text{in } \Omega, \\ u > 0 & \text{in } \Omega, \\ u = 0 & \text{on } \partial\Omega. \end{cases}$$

Here we prove the analog for problem (8.5.8), associated to the operator $\mathcal{L}$. First, we need to recall the following definition.

Definition 8.5.4. A subsolution (supersolution) to (8.5.8) is a positive function $u \in H_0^s(\Omega) \cap L^\infty(\Omega)$ such that $f(x,u) \in L^1(\Omega)$ and

$$\int_\Omega u\mathcal{L}\varphi \leq (\text{respectively } \geq) \int_\Omega f(x,u)\varphi, \quad \forall \varphi \in H_0^s(\Omega),\ \varphi \geq 0. \tag{8.5.10}$$

A solution is both a subsolution and a supersolution.

Now we can state our Brezis–Kamin–Oswald type result.

Theorem 8.5.5. *Let u and v be a sub- and a supersolution, respectively, to* (8.5.8). *Assume that* (8.5.9) *holds true. Then $u \leq v$ a. e. in Ω.*

Remark 8.5.6. Before proving the comparison result that in particular proves the uniqueness, observe that solutions of problem (8.5.8) exist under different conditions. For instance, according to [95], the following hypotheses are sufficient:
(f1) the function $x \longmapsto f(x,\sigma)$ belongs to $L^\infty(\Omega)$ for each $\sigma \geqslant 0$;
(f2) $\frac{f(x,\sigma)}{\sigma} \to a_0(x)$ as $\sigma \to 0$, where $-\infty < -c \leqslant a_0(x) \leqslant \infty$ and $\frac{f(x,\sigma)}{\sigma} \to 0$, as $\sigma \to \infty$, uniformly in Ω;
(f3) there exists a function $\phi \in H_0^s(\Omega) \cap L^\infty(\Omega)$ such that

$$\|\phi\|^2_{H_0^s(\Omega)} - \int_{\{\phi \neq 0\}} a_0(x)|\phi(x)|^2\,dx < 0.$$

Proof. Define the subset $\Omega_0 = \{x \in \Omega \mid u(x) > v(x)\}$. We have to prove that $|\Omega_0| = 0$. Set $\phi = (u^2 - v^2)_+$. Then $\frac{\phi}{u}$ and $\frac{\phi}{v}$ are admissible test functions, thanks to the Picone inequality (8.5.1). Hence

$$\int_\Omega u\mathcal{L}\Big(\frac{\phi}{u}\Big)\,dx - \int_\Omega v\mathcal{L}\Big(\frac{\phi}{v}\Big)\,dx \leq \int_\Omega \Big(\frac{f(x,u(x))}{u(x)} - \frac{f(x,v(x))}{v(x)}\Big)\phi(x)\,dx. \tag{8.5.11}$$

Note that the right hand side is negative due to (8.5.9), so if we show that the left hand side is positive, then $\phi \equiv 0$ and we conclude the proof. To prove that the left hand side is nonnegative, observe first that if we call $u_0 = u\chi_{\Omega_0}$, $v_0 = v\chi_{\Omega_0}$ and $\phi = u_0^2 - v_0^2$, it suffices to prove the following pointwise inequality:

$$\begin{aligned}(v(x) - v(y))&\Big(\frac{u_0^2(x) - v_0^2(x)}{v(x)} - \frac{u_0^2(y) - v_0^2(y)}{v(y)}\Big)\\ &\leq (u(x) - u(y))\Big(\frac{u_0^2(x) - v_0^2(x)}{u(x)} - \frac{u_0^2(y) - v_0^2(y)}{u(y)}\Big).\end{aligned} \tag{8.5.12}$$

By rearranging the terms, (8.5.12) is equivalent to

$$\begin{aligned}&\underbrace{(v(x) - v(y))\Big(\frac{u_0^2(x)}{v(x)} - \frac{u_0^2(y)}{v(y)}\Big)}_{(A)} + \underbrace{(u(x) - u(y))\Big(\frac{v_0^2(x)}{u(x)} - \frac{v_0^2(y)}{u(y)}\Big)}_{(B)}\\ &\quad\leq \underbrace{(u(x) - u(y))(u_0(x) - u_0(y))}_{(C)} + \underbrace{(v(x) - v(y))(v_0(x) - v_0(y))}_{D}.\end{aligned}$$

By symmetry it is sufficient to consider the following three cases:
(i) $x, y \notin \Omega_0$; the inequality is trivial because the terms (A), (B), (C) and (D) are identically zero;
(ii) $x, y \in \Omega_0$; by the pointwise Picone inequality we have $(A) \leq (C)$ and $(B) \leq (D)$;

(iii) $x \in \Omega_0$ and $y \notin \Omega_0$; in this case,

$$\begin{cases} (A) + (B) = u_0^2(x) - u_0^2(x)\frac{v(y)}{u_0(x)} - v_0^2(x)\frac{u(y)}{u_0(x)} + v_0^2(x), \\ (C) + (D) = u_0^2(x) - u_0(x)u(y) + v_0^2(x) - v_0(x)v(y). \end{cases}$$

Therefore $(A) + (B) \leq (C) + (D)$ is equivalent to

$$u(y)\left(u_0(x) - \frac{v_0^2(x)}{u_0(x)}\right) \leq v(y)\left(\frac{u_0^2(x)}{v_0(x)} - v_0(x)\right),$$

which itself is equivalent to

$$\frac{u(y)}{u_0(x)} \leq \frac{v(y)}{v_0(x)},$$

that is, $u(y)v_0(x) \leq v(y)u_0(x)$. This is true because $x \in \Omega_0$ and $y \notin \Omega_0$, and then $u(y) \leq v(y)$ and $v_0(x) \leq u_0(x)$.

So, we conclude the proof. □

Remark 8.5.7. A natural extension to the case of the fractional p-Laplacian can be found using Proposition 8.5.3.

8.6 Nonvariational setting for elliptic problems: weak solutions

When $f \in L^m(\Omega)$ with $1 \leq m < \frac{2N}{N+2s}$ we cannot expect a solution of finite energy. However, as in the local case, by relaxing the meaning of solution we can show the existence of weaker solutions for nonvariational data. Consider the problem

$$\begin{cases} \mathcal{L}u = f & \text{in } \Omega, \\ u = 0 & \text{in } \mathbb{R}^N \setminus \Omega. \end{cases} \tag{8.6.1}$$

In the case of $\Omega = \mathbb{R}^N$ the existence and uniqueness of *renormalized solutions* is obtained in [2] (see also [206]). However we consider a *weak solution* to (8.6.1) in a very elementary context, in bounded domains and making explicit the summability of the solution.

Definition 8.6.1. We define the class of test functions

$$\mathcal{T}(\Omega) = \{\phi \mid \mathcal{L}(\phi) = \psi \text{ in } \Omega,\ \phi = 0 \text{ in } \mathbb{R}^N \setminus \Omega,\ \psi \in C_0^\infty(\Omega)\}. \tag{8.6.2}$$

Note that if $v \in \mathcal{T}(\Omega)$, then, using the results in the previous section, $v \in H_0^s(\Omega) \cap L^\infty(\Omega)$. Moreover, according to the regularity theory developed in [287], if Ω is smooth enough, there exists a constant $\beta > 0$ (that depends only on the structural constants) such that $v \in C^\beta(\Omega)$ (see also [207]).

Definition 8.6.2. We say that $u \in L^1(\Omega)$ is a *weak solution* to (8.6.1) if for $f \in L^1(\Omega)$ we have

$$\int_\Omega u\psi\,dx = \int_\Omega f\phi\,dx$$

for any $\phi \in \mathcal{T}(\Omega)$ with $\psi \in C_0^\infty(\Omega)$.

Remark 8.6.3. We recall that if in the definition of the test function space, we consider $\psi \in L^m(\Omega)$, $m > N/2$, we recover the classical definition of *duality solution* introduced by Stampacchia in [291]. Indeed, using Theorem 8.4.1, ϕ is bounded and consequently all the terms in the above identity make sense.

We also stress that such a definition can be extended to equations with measure data if we have a regularity result that guarantees the continuity, up to the boundary, of the solution of the problem $\mathcal{L}(\phi) = \psi$ in Ω, $\phi = 0$ in $\mathbb{R}^N \setminus \Omega$. Results in this direction can be found in [206].

Now we can state the existence result for L^1-data.

Theorem 8.6.4. *For any f belonging to $L^1(\Omega)$ there exists a unique weak solution to* (8.6.1). *Moreover,*

$$\forall k \geq 0, \quad T_k(u) \in H_0^s(\Omega), \tag{8.6.3}$$

$$u \in L^q(\Omega), \quad \forall q \in \left(1, \frac{N}{N-2s}\right), \tag{8.6.4}$$

and

$$\left|(-\Delta)^{\frac{s}{2}} u\right| \in L^r(\Omega), \quad \forall r \in \left(1, \frac{N}{N-s}\right). \tag{8.6.5}$$

Remark 8.6.5. The philosophy behind the above result follows the ideas that come from the analog in the local case (see, for instance [291] and [69] for the nonlinear case). Indeed, as already observed, we cannot expect solutions to belong to the Sobolev space $H_0^s(\Omega)$ since $L^1(\Omega)$ is not embedded into $H^{-s}(\Omega)$. Anyway, we prove that any truncation at level k does belong to the energy space.

On the other hand, observe that if $s = 1$ we recover the existence and regularity results that are, nowadays, well known for second order elliptic operators in divergence form with L^1-data. In the local case, according to the results mentioned above, weak solutions u are such that $|\nabla u| \in L^r(\Omega)$, $\forall r \in (1, \frac{N}{N-1})$, and $u \in L^q(\Omega)$, $\forall q \in (1, \frac{N}{N-2})$.

Before we begin with the proof, it is interesting to observe that the two estimates (8.6.4) and (8.6.5) are quite different from one another. This is because since u has been defined as zero outside Ω, (8.6.4) turns out to be a global estimate on u. On the other hand, the *a priori* estimate on the $s/2$-derivative of u cannot be extended to a global one. Also, the nonlocal nature of the equation makes $|(-\Delta)^{\frac{s}{2}} u|$ to be different from zero in $\mathbb{R}^N \setminus \Omega$, but since the integro-differential equation is set only in Ω we cannot get summability information outside Ω.

Proof of Theorem 8.6.4. The proof of the above theorem is split into two parts. We first prove uniqueness of a weak solution. We stress that the proof follows the idea of the one by Stampacchia. Subsequently we construct the solution by approximation with solutions of problems with smooth data. We prove suitable *a priori* estimates (that, in particular, imply (8.6.3)–(8.6.5)) and compactness of the approximating sequence in $L^1(\Omega)$.

UNIQUENESS. Assume that u is a weak solution to (8.6.1) with $f = 0$. Then

$$\int_\Omega u\psi\,dx = 0 \quad \text{for any } \psi \in C_0^\infty(\Omega).$$

Therefore $u \equiv 0$.

EXISTENCE. We obtain the solution to (8.6.1) as a limit of solutions to approximated problems.

Consider $f_n \in L^\infty(\Omega)$ such that $f_n \to f$ in $L^1(\Omega)$ and let u_n be the solution to the problem

$$\begin{cases} \mathcal{L}u_n = f_n & \text{in } \Omega, \\ u_n = 0 & \text{in } \mathbb{R}^N \setminus \Omega. \end{cases} \tag{8.6.6}$$

Then we prove the existence of solution in several steps.

Step 1. *There exists a positive constant c, only depending on N, Ω and s, such that*

$$\|u_n\|_{L^q(\Omega)} \le c\|f_n\|_{L^1(\Omega)}, \quad \forall q \in \left(1, \frac{N}{N-2s}\right). \tag{8.6.7}$$

Let us multiply the equation in (8.6.6) by $T_k(u_n)$, for $k \ge 0$, and let us integrate over Ω. Thanks to Proposition 8.3.6 we obtain

$$\lambda\|T_k(u_n)\|^2_{L^{2_s^*}(\Omega)} \le \mathcal{S}^2 k\|f_n\|_{L^1(\Omega)}. \tag{8.6.8}$$

Moreover by using the Sobolev inequality and an elementary estimate we have

$$\lambda k^2|A_{n,k}(u_n)|^{\frac{N-2s}{N}} \le \lambda\|T_k(u_n)\|^2_{L^{2_s^*}(\Omega)} \le \mathcal{S}^2 k\|f_n\|_{L^1(\Omega)},$$

where $A_{n,k}(u_n) = \{x \in \Omega : u_n(x) \ge k\}$. It follows that

$$|A_{n,k}(u_n)| \le c\left(\frac{\|f_n\|_{L^1(\Omega)}}{k}\right)^{\frac{N}{N-2s}}. \tag{8.6.9}$$

It means that u_n is bounded in the Marcinkiewicz space $\mathcal{M}^{\frac{N}{N-2s}}(\Omega)$ and consequently (8.6.7) holds true.

Step 2. *There exists a positive constant c, just depending on q, N, Ω and s, such that*

$$\|(-\Delta)^{\frac{s}{2}}u_n\|_{L^r(\Omega)} \le c\|f_n\|_{L^1(\Omega)}, \quad \forall r \in \left(1, \frac{N}{N-s}\right). \tag{8.6.10}$$

We fix $\lambda > 0$ and, for any positive k, we want to estimate the measure of the following set:

$$\begin{aligned}&\{x \in \Omega : |(-\Delta)^{\frac{s}{2}} u_n| \geq \lambda\}\\&\quad = \{x \in \Omega : |(-\Delta)^{\frac{s}{2}} u_n| \geq \lambda, u_n < k\} \cup \{x \in \Omega : |(-\Delta)^{\frac{s}{2}} u_n| \geq \lambda, u_n \geq k\}.\end{aligned}$$

Consequently,

$$\{x \in \Omega : |(-\Delta)^{\frac{s}{2}} u_n| \geq \lambda\} \subset \{x \in \Omega : |(-\Delta)^{\frac{s}{2}} u_n| \geq \lambda, u_n < k\} \cup A_{n,k}(u_n).$$

Since

$$|\{x \in \Omega : |(-\Delta)^{\frac{s}{2}} u_n| \geq \lambda, u_n < k\}| \leq \frac{1}{\lambda^2} \int_{\{x\in\Omega, u_n<k\}} |(-\Delta)^{\frac{s}{2}} u_n|^2 \, dx,$$

we apply Corollary 8.3.9 and deduce that

$$\begin{aligned}&|(\{x \in \Omega : |(-\Delta)^{\frac{s}{2}} u_n| \geq \lambda, u_n < k\})|\\&\quad \leq \frac{1}{\lambda^2} \int_{\mathbb{R}^N} |(-\Delta)^{\frac{s}{2}} T_k(u_n)|^2 \, dx \leq \frac{k}{\lambda^2} \|f_n\|_{L^1(\Omega)}.\end{aligned}$$

Moreover, using (8.6.9), for every $k > 0$,

$$|(\{x \in \Omega : |(-\Delta)^{\frac{s}{2}} u_n| \geq \lambda\})| \leq \frac{k}{\lambda^2} \|f_n\|_{L^1(\Omega)} + c\left(\frac{\|f_n\|_{L^1(\Omega)}}{k}\right)^{\frac{N}{N-2s}}.$$

Minimizing in k we find that the minimum is achieved at $k = \lambda^{\frac{N-2s}{N-s}} \|f_n\|_{L^1(\Omega)}^{\frac{s}{N-s}}$, thus we have

$$|(\{x \in \Omega : |(-\Delta)^{\frac{s}{2}} u_n| \geq \lambda\})| \leq c\left(\frac{\|f_n\|_{L^1(\Omega)}}{\lambda}\right)^{\frac{N}{N-s}}. \tag{8.6.11}$$

This means that $|(-\Delta)^{s/2} u_n|$ is bounded in the Marcinkiewicz space $\mathcal{M}^{\frac{N}{N-s}}(\Omega)$ and consequently (8.6.10) holds true.

Step 3. *Passing to the limit.*

Before passing to the limit in the equation, we need to determine the a. e. limit of u_n. Using the linearity of the equation, for any m and $n \in \mathbb{N}$, $u_n - u_m$ solves

$$\begin{cases}\mathcal{L}(u_n - u_m) = f_n - f_m & \text{in } \Omega,\\ u_n = 0, \; u_m = 0 & \text{in } \mathbb{R}^N \setminus \Omega.\end{cases}$$

Hence, choosing, for any $k > 0$, $T_k(u_n - u_m)$ as a test function in the weak formulation of the above problem, we deduce, by repeating the computations of step 1, that

$$|(\{x \in \Omega : |u_n - u_m| \geq k\})| \leq c\left(\frac{\|f_n - f_m\|_{L^1(\Omega)}}{k}\right)^{\frac{N}{N-2s}}.$$

Since the right hand side of the above inequality is small for n and m large enough, it follows that $\{u_n\}$ is a Cauchy sequence in measure. Consequently, up to subsequences (not relabeled), it converges in Ω almost everywhere, toward a function u.

By step 1 we also deduce (using the embedding of the $\mathcal{M}^p(\Omega)$ spaces into $L^p(\Omega)$, for all p finite) that u_n also converges to u in $L^q(\Omega)$, for any $1 \le q < \frac{N}{N-2s}$. Note that this is sufficient to pass to the limit in the equation and obtain a weak solution of (8.6.1). Observe that, by the uniqueness, the whole sequence converges to u in $L^q(\Omega)$ and that (8.6.4) holds.

Since (8.6.11) holds, we have

$$\left|\left(\{x \in \Omega : |(-\Delta)^{\frac{s}{2}}(u_n - u_m)| \ge \lambda\}\right)\right| \le c\left(\frac{\|f_n - f_m\|_{L^1(\Omega)}}{\lambda}\right)^{\frac{N}{N-s}},$$

thus $(-\Delta)^{\frac{s}{2}}(u_n)$ is a Cauchy sequence in measure in Ω. Therefore, up to a subsequence, $(-\Delta)^{\frac{s}{2}}(u_n)$ converges a. e. in Ω. □

8.6.1 Calderón–Zygmund type result for weak solutions

In the local case it is well known that the Calderón–Zygmund results are true if the right hand side is $L^m(\Omega)$ for $1 < m < \infty$ (see for instance [186, Chapter 9] and [69]). The existence of a unique weak solution of (8.4.1) is a consequence of Theorem 8.6.4. Thus we can define the operator T as in (8.4.3) and we can get the Calderón–Zygmund type results in such a range of m by using the interpolation Theorem 8.3.13. The statement of our last result concerning solutions of elliptic problems is the following.

Theorem 8.6.6. *Let f belong to $L^m(\Omega)$, with $1 < m < \frac{2N}{N+2s}$. Then there exists a unique weak solution of* (8.4.1). *Moreover there exists a constant c, only depending on N, Ω, m and s, such that*

$$\|u\|_{L^{m_s^{**}}(\Omega)} \le c\|f\|_{L^m(\Omega)} \tag{8.6.12}$$

and

$$\left\|(-\Delta)^{s/2}u\right\|_{L^{m_{s/2}^{**}}(\Omega)} \le c\|f\|_{L^m(\Omega)}, \tag{8.6.13}$$

where

$$\frac{1}{m_t^{**}} = \frac{1}{m} - \frac{2t}{N}, \qquad \forall t \in (0,1).$$

Proof. In order to get the estimate we argue by interpolation, namely, we apply the above interpolation theorem. We define the map T as the inverse of $\mathcal{L}$ (see (8.4.3)), and due to Theorem 8.6.4 we know that it acts between $L^1(\Omega)$ and $L^r(\Omega)$, for any $r < \frac{N}{N-2s}$.

On the other hand, estimate (8.6.12) in the case $(2_s^*)' = m$ tells us that it acts between $L^{(2_s^*)'}(\Omega)$ and $L^{2_s^*}(\Omega)$, too. Consequently

$$\|u_n\|_{L^p(\Omega)} = \|Tf_n\|_{L^p(\Omega)} \le c\|f_n\|_{L^q(\Omega)},$$

where

$$\frac{1}{p} = \frac{1-\theta}{2_s^*} + \frac{\theta}{r} \quad \text{and} \quad \frac{1}{q} = \frac{1-\theta}{(2_s^*)'} + \theta,$$

i. e.,

$$\frac{1}{q} - \frac{1}{p} = \frac{2s}{N}, \quad \text{which implies } \frac{1}{p} = \frac{1}{m_s^{**}}.$$

On the other hand, arguing in the same way with the norms of $|(-\Delta)^{s/2}u_n|$ we deduce (8.6.13). □

8.7 Further results

Consider the minimization problem (8.4.2), where f is a measurable function defined in Ω and the hypotheses about the kernel K hold. In order to have that for all $u \in H_0^s(\Omega)$, $fu \in L^1(\Omega)$, we must assume that $f \in L^{\frac{2N}{N+2s}}(\Omega)$ and, passing to the Euler–Lagrange equation, we want to find the natural conditions to have bounded solutions of the problem

$$\begin{cases} \mathcal{L}u = f & \text{in } \Omega, \\ u = 0 & \text{in } \mathbb{R}^N \setminus \Omega. \end{cases} \tag{8.7.1}$$

From the results in Section 8.4.1 we know that f must be in $L^m(\Omega)$ with $m > \frac{N}{2s}$ in order to have bounded solutions. This condition is natural in the *De Giorgi program*. The solution to the 19th Hilbert problem proposed in [200] was achieved by Ennio De Giorgi in his famous paper [137]. The De Giorgi program has two main steps.

Given a variational solution to problem (8.7.1),

Step 1. *L^2-estimates implies L^∞-estimates;*

Step 2. *bounded solutions of finite energy become C^α-continuous.*

The result by De Giorgi was also obtained by J. F. Nash in [250] by different methods and, later on, was reproved by J. Moser in [245] using a different approach. The De Giorgi proof has been applied in many different settings, for instance in [185] and [109]. The book [188] could be a good reference to see the modern development of the De Giorgi ideas in the local case. For the nonlocal case there are many references and the following is a (not exhaustive) list of them: [109], [124], [136], [207], [208], [273], [288].

In our case, the first step is solved by the methods of Stampacchia and Moser in Section 8.4.1 when $f \in L^m(\Omega)$, $m > \frac{N}{2s}$. That is

the solution $u \in H_0^s(\Omega)$ to problem (8.7.1) *with $f \in L^m(\Omega)$ for $m > \frac{N}{2s}$ is bounded.*

Regarding the second step, the following result holds.

Theorem 8.7.1. *Let $u \in H_0^s(\Omega) \cap L^\infty(\Omega)$ be the solution to problem* (8.7.1)*. Then u is Hölder continuous in Ω. More precisely, there exist α and C, both depending only on N, s and the ellipticity constants, such that if $B_{2r}(x_0) \subset \Omega$, then*

$$\operatorname{osc}_{B_\rho(x_0)}(u) \le C\left(\frac{\rho}{r}\right)^\alpha \left[\|u\|_{L^2(B_{2r}(x_0))} + \|f\|_{L^m(B_{2r}(x_0))}\right].$$

The main difficulty in the proof is to find local estimates for solutions of a nonlocal operator and then find similar results as in the local case. In the references cited above the reader can find detailed proofs of the C^α-regularity. In references [109], [124] and [136] the arguments are extensions of those by De Giorgi. In references [207] and [208] the arguments are close to the Moser techniques, while in [273] the method of Krylov–Safonov is used. In [288] general results are presented and some interesting applications are obtained.

9 The fractional Hardy inequality

9.1 Introduction

In this chapter we will study Hardy's inequality for the fractional $\frac{s}{2}$-Laplacian and the influence of the Hardy potential in semilinear problems. The motivation comes from several points of view, but we take here potential theory and theoretical physics as our main references.

- Assume $0 < s < 1$. Let us try to find a potential $|x|^{-\alpha}$, with $\alpha > 0$, for which there exists $C > 0$ such that

$$C \int_{\mathbb{R}^N} \frac{u^2}{|x|^\alpha}\, dx \le \int_{\mathbb{R}^N} |\xi|^{2s} |\mathcal{F}(u)(\xi)|^2\, d\xi, \quad u \in C_0^\infty(\mathbb{R}^N).$$

  Then, by homogeneity as in the local case $s = 1$, we must have $\alpha = 2s$. By the analysis done in Chapter 8,

$$\|(-\Delta)^{s/2} u\|^2_{L^2(\mathbb{R}^N)} = (2\pi)^{2s} \int_{\mathbb{R}^N} |\xi|^{2s} |\mathcal{F}(u)(\xi)|^2\, d\xi, \quad u \in C_0^\infty(\mathbb{R}^N).$$

  Thus, we have the equivalent inequality in terms of fractional derivatives

$$C' \int_{\mathbb{R}^N} \frac{u^2}{|x|^{2s}}\, dx \le \|(-\Delta)^{s/2} u\|^2_{L^2(\mathbb{R}^N)}, \quad u \in C_0^\infty(\mathbb{R}^N).$$

  Therefore the fractional Hardy inequality is just the extension of the classical Hardy inequality studied in Chapter 2.
- From theoretical physics, the previous inequality can be seen as *the uncertainty principle* with respect to the $2s$-fractional momentum. The inequality plays an important role in the study of the stability of matter from a relativistic point of view, as is explained in the fundamental paper by R. Frank, E. Lieb and R. Seiringer [167]. We recommend the reader to see also the references in this paper.

The goal in this chapter is to prove the following inequality, in which we include the explicit calculation of the optimal constant.

FRACTIONAL HARDY INEQUALITY. Given $u \in C_0^\infty(\mathbb{R}^N)$,

$$\Lambda_{N,2s} \int_{\mathbb{R}^N} \frac{u^2}{|x|^{2s}}\, dx \le \int_{\mathbb{R}^N} |\xi|^{2s} |\hat{u}(\xi)|^2\, d\xi,$$

where $0 < s < 1$, $N > 2s$ and

$$\Lambda_{N,2s} = \pi^{-2s} \frac{\Gamma^2(\frac{N+2s}{4})}{\Gamma^2(\frac{N-2s}{4})}.$$

https://doi.org/10.1515/9783110606270-009

As we said above, this inequality is a cornerstone in the proof of *the relativistic stability of matter* as is described in [167], and it has become a classical tool in quantum mechanics for years. Despite this, the main results from an analysis point of view are more recent. Perhaps the first work in which the optimal constant $\Lambda_{N,2s}$ is obtained is the paper by Herbst in [199] and then by Beckner in [54] and Yafaev in [319].

We will present in this chapter a proof of Hardy's inequality that is inspired in the articles [54] and [295] and, as these, uses classical tools from potential theory. Our approach, however, is more direct and elementary. We have made it also self-contained.

Extensions of the Hardy inequality to the fractional quasilinear framework can be found in the literature; see for instance [168] and the references therein.

9.2 The fractional Hardy inequality

The functional inequality known as the fractional Hardy inequality is the following.

Theorem 9.2.1. *Assume $2s < N$ and $0 < s < 1$. Then the following inequality holds:*

$$\Lambda_{N,2s} \int_{\mathbb{R}^N} \frac{u^2}{|x|^{2s}}\, dx \le \int_{\mathbb{R}^N} |\xi|^{2s} |\mathcal{F}(u)(\xi)|^2\, d\xi, \quad u \in C_0^\infty(\mathbb{R}^N), \tag{9.2.1}$$

where

$$\Lambda_{N,2s} = \pi^{-2s} \frac{\Gamma^2(\frac{N+2s}{4})}{\Gamma^2(\frac{N-2s}{4})}. \tag{9.2.2}$$

Remark 9.2.2. In the complementary case, $N \le 2s$, a similar inequality to (9.2.1) holds on the space $C_0^\infty(\mathbb{R}^N \setminus \{0\})$. See [167]. It is worth pointing out that $C_0^\infty(\mathbb{R}^N \setminus \{0\})$ is dense in $H^s(\mathbb{R}^N)$ if $N > 2s$. If $N \le 2s$ the completion of $C_0^\infty(\mathbb{R}^N \setminus \{0\})$ is $H^s(\mathbb{R}^N \setminus \{0\})$. See [140] and [139].

Remark 9.2.3. Since

$$\|(-\Delta)^{s/2} u\|^2_{L^2(\mathbb{R}^N)} = (2\pi)^{2s} \int_{\mathbb{R}^N} |\xi|^{2s} |\mathcal{F}(u)(\xi)|^2\, d\xi, \quad u \in C_0^\infty(\mathbb{R}^N), \tag{9.2.3}$$

we can write the Hardy inequality as

$$\|(-\Delta)^{s/2} u\|^2_{L^2(\mathbb{R}^N)} \ge \widetilde{\Lambda}_{N,2s} \int_{\mathbb{R}^N} \frac{u^2(x)}{|x|^{2s}}\, dx, \quad u \in C_0^\infty(\mathbb{R}^N), \tag{9.2.4}$$

where $\widetilde{\Lambda}_{N,2s} = 2^{2s} \frac{\Gamma^2(\frac{N+2s}{4})}{\Gamma^2(\frac{N-2s}{4})}$.

The result in Theorem 9.2.1 is a direct consequence of a *family of Hardy inequalities* that have many applications to other areas of analysis, including certain estimates for higher order fractional and local operators. The notation for the Riesz potentials and their corresponding kernels is the same as used in Section 8.1. We will use repeatedly that the Fourier transform is an isomorphism on the Schwartz class. We will also use the following elementary result of measure theory.

Lemma 9.2.4. *Let $d\mu$ be a probability measure and $u \in L^2(d\mu)$, $u \geq 0$. If*

$$\left(\int u(x)\,d\mu\right)^2 = \int u^2(x)\,dx, \tag{9.2.5}$$

then $u(x) = \|u\|_{L^2(d\mu)}$ for $d\mu$-almost x.

Proof. Assume that $\|u\|_{L^2(d\mu)} \neq 0$, otherwise there is nothing to prove. Hence by normalization we can assume that $\|u\|_{L^2(d\mu)} = 1$. Therefore by hypothesis,

$$\int u(x)\,d\mu = \left(\int u^2(x)\,d\mu\right)^{\frac{1}{2}} = 1.$$

Since $0 \leq u(x) \leq \frac{1}{2}u^2(x) + \frac{1}{2}$ for all x, we see that

$$1 = \int u(x)\,d\mu \leq \int \left(\frac{1}{2}u^2(x) + \frac{1}{2}\right)d\mu = 1.$$

Hence, we must have $u(x) = \frac{1}{2}u^2(x) + \frac{1}{2}$ for $d\mu$-almost x, and therefore $u(x) = 1$ for $d\mu$-almost x. □

A more general form of Hardy's inequality is given by the following.

Theorem 9.2.5. *Assume $0 < \alpha < N$. Then*

$$\int_{\mathbb{R}^N} |\mathcal{F}(f)(s)|^2 \frac{ds}{|s|^\alpha} \leq \Lambda_{N,\alpha}^{-1} \int_{\mathbb{R}^N} |x|^\alpha |f(x)|^2\,dx, \quad \textit{for all } f \in \mathcal{S}(\mathbb{R}^N), \tag{9.2.6}$$

where

$$\Lambda_{N,\alpha}^{-1} = I_\alpha(g_{\frac{N+\alpha}{2}})(\omega). \tag{9.2.7}$$

Here ω is an arbitrary unit vector and $g_{\frac{N+\alpha}{2}}$, I_α were defined in (8.1.9).

Moreover, with $c_{N,\frac{N+\alpha}{2}}$ defined in (8.1.8),

$$\Lambda_{N,\alpha}^{-1} = [c_{N,\frac{N+\alpha}{2}}]^2 = \pi^\alpha \left[\frac{\Gamma(\frac{N-\alpha}{4})}{\Gamma(\frac{N+\alpha}{4})}\right]^2 \tag{9.2.8}$$

is the best constant. The identity in (9.2.6) is never attained, except for $f \equiv 0$ a. e.

Proof. We present the proof in three steps.

FIRST STEP. PROOF OF (9.2.7) AND (9.2.8).

Inequality (9.2.6) by the Plancherel theorem is equivalent to

$$\int_{\mathbb{R}^N} \left|I_{\frac{\alpha}{2}}(f)(x)\right|^2 dx \le \Lambda_{N,\alpha}^{-1} \int_{\mathbb{R}^N} |x|^\alpha |f(x)|^2 dx. \tag{9.2.9}$$

Since $I_{\frac{\alpha}{2}}$ is self-adjoint, by the semigroup property the left hand side of (9.2.9) can be written as follows:

$$\begin{aligned}\int_{\mathbb{R}^N} I_{\frac{\alpha}{2}}(f)(x) I_{\frac{\alpha}{2}}(f)(x)\,dx &= \int_{\mathbb{R}^N} I_\alpha(f)(x) f(x)\,dx \\ &= c_{N,\alpha} \int_{\mathbb{R}^N}\int_{\mathbb{R}^N} f(x)\frac{1}{|x-y|^{N-\alpha}} f(y)\,dy\,dx. \end{aligned}\tag{9.2.10}$$

Call $h(x) = f(x)|x|^{\frac{N+\alpha}{2}}$, that is, $f(x) = h(x) g_{\frac{N+\alpha}{2}}(x)$. Then, inequality (9.2.9) is equivalent to

$$c_{N,\alpha} \int_{\mathbb{R}^N}\int_{\mathbb{R}^N} h(x)\frac{|x|^{\frac{N-\alpha}{2}}|y|^{\frac{N-\alpha}{2}}}{|x-y|^{N-\alpha}} h(y)\frac{dy}{|y|^N}\frac{dx}{|x|^N} \le \Lambda_{N,\alpha}^{-1}\int_{\mathbb{R}^N} |h(x)|^2 \frac{dx}{|x|^N}. \tag{9.2.11}$$

We set

$$K(x,y) = c_{N,\alpha}\frac{|x|^{\frac{N-\alpha}{2}}|y|^{\frac{N-\alpha}{2}}}{|x-y|^{N-\alpha}} \quad \text{and} \quad d\mu(x) = \frac{dx}{|x|^N}.$$

Then, the left hand side of (9.2.11) can be estimated as follows:

$$\begin{aligned}&\int_{\mathbb{R}^N}\int_{\mathbb{R}^N} h(x)h(y)K(x,y)\,d\mu(x)\,d\mu(y) \\ &\le \left(\int_{\mathbb{R}^N} |h(x)|^2\,d\mu(x)\right)^{\frac{1}{2}} \left(\int_{\mathbb{R}^N}\left|\int_{\mathbb{R}^N} h(y)K(x,y)\,d\mu(x)\right|^2 d\mu(y)\right)^{\frac{1}{2}} \\ &\le \|h\|_{L^2(d\mu)} \left(\int_{\mathbb{R}^N}\left(\int_{\mathbb{R}^N} |h(y)|^2 K(x,y)\,d\mu(y)\right)\left(\int_{\mathbb{R}^N} K(x,z)\,d\mu(z)\right) d\mu(x)\right)^{\frac{1}{2}} \\ &= \|h\|_{L^2(d\mu)} \left(\int_{\mathbb{R}^N} |h(y)|^2 \left(\int_{\mathbb{R}^N} K(x,y)\left(\int_{\mathbb{R}^N} K(x,z) d\mu(z)\right) d\mu(x)\right) d\mu(y)\right)^{\frac{1}{2}}. \end{aligned}\tag{9.2.12}$$

For the inequalities we have used simply Cauchy–Schwarz and the last identity is just the application of Fubini's theorem. The proof will conclude if we are able to prove that

$$\int_{\mathbb{R}^N} K(x,z)\,d\mu(z) = \Lambda_{N,\alpha}^{-1}, \quad \text{independent of } x. \tag{9.2.13}$$

Note that, by symmetry, (9.2.13) is the same as

$$\int_{\mathbb{R}^N} K(x,y)\,d\mu(x) = \Lambda_{N,\alpha}^{-1}, \quad \text{independent of } y.$$

To prove (9.2.13) we write $x = |x|x'$ with $x' \in S^{N-1}$ and make the change of variable $w = \frac{z}{|x|}$; therefore

$$\int_{\mathbb{R}^N} K(x,z)\,d\mu(z) = c_{N,\alpha} \int_{\mathbb{R}^N} \frac{|x|^{\frac{N-\alpha}{2}}|z|^{\frac{N-\alpha}{2}}}{|x' - \frac{z}{|x|}|^{N-\alpha}|x|^{N-\alpha}} \frac{dz}{|z|^N}$$

$$= c_{N,\alpha} \int_{\mathbb{R}^N} \frac{|w|^{\frac{N-\alpha}{2}}}{|x' - w|^{N-\alpha}} \frac{dw}{|w|^N} = c_{N,\alpha} \int_{\mathbb{R}^N} \frac{1}{|x' - w|^{N-\alpha}} \frac{dw}{|w|^{\frac{N+\alpha}{2}}} = I_\alpha(g_{\frac{N+\alpha}{2}})(x').$$

We need to evaluate $I_\alpha(g_{\frac{N+\alpha}{2}})(x')$, which, clearly, is independent of x' because $g_{\frac{N+\alpha}{2}}$ is a radial function. Now,

$$\mathcal{F}(I_\alpha(g_{\frac{N+\alpha}{2}}))(\xi) = \frac{1}{|\xi|^\alpha}\mathcal{F}(g_{\frac{N+\alpha}{2}})(\xi) = \frac{1}{|\xi|^\alpha} c_{N,\frac{N+\alpha}{2}} \frac{1}{|\xi|^{\frac{N-\alpha}{2}}} = c_{N,\frac{N+\alpha}{2}} g_{\frac{N+\alpha}{2}}(\xi).$$

That means that $g_{\frac{N+\alpha}{2}}$ is invariant for the operator $h \to \mathcal{F}(I_\alpha)(h)$. Therefore,

$$I_\alpha(g_{\frac{N+\alpha}{2}})(x) = c_{N,\frac{N+\alpha}{2}} \mathcal{F}^{-1}(g_{\frac{N+\alpha}{2}})(x) = [c_{N,\frac{N+\alpha}{2}}]^2 \frac{1}{|x|^{\frac{N-\alpha}{2}}}.$$

This completes the proof of (9.2.8). According with the previous estimates and (9.2.12) the inequality (9.2.6) follows.

Note that, by density, inequality (9.2.6) holds for all $f \in L^2(\mathbb{R}^N, |x|^\alpha\,dx)$.

SECOND STEP. OPTIMALITY OF $\Lambda_{N,\alpha}^{-1}$. Proving the fact that $\Lambda_{N,\alpha}$ is sharp is equivalent to proving the following statement.

If for $0 < \alpha < N$ and for every $f \in L^2(\mathbb{R}^N, |x|^\alpha\,dx)$ we have

$$\int_{\mathbb{R}^N} |I_{\frac{\alpha}{2}}(f)(x)|^2\,dx \le C_0 \int_{\mathbb{R}^N} |f(x)|^2|x|^\alpha\,dx, \tag{9.2.14}$$

then

$$C_0 \ge \Lambda_{N,\alpha}^{-1}.$$

Before proving the statement, we perform a formal calculation to motivate the rigorous proof and, as a byproduct, to find a pseudo-minimizer as in the local case. Indeed, consider $f(x) = g_{\frac{N+\alpha}{2}}(x) := g(x)$. Then

$$\mathcal{F}(I_{\frac{\alpha}{2}})(f)(\xi) = \frac{1}{|\xi|^{\frac{\alpha}{2}}} c_{N,\frac{N+\alpha}{2}} \frac{1}{|\xi|^{\frac{N-\alpha}{2}}} = c_{N,\frac{N+\alpha}{2}} \frac{1}{|\xi|^{\frac{N}{2}}}.$$

Hence

$$I_{\frac{\alpha}{2}}(f)(x) = c_{N,\frac{N+\alpha}{2}}\mathcal{F}^{-1}(g_{\frac{N}{2}})(x) = c_{N,\frac{N+\alpha}{2}}g_{\frac{N}{2}}(x) = c_{N,\frac{N+\alpha}{2}}f(x)|x|^{\frac{\alpha}{2}},$$

that is, we obtain the following pointwise identity:

$$\left|I_{\frac{\alpha}{2}}(f)(x)\right|^2 = \Lambda_{N,\alpha}^{-1}\left|f(x)\right|^2|x|^{\alpha}.$$

Since $f(x) = g_{\frac{N+\alpha}{2}}(x) \notin L^2(\mathbb{R}^N, |x|^{\alpha}\,dx)$, we cannot obtain what we want, but we have two consequences. The first one is that $g_{\frac{N+\alpha}{2}}(x)$ behaves really like a candidate to be a minimizer. The problem is that it does not belong to the space that we need. For this reason we will call $g_{\frac{N+\alpha}{2}}(x)$ a *pseudo-minimizer*. The second conclusion is that a complete proof will be obtained doing the same estimates on truncates of our pseudo-minimizer. This is essentially the same kind of argument that we used in Chapter 2, Section 2.3.2 to study the classical Hardy–Leray inequality.

We proceed now to prove the previous statement and then that $\Lambda_{N,\alpha}$ is the best constant in inequality (9.2.6).

For R large enough, take $\psi_R \in \mathcal{S}(\mathbb{R}^N)$ such that

$$\chi_{[\frac{1}{R},R]}(|x|) \leq \psi_R(x) \leq \chi_{[\frac{1}{2R},2R]}(|x|).$$

We will multiply the pseudo-minimizer by the cutoff ψ_R, that is, we consider

$$f(x) = \frac{\psi_R(x)}{|x|^{\frac{N+\alpha}{2}}}.$$

Clearly, we have $f \in \mathcal{S}(\mathbb{R}^N)$. We can compute directly both sides of inequality (9.2.14) for f,

$$\int_{\mathbb{R}^N}\left|f(x)\right|^2|x|^{\alpha}\,dx = \int_{\mathbb{R}^N}\left|\psi(x)\right|^2\frac{1}{|x|^N}\,dx \leq \int_{\frac{1}{2R}<|x|<2R}\frac{dx}{|x|^N} = \omega_{N-1}2\log(2R). \tag{9.2.15}$$

The first term in (9.2.14) is

$$\begin{aligned}
&\int_{\mathbb{R}^N}\left|I_{\frac{\alpha}{2}}(f)(x)\right|^2 dx\\
&\quad= \int_{\mathbb{R}^N} f(x)I_{\alpha}(f)(x)\,dx = \int_{\mathbb{R}^N}\frac{\psi_R(x)}{|x|^{\frac{N+\alpha}{2}}}\left(c_{N,\alpha}\int_{\mathbb{R}^N}\frac{\psi_R(y)}{|y|^{\frac{N+\alpha}{2}}}\frac{1}{|x-y|^{N-\alpha}}\,dy\right)dx\\
&\quad= \int_{\mathbb{R}^N}\frac{\psi_R(x)}{|x|^{N+\alpha}}\left(c_{N,\alpha}\int_{\mathbb{R}^N}\psi_R(y)\frac{|x|^{\frac{N+\alpha}{2}}}{|y|^{\frac{N+\alpha}{2}}}\frac{1}{\left|\frac{x}{|x|}-\frac{y}{|x|}\right|^{N-\alpha}}\frac{dy}{|x|^{N-\alpha}}\right)dx.
\end{aligned} \tag{9.2.16}$$

Making the change of variables $w = \frac{y}{|x|}$ and calling $x' = \frac{x}{|x|}$, we get

$$\int_{\mathbb{R}^N} |I_{\frac{\alpha}{2}}(f)(x)|^2\,dx = \int_{\mathbb{R}^N} \frac{\psi_R(x)}{|x|^{N+\alpha}}|x|^\alpha \Big(c_{N,\alpha} \int_{\mathbb{R}^N} \frac{\psi(|x|w)}{|w|^{\frac{N+\alpha}{2}}} \frac{dw}{|x'-w|^{N-\alpha}} \Big)\,dx$$
$$\geq \int_{\frac{1}{R}<|x|<R} \frac{1}{|x|^N} \Big(c_{N,\alpha} \int_{\frac{1}{R}<|x||w|<R} \frac{1}{|w|^{\frac{N+\alpha}{2}}} \frac{dw}{|x'-w|^{N-\alpha}} \Big)\,dx. \tag{9.2.17}$$

We take $0 < \epsilon < 1$. Note that if $\frac{1}{R^{1-\epsilon}} < |x| < R^{1-\epsilon}$ and $\frac{1}{R^\epsilon} < |w| < R^\epsilon$, then $\frac{1}{R} < |x||w| < R$. Therefore we have from (9.2.17)

$$\int_{\mathbb{R}^N} |I_{\frac{\alpha}{2}}(f)(x)|^2\,dx \geq \int_{\frac{1}{R^{1-\epsilon}}<|x|<R^{1-\epsilon}} \frac{1}{|x|^N} J_{R,\epsilon}(x')\,dx,$$

where

$$J_{R,\epsilon}(x') = c_{N,\alpha} \int_{\frac{1}{R^\epsilon}<|w|<R^\epsilon} \frac{1}{|w|^{\frac{N+\alpha}{2}}} \frac{dw}{|x'-w|^{N-\alpha}}.$$

Since $J_{R,\epsilon}$ is rotation-invariant, it is independent of $x' \in S_{N-1}$. Hence, for $x' = e_1 = (1, 0, \dots, 0)$ we have

$$\int_{\mathbb{R}^N} |I_{\frac{\alpha}{2}}(f)(x)|^2\,dx \geq J_{R,\epsilon}(e_1) \int_{\frac{1}{R^{1-\epsilon}}<|x|<R^{1-\epsilon}} \frac{1}{|x|^N}\,dx = J_{R,\epsilon}(e_1)\omega_{N-1} 2(1-\epsilon)\log R.$$

Therefore, if inequality (9.2.14) holds we obtain

$$J_{R,\epsilon}(e_1) 2(1-\epsilon)\log R \leq C_0 2\log(2R), \quad \text{for all } 0 < \epsilon < 1 \text{ and all } R > 1.$$

Thus

$$J_{R,\epsilon}(e_1) \leq C_0 \frac{1}{(1-\epsilon)} \frac{\log 2R}{\log R}. \tag{9.2.18}$$

Finally, observe that by the monotone convergence theorem,

$$\lim_{R\to\infty} J_{R,\epsilon}(e_1) = I_\alpha(g_{\frac{N+\alpha}{2}})(e_1) = \Lambda_{N,\alpha}^{-1}, \quad \text{for all } \epsilon > 0.$$

Therefore taking limit first in $R \to \infty$ and then in $\epsilon \to 0$ in (9.2.18), we conclude that

$$\Lambda_{N,\alpha}^{-1} \leq C_0.$$

In other words, we have proved that

$$\Lambda_{N,\alpha} = \inf_{\{f\in L^2(\mathbb{R}^N,|x|^\alpha\,dx), f\neq 0\}} \frac{\int_{\mathbb{R}^N} |f(x)|^2|x|^\alpha\,dx}{\int_{\mathbb{R}^N} |I_{\frac{\alpha}{2}}(f)(x)|^2\,dx}.$$

In the next step we will prove that is not a minimum.

THIRD STEP. $\Lambda_{N,\alpha}^{-1}$ IS NO ATTAINED IN $L^2(\mathbb{R}^N, |x|^\alpha\,dx)$. Proving the nonattainability of $\Lambda_{N,\alpha}^{-1}$ is obviously equivalent to proving the following statement.

If for $f \in \mathcal{S}(\mathbb{R}^N)$, $f \geq 0$,

$$\int_{\mathbb{R}^N} \left|I_{\frac{\alpha}{2}}(f)(x)\right|^2 dx = \Lambda_{N,\alpha}^{-1} \int_{\mathbb{R}^N} |x|^\alpha |f(x)|^2\,dx, \tag{9.2.19}$$

then $f \equiv 0$ almost everywhere.

To prove the statement, recall the notation used in the first step,

$$h(x) = f(x)|x|^{\frac{N+\alpha}{2}}, \quad K(x,y) = c_{N,\alpha}\frac{|x|^{\frac{N-\alpha}{2}}|y|^{\frac{N-\alpha}{2}}}{|x-y|^{N-\alpha}} \quad \text{and} \quad d\mu(x) = \frac{dx}{|x|^N}.$$

Then (9.2.19) is equivalent to

$$\left(\int_{\mathbb{R}^N}\left(\int_{\mathbb{R}^N} h(y)K(x,y)\,d\mu(y)\right)^2 d\mu(x)\right)^{\frac{1}{2}} = \Lambda_{N,\alpha}^{-1}\left(\int_{\mathbb{R}^N} |h(y)|^2\,d\mu(y)\right)^{\frac{1}{2}}. \tag{9.2.20}$$

Moreover recall that

$$\int_{\mathbb{R}^N} K(x',y)\,d\mu(y) = \Lambda_{N,\alpha}^{-1}.$$

Let us set

$$A(x) := \left(\int_{\mathbb{R}^N} h(y)K(x,y)\,d\mu(y)\right)^2 \quad \text{and} \quad B(x) := \Lambda_{N,\alpha}^{-1}\int_{\mathbb{R}^N} K(x,y)|h(y)|^2 d\mu(y).$$

Using the Cauchy–Schwarz inequality we have

$$A(x) \leq \left(\int_{\mathbb{R}^N} K(x,y)|h(y)|^2\,d\mu(y)\right)\left(\int_{\mathbb{R}^N} K(x,y)\,d\mu(y)\right) = B(x).$$

Integrating with respect to $d\mu(x)$ in $\mathbb{R}^N$,

$$\int_{\mathbb{R}^N} A(x)\,d\mu(x) \leq \int_{\mathbb{R}^N} B(x)d\mu(x) = \Lambda_{N,\alpha}^{-2}\int_{\mathbb{R}^N} |h(y)|^2 d\mu(y) = \int_{\mathbb{R}^N} A(x)\,d\mu(x),$$

where the last identity follows from (9.2.20). As a consequence we have $A(x) = B(x)$ almost everywhere, that is,

$$\Big(\int_{\mathbb{R}^N} h(y)K(x,y)\,d\mu(y)\Big)^2 = \Lambda_{N,\alpha}^{-1}\int_{\mathbb{R}^N} K(x,y)\big|h(y)\big|^2\,d\mu(y), \quad a.e.$$

Fix $x \in \mathbb{R}^N$ for which the identity above holds. Writing

$$d\bar{\mu}(x) = \Lambda_{N,\alpha}K(x,y)\,d\mu(y),$$

we get a probability measure in $\mathbb{R}^N$ and h satisfies

$$\Big(\int_{\mathbb{R}^N} h(y)d\bar{\mu}(y)\Big)^2 = \Big(\int_{\mathbb{R}^N} \big|h(y)\big|^2 d\bar{\mu}(y)\Big). \tag{9.2.21}$$

By Lemma 9.2.4, we conclude that h is constant and this is equivalent to saying

$$f(x) = \frac{c}{|x|^{\frac{N+\alpha}{2}}},$$

for some constant $c > 0$. The only way in which $f \in L^2(\mathbb{R}^N, |x|^\alpha\,dx)$ is that $c = 0$ and this finishes the proof. □

Remark 9.2.6. Inequality (9.2.6) has several important consequences.
1. The classical Hardy inequality (9.2.1) follows from (9.2.6) taking $0 < s < 1$, $2s < N$, $\alpha = 2s$ and interchanging u and $\mathcal{F}(u)$.
2. If $N > 2$ we can take $s = 1$ and recover the inequality

$$\widetilde{\Lambda}_{N,2}\int_{\mathbb{R}^N} \frac{u^2(x)}{|x|^2}\,dx \le \|(-\Delta)^{1/2}u\|^2_{L^2(\mathbb{R}^N)} \equiv \int_{\mathbb{R}^N} |\nabla u|^2\,dx.$$

Indeed, $\widetilde{\Lambda}_{N,2} = 4\frac{\Gamma^2(\frac{N+2}{4})}{\Gamma^2(\frac{N-2}{4})}$ and $\frac{N+2}{4} = 1 + \frac{N-2}{4}$. Therefore, since $N > 2$ and $\Gamma(\frac{N+2}{4}) = (\frac{N-2}{4})\Gamma(\frac{N-2}{4})$, we finally get

$$\widetilde{\Lambda}_{N,2} = 4\Big(\frac{N-2}{4}\Big)^2.$$

In this way we arrive at the classical inequality

$$\Big(\frac{N-2}{2}\Big)^2\int_{\mathbb{R}^N} \frac{u^2(x)}{|x|^2}\,dx \le \int_{\mathbb{R}^N} |\nabla u|^2\,dx.$$

3. For $0 < s < \frac{N}{2}$, $\alpha = 2s$ and interchanging again u and $\mathcal{F}(u)$ in (9.2.6), we have

$$\int_{\mathbb{R}^N} |f(x)|^2 \frac{dx}{|x|^{2s}} \leq \Lambda_{N,2s}^{-1} \int_{\mathbb{R}^N} |x|^{2s} |\mathcal{F}(f)(x)|^2 \, dx.$$

Setting, as before, for $s < \frac{N}{2}$

$$\mathcal{F}((-\Delta)^{\frac{s}{2}} u)(\xi) = |2\pi\xi|^s \mathcal{F}(u)(\xi),$$

we get the extension to higher fractional powers of the Laplacian,

$$\int_{\mathbb{R}^N} |u(x)|^2 \frac{dx}{|x|^{2s}} \leq \widetilde{\Lambda}_{N,2s}^{-1} \int_{\mathbb{R}^N} |(-\Delta)^{s/2}(u)(x)|^2 \, dx.$$

In particular, for $N > 4$ and $s = 2$ we obtain Hardy's inequality for the biharmonic operator,

$$N^2 \left(\frac{N-4}{4} \right)^2 \int_{\mathbb{R}^N} \frac{u^2(x)}{|x|^4} \, dx \leq \int_{\mathbb{R}^N} |\Delta u(x)|^2 \, dx = \int_{\mathbb{R}^N} u(x) \Delta^2 u(x) \, dx.$$

9.3 Normalizing the constants

To pursue a more standard notation and normalization of the constants, let us modify our initial setup of the Fourier transform so that we can identify the right hand side of (9.2.1) with the L^2-norm of the fractional Laplacian, $(-\Delta)^{s/2}u$. To that end, we consider the following definition.

Definition 9.3.1. Given u in the Schwartz class $\mathcal{S}$, set

$$\mathcal{F}(u)(\xi) = \widehat{u}(\xi) := (2\pi)^{-N/2} \int_{\mathbb{R}^N} u(x) e^{-i\langle x, \xi \rangle} \, dx. \tag{9.3.1}$$

By abuse of notation, we will use again $\mathcal{F}$ and $\widehat{\cdot}$ to denote this new transform. As before, we have Parseval–Plancherel identities

$$\int_{\mathbb{R}^N} u(x)\overline{v(x)} \, dx = \int_{\mathbb{R}^N} \widehat{u}(\xi)\overline{\widehat{v}(\xi)} \, d\xi \quad \text{and} \quad \int_{\mathbb{R}^N} |u(x)|^2 \, dx = \int_{\mathbb{R}^N} |\widehat{u}(\xi)|^2 \, d\xi. \tag{9.3.2}$$

One easily sees now that $\widehat{(-\Delta)}u(\xi) = |\xi|^2 \widehat{u}(\xi)$ and then $\widehat{(-\Delta)^s}u(\xi) = |\xi|^{2s}\widehat{u}(\xi)$. In particular, we have

$$\int_{\mathbb{R}^N} |\xi|^{2s} |\widehat{u}|^2 \, d\xi = \int_{\mathbb{R}^N} |(-\Delta)^{s/2} u(x)|^2 \, dx, \tag{9.3.3}$$

as desired.

In this way, Hardy's inequality reads

$$\int_{\mathbb{R}^N} |\xi|^{2s}|\widehat{u}|^2\,d\xi \geq \Lambda_{N,2s}\int_{\mathbb{R}^N} |x|^{-2s}|u|^2\,dx, \quad \forall u \in C_0^\infty(\mathbb{R}^N), \tag{9.3.4}$$

now with

$$\Lambda_{N,2s} := 2^{2s}\frac{\Gamma^2(\frac{N+2s}{4})}{\Gamma^2(\frac{N-2s}{4})}. \tag{9.3.5}$$

As before, we obtain this as

$$\Lambda_{N,2s}^{-1} = I_{2s}\left(\frac{1}{|\cdot|^{\frac{N+2s}{2}}}\right)(x') = [(2\pi)^{N/2}c_{N,\frac{N+2s}{2}}]^2, \quad x' \in S^{N-1}, \tag{9.3.6}$$

where $c_{N,\alpha}$ denotes the constant that appears in the Riesz kernel of order α, I_α. To be more precise, we define

$$\widehat{I_\alpha u}(\xi) = (\widehat{-\Delta)^{-\alpha/2}u}(\xi) = \frac{1}{|\xi|^\alpha}\widehat{u}(\xi), \tag{9.3.7}$$

that is, the multiplier operator associated to $g_\alpha(\xi) := \frac{1}{|\xi|^\alpha}$ with respect to the new Fourier transform. From this, we get

$$I_\alpha u(x) = c_{N,\alpha}\int_{\mathbb{R}^N}\frac{u(y)}{|x-y|^{N-\alpha}}\,dx = c_{N,\alpha}g_{N-\alpha} * u(x). \tag{9.3.8}$$

As in Section 8.1.1, we now compute the Fourier transform of its kernel, $k_{N,\alpha} = c_{N,\alpha}g_{N-\alpha}$, and obtain

$$c_{N,\alpha} = 2^{-\alpha}\pi^{-N/2}\frac{\Gamma(\frac{N-\alpha}{2})}{\Gamma(\frac{\alpha}{2})}. \tag{9.3.9}$$

From here, the fundamental solution to the equation $(-\Delta)^s u = f$ is given by

$$S(x) = c_{N,2s}\frac{1}{|x|^{N-2s}}. \tag{9.3.10}$$

We now deduce the constant in the pointwise, principal value definition of the fractional Laplacian,

$$(-\Delta)^s u(x) = C_{N,s}\,P.V.\int_{\mathbb{R}^N}\frac{u(x)-u(y)}{|x-y|^{N+2s}}\,dy. \tag{9.3.11}$$

With the new normalization, we have

$$C_{N,s} = -c_{N,-2s} = 2^{2s}\pi^{-\frac{N}{2}}\frac{\Gamma(\frac{N+2s}{2})}{|\Gamma(-s)|}, \tag{9.3.12}$$

which coincides with the one given initially.

Now, formula (8.1.16) in Lemma 8.1.8 becomes

$$\int_{\mathbb{R}^N} |\xi|^{2s} |\hat{u}(\xi)|^2 \, d\xi = \frac{1}{2} C_{N,s} \int_{\mathbb{R}^N} \int_{\mathbb{R}^N} \frac{|u(x) - u(y)|^2}{|x - y|^{N+2s}} \, dx \, dy. \tag{9.3.13}$$

Often we will also write

$$a_{N,s} = \frac{1}{2} C_{N,s} = 2^{2s-1} \pi^{-\frac{N}{2}} \frac{\Gamma(\frac{N+2s}{2})}{|\Gamma(-s)|}. \tag{9.3.14}$$

9.4 Local behavior of solutions of the elliptic equation

The purpose of this section is to analyze the behavior of the radial solutions to the homogeneous problem

$$(-\Delta)^s u = \lambda \frac{u}{|x|^{2s}} \quad \text{in } \mathbb{R}^N \setminus \{0\}, \tag{9.4.1}$$

in a neighborhood of the origin, in order to use this information as a tool for proving the existence and nonexistence results in the forthcoming chapters.

As usual, we will write $\mathcal{F}(u) = \hat{u}$ and we will consider

$$h_s(u) := \int_{\mathbb{R}^N} |\xi|^{2s} |\hat{u}|^2 \, d\xi - \Lambda_{N,2s} \int_{\mathbb{R}^N} \frac{u^2(x)}{|x|^{2s}} \, dx. \tag{9.4.2}$$

According to (8.1.8) and (9.3.9) we can write for $0 < \alpha < N$ that the Fourier transform of the tempered distribution $g_\alpha(x) = |x|^{-\alpha}$ is given by the identity

$$b_\alpha \widehat{g}_\alpha(\xi) = b_{N-\alpha} |\xi|^{-N+\alpha}, \tag{9.4.3}$$

where $b_\alpha = 2^{\frac{\alpha}{2}} \Gamma(\frac{\alpha}{2})$.

As a consequence, if $0 < s < \frac{N}{2}$ and $0 < \alpha < (N - 2s)$, then $(-\Delta)^s |x|^{-\alpha} \in L^1_{\text{loc}}$. Indeed,

$$\widehat{(-\Delta)^s(g_\alpha)}(\xi) = |\xi|^{2s} \widehat{g_\alpha}(\xi) = 2^{N/2-\alpha} \frac{\Gamma(\frac{N-\alpha}{2})}{\Gamma(\frac{\alpha}{2})} \frac{1}{|\xi|^{N-\alpha-2s}} = c_{N,\alpha} \frac{1}{|\xi|^{N-\alpha-2s}},$$

and then

$$(-\Delta)^s(g_\alpha)(x) = c_{N,\alpha} c_{N,N-\alpha-2s} \frac{1}{|x|^{\alpha+2s}}, \tag{9.4.4}$$

which is in $L^1_{\text{loc}}(\mathbb{R}^N)$. Call

$$\Psi_N(\alpha) := \frac{b_{\alpha+2s} b_{N-\alpha}}{b_{N-\alpha-2s} b_\alpha} = 2^{2s} \left(\frac{\Gamma(\frac{(\alpha+2s)}{2}) \Gamma(\frac{(N-\alpha)}{2})}{\Gamma(\frac{(N-\alpha-2s)}{2}) \Gamma(\frac{\alpha}{2})} \right) \tag{9.4.5}$$

and define

$$\lambda = \lambda(\alpha) = \lambda(-\alpha) = \frac{2^{2s}\Gamma(\frac{N+2s+2\alpha}{4})\Gamma(\frac{N+2s-2\alpha}{4})}{\Gamma(\frac{N-2s+2\alpha}{4})\Gamma(\frac{N-2s-2\alpha}{4})}. \tag{9.4.6}$$

That is,

$$\lambda(\alpha) = \Psi_N\left(\frac{N-2s}{2} - \alpha\right).$$

Remark 9.4.1. Note that $\lambda(\alpha) = \lambda(-\alpha) = m_\alpha m_{-\alpha}$, with $m_\alpha = 2^{\alpha+s}\frac{\Gamma(\frac{N+2s+2\alpha}{4})}{\Gamma(\frac{N-2s-2\alpha}{4})}$.

Lemma 9.4.2. *The function $\lambda = \lambda(\alpha)$ defined in (9.4.6) is strictly decreasing and, as a consequence, the following equivalence holds true:*

$$0 < \lambda(\alpha) = \lambda(-\alpha) \le \Lambda_{N,2s} \quad \textit{if and only if } 0 \le \alpha < \frac{N-2s}{2}.$$

Proof. We give an elementary proof of this result. See also [167, 199]. Note that $\lambda(\alpha)$ is a positive continuous function for $0 \le \alpha < \frac{N-2s}{2}$, such that $\lambda(0) = \Lambda_{N,2s}$ and $\lim_{\alpha\to\frac{N-2s}{2}} \lambda(\alpha) = 0$.

We prove that for fixed s, $\lambda(\alpha)$ is a decreasing function. Let us consider the following representation of the Gamma function:

$$\frac{1}{\Gamma(x)} = xe^{\gamma x}\prod_{n=1}^{\infty}\left(1 + \frac{x}{n}\right)e^{-\frac{x}{n}},$$

where γ is the Euler–Mascheroni constant (see for instance [40], [39] for more details). We aim to prove that $\log\frac{1}{\lambda(\alpha)}$ is an increasing function in α. We have

$$\begin{aligned}\log\frac{1}{\lambda(\alpha)} &= \log\frac{1}{2^{2s}}\frac{\frac{1}{\Gamma(\frac{N+2s+2\alpha}{4})}\cdot\frac{1}{\Gamma(\frac{N+2s-2\alpha}{4})}}{\frac{1}{\Gamma(\frac{N-2s+2\alpha}{4})}\cdot\frac{1}{\Gamma(\frac{N-2s-2\alpha}{4})}}\\ &= -2s\cdot\log 2 + \log\frac{(N+2s)^2-4\alpha^2}{(N-2s)^2-4\alpha^2} + 2\gamma s + \sum_{n=1}^{\infty}\left[\log\frac{(\frac{N+4n+2s}{4n})^2 - \frac{\alpha^2}{4n^2}}{(\frac{N+4n-2s}{4n})^2 - \frac{\alpha^2}{4n^2}} - \frac{2s}{n}\right].\end{aligned}$$

Note that the last term is a convergent series in the same way as

$$\prod_{n=1}^{\infty}\left(1+\frac{x}{n}\right)e^{-\frac{x}{n}}$$

is a convergent product. We conclude just by noting that if $a > b$ and $\zeta > 0$, then $\frac{a^2-\zeta^2}{b^2-\zeta^2}$ is an increasing function in ζ. □

Therefore we can formulate the following result, which is the extension to the fractional setting of the existence of two radial solutions to the local homogeneous equation.

Lemma 9.4.3. *Let* $0 < \lambda \le \Lambda_{N,2s}$. *Then* $v_{\pm\alpha} = |x|^{-\frac{N-2s}{2}\pm\alpha}$ *are solutions to*

$$(-\Delta)^s u = \lambda \frac{u}{|x|^{2s}} \quad \text{in } (\mathbb{R}^N \setminus \{0\}), \tag{9.4.7}$$

where α *is obtained by the identity*

$$\lambda = \lambda(\alpha) = \lambda(-\alpha) = \frac{2^{2s}\Gamma(\frac{N+2s+2\alpha}{4})\Gamma(\frac{N+2s-2\alpha}{4})}{\Gamma(\frac{N-2s+2\alpha}{4})\Gamma(\frac{N-2s-2\alpha}{4})}. \tag{9.4.8}$$

Proof. Applying the Fourier transform (as in Section 9.3) yields

$$(-\Delta)^s v_\alpha = \mathcal{F}^{-1}(|\xi|^{2s}\mathcal{F}(v_\alpha)(\xi)) = \frac{2^{2s}\Gamma(\frac{N+2s+2\alpha}{4})\Gamma(\frac{N+2s-2\alpha}{4})}{\Gamma(\frac{N-2s+2\alpha}{4})\Gamma(\frac{N-2s-2\alpha}{4})} v_\alpha(x) == \lambda |x|^{-2s} v_\alpha,$$

with $\lambda = \lambda(\alpha)$ defined in (9.4.8). By Lemma 9.4.3 we conclude that for each $0 < \lambda < \Lambda_{N,2s}$ there exists a unique positive $\alpha \in (0, \frac{N-2s}{2}]$ such that $\lambda = \lambda(\alpha)$, which implies the result. The symmetry with respect to α gives the other solution. □

We will use these results to study the unboundedness of any positive weak supersolution and, moreover, to obtain explicit quantitative information on the growth around the origin when the summability of the datum is good enough.

Remark 9.4.4. Note that we can explicitly construct two positive solutions to the homogeneous problem (9.4.7). We denote

$$\gamma = \frac{N-2s}{2} - \alpha \text{ and } \bar{\gamma} = \frac{N-2s}{2} + \alpha, \quad \text{with } 0 < \gamma \le \frac{N-2s}{2} \le \bar{\gamma} < (N-2s). \tag{9.4.9}$$

Since $N - 2\gamma - 2s = 2\alpha > 0$ and $N - 2\bar{\gamma} - 2s = -2\alpha < 0$, we have $(-\Delta)^{s/2}(|x|^{-\gamma}) \in L^2_{\text{loc}}(\mathbb{R}^N)$, but $(-\Delta)^{s/2}(|x|^{-\bar{\gamma}}) \notin L^2_{\text{loc}}(\mathbb{R}^N)$.

9.5 *Ground state* representation

As we have seen in Section 9.4, if $0 < s < \frac{N}{2}$ and $0 < \alpha < (N-2s)$, we have

$$((-\Delta)^s - \Lambda_{N,2s}|x|^{-2s})|x|^{-\alpha} = \Phi_{N,s}(\alpha)|x|^{-\alpha-2s}, \tag{9.5.1}$$

where $\Lambda_{N,2s}$ is defined in (9.2.2) and

$$\Phi_{N,s}(\alpha) := \frac{b_{\alpha+2s}b_{N-\alpha}}{b_{N-\alpha-2s}b_\alpha} - \Lambda_{N,2s} = 2^{2s}\left(\frac{\Gamma(\frac{(\alpha+2s)}{2})\Gamma(\frac{(N-\alpha)}{2})}{\Gamma(\frac{(N-\alpha-2s)}{2})\Gamma(\frac{\alpha}{2})} - \frac{\Gamma^2(\frac{N+2s}{4})}{\Gamma^2(\frac{N-2s}{4})}\right). \tag{9.5.2}$$

The following result will be used in the proof of Theorem 9.6.1 and will play an important role in the following chapters.

Lemma 9.5.1. *The function $\Phi_{N,s}$ is negative and strictly increasing in the interval $(0, \frac{N-2}{2})$, with $\Phi_{N,s}(\frac{N-2}{2}) = 0$.*

Proof. Note that it is a direct consequence of Lemma 9.4.2, changing $\alpha \to \frac{N-2s}{2}$. □

According to the previous section, for $\lambda = \lambda(\alpha)$, $|x|^{-\alpha}$ is almost a *ground state* for the linear equation in the whole $\mathbb{R}^N$. The idea of the *ground state* representation, which goes back to Jacobi, is useful in many situations. The ground state representation relies on the change $v_\alpha(x) = u(x)|x|^\alpha$. We follow the arguments in [167] to prove the following result.

Lemma 9.5.2. *Let $0 < s < \min\{1, \frac{N}{2}\}$. If $0 < \alpha < \frac{N-2s}{2}$, $u \in C_0^\infty(\mathbb{R}^N \setminus \{0\})$ and $v_\alpha(x) = |x|^\alpha u(x)$, then*

$$\int_{\mathbb{R}^N} |\xi|^{2s}|\hat{u}|^2\, d\xi - (\Lambda_{N,2s} + \Phi_{N,s}(\alpha)) \int_{\mathbb{R}^N} \frac{u^2}{|x|^2}\, dx = a_{N,s} \int_{\mathbb{R}^N}\int_{\mathbb{R}^N} \frac{|v_\alpha(x) - v_\alpha(y)|^2}{|x-y|^{N+2s}} \frac{dx}{|x|^\alpha} \frac{dy}{|y|^\alpha}, \tag{9.5.3}$$

where $a_{N,s}$ is given by (9.3.14).

Proof. Define $w_n(x) = |x|^{-\alpha}\chi(\frac{x}{n})$, where

$$\chi \in C_0^\infty(\mathbb{R}^N), \quad \chi(x) \geq 0, \quad \chi(x) = 1, \text{ if } |x| \leq 1.$$

Note that $w_n \in H^s(\mathbb{R}^N)$; it is clear that $w_n \in L^2(\mathbb{R}^N)$. Moreover $|\xi|^s\widehat{w_n}(\xi) \in L^2(\mathbb{R}^N)$, indeed, since $\hat{\chi} \in \mathcal{S}(\mathbb{R}^N)$ and if $g_\alpha(x) = |x|^{-\alpha}$, then

$$b_\alpha \hat{g}_\alpha(\xi) = b_{N-\alpha}|\xi|^{-N+\alpha}.$$

We obtain

$$\int_{\mathbb{R}^N} |\xi|^{2s}|\widehat{w_n}(\xi)|^2\, d\xi = \int_{B_r(0)} |\xi|^{2s+\alpha-N}\, d\xi + O(r) < \infty.$$

By the Plancherel theorem and the polarization of the norm in $H^s(\mathbb{R}^N)$, we can write

$$\int_{\mathbb{R}^N} |\xi|^{2s}\overline{\hat{f}}(\xi)\hat{g}(\xi)\, d\xi = a_{N,s} \int_{\mathbb{R}^N}\int_{\mathbb{R}^N} \frac{(\overline{f}(x) - \overline{f}(y))(g(x) - g(y))}{|x-y|^{N+2s}}\, dx\, dy. \tag{9.5.4}$$

Applying (9.5.4) to $g(x) = w_n(x)$ and $f(x) = \frac{|u^2(x)|}{w_n(x)}$, a direct calculation gives

$$\begin{aligned}
&\int_{\mathbb{R}^N} |\xi|^{2s}\overline{\hat{f}}(\xi)\hat{g}(\xi)\, d\xi \\
&= a_{N,s} \int_{\mathbb{R}^N}\int_{\mathbb{R}^N} \left(\frac{|u(x)|^2}{w_n(x)} - \frac{|u(y)|^2}{w_n(y)}\right)(w_n(x) - w_n(y)) \frac{dx\, dy}{|x-y|^{N+2s}} \\
&= a_{N,s} \int_{\mathbb{R}^N}\int_{\mathbb{R}^N} \left[(u(x) - u(y))^2 - (\frac{u(x)}{w_n(x)} - \frac{u(y)}{w_n(y)})\right]^2 w_n(x)w_n(y) \frac{dx\, dy}{|x-y|^{N+2s}} \equiv: I_n.
\end{aligned}$$

By noting that $\operatorname{supp}(u) \subset B_1(0)$, $\frac{u(x)}{w_n(x)} = v_\alpha(x)$ independently of n, $0 \notin \operatorname{supp}(u)$ and $v_\alpha \in C_0^\infty(\mathbb{R}^N)$, the dominated convergence theorem proves that

$$\lim_{n\to\infty} I_n = a_{N,s} \int_{\mathbb{R}^N}\int_{\mathbb{R}^N} [(u(x) - u(y))^2 - |v_\alpha(x) - v_\alpha(y)|^2]|x|^{-\alpha}|y|^{-\alpha} \frac{dx\,dy}{|x-y|^{N+2s}}. \quad (9.5.5)$$

The left hand side in (9.5.4) can be evaluated as follows:

$$\int_{\mathbb{R}^N} |\xi|^{2s}\bar{\hat{f}}(\xi)\hat{g}(\xi)\,d\xi = \frac{b_{N-\alpha}}{b_\alpha} \int_{\mathbb{R}^N}\int_{\mathbb{R}^N} |\xi|^{2s}\bar{\hat{f}}(\xi) n^N \hat{\chi}(n(\xi-\eta))|\eta|^{\alpha-N}\,d\xi d\eta \equiv: J_n. \quad (9.5.6)$$

Since $\hat{f} \in \mathcal{S}(\mathbb{R}^N)$ (hence $\hat{f} \in L^p(\mathbb{R}^N)$ for all $1 \le p < \infty$) and $\tau_n(\eta) = n^N\hat{\chi}(n\eta)$ is an approximation to the identity, we have

$$\lim_{n\to\infty} J_n = \frac{b_{\alpha+2s}b_{N-\alpha}}{b_{N-\alpha-2s}b_\alpha} \int_{\mathbb{R}^N} \frac{u^2(x)}{|x|^{2s}}\,dx.$$

Since (9.5.5) holds and $I_n = J_n$ we conclude the proof. □

The continuity of (9.5.3) in α and the fact that $\Phi_{N,s}(\frac{N-2s}{2}) = 0$ allow us to formulate the following.

Corollary 9.5.3. *Let* $0 < s < \min\{1, \frac{N}{2}\}$. *If* $u \in C_0^\infty(\mathbb{R}^N \setminus \{0\})$ *and* $v(x) = |x|^{\frac{N-2s}{2}}u(x)$, *then*

$$\int_{\mathbb{R}^N} |\xi|^{2s}|\hat{u}|^2\,d\xi - \Lambda_{N,2s}\int_{\mathbb{R}^N} \frac{u^2}{|x|^2}\,dx = a_{N,s}\int_{\mathbb{R}^N}\int_{\mathbb{R}^N} \frac{|v(x)-v(y)|^2}{|x-y|^{N+2s}} \frac{dx}{|x|^{\frac{N-2s}{2}}} \frac{dy}{|y|^{\frac{N-2s}{2}}}. \quad (9.5.7)$$

Motivated by the *ground state representation* (9.5.3) it is natural to define $\dot{H}^{s,\alpha}(\mathbb{R}^N)$ as the closure of $C_0^\infty(\mathbb{R}^N)$ with respect to the norm

$$\|\phi\|_{\dot{H}^{s,\alpha}(\mathbb{R}^N)} = \left(\int_{\mathbb{R}^N} \frac{|\phi(x)|^{2_s^*}}{|x|^{\alpha 2_s^*}}\,dx\right)^{\frac{1}{2_s^*}} + \left(\int_{\mathbb{R}^N}\int_{\mathbb{R}^N} \frac{|\phi(x)-\phi(y)|^2}{|x-y|^{N+2s}} \frac{dx}{|x|^\alpha}\frac{dy}{|y|^\alpha}\right)^{\frac{1}{2}}.$$

This definition is based on the weighted Sobolev inequality that we will prove in Sections 10.2 and 10.2.1. Let us define

$$\dot{W}^{s,\alpha}(\mathbb{R}^N) := \{\phi : \mathbb{R}^N \to \mathbb{R} \text{ measurable: } \|\phi\|_{\dot{H}^{s,\alpha}(\mathbb{R}^N)} < +\infty\}.$$

Note that, following, e. g., [139], the space $\dot{W}^{s,\alpha}(\mathbb{R}^N)$ coincides with $\dot{H}^{s,\alpha}(\mathbb{R}^N)$ provided that $N > 2s$. We will come back to the functional study of these spaces in the following chapter.

We state a refined version of Lemma 9.5.2.

Lemma 9.5.4 (Ground state representation). *For $0 < \alpha < (N-2s)/2$ and $2s < N$ it follows that the representation formula (9.5.3) still holds for any $u \in \dot{H}^s(\mathbb{R}^N)$ with $v_\alpha(x) := |x|^\alpha u(x)$.*

Proof. We have proved that (9.5.3) holds for any $u \in C_0^\infty(\mathbb{R}^N \setminus \{0\})$. Since $2s < N$ the space $C_0^\infty(\mathbb{R}^N$ is dense in $\dot{H}^s(\mathbb{R}^N)$. See the details for instance in [139]. We can approximate $u \in \dot{H}^s(\mathbb{R}^N)$ with a sequence of functions $u_n \in C_0^\infty(\mathbb{R}^N \setminus \{0\})$. Then, with simple computations, and since $2s < N$, we can pass to the limit on the left hand side of (9.5.3). In order to pass to the limit also on the right hand side of (9.5.3), one needs to note that, by the representation formula (9.5.3), Cauchy sequences of $C_0^\infty(\mathbb{R}^N \setminus \{0\})$ are sent into Cauchy sequences in the weighted Hilbert space $\dot{H}^{s,\alpha}(\mathbb{R}^N)$. Therefore, the conclusion follows by observing that on the left hand side we have a Cauchy sequence since it is convergent. □

Remark 9.5.5.
1. By the continuity in α we can obtain Corollary 9.5.3 in $\dot{H}^{s,\alpha}(\mathbb{R}^N)$.
2. The *ground state representation* (9.5.7) shows in a different way that $\Lambda_{N,2s}$ is the optimal constant in the Hardy inequality and that it is not attained.
3. As we pointed out above, in the complementary case, $N \le 2s$, similar inequalities to (9.2.1), (9.5.3) and (9.5.7) hold in the space $C_0^\infty(\mathbb{R}^N \setminus \{0\})$. See [167].
   If $N \le 2s$ the completion of $C_0^\infty(\mathbb{R}^N \setminus \{0\})$ is $H^s(\mathbb{R}^N \setminus \{0\})$. See [167] and [139]. Therefore (9.2.1), (9.5.3) and (9.5.7) can be extended to the whole $H^s(\mathbb{R}^N)$ only if $N > 2s$.

9.6 Hardy's inequality with remainder terms

We have seen in Section 9.2 that the fractional Hardy inequality is strict. We will give another approach to this kind of results. In [167] the following result evaluating a remainder term in the Hardy inequality is obtained. This result can be seen as the extension to the fractional framework of the local one obtained in [97]. We will follow the ideas in [167].

Theorem 9.6.1. *Let $0 < s < \min\{1, \frac{N}{2}\}$ and $1 \le q < \frac{2N}{N-2s}$. Then there exists a constant $C_{q,s,N} > 0$ such that*

$$\left(\int_\Omega |u|^q\,dx\right)^{\frac{2}{q}} \le C_{q,s,N}|\Omega|^{2(\frac{1}{q}-\frac{1}{2_s^*})}\left(\int_{\mathbb{R}^N} |\xi|^{2s}|\hat{u}|^2\,d\xi - \Lambda_{N,2s}\int_{\mathbb{R}^N} \frac{|u|^2}{|x|^{2s}}\,dx\right), \quad \forall u \in C_0^\infty(\Omega). \tag{9.6.1}$$

Remark 9.6.2. Note that

$$h_s(u) := \left(\int_{\mathbb{R}^N} |\xi|^{2s}|\hat{u}|^2 d\xi - \Lambda_{N,2s}\int_{\mathbb{R}^N} \frac{|u|^2}{|x|^{2s}}\,dx\right), \quad \forall u \in C_0^\infty(\Omega),$$

is the square of a pre-Hilbert seminorm, associated to the following scalar product

$$\langle u, \psi\rangle := \int_{\mathbb{R}^N} |\xi|^{2s}\hat{u}\overline{\hat{\psi}}d\xi - \Lambda_{N,2s}\int_{\mathbb{R}^N} \frac{u\psi}{|x|^{2s}}\,dx = \langle u, H_s\psi\rangle, \quad \forall u, \psi \in \mathcal{C}_0^\infty(\Omega),$$

where $H_s(\psi) = (-\Delta)^s\psi - \Lambda_{N,2s}\frac{\psi}{|x|^{2s}}$. Notice that the brackets in the last term means the duality product between $H^s(\mathbb{R}^N)$ and its dual $H^{-s}(\mathbb{R}^N)$.

The result in Theorem 9.6.1, in particular, shows that $h_s(u)$ is in fact a Hilbertian norm in a bigger space than $H^s(\mathbb{R}^N)$ and, has a byproduct, that the optimal constant is not attained.

To prove Theorem 9.6.1 we need some previous results.

Lemma 9.6.3. *Let $0 \le \chi \le 1$ be a smooth function in $\mathbb{R}_+$ of compact support with $\chi(r) = 1$ for $r \le 1$. Define*

$$\psi_\lambda(x) := \chi\Big(\frac{|x|}{\lambda}\Big)|x|^{-\alpha} \tag{9.6.2}$$

for $0 < \alpha < \frac{(N-2s)}{2}$ and $\lambda > 0$. Then, $\psi_\lambda \in H^s(\mathbb{R}^N)$ for $0 < s < 1$ and, moreover, for all $\epsilon > 0$ there exists a $\lambda_\epsilon = \lambda_\epsilon(\alpha, N, \chi)$ such that for any $\lambda \ge \lambda_\epsilon$,

$$(((-\Delta)^s - \Lambda_{N,s}|x|^{-2s})\psi_\lambda)(x) \le -(|\Phi_{N,s}(\alpha)| - \epsilon)|x|^{-\alpha-2s}, \quad \forall x \in B, \tag{9.6.3}$$

in the sense of distributions, and with B representing the unit ball in $\mathbb{R}^N$.

Proof. The proof that $\psi_\lambda \in H^s(\mathbb{R}^N)$ is a direct computation that follows from [167, Proposition 4.1].

Consider $\phi \in \mathcal{C}_0^\infty(\Omega)$. Using (9.5.1) we find

$$\langle \psi_\lambda, ((-\Delta)^s - \Lambda_{N,s}|x|^{-2s})\phi\rangle = \Phi_{N,s}(\alpha)\langle |x|^{-\alpha-2s}, \phi\rangle - \Big\langle \Big(1 - \chi\Big(\frac{|x|}{\lambda}\Big)|x|^{-\alpha}\Big), (-\Delta)^s\phi\Big\rangle.$$

Call $\widetilde{\psi}_\lambda(x) = \Big(1 - \chi(\frac{|x|}{\lambda})\Big)|x|^{-\alpha}$. By (9.3.13) we obtain

$$\Big\langle \Big(1 - \chi\Big(\frac{|x|}{\lambda}\Big)|x|^{-\alpha}\Big), (-\Delta)^s\phi\Big\rangle = -2a_{N,s}\int_{\mathbb{R}^N}\int_{\mathbb{R}^N} \frac{\widetilde{\psi}_\lambda(y)\phi(x)}{|x-y|^{N+2s}}\,dy\,dx \ge$$

$$-\sup_{|x|\le 1} 2a_{N,s}|x|^{\alpha+2s}\int_{\mathbb{R}^N} \frac{\widetilde{\phi}_\lambda(y)}{|x-y|^{N+2s}}\,dy \int_{\mathbb{R}^N} \frac{\phi(x)}{|x-y|^{N+2s}}\,dx.$$

Set

$$\rho(\lambda) = \sup_{|x|\le 1} 2a_{N,s}|x|^{\alpha+2s}\int_{\mathbb{R}^N} \frac{\widetilde{\phi}_\lambda(y)}{|x-y|^{N+2s}}\,dy = \sup_{|x|\le \frac{1}{\lambda}} 2a_{N,s}|x|^{\alpha+2s}\int_{\mathbb{R}^N} \frac{(1-\phi(y))}{|y|^\alpha|x-y|^{N+2s}}\,dy.$$

Now

(i) $\rho(\lambda) < \infty$ if $\lambda > 1$;
(ii) $\rho(\lambda) \downarrow 0$ as $\lambda \to \infty$.

Hence, given $\epsilon > 0$ we can choose λ_ε such that $\rho(\lambda_\epsilon) = \epsilon$. Since $\Phi_{N,s}(\alpha) < 0$ by Lemma 9.5.1, we obtain (9.6.3). □

Lemma 9.6.4. *Let $0 \le q < 1$ and $u \in L^q(\mathbb{R}^N)$ be a positive, radially symmetric, decreasing function. Then, denoting by B the unit ball,*

$$\|u\|_q \le \frac{1}{q}|B|^{-\frac{1}{q'}} \int_{\mathbb{R}^N} u(x)|x|^{-\frac{N}{q'}}\,dx. \tag{9.6.4}$$

Proof. If u is the characteristic function of a ball, a simple computation shows that the above is in fact an identity. For general positive, radially symmetric, decreasing u we have for every $t > 0$ that the level set $\{x : u(x) > t\}$ is a ball. We denote by $\chi_t(x)$ its characteristic function. Then,

$$u(x) = \int_0^\infty \chi_t(x)\,dt.$$

Therefore by Minkowski's inequality we conclude that

$$\|u\|_q \le \int_0^\infty \|\chi_t\|_q\,dt = \frac{1}{q}|B|^{-\frac{1}{q'}} \int_0^\infty \int_{\mathbb{R}^N} \chi_t(x)|x|^{-\frac{N}{q'}}\,dx\,dt = \frac{1}{q}|B|^{-\frac{1}{q'}} \int_{\mathbb{R}^N} u(x)|x|^{-\frac{N}{q'}}\,dx,$$

where the last identity is just Fubini's theorem. □

Proof of Theorem 9.6.1. By decreasing rearrangement and scaling, we can assume that Ω is the unit ball centered at the origin, B, and then perform the proof in the radial setting. By Lemmas 9.6.3 and 9.6.4, if u is a symmetric function in B,

$$\|u\|_q \le \frac{1}{q}|B|^{-\frac{1}{q'}} \int_{\mathbb{R}^N} u(x)|x|^{-\frac{N}{q'}}\,dx \le 2\frac{1}{q}|B|^{-\frac{1}{q'}} |\Phi_{N,s}(\alpha)|^{-1} |\langle u, H_s\psi\rangle|.$$

Here $\psi \equiv \psi_{\lambda_\epsilon}$ is as in Lemma 9.6.3 with $\epsilon = \frac{|\Phi_{N,s}(\alpha)|}{2}$. Finally by Cauchy–Schwarz's inequality,

$$|\langle u, H_s\psi\rangle|^2 \le h_s(u)h_s(\psi)$$

and we conclude the proof. □

We also have the following result obtained in [24]. See also [164] for a different proof.

Theorem 9.6.5. *Let $0 < s < 1$ and $N > \max\{2s, 1\}$. Assume that $\Omega \subset \mathbb{R}^N$ is a bounded domain. Then for all $1 \le q < 2$, there exists a positive constant $C = C(\Omega, q, N, s)$ such that for all $u \in C_0^\infty(\Omega)$,*

$$a_{N,s} \int_{\mathbb{R}^N}\int_{\mathbb{R}^N} \frac{(u(x)-u(y))^2}{|x-y|^{N+2s}}\,dx\,dy - \Lambda_{N,2s}\int_{\mathbb{R}^N} \frac{u^2(x)}{|x|^{2s}}\,dx \ge C(\Omega, q, N, s)\int_\Omega\int_\Omega \frac{(u(x)-u(y))^2}{|x-y|^{N+qs}}\,dx\,dy. \tag{9.6.5}$$

Proof. We follow the same strategy as in [167], that is, the *ground state representation*. For simplicity, we set $\alpha = \frac{N-2s}{2}$. Then $w(x) = |x|^{-\alpha}$ and $v(x) = \frac{u(x)}{w(x)}$. Thus,

$$f_1(x,y) := \frac{(v(x)-v(y))^2}{|x-y|^{N+2s}} w(x)w(y) = \frac{((u(x)-u(y)) - \frac{u(y)}{w(y)}(w(x)-w(y)))^2}{|x-y|^{N+2s}} \frac{w(y)}{w(x)}.$$

In the same way, thanks to the symmetry of $f_1(x,y)$, it immediately follows that

$$\frac{(v(x)-v(y))^2}{|x-y|^{N+2s}} w(x)w(y) = \frac{((u(y)-u(x)) - \frac{u(x)}{w(x)}(w(y)-w(x)))^2}{|x-y|^{N+2s}} \frac{w(x)}{w(y)} \equiv f_2(x,y).$$

Then,

$$h_s(u) = \frac{c_{N,2s}}{2}\Big(\int_{\mathbb{R}^N}\int_{\mathbb{R}^N} f_1(x,y)\,dx\,dy + \int_{\mathbb{R}^N}\int_{\mathbb{R}^N} f_2(x,y)\,dx\,dy\Big).$$

Since f_1 and f_2 are positive functions, it follows that

$$h_s(u) \ge \frac{c_{N,2s}}{2}\Big(\int_\Omega\int_\Omega f_1(x,y)\,dx\,dy + \int_\Omega\int_\Omega f_2(x,y)\,dx\,dy\Big).$$

Note that for all $(x,y) \in \Omega \times \Omega$ and $q < 2$, we have $\frac{1}{|x-y|^{N+2s}} \ge \frac{C}{|x-y|^{N+qs}}$ and

$$K(x,y) \equiv \frac{w(x)w(y)}{w^2(x)+w^2(y)} \le \frac{1}{2}.$$

It is clear that $K(x,y)(\frac{w(x)}{w(y)} + \frac{w(y)}{w(x)}) = 1$. Hence,

$$\begin{aligned} f_1(x,y) \ge{}& C_1 K(x,y)\frac{w(y)}{w(x)} \\ &\times \Big[\frac{(u(x)-u(y))^2}{|x-y|^{N+qs}} + \Big(\frac{u(y)}{w(y)}\Big)^2 \frac{(w(x)-w(y))^2}{|x-y|^{N+qs}} \\ &- 2\frac{u(y)}{w(y)}\frac{(u(x)-u(y))(w(x)-w(y))}{|x-y|^{N+qs}}\Big] \end{aligned}$$

and

$$f_2(x,y) \geq C_1 K(x,y)\frac{w(x)}{w(y)} \times \left[\frac{(u(y)-u(x))^2}{|x-y|^{N+qs}} + \left(\frac{u(x)}{w(x)}\right)^2 \frac{(w(y)-w(x))^2}{|x-y|^{N+qs}} - 2\frac{u(x)}{w(x)}\frac{(u(y)-u(x))(w(y)-w(x))}{|x-y|^{N+qs}}\right].$$

Therefore,

$$\begin{aligned} h_s(u) &\geq \frac{C_1}{2}\int_\Omega\int_\Omega K(x,y)\left(\frac{w(y)}{w(x)} + \frac{w(x)}{w(y)}\right)\frac{(u(x)-u(y))^2}{|x-y|^{N+qs}}\,dx\,dy \\ &\quad - C_1\int_\Omega\int_\Omega K(x,y)\left[\frac{u(y)}{w(x)}\frac{(u(x)-u(y))(w(x)-w(y))}{|x-y|^{N+qs}} \right. \\ &\quad \left. + \frac{u(x)}{w(y)}\frac{(u(y)-u(x))(w(y)-w(x))}{|x-y|^{N+qs}}\right]dx\,dy \\ &\geq \frac{C_1}{2}\int_\Omega\int_\Omega \frac{(u(x)-u(y))^2}{|x-y|^{N+qs}}\,dx\,dy - C_1\int_\Omega\int_\Omega g_1(x,y)\,dx\,dy - C_1\int_\Omega\int_\Omega g_2(x,y)\,dx\,dy, \end{aligned}$$

with

$$g_1(x,y) = K(x,y)\frac{u(y)}{w(x)}\left(\frac{(u(x)-u(y))(w(x)-w(y))}{|x-y|^{N+qs}}\right)$$

and

$$g_2(x,y) = K(x,y)\frac{u(x)}{w(y)}\left(\frac{(u(y)-u(x))(w(y)-w(x))}{|x-y|^{N+qs}}\right).$$

Using the fact that $u \in \mathcal{C}_0^\infty(\Omega)$ it follows that $\int_\Omega\int_\Omega |g_i(x,y)|\,dx\,dy < \infty$ for $i = 1,2$.

Since $g_2(x,y) = g_1(y,x)$, we just have to estimate $\int_\Omega\int_\Omega g_2(x,y)\,dx\,dy$. Note that

$$g_2(x,y) = \frac{w(x)u(x)}{w^2(x)+w^2(y)}\frac{(u(y)-u(x))(w(y)-w(x))}{|x-y|^{N+qs}}.$$

Hence, using the Young inequality,

$$\begin{aligned} &\int_\Omega\int_\Omega |g_2(x,y)|\,dx\,dy \\ &\quad\leq \epsilon\int_\Omega\int_\Omega \frac{(u(x)-u(y))^2}{|x-y|^{N+qs}}\,dx\,dy + C(\epsilon)\int_\Omega\int_\Omega \frac{w^2(x)u^2(x)(w(x)-w(y))^2}{(w^2(x)+w^2(y))^2|x-y|^{N+qs}}\,dx\,dy. \end{aligned}$$

We claim that

$$I \equiv \int_\Omega\int_\Omega \frac{w^2(x)u^2(x)(w(x)-w(y))^2}{(w^2(x)+w^2(y))^2|x-y|^{N+qs}}\,dx\,dy \leq Ch_s(u).$$

It is clear that

$$I = \int_\Omega u^2(x)\left[\int_\Omega \frac{(|x|^\alpha - |y|^\alpha)^2|y|^{2\alpha}}{(|x|^{2\alpha} + |y|^{2\alpha})^2|x-y|^{N+qs}}\,dy\right]dx.$$

To compute the above integral, we use a similar argument as in the proof of Theorem 1.1 in [158]. We set $y = \rho y'$ and $x = rx'$. Then, since $\Omega \subset B_R(0)$ for a suitable R, we get

$$I \le \int_\Omega u^2(x)\int_0^R \frac{(r^\alpha - \rho^\alpha)^2\rho^{N+2\alpha-1}}{(r^{2\alpha}+\rho^{2\alpha})^2}\left(\int_{|y'|=1}\frac{dH^{N-1}(y')}{|\rho y' - rx'|^{N+qs}}\right)d\rho\,dx,$$

where dH^{N-1} is the $(N-1)$ dimensional Hausdorff measure on the sphere. Let $\rho = r\sigma$. Then

$$\begin{aligned} I &\le \int_\Omega \frac{u^2(x)}{|x|^{qs}}\int_0^{\frac{R}{r}} \frac{(1-\sigma^\alpha)^2\sigma^{N+2\alpha-1}}{(1+\sigma^{2\alpha})^2}\left(\int_{|y'|=1}\frac{dH^{N-1}(y')}{|\sigma y' - x'|^{N+qs}}\right)d\sigma\,dx \\ &\le \int_\Omega \frac{u^2(x)}{|x|^{qs}}\,dx\int_0^\infty \frac{(1-\sigma^\alpha)^2\sigma^{N+2\alpha-1}}{(1+\sigma^{2\alpha})^2}K(\sigma)d\sigma, \end{aligned}$$

where

$$K(\sigma) = 2\frac{\pi^{\frac{N-1}{2}}}{\Gamma(\frac{N-1}{2})}\int_0^\pi \frac{\sin^{N-2}(\theta)}{(1-2\sigma\cos(\theta)+\sigma^2)^{\frac{N+qs}{2}}}d\theta.$$

Let us analyze $\int_0^\infty \frac{(1-\sigma^\alpha)^2\sigma^{N+2\alpha-1}}{(1+\sigma^{2\alpha})^2}K(\sigma)d\sigma$. We have

$$\int_0^\infty \frac{(1-\sigma^\alpha)^2\sigma^{N+2\alpha-1}}{(1+\sigma^{2\alpha})^2}K(\sigma)d\sigma = \int_0^1 \frac{(1-\sigma^\alpha)^2\sigma^{N+2\alpha-1}}{(1+\sigma^{2\alpha})^2}K(\sigma)d\sigma + \int_1^\infty \frac{(1-\sigma^\alpha)^2\sigma^{N+2\alpha-1}}{(1+\sigma^{2\alpha})^2}K(\sigma)d\sigma.$$

By setting $\xi = \frac{1}{\sigma}$ in the first integral and taking into account that $K(\frac{1}{\xi}) = \xi^{N+qs}K(\xi)$, we get

$$\int_0^\infty \frac{(1-\sigma^\alpha)^2\sigma^{N+2\alpha-1}}{(1+\sigma^{2\alpha})^2}K(\sigma)d\sigma = \int_1^\infty \frac{(\xi^\alpha-1)^2\xi^{qs-1}}{(1+\xi^{2\alpha})^2}K(\xi)d\xi + \int_1^\infty \frac{(\sigma^\alpha-1)^2\sigma^{N+2\alpha-1}}{(1+\sigma^{2\alpha})^2}K(\sigma)d\sigma.$$

Note that $K(\sigma) \backsimeq \sigma^{-N-qs}$ as $\sigma \to \infty$, thus both integrals converge near ∞.

On the other hand and following the computation of [158] (estimates (2.1) to (2.10), see also the definition of the function H in formula (2.6)), we reach $K(\sigma) \le C(\sigma^2 - 1)^{-1-2s}$ as $\sigma \to 1$. Hence, as $\sigma, \xi \to 1^+$,

$$(\sigma^\alpha - 1)^2 K(\sigma) \le C(\sigma-1)^{1-2s} \in L^1(1,2).$$

Thus, combining the above estimates, we get $I \le C\int_\Omega \frac{u^2(x)}{|x|^{qs}}\,dx$.

Since $q < 2$, using the Hölder inequality and (9.6.1), it follows that $I \le Ch_s(\dot{u})$. Then the claim follows.

As a conclusion, we obtain

$$h_s(u) \ge C(\Omega, q, N, s) \int_\Omega \int_\Omega \frac{(u(x) - u(y))^2}{|x - y|^{N+qs}} \, dx \, dy,$$

which finishes the proof. □

9.7 Further results and comments

The Hardy inequality has its counterpart with homogeneity $sp > N$; see for instance [168]. More precisely in this paper the following result is obtained.

Theorem 9.7.1 (Hardy inequality). *Let $N \ge 1$ and $0 < s < 1$. Then for all $u \in \dot{W}^{s,p}(\mathbb{R}^N)$ in case $1 \le p < \frac{N}{s}$ and for all $u \in \dot{W}^{s,p}(\mathbb{R}^N \setminus \{0\})$ in case $p > \frac{N}{s}$,*

$$\int_{\mathbb{R}^N} \int_{\mathbb{R}^N} \frac{|u(x) - u(y)|^p}{|x - y|^{N+ps}} \, dx \, dy \ge \Lambda_{N,s,p} \int_{\mathbb{R}^N} \frac{|u(x)|^p}{|x|^{ps}} \, dx, \tag{9.7.1}$$

with

$$\Lambda_{N,s,p} := 2 \int_0^1 r^{ps-1} |1 - r^{\frac{N-ps}{p}}|^p \Phi_{N,s,p}(r) \, dr, \tag{9.7.2}$$

and

$$\Phi_{N,s,p}(r) = \begin{cases} \omega_{N-2} \int_{-1}^1 \frac{(1-t^2)^{\frac{N-3}{2}}}{(1-2rt+r^2)^{\frac{N+ps}{2}}} \, dt & \text{if } N \ge 2, \\ \left(\frac{1}{(1-r)^{1+ps}} + \frac{1}{(1+r)^{1+ps}}\right) & \text{if } N = 1. \end{cases} \tag{9.7.3}$$

The constant $\Lambda_{N,s,p}$ is optimal. If $p > 1$ the inequality is strict for any function $0 \ne u \in \dot{W}^{s,p}(\mathbb{R}^N)$ in case $1 \le p < \frac{N}{s}$, and for all $0 \ne u \in \dot{W}^{s,p}(\mathbb{R}^N \setminus \{0\})$ in case $p > \frac{N}{s}$.

If $p = 1$, equality holds if and only if u is proportional to a symmetric decreasing function.

In the same paper [168] Frank and Seiringer find a general Hardy inequality and prove for $p \ge 2$ a Hardy inequality with remainder term.

The previous inequality in the above theorem is related to the following integro-differential problem. Defining $f(x, u)$ by means of the difference

$$\int_{\mathbb{R}^N} \frac{|u(x) - u(y)|^{p-2}(u(x) - u(y))}{|x - y|^{N+ps}} \, dy - \lambda \frac{|u(x)|^{p-2} u}{|x|^{ps}} = f(x, u),$$

find lower estimates for f depending on the value of λ with respect to $\Lambda_{N,s,p}$. The function f could also be dependent on ∇u if $s > \frac{1}{2}$.

10 Calderón–Zygmund summability in the fractional setting

10.1 Introduction and statement of the problem

In Chapter 3 we have established how the Hardy potential affects the Calderón–Zygmund summability of the solutions to the linear problem (3.1.1). In this chapter we will study the following problem:

$$\begin{cases} (-\Delta)^s u - \lambda \frac{u}{|x|^{2s}} = f(x) & \text{in } \Omega, \\ u > 0 & \text{in } \Omega, \\ u = 0 & \text{in } \mathbb{R}^N \setminus \Omega, \end{cases} \tag{10.1.1}$$

where $s \in (0, 1)$ is such that $2s < N$, $\Omega \subset \mathbb{R}^N$ is a bounded regular domain containing the origin and f is a measurable function satisfying suitable hypotheses of integrability. The fractional s-Laplacian has been defined in Chapter 8, Sections 8.1 and 8.2 in different settings. Recall that we define the fractional Laplacian as the following integral operator:

$$(-\Delta)^s u(x) := C_{N,s} P.V. \int_{\mathbb{R}^N} \frac{u(x) - u(y)}{|x - y|^{N+2s}} \, dy, \quad s \in (0, 1), \ u \in \mathcal{S}(\mathbb{R}^N), \tag{10.1.2}$$

where $C_{N,s}$ is given by (9.3.12). The asymptotics of the constant $C_{N,s}$ can be found in [138, Section 4].

Due to its second term, problem (10.1.1) is related to the following fractional Hardy inequality, proved in Chapter 9:

$$\int_{\mathbb{R}^N} |\xi|^{2s} |\hat{u}|^2 \, d\xi \ge \Lambda_{N,2s} \int_{\mathbb{R}^N} |x|^{-2s} |u|^2 \, dx \quad \forall u \in C_0^\infty(\mathbb{R}^N), \tag{10.1.3}$$

where

$$\Lambda_{N,2s} := 2^{2s} \frac{\Gamma^2(\frac{N+2s}{4})}{\Gamma^2(\frac{N-2s}{4})} \tag{10.1.4}$$

is optimal and not attained. See Theorem 9.2.1. Moreover,

$$\lim_{s \to 1} \Lambda_{N,2s} = \left(\frac{N-2}{2} \right)^2,$$

the classical Hardy constant.

We can also rewrite (10.1.3) in the form

$$a_{N,s} \int_{\mathbb{R}^N} \int_{\mathbb{R}^N} \frac{|u(x) - u(y)|^2}{|x - y|^{N+2s}} \, dx \, dy \ge \Lambda_{N,2s} \int_{\mathbb{R}^N} \frac{u^2}{|x|^{2s}} \, dx, \quad u \in C_0^\infty(\mathbb{R}^N), \quad a_{N,s} = \frac{C_{N,s}}{2},$$

https://doi.org/10.1515/9783110606270-010

which we will use along this chapter. See also [167]. This critical value $\Lambda_{N,2s}$ will also play a fundamental role concerning the regularity of the solution in terms of the summability of the datum.

In particular, we already know that for $\lambda > \Lambda_{N,2s}$ problem (10.1.1) has no positive supersolution (see for example [46, 171]). Hence, from now on we will assume $0 < \lambda \leq \Lambda_{N,2s}$.

Remark 10.1.1. The behavior of changing-sign solutions in problems involving the Hardy potential is quite different. In fact some obstructions for the existence disappear (see [311, Section 8]). Roughly speaking, the oscillation can produce cancellations near the origin that kill the critical behavior of the potential. Since in this book we are interested in the critical behavior we will focus on nonnegative solutions.

The goal in this chapter will be to obtain the optimal summability of u according to the summability of the datum f and the parameter λ. The case $\lambda = 0$ was studied in Chapter 8, Sections 8.4 and 8.6 (see also [224]), where some Calderón–Zygmund type results were obtained. Since this problem is linear, we will assume, without loss of generality, that the datum f is positive and we will deal with positive solutions. The influence of the Hardy potential in the local case ($s = 1$) was studied in Chapter 3 (see also [80]). The results in the nonlocal case, $0 < s < 1$, are more involved and need a change in the functional setting.

In order to find an analogous optimal condition for problem (10.1.1), we will need a *weak Harnack inequality* for a singular weighted nonlocal operator associated with the *ground state transform* that we have defined in Section 9.5 and we will recall in (10.2.3) below. This study, performed in Section 3, requires the combination of techniques on elliptic operators and the behavior of nonlocal radial integrals studied in Chapter 9, and it provides the precise behavior of the solutions around the origin. This result will be the key in the proofs of existence and regularity in the next sections.

This chapter is organized as follows. In Section 10.2 we precise the meaning of solutions that will be used along this part of the work, with the corresponding functional setting. In Section 10.2.1 we prove some useful tools as a weighted Sobolev inequality, compactness results and certain algebraic inequalities.

In Section 10.3 we prove a weighted singular version of the Harnack inequality. Note that using the *ground state transformation* stated in Lemma 10.2.4, the weak Harnack inequality gives the exact blow-up rate for the positive supersolutions to (10.1.1) near the origin. This result will be the key for the optimality in the results of the following sections. Finally, in Section 10.4 we prove the main results. According to λ and the summability of f, we obtain the optimal summability of the solution u for certain values of the spectral parameter λ. In particular, we see that the local techniques applied in Chapter 3 do not give complete information in this framework, leaving the optimality for certain ranges of λ as an open problem. Some results in this chapter were obtained in [12].

10.2 Functional setting: inequalities with weights

The Sobolev spaces $H^s(\mathbb{R}^N)$, $H^s(\Omega)$, their duals, the definition of the fractional Laplacian and its fundamental properties have been introduced in Chapter 8. In this section we will concentrate on the specific functional tools to study the summability of the solutions to (10.1.1). This new framework is suggested by the meaning of solutions for problem (10.1.1):

(i) *energy solutions* when the variational framework can be used, that is, when the datum belongs to the dual space;
(ii) *weak solutions* for data that are integrable but not in the dual space.

Recall that we also assume $0 \in \Omega$ to avoid the classical situation of a positive bounded potential.

Definition 10.2.1. Assume $0 < \lambda < \Lambda_{N,2s}$. For $f \in H^{-s}(\Omega)$ we say that $u \in H_0^s(\Omega)$ is a *finite energy solution* to (10.1.1) if

$$\langle (-\Delta)^s u, w\rangle - \lambda \int_\Omega \frac{uw}{|x|^{2s}}\,dx = \langle f, w\rangle, \quad \forall w \in H_0^s(\Omega).$$

If $\lambda < \Lambda_{N,2s}$, then existence and uniqueness of a solution $u \in H_0^s(\Omega)$ for all $f \in H^{-s}(\Omega)$ easily follows from the Lax–Milgram theorem.

Remark 10.2.2. If $\lambda = \Lambda_{N,2s}$, the same result holds but in a space $H(\Omega)$ defined as the completion of $C_0^\infty(\Omega)$ with respect to the norm

$$\|\phi\|^2_{H(\Omega)} := a_{N,s} \iint_{D_\Omega} \frac{|\phi(x) - \phi(y)|^2}{|x-y|^{N+2s}}\,dx\,dy - \Lambda_{N,2s} \int_\Omega \frac{\phi^2}{|x|^{2s}}\,dx.$$

Using the Hardy inequality with remainder term obtained in Section 9.6 (see also [19] and the references therein), we get that $H(\Omega)$ is a Hilbert space and $H_0^s(\Omega) \subsetneq H(\Omega) \subsetneq W_0^{s,q}(\Omega)$, for all $q < 2$. This mechanism is similar to the one used in the local case, $s = 1$.

To deal with the case of a general $f \in L^1(\Omega)$, we need to define the notion of *weak solution*, where we only request the regularity needed to give weak sense to the equation. Since the operator is nonlocal, we need to detail the class of test functions to be considered, which is precisely

$$\mathcal{T} := \{\phi : \mathbb{R}^N \to \mathbb{R} \mid (-\Delta)^s \phi = \varphi, \varphi \in L^\infty(\Omega) \cap C^\alpha(\Omega), 0 < \alpha < 1, \phi = 0 \text{ in } \mathbb{R}^N \setminus \Omega\}. \tag{10.2.1}$$

Note that every $\phi \in \mathcal{T}$ belongs in particular to $L^\infty(\Omega)$ (see Chapter 8) and, moreover, is a strong solution to the equation $(-\Delta)^s \phi = \varphi$. See [282] and [287].

Definition 10.2.3. Assume $f \in L^1(\Omega)$. We say that $u \in L^1(\Omega)$ is a weak supersolution (subsolution) of problem (10.1.1) if $\frac{u}{|x|^{2s}} \in L^1(\Omega)$, $u = 0$ in $\mathbb{R}^N \setminus \Omega$, and for all nonnegative $\phi \in \mathcal{T}$, the following inequality holds:

$$\int_\Omega u(-\Delta)^s \phi\, dx - \lambda \int_\Omega \frac{u\phi}{|x|^{2s}}\, dx \geq (\leq) \int_\Omega f\phi\, dx.$$

If u is a super- and subsolution, then we say that u is a weak solution.

Recall that if $u \in C_0^\infty(\mathbb{R}^N)$, in Section 9.5 we have proved, following [167], the *ground state representation.*

Lemma 10.2.4 (Ground state representation). *Let* $0 < \gamma < \frac{N-2s}{2}$. *If* $u \in C_0^\infty(\mathbb{R}^N)$ *and* $v(x) := |x|^\gamma u(x)$, *then*

$$\int_{\mathbb{R}^N} |\xi|^{2s} |\hat{u}(\xi)|^2\, d\xi - (\Lambda_{N,2s} + \Phi_{N,s}(\gamma)) \int_{\mathbb{R}^N} |x|^{-2s} |u(x)|^2\, dx = a_{N,s} \iint_{\mathbb{R}^{2N}} \frac{|v(x) - v(y)|^2}{|x-y|^{N+2s}} \frac{dx}{|x|^\gamma} \frac{dy}{|y|^\gamma},$$

where

$$\Phi_{N,s}(\gamma) = 2^{2s} \left(\frac{\Gamma(\frac{\gamma+2s}{2})\Gamma(\frac{N-\gamma}{2})}{\Gamma(\frac{N-\gamma-2s}{2})\Gamma(\frac{\gamma}{2})} - \frac{\Gamma^2(\frac{N+2s}{4})}{\Gamma^2(\frac{N-2s}{4})} \right)$$

and $a_{N,s}$ *is defined in* (9.3.14).

In particular this representation proves that the constant $\Lambda_{N,2s}$ is optimal and not attained.

Using this representation with $\lambda = \Lambda_{N,2s} + \Phi_{N,s}(\gamma)$, we obtain that if u solves

$$\begin{cases} (-\Delta)^s u - \lambda \frac{u}{|x|^{2s}} = f(x) & \text{in } \Omega, \\ u = 0 & \text{in } \mathbb{R}^N \setminus \Omega, \end{cases}$$

then v satisfies

$$\begin{cases} L_{\gamma,\Omega} v = |x|^{-\gamma} f(x) |x|^{-\gamma} =: g(x) & \text{in } \Omega, \\ v = 0 & \text{in } \mathbb{R}^N \setminus \Omega, \end{cases} \tag{10.2.2}$$

with

$$L_{\gamma,\Omega} v := C_{N,s}\, P.V. \int_{\mathbb{R}^N} \frac{v(x) - v(y)}{|x-y|^{N+2s}} \frac{dy}{|x|^\gamma |y|^\gamma}. \tag{10.2.3}$$

Observe that if $\gamma \to 0$, then $\Phi_{N,s}(\gamma) \to -\Lambda_{N,2s}$ and $\lambda \to 0$. On the other hand, if $\gamma \to \frac{N-2s}{2}$, then $\Phi_{N,s}(\gamma) \to 0$ and $\lambda \to \Lambda_{N,2s}$.

To analyze the behavior and the regularity of u, we deal with the same questions for v. Thus, we need to work in fractional Sobolev spaces with admissible weights. For simplicity, we denote

$$d\mu := \frac{dx}{|x|^{2\gamma}} \quad \text{and} \quad d\nu := \frac{dx\,dy}{|x-y|^{N+2s}|x|^{\gamma}|y|^{\gamma}}. \tag{10.2.4}$$

Remark 10.2.5. Note that $d\mu := \frac{dx}{|x|^{2\gamma}}$ is a positive Radon measure with the doubling property, that is, there exists a positive constant C such that

$$|B_{2r}(x)|_{d\mu} \le C|B_r(x)|_{d\mu}, \quad \text{for each ball } B_r(x) \subset \mathbb{R}^N.$$

This property will be relevant to obtain the Harnack inequality for operators with weights.

For $\Omega \subseteq \mathbb{R}^N$, we define the regional, weighted, fractional Sobolev space $W^{s,\gamma}(\Omega)$ as follows:

$$W^{s,\gamma}(\Omega) := \left\{\phi \in L^2(\Omega, d\mu) \text{ s. t. } \int\limits_\Omega \int\limits_\Omega (\phi(x)-\phi(y))^2\, d\nu < +\infty\right\}.$$

It is clear that $W^{s,\gamma}(\Omega)$ is a Hilbert space endowed with the norm

$$\|\phi\|_{W^{s,\gamma}(\Omega)} := \left(\int\limits_\Omega |\phi(x)|^2\, d\mu + \int\limits_\Omega \int\limits_\Omega (\phi(x)-\phi(y))^2\, d\nu\right)^{\frac{1}{2}}.$$

The following extension lemma can be proved by using the same arguments of [28] (see also [138]).

Lemma 10.2.6. *Let $\Omega \subset \mathbb{R}^N$ be a smooth domain. Then for all $w \in W^{s,\gamma}(\Omega)$, there exists $\tilde{w} \in W^{s,\gamma}(\mathbb{R}^N)$ such that $\tilde{w}_{|\Omega} = w$ and*

$$\|\tilde{w}\|_{W^{s,\gamma}(\mathbb{R}^N)} \le C\|w\|_{W^{s\gamma}(\Omega)},$$

where $C := C(N, s, \Omega, \gamma) > 0$.

If $\Omega = \mathbb{R}^N$, we have the next result.

Lemma 10.2.7. *If $w(x) := |x|^{-\theta}$, with $0 < \theta < (N-2s-2\gamma)$, then there exists a positive constant $C := C(N, s, \gamma, \theta)$ such that*

$$L_{\gamma,\mathbb{R}^N}(w)(x) = C\frac{w(x)}{|x|^{2s+2\gamma}} \quad a.e \text{ in } \mathbb{R}^N.$$

Proof. In $\mathbb{R}^N$, the operator has the form

$$L_{\gamma,\mathbb{R}^N}(w)(x) := C_{N,s}\, P.V. \int\limits_{\mathbb{R}^N} \frac{(w(x)-w(y))}{|x|^{\gamma}|y|^{\gamma}|x-y|^{N+2s}}\, dy.$$

We closely follow the arguments used in [158]. By setting $r := |x|$ and $\rho := |y|$, we have $x = rx'$ and $y = \rho y'$, where $|x'| = |y'| = 1$. Thus,

$$L_{\gamma,\mathbb{R}^N}(w)(x) = \frac{C_{N,s}}{|x|^\gamma} \int_0^{+\infty} \frac{(r^{-\theta} - \rho^{-\theta})\rho^{N-1}}{\rho^\gamma r^{N+2s}} \Bigg(\int_{|y'|=1} \frac{dH^{N-1}(y')}{|x' - \frac{\rho}{r} y'|^{N+2s}} \Bigg) d\rho.$$

Set now $\sigma := \frac{\rho}{r}$. Then

$$L_{\gamma,\mathbb{R}^N}(w)(x) = \frac{C_{N,s} w(x)}{|x|^{2s+2\gamma}} \int_0^{+\infty} (1 - \sigma^{-\theta})\sigma^{N-\gamma-1} \Bigg(\int_{|y'|=1} \frac{dH^{N-1}(y')}{|x' - \sigma y'|^{N+2s}} \Bigg) d\sigma.$$

Define

$$D(\sigma) := \int_{|y'|=1} \frac{dH^{N-1}(y')}{|x' - \sigma y'|^{N+2s}}.$$

Then

$$D(\sigma) = 2\frac{\pi^{\frac{N-1}{2}}}{\Gamma(\frac{N-1}{2})} \int_0^\pi \frac{\sin^{N-2}(\eta)}{(1 - 2\sigma\cos(\eta) + \sigma^2)^{\frac{N+2s}{2}}}\, d\eta. \tag{10.2.5}$$

Thus

$$L_{\gamma,\mathbb{R}^N}(w) = \Lambda_{N,2s,\gamma} \frac{w(x)}{|x|^{2s+2\gamma}},$$

where we define

$$\Lambda_{N,2s,\gamma} = C_{N,s} \int_0^{+\infty} (\sigma^\theta - 1)\sigma^{N-\gamma-\theta-1} D(\sigma)\, d\sigma.$$

In [158] (see also [223]) they proved the behavior of D near $\sigma = 1$ and at $+\infty$, namely, $D(\sigma) \le C(\sigma^2 - 1)^{-1-2s}$ close to $\sigma = 1$ and $D(\sigma) \approx \sigma^{-N-2s}$ as $\sigma \to \infty$. Then we obtain $|\Lambda_{N,2s,\gamma}| < \infty$.

To conclude we just have to show that $\Lambda_{N,2s,\gamma} > 0$. Since $D(\frac{1}{s}) = s^{N+2s} D(s)$ for all $s > 0$, we get

$$\begin{aligned}
\Lambda_{N,2s,\gamma} &= \int_0^1 (\sigma^\theta - 1)\sigma^{N-\gamma-\theta-1} D(\sigma)\, d\sigma + \int_1^\infty (\sigma^\theta - 1)\sigma^{N-\gamma-\theta-1} D(\sigma)\, d\sigma \\
&= -\int_1^\infty (\xi^\theta - 1)\xi^{2s+\gamma-1} D(\xi)\, d\xi + \int_1^\infty (\sigma^\theta - 1)\sigma^{N-\gamma-\theta-1} D(\sigma)\, d\sigma \\
&= \int_1^\infty D(\sigma)(\sigma^\theta - 1)(\sigma^{N-\gamma-\theta-1} - \sigma^{2s+\gamma-1})\, d\sigma.
\end{aligned}$$

Since $0 < \theta < N - 2s - 2\gamma$, by formulas (2.5) and (2.6) in [158], we have $D(\sigma) > 0$ in $(1, \infty)$, thus $\Lambda_{N,2s,\gamma} > 0$. □

10.2.1 Weighted Sobolev inequalities and applications

In [3], the authors proved a weighted Sobolev inequality for general $1 < p$. Here we will prove the result in the Hilbertian case, which is the case that we will use in this book. To be more precise, we will prove the following weighted Sobolev inequality.

Theorem 10.2.8. *Given $0 < s < 1$, $0 < \beta < \frac{N-2s}{2}$ and $p = 2_s^* = \frac{2N}{N-2s}$, there exists a constant $C_0 = C(N,s,\beta)$ so that*

$$\left(\int_{\mathbb{R}^N}\left(\frac{|u(x)|}{|x|^\beta}\right)^p dx\right)^{1/p} \le C_0\left(\iint_{\mathbb{R}^N\times\mathbb{R}^N}\frac{|u(x)-u(y)|^2}{|x-y|^{N+2s}}\frac{dy}{|y|^\beta}\frac{dx}{|x|^\beta}\right)^{1/2}, \tag{10.2.6}$$

for all u, say, in the Schwartz class.

Proof. From the usual, unweighted, Sobolev inequality we only need to prove

$$\iint_{\mathbb{R}^N\times\mathbb{R}^N}\frac{\left|\frac{u(x)}{|x|^\beta}-\frac{u(y)}{|y|^\beta}\right|^2}{|x-y|^{N+2s}}\,dy\,dx \le C\iint_{\mathbb{R}^N\times\mathbb{R}^N}\frac{|u(x)-u(y)|^2}{|x-y|^{N+2s}}\frac{dy}{|y|^\beta}\frac{dx}{|x|^\beta}, \tag{10.2.7}$$

for suitable constant C independent of u. In order to do this, let us consider the following algebraic identities with $A = u(x)$, $B = u(y)$, $a = |x|^\beta$ and $b = |y|^\beta$:

$$\frac{A}{a}-\frac{B}{b} = \begin{cases}(A-B)\frac{1}{a}+B(\frac{1}{a}-\frac{1}{b}),\\ (A-B)\frac{1}{b}+A(\frac{1}{a}-\frac{1}{b}).\end{cases}$$

Hence, for every $t \in [0,1]$ we have

$$\left|\frac{A}{a}-\frac{B}{b}\right| \le |A-B|\left(\frac{t}{a}+\frac{1-t}{b}\right)+(t|B|+(1-t)|A|)\left|\frac{1}{a}-\frac{1}{b}\right|.$$

Choosing $t = \frac{a}{a+b}$ and squaring both sides, we get

$$\left|\frac{A}{a}-\frac{B}{b}\right|^2 \le C\left(|A-B|^2\left(\frac{1}{a+b}\right)^2+\left(\frac{|B|}{b}\right)^2\left|\frac{b-a}{a+b}\right|^2+\left(\frac{|A|}{a}\right)^2\left|\frac{b-a}{a+b}\right|^2\right).$$

Integrating in $L^2(\mathbb{R}^N\times\mathbb{R}^N)$ against the kernel $K(x,y) = \frac{1}{|x-y|^{N+2s}}$ and using symmetry we have

$$\begin{aligned}\iint_{\mathbb{R}^N\times\mathbb{R}^N}\left|\frac{u(x)}{|x|^\beta}-\frac{u(y)}{|y|^\beta}\right|^2 K(x,y)\,dy\,dx &\le C\iint_{\mathbb{R}^N\times\mathbb{R}^N}\frac{|u(x)-u(y)|^2}{|x-y|^{N+2s}}\frac{1}{(|x|^\beta+|y|^\beta)^2}\,dy\,dx\\ &\quad + 2C\iint_{\mathbb{R}^N\times\mathbb{R}^N}\frac{|u(x)|^2}{|x|^{2\beta}}\left(\frac{|x|^\beta-|y|^\beta}{|x|^\beta+|y|^\beta}\right)^2 K(x,y)\,dy\,dx.\end{aligned}$$

Now, using that $\frac{1}{(|x|^\beta+|y|^\beta)^2} \le \frac{1}{4|x|^\beta|y|^\beta}$ we observe that the first term in the previous sum is controlled by the expression on the right hand side of (10.2.7).

In order to estimate the second term in the sum we make use of the following identity:

$$\int_{\mathbb{R}^N} \frac{1}{|x|^\beta}\left(\frac{|x|^\beta - |y|^\beta}{|x|^\beta + |y|^\beta}\right)^2 \frac{1}{|x-y|^{N+2s}}\,dy = C_3 \int_{\mathbb{R}^N} \frac{|x|^\beta - |y|^\beta}{|x|^\beta|y|^\beta}\frac{1}{|x-y|^{N+2s}}\,dy, \tag{10.2.8}$$

for some $C_3 > 0$. This can be easily seen by homogeneity and the change of variables $z = y/|x|$ in both integrals. This gives, aside from the constants, the identities for the second term

$$\begin{aligned}
&\iint_{\mathbb{R}^N\times\mathbb{R}^N} \frac{|u(x)|^2}{|x|^{2\beta}}\left(\frac{|x|^\beta - |y|^\beta}{|x|^\beta + |y|^\beta}\right)^2 \frac{1}{|x-y|^{N+2s}}\,dy\,dx\\
&= C_3 \iint_{\mathbb{R}^N\times\mathbb{R}^N} \frac{|u(x)|^2}{|x|^\beta}\frac{|x|^\beta - |y|^\beta}{|x|^\beta|y|^\beta}\frac{1}{|x-y|^{N+2s}}\,dy\,dx\\
&= C_3\frac{1}{2} \iint_{\mathbb{R}^N\times\mathbb{R}^N} \left(\frac{|u(x)|^2}{|x|^\beta} - \frac{|u(y)|^2}{|y|^\beta}\right)(|x|^\beta - |y|^\beta)\frac{1}{|x-y|^{N+2s}}\frac{dy}{|y|^\beta}\frac{dx}{|x|^\beta}\\
&\le C_3\frac{1}{2} \iint_{\mathbb{R}^N\times\mathbb{R}^N} \frac{|u(x)-u(y)|^2}{|x-y|^{N+2s}}\frac{dy}{|y|^\beta}\frac{dx}{|x|^\beta}.
\end{aligned}$$

The last inequality follows from the trivial pointwise estimate

$$\left(\frac{A^2}{a} - \frac{B^2}{b}\right)(a-b) = A^2 + B^2 - A^2\frac{b}{a} - B^2\frac{a}{b} \le A^2 + B^2 - 2AB = (A-B)^2.$$

This finishes the proof of the theorem. □

Remark 10.2.9. Note that inequality (10.2.6) can be read as a weighted Caffarelli–Kohn–Nirenberg inequality.

Observe that the argument used here is reminiscent of the one used to prove Picone's lemma replacing $K(x,y) = \frac{1}{|x-y|^{N+2s}}$ by the symmetric kernel

$$K_\beta(x,y) = \frac{1}{|x-y|^{N+2s}}\frac{1}{|y|^\beta}\frac{1}{|x|^\beta}.$$

One can check that symmetry is actually the only requirement needed for such a lemma to work.

We define the space $W_0^{s,\gamma}(\Omega)$ as the completion of $\mathcal{C}_0^\infty(\Omega)$ with respect to the norm of $W^{s,\gamma}(\Omega)$. We obtain the following Sobolev inequality in $W_0^{s,\gamma}(\Omega)$.

Proposition 10.2.10. *Let Ω be a bounded Lipschitz domain and suppose that the hypotheses of Theorem 10.2.8 hold. Then, for all $v \in W_0^{s,\gamma}(\Omega)$, there exists a positive constant $v \in$, there exists a positive constant $S = S(N,s,\gamma,\Omega)$ such that*

$$a_{N,s} \int_\Omega \int_\Omega \frac{|v(x)-v(y)|^2}{|x-y|^{N+2s}} \frac{dx}{|x|^\gamma} \frac{dy}{|y|^\gamma} \geq S\Big(\int_\Omega \frac{|v(x)|^{2_s^*}}{|x|^{2_s^*\gamma}} \Big)^{\frac{2}{2_s^*}}. \tag{10.2.9}$$

Proof. Let $v \in W_0^{s,\gamma}(\Omega)$ and define $\tilde{v}$ to be the extension of v to $\mathbb{R}^N$ given in Lemma 10.2.6. Then using Theorem 10.2.8 and an argument of density, we get

$$C\|v\|_{W_0^{s,\gamma}(\Omega)} \geq a_{N,s} \int_{\mathbb{R}^N} \int_{\mathbb{R}^N} \frac{|\tilde{v}(x)-\tilde{v}(y)|^2}{|x-y|^{N+2s}} \frac{dx\,dy}{|x|^\gamma |y|^\gamma} \geq S(N,s,\gamma)\Big(\int_{\mathbb{R}^N} \frac{|\tilde{v}(x)|^{2_s^*}}{|x|^{2_s^*\gamma}}\, dx \Big)^{\frac{2}{2_s^*}}.$$

Since $\tilde{v}_{|\Omega} = v$ we reach

$$a_{N,s} \int_\Omega \int_\Omega \frac{|v(x)-v(y)|^2}{|x-y|^{N+2s}} \frac{dx\,dy}{|x|^\gamma |y|^\gamma} \geq S(N,s,\gamma,\Omega)\Big(\int_\Omega \frac{|v(x)|^{2_s^*}}{|x|^{2_s^*\gamma}}\, dx \Big)^{\frac{2}{2_s^*}}$$

and the result follows. □

If Ω is a bounded regular domain, then as in the case $\gamma = 0$, we have the next Poincaré inequality.

Theorem 10.2.11. *There exists a positive constant $C := C(\Omega, N, s, \gamma)$ such that for all $\phi \in W_0^{s,\gamma}(\Omega)$, we have*

$$C \int_\Omega \phi^2(x)\, d\mu \leq \int_\Omega \int_\Omega (\phi(x) - \phi(y))^2\, d\nu.$$

Proof. By density it is sufficient to consider $\phi \in C_0^\infty(\Omega)$. If $\phi \equiv 0$, the inequality follows trivially. Thus, let us define

$$\lambda_1(\Omega) := \inf_{\{\phi \in C_0^\infty(\Omega), \phi \not\equiv 0\}} \frac{\int_\Omega \int_\Omega (\phi(x)-\phi(y))^2\, d\nu}{\int_\Omega \phi^2(x)\, d\mu}.$$

Hence, to prove the lemma we need to check that $\lambda_1(\Omega) > 0$. We argue by contradiction, that is, let us suppose $\lambda_1(\Omega) = 0$. Then there exists a sequence $\{\phi_n\}_{n\in\mathbb{N}} \subset C_0^\infty(\Omega)$ such that

$$\int_\Omega \phi_n^2(x)\, d\mu = 1 \quad \text{and} \quad \int_\Omega \int_\Omega (\phi_n(x) - \phi_n(y))^2\, d\nu \to 0 \text{ as } n \to \infty.$$

It is clear that $\|\phi_n\|_{W_0^{s,\gamma}(\Omega)} \leq C$ and, hence, there exists a function $\bar{\phi} \in W_0^{s,\gamma}(\Omega)$ such that $\phi_n \rightharpoonup \bar{\phi}$ weakly in $W_0^{s,\gamma}(\Omega)$.

From the Sobolev inequality (10.2.9) it follows that

$$\int_\Omega \frac{|\phi_n|^{2_s^*}}{|x|^{2_s^*\gamma}}\,dx \le C(N,s,\Omega,\gamma)\|\phi_n\|_{W_0^{s,\gamma}(\Omega)} \le \bar{C},$$

with $\bar{C}$ independent of n. Then, from [28] (see also [138]) we have $\phi_n \to \bar{\phi}$ strongly in $L^2(\Omega)$. Hence, combining the estimates above and using Vitali's lemma we obtain, up to a subsequence,

$$\phi_n \to \bar{\phi} \quad \text{strongly in } L^2(\Omega, d\mu),$$

and thus,

$$\int_\Omega \bar{\phi}^2(x)\,d\mu = 1. \tag{10.2.10}$$

Since $\|\bar{\phi}\|_{W_0^{s,\gamma}(\Omega)} \le \|\phi_n\|_{W_0^{s,\gamma}(\Omega)}$ and

$$\int_\Omega\int_\Omega (\phi_n(x) - \phi_n(y))^2\,d\nu \to 0 \quad \text{as } n \to \infty,$$

we get

$$\phi_n \to \bar{\phi} \quad \text{strongly in } W_0^{s,\gamma}(\Omega), \quad \text{thus } \int_\Omega\int_\Omega (\bar{\phi}(x) - \bar{\phi}(y))^2\,d\nu = 0.$$

Hence $\bar{\phi} \equiv C$. Now, since $\bar{\phi} \in W_0^{s,\gamma}(\Omega)$, necessarily $\bar{\phi} \equiv 0$, a contradiction with (10.2.10). □

As a direct application of Theorem 10.2.11 we obtain that if Ω is a bounded regular domain, then every $w \in W_0^{s,\gamma}(\Omega)$ satisfies

$$\|\tilde{w}\|_{W_0^{s,\gamma}(\Omega)} \le C\Big(\int_\Omega\int_\Omega |w(x) - w(y)|^2\,d\nu\Big)^{\frac{1}{2}},$$

where $C := C(N,s,\Omega,\gamma) > 0$ and $\tilde{w}$ is the extension of w given in Lemma 10.2.6. As a consequence we reach that if Ω is a bounded domain, we can consider $W_0^{s,\gamma}(\Omega)$ with the equivalent norm

$$|||\phi|||_{W_0^{s,\gamma}(\Omega)} := \Big(\int_\Omega\int_\Omega |\phi(x) - \phi(y)|^2\,d\nu\Big)^{\frac{1}{2}}.$$

We state now a weighted version of the Poincaré–Wirtinger inequality.

Theorem 10.2.12. *Let $w \in W_0^{s,\gamma}(\Omega)$ and assume that ψ is a radial decreasing function such that* $\operatorname{supp}\psi \subset B_1$ *and* $0 \lneqq \psi \le 1$. *Define*

$$W_\psi := \frac{\int_{B_1} w(x)\psi(x)\,d\mu}{\int_{B_1} \psi(x)\,d\mu}.$$

Then, there exists $C := C(N,s,\psi) > 0$ such that

$$\int_{B_1} (w(x) - W_\psi)^2 \psi(x)\,d\mu \le C \int_{B_1}\int_{B_1} (w(x) - w(y))^2 \min\{\psi(x), \psi(y)\}\,dv.$$

Proof. Consider $\Psi(x) := \frac{\psi(x)}{|x|^{2\gamma}}$, which is a radial decreasing function. Then using [146, Corollary 6] we get

$$\int_{B_1} (w(x) - \bar{W}_\Psi)^2 \Psi(x)\,dx \le C \int_{B_1}\int_{B_1} \frac{(w(x) - w(y))^2}{|x-y|^{N+2s}} \min\{\Psi(x), \Psi(y)\}\,dx\,dy,$$

where

$$\bar{W}_\Psi = \frac{\int_{B_1} w(x)\Psi(x)\,dx}{\int_{B_1} \Psi(x)\,dx}.$$

Then, we get

$$\int_{B_1} (w(x) - \bar{W}_\Psi)^2 \Psi(x)\,dx = \int_{B_1} (w(x) - W_\psi)^2 \psi(x)\,d\mu,$$

and

$$\begin{aligned}&\int_{B_1}\int_{B_1} \frac{(w(x) - w(y))^2}{|x-y|^{N+2s}} \min\{\Psi(x), \Psi(y)\}\,dx\,dy \\ &\quad = \int_{B_1}\int_{B_1} \frac{(w(x) - w(y))^2}{|x-y|^{N+2s}} \min\left\{\frac{\psi(x)}{|x|^{2\gamma}}, \frac{\psi(y)}{|y|^{2\gamma}}\right\}\,dx\,dy.\end{aligned}$$

Hence, to finish we just have to show that

$$\min\left\{\frac{\psi(x)}{|x|^{2\gamma}}, \frac{\psi(y)}{|y|^{2\gamma}}\right\} \le \frac{\min\{\psi(x), \psi(y)\}}{|x|^\gamma |y|^\gamma} \quad \text{in } B_1 \times B_1.$$

Without loss of generality we can assume that $|x| \ge |y|$.

Define $H(s) := \frac{\psi(s)}{s^{2\gamma}}$, which is again a decreasing function in $(0,1)$. Let $s_1 := |x|$ and $s_2 := |y|$. Then

$$\min\left\{\frac{\psi(x)}{|x|^{2\gamma}}, \frac{\psi(y)}{|y|^{2\gamma}}\right\} = H(s_1).$$

Using that ψ is decreasing, we obtain $\psi(s_1) \le \psi(s_2)$. Thus

$$\frac{\min\{\psi(x), \psi(y)\}}{|x|^\gamma |y|^\gamma} = \frac{\psi(s_1)}{s_1^\gamma s_2^\gamma}.$$

Since $s_2 \le s_1 \le 1$, we conclude that $H(s_1) \le \frac{\psi(s_1)}{s_1^\gamma s_2^\gamma}$ and the result follows. □

Finally, we define

$$W^{s,\gamma}_{\mathrm{loc}}(\Omega) := \left\{ u \in L^2_{\mathrm{loc}}(\Omega, d\mu) \text{ s. t. } \forall \Omega_1 \subset\subset \Omega, \int_{\Omega_1}\int_{\Omega_1} \frac{|u(x)-u(y)|^2}{|x-y|^{N+2s}} \frac{dx}{|x|^\gamma} \frac{dy}{|y|^\gamma} < +\infty \right\}.$$

Remark 10.2.13. The definition for $\gamma = 0$ is similar. In this case we will denote the associated space simply as $H^s_{\mathrm{loc}}(\Omega)$.

We consider the following natural definition.

Definition 10.2.14. Let $v \in W^{s,\gamma}_{\mathrm{loc}}(\Omega)$. We say that v is a supersolution to problem (10.2.2) if

$$\frac{1}{2} \iint_{D_{\Omega_1}} (v(x)-v(y))(\varphi(x)-\varphi(y))\, d\nu \ge \int_{\Omega_1} g\varphi \, dx \tag{10.2.11}$$

for every nonnegative $\varphi \in W^{s,\gamma}_0(\Omega_1))$ and every $\Omega_1 \subset\subset \Omega$.

Consider the operator

$$L_{\gamma,\Omega}(w)(x) := C_{N,s}\, P.V. \int_\Omega (w(x)-w(y))K(x,y)\, dy, \tag{10.2.12}$$

where

$$K(x,y) := \frac{1}{|x|^\gamma |y|^\gamma |x-y|^{N+2s}}.$$

Note that Picone's inequality in Section 8.5 is based on a pointwise inequality and is valid for any positive symmetric kernel. Therefore we can formulate the following result.

Theorem 10.2.15 (Picone type inequality). *Let $w \in W^{s,\gamma}_0(\Omega)$ be such that $w > 0$ in Ω, and assume that $L_{\gamma,\Omega}(w) = \nu$ with $\nu \in L^1_{\mathrm{loc}}(\mathbb{R}^N)$ and $\nu \gneqq 0$. Then for all $v \in W^{s,\gamma}_0(\Omega)$ we have*

$$a_{N,s} \int_\Omega \int_\Omega \frac{|v(x)-v(y)|^2}{|x-y|^{N+2s}} \frac{dx\, dy}{|x|^\gamma |y|^\gamma} \ge \left\langle L_{\gamma,\Omega}(w), \frac{v^2}{w} \right\rangle, \tag{10.2.13}$$

where $L_{\gamma,\Omega} w$ is defined by (10.2.12).

The proof is the same as the one in Section 8.5 (see also [224]). As a consequence we get the next weighted Hardy inequality.

Theorem 10.2.16. *There exists a positive constant $C(N,s,\gamma)$ such that for all $\phi \in C_0^\infty(\mathbb{R}^N)$, we have*

$$C\int\limits_{\mathbb{R}^N} \frac{\phi^2(x)}{|x|^{2s+2\gamma}}\,dx \le \int\limits_{\mathbb{R}^N}\int\limits_{\mathbb{R}^N} (\phi(x)-\phi(y))^2\,dv.$$

Proof. Let $\phi \in C_0^\infty(\mathbb{R}^N)$ and define $w(x) := |x|^{-\theta}$, with $0 < \theta < \frac{N-2s-2\gamma}{2}$. Then, by Lemma 10.2.7,

$$L_{\gamma,\mathbb{R}^N}(w)(x) = C\frac{w(x)}{|x|^{2s+2\gamma}} \quad \text{a. e. in } \mathbb{R}^N.$$

Using (10.2.13) it follows that

$$a_{N,s}\int\limits_{\mathbb{R}^N}\int\limits_{\mathbb{R}^N} \frac{|\phi(x)-\phi(y)|^2}{|x-y|^{N+2s}}\frac{dx\,dy}{|x|^\gamma|y|^\gamma} \ge \left\langle L_{\gamma,\mathbb{R}^N}(w), \frac{\phi^2}{w}\right\rangle = C(N,s,\gamma)\int\limits_{\mathbb{R}^N} \frac{\phi^2(x)}{|x|^{2s+2\gamma}}\,dx.$$

Hence we conclude the proof. □

Remark 10.2.17. It should be interesting from the analytical point of view to find the value of the optimal constant.

If Ω is a bounded domain, we have a similar result.

Theorem 10.2.18. *There exists a positive constant $C(\Omega,N,s,\gamma)$ such that for all $\phi \in C_0^\infty(\Omega)$, we have*

$$C\int\limits_{\Omega} \frac{\phi^2(x)}{|x|^{2s+2\gamma}}\,dx \le \int\limits_{\Omega}\int\limits_{\Omega} (\phi(x)-\phi(y))^2\,dv.$$

Proof. Let $\phi \in C_0^\infty(\Omega)$ and define $\tilde\phi$ to be the extension of ϕ to $\mathbb{R}^N$ given in Lemma 10.2.6. Then from Theorem 10.2.16, we get

$$\int\limits_{\mathbb{R}^N}\int\limits_{\mathbb{R}^N} \frac{|\tilde\phi(x)-\tilde\phi(y)|^2}{|x-y|^{N+2s}}\frac{dx\,dy}{|x|^\gamma|y|^\gamma} \ge C(N,s,\gamma)\int\limits_{\mathbb{R}^N} \frac{\tilde\phi^2(x)}{|x|^{2s+2\gamma}}\,dx.$$

Now, using the fact that $\tilde\phi_{|\Omega} = \phi$ and combining the results of Lemma 10.2.6 and Theorem 10.2.11, we reach the desired result. □

In the case of nonzero boundary conditions, we obtain the following version of the Hardy inequality.

Theorem 10.2.19. *There exists a positive constant $C(\Omega, N, s, \gamma)$ such that for all $\phi \in W_0^{s,\gamma}(\Omega)$, we have*

$$C\int_{\Omega} \frac{\phi^2(x)}{|x|^{2s+2\gamma}}\,dx \le \int_{\Omega}\int_{\Omega} (\phi(x)-\phi(y))^2\,d\nu + \int_{\Omega} \phi^2(x)\,d\mu.$$

Proof. Fix $\phi \in W_0^{s,\gamma}(\Omega)$ and define $\tilde{\phi}$ as the extension of ϕ to $\mathbb{R}^N$ given in Lemma 10.2.6. Then from Theorem 10.2.16, we get

$$\int_{\mathbb{R}^N}\int_{\mathbb{R}^N} \frac{|\tilde{\phi}(x)-\tilde{\phi}(y)|^2}{|x-y|^{N+2s}}\frac{dx\,dy}{|x|^{\gamma}|y|^{\gamma}} \ge C(N,s,\gamma)\int_{\mathbb{R}^N} \frac{\tilde{\phi}^2(x)}{|x|^{2s+2\gamma}}\,dx \ge C(N,s,\gamma)\int_{\Omega} \frac{\phi^2(x)}{|x|^{2s+2\gamma}}\,dx.$$

Since $\|\tilde{\phi}\|_{W^{s,\gamma}(\mathbb{R}^N)} \le C(\Omega)\|\phi\|_{W^{s,\gamma}(\Omega)}$, the result follows. □

Moreover, in the particular case $\phi \in W^{s,\gamma}(B_R)$, as an application of Theorem 10.2.19 we can prove the following inequality.

Theorem 10.2.20. *Let $R > 0$ and $\phi \in W^{s,\gamma}(B_R)$. Then, there exists $C = C(N, s, R, \gamma) > 0$ such that*

$$C\left(\int_{B_R} \frac{|\phi|^{2_s^*}}{|x|^{2_s^*\gamma}}\,dx\right)^{\frac{2}{2_s^*}} \le \int_{B_R}\int_{B_R} (\phi(x)-\phi(y))^2\,d\nu + R^{-2s}\int_{B_R} \phi^2\,d\mu. \tag{10.2.14}$$

Proof. We prove the result for $R = 1$, and then (10.2.14) follows by a scaling argument. We set $\phi_1(x) := \frac{\phi(x)}{|x|^{\gamma}}$. Then we have

$$C\left(\int_{B_1} |\phi_1|^{2_s^*}\,dx\right)^{\frac{2}{2_s^*}} \le \int_{B_1}\int_{B_1} \frac{(\phi_1(x)-\phi_1(y))^2}{|x-y|^{N+2s}}\,dx\,dy + \int_{B_1} \phi_1^2\,dx. \tag{10.2.15}$$

See [28] for more details.

To finish the proof we just have to estimate the term

$$\int_{B_1}\int_{B_1} \frac{(\phi_1(x)-\phi_1(y))^2}{|x-y|^{N+2s}}\,dxdy.$$

Since

$$(\phi_1(x)-\phi_1(y))^2 = \frac{(\phi(x)-\phi(y))^2}{|x|^{\gamma}|y|^{\gamma}} + \left(\frac{\phi^2(x)}{|x|^{\gamma}} - \frac{\phi^2(y)}{|y|^{\gamma}}\right)\left(\frac{1}{|x|^{\gamma}} - \frac{1}{|y|^{\gamma}}\right),$$

it follows that

$$\int_{B_1}\int_{B_1} \frac{(\phi_1(x)-\phi_1(y))^2}{|x-y|^{N+2s}}\,dxdy \le \int_{B_1}\int_{B_1} \frac{(\phi(x)-\phi(y))^2}{|x|^{\gamma}|y|^{\gamma}|x-y|^{N+2s}}\,dxdy + \int_{B_1} \frac{\phi^2(x)}{|x|^{\gamma}} L_{0,B_1}(|x|^{-\gamma})\,dx.$$

Proceeding as in the proof of Lemma 10.2.7, since $0 < \gamma < \frac{N-2s}{2}$, we realize that $L_{0,B_1}(|x|^{-\gamma}) \le \frac{C}{|x|^{\gamma+2s}}$, and hence

$$\int_{B_1}\int_{B_1} \frac{(\phi_1(x)-\phi_1(y))^2}{|x-y|^{N+2s}}\,dxdy \le \int_{B_1}\int_{B_1} \frac{(\phi(x)-\phi(y))^2}{|x|^\gamma |y|^\gamma |x-y|^{N+2s}}\,dxdy + C\int_{B_1} \frac{\phi^2}{|x|^{2s+2\gamma}}\,dx.$$

Finally, using Theorem 10.2.19 and substituting $\phi(x) = |x|^\gamma \phi_1(x)$, we reach (10.2.14). □

Remark 10.2.21. By interpolation of the Sobolev and Hardy weighted inequalities, we can arrive at a family of fractional Caffarelli–Kohn–Nirenberg type inequalities. See [3]. We will not go into more detail in this direction.

Remark 10.2.22. As a consequence of the weighted Picone inequality, we have the next comparison principle, which extends to the weighted fractional framework the classical one obtained by Brezis and Kamin in [90] for the local case and the extension to the nonlocal case studied in Section 8.5.

Lemma 10.2.23. *Let Ω be a bounded domain and let f be a nonnegative continuous function such that $f(x,\sigma) > 0$ if $\sigma > 0$ and $\frac{f(x,\sigma)}{\sigma}$ is decreasing. Let $u, v \in W_0^{s,\gamma}(\Omega)$ be such that $u, v > 0$ in Ω and*

$$\begin{cases} L_{\gamma,\Omega}(u) \ge f(x,u) & \text{in } \Omega, \\ L_{\gamma,\Omega}(v) \le f(x,v) & \text{in } \Omega. \end{cases}$$

Then, $u \ge v$ in Ω.

Proof. Since the proof relies on pointwise inequalities, the result follows in a similar way to the constant coefficients case, obtained in Section 8.5. □

This result will be applied in the next chapter.

10.2.2 Some compactness results

Here, we will need the next compactness result.

Lemma 10.2.24. *Let $\{u_n\}_{n\in\mathbb{N}} \subset W_0^{s,\gamma}(\Omega)$ be an increasing sequence of nonnegative functions such that $L_{\gamma,\Omega}(u_n) \ge 0$. Assume that $u_n \rightharpoonup u$ weakly in $W_0^{s,\gamma}(\Omega)$. Then, $u_n \to u$ strongly in $W_0^{s,\gamma}(\Omega)$.*

Proof. Using the hypothesis $L_{\gamma,\Omega}(u_n) \ge 0$, we have $\langle L_{\gamma,\Omega}(u_n), w_n\rangle \le 0$, where $w_n := u_n - u$. Thus

$$\int_\Omega\int_\Omega \frac{(u_n(x)-u_n(y))(w_n(x)-w_n(y))}{|x-y|^{N+2s}} \frac{dx\,dy}{|x|^\gamma |y|^\gamma} \le 0.$$

Since

$$(u_n(x) - u_n(y))(w_n(x) - w_n(y)) = (u_n(x) - u_n(y))^2 - (u_n(x) - u_n(y))(u(x) - u(y))$$

by the Young inequality we conclude that

$$\int_\Omega \int_\Omega \frac{(u_n(x) - u_n(y))^2}{|x - y|^{N+2s}} \frac{dx\,dy}{|x|^\gamma |y|^\gamma} \le \int_\Omega \int_\Omega \frac{(u(x) - u(y))^2}{|x - y|^{N+2s}} \frac{dx\,dy}{|x|^\gamma |y|^\gamma}.$$

Therefore $\|u_n\|_{W_0^{s,\gamma}(\Omega)} \le \|u\|_{W_0^{s,\gamma}(\Omega)}$ and hence $u_n \to u$ strongly in $W_0^{s,\gamma}(\Omega)$. □

Likewise, we have the following local version of Lemma 10.2.24.

Lemma 10.2.25. *Let $\{u_n\}_{n\in\mathbb{N}} \subset W_0^{s,\gamma}(\Omega)$ be an increasing sequence of nonnegative functions such that $L_{\gamma,\Omega}(u_n) \ge 0$. Assume that $\{u_n\}_{n\in\mathbb{N}}$ is uniformly bounded in $W_{\mathrm{loc}}^{s,\gamma}(\Omega)$. Then there exists $u \in W_{\mathrm{loc}}^{s,\gamma}(\Omega)$ such that $u_n \to u$ strongly in $W_{\mathrm{loc}}^{s,\gamma}(\Omega)$.*

Proof. By using a straightforward modification of Lemma 5.3 in [138] and multiplying the sequence $\{u_n\}_{n\in\mathbb{N}}$ by a Lipschitz cutoff function ψ such that $\psi \equiv 1$ on $\Omega' \subset \Omega$, we can apply Lemma 10.2.24 to conclude. □

Remark 10.2.26. Note that Lemma 10.2.24 and Lemma 10.2.25 also hold for $\gamma = 0$, i. e., for the spaces $H_0^s(\Omega)$ and $H_{\mathrm{loc}}^s(\Omega)$.

10.2.3 Some numerical inequalities

The next elementary algebraic inequality in [208] will be used in some arguments. See [2]. For the reader's convenience we give a complete proof here.

Lemma 10.2.27. *Let $s_1, s_2 \ge 0$ and $a > 0$. Then*

$$(s_1 - s_2)(s_1^a - s_2^a) \ge \frac{4a}{(a+1)^2}\left(s_1^{\frac{a+1}{2}} - s_2^{\frac{a+1}{2}}\right)^2. \tag{10.2.16}$$

Proof. Since $\frac{4a}{(a+1)^2} \le 1$ for $0 \le a$, if $s_1 = 0$ or $s_2 = 0$ the inequality trivially follows. Hence, we can assume $s_1 > s_2 > 0$. Thus, setting $x := \frac{s_2}{s_1}$, (10.2.16) is equivalent to

$$(1 - x)(1 - x^a) \ge \frac{4a}{(a+1)^2}\left(1 - x^{\frac{a+1}{2}}\right)^2 \quad \text{for all } x \in (0,1). \tag{10.2.17}$$

We set

$$h(x) := (1 - x)(1 - x^a)(a+1)^2 - 4a\left(1 - x^{\frac{a+1}{2}}\right)^2,$$

and then we just have to show that $h(x) \ge 0$ for all $x \in (0,1)$. Moreover, h can be written as

$$h(x) = (a-1)^2\left(1 - x^{\frac{a+1}{2}}\right)^2 - (a+1)^2\left(x^{\frac{1}{2}} - x^{\frac{a}{2}}\right)^2.$$

First, we assume $a > 1$. We claim that

$$(a-1)(1-x^{\frac{a+1}{2}}) \geq (a+1)(x^{\frac{1}{2}} - x^{\frac{a}{2}}).$$

In fact, let us define

$$h_1(x) := (a-1)(1-x^{\frac{a+1}{2}}) - (a+1)(x^{\frac{1}{2}} - x^{\frac{a}{2}}),$$

so that $h_1'(x) = \frac{(a+1)}{2}(-(a-1)x^{\frac{a-1}{2}} - x^{-\frac{1}{2}} + ax^{\frac{a-2}{2}})$.

Using the Young inequality, we obtain

$$x^{\frac{a}{2}-1} \leq \frac{a-1}{a}x^{\frac{a-1}{2}} + \frac{1}{a}x^{-\frac{1}{2}}.$$

Thus $h_1'(x) \leq 0$ and hence $h_1(x) \geq h_1(1) = 0$. Therefore $h(x) \geq 0$ and the result follows in this case.

Consider now the case $a < 1$. We prove the result if we show

$$(1-a)(1-x^{\frac{a+1}{2}}) \geq (a+1)(x^{\frac{a}{2}} - x^{\frac{1}{2}}).$$

Defining

$$h_2(x) := (1-a)(1-x^{\frac{a+1}{2}}) - (a+1)(x^{\frac{a}{2}} - x^{\frac{1}{2}})$$

and using again the Young inequality we obtain $h_2'(x) \leq 0$ for all $x \in (0,1)$. Thus $h_2(x) \geq h_2(1) = 0$. Hence $h(x) \geq 0$ and we conclude the proof. □

Lemma 10.2.28. *For $a > 0$, $b > 0$ the following identities hold:*

$$\frac{(a-b)^2}{ab} = (a-b)\left(\frac{1}{b} - \frac{1}{a}\right) = (\log a - \log b)^2 + 2\sum_{k=2}^{\infty} \frac{(\log a - \log b)^{2k}}{(2k)!}. \tag{10.2.18}$$

As a consequence

$$\frac{(a-b)^2}{ab} = (a-b)\left(\frac{1}{b} - \frac{1}{a}\right) \geq (\log a - \log b)^2. \tag{10.2.19}$$

Proof. The first identity is evident. With respect to the second, without any loss of generality we can assume that $a > b > 0$. For $x = \frac{a}{b} > 1$ the second identity is equivalent to

$$(x-1)^2 = 2x \sum_{k=1}^{\infty} \frac{(\log x)^{2k}}{2k!}.$$

Calling $t = \log x \in [0, \infty)$ the identity above is equivalent to

$$(e^t - 1)^2 = 2e^t \sum_{k=1}^{\infty} \frac{t^{2k}}{2k!}.$$

The identity follows by noting that the functions $f(t) = (e^t - 1)^2$ and $g(t) = 2e^t \sum_{k=1}^{\infty} \frac{t^{2k}}{2k!}$ verify $f(0) = g(0) = 0$ and

$$g'(t) = 2e^t \sum_{k=1}^{\infty} \frac{t^{2k}}{2k!} + 2e^t \sum_{k=1}^{\infty} \frac{t^{2k-1}}{(2k-1)!} \equiv 2e^t(e^t - 1) = f'(t).$$

Inequality (10.2.19) is an immediate consequence. □

10.3 Weak Harnack inequality and local behavior of nonnegative supersolutions

Consider $f \in L^m(\Omega)$, $m \geq \frac{2N}{N+2s}$ and u, the energy solution to problem (10.1.1) with $0 < \lambda < \Lambda_{N,2s}$. By setting $v(x) := |x|^\gamma u(x)$, where γ is defined in (9.4.9), it follows that v solves

$$\begin{cases} L_{\gamma,\Omega} v(x) = |x|^{-\gamma} f(x) & \text{in } \Omega, \\ v = 0 & \text{in } \mathbb{R}^N \setminus \Omega, \end{cases} \tag{10.3.1}$$

with $0 < \gamma < \frac{N-2s}{2}$ and $L_{\gamma,\Omega}$ defined in (10.2.3). Hence, to study the behavior of u near the origin, we may deal with the same question for v. More precisely, we want to prove that the weighted operator $L_\gamma v$ satisfies a suitable weak Harnack inequality. Note that the natural functional framework for the new equation of v is the space $W^s_\gamma(\mathbb{R}^N)$ defined in Section 10.2. The statement of the result that we want to prove is the following.

Theorem 10.3.1 (Weak Harnack inequality). *Let $r > 0$ such that $B_{2r} \subset \Omega$. Assume that $f \geq 0$ and let $v \in W^{s,\gamma}(\mathbb{R}^N)$ be a supersolution to* (10.3.1) *with $v \gneqq 0$ in $\mathbb{R}^N$. Then, for every $q < \frac{N}{N-2s}$ there exists a positive constant $C = C(N, s, \gamma)$ such that*

$$\left(\int_{B_r} v^q \, d\mu \right)^{\frac{1}{q}} \leq C \inf_{B_{\frac{3}{2}r}} v.$$

Note that in the nonlocal framework we must have $v \geq 0$ in the whole space. Otherwise it is possible to find counterexamples (see the work by Kassmann [207] and the references therein).

The proof follows classical arguments by Moser and Krylov–Safonov (see [150] for the local case with weights). For the nonlocal case we have the precedents of [135, 215], where the kernel is comparable to a fractional Laplacian and the operator considered is of fractional p-Laplacian type.

Remark 10.3.2. The *ground state representation* has been considered for the case of the fractional p-Laplacian, $p > 2$ and the corresponding Hardy's inequality in [168].

When $p > 2$ the identity of Lemma 10.2.4 becomes an inequality. However, the corresponding operator, that is,

$$(-\Delta)^s_{p,\gamma} u(x) := P.V. \int_{\mathbb{R}^N} \frac{|u(x)-u(y)|^{p-2}(u(x)-u(y))}{|x-y|^{N+ps}} \frac{dy}{|x|^\gamma |y|^\gamma},$$

with $0 \le \gamma < \frac{N-ps}{2}$, $1 < p < N$, $s \in (0,1)$ and $ps < N$, is studied in [1, 2, 3]. In these references, the authors prove the existence of solution for general data using suitable approximation arguments. They extend the results of [218] using a different approach and they also get a weak Harnack inequality partially extending the results of [135]. The behavior of the radial solution in the local case for $p \neq 2$ is quite involved (see [10, Section 3.2]). For the fractional p-Laplacian this seems to be an open problem.

Since the kernel defined in (10.2.3) is singular we have to check the arguments step by step. That is, our result can be seen as the fractional counterpart of [150].

Note that it is enough to consider the case $B_r(x_0) = B_r(0)$. For simplicity of typing, we will write B_r instead of $B_r(0)$. We start by proving the following estimate.

Lemma 10.3.3. *Let $R > 0$ such that $B_R \subset \Omega$, and assume that $v \in W^{s,\gamma}(\mathbb{R}^N)$ with $v \gneqq 0$ is a supersolution to (10.3.1). Let $k > 0$ and suppose that for some $\sigma \in (0,1]$ we have*

$$|B_r \cap \{v \ge k\}|_{d\mu} \ge \sigma |B_r|_{d\mu} \tag{10.3.2}$$

with $0 < r < \frac{R}{16}$. Then there exists a positive constant $C = C(N,s,\gamma)$ such that

$$|B_{6r} \cap \{v \le 2\delta k\}|_{d\mu} \le \frac{C}{\sigma \log(\frac{1}{2\delta})} |B_{6r}|_{d\mu}$$

for all $\delta \in (0, \frac{1}{4})$.

Proof. Without loss of generality we can assume that $v > 0$ in B_R (otherwise we can deal with $v + \varepsilon$ and let $\varepsilon \to 0$ at the end). Let $\psi \in W^{s,\gamma}_0(B_R)$ be such that $0 \le \psi \le 1$, $\operatorname{supp} \psi \subset B_{7r}$, $\psi = 1$ in B_{6r} and $|\nabla \psi| \le \frac{C}{r}$.

Using $\psi^2 v^{-1}$ as a test function in (10.3.1), it follows that

$$\int_{\mathbb{R}^N}\int_{\mathbb{R}^N} (v(x)-v(y))(\psi^2(x)v^{-1}(x) - \psi^2(y)v^{-1}(y))\, d\nu \ge 0.$$

Thus

$$\begin{aligned} 0 \le & \int_{B_{8r}}\int_{B_{8r}} (v(x)-v(y))\left(\frac{\psi^2(x)}{v(x)} - \frac{\psi^2(y)}{v(y)}\right) d\nu \\ & + 2 \int_{\mathbb{R}^N \setminus B_{8r}} \int_{B_{8r}} (v(x)-v(y)) \frac{\psi^2(x)}{v(x)}\, d\nu. \end{aligned} \tag{10.3.3}$$

Denote $x = |x|x'$ and $y = \rho y'$, where $|x'| = |y'| = 1$ and $\rho := |y|$. We have

$$\int_{\mathbb{R}^N\setminus B_{8r}}\int_{B_{8r}} (v(x)-v(y))\frac{\psi^2(x)}{v(x)}\,dv \le \int_{\mathbb{R}^N\setminus B_{8r}}\int_{B_{8r}} \psi^2(x)\,dv$$

$$\le \int_{B_{7r}} \frac{\psi^2(x)}{|x|^{\gamma}} \int_{8r}^{\infty} \frac{\rho^{N-\gamma-1}}{|x|^{N+2s}}\bigg(\int_{\mathbb{S}^{N-1}} \frac{dy'}{|\frac{\rho}{|x|}y'-x'|^{N+2s}}\bigg)\,d\rho\,dx.$$

Setting here $\tau := \frac{\rho}{|x|}$,

$$\int_{\mathbb{R}^N\setminus B_{8r}}\int_{B_{8r}} (v(x)-v(y))\frac{\psi^2(x)}{v(x)}\,dv \le C\int_{B_{7r}} \frac{\psi^2(x)}{|x|^{2\gamma+2s}} \int_{\frac{8}{7}}^{\infty} \tau^{N-\gamma-1}\bigg(\int_{\mathbb{S}^{N-1}} \frac{dy'}{|\tau y'-x'|^{N+2s}}\bigg)\,d\tau\,dx$$

$$\le C\int_{B_{7r}} \frac{\psi^2(x)}{|x|^{2\gamma+2s}} \int_{\frac{8}{7}}^{\infty} \tau^{N-\gamma-1}D(\tau)\,d\tau\,dx,$$

where

$$D(\tau) := 2\frac{\pi^{\frac{N-1}{2}}}{\Gamma(\frac{N-1}{2})}\int_0^{\pi} \frac{\sin^{N-2}(\theta)}{(1-2\sigma\cos(\theta)+\tau^2)^{\frac{N+2s}{2}}}\,d\theta.$$

Considering the behavior of D near 0 and 1 and at ∞ as in [158] (see also the proof of Lemma 10.2.7), we obtain

$$\int_{\frac{8}{7}}^{\infty} \tau^{N-\gamma-1}D(\tau)\,d\tau \le C,$$

and therefore we conclude that

$$\int_{\mathbb{R}^N\setminus B_{8r}}\int_{B_{8r}} (v(x)-v(y))\frac{\psi^2(x)}{v(x)}\,dv \le Cr^{N-2s-2\gamma}. \tag{10.3.4}$$

Moreover

$$\int_{B_{8r}}\int_{B_{8r}} (v(x)-v(y))\bigg(\frac{\psi^2(x)}{v(x)}-\frac{\psi^2(y)}{v(y)}\bigg)\,dv$$

$$= \int_{B_{6r}}\int_{B_{6r}} (v(x)-v(y))\bigg(\frac{\psi^2(x)}{v(x)}-\frac{\psi^2(y)}{v(y)}\bigg)\,dv$$

$$+ \iint_{B_{8r}\times B_{8r}\setminus B_{6r}\times B_{6r}} (v(x)-v(y))\bigg(\frac{\psi^2(x)}{v(x)}-\frac{\psi^2(y)}{v(y)}\bigg)\,dv$$

$$\le \int_{B_{6r}}\int_{B_{6r}} (v(x)-v(y))\bigg(\frac{1}{v(x)}-\frac{1}{v(y)}\bigg)\,dv + Cr^{N-2s-2\gamma}, \tag{10.3.5}$$

where the last inequality follows as a consequence of $\psi \equiv 1$ in B_{6r} and the fact that the integral in $(B_{8r} \times B_{8r}) \setminus (B_{6r} \times B_{6r})$ can be estimated in the same way as (10.3.4).

If $(x,y) \in (B_{6r} \times B_{6r})$

$$(v(x) - v(y))\left(\frac{\psi^2(x)}{v(x)} - \frac{\psi^2(y)}{v(y)}\right) = (v(x) - v(y))\left(\frac{1}{v(x)} - \frac{1}{v(y)}\right)$$
$$\leq -(\log(v(x)) - \log(v(y)))^2, \tag{10.3.6}$$

where the last inequality is a consequence of (10.2.19). Hence from (10.3.6) we deduce

$$\int_{B_{6r}}\int_{B_{6r}} (v(x) - v(y))\left(\frac{1}{v(x)} - \frac{1}{v(y)}\right) dv \leq -\int_{B_{6r}}\int_{B_{6r}} (\log(v(x)) - \log(v(y)))^2 \, dv, \tag{10.3.7}$$

and thus, putting together (10.3.3), (10.3.4), (10.3.5) and (10.3.7), it follows that

$$\int_{B_{6r}}\int_{B_{6r}} (\log(v(x)) - \log(v(y)))^2 \, dv \leq Cr^{N-2s-2\gamma}. \tag{10.3.8}$$

Let $\delta \in (0, 1/4)$. We set $w(x) := \min\{\log(\frac{1}{2\delta}), \log(\frac{k}{v})\}_+$. Now, since w is a truncation of $\log(\frac{k}{v})$, from (10.3.8) we obtain

$$\int_{B_{6r}}\int_{B_{6r}} (w(x) - w(y))^2 \, dv \leq Cr^{N-2s-2\gamma}.$$

Call

$$\langle w \rangle_{B_{6r}} := \frac{1}{|B_{6r}|_{d\mu}} \int_{B_{6r}} w(x) \, d\mu.$$

Thus, using the Hölder and Poincaré–Wirtinger inequalities (Theorem 10.2.12),

$$\int_{B_{6r}} \left|w(x) - \langle w \rangle_{B_{6r}}\right| d\mu \leq C|B_{6r}|_{d\mu}. \tag{10.3.9}$$

Note that $\{x \in \Omega : w(x) = 0\} = \{x \in \Omega : v(x) \geq k\}$, and then from (10.3.2) we have

$$\left|B_{6r} \cap \{w = 0\}\right|_{d\mu} \geq \frac{\sigma}{6^{N-2\gamma}} |B_{6r}|_{d\mu}.$$

As a consequence of this, we can see that

$$\left|B_{6r} \cap \left\{w = \log\left(\frac{1}{2\delta}\right)\right\}\right|_{d\mu} \leq \frac{6^{N-2\gamma}}{\sigma \log(\frac{1}{2\delta})} \int_{B_{6r}} \left|w(x) - \langle w \rangle_{B_{6r}}\right| d\mu,$$

and, hence, we conclude the result by applying (10.3.9) and the fact that

$$\{B_{6r} \cap \{v \leq 2\delta k\}\} = \left\{B_{6r} \cap \left\{w = \log\left(\frac{1}{2\delta}\right)\right\}\right\}.$$

□

As a consequence we have the next estimate on $\inf_{B_{4r}} v$.

Lemma 10.3.4. *Assume that the hypotheses of Lemma* 10.3.3 *are satisfied. Then, there exists* $\delta \in (0, \frac{1}{2})$, *depending only on* N, s, σ *and* γ, *such that*

$$\inf_{B_{4r}} v \geq \delta k. \tag{10.3.10}$$

Proof. We set $w := (l - v)_+$, where $l \in (\delta k, 2\delta k)$, and we let $\psi \in W_0^{s,\gamma}(B_\rho)$ with $r \leq \rho < 6r$. Using $w\psi^2$ as a test function in (10.3.1) and using similar arguments to those in [135, Lemma 3.2], we reach

$$\begin{aligned}\int_{B_\rho}\int_{B_\rho} &(w(x)\psi(x) - w(y)\psi(y))^2\,dv \\ &\leq C_1 \int_{B_\rho}\int_{B_\rho} \max\{w(x), w(y)\}^2 (\psi(x) - \psi(y))^2\,dv \\ &\quad + l^2 |B_\rho \cap \{v < l\}|_{d\mu} \times \sup_{\{x \in \operatorname{supp}(\psi)\}} \int_{\mathbb{R}^N \setminus B_\rho} \frac{dy}{|x - y|^{N+2s}}.\end{aligned} \tag{10.3.11}$$

We define now the sequences $\{l_j\}_{j\in\mathbb{N}}$, $\{\rho_j\}_{j\in\mathbb{N}}$ and $\{\bar{\rho}_j\}_{j\in\mathbb{N}}$ by setting

$$l_j := \delta k + 2^{-j-1}\delta k, \quad \rho_j := 4r + 2^{1-j} r, \quad \bar{\rho}_j := \frac{\rho_j + \rho_{j+1}}{2}.$$

Likewise, let us denote

$$w_j := (l_j - v)_+, \quad B_j := B_{\rho_j},$$

and let $\psi_j \in C_0^\infty(B_{\bar{\rho}_j})$ be such that $0 \leq \psi \leq 1$, $\psi \equiv 1$ in B_{j+1} and $|\nabla \psi_j| \leq 2^{j+3}/r$. Using the Sobolev inequality stated in Proposition 10.2.10 we obtain

$$C(N, s, \gamma) \Big(\int_{B_j} \frac{|w_j \psi_j|^{2_s^*}}{|x|^{\gamma 2_s^*}}\,dx \Big)^{\frac{2}{2_s^*}} \leq \int_{B_j}\int_{B_j} (w_j(x)\psi_j(x) - w_j(y)\psi_j(y))^2\,dv.$$

Hence, using the facts that

$$w_j \psi_j \geq (l_j - l_{j+1}) \quad \text{in } B_{j+1} \cap \{v < l_{j+1}\}$$

and

$$|x|^{-2_s^*\gamma} \geq \bar{C} r^{-(2_s^* - 2)\gamma} |x|^{-2\gamma} \quad \text{in } B_j,$$

with $\bar{C}$ independent of j, it follows that

$$\Big(\int_{B_j} \frac{|w_j \psi_j|^{2_s^*}}{|x|^{\gamma 2_s^*}}\,dx \Big)^{\frac{2}{2_s^*}} \geq \frac{C}{r^{(2_s^* - 2)\gamma}} (l_j - l_{j+1})^2 |B_{j+1} \cap \{v < l_{j+1}\}|_{d\mu}^{\frac{2}{2_s^*}}.$$

Since $|B_{j+1}|_{d\mu} = Cr^{N-2\gamma}$, we have

$$\frac{r^{(2_s^*-2)\gamma}}{(|B_{j+1}|_{d\mu})^{\frac{2}{2^*}}} = Cr^{-(N-2s-2\gamma)}r^{4s\gamma(\frac{1}{N-2s}-\frac{1}{N})} \le C(N,s,\gamma,\Omega)r^{-(N-2s-2\gamma)}.$$

Hence we get

$$\begin{aligned}(l_j - l_{j+1})^2\Big(\frac{|B_{j+1}\cap\{v<j+1\}|_{d\mu}}{|B_{j+1}|_{d\mu}}\Big)^{\frac{2}{2_s^*}}\\ \le C(N,s,\gamma,\Omega)r^{-(N-2s-2\gamma)}\int_{B_j}\int_{B_j}(w_j(x)\psi_j(x)-w_j(y)\psi_j(y))^2\,d\nu.\end{aligned}$$

Applying (10.3.11) to w_j, we conclude that

$$\begin{aligned}&(l_j - l_{j+1})^2\Big(\frac{|B_{j+1}\cap\{v<j+1\}|_{d\mu}}{|B_{j+1}|_{d\mu}}\Big)^{\frac{2}{2_s^*}}\\ &\quad\le \frac{C(N,s,\gamma)}{r^{(N-2s-2\gamma)}}\Big(C_1\int_{B_j}\int_{B_j}\max\{w_j(x),w_j(y)\}^2(\psi_j(x)-\psi_j(y))^2\,d\nu\\ &\qquad + l_j^2|B_j\cap\{v<l_j\}|_{d\mu}\sup_{\{x\in\operatorname{supp}(\psi_j)\}}\int_{\mathbb{R}^N\setminus B_j}\frac{dy}{|x-y|^{N+2s}}\Big).\end{aligned} \tag{10.3.12}$$

We have

$$\begin{aligned}&\int_{B_j}\int_{B_j}\max\{w_j(x),w_j(y)\}^2(\psi_j(x)-\psi_j(y))^2\,d\nu\\ &\quad\le l_j^2\|\nabla\psi_j\|^2_{L^\infty(B_j)}\int_{B_j\cap\{v<l_j\}}\frac{dx}{|x|^\gamma}\int_{B_j}\frac{|x-y|^{2-2s}}{|x-y|^N}\frac{dy}{|y|^\gamma}\\ &\quad\le C2^{2j}l_j^2r^{-2s}\int_{B_j\cap\{v<l_j\}}\frac{dx}{|x|^{2\gamma}} = C2^{2j}l_j^2r^{-2s}|B_j\cap\{v<l_j\}|_{d\mu}.\end{aligned} \tag{10.3.13}$$

Now, estimating the term

$$\sup_{\{x\in\operatorname{supp}(\psi_j)\}}\int_{\mathbb{R}^N\setminus B_j}\frac{dy}{|x-y|^{N+2s}}$$

as in [135, Lemma 3.2] and considering (10.3.12) and (10.3.13), we obtain

$$\begin{aligned}(l_j - l_{j+1})^2\Big(\frac{|B_{j+1}\cap\{v<j+1\}|_{d\mu}}{|B_{j+1}|_{d\mu}}\Big)^{\frac{2}{2_s^*}} &\le 2^{j(2+2s+N)}l_j^2\frac{C(N,s,\gamma)}{r^{(N-2s-2\gamma)}}r^{-2s}|B_j\cap\{v<l_j\}|_{d\mu}\\ &\le \tilde{C}2^{j(2+2s+N)}l_j^2\frac{|B_j\cap\{v<j\}|_{d\mu}}{|B_j|_{d\mu}},\end{aligned}$$

where $\tilde{C} = \tilde{C}(N,s,\gamma)$ but independent of j and r.

Defining $A_j := \frac{|B_j \cap \{v<j\}|_{d\mu}}{|B_j|_{d\mu}}$ and following as in [135], we get the result. □

Now, we need to obtain a kind of *reverse Hölder inequality* for v.

Lemma 10.3.5. *Let $r > 0$ such that $B_{3r/2} \subset \Omega$ and suppose that v is a supersolution to (10.3.1). Then, for every $0 < \alpha_1 < \alpha_2 < \frac{N}{N-2s}$, we have*

$$\left(\frac{1}{|B_r|_{d\mu}} \int_{B_r} v^{\alpha_2}\, d\mu\right)^{\frac{1}{\alpha_2}} \le C\left(\frac{1}{|B_{3r/2}|_{d\mu}} \int_{B_{3r/2}} v^{\alpha_1}\, d\mu\right)^{\frac{1}{\alpha_1}}, \tag{10.3.14}$$

with $C = C(N, s, \gamma, \alpha_1, \alpha_2) > 0$.

Proof. Let $q \in (1,2)$ and $d > 0$. Set $\tilde{v} := (v + d)$, and assume that $\psi \in W^s_{\gamma,0}(\Omega)$ is such that $\operatorname{supp}(\psi) \subset B_{\tau r}$, $\psi = 1$ in $B_{\tau' r}$ and $|\nabla\psi| \le \frac{C}{(\tau-\tau')r}$, where $\frac{1}{2} \le \tau' < \tau < \frac{3}{2}$. Then using $\tilde{v}^{1-q}\psi^2$ as a test function in (10.2.11), we obtain

$$0 \le \int_{B_{\tau r}} \int_{B_{\tau r}} (\tilde{v}(x) - \tilde{v}(y))\left(\frac{\psi^2(x)}{\tilde{v}^{q-1}(x)} - \frac{\psi^2(y)}{\tilde{v}^{q-1}(y)}\right) d\nu + 2 \int_{\mathbb{R}^N \setminus B_{\tau r}} \int_{B_{\tau r}} (\tilde{v}(x) - \tilde{v}(y)) \frac{\psi^2(x)}{\tilde{v}^{q-1}(x)}\, d\nu.$$

Since $|x| < |y|$ in $B_{\tau r} \times (\mathbb{R}^N \setminus B_{\tau r})$, using the positivity of $\tilde{v}$ it follows that

$$\int_{\mathbb{R}^N \setminus B_{\tau r}} \int_{B_{\tau r}} (\tilde{v}(x) - \tilde{v}(y)) \frac{\psi^2(x)}{\tilde{v}^{q-1}(x)}\, d\nu \le \left(\int_{B_{\tau r}} \tilde{v}^{2-q}\psi^2\, d\mu\right)\left(\sup_{\{x \in \operatorname{supp}(\psi)\}} \int_{\mathbb{R}^N \setminus B_{\tau r}} \frac{dy}{|x-y|^{N+2s}}\right).$$

Furthermore, by the pointwise inequality of [155, Lemma 3.3-(i)], there exist positive constants C_1 and C_2, depending on q, such that

$$\begin{aligned}
&\int_{B_{\tau r}} \int_{B_{\tau r}} (\tilde{v}(x) - \tilde{v}(y))\left(\frac{\psi^2(x)}{\tilde{v}^{q-1}(x)} - \frac{\psi^2(y)}{\tilde{v}^{q-1}(y)}\right) d\nu \\
&\quad \le -C_1 \int_{B_{\tau r}} \int_{B_{\tau r}} \left(\tilde{v}^{\frac{2-q}{2}}(x)\psi(x) - \tilde{v}^{\frac{2-q}{2}}(y)\psi(y)\right)^2 d\nu \\
&\quad\quad + C_2 \int_{B_{\tau r}} \int_{B_{\tau r}} \left((\tilde{v}^{2-q}(x) + \tilde{v}^{2-q}(y))(\psi(x) - \psi(y)\right)^2 d\nu.
\end{aligned}$$

By symmetry we have

$$\int_{B_{\tau r}} \int_{B_{\tau r}} \left((\tilde{v}^{2-q}(x) + \tilde{v}^{2-q}(y))(\psi(x) - \psi(y)\right)^2 d\nu = 2 \int_{B_{\tau r}} \int_{B_{\tau r}} \left(\tilde{v}^{2-q}(x)(\psi(x) - \psi(y)\right)^2 d\nu,$$

and we obtain

$$\int_{B_{\tau r}} \int_{B_{\tau r}} \left((\tilde{v}^{2-q}(x) + \tilde{v}^{2-q}(y))(\psi(x) - \psi(y)\right)^2 d\nu \le \frac{Cr^{-2s}}{(\tau - \tau')^2} \int_{B_{\tau r}} \tilde{v}^{2-q}\, d\mu.$$

See [11, Lemma 4.6].

Since

$$\sup_{\{x\in\operatorname{Supp}(\psi)\}} \int_{\mathbb{R}^N\setminus B_{\tau r}} \frac{dy}{|x-y|^{N+2s}} \le Cr^{-2s},$$

combining the estimates above we reach

$$\int_{B_{\tau r}}\int_{B_{\tau r}} \left(\tilde{v}^{\frac{2-q}{2}}(x)\psi(x) - \tilde{v}^{\frac{2-q}{2}}(y)\psi(y)\right)^2 d\nu \le \frac{Cr^{-2s}}{(\tau-\tau')^2}\int_{B_{\tau r}} \tilde{v}^{2-q}\, d\mu.$$

Hence, from the previous inequality and the Sobolev inequality in Proposition 10.2.10, we get

$$\begin{aligned}\left(\frac{1}{|B_{\tau' r}|_{d\mu}}\int_{B_{\tau' r}} \tilde{v}^{\frac{(2-q)N}{N-2s}}\, d\mu\right)^{\frac{N-2s}{N}} &\le \left(\frac{1}{|B_{\tau' r}|_{d\mu}}\int_{B_{\tau r}} (\tilde{v}^{\frac{2-q}{2}}\psi)^{2_s^*}\, d\mu\right)^{\frac{N-2s}{N}}\\ &\le \frac{C}{|B_{\tau r}|_{d\mu}(\tau-\tau')^2}\int_{B_{\tau r}} \tilde{v}^{2-q}\, d\mu.\end{aligned}$$

Since $q \in (1,2)$ is arbitrary and $\frac{N}{N-2s} > 1$ by using the Hölder inequality we obtain the estimate (10.3.14) for $\tilde{v} = v + d$ with α_1 and α_2 in the hypotheses. Finally letting $d \to 0$ and by the *monotone convergence theorem* we conclude the proof. □

In order to obtain the weak Harnack inequality, we need to prove the following estimate as in the De Giorgi seminal work [137].

Lemma 10.3.6. *Let $r > 0$ such that $B_r \subset \Omega$. Assume that v is a supersolution to* (10.3.1). *Then, there exists a constant $\eta \in (0,1)$ depending only on N, s and γ such that*

$$\left(\frac{1}{|B_r|_{d\mu}}\int_{B_r} v^{\eta}\, d\mu\right)^{\frac{1}{\eta}} \le C\inf_{B_r} v.$$

To prove Lemma 10.3.6 (see [213], [135, Lemma 4.1] and the pioneering work [137]) we use the next covering result in the spirit of Krylov–Safonov theory; see [216]. Recall that in general, we say that a measure ω in a metric space has the *doubling property* if there exists a constant $C > 0$ such that $\omega(B_{2r}(x)) \le C\omega(B_r(x))$ for all x and all $r > 0$. Note that we are working here with μ, a doubling measure on bounded domains of $\mathbb{R}^N$.

Lemma 10.3.7. *Assume that $E \subset B_r(x_0)$ is a measurable set. For $\delta \in (0,1)$, we define*

$$[E]_\delta := \bigcup_{\rho>0}\{B_{3\rho}(x)\cap B_r(x_0), x \in B_r(x_0) : |E\cap B_{3\rho}(x)|_{d\mu} > \delta|B_\rho(x)|_{d\mu}\}.$$

Then, either

1. $|[E]_\delta|_{d\mu} \geq \frac{\tilde{C}}{\delta}|E|_{d\mu}$, *or*
2. $[E]_\delta = B_r(x_0)$;

where $\tilde{C}$ is the doubling constant of $d\mu$.

Proof. Define the maximal operator

$$\begin{cases} M : B_r(x_0) & \to & \mathbb{R}, \\ x & \to & M(x) = \sup\limits_{\rho>0,\text{ so that } |x-x_0|\leq\min\{r,3\rho\}} \dfrac{|E \cap B_{3\rho}(x)|_{d\mu}}{|B_\rho(x)|_{d\mu}}. \end{cases} \tag{10.3.15}$$

It is clear that

$$[E]_\delta = \{x \in B_r(x_0), M(x) > \delta\}, \quad \forall \delta \in (0,1).$$

Assume that alternative 2 does not hold, that is, $B_\rho(x) \setminus [E]_\delta \neq \emptyset$. The set $[E]_\delta$ is open by definition. We consider a covering of $[E]_\delta$ by balls $B_{r_x}(x)$ such that $x \in [E]_\delta$ and $r_x = \frac{1}{2}\operatorname{dist}(x, B_\rho(x) \setminus [E]_\delta)$. By the Besicovitch–Vitali theorem (see [192, Chapter 1] and [117, Chapter 3]) there are countably many pairwise disjoint balls

$$\{B_{r_i}(x_i)\}_{i\in\mathbb{N}}, \quad r_i \equiv r_{x_i}, \text{ such that } [E]_\delta \subset \bigcup_{i\in\mathbb{N}} B_{5r_i}(x_i).$$

Then

$$B_{5r_i}(x_i) \cap (B_\rho(x) \setminus [E]_\delta) \neq \emptyset, \quad i = 1,2,3\ldots,$$

and there exists a point $y_i \in B_{5r_i}(x_i) \cap (B_\rho(x)\setminus[E]_\delta)$, that is, $M(y_i) \leq \delta$ for all $i = 1,2,3\ldots$. Since $x_i \in B_{5r_i}(y_i)$ we get

$$|E \cap B_{5r_i}(x_i)|_{d\mu} \leq \delta |B_{\frac{5}{3}r_i}(x_i)|_{d\mu} \leq C\delta |B_{r_i}(x_i)|_{d\mu},$$

where C is the constant of the doubling property. For y, a density point of E, we have

$$\liminf_{\rho\to 0} \frac{|E \cap B_{3\rho}(y)|_{d\mu}}{|B_\rho(y)|_{d\mu}} \geq \limsup_{\rho\to 0} \frac{|E \cap B_\rho(y)|_{d\mu}}{|B_\rho(y)|_{d\mu}} = 1 > \delta.$$

By the Lebesgue theorem $d\mu$-almost every point of E is a density point, and hence $d\mu$-almost every point of E is in $[E]_\delta$, for every $0 < \delta < 1$. Therefore

$$\begin{aligned} |E|_{d\mu} = |E \cap [E]_\delta|_{d\mu} &\leq \sum_{i=1}^{\infty} |E \cap B_{5r_i}(x_i)|_{d\mu} \\ &\leq C\delta \sum_{i=1}^{\infty} |E \cap B_{r_i}(x_i)|_{d\mu} \leq C\delta \sum_{i=1}^{\infty} |B_{r_i}(x_i)|_{d\mu} = C\delta |[E]_\delta|_{d\mu}, \end{aligned}$$

which implies alternative 1, as we wanted to prove. □

We next prove Lemma 10.3.6.

Proof of Lemma 10.3.6. Note that, for any $\eta > 0$,

$$\frac{1}{|B_r|_{d\mu}} \int_{B_r} v^\eta \, d\mu = \eta \int_0^\infty t^{\eta-1} \frac{|B_r \cap \{v > t\}|_{d\mu}}{|B_r|_{d\mu}} \, dt. \tag{10.3.16}$$

Then, for $t > 0$ and $i \in \mathbb{N}$, we set $A_t^i := \{x \in B_r : v(x) > t\delta^i\}$, where δ is given by Lemma 10.3.4. Note that $A_t^{i-1} \subset A_t^i$.

Let $\rho > 0$ and $x \in B_r$ such that $B_{3\rho}(x) \cap B_r \subset [A_t^{i-1}]_{\bar{\delta}}$. Thus,

$$|A_t^{i-1} \cap B_{3\rho}(x)|_{d\mu} > \bar{\delta}|B_\rho|_{d\mu} = \frac{\bar{\delta}}{3^{N-2\gamma}}|B_{3\rho}|_{d\mu}.$$

Hence, using Lemma 10.3.4, we reach

$$v(x) > \delta(t\delta^{i-1}) = t\delta^i \quad \text{for all } x \in B_r,$$

and therefore $[A_t^{i-1}]_{\bar{\delta}} \subset A_t^i$, being $[A_t^{i-1}]_{\bar{\delta}}$ as in Lemma 10.3.7. This fact, together with Lemma 10.3.7, allows us to deduce that

$$A_t^i = B_r \quad \text{or} \quad |A_t^i|_{d\mu} \geq \frac{\tilde{C}}{\bar{\delta}}|A_t^{i-1}|_{d\mu}. \tag{10.3.17}$$

Thus, if for some $m \in \mathbb{N}$ we have

$$|A_t^0|_{d\mu} > \left(\frac{\bar{\delta}}{\tilde{C}}\right)^m |B_r|_{d\mu}, \tag{10.3.18}$$

then $A_t^m = B_r$. If not, it follows from (10.3.17) that

$$|A_t^m|_{d\mu} \geq \frac{\tilde{C}}{\bar{\delta}}|A_t^{m-1}|_{d\mu}.$$

Since $A_t^{i-1} \subset A_t^m \subsetneq B_r$ for all $i \leq m$, the second point of the alternative (10.3.17) holds for A_t^{i-1} and then

$$|A_t^{m-1}|_{d\mu} \geq \frac{\tilde{C}}{\bar{\delta}}|A_t^{m-2}|_{d\mu} \cdots \geq \left(\frac{\tilde{C}}{\bar{\delta}}\right)^{m-1} |A_t^0|_{d\mu} > \left(\frac{\tilde{C}}{\bar{\delta}}\right)^{-1} |B_r|_{d\mu}.$$

Thus $|A_t^m|_{d\mu} > |B_r|_{d\mu}$, a contradiction with the fact that $A_t^m \subsetneq B_r$. Hence $A_t^m = B_r$.

It is clear that (10.3.18) holds if

$$m > \frac{1}{\log(\frac{\bar{\delta}}{\tilde{C}})} \log\left(\frac{|A_t^0|_{d\mu}}{|B_r|_{d\mu}}\right), \tag{10.3.19}$$

and consequently, fixing m to be the smallest integer such that (10.3.19) holds, we have $m \geq 1$ and

$$0 \leq m - 1 \leq \frac{1}{\log(\frac{\delta}{\tilde{C}})} \log\left(\frac{|A_t^0|_{d\mu}}{|B_r|_{d\mu}} \right).$$

Thus, using the fact that $\delta \in (0, \frac{1}{2})$, it can be checked that

$$\inf_{B_r} v > t\delta^m => t\delta\left(\frac{|A_t^0|_{d\mu}}{|B_r|_{d\mu}} \right)^{\frac{1}{\beta}},$$

with $\beta := \frac{\log(\frac{\delta}{\tilde{C}})}{\log(\delta)}$.

Set now $\xi := \inf_{B_r} v$. Then,

$$\frac{|B_r \cap \{v > t\}|_{d\mu}}{|B_r|_{d\mu}} = \frac{|A_t^0|_{d\mu}}{|B_r|_{d\mu}} \leq \tilde{C}\delta^{-\beta} t^{-\beta} \xi^{\beta}.$$

Going back to (10.3.16), we have

$$\begin{aligned} \frac{1}{|B_r|_{d\mu}} \int_{B_r} v^\eta \, d\mu &\leq \eta \int_0^a t^{\eta-1} \, dt + \eta\tilde{C} \int_a^\infty t^{\eta-1}\delta^{-\beta} t^{-\beta} \xi^\beta \, dt \\ &= a^\eta - \eta\tilde{C}\delta^{-\beta}\xi^\beta \frac{a^{\eta-\beta}}{\eta-\beta}. \end{aligned}$$

Choosing $a := \xi$ and $\eta := \frac{\beta}{2}$, we reach the result. □

After this result, we can prove the weighted weak Harnack inequality in Theorem 10.3.1.

Proof. Using Lemma 10.3.6 we obtain

$$\left(\frac{1}{|B_r|_{d\mu}} \int_{B_r} v^\eta \, d\mu \right)^{\frac{1}{\eta}} \leq C \inf_{B_r} v$$

for some $\eta \in (0, 1)$. Fixing $1 \leq q < \frac{N}{N-2s}$, by Lemma 10.3.5 with $\alpha_1 = \eta$ and $\alpha_2 = q$, it follows that

$$\left(\frac{1}{|B_r|_{d\mu}} \int_{B_r} v^q \, d\mu \right)^{\frac{1}{q}} \leq C \left(\frac{1}{|B_{\frac{3}{2}r}|_{d\mu}} \int_{B_{\frac{3}{2}r}} v^\eta \, d\mu \right)^{\frac{1}{\eta}}.$$

Hence

$$\left(\frac{1}{|B_r|_{d\mu}} \int_{B_r} v^q \, d\mu \right)^{\frac{1}{q}} \leq C \inf_{B_{\frac{3}{2}r}} v$$

and we conclude the proof. □

As a consequence of the previous Harnack inequality, we get much information about the behavior of the supersolutions to (10.1.1) around the origin. In particular, we see that any of them must be unbounded, even if $f \in L^\infty(\Omega)$.

Lemma 10.3.8. *Let $\lambda \le \Lambda_{N,2s}$. Assume that u is a nonnegative function defined in Ω such that $u \not\equiv 0$, $u \in L^1(\Omega)$, $\frac{u}{|x|^{2s}} \in L^1(\Omega)$ and $u \ge 0$ in $\mathbb{R}^N \setminus \Omega$. If u satisfies $(-\Delta)^s u - \lambda \frac{u}{|x|^{2s}} \ge 0$ in the weak sense in Ω, then there exist $\delta > 0$ and a constant $C = C(N, \delta, \gamma)$ such that for each ball $B_\delta(0) \subset\subset \Omega$,*

$$u \ge C|x|^{-\gamma} \quad \text{in } B_\delta(0),$$

where $\gamma = \frac{N-2s}{2} - \alpha$ is defined in (9.4.9). In particular, for δ conveniently small, we can assume that $u > 1$ in $B_\delta(0)$.

Proof. Considering $v := |x|^\gamma u$, $v \gneqq 0$ and satisfies $L_\gamma v \ge 0$, with L_γ defined in (10.2.3). Hence using the weak Harnack inequality in Theorem 10.3.1, we conclude that $\inf_{B_r(0)} v \ge C$. Thus $u(x) \ge C|x|^{-\gamma}$ in $B_r(0)$ and the result follows. □

10.4 Optimal summability in the presence of Hardy potential

In this section we analyze the question of the optimal summability of the solution to the problem

$$\begin{cases} (-\Delta)^s u - \lambda \frac{u}{|x|^{2s}} = f & \text{in } \Omega, \\ u = 0 & \text{in } \mathbb{R}^N \setminus \Omega, \end{cases} \tag{10.4.1}$$

with $0 < \lambda < \Lambda_{N,2s}$ in terms of the summability of f.

We try to obtain the corresponding results to those in Chapter 3 for fractional elliptic operators. In this fractional framework the results are more involved and up to now the optimality is not well understood.

10.4.1 Regularity of energy solutions

Along this subsection we will assume that $f \in L^m(\Omega)$ with $m \ge \frac{2N}{N+2s}$, and thus the corresponding solution, u, belongs to $H_0^s(\Omega)$.

We have seen in Chapter 8 that in the unperturbed problem, $\lambda = 0$, if $f \in L^m(\Omega)$ with $m > \frac{N}{2s}$, then, as in the classical case, $u \in L^\infty(\Omega)$. However, as a consequence of Lemma 10.3.8, this feature is no longer true for $\lambda > 0$, and actually $u(x) \ge C|x|^{-\gamma}$ in a neighborhood of the origin. Hence, the first natural question here is whether this rate is exactly the rate of growth of u, and the answer is yes for regular data, as the following theorem shows.

Theorem 10.4.1. *Let $f \in L^m(\Omega), m > \frac{N}{2s}$. Let us consider $u \in H_0^s(\Omega)$, the unique energy solution to problem* (10.4.1), *with $\lambda < \Lambda_{N,2s}$. Then $u(x) \le C|x|^{-\gamma}$ in $\mathbb{R}^N$.*

Proof. Since the problem is linear, without loss of generality we can assume $f \ge 0$. Defining $v(x) := |x|^{\gamma} u(x)$, it can be checked that it solves

$$\begin{cases} L_{\gamma,\Omega}(v) = |x|^{-\gamma} f & \text{in } \Omega, \\ v = 0 & \text{in } \mathbb{R}^N \setminus \Omega, \end{cases} \tag{10.4.2}$$

where the operator $L_{\gamma,\Omega}(v)$ was defined by (10.2.3). Consider now $G_k(v(x))$, specified in (8.3.12), with $k > 0$ as test function in (10.4.2). Hence,

$$a_{N,s} \iint_{D_\Omega} \frac{(v(x) - v(y))(G_k(v(x)) - G_k(v(y)))}{|x - y|^{N+2s}} \frac{dx}{|x|^{\gamma}} \frac{dy}{|y|^{\gamma}} = \int_{\Omega} |x|^{-\gamma} f G_k(v)\, dx.$$

Since for any $\sigma \in \mathbb{R}$, $\sigma = T_k(\sigma) + G_k(\sigma)$, we have

$$\begin{aligned} &(v(x) - v(y))(G_k(v(x)) - G_k(v(y))) \\ &\quad = (G_k(v(x)) - G_k(v(y)))^2 + (T_k(v(x)) - T_k(v(y)))(G_k(v(x)) - G_k(v(y))). \end{aligned}$$

Moreover, by Proposition 8.3.6, we know that

$$(T_k(v(x)) - T_k(v(y)))(G_k(v(x)) - G_k(v(y))) \ge 0,$$

and therefore

$$a_{N,s} \iint_{D_\Omega} \frac{|G_k(v(x)) - G_k(v(y))|^2}{|x - y|^{N+2s}} \frac{dx}{|x|^{\gamma}} \frac{dy}{|y|^{\gamma}} \le \int_{\Omega} f G_k(v(x)) \frac{dx}{|x|^{\gamma}}.$$

Let us denote $A_k := \{x \in \Omega : v(x) \ge k\}$. Applying the weighted Sobolev inequality on the left hand side we obtain

$$\mathcal{S} \|G_k(v)\|^2_{L^{2_s^*}(\Omega, |x|^{-\gamma} dx)} \le \int_{A_k} f G_k(v(x)) \frac{dx}{|x|^{\gamma}},$$

and using Hölder's inequality on the right hand side,

$$\left| \int_{A_k} f G_k(v(x)) \frac{dx}{|x|^{\gamma}} \right| \le \|f\|_{L^m(\Omega)} \|G_k(v)\|_{L^{2_s^*}(\Omega, |x|^{-\gamma} dx)} |A_k|^{1 - \frac{1}{2_s^*} - \frac{1}{m}}.$$

Thus we have

$$\|G_k(v)\|_{L^{2_s^*}(\Omega, |x|^{-\gamma} dx)} \le \mathcal{S}^{-1} \|f\|_{L^m(\Omega)} |A_k|^{1 - \frac{1}{2_s^*} - \frac{1}{m}}.$$

On the other hand, since Ω is bounded, there exists a constant $c > 0$ such that

$$\|G_k(v)\|_{L^{2_s^*}(\Omega,|x|^{-\gamma}\,dx)} \ge c\|G_k(v)\|_{L^{2_s^*}(\Omega)}.$$

Moreover, for any $z > k$, we have $A_z \subset A_k$ and $G_k(s)\chi_{A_z} \ge (z-k)$ for every $s \in \mathbb{R}$ and thus

$$(z-k)|A_z|^{\frac{1}{2_s^*}} \le \frac{1}{c\mathcal{S}}\|f\|_{L^m(\Omega)}|A_k|^{1-\frac{1}{2_s^*}-\frac{1}{m}}.$$

Manipulating the inequality above, we deduce that

$$|A_z| \le \frac{\|f\|_{L^m(\Omega)}^{2_s^*}|A_k|^{2_s^*(1-\frac{1}{2_s^*}-\frac{1}{m})}}{(c\mathcal{S})^{2_s^*}(z-k)^{2_s^*}}.$$

Hence we apply Lemma 8.4.2 with the choice $\psi(s) := |A_s|$. Since $m > \frac{N}{2s}$ we have

$$2_s^*\left(1-\frac{1}{2_s^*}-\frac{1}{m}\right) > 1.$$

Consequently, there exists k_0 such that $\psi(k) \equiv 0$ for any $k \ge k_0$ and thus

$$\operatorname*{ess\,sup}_{\Omega} v \le k_0. \qquad \square$$

We next try to obtain conditions on λ, under which the Calderón–Zygmund summability holds for the rest of the Lebesgue spaces contained in the dual of $H_0^s(\Omega)$, that is, when $f \in L^m(\Omega)$, where $\frac{2N}{N+2s} \le m < \frac{N}{2s}$. Then we have the following result.

Theorem 10.4.2. *Let f be a positive function $f \in L^m(\Omega)$, with $\frac{2N}{N+2s} \le m < \frac{N}{2s}$. If*

$$\lambda < \Lambda_{N,2s}\frac{4N(m-1)(N-2ms)}{m^2(N-2s)^2}, \tag{10.4.3}$$

then there exists a constant $c = c(N,m,s) > 0$ such that the unique energy solution of problem (10.4.1) *verifies*

$$\|u\|_{L^{m_s^{**}}(\Omega)} \le c\|f\|_{L^m(\Omega)}, \quad \text{where } m_s^{**} = \frac{mN}{N-2ms}. \tag{10.4.4}$$

Proof. Since $f \in H^{-s}(\Omega)$, the existence and uniqueness of an energy solution is a direct consequence of the Lax–Milgram theorem applied in $H_0^s(\Omega)$. To study the regularity of the solution, for every $k \in \mathbb{N}$, we consider $u_k \in L^\infty(\mathbb{R}^N) \cap H_0^s(\Omega)$, the solution to the following approximated problem:

$$\begin{cases} (-\Delta)^s u_k - \lambda\frac{u_{k-1}}{|x|^{2s}+\frac{1}{k}} = f_k(x) & \text{in } \Omega, \\ u_k = 0 & \text{in } \mathbb{R}^N \setminus \Omega, \end{cases} \tag{10.4.5}$$

where $f_k(x) := \min\{f(x), k\}$ and $u_0 = 0$.

In this way we obtain the following properties:

(i) $\{u_k\}$ is an increasing sequence;
(ii) each u_k is bounded;
(iii) $u_k \to u$, the unique solution to problem (10.4.1), in $L^p(\Omega)$, for every $1 \le p \le 2_s^*$.

Define $\beta = \frac{2_s^*}{2m'-2_s^*} = \frac{N(m-1)}{N-2ms} \ge 1$. Then $\beta m' = \frac{(\beta+1)}{2}2_s^*$. Since u_k is bounded we can take u_k^β as a test function in (10.4.1), obtaining

$$a_{N,s}\iint_{D_\Omega}\frac{(u_k(x)-u_k(y))(u_k^\beta(x)-u_k^\beta(y))}{|x-y|^{N+2s}}\,dx\,dy \le \lambda\int_\Omega \frac{u_k^{\beta+1}}{|x|^{2s}}\,dx + \int_\Omega f_k u_k^\beta\,dx.$$

By Hölder's inequality,

$$\int_\Omega f_k u_k^\beta\,dx \le \|f_k\|_{L^m(\Omega)}\Big(\int_\Omega u_k^{m'\beta}\,dx\Big)^{\frac{1}{m'}} \le \|f\|_{L^m(\Omega)}\Big(\int_\Omega u_k^{\frac{(\beta+1)2_s^*}{2}}\,dx\Big)^{\frac{\beta}{\beta+1}\frac{2}{2_s^*}}.$$

Now, by the algebraic inequality in (10.2.16), we get

$$(u_k(x)-u_k(y))(u_k^\beta(x)-u_k^\beta(y)) \ge \frac{4\beta}{(\beta+1)^2}\big(u_k^{\frac{\beta+1}{2}}(x)-u_k^{\frac{\beta+1}{2}}(y)\big)^2,$$

and, hence, using Hardy's inequality again, we conclude that

$$a_{N,s}\Big(\frac{4\beta}{(\beta+1)^2}-\frac{\lambda}{\Lambda_{N,2s}}\Big)\iint_{D_\Omega}\frac{(u_k^{\frac{\beta+1}{2}}(x)-u_k^{\frac{\beta+1}{2}}(y))^2}{|x-y|^{N+2s}}\,dx\,dy \le \|f\|_{L^m(\Omega)}\Big(\int_\Omega u_k^{\frac{(\beta+1)2_s^*}{2}}\,dx\Big)^{\frac{\beta}{\beta+1}\frac{2}{2_s^*}}.$$

On the other hand, hypothesis (10.4.3) is equivalent to

$$\Big(\frac{4\beta}{(\beta+1)^2}-\frac{\lambda}{\Lambda_{N,2s}}\Big) > 0.$$

Thus, by the Sobolev inequality, we reach

$$\Big(\int_\Omega u_k^{\frac{(\beta+1)2_s^*}{2}}\,dx\Big)^{\frac{2}{2_s^*(\beta+1)}} \le C\|f\|_{L^m(\Omega)}.$$

Furthermore, $\frac{(\beta+1)2_s^*}{2} = \frac{mN}{N-2sm} = m_s^{**}$, and therefore passing to the limit we conclude the proof. □

Remark 10.4.3. Note that, making $s \to 1$, the condition over λ becomes

$$\lambda < \frac{N(m-1)(N-2m)}{m^2},$$

the optimal curve obtained in Chapter 3 for the local case (see [80]).

10.4.2 About the optimality of the regularity results

For simplicity, we set

$$J_s(m) := \Lambda_{N,2s} \frac{4N(m-1)(N-2ms)}{m^2(N-2s)^2}. \tag{10.4.6}$$

In Chapter 3 we have proved that, in the local case, condition (10.4.3) (with $s = 1$) is optimal. In particular, if $\lambda > J_1(m)$ and $\Omega = B_1(0)$, there exists a suitable radial function $f \in L^m(\Omega)$ such that the solution u does not belong to $L^{m^{**}}(\Omega)$. This proves the optimality of the curve $\lambda = J_1(m)$.

In the nonlocal case the situation is more delicate. Indeed, we will see that in this case, the previous example does not provide the optimality of the curve $J_s(m)$. It proves that the m_s^{**} summability does not hold above a curve, which we will call $P_s(m)$, which is in general above $J_s(m)$. Thus, *to the best of our knowledge, the optimality of $J_s(m)$ for every $\frac{2N}{N+2s} \le m < \frac{N}{2s}$ remains open.*

In order to define such curve $P_s(m)$, let us first recall that for all $\lambda \in (0, \Lambda_{N,2s}]$ there exists a unique nonnegative constant $\alpha \in [0, \frac{N-2s}{2})$ such that

$$\lambda = \lambda(\alpha) = \frac{2^{2s}\Gamma(\frac{N+2s+2\alpha}{4})\Gamma(\frac{N+2s-2\alpha}{4})}{\Gamma(\frac{N-2s+2\alpha}{4})\Gamma(\frac{N-2s-2\alpha}{4})}, \tag{10.4.7}$$

and $\lambda(\alpha)$ is a decreasing function of α (see Section 9.4 and [11]). Hence, for $m \in [\frac{2N}{N+2s}, \frac{N}{2s})$ fixed, we consider

$$\alpha_0(m) := \frac{N+2s}{2} - \frac{N}{m} \quad \text{and} \quad P_s(m) := \lambda(\alpha_0(m)) \tag{10.4.8}$$

given by (10.4.7). Then we have the following result.

Lemma 10.4.4. *Assume that $f(x) = \frac{1}{|x|^\nu}$ with $\nu = \frac{N-\varepsilon}{m}$, for some $\varepsilon > 0$. Let $\lambda_1 \in (0, \Lambda_{N,2s}]$ be such that $\lambda_1 \ge P_s(m)$. If u is the unique solution of (10.4.1) with $\lambda = \lambda_1$, then $u \notin L^{m_s^{**}}(B_1(0))$.*

Proof. Note that $f \in L^m(B_1(0))$. From (10.4.7) we know that $\lambda_1 = \lambda(\alpha_1)$ for some $\alpha_1 \in [0, \frac{N-2s}{2})$. Since $\lambda_1 \ge P_s(m)$, using the fact that $\lambda(\alpha)$ is a decreasing function we reach

$$\alpha_1 \le \alpha_0(m) = \frac{N+2s}{2} - \frac{N}{m}.$$

Define

$$v(x) := C\left(\frac{1}{|x|^\gamma} - \frac{1}{|x|^{\nu-2s}}\right), \tag{10.4.9}$$

where $\gamma = \frac{N-2s}{2} - \alpha_1$. Since $\gamma \ge \frac{N}{m} - 2s$, using the fact that $\nu < \frac{N}{m}$ it follows that $\gamma > \nu - 2s$. Thus, $v \ge 0$ in $B_1(0)$. Hence, choosing a suitable positive constant C, we reach

$$(-\Delta)^s v - \lambda_1 \frac{v}{|x|^{2s}} = f \quad \text{in } B_1(0).$$

Since $v \le 0$ in $\mathbb{R}^N \setminus B_1(0)$, by comparison it follows that $v \le u$, where u is the unique solution of (10.4.1) for $\lambda = \lambda_1$. Since $\gamma \ge \frac{N}{m} - 2s$, we have $v \notin L^{m_s^{**}}(B_1(0))$ and we conclude. □

As a direct application of this lemma we can prove the optimality of the curve $J_s(m)$ in a particular case.

Lemma 10.4.5. *Assume that the hypotheses of Lemma* 10.4.4 *hold and let u be the solution of problem* (10.4.1). *If* $m = \frac{2N}{N+2s}$ *and* $\lambda \ge J_s(m)$, *then* $u \notin L^{m_s^{**}}(B_1(0))$.

Proof. Note that $P_s(m)$, defined in (10.4.8), can be rewritten as

$$P_s(m) = \frac{2^{2s}\Gamma(\frac{N+2s}{2} - \frac{N}{2m})\Gamma(\frac{N}{2m})}{\Gamma(\frac{N}{2} - \frac{N}{2m})\Gamma(\frac{N}{2m} - s)}, \tag{10.4.10}$$

and in the particular case of $m = \frac{2N}{N+2s}$, it satisfies

$$J_s(m) = P_s(m).$$

Hence $\lambda \ge P_s(m)$ by hypothesis, and we conclude applying Lemma 10.4.4. □

Remark 10.4.6. Note that in the local case, $s = 1$, $\alpha = \sqrt{\Lambda_{N,2} - \lambda}$ and $P_1(m) = J_1(m)$ for all $m \in [\frac{2N}{N+2}, \frac{N}{2})$, and thus Lemma 10.4.4 holds in the whole range.

Next we show that for radial functions the result in Lemma 10.4.4 cannot be improved. In other words the optimality of the curve $J_s(m)$ cannot be proved with radial functions. We start by proving the following result (for the properties of the Gamma function we refer to [27]).

Lemma 10.4.7. *Assume that* $s \in (0, 1)$ *and* $m \in [\frac{2N}{N+2s}, \frac{N}{2s})$. *Then*

$$J_s(m) \le P_s(m) \tag{10.4.11}$$

and equality holds in (10.4.11) *if and only if* $m = \frac{2N}{N+2s}$.

Proof. Using (10.4.10) and the definition of $\Lambda_{N,2s}$ (see Chapter 9 and (10.1.4)), it easily follows that (10.4.11) is equivalent to

$$\begin{aligned} D(m) &:= \frac{m^2}{(m-1)(N-2sm)} \frac{\Gamma(\frac{N+2s}{2} - \frac{N}{2m})\Gamma(\frac{N}{2m})}{\Gamma(\frac{N}{2} - \frac{N}{2m})\Gamma(\frac{N}{2m} - s)} \\ &\ge \frac{4N}{(N-2s)^2} \frac{\Gamma^2(\frac{N+2s}{4})}{\Gamma^2(\frac{N-2s}{4})} =: \Theta(N, s). \end{aligned} \tag{10.4.12}$$

Furthermore, as we saw in the proof of Lemma 10.4.5,

$$D\Big(\frac{2N}{N+2s}\Big) = \Theta(N, s),$$

and thus (10.4.12) holds if we prove that D is an increasing function. Let us denote $D(m) = D_1(m)D_2(m)$, where

$$D_1(m) := \frac{\Gamma(\frac{N+2s}{2} - \frac{N}{2m})\Gamma(\frac{N}{2m})}{\Gamma(\frac{N}{2} - \frac{N}{2m})\Gamma(\frac{N}{2m} - s)} \quad \text{and} \quad D_2(m) := \frac{m^2}{(m-1)(N-2sm)}.$$

On the other hand, it is known that for $t > 0$ one has $\Gamma'(t) = \psi(t)\Gamma(t)$, where $\psi(t)$, the so-called *digamma function*, is given by

$$\psi(t) := -\frac{1}{t} - C_0 + t\sum_{n=1}^{\infty}\frac{1}{n(n+t)},$$

with C_0 being the Euler constant. Hence, it follows that

$$D_1'(m) = \frac{N}{2m^2}D_1(m)K(m),$$

where

$$K(m) = [\psi(a) - \psi(b) + \psi(c) - \psi(d)]$$

and

$$a := \frac{N+2s}{2} - \frac{N}{2m}, \quad b := \frac{N}{2} - \frac{N}{2m}, \quad c := \frac{N}{2m} - s, \quad d := \frac{N}{2m}.$$

Thus

$$D'(m) = D_1(m)\frac{m}{(m-1)(N-2sm)}\left(\frac{(m-2)N+2sm}{(m-1)(N-2sm)} + \frac{N}{2m}K(m)\right).$$

The first two terms here are positive, and then to analyze the sign of D' we have to study the function

$$H(m) := \left(\frac{(m-2)N+2sm}{(m-1)(N-2sm)} + \frac{N}{2m}K(m)\right).$$

By definition, there holds

$$\begin{aligned} K(m) = &-\frac{4sm^3}{N}\frac{(m-2)N+2sm}{(m-1)(N-2sm)((m-1)N+2sm)} \\ &+ s\sum_{n=1}^{\infty}\left(\frac{1}{(n+a)(n+b)} - \frac{1}{(n+c)(n+d)}\right), \end{aligned}$$

and noting that $a = b + s$ and $d = c + s$, we have

$$\frac{1}{(n+a)(n+b)} = \frac{1}{s}\left(\frac{1}{n+b} - \frac{1}{n+b+s}\right)$$

and

$$\frac{1}{(n+c)(n+d)} = \frac{1}{s}\left(\frac{1}{n+c} - \frac{1}{n+c+s}\right).$$

Thus,

$$\begin{aligned}\frac{1}{(n+a)(n+b)} - \frac{1}{(n+c)(n+d)} &= \frac{1}{s}\left(\left(\frac{1}{n+b} - \frac{1}{n+c}\right) + \left(\frac{1}{n+c+s} - \frac{1}{n+b+s}\right)\right)\\ &= -\frac{b-c}{s}\left(\frac{1}{(n+c)(n+b)} - \frac{1}{(n+c+s)(n+b+s)}\right).\end{aligned}$$

Hence $H(m) = ((m-2)N + 2sm)H_1(m)$, where

$$H_1(m) := \frac{1}{(m-1)N + 2sm} - \frac{N}{4m^2}\sum_{n=1}^{\infty}\left[\frac{1}{(n+c)(n+b)} - \frac{1}{(n+c+s)(n+b+s)}\right].$$

Using the fact that $m \in (\frac{2N}{N+2s}, \frac{N}{2s})$, we obtain $(m-2)N + 2sm > 0$ and therefore the sign of H is the sign of H_1. On the other hand, since $s \in (0,1)$, we have

$$\sum_{n=1}^{\infty}\left[\frac{1}{(n+c)(n+b)} - \frac{1}{(n+c+s)(n+b+s)}\right] \leq \frac{1}{(1+c)(1+b)},$$

whence we conclude that

$$\begin{aligned}H_1(m) &\geq \frac{1}{(m-1)N + 2sm} - \frac{N}{4m^2}\frac{1}{(1+c)(1+b)}\\ &= \frac{1}{(m-1)N + 2sm} - \frac{N}{(N + 2m - 2sm)((m-1)N + 2m)}.\end{aligned}$$

Now, using the fact that $s \in (0,1)$, we get $H_1(m) > 0$ and the result follows. □

As a consequence we get the next regularity result.

Proposition 10.4.8. *Let $s \in (0,1)$. Assume that $m \in (\frac{2N}{N+2s}, \frac{N}{2s})$, $s < 1$ and $f(x) = \frac{1}{|x|^\nu}$ with $\nu = \frac{N-\varepsilon}{m}$ for some small $\varepsilon > 0$. Then there exists $\tilde{\lambda} \in (J_s(m), P_s(m))$ such that the solution u of problem (10.4.1) with $\lambda = \tilde{\lambda}$ satisfies $u \in L^{m_s^{**}}(B_1(0))$.*

Proof. Fix $\delta > 0$ small enough so that if $\alpha \in (\alpha_0(m), \alpha_0(m) + \delta)$, then $\gamma := \frac{N-2s}{2} - \alpha$ satisfies $\gamma > \nu - 2s$ and $\gamma m_s^{**} < N$.

Define $\tilde{\lambda} = \lambda(\alpha)$. Since $\alpha > \alpha_0$, we have $\tilde{\lambda} < P_s(m)$. Thus, due to the continuity of λ as a function of α, choosing δ small enough and applying Lemma 10.4.7 we deduce

$$J_s(m) < \tilde{\lambda} < P_s(m).$$

Let u be the unique solution of problem (10.4.1) with $\lambda = \tilde{\lambda}$ and consider the function v defined in (10.4.9). Since $\gamma > \nu - 2s$, by similar arguments as in the proof of Lemma 10.4.4 we conclude that $v \leq u$. By setting $w := u - v$, it follows that

$$\begin{cases} (-\Delta)^s w - \lambda\frac{w}{|x|^{2s}} = 0 & \text{in } \Omega,\\ w = -v \geq 0 & \text{in } \mathbb{R}^N \setminus \Omega.\end{cases}$$

Thus, by Lemma 10.3.8 and Theorem 10.4.1, $w \simeq |x|^{-\gamma}$ close to the origin. Since $v \simeq |x|^{-\gamma}$, also $u \simeq |x|^{-\gamma}$. By the definition of γ, we obtain $\gamma m_s^{**} < N$, thus $u \in L^{m_s^{**}}(B_1(0))$ and then we conclude the proof. □

10.4.3 Nonvariational setting: weak solutions

In this subsection we consider $f \in L^m(\Omega)$, with $1 < m < \frac{2N}{N+2s}$.

Theorem 10.4.9. *Assume* $1 < m < \frac{2N}{N+2s}$ *and* $\lambda < J_s(m)$ *defined in* (10.4.6). *Then problem* (10.4.1) *has a unique weak solution u, which satisfies*

$$\|u\|_{L^{m_s^{**}}(\Omega)} \le c\|f\|_{L^m(\Omega)}, \quad \text{where } m_s^{**} = \frac{mN}{N-2ms}. \tag{10.4.13}$$

Moreover, $u \in W_0^{s_1,m_s^*}(\Omega)$, *for all* $s_1 < s$, *with* $m_s^* = \frac{mN}{N-ms}$.

Proof. Let $\{f_n\}_{n\in\mathbb{N}} \subset L^\infty(\Omega)$ be such that $0 \le f_n \le f$ and $f_n \uparrow f$ strongly in $L^m(\Omega)$. Define u_n to be the unique positive solution to the approximated problem

$$\begin{cases} (-\Delta)^s u_n - \lambda \frac{u_n}{|x|^{2s}+\frac{1}{n}} = f_n & \text{in } \Omega, \\ u_n = 0 & \text{in } \mathbb{R}^N \setminus \Omega. \end{cases} \tag{10.4.14}$$

Then $\{u_n\}_{n\in\mathbb{N}}$ is monotone in n. As in the proof of Theorem 10.4.2, we use u_n^β as a test function in (10.4.14) with

$$0 < \beta := \frac{N(m-1)}{N-2ms} < 1$$

(actually we have to test with $(u_n+\delta)^\beta$, $\delta > 0$ and make $\delta \to 0$ at the end, but to simplify we will drop this parameter here). Then,

$$a_{N,s} \iint_{D_\Omega} \frac{(u_n(x)-u_n(y))(u_n^\beta(x)-u_n^\beta(y))}{|x-y|^{N+2s}}\,dx\,dy \le \lambda \int_\Omega \frac{u_n^{\beta+1}}{|x|^{2s}}\,dx + \int_\Omega f_n u_n^\beta\,dx. \tag{10.4.15}$$

By the fact that $m'\beta = \frac{(\beta+1)2_s^*}{2}$ and Hölder's inequality,

$$\int_\Omega f_n u_n^\beta\,dx \le \|f\|_{L^m(\Omega)} \Big(\int_\Omega u_n^{\frac{(\beta+1)2_s^*}{2}}\,dx \Big)^{\frac{\beta}{\beta+1}\frac{2}{2_s^*}}. \tag{10.4.16}$$

Now, using Lemma 10.2.27 and the Hardy and Sobolev inequalities, it follows that

$$a_{N,s}\Big(\frac{4\beta}{(\beta+1)^2} - \frac{\lambda}{\Lambda_{N,2s}}\Big)\Big(\iint_{D_\Omega} \frac{(u_n^{\frac{\beta+1}{2}}(x)-u_n^{\frac{\beta+1}{2}}(y))^2}{|x-y|^{N+2s}}\,dx\,dy\Big)^{\frac{1}{\beta+1}} \le C(N,s)\|f\|_{L^m(\Omega)}.$$

Since $\lambda < J_s(m)$, noting that $m_s^{**} = 2_s^* \frac{\beta+1}{2}$ and applying the Sobolev inequality, (10.4.13) follows.

Furthermore, in particular

$$\int_\Omega u_n^{\frac{(\beta+1)2_s^*}{2}} dx \le C_1 \quad \text{and} \quad \int_\Omega \frac{u_n^{\beta+1}}{|x|^{2s}} dx \le C_2, \tag{10.4.17}$$

where C_1 and C_2 are independent of n. Since $\frac{(\beta+1)2_s^*}{2} > 1$, the Lebesgue theorem implies that $u_n \uparrow u$ a. e. and strongly in $L^\sigma(\Omega)$ for all $1 \le \sigma \le \frac{(\beta+1)2_s^*}{2}$ and $\frac{u_n}{|x|^{2s}} \uparrow \frac{u}{|x|^{2s}}$ strongly in $L^1(\Omega)$. Therefore, u is a weak solution of (10.4.1).

Moreover, using Fatou's lemma, (10.4.16) and (10.4.17) in (10.4.15) we obtain

$$\iint_{D_\Omega} \frac{(u(x)-u(y))(u^\beta(x)-u^\beta(y))}{|x-y|^{N+2s}} dx\, dy \le C. \tag{10.4.18}$$

In addition, we know that u is the unique weak solution. Indeed if u_2 is another solution of (10.4.1) with the regularities above, then setting $w := u_2 - u$, we conclude that

$$\begin{cases} (-\Delta)^s w - \lambda \frac{w}{|x|^{2s}} = 0 & \text{in } \Omega, \\ w = 0 & \text{in } \mathbb{R}^N \setminus \Omega. \end{cases}$$

By testing with $\phi \in \mathcal{T}$ defined in (10.2.1), with $(-\Delta)^s \phi = \varphi > 0$, we obtain that $w = 0$, and hence $u_2 = u$. To finish, we prove the regularity of the *fractional gradient*. Fix $s_1 < s$ and let $q = m_s^* < 2$. Call

$$d\sigma := \begin{cases} \left(\frac{u^\beta(x)-u^\beta(y)}{u(x)-u(y)}\right) dx\, dy & \text{if } u(x) \ne u(y), \\ 0 & \text{if } u(x) = u(y), \end{cases}$$

and note that $d\sigma$ is positive. Therefore, by Hölder's inequality

$$\begin{aligned}
&\int_\Omega \int_\Omega \frac{|u(x)-u(y)|^q}{|x-y|^{N+qs_1}} dx\, dy \\
&= \int_\Omega \int_\Omega \frac{|u(x)-u(y)|^q}{|x-y|^{N+qs_1}} \left(\frac{u^\beta(x)-u^\beta(y)}{u(x)-u(y)}\right) \times \left(\frac{u(x)-u(y)}{u^\beta(x)-u^\beta(y)}\right) dx\, dy \\
&\le \left(\int_\Omega \int_\Omega \frac{|u(x)-u(y)|^2}{|x-y|^{N+2s}} d\sigma\right)^{\frac{q}{2}} \times \left(\int_\Omega \int_\Omega \left(\frac{u(x)-u(y)}{u^\beta(x)-u^\beta(y)}\right)^{\frac{2}{2-q}} \frac{d\sigma}{|x-y|^{N-\theta}}\right)^{\frac{2-q}{2}} \\
&= \left(\int_\Omega \int_\Omega \frac{(u(x)-u(y))(u^\beta(x)-u^\beta(y))}{|x-y|^{N+2s}} dx\, dy\right)^{\frac{q}{2}} \\
&\quad \times \left(\int_\Omega \int_\Omega \left(\frac{u(x)-u(y)}{u^\beta(x)-u^\beta(y)}\right)^{\frac{q}{2-q}} \frac{dx\, dy}{|x-y|^{N-\theta}}\right)^{\frac{2-q}{2}},
\end{aligned}$$

where $\theta := \frac{2(s-s_1)}{2-q}$. The first term is bounded by (10.4.18), and the second one can be estimated as follows. Since $\beta < 1$, we have

$$0 \le \frac{u(x)-u(y)}{u^\beta(x)-u^\beta(y)} \le \frac{1}{\beta}(u^{1-\beta}(x)+u^{1-\beta}(y)),$$

and hence

$$\begin{aligned}\int_\Omega\int_\Omega\Big(\frac{u(x)-u(y)}{u^\beta(x)-u^\beta(y)}\Big)^{\frac{q}{2-q}}\frac{dx\,dy}{|x-y|^{N-\theta}} &\le \frac{C}{\beta}\int_\Omega\int_\Omega(u^{\frac{(1-\beta)q}{2-q}}(x)+u^{\frac{(1-\beta)q}{2-q}}(y))\frac{dx\,dy}{|x-y|^{N-\theta}}\\ &\le \frac{2C}{\beta}\int_\Omega u^{\frac{(1-\beta)q}{2-q}}(x)\Big(\int_\Omega\frac{dy}{|x-y|^{N-\theta}}\Big)\,dx.\end{aligned}$$

Note that

$$\sup_{\{x\in\Omega\}}\Big(\int_\Omega\frac{dy}{|x-y|^{N-\theta}}\Big) \le C,$$

hence

$$\int_\Omega\int_\Omega\Big(\frac{u(x)-u(y)}{u^\beta(x)-u^\beta(y)}\Big)^{\frac{q}{2-q}}\frac{dx\,dy}{|x-y|^{N-\theta}} \le C_1\int_\Omega u^{\frac{(1-\beta)q}{2-q}}(x)\,dx.$$

Now, since $\frac{(1-\beta)q}{2-q} = m_s^{**}$, the result follows using (10.4.13). □

10.4.4 A necessary and sufficient condition for solvability

As in the local problem studied in Chapter 3, in the case where no condition is imposed on λ, an additional condition on f is needed.

The next theorem gives a necessary and sufficient condition to ensure the existence of a weak solution.

Theorem 10.4.10. *Let $\lambda \le \Lambda_{N,2s}$ and suppose that $f \in L^1(\Omega)$, $f \ge 0$. Then u is a positive weak solution to problem* (10.4.1) *if and only if f satisfies*

$$\int_{B_r(0)} |x|^{-\gamma} f\,dx < +\infty,$$

for some $B_r(0) \subset\subset \Omega$ with γ defined in Lemma 10.2.4.

Moreover, if u is the unique weak solution to (10.4.1), *then*

$$\forall k \ge 0,\quad T_k(u) \in H_0^s(\Omega),\quad u \in L^q(\Omega),\quad \forall q \in \Big(1, \frac{N}{N-2s}\Big), \tag{10.4.19}$$

$$\left|(-\Delta)^{\frac{s}{2}}(u)\right| \in L^r(\Omega), \quad \forall r \in \left(1, \frac{N}{N-s}\right), \tag{10.4.20}$$

and $u \in W_0^{s_1,q_1}(\Omega)$ *for all* $s_1 < s$ *and for all* $q_1 < \frac{N}{N-s}$.

Proof. Necessary condition: Consider u a positive weak solution to problem (10.4.1) and consider $\varphi_n \in L^\infty(\Omega) \cap H_0^s(\Omega)$, the positive solution to

$$\begin{cases} (-\Delta)^s \varphi_n = \lambda \frac{\varphi_{n-1}}{|x|^{2s}+\frac{1}{n}} + 1 & \text{in } \Omega, \\ \varphi_n = 0 & \text{in } \mathbb{R}^N \setminus \Omega, \end{cases}$$

where

$$\begin{cases} (-\Delta)^s \varphi_0 = 1 & \text{in } \Omega, \\ \varphi_0 = 0 & \text{in } \mathbb{R}^N \setminus \Omega. \end{cases}$$

Then it is easy to check that $\varphi_0 \le \varphi_1 \le \varphi_{n-1} \le \varphi_n \le \varphi$, where φ is the pointwise limit. Then,

$$\begin{cases} (-\Delta)^s \varphi = \lambda \frac{\varphi}{|x|^{2s}} + 1 & \text{in } \Omega, \\ \varphi > 0 & \text{in } \Omega, \\ \varphi = 0 & \text{in } (\mathbb{R}^N \setminus \Omega). \end{cases} \tag{10.4.21}$$

Taking φ_n as a test function in (10.4.1), we get

$$\int_\Omega f\varphi_n \, dx \le \int_\Omega u \, dx = C < \infty.$$

Hence, $\{f\varphi_n\}_{n\in\mathbb{N}}$ is an increasing sequence uniformly bounded in $L^1(\Omega)$. Applying the monotone convergence theorem and Lemma 10.3.8 we obtain

$$\tilde{C} \int_{B_r(0)} |x|^{-\gamma} f \, dx \le \int_\Omega f\varphi \, dx \le C.$$

Sufficient condition: Assume that

$$\int_{B_r(0)} |x|^{-\gamma} f \, dx < +\infty,$$

for some $B_r(0) \subset\subset \Omega$. Let us consider the sequence of energy solutions $u_n \in L^\infty(\Omega) \cap H_0^s(\Omega)$ to the following approximated problems:

$$\begin{cases} (-\Delta)^s u_n = \lambda \frac{u_{n-1}}{|x|^{2s}+\frac{1}{n}} + f_n & \text{in } \Omega, \\ u_n(x) = 0 & \text{in } \mathbb{R}^N \setminus \Omega, \end{cases} \tag{10.4.22}$$

where

$$\begin{cases} (-\Delta)^s u_0 = f_1 & \text{in } \Omega, \\ u_0(x) = 0 & \text{in } (\mathbb{R}^N \setminus \Omega), \end{cases}$$

with $f_n = T_n(f)$ and $u_0 \le u_1 \le u_{n-1} \le u_n$ in $\mathbb{R}^N$. Since $f_n \ge 0$, $u_n(x) \ge 0$ in Ω. Take $\varphi \in H_0^s(\Omega)$, the positive energy solution to (10.4.21), as a test function in (10.4.22). As a consequence of Lemma 10.4.1, it follows that

$$\int_\Omega u_n\,dx \le \int_\Omega f\varphi\,dx \le \tilde{C}\int_\Omega f|x|^{-\gamma} \le C.$$

Hence, since the sequence $\{u_n\}_{n\in\mathbb{N}}$ is increasing, we can define $u := \lim_{n\to\infty} u_n$, and conclude that $u \in L^1(\Omega)$. We claim that $\frac{u}{|x|^{2s}} \in L^1(\Omega)$. Indeed, let ψ be the unique bounded positive solution to the problem

$$\begin{cases} (-\Delta)^s \psi = 1 & \text{in } \Omega, \\ \psi = 0 & \text{in } \mathbb{R}^N \setminus \Omega. \end{cases}$$

Then $\psi \ge C$ in $B_r(0)$. By using ψ as a test function in (10.4.22),

$$\int_\Omega \frac{u_{n-1}}{|x|^{2s} + \frac{1}{n}}\,dx \le \frac{1}{C}\int_{B_r(0)} \frac{\psi u_{n-1}}{|x|^{2s} + \frac{1}{n}}\,dx + C(r)\int_{\Omega\setminus B_r(0)} u_n\,dx \le C,$$

and thus

$$\lambda\frac{u_{n-1}}{|x|^{2s} + \frac{1}{n}} + f_n \nearrow \lambda\frac{u}{|x|^{2s}} + f \quad \text{strongly in } L^1(\Omega).$$

Testing with $T_k(u_n)$ in (10.4.22) and by the previous estimates, we easily get $T_k(u_n) \rightharpoonup T_k(u)$ weakly in $H_0^s(\Omega)$. Since the sequence $\{\lambda\frac{u_{n-1}}{|x|^{2s}+\frac{1}{n}} + f_n\}_{n\in\mathbb{N}}$ converges strongly in $L^1(\Omega)$, by the results of [224], we reach that $u \in L^\sigma(\Omega)$ for all $\sigma < \frac{N}{N-2s}$ and $|(-\Delta)^{\frac{s}{2}}(u)| \in L^r(\Omega)$ for all $r \in (1, \frac{N}{N-s})$. Moreover, according to Theorem 5 (C) of Chapter 5 in [293], we conclude that $u \in W_0^{s_1,q_1}(\Omega)$ for all $s_1 < s$ and for all $q_1 < \frac{N}{N-s}$. See also [2, 218] for a simple proof. □

Remark 10.4.11. As a consequence of this result, together with the weak Harnack inequality, one can easily prove nonexistence for $\lambda > \Lambda_{N,2s}$.

10.5 Further results and comments

1. An intellectual challenge here is to investigate what happens in the context of variational solutions in the interval

$$J_s(m) < \lambda < P_s(m),$$

defined in Section 10.4.2. Independently of the result, we believe that the techniques needed to attack this problem would be very interesting.
2. It would also be interesting to find negative results to the summability for $\lambda > J_s(m)$ if $m \in (1, \frac{2N}{N+2s})$.
3. Another direction of research comes from studying the analysis of other kinds of nonlocal elliptic operators. However, there is no substantial difference for those operators with the kernels of Definition 8.3.2.

11 Fractional semilinear elliptic problems

11.1 Introduction

During the last 20 years much attention has been devoted to understand the role of the Hardy–Leray potential in the solvability of sublinear and quasilinear elliptic problems. References to study the evolution of the research on elliptic problems involving the *inverse square potential* can be found in the papers [13, 14, 15, 17, 87, 88, 89, 99, 145, 184, 179, 181, 257, 311], among others.

In this chapter we will study the effect of the extension of the Hardy inequality studied in Chapter 9 in the semilinear problem involving the fractional Hardy potential.

There exist many recent results related to the Dirichlet problem for the fractional Laplacian with semilinear perturbations; see [44, 279, 280, 281, 283]. M. M. Fall, in [151], has extended for the nonlocal case some results given by Brezis–Dupaigne–Tesei in [89]. The problem he analyzes is

$$(P_0) = \begin{cases} (-\Delta)^s u - \lambda \dfrac{u}{|x|^{2s}} = u^p & \text{in } \Omega, \\ u > 0 & \text{in } \Omega, \\ u = 0 & \text{in } \mathbb{R}^N \setminus \Omega, \end{cases}$$

for which he finds the critical power p to have solvability. In his proof, he uses the extension given by L. Caffarelli and L. Silvestre in [105]. We will also discuss the problem (P_0) but without such extension.

In the spirit of the classical paper by H. Brezis and L. Nirenberg [94], along this chapter we will analyze the solvability of the Dirichlet problem for the nonlocal operator $(-\Delta)^s$ and a zero order convex-concave term. In particular, we will analyze the existence of nontrivial solutions for the following problem with the presence of the Hardy potential:

$$(P_\mu) = \begin{cases} (-\Delta)^s u - \lambda \frac{u}{|x|^{2s}} = u^p + \mu u^q & \text{in } \Omega, \\ u > 0 & \text{in } \Omega, \\ u = 0 & \text{in } \mathbb{R}^N \setminus \Omega, \end{cases} \tag{11.1.1}$$

where Ω is a bounded Lipschitz domain of $\mathbb{R}^N$ satisfying the exterior ball condition and $0 \in \Omega$, $s \in (0,1)$, $N > 2s$, $\mu > 0$, $0 < q < 1$, $\lambda < \Lambda_{N,2s}$. Here $p > 1$ but smaller than an upper bound that we will explain right after. The case $q = 0$ can be extended in an easy way to the case with a source term, $f(x)$, with suitable integrability.

We recall that a domain Ω satisfies the exterior ball condition if there exists a positive radius ρ^* such that all the points on $\partial\Omega$ can be touched by some exterior ball of radius ρ^*. The hypotheses that we have presented above are supposed to hold in

https://doi.org/10.1515/9783110606270-011

what follows. We remark again that we will study directly the nonlocal operator, without the harmonic extension in an extra space dimension as in the Caffarelli–Silvestre argument.

The organization of the chapter is as follows. Section 11.2 is devoted to the development of some important tools for the fractional Laplacian: the maximum principle (L. Silvestre [287]) and the Pohozaev type inequality (X. Ros-Oton and J. Serra [274]). In Section 11.3 we deal with the existence of at least one solution for the whole range $1 < p < p(\lambda, s)$, where $p(\lambda, s)$ is the threshold for the existence of a positive radial weak solution of the equation in the whole $\mathbb{R}^N$ (see (11.2.1)). Proceeding by a monotonicity argument, if $1 < p \le 2_s^* - 1$ we will reach a finite energy solution, while if $2_s^* - 1 < p < p(\lambda, s)$ we find a weak solution. In Section 11.4.1, we prove the existence of a second solution in the subcritical case, that is, for $1 < p < 2_s^* - 1$, taking advantage of the corresponding variational framework. For μ small, we find two solutions using first a minimization procedure and later applying the mountain pass theorem in [37]. For μ large, we will need to prove that the solution obtained in Section 2 is a local minimum, using an argument due to Alama (see [32]). This generalization is not immediate because, due to the nonlocal behavior of our operator, the bounded support of the test functions is not preserved.

In Section 11.5 we will study the nonexistence of solution for the case $p \ge p(\lambda, s)$. In fact, using the nonlocal version of Lemma 3.2 in [87] (see Lemma 11.5.3 below), we will prove that the solutions to the truncated problems blow up for every x in Ω. Finally in Section 11.6 the problem with nonlinearities singular on the boundary is analyzed.

11.2 Preliminaries

In the first place, let us guess the candidate to be an upper bound for p. To that end, we look for a radial solution to the problem in the whole $\mathbb{R}^N$:

$$(-\Delta)^s w - \lambda \frac{w}{|x|^{2s}} = w^p.$$

If we choose $w = A|x|^{\frac{2s-N}{2}+\beta}$, with A a positive constant, we have

$$A\gamma_\beta |x|^{-2s+\frac{2s-N}{2}+\beta} - \lambda A|x|^{-2s+\frac{2s-N}{2}+\beta} = A^p |x|^{(\frac{2s-N}{2}+\beta)p},$$

where

$$\gamma_\beta := \frac{2^{2s}\Gamma(\frac{N+2s+2\beta}{4})\Gamma(\frac{N+2s-2\beta}{4})}{\Gamma(\frac{N-2s-2\beta}{4})\Gamma(\frac{N-2s+2\beta}{4})}$$

was calculated in Chapter 9. Hence, in order to have homogeneity, we need

$$\frac{2s-N}{2} + \beta = \frac{-2s}{p-1} \quad \text{and, therefore, the equation becomes } \gamma_\beta - \lambda = A^{p-1}.$$

Since $A > 0$, we need $\gamma_\beta - \lambda > 0$. Note that the map

$$\gamma : \left[0, \frac{N-2s}{2}\right) \mapsto (0, \Lambda_{N,2s}], \quad \beta \mapsto \gamma_\beta$$

by Lemma 9.5.1 is decreasing (see also [167]). Therefore there exists a unique α_λ such that $\gamma_{\alpha_\lambda} = \lambda$.

Thus $\gamma_\beta - \lambda > 0$ is equivalent to $\alpha_\lambda > \beta$, that is, $\alpha_\lambda > \frac{-2s}{p-1} + \frac{N-2s}{2}$ and, consequently,

$$p < \frac{N + 2s - 2\alpha_\lambda}{N - 2s - 2\alpha_\lambda} := p(\lambda, s). \tag{11.2.1}$$

Therefore, for $p < p(\lambda, s)$ we will be able to construct a radial supersolution for the Dirichlet problem (11.1.1), just modifying the w found above. Hence this bound for p appears as the threshold for the existence also for the Dirichlet problem. In other words, we are guessing that $p(\lambda, s)$ is the threshold to have the solvability of problem (11.1.1). This chapter is devoted to prove that this is true.

As usual, we start by detailing the meaning of solution that will be considered in this context. Observe that if $p \le 2^* - 1$ the variational setting is admissible, that is, we can consider variational solutions.

In order to describe correctly the energy formulation of problem (11.1.1) and taking into account that we are looking for positive solutions, in what follows we consider the following Dirichlet problem:

$$(P_\mu^+) = \begin{cases} (-\Delta)^s u - \lambda \frac{u_+}{|x|^{2s}} = \mu u_+^q + u_+^p & \text{in } \Omega, \\ u = 0 & \text{in } \mathbb{R}^N \setminus \Omega. \end{cases} \tag{11.2.2}$$

Definition 11.2.1. We say that $u \in H_0^s(\Omega)$ is an *energy* solution of (11.2.2) if, for every $\varphi \in H_0^s(\Omega)$, the following conditions hold:

$$\int_\Omega u_+^p \varphi \, dx < \infty \tag{11.2.3}$$

and

$$\begin{aligned} a_{N,s} \int_Q \frac{(u(x) - u(y))(\varphi(x) - \varphi(y))}{|x - y|^{N+2s}} \, dx \, dy - \lambda \int_\Omega \frac{u_+ \varphi}{|x|^{2s}} \, dx \\ = \int_\Omega u_+^p \varphi \, dx + \mu \int_\Omega u_+^q \varphi \, dx. \end{aligned} \tag{11.2.4}$$

If in (11.2.4) we replace equality with $\ge$ (respectively, $\le$) and $u \ge 0$ ($\le$) in $\mathbb{R}^N \setminus \Omega$, we will say that u is a supersolution (subsolution) of (11.2.2).

As we said, problem (11.2.2) is variational in nature if $0 < p \leq 2_s^* - 1$, i. e., solutions in the sense of Definition 11.2.1 correspond to critical points of the functional $\mathcal{J}_\mu : H_0^s(\Omega) \to \mathbb{R}$ given by

$$\mathcal{J}_\mu(u) := \frac{1}{2}\left(a_{N,s}\int\limits_Q \frac{|u(x)-u(y)|^2}{|x-y|^{N+2s}}\,dx\,dy - \lambda\int\limits_\Omega \frac{(u_+)^2}{|x|^{2s}}\,dx\right) - \frac{1}{p+1}\int\limits_\Omega u_+^{p+1}\,dx - \frac{\mu}{q+1}\int\limits_\Omega u_+^{q+1}\,dx.$$

That is, problem (11.2.2) represents the Euler–Lagrange equation associated to the functional. Observe that, since $p \leq 2_s^* - 1$, we have $u_+^p \in L^{\frac{2N}{N+2s}}(\Omega)$. In particular, (11.2.3) is satisfied. Hence, if $\mu = 0$ and $p < 2_s^* - 1$ it is possible to find a variational solution using the classical mountain pass lemma introduced by A. Ambrosetti and P. Rabinowitz in [37] (see Section 11.4.1).

To prove that the solutions to problem (11.2.2) (and to (11.1.1)) are positive, we will need a maximum principle for the fractional Laplacian. Section 11.2.2 is dedicated to presenting the interesting and beautiful proof by L. Silvestre [287, Proposition 2.17].

In the complementary interval, $p \geq 2_s^* - 1$, with $\mu = 0$ and Ω being a star-shaped domain, one can prove that the only solution in $H_0^s(\Omega)$ is the trivial one. This result follows by an argument of local Pohozaev type (see [153, Corollary 1.3]) by using the extension of Caffarelli–Silvestre. However, we prefer to use the direct Pohozaev identity obtained by X. Ros-Oton and J. Serra in [274], which will be sketchy explained in Section 11.2.3 .

It is important to remark that in the local case, despite the result by Pohozaev, the deep ideas by Brezis and Nirenberg in [94] show that a subcritical semilinear perturbation removes the obstruction for the existence if $p = 2_s^* - 1$. This local result motivates the term u^q, $q < 1$, in our work in the fractional case.

In what follows we will omit the term energy when we refer to solutions that satisfy Definition 11.2.1.

If $p > 2_s^* - 1$, we are concerned with a supercritical problem and there is no possible variational formulation. In this supercritical context we give the following definition.

Definition 11.2.2. We say that $u \in L^1(\Omega)$ is a *weak* solution of (P_μ) if $u \geq 0$ a. e., $u = 0$ in $\mathbb{R}^N \setminus \Omega$, and it satisfies

$$\int\limits_\Omega \left(\lambda\frac{u}{|x|^{2s}} + u^p + \mu u^q\right)\delta^s\,dx < \infty \tag{11.2.5}$$

and

$$\int\limits_{\mathbb{R}^N} u(-\Delta)^s\varphi\,dx = \int\limits_\Omega \left(\lambda\frac{u}{|x|^{2s}} + u^p + \mu u^q\right)\varphi\,dx,$$

for all $\varphi \in C^{2s+\beta}(\Omega) \cap C^s(\overline{\Omega})$, $\beta > 0$, with $\varphi = 0$ in $\mathbb{R}^N \setminus \Omega$ and $\delta(x) := \text{dist}(x, \partial\Omega)$.

Note that the right hand side makes sense because, since $\varphi \in C^s(\overline{\Omega})$, it follows that $|\varphi(x)| \leq C\delta^s(x)$. See also [287, Proposition 2.4].

It is interesting to remark that for the existence results, the size of the biggest exponent, the smallest power q (that introduces an appropriate behavior of the nonlinearity that allows the use of monotonicity arguments) and the size of λ are relevant. Also, as an application of the weak setting, we will study a semilinear problem that is singular at the boundary. More precisely, we will consider the problem

$$\begin{cases} (-\Delta)^s u - \lambda \frac{u}{|x|^{2s}} = \frac{h(x)}{u^\sigma} & \text{in } \Omega, \\ u > 0 & \text{in } \Omega, \\ u = 0 & \text{in } \mathbb{R}^N \setminus \Omega. \end{cases} \tag{11.2.6}$$

The local case ($s = 1$) with $\lambda = 0$ was studied in [79]. Here the authors proved that for all $h \in L^1(\Omega)$, there exists at least one distributional solution. Regularity is obtained according to the regularity of h and the value of σ. The fractional case when $\lambda = 0$ was studied in [45] (see also the recent paper [163] for some extensions of [46]).

11.2.1 A convergence tool by Brezis–Lieb

For completeness we include the following result by H. Brezis and E. Lieb [91].

Theorem 11.2.3. *Assume $1 \leq p < \infty$. Let $\{f_n\}_{n\in\mathbb{N}} \subset L^p(\Omega)$ be a sequence such that:*
(a) *$f_n(x) \to f(x)$ almost everywhere as $n \to \infty$;*
(b) *$\|f_n\| \leq M < \infty$.*

Then $f \in L^p(\Omega)$ and

$$\lim_{n\to\infty} \int_\Omega (|f_n|^p - |f_n - f|^p)\,dx = \int_\Omega |f|^p\,dx.$$

Moreover, if $\|f_n\|_p \to \|f\|_p$ as $n \to \infty$, then

$$\|f_n - f\|_p \to 0 \quad \text{as } n \to \infty.$$

Proof. Note that Fatou's lemma gives

$$\int_\Omega |f|^p\,dx \leq \liminf_{n\to\infty} \int_\Omega |f_n|^p\,dx \leq M.$$

Hence, $f \in L^p(\Omega)$. Fixing $p \in [1, \infty)$ we have

$$\lim_{t\to\infty} \frac{|t+1|^p - |t|^p - 1}{|t|^p} = 0,$$

and hence for all $\epsilon > 0$ there exists $C_\epsilon > 0$ such that

$$|t+1|^p - |t|^p - 1 \le \epsilon |t|^p + C_\epsilon, \quad \forall t \in \mathbb{R}.$$

Therefore by an elementary argument, for all $\epsilon > 0$ there exists $C_\epsilon > 0$ such that

$$\big||a+b|^p - |a|^p - |b|^p\big| \le \epsilon |a|^p + C_\epsilon |b|^p, \quad \text{for all } a, b \in \mathbb{R}.$$

We use the previous inequality with $a = f_n - f$ and $b = f$. Then, for $\epsilon > 0$ and the corresponding C_ϵ,

$$\phi_n(x) =: \big(\big||f_n|^p - |f|^p - |f_n - f|^p\big| - \epsilon |f_n - f|^p\big) \le C_\epsilon |f|^p.$$

By hypothesis (a), $\phi_n(x) \to 0$ almost everywhere as $n \to \infty$ and $0 \le \phi_n \le C_\epsilon |f|^p$. Hence, by the dominated convergence theorem, $\lim_{n\to\infty} \int_\Omega \phi_n = 0$. Since

$$\begin{aligned}
&\big||f_n|^p - |f|^p - |f_n - f|^p\big| \le \phi_n + \epsilon |f_n - f|^p, \\
&\limsup_{n\to\infty} \int_\Omega \big||f_n|^p - |f|^p - |f_n - f|^p\big|\, dx \le \epsilon \int_\Omega |f_n - f|^p\, dx \le \epsilon (2M)^p, \quad \text{for all } \epsilon > 0,
\end{aligned}$$

which implies the result. □

11.2.2 Maximum principle for the fractional Laplacian

We follow in this section the ideas in [287].

Consider the fundamental solution of the s-fractional Laplacian defined in (9.3.10),

$$S_{N,s}(x) = \frac{c_{N,2s}}{|x|^{N-2s}},$$

where $c_{N,2s}$ is defined in (9.3.9).

Consider a paraboloid $P_{N,s}(x)$ tangent from below to $S_{N,s}$ on $|x| = 1$ and define

$$\Psi(x) = \begin{cases} P_{N,s}(x) & \text{if } |x| \le 1 \\ S_{N,s}(x) & \text{if } |x| \ge 1. \end{cases}$$

Hence $\Psi \in C^{1,1}(\mathbb{R}^N)$. For $\lambda > 0$ define

$$\Psi_\lambda(x) = \frac{1}{\lambda^{N-2s}} \Psi\Big(\frac{x}{\lambda}\Big) \in C^{1,1}(\mathbb{R}^N).$$

By Proposition 8.2.11, $(-\Delta)^s \Psi$ is a continuous function. Moreover we have the following properties of Ψ:

1. We have $(-\Delta)^s\Psi(x) \geq 0$. By the regularity of Ψ we can write

$$(-\Delta)^s\Psi(x_0) = C_{N,s}\, p.v. \int_{\mathbb{R}^N} \frac{\Psi(x_0) - \Psi(y)}{|x_0 - y|^{N+2s}}\, dy.$$

- If $|x_0| > 1$, $\Psi(x_0) = S_{N,s}(x_0)$ and for $y \neq x_0$, $\Psi(y) \leq S_{N,s}(y)$. Then

$$(-\Delta)^s\Psi(x_0) = C_{N,s} \int_{\mathbb{R}^N} \frac{\Psi(x_0) - \Psi(y)}{|x_0 - y|^{N+2s}}\, dy \geq C_{N,s} \int_{\mathbb{R}^N} \frac{S_{N,s}(x_0) - S_{N,s}(y)}{|x_0 - y|^{N+2s}}\, dy = 0.$$

- If $0 < |x| \leq 1$ we choose a point x_1 in such a way that $S_{N,s}(x_0 - x_1) < \Psi(x_0)$ and $\nabla S_{N,s}(x_0 - x_1) = \nabla\Psi(x_0)$. Then we choose $\delta > 0$ such that $S_{N,s}(x - x_1) + \delta$ touches $\Psi(x)$ from above at x_0. Therefore

$$\begin{aligned}(-\Delta)^s\Psi(x_0) &= C_{N,s} \int_{\mathbb{R}^N} \frac{\Psi(x_0) - \Psi(y)}{|x_0 - y|^{N+2s}}\, dy \\ &\geq C_{N,s} \int_{\mathbb{R}^N} \frac{S_{N,s}(x_0 - x_1) - S_{N,s}(y - x_1)}{|x_0 - y|^{N+2s}}\, dy = 0.\end{aligned}$$

- If $x_0 = 0$ then Ψ attains its maximum at x_0, that is, $\Psi(x_0) - \Psi(y) \geq 0$ that implies $(-\Delta)^s\Psi(x_0) \geq 0$.

2. We have $(-\Delta)^s\Psi(x) \in L^1(\mathbb{R}^N)$. Moreover

$$\|(-\Delta)^s\Psi\|_1 = 1.$$

We compute the L^1-norm of $(-\Delta)^s\Psi(x)$. Consider a cutoff function $\eta \in C_0^\infty(\mathbb{R}^N)$ such that

$$\eta(x) \leq 1 \quad \text{for all } x \in \mathbb{R}^N,\ \operatorname{supp}(\eta) \subset B_2 \text{ and } \eta(x) = 1,\ x \in B_1.$$

Define $\eta_R(x) = \eta(\frac{x}{R})$.
Since $\Psi(x) - S_{N,s}(x) \in L^1(\mathbb{R}^N)$ is compact supported and $(-\Delta)^s\eta_R(x) \to 0$ as $R \to \infty$ on compact sets, we find

$$\begin{aligned}\int_{\mathbb{R}^N} (-\Delta)^s\Psi(x)\, dx - 1 &= \lim_{R\to\infty} \int_{\mathbb{R}^N} ((-\Delta)^s\Psi(x) - (-\Delta)^s S_{N,s}(x))\eta_R(x)\, dx \\ &= \lim_{R\to\infty} \int_{\mathbb{R}^N} (\Psi(x) - S_{N,s}(x))(-\Delta)^s\eta_R(x)\, dx = 0.\end{aligned}$$

Define

$$\gamma_\lambda(x) = (-\Delta)^s\Psi_\lambda(x).$$

By the previous remarks γ_λ is a compact supported function and

$$\int_{\mathbb{R}^N} \gamma_1(x)\,dx = 1.$$

On the other hand by a simple scaling we find that

$$\gamma_\lambda(x) = \lambda^{-N}\gamma\left(\frac{x}{\lambda}\right).$$

Thus,

$$\{\gamma_\lambda(x)\}_{\lambda>0} \quad \text{is an approximation to the identity.}$$

In particular if $u \in \mathcal{L}^s$, where we recall

$$\mathcal{L}^s := \left\{u : \mathbb{R}^N \to \mathbb{R} \text{ measurable: } \int_{\mathbb{R}^N} \frac{|u(x)|}{(1+|x|^{N+2s})}\,dx < \infty\right\}$$

is a Banach space endowed with the norm

$$\|u\|_{\mathcal{L}^s} := \int_{\mathbb{R}^N} \frac{|u(x)|}{(1+|x|^{N+2s})}\,dx,$$

we have

$$\gamma_\lambda * u(x) \to u(x), \quad \lambda \to 0, \text{ a. e.}$$

The next result plays the same role as the mean inequality in the classical case.

Proposition 11.2.4. *Let $u \in \mathcal{L}^s$ be such that $(-\Delta)^s u(x) \geq 0$ in an open set Ω. Then u is lower semicontinuous in Ω and*

$$u(x_0) \geq \int_{\mathbb{R}^N} u(x)\gamma_\lambda(x - x_0)\,dx, \tag{11.2.7}$$

for any $x_0 \in \Omega$ and $\lambda \leq \operatorname{dist}(x_0, \partial\Omega)$.

Proof. Call $r = \operatorname{dist}(x_0, \partial\Omega)$ and take $r > \lambda_1 > \lambda_2 > 0$. Then $\Psi_{\lambda_2} - \Psi_{\lambda_1} \geq 0$ is a $\mathcal{C}^{1,1}$-function supported in B_r. Since $(-\Delta)^s u(x) \geq 0$ in $B_r(x_0)$ we get

$$0 \leq \langle (-\Delta)^s u(x), \Psi_{\lambda_2}(x - x_0) - \Psi_{\lambda_1}(x - x_0)\rangle = C_{N,s}\int_{\mathbb{R}^N} u(x)(\gamma_{\lambda_2}(x - x - 0) - \gamma_{\lambda_1}(x - x_0))\,dx,$$

and hence

$$\gamma_{\lambda_2} * u(x_0) \geq \gamma_{\lambda_1} * u(x_0).$$

For each $\lambda > 0$, $\gamma_\lambda * u$ is a continuous function and is an increasing family in λ. Given that the family γ_λ is an approximation of the identity u is the increasing limit of continuous functions and hence is lower semicontinuous up to a set of zero measure. Taking limits for $\lambda_2 \to 0$,

$$u(x_0) \geq \int_{\mathbb{R}^N} u(x)\gamma_\lambda(x - x_0)\,dx. \qquad \square$$

Proposition 11.2.4 is the main tool to prove the following maximum principle. Note that the setting in which we find the result is very general.

Proposition 11.2.5. *Assume that Ω is a bounded open set in $\mathbb{R}^N$. Let u be lower semicontinuous in $\overline{\Omega}$ and such that*
(i) $(-\Delta)^s u \geq 0$ *in* Ω;
(ii) $u \geq 0$ *in* $\mathbb{R}^N \setminus \Omega$.

Then $u \geq 0$ in $\mathbb{R}^N$.

Moreover if $u(x) = 0$ in a point $x \in \Omega$, then $u = 0$ in $\mathbb{R}^N$.

Proof. Since, by hypothesis, $u \geq 0$ in $\mathbb{R}^N$, if by contradiction we suppose that there exists a point $y \in \mathbb{R}^N$ such that $u(y) < 0$, then $y \in \Omega$. By the hypothesis of lower semicontinuity u attains its minimum in $x_0 \in \overline{\Omega}$. If we assume that $x_0 \in \Omega$ and $u(x_0) < 0$, by Proposition 11.2.4 there is a $\lambda > 0$ such that

$$u(x_0) \geq \int_{\mathbb{R}^N} u(x)\gamma_\lambda(x - x_0)\,dx.$$

Since γ_λ is positive and has integral equal to 1, we have

$$0 \geq \int_{\mathbb{R}^N} (u(x) - u(x_0))\gamma_\lambda(x - x_0)\,dx,$$

which is a contradiction with the fact that the second member is positive.

If there exists $x_0 \in \Omega$ such that $u(x_0) = 0$, we find

$$0 \geq \int_{\mathbb{R}^N} u(x)\gamma_\lambda(x - x_0)\,dx,$$

which, jointly with the nonnegativity of u, implies

$$0 = \int_{\mathbb{R}^N} u(x)\gamma_\lambda(x - x_0)\,dx.$$

Since $\gamma_\lambda(y) > 0$ we conclude that $u = 0$. $\square$

11.2.3 The Pohozaev identity for the fractional Laplacian

The classical paper by S. Pohozaev [261] studies solutions u to the problem

$$\begin{cases} -\Delta u = f(u) & \text{in } \Omega, \\ u = 0 & \text{on } \partial\Omega, \end{cases}$$

where f is a continuous function and Ω is a smooth star-shaped bounded domain in $\mathbb{R}^N$. Assuming a suitable regularity of the solution u and by using the multiplier $\langle x, \nabla u(x)\rangle$, i. e.,

$$\int_{\Omega} \langle x, \nabla u(x)\rangle(-\Delta u(x))\,dx = \int_{\Omega} \langle x, \nabla u(x)\rangle f(u(x))\,dx,$$

an integration by parts gives

$$\int_{\Omega} \langle x, \nabla u(x)\rangle(-\Delta u(x))\,dx = \frac{2-N}{2}\int_{\Omega} |\nabla u(x)|^2\,dx - \frac{1}{2}\int_{\partial\Omega} |\nabla u(x)|^2 \langle x, \nu\rangle\,d\sigma$$

and

$$\int_{\Omega} \langle x, \nabla u(x)\rangle f(u(x))\,dx = N\int_{\Omega} F(u(x))\,dx, \quad \text{where } F(u) = \int_0^u f(s)\,ds.$$

On the other hand, multiplying the equation by u and integrating by parts we find that

$$\int_{\Omega} |\nabla u|^2\,dx = \int_{\Omega} uf(u)\,dx,$$

which brings the final form of the Pohozaev identity

$$\frac{N-2}{2}\int_{\Omega} uf(u)\,dx - N\int_{\Omega} F(u)\,dx = \frac{1}{2}\int_{\partial\Omega} |\nabla u(x)|^2 \langle x, \nu\rangle\,d\sigma.$$

If we try to find similar identities for solutions to the problem

$$(P_s)\begin{cases} (-\Delta)^s u = f(u) & \text{in } \Omega, \\ u = 0 & \text{in } \mathbb{R}^N \setminus \Omega, \end{cases} \tag{11.2.8}$$

by using the multiplier of Pohozaev, the first difficulty is to calculate the integral

$$\int_{\Omega} \langle x, \nabla u(x)\rangle(-\Delta)^s u(x)\,dx$$

and find the local behavior of the solution on the boundary $\partial\Omega$. The solution to this problem is a deep result found by X. Ros-Oton and J. Serra in [274]. We will present an approach to this in a sketchy way, with the recommendation for the reader to read the original paper for more details.

Regularity up to the boundary

There are several key points in the research of the fractional Pohozaev identity. The first one is the paper [273], in which Ros-Oton and Serra give the precise behavior of a solution on the boundary.

As usual, we denote $\delta(x) = d(x, \partial\Omega)$.

Let $\beta > 0$ be so that $\beta \notin \mathbb{N}$. Denote by $[v]_{C^{\beta}(O)}$ the classical seminorm of Hölder spaces in an open set $\mathcal{O}$. That is, if k is the integral part of β and $\beta' = \beta - k$, then

$$[v]_{C^{\beta}(\mathcal{O})} = [v]_{C^{k,\beta'}(\mathcal{O})} = \sup_{x,y\in\mathcal{O}, x\neq y} \frac{|D^k u(x) - D^k u(y)|}{|x-y|^{\beta'}}.$$

We are in a position to formulate the regularity result.

Assume in problem (11.2.8) *that Ω is a bounded and $C^{1,1}$-domain and f is a locally Lipschitz function. If u is a bounded solution to* (11.2.8) *and $g = f(u)$, then $u \in C^s(\mathbb{R}^N)$ and*

$$\|u\|_{C^s(\mathbb{R}^N)} \leq C\|g\|_{L^\infty(\Omega)},$$

where $C = C(\Omega, s)$.

Moreover, there exists $\alpha = \alpha(\Omega, s) > 0$, $\alpha < \min\{s, (1-s)\}$, such that $\frac{u}{\delta^s} \in C^\alpha(\overline{\Omega})$ and

$$\left\|\frac{u}{\delta^s}\right\|_{C^s(\mathbb{R}^N)} \leq C\|g\|_{L^\infty(\Omega)}, \quad C = C(\Omega, s).$$

The last inequality tells the precise behavior of u at the boundary and gives some kind of Hopf lemma for the fractional Laplacian.

The following extra interior regularity of u, a bounded solution to (11.2.8), *is also obtained:*

(i) If $\beta \in [s, 1+2s)$, $u \in C^\beta(\Omega)$ and

$$[u]_{C^\beta(\{x\in\Omega|\delta(x)>\rho\})} \leq C\rho^{s-\beta}.$$

(ii) *If $\beta \in [\alpha, s+\alpha]$,*

$$\left[\frac{u}{\delta^s}\right]_{C^\beta(\{x\in\Omega|\delta(x)>\rho\})} \leq C\rho^{\alpha-\beta}.$$

The constant C depends on $\beta, f, \|u\|_\infty, \Omega$ and s.

The precise statement of the previous regularity results and the proofs can be seen in [273].

Key estimates to prove the fractional Pohozaev identity

Lemma 11.2.6. *Let Ω be a bounded and $C^{1,1}$-domain, and let u be a function such that $u = 0$ in $\mathbb{R}^N \setminus \Omega$ that satisfies the same regularity as the bounded solution of prob-*

lem (11.2.8). *Then, there exists a* $C^\alpha(\mathbb{R}^N)$*-extension* v *of* $\frac{u}{\delta^s}|_\Omega$ *such that*

$$(-\Delta)^{\frac{s}{2}}u(x) = \frac{\Gamma(1+s)\sin(\frac{s\pi}{2})}{\pi}\left[\log_- \delta(x) + \frac{\pi}{\tan(\frac{s\pi}{2})}\chi_\Omega(x)\right]v(x) + h(x),$$

where $h \in C^\alpha(\mathbb{R}^N)$. *Moreover, if* $\beta \in (0, 1+s)$,

$$[(-\Delta)^{\frac{s}{2}}u]_{C^\beta(\{x\in\mathbb{R}^N|\delta(x)>\rho\})} \leq C\rho^{-\beta} \quad \text{for all } \rho \in (0.1),$$

with C *independent of* ρ.

We also need the following technical result.

Lemma 11.2.7. *Let* $a, b \in \mathbb{R}$ *and*

$$\phi(t) = a\log_- |t-1| + b\chi_{[0,1]}(t) + h(t),$$

where $\log_- |\tau| = \min\{0, \log|\tau|\}$ *and* h *is a real function satisfying that there exist* $\alpha, \gamma \in (0,1)$ *and* $C_0 > 0$ *such that:*
(i) $\|h\|_{C^\alpha[0,\infty)} \leq C_0$;
(ii) *for all* $\beta \in [\gamma, \gamma+1]$, $\|h\|_{C^\beta(0,1-\rho\cup(1+\rho,2))} \leq C_0\rho^{-\beta}$, $\rho \in (0,1)$;
(iii) $|h'(t) \leq C_0 t^{-2-\gamma}$ *and* $|h''(t) \leq C_0 t^{-3-\gamma}$, *for all* $t > 2$.

Then

$$\frac{d}{dt}\bigg|_{t=1^+} \int_0^\infty \phi(\lambda t)\phi\left(\frac{t}{\lambda}\right) dt = a^2\pi^2 + b^2.$$

See the proofs of the two previous lemmas in [274].

The fractional Pohozaev identity by Ros-Oton and Serra in strictly star-shaped domains

Next we formulate the fractional Pohozaev in star-shaped domains with respect to the origin. The reader can also find in the paper [274] the proof of the identity in general $C^{1,1}$-domains.

Theorem 11.2.8. *Let* $\Omega \subset \mathbb{R}^N$ *be a bounded star-shaped and* $C^{1,1}$*-domain,* $f : \mathbb{R} \to \mathbb{R}$ *a locally Lipschitz and* u *a bounded solution to problem* (11.2.8). *Then* $\frac{u}{\delta^s}$ *has a continuous extension to* $\overline{\Omega}$, *that is,* $C^\alpha(\overline{\Omega})$, *for some* $\alpha \in (0,1)$. *Moreover, the following identity holds:*

$$(2s-N)\int_\Omega uf(u)\,dx + 2N\int_\Omega F(u) = \Gamma(1+s)^2 \int_{\partial\Omega} \left(\frac{u}{\delta^s}\right)^2 \langle x, \nu\rangle\, d\sigma,$$

where ν *is the unit outward normal to* $\partial\Omega$ *at* x, $\delta(x)$ *is the distance to the boundary,* Γ *is the Gamma function and* $F(u) = \int_0^u f(\tau)\,d\tau$.

Proof. The basic idea is to use the same Pohozaev multiplier as in the local case, $\langle x, \nabla u(x)\rangle$. In the fractional case the integration by parts must be carefully done. This integration by parts will be done step by step.

STEP 1. *Define* $u_\lambda(x) = u(\lambda x)$. *Then*

$$\int_\Omega \langle x, \nabla u(x)\rangle(-\Delta)^s u(x)\,dx = \frac{d}{d\lambda}\bigg|_{\lambda=1^+} \int_\Omega u_\lambda(x)(-\Delta)^s u(x)\,dx. \tag{11.2.9}$$

By the regularity results, in particular we know that $(-\Delta)^s u(x)$ is pointwise defined and bounded. Call $w(x) = (-\Delta)^s u(x)$ and note that if $\lambda > 1$, then $\frac{1}{\lambda}\Omega \subset \Omega$, and hence $\operatorname{supp} u_\lambda = \frac{1}{\lambda}\Omega$. Then

$$\frac{d}{d\lambda}\bigg|_{\lambda=1^+} \int_\Omega u_\lambda(x)(-\Delta)^s u(x)\,dx = \lim_{\lambda\downarrow 1} \int_\Omega \frac{u(\lambda x) - u(x)}{\lambda - 1} w(x)\,dx.$$

Making the change of variables $y = \lambda x$, we obtain

$$\begin{aligned}\lim_{\lambda\downarrow 1} \int_\Omega \frac{u(\lambda x) - u(x)}{\lambda - 1} w(x)\,dx &= \lim_{\lambda\downarrow 1} \lambda^{-N} \int_{\lambda\Omega} \frac{u(y) - u(\frac{y}{\lambda})}{\lambda - 1} w\Big(\frac{y}{\lambda}\Big)\,dy \\ &= \lim_{\lambda\downarrow 1} \int_{\lambda\Omega} \frac{u(y) - u(\frac{y}{\lambda})}{\lambda - 1} w\Big(\frac{y}{\lambda}\Big)\,dy + \lim_{\lambda\downarrow 1} \int_{\lambda\Omega\setminus\Omega} \frac{-u(\frac{y}{\lambda})}{\lambda - 1} w\Big(\frac{y}{\lambda}\Big)\,dy.\end{aligned}$$

Consider $\beta = 1$ in the regularity result. Then $|\nabla u(x)| \le \delta(x)^{s-1} \le \lambda^{1-s}\delta(x)^{s-1}$ for all x in the segment joining y and $\frac{y}{\lambda}$. Observe that $\delta(x)^{s-1}$ is integrable. Therefore we can apply the dominated convergence theorem to find

$$\lim_{\lambda\downarrow 1} \int_{\lambda\Omega} \frac{u(y) - u(\frac{y}{\lambda})}{\lambda - 1} w\Big(\frac{y}{\lambda}\Big)\,dy = \int_\Omega \langle x, \nabla u(x)\rangle(-\Delta)^s u(x)\,dx.$$

To finish the first step we have to prove that

$$\lim_{\lambda\downarrow 1} \int_{\lambda\Omega\setminus\Omega} \frac{-u(\frac{y}{\lambda})}{\lambda - 1} w\Big(\frac{y}{\lambda}\Big)\,dy = 0.$$

But this follows immediately from the facts that $|\lambda\Omega\setminus\Omega| \le C(\lambda-1)$ and since $u \in C^s(\mathbb{R}^N)$ and $u = 0$ in $\mathbb{R}^N \setminus \Omega$, we have $\|u\|_{L^\infty(\lambda\Omega\setminus\Omega)} \to 0$ as $\lambda \to 1^+$.

STEP 2. *Under the general hypotheses the following identity holds:*

$$\int_\Omega \langle x, \nabla u(x)\rangle(-\Delta)^s u(x)\,dx = \frac{(2s - N)}{2} \int_\Omega u(-\Delta)^s u\,dx - \frac{\Gamma(1+s)^2}{2} \int_{\partial\Omega} \Big(\frac{u}{\delta^s}\Big)^2 \langle x, \nu\rangle\,d\sigma.$$

Note that with the change of variables $y = \sqrt{\lambda}x$, we have

$$\int_{\Omega} u_\lambda(x)(-\Delta)^s u(x)\,dx = \int_{\mathbb{R}^N} (-\Delta)^{\frac{s}{2}} u_\lambda(x)(-\Delta)^{\frac{s}{2}} u(x)\,dx = \lambda^{\frac{2s-N}{2}} \int_{\mathbb{R}^N} w_{\sqrt{\lambda}} w_{\frac{1}{\sqrt{\lambda}}}\,dx.$$

Therefore, according to the first step

$$\begin{aligned}\int_{\Omega} \langle x, \nabla u(x)\rangle(-\Delta)^s u(x)\,dx &= \frac{d}{d\lambda}\bigg|_{\lambda=1^+} \left\{\lambda^{\frac{2s-N}{2}} \int_{\mathbb{R}^N} w_{\sqrt{\lambda}} w_{\frac{1}{\sqrt{\lambda}}}\,dx\right\}\\ &= \frac{(2s-N)}{2} \int_{\mathbb{R}^N} w^2\,dx + \frac{d}{d\lambda}\bigg|_{\lambda=1^+} I_{\sqrt{\lambda}}\\ &= \frac{(2s-N)}{2} \int_{\mathbb{R}^N} u(-\Delta)^s u\,dx + \frac{1}{2}\frac{d}{d\lambda}\bigg|_{\lambda=1^+} I_\lambda,\end{aligned}$$

where

$$I_\lambda = \int_{\mathbb{R}^N} w_{\sqrt{\lambda}} w_{\frac{1}{\sqrt{\lambda}}}\,dx. \tag{11.2.10}$$

So, to finish step 2 we need to evaluate the derivative of I_λ at $t = 1$ on the right.

Since Ω is a star-shaped and $C^{1,1}$-domain we can define the radial projection of $\mathbb{S}^{N-1}$ into $\partial\Omega$. More precisely, if $x \in \partial\Omega$, then $\theta = \frac{x}{|x|}$ and since Ω is star-shaped with respect to the origin there exists a unique $\rho(\theta) > 0$ such that $\rho(\theta)\theta \in \partial\Omega$. Then we define the radial projection by

$$\begin{aligned} f : \mathbb{S}^{N-1} &\to \partial\Omega,\\ \theta = \tfrac{x}{|x|} &\to x = f(\theta) = r(\theta)\theta.\end{aligned}$$

Writing the integral (11.2.10) in spherical coordinates and performing the change of variables $t = \frac{r}{r(\theta}$ we get

$$\begin{aligned}\frac{d}{d\lambda}\bigg|_{\lambda=1^+} I_\lambda &= \frac{d}{d\lambda}\bigg|_{\lambda=1^+} \int_{\mathbb{S}^{N-1}} d\theta \int_0^\infty r^{N-1} w(\lambda r\theta) w\Big(\frac{r}{\lambda}\theta\Big) dr\\ &= \frac{d}{d\lambda}\bigg|_{\lambda=1^+} \int_{\mathbb{S}^{N-1}} \int_0^\infty (r(\theta)t)^{N-1} w(\lambda r(\theta)t\theta) w\Big(\frac{r(\theta)t}{\lambda}\theta\Big)\,dt\\ &= \frac{d}{d\lambda}\bigg|_{\lambda=1^+} \int_{\mathbb{S}^{N-1}} \Big\langle \frac{x}{|x|}, \nu\Big\rangle\,d\sigma(x) \int_0^\infty t^{N-1} w(\lambda tx) w\Big(\frac{tx}{\lambda}\Big)\,dt,\end{aligned}$$

where the identity

$$r(\theta)^{N-1}\,d\theta = \frac{1}{r(\theta)}\Big\langle \frac{x}{|x|}, \nu\Big\rangle\,d\sigma(x)$$

follows from the change of variables associated to the radial projection on $\partial\Omega$. For $x_0 \in \partial\Omega$ fixed and recalling the meaning of the function w, we define

$$\phi(t) = t^{\frac{N-1}{2}}(-\Delta)^{\frac{s}{2}}u(tx_0).$$

By Lemma 11.2.6 we find

$$\phi(t) = c_1[\log_- \delta(tx_0) + c_2\chi_{[0,1]}]v(tx_0) + h_0(t),$$

where v is a $C^\alpha(\mathbb{R}^N)$-extension of $\frac{u}{\delta^s}|_\Omega$ and h_0 is a $C^\alpha([0,\infty))$-function.

By the regularity and the geometry of the domain Ω the authors in [274] are able to prove that the function $\phi(t)$ satisfies the hypotheses of Lemma 11.2.7. Therefore

$$\frac{d}{d\lambda}\bigg|_{\lambda=1^+} \int_0^\infty \phi(\lambda t)\phi\Big(\frac{t}{\lambda}\Big)\,dt = (v(x_0))^2 c_1^2(\pi^2 + c_2^2)$$

and then

$$\frac{d}{d\lambda}\bigg|_{\lambda=1^+} \int_0^\infty t^{N-1}w(\lambda t x_0)w\Big(\frac{t}{\lambda}x_0\Big)\,dt = (v(x_0))^2 c_1^2(\pi^2 + c_2^2).$$

Now, taking into account that v is a $C^\alpha(\mathbb{R}^N)$-extension of $\frac{u}{\delta^s}|_\Omega$, we find that for $x_0 \in \partial\Omega$

$$\frac{u}{\delta^s}(x_0) = \lim_{x\to x_0, x\in\Omega} \frac{u(x)}{\delta^s(x)}.$$

We conclude that

$$\begin{aligned}\frac{d}{d\lambda}\bigg|_{\lambda=1^+} I_\lambda &= \frac{d}{d\lambda}\bigg|_{\lambda=1^+} \int_{\mathbb{S}^{N-1}} \Big\langle \frac{x_0}{|x_0|}, \nu\Big\rangle\,d\sigma(x) \int_0^\infty t^{N-1}w(\lambda t x_0)w\Big(\frac{tx_0}{\lambda}\Big)\,dt \\ &= c_1^2(\pi^2 + c_2^2) \int_{\partial\Omega} \Big\langle \frac{x_0}{|x_0|}, \nu\Big\rangle \Big(\frac{u}{\delta^s}(x_0)\Big)^2\,d\sigma(x_0).\end{aligned}$$

By Lemma 11.2.6 we have

$$c_1 = \frac{\Gamma(1+s)\sin(\frac{s\pi}{2})}{\pi}, \quad \text{and} \quad c_2 = \frac{\pi}{\tan(\frac{s\pi}{2})},$$

and then

$$c_1^2(\pi^2 + c_2^2) = (\Gamma(1+s))^2,$$

which finishes step 2.

STEP 3. Consider u, a bounded solution to problem (11.2.8). Multiplying the equation by $\langle x, \nabla u(x)\rangle$ and integrating in Ω we get

$$\int_\Omega \langle x, \nabla u(x)\rangle (-\Delta)^s u(x)\,dx = \int_\Omega \langle x, \nabla u(x)\rangle f(u(x))\,dx = \int_\Omega \langle x, \nabla F(u(x))\rangle\,dx.$$

Integrating by parts, since $u = 0$ in $\mathbb{R}^N \setminus \Omega$,

$$\int_\Omega \langle x, \nabla F(u(x))\rangle\,dx = -N\int_\Omega F(u(x))\,dx.$$

Moreover multiplying the equation by u and integrating we have

$$\int_\Omega u(x)(-\Delta)^s u(x)\,dx = \int_\Omega u(x) f(u(x))\,dx.$$

Hence summarizing the results in the previous steps and the last calculation we have

$$\begin{aligned}
-N\int_\Omega F(u(x))\,dx &= \int_\Omega \langle x, \nabla u(x)\rangle (-\Delta)^s u(x)\,dx \\
&= \frac{(2s-N)}{2}\int_\Omega u(-\Delta)^s u\,dx - \frac{\Gamma(1+s)^2}{2}\int_{\partial\Omega}\left(\frac{u}{\delta^s}\right)^2 \langle x, \nu\rangle\,d\sigma \\
&= \frac{(2s-N)}{2}\int_\Omega u(x) f(u(x))\,dx - \frac{\Gamma(1+s)^2}{2}\int_{\partial\Omega}\left(\frac{u}{\delta^s}\right)^2 \langle x, \nu\rangle\,d\sigma. \qquad \square
\end{aligned}$$

Pohozaev with the Hardy potential

For the application to solutions of problems involving power nonlinearities of the type

$$(-\Delta u)^s - \lambda\frac{u}{|x|^{2s}} = f(u), \quad u \geq 0 \quad x \in \Omega, \quad u = 0 \quad x \in \mathbb{R}^N \setminus \Omega,\ 0 < \lambda < \Lambda_{N,2s}, \tag{11.2.11}$$

we need to adapt the arguments above. We will begin with a solution $u \in H_0^s(\Omega)$ that satisfies the same interior regularity as the solutions to (11.2.8) referring to $\Omega \setminus \{0\}$ and (ii) assume that the function $\frac{u}{\delta^s}$ has a continuous extension to $\overline{\Omega \setminus B_r(0)}$, that is, $\mathcal{C}^\alpha(\overline{\Omega} \setminus B_r(0))$, for some $\alpha \in (0,1)$. Since $u \in H_0^s(\Omega)$, $f(u) \in L^m(\Omega)$ with $m \geq \frac{2N}{N+2s}$, and then $uf(u) \in L^1(\Omega)$. The first difference is the term involving the Hardy potential. The second difference is that the positive solutions to problem (11.2.11) are in general unbounded. However, we have extra information about the precise behavior of the solution in $B_r(0)$ given in Theorem 10.4.1.

To handle the term involving the Hardy potential we will proceed as follows. First note that

$$\int_\Omega \langle x, \nabla u\rangle \lambda\frac{u}{|x|^{2s}}\,dx = \lim_{\rho\to 0}\int_{\Omega\setminus B_\rho(0)} \frac{\lambda}{2}\left\langle \frac{x}{|x|^{2s}} \nabla u^2\right\rangle dx.$$

This last term is calculated as follows. First we compute the integral in the perforated domain,

$$\begin{aligned}\int_{\Omega\setminus B_\rho(0)} \frac{u}{|x|^{2s}}\langle x, \nabla u\rangle &= -\frac{1}{2}\int_{\Omega\setminus B_\rho(0)} u^2 \operatorname{div}\Big(\frac{x}{|x|^{2s}}\Big) + \frac{1}{2}\int_{\partial(\Omega\setminus B_\rho(0))} \frac{u^2}{|x|^{2s}}\langle \nu, x\rangle\, d\sigma \\ &= -\frac{N-2s}{2}\int_{\Omega\setminus B_\rho(0)} \frac{u^2}{|x|^{2s}} + \frac{1}{2}\int_{\partial(\Omega\setminus B_\rho(0))} \frac{u^2}{|x|^{2s}}\langle \nu, x\rangle\, d\sigma \\ &= -\frac{N-2s}{2}\int_{\Omega\setminus B_\rho(0)} \frac{u^2}{|x|^{2s}} + \frac{1}{2}\int_{\partial B_\rho(0)} \frac{u^2}{|x|^{2s}}\langle \nu, x\rangle\, d\sigma,\end{aligned}$$

using the boundary condition. Taking into account the local behavior of u, the last integral can be estimated as

$$\frac{1}{2}\int_{\partial B_\rho(0)} \frac{u^2}{|x|^{2s}}\langle \nu, x\rangle\, d\sigma \le C\frac{\rho}{\rho^{2\gamma+2s}}\rho^{N-1},$$

and we have $N - 2\gamma - 2s > 0$. Therefore

$$\int_\Omega \langle x, \nabla u\rangle \lambda \frac{u}{|x|^{2s}}\, dx = \lim_{\rho\to 0}\int_{\Omega\setminus B_\rho(0)} \frac{\lambda}{2}\Big\langle \frac{x}{|x|^{2s}}\nabla u^2\Big\rangle\, dx = -\frac{N-2s}{2}\int_\Omega \frac{u^2}{|x|^{2s}}\, dx,$$

by the monotone convergence theorem.

Since the zero order term can be evaluated as in the case $\lambda = 0$, it remains to calculate

$$\int_\Omega \langle x, \nabla u(x)\rangle (-\Delta)^s u(x)\, dx.$$

We will proceed by approximation of the solution u by bounded functions. Call $h(x) = f(u(x)) > 0$. We consider w_0 the solution to $(-\Delta)^s w_0 = T_{\frac{1}{2}}(h(x))$ with $w_0 = 0$ in $\mathbb{R}^N \setminus \Omega$ and by recurrence we define w_k, $k = 1, 2, \ldots$, the solution to

$$\begin{cases} (-\Delta)^s w_k = \lambda \frac{w_{k-1}}{|x|^{2s}} + h_k(x) & \text{in } \Omega, \\ w_k = 0 & \text{in } \mathbb{R}^N \setminus \Omega, \end{cases} \tag{11.2.12}$$

where $h_k = T_k(h)$. According to Corollary 8.3.12, we easily find

$$w_0 \le w_1 \le \cdots \le w_{k-1} \le w_k \le \cdots \le u.$$

Moreover, testing in problem (11.2.12) with w_k we have

$$\int_\Omega (-\Delta)^s w_k w_k\,dx = \lambda \int_\Omega \frac{w_{k-1}}{|x|^{2s}} w_k\,dx + \int_\Omega h_k w_k\,dx$$
$$\leq \lambda \int_\Omega \frac{w_k^2}{|x|^{2s}}\,dx + \int_\Omega h_k u\,dx$$
$$\leq \lambda \int_\Omega \frac{u^2}{|x|^{2s}}\,dx + \int_\Omega f(u)u\,dx < \infty.$$

Then $w_k \rightharpoonup v$ weakly in $H_0^s(\Omega)$ and it is standard to see that v solves the problem

$$(-\Delta)^s v - \lambda \frac{v}{|x|^{2s}} = h(x) \text{ in } \Omega, \quad v = 0 \text{ in } \mathbb{R}^N \setminus \Omega.$$

By the uniqueness of the linear problem we conclude that $u = v$. We will prove that $w_k \to u$ strongly in $H_0^s(\Omega)$. Indeed,

$$\|(u-w_k)\|^2_{H_0^s(\Omega)} = \int_\Omega (-\Delta)^s(u-w_k)(u-w_k)\,dx = \lambda \int_\Omega \frac{u-w_{k-1}}{|x|^{2s}}(u-w_k)\,dx + \int_\Omega (h-h_k)(u-w_k)\,dx.$$

The second member converges pointwise to zero and it is uniformly bounded by the function

$$F(x) = 4\left(\lambda \frac{u^2}{|x|^{2s}} + f(u)u\right) \in L^1(\Omega),$$

and hence by the dominated convergence theorem we conclude that

$$\|(u-w_k)\|_{H_0^s(\Omega)} \to 0 \quad \text{as } k \to \infty.$$

Applying the estimate by Ros-Oton and Serra to each w_k we find

$$\int_\Omega \langle x, \nabla w_k(x)\rangle (-\Delta)^s w_k(x)\,dx = \frac{2s-N}{2}\|w_k\|^2_{H_0^s(\Omega)} - \frac{\Gamma(1+s)^2}{2}\int_{\partial\Omega} \left(\frac{w_k}{\delta^s}\right)^2 \langle x, \nu\rangle\,d\sigma$$

and passing to the limit when $k \to \infty$

$$\int_\Omega \langle x, \nabla u(x)\rangle (-\Delta)^s u(x)\,dx = \frac{2s-N}{2}\|u\|^2_{H_0^s(\Omega)} - \frac{\Gamma(1+s)^2}{2}\int_{\partial\Omega} \left(\frac{u}{\delta^s}\right)^2 \langle x, \nu\rangle\,d\sigma.$$

Multiplying equation (11.2.11) by u, integrating and doing the same calculations as in the case $\lambda = 0$, we find the Pohozaev identity with Hardy potential

$$(2s-N)\int_\Omega uf(u)\,dx + 2N\int_\Omega F(u) = \Gamma(1+s)^2 \int_{\partial\Omega} \left(\frac{u}{\delta^s}\right)^2 \langle, x, \nu\rangle\,d\sigma, \tag{11.2.13}$$

where, as above, $F(u) = \int_0^u f(t)\,dt$.

As an immediate consequence of (11.2.13) we find that problem (11.2.11) in a star-shaped domain has no positive solution if $f(u) = u^p$ with $p \geq 2_s^* - 1$.

Assume $0 < \lambda \leq \Lambda_{N,2s}$. If $f(u) = u^p$ with $0 < p < 2_s^* - 1$ we can find a positive solution to (11.2.13) by application of classical results of *critical point theory*, minimization if $q < 1$ and *mountain pass* arguments if $2_s^* - 1 > p > 1$. The case $p = 1$ corresponds to finding the eigenvalues of the coercive operator, with compact inverse, $(-\Delta)^s(\cdot) - \lambda \frac{1}{|x|^{2s}}(\cdot)$.

11.3 Existence of minimal solutions for $1 < p < p(\lambda, s)$

We will study the existence of solution to problem (11.1.1). We use the comparison results obtained in Section 8.3.3 and for the supercritical case with respect to the Sobolev exponent we will need the following version of the results for weak solutions in the sense of Definition 8.6.2. This result is the extension of [88, Lemma 1] to the nonlocal framework.

Lemma 11.3.1. *Let $f \in L^1(\Omega, \delta^s(x)\,dx)$, where $\delta(x) = \operatorname{dist}(x, \partial\Omega)$. Then there exists a unique $v \in L^1(\Omega)$, which is a weak solution of the problem*

$$\begin{cases} (-\Delta)^s v = f & \text{in } \Omega, \\ v = 0 & \text{in } \mathbb{R}^N \setminus \Omega \end{cases}$$

and there exists a constant $C > 0$ such that

$$\|v\|_{L^1(\Omega)} \leq C\|f\|_{L^1(\Omega, \delta^s(x)\,dx)}.$$

Moreover if $f \geq 0$ a. e. in Ω, then $v \geq 0$ a. e. in Ω.

Proof. The uniqueness and the existence are obtained like in Theorem 8.6.4. By the linearity of the problem we can assume without loss of generality that $f \geq 0$. We prove that

$$v \geq 0 \quad \text{a. e. in } \Omega. \tag{11.3.1}$$

Indeed, given $\psi \in C_0^\infty(\Omega)$, $\psi \geq 0$, define φ_ψ as the solution of

$$\begin{cases} (-\Delta)^s \varphi_\psi = \psi \geq 0 & \text{in } \Omega, \\ \varphi_\psi = 0 & \text{in } \mathbb{R}^N \setminus \Omega. \end{cases}$$

Since $\varphi_\psi \geq 0$ in $\mathbb{R}^N$, considering the definition of weak solution (Definition 11.2.2) associated to this problem we obtain

$$\int_\Omega v\phi(x)\,dx \geq 0.$$

Hence $v \geq 0$ a. e. in Ω. With this observation the estimate follows immediately. Consider

$$\begin{cases} (-\Delta)^s \phi = 1 & \text{in } \Omega, \\ \phi = 0 & \text{in } \mathbb{R}^N \setminus \Omega. \end{cases}$$

Then

$$\begin{aligned} \|v\|_{L^1(\Omega)} &= \int_\Omega v\,dx = \int_\Omega v(-\Delta)^s(\phi)\,dx = \int_\Omega (-\Delta)^s(v)\phi\,dx = \int_\Omega f\phi\,dx \leq C\int_\Omega f\delta^s\,dx \\ &= \|f\|_{L^1(\Omega,\delta^s\,dx)}, \end{aligned}$$

according to the regularity until the boundary obtained in [273]. □

The first important result in this section is the following.

Theorem 11.3.2. *Consider* $1 < p < p(\lambda, s)$. *Let* M *be defined by*

$$M = \sup\{\mu > 0 : \textit{problem } (11.1.1) \textit{ has a solution}\}. \tag{11.3.2}$$

Then $0 < M < \infty$.

Proof. We use the same kind of ideas as in [65], that is, we have to construct a subsolution and a supersolution to problem (11.2.2) for μ small enough, in such a way that pointwise the subsolution is less than or equal to the supersolution.

STEP 1. FINDING A SUBSOLUTION. For the subsolution, consider the eigenvalue problem

$$(P1) = \begin{cases} (-\Delta)^s \varphi_1 = \lambda_{1,(-\Delta)^s}\varphi_1 & \text{in } \Omega, \\ \varphi_1 = 0 & \text{in } \mathbb{R}^N \setminus \Omega. \end{cases} \tag{11.3.3}$$

By the regularity theory in Chapter 8 (see also [283, Proposition 4]), the nonnegative function φ_1 belongs to $H_0^s(\Omega) \cap L^\infty(\Omega)$. Hence, taking t small enough, we have, in Ω,

$$(-\Delta)^s(t\varphi_1) = \lambda_{1,(-\Delta)^s} t\varphi_1 \leq \mu(t\varphi_1)^q \leq \mu(t\varphi_1)^q + (t\varphi_1)^p + \lambda\frac{t\varphi_1}{|x|^{2s}}.$$

Thus, $\underline{u} := t\varphi_1$ is a subsolution of (P_μ).

STEP 2. FINDING A SUPERSOLUTION. To build the supersolution, we need to deal with the subcritical, critical and supercritical cases separately.

(i) Subcritical and critical case: $1 < p \leq 2_s^* - 1$.
 We look for a supersolution of the form $w(x) := A|x|^{-\beta}$, where $A > 0$ and β is a positive real parameter that verifies

$$\beta < \frac{N-2s}{2}.$$

Notice that $(-\Delta)^s w = C|x|^{-\beta-2s}$. Since $p \le 2_s^* - 1$, we can choose β satisfying that

$$p\beta < \beta + 2s \tag{11.3.4}$$

and

$$\beta(p+1) < N. \tag{11.3.5}$$

The condition (11.3.4) and an appropriate choice of A clearly imply that

$$(-\Delta)^s w - \lambda \frac{w}{|x|^{2s}} \ge w^p, \quad \text{in } \Omega. \tag{11.3.6}$$

Let $\eta := \inf_\Omega w > 0$. For μ small enough, taking $\overline{u} := C_1 w$ with $0 < C_1 < 1$ a suitable constant such that

$$\eta^{p-q} \ge \mu \frac{1}{C^{1-q}(1 - C^p)}, \tag{11.3.7}$$

it follows that

$$(-\Delta)^s \overline{u} - \lambda \frac{\overline{u}}{|x|^{2s}} \ge \overline{u}^p + \mu \overline{u}^q.$$

Thus, we have obtained a supersolution of (P_μ^+) for $1 < p \le 2_s^* - 1$. Moreover, by (11.3.5) we also have

$$\overline{u} \in L^{p+1}(\Omega) \quad \text{and} \quad \frac{\overline{u}^2}{|x|^{2s}} \in L^1(\Omega). \tag{11.3.8}$$

Choosing the parameter t of $\underline{u} := t\varphi_1$ small enough, it yields $\overline{u} \ge \underline{u}$. Now, since $p \le 2_s^* - 1$, we can build a nonnegative sequence $\{u_k\}$ in $H_0^s(\Omega)$ of solutions to the iterated problems

$$(P_k) = \begin{cases} (-\Delta)^s u_k = \lambda \frac{u_{k-1}}{|x|^{2s}} + u_{k-1}^p + \mu u_{k-1}^q & \text{in } \Omega, \\ u_k = 0 & \text{in } \mathbb{R}^N \setminus \Omega, \end{cases}$$

for $k \ge 1$ and $u_0 := \underline{u}$. By Corollary 8.3.12 it can be easily checked that

$$\underline{u} \le u_1 \le \cdots \le \overline{u}.$$

Hence, we can define $u_\mu := \lim_{k\to\infty} u_k$ in $L^1(\Omega)$. Moreover, by (11.3.8),

$$\begin{aligned} \|(-\Delta)^{s/2} u_k\|_{L^2(\mathbb{R}^N)}^2 &= \lambda \int_\Omega \frac{u_k u_{k-1}}{|x|^{2s}}\, dx + \int_\Omega u_k u_{k-1}^p\, dx + \mu \int_\Omega u_k u_{k-1}^q\, dx \\ &\le \lambda \int_\Omega \frac{\overline{u}^2}{|x|^{2s}}\, dx + \int_\Omega \overline{u}^{p+1}\, dx + \mu \int_\Omega \overline{u}^{q+1}\, dx \le C. \end{aligned} \tag{11.3.9}$$

Therefore, up to a subsequence, we know that $u_k \rightharpoonup u_\mu$ in $H_0^s(\Omega)$. In fact, by monotony, the whole sequence converges weakly. Hence, since $p \le 2_s^* - 1$, we can pass to the limit in the iterated problems to conclude that $u_\mu \ge 0$ is a minimal energy solution of (11.2.2) and, obviously, of (11.1.1).

(ii) Supercritical case: $2_s^* - 1 < p < p(\lambda, s)$.
Firstly, as we explained in the introduction, if $p < p(s,\lambda)$, $p(s,\lambda)$ as in (11.2.1), there exists a radial function $u(x) := A|x|^{\frac{-2s}{p-1}}$, with A a positive constant, satisfying

$$(-\Delta)^s u - \lambda \frac{u}{|x|^{2s}} = u^p \quad \text{in } \mathbb{R}^N.$$

Since $p > 2_s^* - 1 > \frac{N}{N-2s}$,

$$u \in L^p_{\text{loc}}(\mathbb{R}^N) \quad \text{and} \quad \frac{u}{|x|^{2s}} \in L^1_{\text{loc}}(\mathbb{R}^N). \tag{11.3.10}$$

Again, taking $\overline{u} = C_1 u$, with $C_1 > 0$ a suitable constant (see (11.3.7)), we get

$$\begin{cases} (-\Delta)^s \overline{u} - \lambda \frac{\overline{u}}{|x|^{2s}} \geq \overline{u}^p + \mu \overline{u}^q & \text{in } \Omega, \\ \overline{u} > 0 & \text{in } \mathbb{R}^N \setminus \Omega. \end{cases}$$

Moreover by (11.3.10), in particular $\overline{u} \geq 0$ satisfies (11.2.5). Hence, by Lemma 11.3.1 we can define $\{u_k\}$ to be the weak solutions to (P_k), and we will prove by induction that, in fact,

$$u_k \in L^1(\mathbb{R}^N) \quad \text{and} \quad 0 \leq \underline{u} \leq u_{k-1} \leq u_k \leq \overline{u} \text{ a. e. in } \Omega, \text{ for every } k \in \mathbb{N}.$$

For $\underline{u}$ there is nothing to prove. Suppose the result is true up to order $k-1$, that is,

$$\underline{u} \leq u_{j-1} \leq u_j \leq \overline{u} \quad \text{for } j \leq k-1 \text{ a. e. in } \Omega.$$

Then

$$(-\Delta)^s u_k = \lambda \frac{u_{k-1}}{|x|^{2s}} + u_{k-1}^p + \mu u_{k-1}^q \leq \lambda \frac{\overline{u}}{|x|^{2s}} + \overline{u}^p + \mu \overline{u}^q.$$

Hence, by (11.3.10), since the right hand side is in $L^1(\Omega)$, by Lemma 11.3.1 u_k, in particular, is integrable too. Moreover, by the induction hypothesis,

$$(-\Delta)^s (u_k - u_{k-1}) = \lambda \frac{(u_{k-1} - u_{k-2})}{|x|^{2s}} + (u_{k-1}^p - u_{k-2}^p) + \mu(u_{k-1}^q - u_{k-2}^q) \geq 0$$

and

$$(-\Delta)^s (\overline{u} - u_k) = \lambda \frac{(\overline{u} - u_{k-1})}{|x|^{2s}} + (\overline{u}^p - u_{k-1}^p) + \mu(\overline{u}^q - u_{k-1}^q) \geq 0.$$

Therefore, by (11.3.1), we have

$$0 \leq \underline{u} \leq u_{k-1} \leq u_k \leq \overline{u} \quad \text{a. e. in } \Omega.$$

By the monotone convergence argument we conclude that $\{u_k\}$ converges in $L^1(\mathbb{R}^N)$ to a weak nonnegative solution u_μ of (P_μ) for $2_s^* - 1 < p < p(\lambda, s)$.

Therefore, for μ small enough, we have built a minimal solution in the subcritical, critical and supercritical cases. That is, we have proved that $M > 0$.

To finish the proof we need to check that $M < \infty$. Consider $1 < p \le 2_s^* - 1$ and the eigenvalue problem with the Hardy potential given by

$$(P2) = \begin{cases} (-\Delta)^s \phi_1 - \lambda \frac{\phi_1}{|x|^{2s}} = \lambda_1 \phi_1 & \text{in } \Omega, \\ \phi_1 = 0 & \text{in } \mathbb{R}^N \setminus \Omega. \end{cases}$$

Note that since $\lambda < \Lambda_{N,2s}$, this problem is well defined and, following the same ideas as in the proof of assertion (*b*) of [281, Lemma 9], $\phi_1 \in H_0^s(\Omega)$. Suppose that u is a solution to problem (11.2.2). By the maximum principle we know that $u > 0$ in Ω. Then taking ϕ_1 as a test function in this problem we get

$$\begin{aligned} a_{N,s} \int_Q \frac{(\phi_1(x) - \phi_1(y))(u(x) - u(y))}{|x-y|^{N+2s}}\, dx\, dy - \lambda \int_\Omega \frac{\phi_1 u}{|x|^{2s}}\, dx \\ = \int_\Omega u^p \phi_1\, dx + \mu \int_\Omega u^q \phi_1\, dx. \end{aligned}$$

Using now that ϕ_1 is a solution of $(P2)$ it follows that

$$\int_\Omega (u^p + \mu u^q)\phi_1\, dx = \lambda_1 \int_\Omega u\phi_1\, dx. \tag{11.3.11}$$

If $2_s^* - 1 < p < p(\lambda, s)$, we consider $\varphi_1 \ge 0$, a solution to (P_1), as a test function in (11.1.1). Then

$$\begin{aligned} \int_\Omega u(-\Delta)^s \varphi_1\, dx &= \int_\Omega \left(\lambda \frac{u}{|x|^{2s}} + u^p + \mu u^q\right)\varphi_1\, dx \\ &\ge \int_\Omega (u^p + \mu u^q)\varphi_1\, dx. \end{aligned} \tag{11.3.12}$$

Moreover, since φ_1 is also an energy bounded solution of the eigenvalue problem (11.3.3), φ_1 is also a classical solution (see [273, Remark 2.1]). Hence from (11.3.12) we also get

$$\lambda_1 \int_\Omega u\varphi_1\, dx \ge \int_\Omega (u^p + \mu u^q)\varphi_1\, dx. \tag{11.3.13}$$

Since there exist structural positive constants c_0, c_1 such that

$$t^p + \mu t^q > c_0 \mu^{c_1} t, \quad \text{for every } t > 0,$$

we obtain from (11.3.11) and (11.3.13) that $c_0 \mu^{c_1} < \lambda_1$. This implies that $M < \infty$ for $p < p(\lambda, s)$. □

Notice that the previous result is true in the local problem and is independent of the Hardy potential term. This is motivated by the structure of the nonlinearity, with a concave part and a convex part, and holds for a larger class of second order operators for which the *concave-convex* structure must be well understood. For the p-laplacian case see [180].

The next result shows that the set of parameters μ for which there exists a solution is in fact the whole interval $(0, M]$.

Proposition 11.3.3. *Problem* (11.1.1) *has at least one positive solution for every* $0 < \mu < M$. *In fact, the sequence* $\{u_\mu\}$ *of minimal solutions is increasing with respect to* μ. *If* $\mu = M$ *problem* (11.1.1) *admits at least one weak solution.*

Proof. Since $M > 0$, we can find a solution for a value of μ as close as we want to M. Denote this value by $\overline{\mu}$ and the associated minimal solution by $u_{\overline{\mu}}$. Then, for all $\mu < \overline{\mu}$, $u_{\overline{\mu}}$ is a supersolution for the problems (11.1.1) with $\mu < \overline{\mu}$. Furthermore, for every μ, the solution to problem (11.3.3) can be modified if necessary to a subsolution to (11.1.1) in such a way that it is pointwise less than or equal to $u_{\overline{\mu}}$. Following the same procedure as in the previous proof, we conclude that there exists a solution u_μ for all $\mu \in (0, \overline{\mu})$, and therefore for all $\mu \in (0, M)$.

For the case $\mu = M$, the idea consists of passing to the limit when $\mu_n \nearrow M$ on the sequence $\{u_n\} = \{u_{\mu_n}\}$, where u_{μ_n} is the minimal solution of (11.2.2) with $\mu = \mu_n$. Since we are considering minimal solutions $u_n \le u_{n+1}$ for all $n \in \mathbb{N}$.

Consider the solution φ_1 to the eigenvalue problem (11.3.3) as a test function in (11.2.2). Since $\operatorname{supp} u_n \subseteq \Omega$ we get

$$\lambda_1 \int_\Omega u_n \varphi_1 \, dx = \lambda \int_\Omega \frac{u_n \varphi_1}{|x|^{2s}} \, dx + \int_\Omega u_n^p \varphi_1 \, dx + \mu \int_\Omega u_n^q \varphi_1 \, dx. \tag{11.3.14}$$

By Young's inequality again and by Hopf's lemma (see [273, Lemma 3.2]) we can conclude that

$$\lambda \int_\Omega \frac{u_n \delta^s}{|x|^{2s}} \, dx + \int_\Omega u_n^p \delta^s \, dx + \mu \int_\Omega u_n^q \delta^s \, dx \le C, \tag{11.3.15}$$

where C is a constant independent of n and $\delta(x) := \operatorname{dist}(x, \partial\Omega), x \in \Omega$. Let now ξ_1 be the solution to the linear problem

$$\begin{cases} (-\Delta)^s \xi_1 = 1 & \text{in } \Omega, \\ \xi_1 = 0 & \text{in } \mathbb{R}^N \setminus \Omega. \end{cases}$$

Using ξ_1 as a test function of (11.2.2), since $\operatorname{supp}(u_n) \subseteq \Omega$, by [273, Proposition 1.1] and (11.3.15), we obtain

$$\int_\Omega u_n \, dx = \lambda \int_\Omega \frac{u_n \xi_1}{|x|^{2s}} \, dx + \int_\Omega u_n^p \xi_1 \, dx + \mu \int_\Omega u_n^q \xi_1 \, dx$$

$$\le C\Big(\lambda \int_\Omega \frac{u_n \delta^s}{|x|^{2s}}\,dx + \int_\Omega u_n^p \delta^s\,dx + \mu \int_\Omega u_n^q \delta^s\,dx\Big)$$
$$\le C.$$

Hence, $\{u_n\}$ converges in $L^1(\Omega)$ to a limit $u_M \ge 0$. Then, since $\{u_n\}$ is an increasing sequence that is uniformly bounded in $L^1(\Omega)$, by the monotone convergence theorem, taking the limit when $n \to \infty$, we conclude that u_M is actually a weak solution of (P_M). □

Remark 11.3.4. The results obtained in this section can be easily translated for the case of $q = 0$, that is, considering a function f with appropriate growth conditions instead of the concave term u^q. In particular, one could obtain the same existence results given in Section 4.4 in the nonlocal framework.

11.4 Existence of at least two nontrivial variational solutions if $1 < p \le 2_s^* - 1$

Problem (11.1.1) is the Euler–Lagrange equation of the energy functional

$$\mathcal{J}_\mu(u) := a_{N,s} \int_Q \frac{|u(x) - u(y)|^2}{|x - y|^{N+2s}}\,dx\,dy$$
$$- \frac{\lambda}{2} \int_\Omega \frac{(u_+)^2}{|x|^{2s}}\,dx - \frac{1}{p+1} \int_\Omega u_+^{p+1}\,dx - \frac{\mu}{q+1} \int_\Omega u_+^{q+1}\,dx, \tag{11.4.1}$$

which is well defined in $H_0^s(\Omega)$ if $1 < p \le 2_s^* - 1$.

There are two different cases for the verification of the Palais–Smale condition:

(i) the subcritical case, $1 < p < 2_s^* - 1$, in which the compactness follows from the Rellich theorem;
(ii) the critical case, $p = 2_s^* - 1$. In this case the verification of the Palais–Smale condition is more involved and it follows from the *concentration compactness* arguments by P. L. Lions in [233, 234].

Both cases will be studied in the following paragraphs.

11.4.1 Subcritical case

Consider $1 < p < 2_s^* - 1$ and $0 < \lambda < \Lambda_{N,2s}$ hereafter in this section. Taking advantage of the variational structure of (11.1.1) we will prove the existence of, at least, two positive solutions as critical points of the functional $\mathcal{J}_\mu(u)$ defined in (11.4.1).

We begin with the parameter μ sufficiently small. In this situation a local minimization argument allows to find the first solution, and the *mountain pass theorem* guarantees the existence of the second one. This behavior in the local case and for the p-Laplacian operator and $\lambda = 0$ was first proved in [177] even for the critical Sobolev exponent. The proof of the existence of two positive solutions in the whole interval $(0, M)$ was proved by Ambrosetti, Brezis and Cerami in the case of the Laplacian in [35]. For radial solutions and the p-Laplacian operator the reader is referred to [36] and for the general case to [180]. In this section we follow the reference [46].

In order to use the mountain pass theorem, we have to check: (i) the mountain pass geometry and (ii) the compactness of the functional, that is, the Palais–Smale condition.

Proposition 11.4.1. *There exists a $\mu_0 > 0$ such that for any $0 < \mu < \mu_0$, there exist $\alpha > 0$ and $\beta > 0$ such that:*
(a) *$\mathcal{J}_\mu(u) \geq \beta$, for any $u \in H_0^s(\Omega)$ with $\|u\|_{H_0^s(\Omega)} = \alpha$ and μ small enough;*
(b) *there exists $u_1 \in H_0^s(\Omega)$ positive such that $\|u_1\|_{H_0^s(\Omega)} > \alpha$ and $\mathcal{J}_\mu(u_1) < \beta$.*

Proof.
(a) Since $q + 1 < p + 1 < 2_s^*$, by (9.2.4) and Theorem 8.2.7, it can be checked that

$$\mathcal{J}_\mu(u) \geq g(\|u\|_{H_0^s(\Omega)}),$$

where $g(t) = c_1 t^2 - c_2 t_+^{p+1} - \mu c_3 t_+^{q+1}$, for some positive constants c_1, c_2 and c_3. Then, choosing μ small enough, there exists $\alpha > 0$ such that if $\beta := g(\alpha) > 0$, therefore $\mathcal{J}_\mu(u) \geq \beta$ for $u \in H_0^s(\Omega)$ with $\|u\|_{H_0^s(\Omega)} = \alpha$.
(b) Fix $u_0 \in H_0^s(\Omega)$ positive such that $\|u_0\|_{H_0^s(\Omega)} = 1$ and take $t > 0$. Since $p > 1$, it is clear that

$$\lim_{t\to\infty} \mathcal{J}_\mu(tu_0) = -\infty.$$

Therefore, there exists t_0 large enough, such that, defining $u_1 := t_0 u_0$, it follows that $\|u_1\|_{H_0^s(\Omega)} > \alpha$ and $\mathcal{J}_\mu(u_1) < \beta$. □

By a similar argument, we obtain

$$\lim_{t\to 0^+} \mathcal{J}_\mu(tu_0) = 0^-. \tag{11.4.2}$$

Finally, we need to check that the functional satisfies the Palais–Smale condition. First we prove the following.

Proposition 11.4.2. *Let $\{u_n\}$ be a bounded sequence in $H_0^s(\Omega)$ such that $\mathcal{J}_\mu'(u_n) \to 0$ in $H^{-s}(\Omega)$ as $n \to \infty$.*
Then there exists $u \in H_0^s(\Omega)$ such that, up to a subsequence, $\|u_n - u\|_{H_0^s(\Omega)} \to 0$ as $n \to \infty$.

Proof. The condition $\mathcal{J}_\mu'(u_n) \to 0$ in $H^{-s}(\Omega)$ is equivalent to saying that

$$(-\Delta)^s u_n - \lambda \frac{(u_n)_+}{|x|^{2s}} - \mu (u_n)_+^q - (u_n)_+^p = \chi_n,$$

with $\chi_n \to 0$ as $n \to \infty$ in $H^{-s}(\Omega)$.

Since $\{u_n\}$ is uniformly bounded in the Hilbert space $H_0^s(\Omega)$ with the norm $\|\cdot\|_{H_0^s(\Omega)}$, there exists $C > 0$ such that

$$\|u_n\|_*^2 := \left(a_{N,s}\|u_n\|^2_{H_0^s(\Omega)} - \lambda \int_\Omega \frac{(u_n)_+^2}{|x|^{2s}}\, dx \right) \le a_{N,s}\|u_n\|^2_{H_0^s(\Omega)} \le C.$$

Since, by (9.2.4), $\|\cdot\|_*^2$ is a norm equivalent to $\|\cdot\|^2_{H_0^s(\Omega)}$, we consider $H_0^s(\Omega)$ endowed with this norm and hence up to a subsequence, there exists $u \in H_0^s(\Omega)$ such that

$$u_n \rightharpoonup u \quad \text{weakly in } H_0^s(\Omega) \text{ with the norm } \|\cdot\|_*, \tag{11.4.3}$$

$$u_n \to u \quad \text{in } L^r(\mathbb{R}^N), 1 \le r < 2_s^*, u_n \to u \text{ a. e. in } \mathbb{R}^N; \tag{11.4.4}$$

see Section 8.2. Therefore, since $\mathcal{J}_\mu'(u_n) \to 0$ in $H^{-s}(\Omega)$ and $q+1 < p+1 < 2_s^*$, by (11.4.4) it follows that

$$\begin{aligned}
&\lim_{n\to\infty}\left(a_{N,s} \int_Q \frac{(u_n(x)-u_n(y))^2}{|x-y|^{N+2s}}\, dx\, dy - \lambda \int_\Omega \frac{(u_n)_+^2}{|x|^{2s}}\, dx \right)\\
&= \lim_{n\to\infty}\left(\mu \int_\Omega (u_n)_+^{q+1}\, dx + \int_\Omega (u_n)_+^{p+1}\, dx \right)\\
&= \mu \int_\Omega u_+^{q+1}\, dx + \int_\Omega u_+^{p+1}\, dx\\
&=: f_\mu(u).
\end{aligned} \tag{11.4.5}$$

Similarly we obtain

$$\lim_{n\to\infty}\left(a_{N,s} \int_Q \frac{(u_n(x)-u_n(y))(u(x)-u(y))}{|x-y|^{N+2s}}\, dx\, dy - \lambda \int_\Omega \frac{(u_n)_+ u}{|x|^{2s}}\, dx \right) = f_\mu(u). \tag{11.4.6}$$

Hence, from (11.4.3), (11.4.5) and (11.4.6) we get

$$\lim_{n\to\infty} \|u_n\|_*^2 = \|u\|_*^2.$$

Consequently, by (11.4.3), we conclude that $\lim_{n\to\infty} \|u_n - u\|_*^2 = 0$. Finally, by the equivalence of norms, we conclude that

$$\lim_{n\to\infty} \|u_n - u\|^2_{H_0^s(\Omega)} = 0. \qquad \square$$

Now we can prove the Palais–Smale condition.

Proposition 11.4.3 (Palais–Smale condition). *Let u_n be a sequence in $H_0^s(\Omega)$, and let $c \in \mathbb{R}$ such that*

$$\mathcal{J}_\mu(u_n) \to c, \tag{11.4.7}$$

$$\mathcal{J}'_\mu(u_n) \to 0 \quad \text{in } H^{-s}(\Omega). \tag{11.4.8}$$

Then, up to a subsequence, there exists u in $H_0^s(\Omega)$ such that

$$\lim_{n\to\infty} \|u_n - u\|^2_{H_0^s(\Omega)} = 0.$$

Proof. By (11.4.7) and (11.4.8) it follows that

$$\mathcal{J}_\mu(u_n) - \frac{1}{p+1}\langle \mathcal{J}'_\mu(u_n), u_n\rangle = c + o(1).$$

Hence, since $p+1 > 2 > q+1$, by (9.2.4) and the Sobolev inequality, Theorem 8.2.7, we obtain

$$\begin{aligned} c + o(1) &= a_{N,s}\Big(\frac{1}{2} - \frac{1}{p+1}\Big)\|u_n\|^2_{H_0^s(\Omega)} \\ &\quad - \lambda\Big(\frac{1}{2} - \frac{1}{p+1}\Big)\int_\Omega \frac{(u_n)_+^2}{|x|^{2s}}\,dx - \mu\Big(\frac{1}{q+1} - \frac{1}{p+1}\Big)\int_\Omega (u_n)_+^{q+1}\,dx \\ &\geq C_1\|u_n\|^2_{H_0^s(\Omega)} - C_2\|u_n\|^{q+1}_{H_0^s(\Omega)}, \end{aligned}$$

with C_1 and C_2 positive constants. Therefore $\|u_n\|_{H_0^s(\Omega)} \leq C$. Applying Proposition 11.4.2 we conclude the strong convergence in the space $H_0^s(\Omega)$. □

Now we can already state the following multiplicity theorem for μ sufficiently small.

Theorem 11.4.4. *For μ small enough, the problem (P_μ) has at least two solutions.*

Proof. We construct the first one by minimization following the construction done in [177]. Indeed, as we saw in Proposition 11.4.1, there exists $\alpha > 0$ such that $\mathcal{J}_\mu(u) \geq \beta > 0$ for all $u \in H_0^s(\Omega)$ with $\|u\|_{H_0^s(\Omega)} = \alpha$. Thus we can choose

$$\alpha_1 = \Big\{\inf_{\alpha\in\mathbb{R}} \alpha : \mathcal{J}_\mu(u) > 0 \text{ for all } u \in H_0^s(\Omega) \text{ with } \|u\|_{H_0^s(\Omega)} = \alpha\Big\}.$$

We know that $\alpha_1 > 0$, because near the origin the functional is negative. We choose now $\alpha_2 > \alpha_1$ so close that $g(\|u\|_{H_0^s(\Omega)})$, the lower radial bound of $\mathcal{J}_\mu(u)$, is nondecreasing for u with $\alpha_1 \leq \|u\|_{H_0^s(\Omega)} \leq \alpha_2$. We define now a smooth function τ as

$$\tau(t) := \begin{cases} 1 & t \leq \alpha_1, \\ 0 & t \geq \alpha_2, \end{cases}$$

and we consider the truncated functional

$$\mathcal{J}_2(u) := \frac{1}{2} a_{N,s} \|u\|^2_{H_0^s(\Omega)} - \frac{\lambda}{2} \int_\Omega \frac{u_+^2}{|x|^{2s}}\,dx - \frac{\tau(\|u\|_{H_0^s(\Omega)})}{p+1} \int_\Omega u_+^{p+1}\,dx - \frac{\mu}{q+1} \int_\Omega u_+^{q+1}\,dx.$$

By definition,

$$\mathcal{J}_2(u) = \mathcal{J}_\mu(u) \quad \text{whenever } \|u\|_{H_0^s(\Omega)} \le \alpha_1$$

and

$$\mathcal{J}_2(u) = \frac{1}{2} a_{N,s} \|u\|^2_{H_0^s(\Omega)} - \frac{\lambda}{2} \int_\Omega \frac{u_+^2}{|x|^{2s}}\,dx - \frac{\mu}{q+1} \int_\Omega u_+^{q+1}\,dx$$

whenever $\|u\|_{H_0^s(\Omega)} \ge \alpha_2$. Note that, by the Sobolev embedding theorem, Theorem 8.2.7, since $q+1 < 2$ the functional $\mathcal{J}_2$ is coercive. The lower semicontinuity is given because $H_0^s(\Omega)$ is a Hilbert space. Therefore there exists u_1, a minimum of $\mathcal{J}_2$ with negative energy, that is also a minimum of $\mathcal{J}_\mu$. Then, we have already found a first solution to (11.1.1).

For the second one, we have proved that if μ is small enough the functional $\mathcal{J}_\mu$ has the suitable geometry, Proposition 11.4.1, and satisfies the Palais–Smale condition, Proposition 11.4.3. Then if we consider

$$\Gamma := \{\gamma \in C^0([0,1], X_0^s(\Omega)) : \gamma(0) = 0, \gamma(1) = u_1\},$$

where u_1 is the local minimum found above, and

$$C := \inf_{\gamma \in \Gamma} \sup_{t \in [0,1]} \mathcal{J}(\gamma(t)),$$

the Ambrosetti–Rabinowitz mountain pass theorem (see Section 4.3.3 and [37]) gives us a solution $u \in H_0^s(\Omega)$ satisfying

$$\mathcal{J}(u) = C \ge \beta > 0.$$

Here β is specified in Proposition 11.4.1. Note that this solution and the one obtained before are different because the previous one had negative energy. Therefore, for μ small enough, problems (P_μ^+) and, consequently, (P_μ) have at least two solutions. □

Now we will prove that in fact problem (P_μ) has two solutions for every $\mu \in (0, M)$. With this purpose we will generalize a result by Alama in [32] to the fractional Laplacian to check that the minimal solution obtained is a local minimum, which will allow us to apply the mountain pass lemma. Due to the involved computations we provide a detailed proof.

Proposition 11.4.5. *Let* $p \in (1, 2_s^* - 1)$. *Then for* $0 < \mu < M$, *where* M *is defined in* (11.3.2), *problem* (P_μ) *has at least two solutions.*

Proof. Let $\mu_0 \in (0, M)$ and take μ_1 such that $\mu_0 < \mu_1 < M$. Then, by Proposition 11.3.3, we can consider w_{μ_0} and w_{μ_1}, the minimal solutions to problems (P_{μ_0}) and (P_{μ_1}), respectively, which satisfy $w_{\mu_0} < w_{\mu_1}$. Now we define

$$W = \{w \in H_0^s(\Omega) : 0 \le w \le w_{\mu_1}\}.$$

Since W is a closed convex set of $H_0^s(\Omega)$, and $\mathcal{J}_{\mu_0}$ is bounded from below and semicontinuous in W, therefore there exists $\underline{w} \in W$ such that $\mathcal{J}_{\mu_0}(\underline{w}) = \inf_{w \in W} \mathcal{J}_{\mu_0}(w)$. Let $w_0 \in H_0^s(\Omega)$ be the positive solution to

$$\begin{cases} (-\Delta)^s w_0 - \lambda \frac{w_0}{|x|^{2s}} = \mu_0 w_0^q & \text{in } \Omega, \\ w_0 = 0 & \text{in } \mathbb{R}^N \setminus \Omega. \end{cases}$$

The uniqueness of this solution is a consequence of the extension of the Brezis–Kamin result to the nonlocal setting, obtained in Theorem 8.5.5. Note that, since $\lambda < \Lambda_{N,2s}$, the existence is given by minimization. Then, for $0 < \varepsilon << \mu_0$, we can affirm that $\mathcal{J}_{\mu_0}(\varepsilon w_0) < 0$ because the term with power $q + 1$ dominates over the quadratic terms. Taking ε small enough, since $\varepsilon w_0 \in W$, we get $\underline{w} \neq 0$ and $\mathcal{J}_{\mu_0}(\underline{w}) < 0$. Following the ideas in [177] and [298, Theorem 2.4, page 17], adapted to the nonlocal framework, also $\underline{w}$ is a solution to problem (11.1.1) with $\mu = \mu_0$.

Hence, we have two possible cases.

- If $\underline{w} \neq w_{\mu_0}$, then we have finished because we have found two different solutions.
- Otherwise, if $\underline{w} = w_{\mu_0}$ and we prove that $\underline{w}$ is a local minimum of $\mathcal{J}_{\mu_0}$, then we obtain a second solution applying the classical *mountain pass theorem* or its refinement given in [183].

Therefore our goal now is to prove that $\underline{w}$ is a local minimum of the functional $\mathcal{J}_{\mu_0}$.

Let us argue by contradiction. Suppose $\underline{w}$ is not a local minimum of $\mathcal{J}_{\mu_0}$ in the space $H_0^s(\Omega)$. Then there exists a sequence $\{v_n\} \subseteq H_0^s(\Omega)$ such that

$$\|v_n - \underline{w}\|_{H_0^s(\Omega)} \to 0 \quad \text{and} \quad \mathcal{J}_{\mu_0}(v_n) < \mathcal{J}_{\mu_0}(\underline{w}). \tag{11.4.9}$$

Call $\overline{w} := w_{\mu_1}$, the minimal solution associated to μ_1. Define

$$w_n := (v_n - \overline{w})^+$$

and

$$z_n(x) := \begin{cases} 0, & v_n(x) \le 0, \\ v_n(x), & 0 \le v_n(x) \le \overline{w}(x), \\ \overline{w}(x), & \overline{w}(x) \le v_n(x). \end{cases}$$

Note that w_n and z_n belong to the energy space $H_0^s(\Omega)$. Consider now the sets given by

$$T_n := \{x : z_n(x) = v_n(x)\}, \quad \tilde{T}_n = T_n \cap \Omega$$

and

$$S_n := \{x : v_n(x) \ge \overline{w}(x)\}, \quad \tilde{S}_n = S_n \cap \Omega.$$

Note that

$$z_n(x) = \overline{w}(x), \quad x \in S_n, \tag{11.4.10}$$

$$z_n(x) = v_n^+(x), \quad x \in S_n^c := \mathbb{R}^n \setminus S_n. \tag{11.4.11}$$

Define now

$$F_{\mu_0}(t) = \frac{1}{p+1} t_+^{p+1} + \frac{\mu_0}{q+1} t_+^{q+1}. \tag{11.4.12}$$

Thus,

$$\int_\Omega F_{\mu_0}(v_n) = \int_{\tilde{T}_n} F_{\mu_0}(v_n) + \int_{\tilde{S}_n} F_{\mu_0}(v_n) = \int_{\tilde{T}_n} F_{\mu_0}(z_n) + \int_{\tilde{S}_n} F_{\mu_0}(v_n).$$

For simplicity, let us denote

$$V_n(x,y) := \frac{(v_n(x) - v_n(y))^2}{|x-y|^{N+2s}},$$

$$V_n^+(x,y) := \frac{(v_n^+(x) - v_n^+(y))^2}{|x-y|^{N+2s}}, \quad V_n^-(x,y) := \frac{(v_n^-(x) - v_n^-(y))^2}{|x-y|^{N+2s}},$$

$$Z_n(x,y) := \frac{(z_n(x) - z_n(y))^2}{|x-y|^{N+2s}} \quad \text{and} \quad W_n(x,y) := \frac{(w_n(x) - w_n(y))^2}{|x-y|^{N+2s}}.$$

Hence, we have

$$\begin{aligned}
\int_{\mathbb{R}^N}\int_{\mathbb{R}^N} V_n(x,y)\,dx\,dy &= \int_{\mathbb{R}^N}\int_{\mathbb{R}^N} V_n^+(x,y)\,dx\,dy + \int_{\mathbb{R}^N}\int_{\mathbb{R}^N} V_n^-(x,y)\,dx\,dy \\
&\quad + 2\int_{\mathbb{R}^N}\int_{\mathbb{R}^N} \frac{(v_n^+(x) - v_n^+(y))(-v_n^-(x) + v_n^-(y))}{|x-y|^{N+2s}}\,dx\,dy \\
&= \int_{\mathbb{R}^N}\int_{\mathbb{R}^N} V_n^+(x,y)\,dx\,dy + \int_{\mathbb{R}^N}\int_{\mathbb{R}^N} V_n^-(x,y)\,dx\,dy \\
&\quad + 4\int_{\mathbb{R}^N}\int_{\mathbb{R}^N} \frac{v_n^+(x)v_n^-(y)}{|x-y|^{N+2s}}\,dx\,dy.
\end{aligned} \tag{11.4.13}$$

Also it is clear that

$$\int_{\Omega} \frac{v_n^2(x)}{|x|^{2s}}\,dx = \int_{\Omega} \frac{(v_n^+(x))^2}{|x|^{2s}}\,dx + \int_{\Omega} \frac{(v_n^-(x))^2}{|x|^{2s}}\,dx. \tag{11.4.14}$$

Then from (11.4.12), (11.4.13) and (11.4.14) it follows that

$$\begin{aligned}
\mathcal{J}_{\mu_0}(v_n) &= a_{N,s} \int_{\mathbb{R}^N}\int_{\mathbb{R}^N} V_n(x,y)\,dx\,dy - \frac{\lambda}{2}\int_{\Omega} \frac{v_n^2(x)}{|x|^{2s}}\,dx \\
&\quad - \int_{\Omega} F_{\mu_0}(v_n)(x)\,dx \\
&\geq a_{N,s}\Big(\int_{\mathbb{R}^N}\int_{\mathbb{R}^N} V_n^+(x,y)\,dx\,dy + \int_{\mathbb{R}^N}\int_{\mathbb{R}^N} V_n^-(x,y)\,dx\,dy\Big) \\
&\quad - \frac{\lambda}{2}\int_{\Omega} \frac{(v_n^+(x)^2)}{|x|^{2s}}\,dx - \frac{\lambda}{2}\int_{\Omega} \frac{(v_n^-(x))^2}{|x|^{2s}}\,dx \\
&\quad - \int_{\tilde{T}_n} F_{\mu_0}(z_n)(x)\,dx - \int_{\tilde{S}_n} F_{\mu_0}(v_n)(x)\,dx.
\end{aligned} \tag{11.4.15}$$

From (11.4.11) we get

$$\begin{aligned}
\int_{\mathbb{R}^N}\int_{\mathbb{R}^N} V_n^+(x,y)\,dx\,dy &= \int_{S_n}\int_{S_n} V_n^+(x,y)\,dx\,dy \\
&\quad + \int_{S_n^c}\int_{S_n^c} Z_n(x,y)\,dx\,dy \\
&\quad + 2\int_{S_n}\int_{S_n^c} V_n^+(x,y)\,dx\,dy.
\end{aligned} \tag{11.4.16}$$

Also, since

$$\begin{aligned}
\int_{\mathbb{R}^N}\int_{\mathbb{R}^N} Z_n(x,y)\,dx\,dy &= \int_{S_n}\int_{S_n} Z_n(x,y)\,dx\,dy \\
&\quad + \int_{S_n^c}\int_{S_n^c} Z_n(x,y)\,dx\,dy + 2\int_{S_n}\int_{S_n^c} Z_n(x,y)\,dx\,dy,
\end{aligned}$$

we obtain from (11.4.16)

$$\begin{aligned}
\int_{\mathbb{R}^N}\int_{\mathbb{R}^N} V_n^+(x,y)\,dx\,dy &= \int_{S_n}\int_{S_n} V_n^+(x,y)\,dx\,dy + 2\int_{S_n}\int_{S_n^c} V_n^+(x,y)\,dx\,dy \\
&\quad + \int_{\mathbb{R}^N}\int_{\mathbb{R}^N} Z_n(x,y)\,dx\,dy - 2\int_{S_n}\int_{S_n^c} Z_n(x,y)\,dx\,dy
\end{aligned}$$

$$-\int_{S_n}\int_{S_n} Z_n(x,y)\,dx\,dy. \tag{11.4.17}$$

Moreover, using the same argument,

$$\begin{aligned}\int_\Omega \frac{(v_n^+(x))^2}{|x|^{2s}}\,dx &= \int_{\tilde{S_n}} \frac{(v_n^+(x))^2}{|x|^{2s}}\,dx + \int_{(\tilde{S_n})^c} \frac{z_n^2(x)}{|x|^{2s}}\,dx \\ &= \int_{\tilde{S_n}} \frac{(v_n^+(x))^2}{|x|^{2s}}\,dx + \int_\Omega \frac{z_n^2(x)}{|x|^{2s}}\,dx - \int_{\tilde{S_n}} \frac{z_n^2(x)}{|x|^{2s}}\,dx. \end{aligned} \tag{11.4.18}$$

Therefore from (11.4.15), by the Hardy inequality given in (9.3.4), (11.4.17) and (11.4.18), we get

$$\begin{aligned}\mathcal{J}_{\mu_0}(v_n) &\ge C_1\|v_n^-\|^2_{H_0^s(\Omega)} + \mathcal{J}_{\mu_0}(z_n) \\ &\quad + a_{N,s}\int_{S_n}\int_{S_n} \frac{(v_n^+(x)-v_n^+(y))^2-(z_n(x)-z_n(y))^2}{|x-y|^{N+2s}}\,dx\,dy \\ &\quad + a_{N,s}\int_{S_n}\int_{S_n^c} \frac{(v_n^+(x)-v_n^+(y))^2-(z_n(x)-z_n(y))^2}{|x-y|^{N+2s}}\,dx\,dy \\ &\quad - \frac{\lambda}{2}\int_{\tilde{S_n}} \frac{(v_n^+(x))^2 - z_n^2(x)}{|x|^{2s}}\,dx \\ &\quad + \int_{\tilde{S_n}} F_{\mu_0}(z_n) - F_{\mu_0}(v_n)\,dx, \end{aligned} \tag{11.4.19}$$

where

$$C_1 = a_{N,s}\left(1 - \frac{\lambda}{\Lambda_{N,2s}}\right).$$

As $w_n(x) = v_n(x) - \overline{w}(x)$ when $x \in S_n$, using (11.4.10) we obtain

$$\begin{aligned}&\int_{S_n}\int_{S_n} \frac{(v_n^+(x)-v_n^+(y))^2-(z_n(x)-z_n(y))^2}{|x-y|^{N+2s}}\,dx\,dy \\ &= \int_{S_n}\int_{S_n} \frac{(w_n(x)+\overline{w}(x)-w_n(y)-\overline{w}(y))^2-(\overline{w}(x)-\overline{w}(y))^2}{|x-y|^{N+2s}}\,dx\,dy \\ &= \int_{S_n}\int_{S_n} \frac{(w_n(x)-w_n(y))^2}{|x-y|^{N+2s}}\,dx\,dy \end{aligned}$$

$$+\int_{S_n}\int_{S_n}\frac{2(w_n(x)-w_n(y))(\overline{w}(x)-\overline{w}(y))}{|x-y|^{N+2s}}\,dx\,dy \tag{11.4.20}$$

and

$$\begin{aligned}\int_{\tilde{S}_n}\frac{(v_n^+(x))^2-z_n^2(x)}{|x|^{2s}}\,dx&=\int_{\tilde{S}_n}\frac{(w_n(x)+\overline{w}(x))^2-\overline{w}(x)^2}{|x|^{2s}}\,dx\\&=\int_{\tilde{S}_n}\frac{w_n^2(x)+2w_n\overline{w}(x)}{|x|^{2s}}\,dx.\end{aligned}\tag{11.4.21}$$

Also from (11.4.10) and (11.4.11) it follows that

$$\begin{aligned}&\int_{S_n}\int_{S_n^c}\frac{(v_n^+(x)-v_n^+(y))^2-(z_n(x)-z_n(y))^2}{|x-y|^{N+2s}}\,dx\,dy\\&\quad=\int_{S_n}\int_{S_n^c}\frac{(v_n^+(x)-w_n(y)-\overline{w}(y))^2-(v_n^+(x)-\overline{w}(y))^2}{|x-y|^{N+2s}}\,dx\,dy\\&\quad=\int_{S_n}\int_{S_n^c}\frac{w_n^2(y)-2w_n(y)(v_n^+(x)-\overline{w}(y))}{|x-y|^{N+2s}}\,dx\,dy.\end{aligned}\tag{11.4.22}$$

Furthermore, since $\operatorname{supp} w_n=S_n$, we have

$$\begin{aligned}\int_{\mathbb{R}^N}\int_{\mathbb{R}^N}W_n(x,y)\,dx\,dy&=\int_{S_n}\int_{S_n}W_n(x,y)\,dx\,dy\\&\quad+2\int_{S_n}\int_{S_n^c}\frac{w_n^2(y)}{|x-y|^{N+2s}}\,dx\,dy.\end{aligned}\tag{11.4.23}$$

Hence, by using (11.4.20), (11.4.21), (11.4.22) and (11.4.23), from (11.4.19) we get

$$\begin{aligned}\mathcal{J}_{\mu_0}(v_n)\ge{}&C_1\|v_n^-\|^2_{H_0^s(\Omega)}+a_{N,s}\|w_n\|^2_{H_0^s(\Omega)}+\mathcal{J}_{\mu_0}(z_n)\\&+C_{N,s}\int_{S_n}\int_{S_n}\frac{(w_n(x)-w_n(y))(\overline{w}(x)-\overline{w}(y))}{|x-y|^{N+2s}}\,dx\,dy\\&-2C_{N,s}\int_{S_n}\int_{S_n^c}\frac{w_n(y)(v_n^+(x)-\overline{w}(y))}{|x-y|^{N+2s}}\,dx\,dy\\&-\frac{\lambda}{2}\int_{\tilde{S}_n}\frac{w_n^2(x)+2w_n\overline{w}(x)}{|x|^{2s}}\,dx\end{aligned}$$

$$+ \int_{\tilde{S}_n} F_{\mu_0}(z_n) - F_{\mu_0}(v_n)\,dx. \tag{11.4.24}$$

Since $v_n^+(x) \leq \overline{w}(x)$, for $x \in S_n^c$, using that $\operatorname{supp} w_n = S_n$, from (11.4.24) it follows that

$$\begin{aligned} \mathcal{J}_{\mu_0}(v_n) \geq{}& C_1\|v_n^-\|^2_{H_0^s(\Omega)} + a_{N,s}\|w_n\|^2_{H_0^s(\Omega)} + \mathcal{J}_{\mu_0}(z_n) \\ &+ C_{N,s}\int_{\mathbb{R}^N}\int_{\mathbb{R}^N} \frac{(w_n(x)-w_n(y))(\overline{w}(x)-\overline{w}(y))}{|x-y|^{N+2s}}\,dx\,dy \\ &- \frac{\lambda}{2}\int_{\tilde{S}_n} \frac{w_n^2(x) + 2w_n\overline{w}(x)}{|x|^{2s}}\,dx \\ &+ \int_{\tilde{S}_n} F_{\mu_0}(z_n) - F_{\mu_0}(v_n)\,dx. \end{aligned} \tag{11.4.25}$$

As $\overline{w}$ is a supersolution of (P_{μ_0}), testing in that problem with the function w_n, we obtain

$$\begin{aligned} &a_{N,s}\int_{\mathbb{R}^n}\int_{\mathbb{R}^n} \frac{(w_n(x)-w_n(y))(\overline{w}(x)-\overline{w}(y))}{|x-y|^{N+2s}}\,dx\,dy - \lambda\int_{\Omega}\frac{w_n\overline{w}(x)}{|x|^{2s}}\,dx \\ &\geq \int_{\tilde{S}_n} w_n F'_{\mu_0}(\overline{w})\,dx. \end{aligned} \tag{11.4.26}$$

Then from (11.4.25), (9.3.4) and (11.4.26) we have

$$\begin{aligned} \mathcal{J}_{\mu_0}(v_n) \geq{}& C_1\|v_n^-\|^2_{H_0^s(\Omega)} + a_{N,s}\|w_n\|^2_{H_0^s(\Omega)} - \frac{\lambda}{2}\int_{\tilde{S}_n}\frac{w_n^2(x)}{|x|^{2s}}\,dx + \mathcal{J}_{\mu_0}(z_n) \\ &+ \int_{\tilde{S}_n} (F_{\mu_0}(\overline{w}) - F_{\mu_0}(\overline{w}+w_n) + w_n F'_{\mu_0}(\overline{w}))(x)\,dx \\ \geq{}& a_{N,s}\|w_n\|^2_{H_0^s(\Omega)} - \frac{\lambda}{2}\int_{\Omega}\frac{w_n^2(x)}{|x|^{2s}}\,dx + \mathcal{J}_{\mu_0}(z_n) \\ &+ \int_{\tilde{S}_n} \frac{1}{p+1}(\overline{w}^{p+1} - (\overline{w}+w_n)^{p+1})(x) + w_n\overline{w}^p(x)\,dx \\ &+ \mu_0\int_{\tilde{S}_n} \frac{(w_{\mu_1}^{q+1} - (w_{\mu_1}+w_n)^{q+1})(x)}{q+1} + w_n w_{\mu_1}^q(x)\,dx. \end{aligned} \tag{11.4.27}$$

Since $0 < q+1 < 2$ it follows that

$$0 \leq \frac{1}{q+1}[(w_{\mu_1}+w_n)^{q+1} - w_{\mu_1}^{q+1}] - w_n w_{\mu_1}^q \leq \frac{q}{2}\frac{w_n^2}{w_{\mu_1}^{1-q}}. \tag{11.4.28}$$

Using that w_{μ_1} is a solution of (P_{μ_1}), since

$$|w_n(x) - w_n(y)|^2 \geq (w_{\mu_1}(x) - w_{\mu_1}(y))\left(\frac{w_n^2}{w_{\mu_1}}(x) - \frac{w_n^2}{w_{\mu_1}}(y)\right),$$

we obtain

$$\begin{aligned} C_{N,s}\|w_n\|^2_{H_0^s(\Omega)} &\geq \int_\Omega \left[\lambda \frac{w_{\mu_1}(x)}{|x|^{2s}} + \mu_1 w_{\mu_1}^q(x)\right] \frac{w_n^2}{w_{\mu_1}}(x)\,dx \\ &\geq \lambda \int_\Omega \frac{w_n^2(x)}{|x|^{2s}}\,dx + \mu_0 \int_\Omega \frac{w_n^2}{w_{\mu_1}^{1-q}}(x)\,dx. \end{aligned} \tag{11.4.29}$$

Hence from (11.4.28) and (11.4.29) we get

$$\begin{aligned} &\frac{q}{2}\left(C_{N,s}\|w_n\|^2_{X_0^s(\Omega)} - \lambda \int_\Omega \frac{w_n^2}{|x|^{2s}}\,dx\right) \\ &\geq \mu_0 \int_{\tilde{S}_n} \frac{((w_{\mu_1} + w_n)^{q+1} - w_{\mu_1}^{q+1})(x)}{q+1} - w_n w_{\mu_1}^q(x)\,dx. \end{aligned} \tag{11.4.30}$$

Moreover, since $p + 1 > 2$ we have

$$0 \leq \frac{1}{p+1}[(\overline{w} + w_n)^{p+1} - \overline{w}^{p+1}] - \overline{w}^p w_n \leq C(p)(\overline{w}^{p-1} w_n^2 + w_n^{p+1}). \tag{11.4.31}$$

Therefore, from (11.4.27), by (11.4.30) and (11.4.31) it follows that

$$\begin{aligned} \mathcal{J}_{\mu_0}(v_n) &\geq (1-q)\left(a_{N,s}\|w_n\|^2_{H_0^s(\Omega)} - \frac{\lambda}{2}\int_\Omega \frac{w_n^2(x)}{|x|^{2s}}\,dx\right) \\ &\quad + \mathcal{J}_{\mu_0}(z_n) + \int_{\tilde{S}_n} C(p)(-\overline{w}^{p-1} w_n^2 - w_n^{p+1})(x)\,dx, \\ &\geq C_2\|w_n\|^2_{X_0^s(\Omega)} + \mathcal{J}_{\mu_0}(z_n) + \int_{\tilde{S}_n} C(p)(-\overline{w}^{p-1} w_n^2 - w_n^{p+1})(x)\,dx, \end{aligned} \tag{11.4.32}$$

where $C_2 = (1-q)C_1 > 0$. What remains to prove now is

$$\lim_{n\to\infty} |\tilde{S}_n| = 0. \tag{11.4.33}$$

Let $\epsilon, \delta > 0$, and define

$$\begin{aligned} A_n &= \{x \in \Omega : v_n(x) \geq w_{\mu_1}(x) \text{ and } w_{\mu_1} > \underline{w} + \delta\}, \\ B_n &= \{x \in \Omega : v_n(x) \geq w_{\mu_1}(x) \text{ and } w_{\mu_1} \leq \underline{w} + \delta\}. \end{aligned}$$

Since

$$0 = \left|\{x \in \Omega : w_{\mu_1}(x) < \underline{w}\}\right| = \left|\bigcap_{j=1}^{\infty}\left\{x \in \Omega : w_{\mu_1}(x) < \underline{w} + \frac{1}{j}\right\}\right|$$

$$= \lim_{j\to\infty}\left|\left\{x \in \Omega : w_{\mu_1}(x) < \underline{w} + \frac{1}{j}\right\}\right|, \tag{11.4.34}$$

for j_0 large enough and $\delta < \frac{1}{j_0}$ we have $\left|\{x \in \Omega : w_{\mu_1}(x) < \underline{w} + \delta\}\right| \le \frac{\epsilon}{2}$.

Therefore $|B_n| \le \frac{\epsilon}{2}$. Moreover, by (11.4.9),

$$\lim_{n\to\infty} \|v_n - \underline{w}\|_{H_0^s(\Omega)} = 0.$$

Then, by Theorem 8.2.7 and the Hölder inequality, it follows that

$$\lim_{n\to\infty} \|v_n - \underline{w}\|_{L^2(\Omega)} = 0.$$

That is, for $n \ge n_0$ large enough we get

$$\frac{\delta^2\varepsilon}{2} \ge \int_\Omega |v_n - \underline{w}|^2\,dx \ge \int_{A_n} |v_n - \underline{w}|^2\,dx \ge \delta^2|A_n|.$$

Therefore $|A_n| \le \frac{\epsilon}{2}$, for $n \ge n_0$. Since $\widetilde{S}_n \subset B_n \cup A_n$ we conclude that $|\widetilde{S}_n| \le \varepsilon$ for $n \le n_0$. Hence (11.4.33) follows.

Then, by (11.4.33), the Sobolev inequality and Theorem 8.2.7, we obtain

$$\int_{\widetilde{S}_n} w_n^{p+1}(x) + \overline{w}^{p-1}w_n^2(x)\,dx \le o(1)\big(\|w_n\|^2_{H_0^s(\Omega)} + \|w_n\|^{p+1}_{H_0^s(\Omega)}\big).$$

Therefore from (11.4.32) we conclude that

$$\mathcal{J}_{\mu_0}(v_n) \ge C_2\|w_n\|^2_{H_0^s(\Omega)} + \mathcal{J}_{\mu_0}(z_n) - o(1)\big(\|w_n\|^2_{H_0^s(\Omega)} + \|w_n\|^{p+1}_{H_0^s(\Omega)}\big).$$

Hence, for n large enough, since $z_n \in W$ and $\underline{w}$ was the infimum of $\mathcal{J}_{\mu_0}$ over W, this implies

$$\mathcal{J}_{\mu_0}(v_n) \ge \mathcal{J}_{\mu_0}(z_n) \ge \mathcal{J}_{\mu_0}(\underline{w}),$$

which is a contradiction with (11.4.9). Hence $\underline{w}$ is a minimum. □

Remark 11.4.6. If $\lambda = \Lambda_{N,2s}$ and $1 < p < 2_s^* - 1$, using the improved Hardy–Sobolev inequality given in section 9.6, (see too [24] and [168]), we could try to prove the existence of at least two positive solutions in a suitable bounded interval $\mu \in (0, M)$.

For the local case these kinds of results were obtained in [181] even in the framework of the Caffarelli–Kohn–Nirenberg inequalities in [108].

11.4.2 The critical problem: $p = 2_s^* - 1$

In Section 4.6 we have described the results about the critical problem with the Hardy potential for the Laplacian in the whole $\mathbb{R}^N$ obtained by S. Terracini in [304]. In the article [140] the corresponding problem for the fractional Laplacian is studied. More precisely, consider the problem

$$(-\Delta)^s u = \lambda \frac{u}{|x|^{2s}} + u^{2_s^*-1}, \quad u \in \dot{H}^s(\mathbb{R}^N), \tag{11.4.35}$$

with $N > 2s$, $0 < s < 1$, $2_s^* = \frac{2N}{N-2s}$ and $0 < \lambda < \Lambda_{N,2s}$. The main results are the following.

Theorem 11.4.7. *Let $0 \le \lambda < \Lambda_{N,2s}$. Then problem* (11.4.35) *has a positive solution $u \in \dot{H}^s(\mathbb{R}^N)$ which is a minimizer of the problem*

$$S_{N,s}(\lambda) := \inf_{u \in \dot{H}^s(\mathbb{R}^N)\setminus\{0\}} Q_\lambda(u), \tag{11.4.36}$$

where

$$Q_\lambda(u) := \frac{\left(a_{N,s} \int_{\mathbb{R}^N}\int_{\mathbb{R}^N} \frac{|u(x)-u(y)|^2}{|x-y|^{N+2s}}\,dx\,dy - \lambda \int_{\mathbb{R}^N} \frac{u^2}{|x|^{2s}}\,dx\right)^{\frac{2_s^*}{2}}}{\int_{\mathbb{R}^N} |u|^{2_s^*}\,dx}. \tag{11.4.37}$$

Moreover:

(1) *If $u \in \dot{H}^s(\mathbb{R}^N)$ is a positive solution to* (11.4.35), *then u is radial and radially decreasing with respect to the origin. Namely, there exists some strictly decreasing function $v : (0, +\infty) \to (0, +\infty)$ such that*

$$u(x) = v(r), \quad r = |x|.$$

(2) *If $u \in \dot{H}^s(\mathbb{R}^N)$ is a positive solution to* (11.4.35), *then there exist two positive constants c and C such that*

$$\frac{c}{(|x|^{1-\eta_\lambda}(1+|x|^{2\eta_\lambda}))^{\frac{N-2s}{2}}} \le u(x) \le \frac{C}{(|x|^{1-\eta_\lambda}(1+|x|^{2\eta_\lambda}))^{\frac{N-2s}{2}}}, \quad x \in \mathbb{R}^N \setminus \{0\}, \tag{11.4.38}$$

where

$$\eta_\lambda = 1 - \frac{2\alpha_\lambda}{N-2s}, \tag{11.4.39}$$

$\alpha_\lambda \in (0, \frac{(N-2s)}{2})$ verifies the equation $\lambda(\alpha) = \Psi_N(\frac{N-2s}{2} - \alpha)$ and $\Psi_{s,N}(t)$ is defined in (9.4.5).

(3) *There exists a function $\rho(\lambda, N, s)$, $(1 - \frac{\lambda}{\Lambda_{N,s}}) \le \rho(\lambda, N, s) < 1$, such that*

$$S_{N,s}(\lambda) = \rho(\lambda, N, s) S_{N,s}.$$

We refer the reader to [140] for the proof and some related details.

Remark 11.4.8. Note that in Theorem 11.4.7 the uniqueness of solutions up to dilations remains open.

Now we come back to the problem under study,

$$\begin{cases} (-\Delta)^s u = \lambda \frac{u}{|x|^{2s}} + u^{2_s^*-1} + \theta u^q, & x \in \Omega, \\ u(x) \ge 0, & x \in \Omega, \\ u(x) = 0, & x \in \mathbb{R}^N \setminus \Omega, \end{cases} \tag{11.4.40}$$

with $N > 2s$, $0 < s < 1$, $2_s^* = \frac{2N}{N-2s}$, $0 < q < 1$, $\theta > 0$ and $0 < \lambda < \Lambda_{N,2s}$. See [284] for a related result in the whole $\mathbb{R}^N$. The solutions in $H_0^s(\Omega)$ (11.4.40) are critical points of the energy functional,

$$\begin{aligned} \mathcal{J}_\theta(u) &:= a_{N,s} \int_Q \frac{|u(x) - u(y)|^2}{|x - y|^{N+2s}} \, dx \, dy \\ &\quad - \frac{\lambda}{2} \int_\Omega \frac{(u_+)^2}{|x|^{2s}} \, dx - \frac{1}{2_s^*} \int_\Omega u_+^{2_s^*} \, dx - \frac{\theta}{q+1} \int_\Omega u_+^{q+1} \, dx, \end{aligned} \tag{11.4.41}$$

which is well defined in $H_0^s(\Omega)$. The existence of a minimal nontrivial solution has been proved in Section 11.3.

Remark 11.4.9. If u is a weak solution to (11.4.41), then by Lemma 10.3.8, for each ball $B_\delta(0) \subset\subset \Omega$,

$$u \ge C(N, \delta, \gamma)|x|^{-\gamma} \quad \text{in } B_\delta(0),$$

where $\gamma = \frac{N-2s}{2} - \alpha(\lambda)$ is defined in (9.4.9). By using the ground state transformation in problem (11.4.40), that is, by putting $u = |x|^\gamma v$ with γ defined in (9.4.9), we arrive at equation (10.3.1) with a second member, that is,

$$f_\theta(v) = \left(\frac{v^{2_s^*-1}}{|x|^{2_s^*\gamma}} + \theta \frac{v^q}{|x|^{(q+1)\gamma}} \right).$$

One can prove that any variational positive solution to the transformed problem is bounded. The idea is to use the fractional Caffarelli–Kohn–Nirenberg inequalities presented in Section 10.2 and then use the Moser type argument introduced by N. Trudinger to study the Yamabe problem in [308]. The details are left to the interested reader. These arguments were used in [256] for the p-Laplacian case and in [44] for problem (11.4.40) with $\lambda = 0$.

Note that positive solutions to problem (11.4.40) are unbounded in any neighborhood of the origin, but with the previous observations they are bounded in the complement set to any ball centered at 0.

We will use the following concentration compactness, which can be found in the P. L. Lions article [234]. As is customary, δ_x means the *Dirac delta* concentrated at the point x.

Lemma 11.4.10. *Assume $\{u_n\}_{n\in\mathbb{N}} \subset H_0^s(\Omega)$ verifies $u_n \to u$ a.e., and $u_n \to u$ in $L^\alpha(\Omega)$, $\alpha \in [1, 2_s^*)$. If*
(i) $|(-\Delta)^{\frac{s}{2}} u_n|^2 \rightharpoonup d\mu$;
(ii) $|u_n|^{2_s^*} \rightharpoonup d\nu$;
(iii) $\frac{u_n^2}{|x|^2} \rightharpoonup d\gamma$;

where $d\mu$, $\delta\nu$ and $d\gamma$ are Radon measures, then there exists an index set $\mathcal{J}$ at most countable such that
1. $d\mu \geq |(-\Delta)^{\frac{s}{2}} u|^2 + \sum_{j\in\mathcal{J}} \mu_j \delta_{x_j} + \mu_0\delta_0$;
2. $d\nu = |u|^{2^*} + \sum_{j\in\mathcal{J}} \nu_j \delta_{x_j} + \nu_0\delta_0$ *with* $S\nu_j^{\frac{2}{2_s^*}} \leq \mu_j$ *for all* $j \in \mathcal{J} \cup \{0\}$, *where* S *is the optimal Sobolev constant;*
3. $\rightharpoonup d\gamma = \frac{u^2}{|x|^2} + \gamma_0\delta_0$, *with* $\gamma_0\Lambda_{N,s} \leq \mu_0$.

The three main steps of the proof are the following.

STEP 1. *The local Palais–Smale condition.* Here the main difference with respect to the subcritical case is the lack of compacteness of the critical term. That is, passing to the limit in the critical term cannot be done directly since the Sobolev embedding is not compact. We need to have a level of energy strictly less than the one given by the concentration compactness by P. L. Lions in [233, 234]. The following lemmas are needed to find the first level of concentration. See Lemmas 2.8 and 2.9 in [44]. These results are the extension to the fractional framework of well-known results for the local case. See for instance [177].

Lemma 11.4.11. *Let $\phi \in C_0^\infty(\mathbb{R}^N)$ be a radial nonincreasing test function and $\phi_\epsilon(x) = \phi(\frac{x}{\epsilon})$. Assume $\{u_n\}_{n\in\mathbb{N}} \subset H_0^s(\Omega)$ verifying the hypotheses of Lemma 11.4.10. Then:*
1. $\lim_{\epsilon\to 0} \lim_{n\to\infty} |\int_{\mathbb{R}^N} u_n (-\Delta)^{\frac{s}{2}} \phi_\epsilon (-\Delta)^{\frac{s}{2}} u_n \, dx| = 0$;
2. $\lim_{\epsilon\to 0} \lim_{n\to\infty} |\int_{\mathbb{R}^N} (-\Delta)^{\frac{s}{2}} u_n \int_{\mathbb{R}^N} \frac{(\phi_\epsilon(x)-\phi_\epsilon(y))(u_n(x)-u_n(y))}{|x-y|^{N+2s}} \, dy \, dx| = 0$.

Lemma 11.4.12. *Let $\{u_n\}_{n\in\mathbb{N}} \subset H_0^s(\Omega)$ be a Palais–Smale sequence for $\mathcal{J}_\theta$, namely,*

$$\mathcal{J}_\theta(u_n) \to c < \infty, \mathcal{J}_\theta'(u_n) \to 0.$$

Let $M = M(\Omega, s, N, q)$ be the constant defined by the identity

$$\min_{t>0} f(t) = f(t_0) = -M\theta^{\frac{2_s^*}{2_s^*-q}},$$

with $f(t) = \frac{s}{N}t^{2_s^*} - (\frac{1}{q} - \frac{1}{2})\theta|\Omega|^{\frac{2_s^*-q}{2_s^*}} t^q$. *If*

$$c < c^* = \frac{s}{N}S_{N,s}(\lambda)^{\frac{N}{2s}} - M\theta^{\frac{2_s^*}{2_s^*-q}} = \frac{s}{N}(S_{N,s}\rho(\lambda, N, s))^{\frac{N}{2s}} - M\theta^{\frac{2_s^*}{2_s^*-q}},$$

then $\{u_n\}_{n\in\mathbb{N}}$ *has a strongly convergent subsequence in* $H_0^s(\Omega)$.

Proof. Assume that $\{u_n\}_{n\in\mathbb{N}}$ verifies the hypotheses. Then the sequence is bounded in $H_0^s(\Omega)$. To see this fact, it is sufficient to note that applying Hardy's, Hölder's and Sobolev's inequalities, it follows that

$$\begin{aligned} c &= \lim_{n\to\infty}\left(\mathcal{J}_\theta(u_n) - \frac{1}{2_s^*}\langle \mathcal{J}_\theta'(u_n), u_n\rangle\right) \\ &\ge \left(\frac{1}{2} - \frac{1}{2_s^*}\right)\left(1 - \frac{\lambda}{\Lambda_{N,2s}}\right)a_{N,s}\|u\|^2_{H_0^s(\Omega)} - \left(\frac{1}{q+1} - \frac{1}{2_s^*}\right)K\|u\|^{\frac{q+1}{2}}_{H_0^s(\Omega)}, \end{aligned}$$

where $K = K(s, N, \Omega, q)$. This implies the boundedness. As a consequence we find a subsequence, relabeled to $\{u_n\}_{n\in\mathbb{N}}$, that verifies

$$\left[\begin{array}{l} u_n \rightharpoonup u \quad \text{as } n \to \infty \text{ weakly in } H_0^s(\Omega), \\ \|u_n - u\|_{L^r(\Omega)} \to 0 \quad \text{as } n \to \infty \text{ if } 1 \le r < 2_s^* - 1, \\ u_n(x) \to u(x) \quad \text{as } n \to \infty \text{ almost everywhere in } \Omega \text{ and} \\ \quad \text{the hypotheses of Lemma 11.4.10.} \end{array}\right. \tag{11.4.42}$$

Hence we have, in particular, the representation of the measures of Lemma 11.4.10.

Claim. *The set* J *is finite and if* $j \ne 0$, *either* $\nu_j = 0$ *or* $\nu_j \ge S_{N,s}^{\frac{N}{2s}}$.

Consider a positive function $\phi(x) \in \mathcal{C}_0^\infty(\mathbb{R}^N)$ such that $\phi(x) = 1$ in $B_1(0)$ and $\phi(x) = 0$ in $\mathbb{R}^N \setminus B_2(0)$ and define

$$\phi_{\epsilon,j}(x) = \phi\left(\frac{x - x_j}{\epsilon}\right), \quad \epsilon > 0, j \in J.$$

Note that we have

$$(-\Delta)^{\frac{s}{2}}\phi_{\epsilon,j}(x) = \epsilon^{-s}(-\Delta)^{\frac{s}{2}}\phi\left(\frac{x - x_j}{\epsilon}\right). \tag{11.4.43}$$

Since $\mathcal{J}_\theta'(u_n) \to 0$ in H^{-s} we have, using the Kato inequality,

$$\begin{aligned} 0 &= \lim_{n\to\infty}\langle \mathcal{J}_\theta'(u_n), (\phi_{\epsilon,j}u_n)\rangle \ge \lim_{n\to\infty}\langle \mathcal{J}_\theta'((u_n)_+), (\phi_{\epsilon,j}(u_n)_+)\rangle \\ &= \int_{\mathbb{R}^N}(\phi_{\epsilon,j}(u_n)_+)(-\Delta)^s((u_n)_+)\,dx - \int_{B_{2\epsilon}(x_j)}\lambda\phi_{\epsilon,j}\frac{(u_n)_+^2}{|x|^{2s}}\,dx - \theta\int_{B_{2\epsilon}(x_j)}(u_n)_+^q - \int_{B_{2\epsilon}(x_j)}\phi_{\epsilon,j}(u_n)_+^{2_s^*}. \end{aligned}$$

Hence

$$\lim_{n\to\infty}\Big(\int_{\mathbb{R}^N}(u_n)_+(-\Delta)^{\frac{s}{2}}(u_n)_+(-\Delta)^{\frac{s}{2}}\phi_{\epsilon,j}\,dx$$
$$-2a_{N,s}\int_{\mathbb{R}^N}(-\Delta)^{\frac{s}{2}}(u_n)_+\int_{\mathbb{R}^N}\frac{(\phi_{\epsilon,j}(x)-\phi_{\epsilon,j}(y))((u_n)_+(x)-(u_n)_+(y))}{|x-y|^{N+2s}}\,dy\,dx\Big)$$
$$\leq\lim_{n\to\infty}\Big(\lambda\int_{B_{2\epsilon}(x_j)}\phi_{\epsilon,j}\frac{(u_n)_+^2}{|x|^{2s}}\,dx+\theta\int_{B_{2\epsilon}(x_j)}(u_n)_+^q\,dx+\int_{B_{2\epsilon}(x_j)}\phi_{\epsilon,j}(u_n)_+^{2_s^*}-\int_{B_{2\epsilon}(x_j)}(-\Delta)^{\frac{s}{2}}\phi_{\epsilon,j}\Big).\tag{11.4.44}$$

Suppose first that $j\neq 0$. Therefore

$$\lim_{\epsilon\to 0}\lim_{n\to\infty}\Big(\int_{\mathbb{R}^N}(u_n)_+(-\Delta)^{\frac{s}{2}}(u_n)_+(-\Delta)^{\frac{s}{2}}\phi_{\epsilon,j}\,dx$$
$$-2\int_{\mathbb{R}^N}(-\Delta)^{\frac{s}{2}}(u_n)_+a_{N,s}\int_{\mathbb{R}^N}\frac{(\phi_{\epsilon,j}(x)-\phi_{\epsilon,j}(y))((u_n)_+(x)-(u_n)_+(y))}{|x-y|^{N+2s}}\,dy\,dx\Big)$$
$$\leq\lim_{\epsilon\to 0}\Big(\lambda\int_{B_{2\epsilon}(x_j)}\phi_{\epsilon,j}\frac{(u)_+^2}{|x|^{2s}}\,dx+\theta\int_{B_{2\epsilon}(x_j)}(u)_+^q\,dx+\int_{B_{2\epsilon}(x_j)}\phi_{\epsilon,j}\,d\nu-\int_{B_{2\epsilon}(x_j)}\phi_{\epsilon,j}\,d\mu\Big).$$

Since by Lemma 11.4.11 the left hand is zero we obtain

$$0\leq\lim_{\epsilon\to 0}\Big(\lambda\int_{B_{2\epsilon}(x_j)}\phi_{\epsilon,j}\frac{(u)_+^2}{|x|^{2s}}\,dx+\theta\int_{B_{2\epsilon}(x_j)}(u)_+^q\,dx+\int_{B_{2\epsilon}(x_j)}\phi_{\epsilon,j}\,d\nu-\int_{B_{2\epsilon}(x_j)}\phi_{\epsilon,j}\,d\mu\Big)=\nu_j-\mu_j.$$

Now, by the P. L. Lions lemma, Lemma 11.4.10, $S\nu_j^{\frac{2}{2_s^*}}\leq\mu_j$ and the measure $d\nu$ is finite, thus the claim follows.

We study next the concentration at the pole of the Hardy potential. We consider the test functions $\phi_{\epsilon,0}$ as above. We have

$$Q(u_n\phi_{\epsilon,0})\geq S_{N,s}(\lambda),$$

where Q is defined in (11.4.37) and $S_{N,s}(\lambda)$ by (11.4.36). That is,

$$a_{N,s}\int_{\mathbb{R}^N}\int_{\mathbb{R}^N}\frac{|u_n(x)\phi_{\epsilon,0}(x)-u_n(y)\phi_{\epsilon,0}(y)|^2}{|x-y|^{N+2s}}\,dx\,dy-\frac{\lambda}{2}\int_{\mathbb{R}^N}\frac{(u_n\phi_{\epsilon,0})^2}{|x|^{2s}}\,dx$$
$$\geq S_{N,s}(\lambda)\Big(\int_{\mathbb{R}^N}|u_n\phi_{\epsilon,0}|^{2_s^*}\,dx\Big)^{\frac{2}{2_s^*}}.\tag{11.4.45}$$

From the nonlocal Leibniz formula in (8.1.17), with

$$B_s(u_n, \phi_{\epsilon,0})(x) = C_{N,\frac{s}{2}} \int_{\mathbb{R}^N} \frac{(u_n(x) - u_n(y))(\phi_{\epsilon,0}(x) - \phi_{\epsilon,0}(y))}{|x-y|^{N+s}}\, dy,$$

we find

$$\begin{aligned}
&\|(-\Delta)^{\frac{s}{2}}(u_n\phi_{\epsilon,0})\|_2^2 \\
&= \|\phi_{\epsilon,0}(-\Delta)^{\frac{s}{2}}(u_n) + u_n(-\Delta)^{\frac{s}{2}}(\phi_{\epsilon,0}) - B_s(u_n,\phi_{\epsilon,0})\|_2^2 \\
&= \|(-\Delta)^{\frac{s}{2}}(u_n)\|_2^2 + \|u_n(-\Delta)^{\frac{s}{2}}(\phi_{\epsilon,0})\|_2^2 + \|B_s(u_n,\phi_{\epsilon,0})\|_2^2 \\
&\quad + 2\int_{\mathbb{R}^N} \phi_{\epsilon,0}(-\Delta)^{\frac{s}{2}}(u_n)u_n(-\Delta)^{\frac{s}{2}}(\phi_{\epsilon,0})\, dx \\
&\quad + 2\int_{\mathbb{R}^N} \phi_{\epsilon,0}(x)(-\Delta)^{\frac{s}{2}}(u_n)(x)B_s(u_n,\phi_{\epsilon,0})\, dx \\
&\quad + 2\int_{\mathbb{R}^N} u_n(x)(-\Delta)^{\frac{s}{2}}(\phi_{\epsilon,0})(x)B_s(u_n,\phi_{\epsilon,0})\, dx \\
&\equiv I_1 + I_2 + I_3 + I_4 + I_5 + I_6.
\end{aligned} \tag{11.4.46}$$

We have to estimate $\lim_{\epsilon\to 0}\lim_{n\to\infty} I_j$, $j = 1,2,3,4,5,6$.

We have

$$\lim_{\epsilon\to 0}\lim_{n\to\infty} I_1 = \lim_{\epsilon\to 0}\int_{\mathbb{R}^N} \phi_{\epsilon,0}^2(x)\, d\mu = \mu_0.$$

Next we deal with I_2:

$$\begin{aligned}
&\lim_{\epsilon\to 0}\lim_{n\to\infty} I_2 \\
&= \lim_{\epsilon\to 0}\lim_{n\to\infty}\int_{\mathbb{R}^N} u_n^2(x)|(-\Delta)^{\frac{s}{2}}(\phi_{\epsilon,0}(x))|^2\, dx \\
&\leq 2\lim_{\epsilon\to 0}\lim_{n\to\infty}\int_{\Omega} |u_n(x) - u(x)|^2|(-\Delta)^{\frac{s}{2}}(\phi_{\epsilon,0}(x))|^2\, dx \\
&\quad + 2\lim_{\epsilon\to 0}\lim_{n\to\infty}\int_{\Omega} |u(x)|^2|(-\Delta)^{\frac{s}{2}}(\phi_{\epsilon,0}(x))|^2\, dx \\
&\leq A_1\lim_{\epsilon\to 0}\lim_{n\to\infty}\epsilon^{-s}\|u_n - u\|_2^2 + A_2\lim_{\epsilon\to 0}\lim_{n\to\infty}\int_{\mathbb{R}^N} u^2(x)|(-\Delta)^{\frac{s}{2}}(\phi_{\epsilon,0}(x))|^2\, dx \\
&= A_2\lim_{\epsilon\to 0}\int_{\mathbb{R}^N} u^2(x)|(-\Delta)^{\frac{s}{2}}(\phi_{\epsilon,0}(x))|^2\, dx \equiv J_1.
\end{aligned}$$

Since $u^2 \in L^{\frac{N}{N-2s}}(\Omega)$,

$$J_1 \le A_3 \lim_{\epsilon\to 0}\big(\|u\|_{L^{\frac{N}{N-2s}}(\Omega)} \|(-\Delta)^{\frac{s}{2}}(\phi_{\epsilon,0})\|_{L^{\frac{N}{2s}}(\Omega)}\big)$$
$$= A_4 \|u\|_{L^{\frac{N}{N-2s}}} \lim_{\epsilon\to 0}\|(-\Delta)^{\frac{s}{2}}(\phi_{\epsilon,0})\|_{L^{\frac{N}{2s}}(\Omega)} = A_4\|u\|_{L^{\frac{N}{N-2s}}} \lim_{\epsilon\to 0} \epsilon^s = 0,$$

and therefore

$$\lim_{\epsilon\to 0}\lim_{n\to\infty} I_2 = 0.$$

To prove that

$$\lim_{\epsilon\to 0}\lim_{n\to\infty} I_3 = 0,$$

we follow the arguments as in the proof of Lemma 2.9 of [44]. The triangular inequality gives us

$$\|B_s(u_n, \phi_{\epsilon,0})\|_2 \le \|B_s(u_n - u, \phi_{\epsilon,0})\|_2 + \|B_s(u, \phi_{\epsilon,0})\|_2$$

and we evaluate separately both terms. By the Cauchy–Schwarz inequality we have

$$\|B_s(u_n - u, \phi_{\epsilon,0})\|_2^2 \le \int_{\mathbb{R}^N} B_s(u_n - u, u_n - u)(x) B_s(\phi_{\epsilon,0}, \phi_{\epsilon,0})(x)\,dx,$$

and it is easy to check that

$$B_s(\phi_{\epsilon,0}, \phi_{\epsilon,0})(x) = \epsilon^{-s} B_s(\phi_0, \phi_0)\Big(\frac{x}{\epsilon}\Big) \le C\frac{\epsilon^{-s}}{1 + |\frac{x}{\epsilon}|^{N+s}} \equiv C\epsilon^{-s}\psi_\epsilon(x) \le C\epsilon^{-s}, \tag{11.4.47}$$

where $\psi_\epsilon(x) := \frac{1}{1+|\frac{x}{\epsilon}|^{N+s}}$. Hence

$$\begin{aligned}
&\|B_s(u_n - u, \phi_{\epsilon,0})\|_2^2 \\
&\quad \le C\epsilon^{-s} \int_{\mathbb{R}^N} B_s(u_n - u, u_n - u)(x)\,dx = C\epsilon^{-s}\int_{\mathbb{R}^N} (u_n - u)(-\Delta)^{\frac{s}{2}}(u_n - u)\,dx \\
&\quad \le C\epsilon^{-s}\|u_n - u\|_2 \|(-\Delta)^{\frac{s}{2}}(u_n - u)\|_2 \le C_1\epsilon^{-s}\|u_n - u\|_2.
\end{aligned} \tag{11.4.48}$$

In a similar way, using (11.4.47), we find

$$\|B_s(u, \phi_{\epsilon,0})\|_2^2 \le \int_{\mathbb{R}^N} B_s(u,u)(x) B_s(\phi_{\epsilon,0}, \phi_{\epsilon,0})(x)\,dx \le C\epsilon^{-s}\int_{\mathbb{R}^N} B_s(u,u)(x)\psi_\epsilon(x)\,dx.$$

We have the following elementary identity:

$$\int_{\mathbb{R}^N} u^2(x)(-\Delta)^{\frac{s}{2}}\psi_\epsilon(x)\,dx = \int_{\mathbb{R}^N} \psi_\epsilon(x)(-\Delta)^{\frac{s}{2}}(u^2)(x)\,dx$$

$$= \int_{\mathbb{R}^N} \psi_\epsilon(x)\{2u(x)(-\Delta)^{\frac{s}{2}}u(x) - B_s(u,u)\}\,dx.$$

Therefore

$$\|B_s(u,\phi_{\epsilon,0})\|_2^2 \le C\epsilon^{-s}\int_{\mathbb{R}^N}\{-u^2(x)(-\Delta)^{\frac{s}{2}}\psi_\epsilon(x) + 2\psi_\epsilon(x)u(x)(-\Delta)^{\frac{s}{2}}u(x)\}\,dx.$$

As in the estimate of I_2,

$$C\epsilon^{-s}\left|\int_{\mathbb{R}^N} -u^2(x)(-\Delta)^{\frac{s}{2}}\psi_\epsilon(x)\,dx\right| \le C_1\epsilon^{-s}\|u^2\|_{L^{\frac{N}{N-2s}}}\|(-\Delta)^{\frac{s}{2}}\psi_\epsilon(x)\|_{L^{\frac{N}{2s}}} \le C_1\|u^2\|_{L^{\frac{N}{N-2s}}}\epsilon^s. \tag{11.4.49}$$

Since $u^2 \in L^{\frac{N}{N-2s}}(\Omega)$, for all $\eta > 0$ there exists $g \in C_0^\infty(\Omega)$ such that

$$\|u^2 - g\|_{L^{\frac{N}{N-2s}}(\Omega)} < \eta.$$

Then the remaining term can be estimated as follows:

$$\begin{aligned}
&2C\epsilon^{-s}\int_{\mathbb{R}^N}\psi_\epsilon(x)u(x)(-\Delta)^{\frac{s}{2}}u(x)\,dx\\
&\le 2C\epsilon^{-s}\|(-\Delta)^{\frac{s}{2}}u\|_2\|\psi_\epsilon(x)u\|_2\\
&\le C_1\epsilon^{-s}\|\psi_\epsilon(x)u\|_2 = C_1\epsilon^{-s}\left(\int_{\mathbb{R}^N}[(u^2-g)\psi_\epsilon^2 + g\psi_\epsilon^2]\,dx\right)^{\frac{1}{2}}\\
&\le C_2\epsilon^{-s}[\|u^2-g\|_{L^{\frac{N}{N-2s}}}\|\psi_\epsilon^2\|_{L^{\frac{N}{2s}}} + \|g\|_\infty\|\psi_\epsilon\|_2^2]^{\frac{1}{2}} \le \epsilon^{-s}[C_3\epsilon^{2s}\eta + C_4\epsilon^N]^{\frac{1}{2}}\\
&\le [C_3\eta + C_4\epsilon^{N-2s}]^{\frac{1}{2}}.
\end{aligned} \tag{11.4.50}$$

Putting together (11.4.48), (11.4.49) and (11.4.50) we find that

$$0 \le \lim_{\epsilon\to 0}\lim_{n\to\infty} I_3 \le C_5\eta^{\frac{1}{2}},$$

and since η is an arbitrary positive number, we conclude. The limits of I_4, I_5 and I_6 are zero as an easy consequence of the previous analysis on the behavior of the three first terms and the Cauchy–Schwarz inequality.

As a consequence we find that taking limits in (11.4.45),

$$\mu_0 - \lambda\gamma_0 \ge S_{N,s}(\lambda)\nu_0^{\frac{2}{2_s^*}} \quad \text{with } \gamma_0\Lambda_{N,2s} \le \mu_0 \text{ and } S_{N,s}\nu_0^{\frac{2}{2_s^*}} \le \mu_0.$$

On the other hand, by the hypotheses,

$$\lim_{\epsilon\to 0}\lim_{n\to\infty}\langle \mathcal{J}_\theta'(u_n), u_n\phi_\epsilon\rangle = 0,$$

and therefore

$$\mu_0 - \lambda\gamma_0 \leq \nu_0.$$

As a consequence, either $\nu_0 = 0$ or $\nu_0 \geq S_{N,s}(\lambda)^{\frac{N}{2s}}$.

Assuming that $u \in H_0^s(\Omega)$ and the weak limit of the sequence and that there exists a concentration point, by Lemma 11.4.10 we find

$$c = \lim_{n\to\infty} \mathcal{J}_\theta(u_n) \geq \mathcal{J}_\theta(u) + \left(\frac{1}{2} - \frac{1}{2_s^*}\right)\sum \nu_j \geq \mathcal{J}_\theta(u) + \frac{s}{N}S_{N,s}(\lambda)^{\frac{N}{2s}},$$

that is, since $c < c^* = \frac{s}{N}S_{N,s}(\lambda)^{\frac{N}{2s}} - M\theta^{\frac{2_s^*}{2_s^*-q}}$, $\mathcal{J}_\theta(u) < 0$, and in particular $u \neq 0$. As a consequence,

$$\begin{aligned} c &= \lim_{n\to\infty}\left(\mathcal{J}_\theta(u_n) - \frac{1}{2}\langle \mathcal{J}_\theta'(u_n), u_n\rangle\right) = \lim_{n\to\infty}\left(\frac{s}{N}\int_\Omega |u_n|^{2_s^*}\,dx - \left(\frac{1}{q}-\frac{1}{2}\right)\theta\int_\Omega u_n^q\,dx\right) \\ &\geq \frac{s}{N}\int_\Omega |u|^{2_s^*}\,dx + \frac{s}{N}S_{N,s}(\lambda)^{\frac{N}{2s}} - \left(\frac{1}{q}-\frac{1}{2}\right)\theta\int_\Omega u^q\,dx. \end{aligned}$$

By Hölder's inequality

$$\left(\frac{1}{q}-\frac{1}{2}\right)\theta\int_\Omega u^q\,dx \leq \left(\frac{1}{q}-\frac{1}{2}\right)\theta|\Omega|^{\frac{2_s^*-q}{2_s^*}}\|u\|_{2_s^*}^q.$$

Note that the function in the statement,

$$f(t) = \frac{s}{N}t^{2_s^*} - \left(\frac{1}{q}-\frac{1}{2}\right)\theta|\Omega|^{\frac{2_s^*-q}{2_s^*}}t^q,$$

attains its negative minimum in $\mathbb{R}_+$ at $t_0 = [(\frac{N-2s}{2s})\theta(\frac{2-q}{2})]^{\frac{1}{2_s^*-q}}|\Omega|^{\frac{1}{2_s^*}}$, that is,

$$f(t) \geq f(t_0) = -M\theta^{\frac{2_s^*}{2_s^*-q}} \quad \text{for all } t > 0, \text{ where } M = M(\Omega, s, N, q).$$

Hence

$$c \geq \frac{s}{N}S_{N,s}(\lambda)^{\frac{N}{2s}} - M\theta^{\frac{2_s^*}{2_s^*-q}}$$

and we arrive at a contradiction with the hypothesis on c. Therefore we have

$$\|u_n\|_{2_s^*} \to \|u\|_{2_s^*} \quad \text{as } n \to \infty$$

and since $u_n \to u$ almost everywhere, by the result of Brezis–Lieb in Theorem 11.2.3 we conclude that

$$\|u_n - u\|_{2_s^*} \to 0 \quad \text{as } n \to \infty.$$

Therefore, using the continuity of the fractional Laplacian from $H^{-s}(\Omega)$ to $H_0^s(\omega)$ we find the Palais–Smale condition. □

Corollary 11.4.13. *For all $\lambda \in [0, \Lambda_{N,2s})$, there exists a $\theta_0(\lambda) > 0$ such that for every $0 < \theta < \theta_0(\lambda)$ there exists $u_0 \in H_0^s(\Omega)$ which is a local minimum of $\mathcal{J}_\theta(u)$.*

Proof. We estimate from above the functional $\mathcal{J}_\theta(u)$ in the following elementary way. Call

$$q_\lambda(u) = \left(a_{N,s}\|u\|_{H^s}^2 - \lambda \int_\Omega \frac{|u|^2}{|x|^{2s}}\,dx\right).$$

Recall that $q_\lambda(u)^{\frac{1}{2}} := \||u\||_{H_0^s(\Omega)}^2$ is an equivalent norm in $H_0^s(\Omega)$. Since $S_{n,s}(\lambda)$ is defined by (11.4.36), we have

$$S_{N,s}^{\frac{2}{2_s^*}}\|u\|_{2_s^*}^2 \le q_\lambda(u).$$

Therefore,

$$\mathcal{J}_\theta(u) \ge \frac{1}{2}q_\lambda(u) - \frac{1}{2_s^* S_{N,s}(\lambda)}q_\lambda(u)^{\frac{2_s^*}{2}} - \frac{\theta}{q+1}|\Omega|^{(1-\frac{q+1}{2^*-s})}\frac{1}{S_{N,s}}q_\lambda^{\frac{q+1}{2}}.$$

Putting $r = q_\lambda(u)^{\frac{1}{2}}$ we have a *radial minorant* of $\mathcal{J}_\theta(u)$ in the Sobolev space, given by

$$\mathcal{J}_\theta(u) \ge \frac{1}{2}r^2 - \frac{1}{2_s^* S^{\frac{2_s^*}{2}}}r^{2_s^*} - C(N,s,q,|\Omega|)\frac{\theta}{q+1}r^{q+1} := h_{\lambda,\theta}(r).$$

For all $0 < \lambda < \Lambda_{N,2s}$ and θ small enough it is easy to check that there exist $0 < r_{\min} < r_0 < r_{\max} < r_1$ such that:

- $h_{\lambda,\theta}(r_{\min}) \le h_{\lambda,\theta}(r) < 0$ for all $0 < r < r_0$;
- $h_{\lambda,\theta}(r_{\max}) \ge\le h_{\lambda,\theta}(r)$ for all $0 < r < r_1$;
- $h_{\lambda,\theta}(r_0) = h_{\lambda,\theta}(r_1) = 0$, $h_{\lambda,\theta}(r) < 0$ if $r > r_1$ and $\lim_{r\to\infty} h_{\lambda,\theta}(r) = -\infty$.

Consider $\tau : \mathbb{R}_+ \to [0,1]$, a nonincreasing function, $\tau \in \mathcal{C}^\infty(\mathbb{R}_+)$ such that $\tau(r) = 1$ if $0 < r \le r_0$ and $\tau(r) = 0$ if $r > r_1$.

Let $\psi(u) := \tau(q_\lambda(u))$ and consider the perturbed functional

$$F(u) = \frac{1}{2}q_\lambda(u) - \frac{1}{2_s^*}\int_\Omega \psi(u)|u|^{2_s^*}\,dx - \frac{\theta}{q+1}\int |u|^{q+1}\,dx.$$

It is clear that F has the same regularity as $\mathcal{J}_\theta$, that is, $F \in C^1(H_0^s(\Omega), \mathbb{R})$. Moreover, (i) if $F(u) \le 0$, then $q_\lambda^{\frac{1}{2}} \le r_0$ and $F(u) = \mathcal{J}_\theta(u)$ in a neighborhood of u; (ii) there exists a $\theta_1 > 0$ such that for all $0 < \theta < \theta_1$, $\frac{s}{N}S_{N,s}(\lambda)^{\frac{N}{2s}} - M\theta^{\frac{2_s^*}{2_s^*-q}} > 0$ and (iii) $F(u)$ is bounded from below.

Let $c_0 = \inf_{u\in H_0^s(\Omega)} F(u)$. Then $c_0 < 0$ and any Palais–Smale sequence to this level has a strongly convergent subsequence. So there exists a $u_0 \in H_0^s(\Omega)$ such that $c_0 = F(u_0)$ is a minimum. Since $F(u) = \mathcal{J}_\theta(u)$ in a neighborhood of u_0, u_0 is a local minimum for $\mathcal{J}_\theta$ and $\mathcal{J}_\theta(u_0) = c_0 < 0$. □

It is possible to read the result in Lemma 11.4.12 in a more accurate way to find the local Palais–Smale condition in order to have a second positive solution (see [178]).

Lemma 11.4.14. *Let $\{u_n\}_{n\in\mathbb{N}} \subset H_0^s(\Omega)$ be a Palais–Smale sequence for $\mathcal{J}_\theta$, namely,*

$$\mathcal{J}_\theta(u_n) \to c < \infty, \mathcal{J}_\theta'(u_n) \to 0.$$

Suppose $\theta > 0$ is small enough and take $u_0 \in H_0^s(\Omega)$ such that $c_0 = \mathcal{J}_\theta(u_0) < 0$ is a local minimum. If c satisfies

$$c < c^* = c_0 + \frac{s}{N} S_{N,s}(\lambda)^{\frac{N}{2s}} = \mathcal{J}_\theta(u_0) + \frac{s}{N} S_{N,s}(\lambda)^{\frac{N}{2s}},$$

then $\{u_n\}_{n\in\mathbb{N}}$ has a strongly convergent subsequence in $H_0^s(\Omega)$.

Proof. We follow the arguments in [178]. First, note that as in the subcritical case $\{u_j\}$ is bounded in $H_0^s(\Omega)$.

Then we obtain a subsequence, again relabeled by $\{u_j\}$, such that it verifies the properties in Lemma 11.4.10 and Lemma 11.4.11.

For $0 < \lambda < \Lambda_{N,2s}$, recall that

$$|||u|||^2_{H_0^s(\Omega)} := a_{N,s} \|u\|^2_{H_0^s(\Omega)} - \lambda \int_\Omega \frac{u^2}{|x|^{2s}}$$

by the Hardy inequality (9.2.1) is an equivalent norm in $H_0^s(\Omega)$. With this notation we can handle simultaneously the principal term and the term associated to the Hardy potential.

By using again Lemma 11.4.10, it is easy to prove that the regular part verifies

$$\langle \mathcal{J}_\theta'(u), u \rangle \leq 0.$$

Assuming the existence of a nontrivial singular part, we get a contradiction with the hypothesis $c < c_0 + \frac{s}{N} S_{N,s}(\lambda)^{\frac{N}{2s}}$.

Recall that u is the weak limit of the sequence u_j. Then

$$\begin{aligned} c_0 + \frac{s}{N} S_{N,s}(\lambda)^{\frac{N}{2s}} &> \lim_n \mathcal{J}_\theta(u_n) = \lim_n \mathcal{J}_\theta(u_n) - \frac{1}{2_s^*} \langle \mathcal{J}_\theta'(u_n), u_n \rangle \\ &= \lim_n \frac{s}{N} |||u_n|||^2_{H_0^s(\Omega)} - \theta \left(\frac{1}{q} - \frac{1}{2_s^*} \right) \int_\Omega |u_n|^q \, dx \\ &\geq \frac{s}{N} |||u|||^2_{H_0^s(\Omega)} - \theta C_q \left(\frac{1}{q} - \frac{1}{2_s^*} \right) |||u|||^{q/2}_{H_0^s(\Omega)}. \end{aligned}$$

In particular, as $c_0 < 0$ we obtain

$$\frac{s}{N} |||u|||^2_{H_0^s(\Omega)} - \theta C_q \left(\frac{1}{q} - \frac{1}{2_s^*} \right) |||u|||^{q/2}_{H_0^s(\Omega)} < \frac{s}{N} S_{N,s}(\lambda)^{\frac{N}{2s}}.$$

Therefore, $|||u|||_{H_0^s(\Omega)} < \alpha(\theta) S_{N,s}(\lambda)^{\frac{N}{4s}}$, where $\alpha(\theta) \to 1$ as $\theta \to 0$.

Taking $g(t) = (t\frac{u}{|||u|||_{H_0^s(\Omega)}})$ we have:

(i) $g'(|||u|||_{H_0^s(\Omega)}) =< \mathcal{J}_\theta'(u), \frac{u}{|||u|||_{H_0^s(\Omega)}} >\le 0$;

(ii) $g(|||u|||_{H_0^s(\Omega)}) \ge f(|||u|||_{H_0^s(\Omega)})$, where $f(x) = \frac{1}{2}x^2 - \frac{1}{2_s^*} S_{N,s}(\lambda)^{-\frac{2_s^*}{2s}} x^{2_s^*} - \frac{\theta}{q} C_q x^q$.

The function f is a *radial minorant* of the functional $\mathcal{J}_\theta$ in the Sobolev space $H_0^s(\Omega)$ (see [177] for some applications of this argument in a different setting).

Moreover $f(\alpha(\theta) S_{N,s}(\lambda)^{\frac{N}{4s}}) \to \frac{s}{N} S_{N,s}(\lambda)^{N/2s}$ as $\theta \to 0$; therefore, for θ small enough we have $f(\alpha(\theta) S_{N,s}(\lambda)^{\frac{N}{4s}}) > 0$. Consider the set A defined by

$$A = \{t \ge 0 \mid \quad g'(t) \le 0\}.$$

Consider I_1, the connected component of A containing $t = 0$, and I_2, the intersection of A with the set where $g(t) \ge 0$. Then $|||u|||_{H_0^s(\Omega)} \in I_1 \cup I_2$. In any case,

$$\mathcal{J}_\theta(u) = g(|||u|||_{H_0^s(\Omega)}) \ge c_o.$$

Then, if there are K Dirac masses we have

$$c_0 + \frac{s}{N} S_{N,s}(\lambda)^{\frac{N}{2s}} > \lim_n \mathcal{J}_\theta(u_n) \ge \mathcal{J}_\theta(u) + K\frac{s}{N} S_{N,s}(\lambda)^{\frac{N}{2s}} \ge c_0 + K\frac{s}{N} S_{N,s}(\lambda)^{\frac{N}{2s}}.$$

Therefore $K = 0$ and we get strong convergence. □

STEP 2. *The mountain pass lemma.* The geometry is the same as in the subcritical case, so that for $\theta > 0$ small enough we find a local minimum. Therefore the application of the *mountain pass lemma* in order to obtain a second positive solution depends of the existence of a Palais–Smale level verifying the smallness stated in Lemma 11.4.12.

STEP 3. *The energy estimate.* This step is just to construct a Palais–Smale sequence verifying the hypotheses of Lemma 11.4.14. At this point the sublinear term and the truncation of the minimizers are crucial in order to break down the energy under the critical level $c^* = c_0 + \frac{s}{N} S_{N,s}(\lambda)^{\frac{N}{2s}}$.

Following [35], the idea is to take $v_\varepsilon = u_0 + tu_\varepsilon$, where

$$u_\varepsilon = \rho(x) U_\varepsilon(x), \quad \rho(x) \text{ being a cutoff function centered at } 0, \tag{11.4.51}$$

and U_ε is a minimizer of (11.4.37), for which we have found the behavior given in (11.4.38). To complete the proof we need to estimate carefully the energy of this function v_ε.

In the fractional case we have an extra difficulty given that we do not have the explicit expression of the minimizer but only the precise behavior, so the determination of his s-Laplacian cannot be done directly. At this point we will

use the strategy developed by Luis Caffarelli and Luis Silvestre in [105], which we recall briefly.
Consider the Poisson kernel

$$p_s(x,y) = \frac{1}{y^N} k\left(\frac{x}{y}\right), \quad \text{where } k(x) = \frac{\alpha_{N,s}}{(1+|x|^2)^{\frac{N+2s}{2}}} \tag{11.4.52}$$

and $\alpha_{N,s} > 0$ is chosen in order to have

$$\int_{\mathbb{R}^N} k(x)\,dx = 1.$$

Given $u \in H^s(\mathbb{R}^N)$, define

$$U(x,y) =: \int_{\mathbb{R}^N} p_s(x-z)u(z)dz. \tag{11.4.53}$$

Then one can check (see [105]) that U solves the following elliptic problem:

$$\begin{cases} -\operatorname{div}(y^{1-2s}\nabla U) = 0 & \text{in } \mathbb{R}^{N+1}_+, \\ U(x,0) = u(x) & \text{on } \mathbb{R}^N, \end{cases} \tag{11.4.54}$$

and moreover the Dirichlet–Neumann transformation gives

$$\lim_{y\to 0} y^{1-2s}\frac{\partial U}{\partial y}(x,y) = -d_s(-\Delta)^s u(x), \quad d_s = 2^{2s-1}\frac{\Gamma(s)}{\Gamma(1-s)}.$$

The last identity establishes the connection between the problem for the s-Laplacian in $\mathbb{R}^N$ and the weighted local problem (11.4.54) in $\mathbb{R}^{N+1}_+$.
If we consider the weighted Sobolev space, $X^s(\mathbb{R}^{N+1}_+)$, as the completion of $\mathcal{C}_0^\infty(\overline{\mathbb{R}^{N+1}_+})$ with respect to the seminorm

$$\|U\|_{X^s} := \left(d_s \int_{\mathbb{R}^{N+1}_+} y^{1-2s}|\nabla U(x,y)|^2\,dx\,dy\right)^{\frac{1}{2}},$$

then we have

$$\|U\|_{X^s} = \|u\|_{H^s}. \tag{11.4.55}$$

Since we know the behavior given by (11.4.38) of the minimizers u_ϵ of (11.4.37), the isometric identity (11.4.55), it is sufficient to evaluate the behavior of $\|U_\epsilon\|_{X^s}$ and then $\|u_\epsilon\|_{H^s}$, where, as above, $u_\varepsilon = \rho(x)U_\varepsilon$, with $\rho(x)$ being a cutoff function centered at the point 0.

Lemma 11.4.15. *Let μ_λ be defined by* (11.4.39) *and $N > 2s$. The following estimates hold:*

(i) $\|u_\epsilon\|^2_{H_0^s(\Omega)} = \|U_1\|^2_{H^s(\mathbb{R}^N)} + \mathcal{O}(\epsilon^{\mu_\lambda(N-2s)})$;
(ii) $\int_\Omega \frac{u_\epsilon^2}{|x|^2 s}\,dx = \int_{\mathbb{R}^N} \frac{U_1^2}{|x|^2 s}\,dx + \mathcal{O}(\epsilon^{\mu_\lambda(N-2s)})$;
(iii) *letting* $0 < q \le \frac{N+2s}{N-2s}$,

$$\int_\Omega |u_\epsilon|^{q+1}\,dx \ge \begin{cases} C\epsilon^{\frac{N-2s}{2}(q+1)\mu_\lambda}, & 0 < q < q_\lambda = \frac{2N}{(N-2s)}\frac{1}{(1+\mu_\lambda)} - 1, \\ C\epsilon^{N-\frac{N-2s}{2}(q+1)}\log(\frac{1}{\epsilon}), & q = q_\lambda = \frac{2N}{(N-2s)}\frac{1}{(1+\mu_\lambda)} - 1, \\ C\epsilon^{N-\frac{N-2s}{2}(q+1)}, & q > q_\lambda = \frac{2N}{(N-2s)}\frac{1}{(1+\mu_\lambda)} - 1; \end{cases}$$

(iv) *moreover,* $\int_\Omega |u_\epsilon|^{2_s^*}\,dx = \int_{\mathbb{R}^N} |U_1|^{2_s^*}\,dx + \mathcal{O}(\epsilon^{\mu_\lambda N})$.

Proof. See [284] to find the details. □

We are looking for an estimate of the type

$$\sup_{t>0} \mathcal{J}_\theta(u_0 + tu_\varepsilon) < \mathcal{J}_\theta(u_0) + \frac{s}{N} S_{N,s}(\lambda)^{N/2s},$$

for $\epsilon > 0$ small. Then for this value of ϵ we can choose $v_\epsilon = u_0 + tu_\varepsilon$, with t large enough, such that $\mathcal{J}_\theta(v_\epsilon) < c_0$.

Given $v \in H_0^s(\Omega)$ such that $\mathcal{J}_\theta(v) < c_0$ there exists a sequence $\{u_j\} \in H_0^s(\Omega)$ such that

$$\text{(i)}\quad \mathcal{J}'_\theta(u_j) \to 0,$$
$$\text{(ii)}\quad \mathcal{J}_\theta(u_j) \to c = \inf_{\gamma\in\mathcal{C}} \sup_{t\in[0,1]} \mathcal{J}_\theta(\gamma(t)),$$

where $\mathcal{C} = \{\gamma \in C([0,1], H_0^s(\Omega)) \mid \gamma(0) = u_0, \gamma(1) = v\}$. Therefore, to obtain a critical point we must find $v \in H_0^s(\Omega)$ such that $c < c_0 + \frac{s}{N} S_{N,s}(\lambda)^{N/2s}$.

Lemma 11.4.16. *Suppose $\theta > 0$ is small enough and take $u_0 \in H_0^s(\Omega)$ such that $c_0 = \mathcal{J}_\theta(u_0) < 0$ is a local minimum. Let u_ϵ be defined in* (11.4.51). *Then*

$$\sup_{t>0} \mathcal{J}_\theta(u_0 + tu_\varepsilon) < \mathcal{J}_\theta(u_0) + \frac{s}{N} S_{N,s}(\lambda)^{N/2s}.$$

Proof. Note that by the scalar product in $H_0^s(\Omega)$, we have

$$\|u_0 + tu_\epsilon\|^2_{H^s} = \|u_0\|^2_{H^s} + t^2\|u_\epsilon\|^2_{H^s} + 2t \int_{\mathbb{R}^N}\int_{\mathbb{R}^N} \frac{(u_0(x) - u_0(y))(u_\epsilon(x) - u_\epsilon(y))}{|x-y|^{N+2s}}\,dx\,dy,$$

$$\int_\Omega \frac{|u_0 + tu_\epsilon|^2}{|x|^{2s}}\,dx = \int_\Omega \frac{|u_0|^2}{|x|^{2s}}\,dx + t^2 \int_\Omega \frac{|u_\epsilon|^2}{|x|^{2s}}\,dx + 2t \int_\Omega \frac{u_0 u_\epsilon}{|x|^{2s}}\,dx. \tag{11.4.56}$$

By using the elementary inequality for the critical exponent $2_s^* = \frac{2N}{N-2s}$,

$$(1+\tau)^{2_s^*} \geq 1 + \tau^{2_s^*} + 2_s^*\tau + 2_s^*\tau^{2_s^*-1} + C\tau^\gamma, \quad \tau \geq 0,$$

valid for $\gamma \in (1, 2_s^* - 1]$, we can estimate from below the critical term. Indeed,

$$\int_\Omega |u_0 + tu_\epsilon|^{2_s^*}\,dx \geq \int_\Omega |u_0|^{2_s^*}\,dx + t^{2_s^*}\int_\Omega |u_\epsilon|^{2_s^*}\,dx$$
$$+ 2_s^* t\int_\Omega |u_0|^{2_s^*-1}u_\epsilon\,dx + 2_s^* t\int_\Omega u_0 u_\epsilon^{2_s^*-1} + C\int_\Omega u_0^{2_s^*-\gamma}u_\epsilon^\gamma. \tag{11.4.57}$$

For $0 < q < 1$, the mean value theorem provides the inequality

$$\int_\Omega (u_0 + tu_\varepsilon)^{q+1}\,dx \geq \int_\Omega u_0^{q+1}\,dx + (q+1)t\int_\Omega u_0^q u_\varepsilon\,dx. \tag{11.4.58}$$

Since u_0 is a local minimum that is, in particular, a solution of problem (11.4.40), by inequalities (11.4.56), (11.4.57) and (11.4.58), we find

$$\mathcal{J}_\theta(u_0 + tu_\varepsilon) \leq \mathcal{J}_\theta(u_0) + \frac{t^2}{2}\left(a_{N,s}\|u_\epsilon\|_{H^s}^2 - \lambda\int_\Omega \frac{|u_\epsilon|^2}{|x|^{2s}}\,dx\right) - \frac{1}{2_s^*}t^{2_s^*}\int_\Omega |u_\epsilon|^{2_s^*}\,dx$$
$$+ 2_s^* t\int_\Omega u_0 u_\epsilon^{2_s^*-1} + C\int_\Omega u_0^{2_s^*-\gamma}u_\epsilon^\gamma.$$

We have to estimate the last two terms. Since $2_s^* - 1 > q_\lambda$ and u_0 is bounded in the complementary of any ball centered at the origin, we find

$$\int_\Omega u_0 u_\epsilon^{2_s^*-1}\,dx \geq C\epsilon^{\frac{N-2s}{2}\mu_\lambda},$$

according to Lemma 11.4.15. Taking a $\gamma \in (\frac{N}{N-2s}, \frac{N+2s}{N-2s})$ we can have $\gamma < q_\lambda$ and then, again by the boundedness of u_0 outside of the origin,

$$\int_\Omega u_0^{2_s^*-\gamma}u_\epsilon^\gamma \geq C\epsilon^{N-\frac{(N-2s)}{2}\gamma},$$

by Lemma 11.4.15.

Therefore we obtain

$$\mathcal{J}_\theta(u_0 + tu_\varepsilon) \leq \mathcal{J}_\theta(u_0) + t^2\frac{1}{2}\left(a_{N,s}\|U_1\|_{H^s}^2 - \lambda\int_\Omega \frac{|U_1|^2}{|x|^{2s}}\,dx\right) - \frac{1}{2_s^*}t^{2_s^*}\int_\Omega |U_1|^{2_s^*}\,dx$$
$$- Ct^{2_s^*-1}\epsilon^{\frac{N-2s}{2}\mu_\lambda} + o(\epsilon^{\frac{N-2s}{2}\mu_\lambda}).$$

To arrive to the strict inequality let us consider the function

$$h_\epsilon(t) := t^2 \frac{1}{2}\left(a_{N,s}\|U_1\|^2_{H^s} - \lambda \int_\Omega \frac{|U_1|^2}{|x|^{2s}}\, dx \right) - \frac{1}{2_s^*} t^{2_s^*} \int_\Omega |U_1|^{2_s^*}\, dx - Ct^{2_s^*-1} \epsilon^{\frac{N-2s}{2}\mu_\lambda}.$$

Putting, as above, $q_\lambda(u) = \frac{1}{2}(a_{N,s}\|U_1\|^2_{H^s} - \lambda \int_\Omega \frac{|U_1|^2}{|x|^{2s}}\, dx)$, we can write

$$h_\epsilon(t) = \frac{t^2}{2} q_\lambda(U_1) - \frac{t^{2_s^*}}{2_s^*} \int_\Omega |U_1|^{2_s^*}\, dx - Ct^{2_s^*-1} \epsilon^{\frac{N-2s}{2}\mu_\lambda}.$$

For $\epsilon = 0$, $h_0(t)$ achieves a maximum at $t_0 = [\frac{q_\lambda(U_1)}{\|U_1\|^{2_s^*}_{2_s^*}}]^{\frac{1}{2_s^*-2}}$ and moreover, since U_1 is a minimizer of (11.4.37),

$$h_0(t_0) = \frac{s}{N} S_{N,s}(\lambda)^{\frac{N}{2s}}.$$

For all $t > 0$, $\epsilon > 0$ we have $h_\epsilon(t) < h_0(t) \leq \frac{s}{N} S_{N,s}(\lambda)^{\frac{N}{2s}}$, and hence

$$\max_{t>0} h_\epsilon(t) = h_\epsilon(t_\epsilon) < h_0(t_\epsilon) \leq h_0(t_0) = \frac{s}{N} S_{N,s}(\lambda)^{\frac{N}{2s}}.$$

Note that the point t_ϵ where h_ϵ achieves its maximum satisfies $0 < t_\epsilon < t_0$ and $t\epsilon \to t_0$ as $\epsilon \to 0$. Therefore we write $t_\epsilon = x_\epsilon t_0$ with $x\epsilon \to 1$ as $\epsilon \to 0$.

To finish the proof we have to prove that the error term is negligible.

Since we have $h'_\epsilon(t_\epsilon) = 0$, we get

$$t_0 x_\epsilon q_\lambda(U_1) - t_0^{2_s^*-1} x_\epsilon^{2_s^*-1} \|U_1\|^{2_s^*}_{2_s^*} = C(2_s^* - 1) t_0^{2_s^*-2} x_\epsilon^{2_s^*-2} \epsilon^{\mu_\lambda \frac{N-2s}{2}}.$$

Including the precise value of t_0 and after some computations we arrive at

$$1 - x_\epsilon^{2_s^*-2} = A x_\epsilon^{2_s^*-3} \varepsilon^{\mu_\lambda \frac{N-2s}{2}},$$

where $A = A(N, s, \lambda, \|U_1\|_{2_s^*})$. By Taylor's expansion,

$$(1 - x_\varepsilon)(2_s^* - 2) x_\varepsilon^{2_s^*-3} + o(1 - x_\varepsilon) = A x_\varepsilon^{2_s^*-3} \varepsilon^{\mu_\lambda \frac{N-2s}{2}}.$$

Therefore, $1 - x_\varepsilon = M\varepsilon^{\mu_\lambda \frac{N-2s}{2}} + o(\varepsilon^{\mu_\lambda \frac{N-2s}{2}})$, for $M = \frac{A}{2_s^*-2}$.

Finally, this identity allows us to prove

$$h_\varepsilon(t_\varepsilon) = \frac{s}{N} S^{N/2s} - Ct_0^{2_s^*-1} \varepsilon^{\mu_\lambda \frac{N-2s}{2}} + o(\varepsilon^{\mu_\lambda \frac{N-2s}{2}}),$$

which shows the strict inequality. □

An immediate consequence of the previous results is the following result.

Theorem 11.4.17. *For every $\lambda \in [0, \Lambda_{N,2s})$, there exists a $\theta_0(\lambda)$ such that for all $0 < \theta < \theta_0(\lambda)$, problem* (11.4.40) *has at least two positive variational solutions of finite energy.*

Proof. From Corollary 11.4.13, for $\theta < \theta_\lambda$ the functional $\mathcal{J}_\theta$ attains a local minimum in u_0 in such a way that $\mathcal{J}_\theta(u_0) < 0$. Then by Lemma 11.4.16 we find a Palais–Smale sequence under the concentration level and then by construction we find a mountain pass critical point that allows us to conclude the proof. □

Remark 11.4.18. The result of existence of at least two positive solutions in the whole range of θ for which we find a minimal solution will be considered in a forthcoming work, with slightly different techniques.

11.5 Nonexistence for $p \geq p(\lambda, s)$: complete blow-up

The main objective in this section is to prove that for $p \geq p(\lambda, s)$ problem (11.1.1) has no weak solutions. To prove the nonexistence result we will use in a suitable way the Picone inequality of Lemma 8.5.2.

Theorem 11.5.1. *Let Ω be a bounded domain of $\mathbb{R}^N$, such that $0 \in \Omega$. Assume that $0 < \lambda \leq \Lambda_{N,2s}$ and $p \geq p(\lambda, s)$. Then, there does not exist any positive weak solution in the sense of Definition* 11.2.2 *to problem* (11.1.1).

Proof. We argue by contradiction. Assume that there exists $u > 0$, a weak solution to problem (11.1.1). We can approximate u by a sequence of approximate solutions, for instance, the solutions $u_k \in H_0^s(\Omega)$ to the sequence of problems

$$(P_k) = \begin{cases} (-\Delta)^s u_k - \lambda \frac{u_k}{|x|^{2s}} = T_k(u)^p + \mu T_k(u)^q & \text{in } \Omega, \\ u_k = 0 & \text{in } \mathbb{R}^N \setminus \Omega. \end{cases}$$

(i) We consider first the case $\lambda < \Lambda_{N,2s}$. By the Hardy inequality and a slight modification of the weak maximum principle proved in Theorem 8.3.11 we obtain

$$u_1 < u_2 < \cdots < u_k < \cdots < u,$$

and then we can conclude that $u(x) = \lim_{k\to\infty} u_k(x)$ pointwise and in $L^p(\Omega)$.

(a) Assume $p > p(\lambda, s)$. By Lemma 10.3.8 there exists $C > 0$ such that

$$u(x) \geq \frac{C}{|x|^\gamma}, \quad \text{for all } x \in B_r(0),$$

with $\gamma = \frac{N-2s-2\alpha_\lambda}{2}$. By the Picone inequality applied to u_k and to any $\phi \in C_0^\infty(B_r(0))$ we find

$$\int\limits_{B_r(0)} \frac{T_k(u)^p}{u_k}\phi^2\,dx \leq \int\limits_{B_r(0)} \frac{(-\Delta)^s u_k}{u_k}\phi^2 \leq \|\phi\|^2_{H_0^s(\Omega)},$$

and therefore passing to the limit as $k \to \infty$ and by the local estimate of u, it follows that

$$C \int_{B_r(0)} \frac{\phi^2}{|x|^{(p-1)\gamma}}\, dx \le \int_{B_r(0)} u^{p-1}\phi^2\, dx \le \|\phi\|^2_{H_0^s(B_r(0))}.$$

Since $p > p(\lambda, s)$,

$$(p-1)\gamma > (p(\lambda,s)-1)\gamma = \left(\frac{N+2s-2\alpha_\lambda}{N-2s-2\alpha_\lambda} - 1\right)\gamma = \left(\frac{4s}{N-2s-2\alpha_\lambda}\right)\frac{N-2s-2\alpha_\lambda}{2}$$
$$= 2s.$$

But then, the inequality $(p-1)\gamma > 2s$ is a contradiction with the Hardy inequality.

(b) If $p = p(\lambda, s)$ the argument above does not give a contradiction in an easy way. The idea is to obtain a stronger singularity to reach the contradiction. To be systematic, the calculations are divided into several steps.

STEP 1. *For* $0 < s < 1$, $(-\Delta)^s(\log(\frac{1}{|x|})) = \frac{a_{N,s}C_0}{|x|^{2s}}$ *with*

$$C_0 = 2\int_{|y|>1} \frac{\log|y|\left(1 - \frac{1}{|y|^{N+2s}}\right)}{|x'-y|^{N+2s}}\, dy > 0.$$

Note that

$$\int_{\mathbb{R}^N} \left(\log\frac{1}{|x|} - \log\frac{1}{|y|}\right)\frac{dy}{|x-y|^{N+2s}} = \frac{1}{|x|^{N+2s}}\int_{\mathbb{R}^N} \log\frac{|y|}{|x|}\frac{dy}{|x' - \frac{y}{|x|}|^{N+2s}},$$

where $x' = \frac{x}{|x|} \in \mathbb{S}^{N-1}$. By the change of variable $z = \frac{|y|}{|x|}$ we obtain

$$\int_{\mathbb{R}^N} \left(\log\frac{1}{|x|} - \log\frac{1}{|y|}\right)\frac{dy}{|x-y|^{N+2s}} = \frac{C_0}{|x|^{2s}},$$

where

$$C_0 = \int_{\mathbb{R}^N} \log|y|\frac{dy}{|x'-y|^{N+2s}},$$

which is constant on $x' =\in \mathbb{S}^{N-1}$. We check next the sign of C_0. We have

$$C_0 = \int_{\mathbb{R}^N} \log|z|\frac{dz}{|x'-y|^{N+2s}} = \int_{|y|>1} \log|y|\frac{dz}{|x'-y|^{N+2s}} + \int_{|y|\le 1} \log|y|\frac{dy}{|x'-y|^{N+2s}}.$$

The first term is positive. With respect to the second term, changing to polar coordinates we obtain

$$\begin{aligned}\int\limits_{|y|\le 1} \log|y| \frac{dy}{|x'-y|^{N+2s}} &= \int\limits_{\mathbb{S}^{N-1}} \int\limits_0^1 \log t \frac{t^{N-1}}{|x'-ty'|^{N+2s}}\, dt\, dy' \\ &= \int\limits_{\mathbb{S}^{N-1}} \int\limits_0^1 \log t \frac{1}{t^{2s}|\frac{x'}{t}-y'|^{N+2s}} \frac{dt}{y}\, dy' \\ &= \int\limits_{\mathbb{S}^{N-1}} \int\limits_1^\infty \log \frac{1}{r} \frac{r^{2s-N}}{|x'-ry'|^{N+2s}}\, dr\, dy' \\ &= -\int\limits_{|y|\le 1} \log|y| \frac{dy}{|x'-y|^{N+2s}}.\end{aligned}$$

Therefore $C_0 = 2\int_{|y|>1} \frac{\log|y|(1-\frac{1}{|y|^{N+2s}})}{|x'-y|^{N+2s}}\, dy > 0$.

Step 2. *If* $0 < s < 1$ *and* $\log_+(\frac{1}{|x|}) = \max\{\log\frac{1}{|x|}, 0\}$, *then*

$$(-\Delta)^s\left(\log_+\left(\frac{1}{|x|}\right)\right) \le \begin{cases} 0 & \text{if } |x| > 1 \\ \frac{C_0}{|x|^{2s}} & \text{if } |x| < 1. \end{cases}$$

It is sufficient that the function $\psi(u) = u_+$ is convex and Lipschitz, in fact, $\psi'(u) = \chi_{(0,\infty)}$. Hence

$$(-\Delta)^s\left(\log_+\left(\frac{1}{|x|}\right)\right) \le \psi'\left(\log\frac{1}{|x|}\right)(-\Delta)^s\left(\log\frac{1}{|x|}\right) = \frac{C_0}{|x|^{2s}}\chi_{(0,1)}.$$

Step 3. According to (8.1.17), if two functions verify that $u(x) \le u(y)$ *if and only if* $v(x) \le v(y)$, then

$$(-\Delta)^s(uv) \le u(-\Delta)^s v + v(-\Delta)^s u.$$

Note that the condition holds if both functions are radial monotone. Then for

$$\overline{w}(x) = \frac{1}{|x|^\gamma}\left(\log_+\frac{1}{|x|}\right), \quad \gamma = \frac{N-2s-2\alpha_\lambda}{2},$$

we find

$$\begin{aligned}(-\Delta)^s\overline{w}(x) &\le (-\Delta)^s\left(\frac{1}{|x|^\gamma}\right)\log_+\frac{1}{|x|} + (-\Delta)^s\left(\log_+\frac{1}{|x|}\right)\frac{1}{|x|^\gamma} \\ &\le \frac{\lambda}{|x|^{2s}}\overline{w}(x) + \frac{C_0}{|x|^{\gamma+2s}}\chi_{\{|y|<1\}}.\end{aligned}$$

Hence we conclude that

$$(-\Delta)^s\overline{w}(x) - \frac{\lambda}{|x|^{2s}}\overline{w}(x) \le \frac{C_0}{|x|^{\gamma+2s}}\chi_{\{|y|<1\}}. \tag{11.5.1}$$

Up to a scaling it can be assumed that $B_1(0) \subset \Omega$. If there exists a positive solution, u, to problem (11.1.1), then Theorem 10.4.10 implies, in particular, that

$$\int_{B_1(0)} \frac{u^p}{|x|^\gamma}\, dx < \infty,$$

and, by Hölder's inequality,

$$\int_{B_1(0)} \frac{u}{|x|^{2s+\gamma}}\, dx \leq \Big(\int_{B_1(0)} \frac{u^p}{|x|^\gamma}\, dx\Big)^{\frac{1}{p}} \Big(\int_{B_1(0)} \frac{dx}{|x|^{\gamma+2s\frac{p}{p-1}}}\Big)^{\frac{p-1}{p}}.$$

Since $p = p(\lambda, s)$ and $\lambda < \Lambda_{N,2s}$,

$$\gamma + 2s\frac{p}{p-1} = N - 2\alpha_\lambda < N,$$

which shows that it is sufficient to assume the weighted L^p-integrability of u to conclude that

$$\int_{B_1(0)} \frac{u}{|x|^{2s+\gamma}}\, dx < \infty.$$

From Lemma 10.3.8 we have

$$u(x) \geq \frac{C_1}{|x|^\gamma}, \quad \text{for all } x \in B_1(0),$$

for some constant $C_1 > 0$. Hence

$$u^p(x) \geq \frac{C_1^p}{|x|^{p\gamma}} \geq \frac{C_1^p}{|x|^{2s+\gamma}}, \quad \text{for all } x \in B_1(0),$$

since $p\gamma > 2s + \gamma$.
Consider $w(x) = A\overline{w}$ with $A > 0$ such that $AC_0 < C_1^p$. Then

$$(-\Delta)^s w(x) - \frac{\lambda}{|x|^{2s}} w(x) \leq \frac{AC_0}{|x|^{\gamma+2s}} \leq \frac{C_1^p}{|x|^{2s+\gamma}} \leq u^p(x) \leq (-\Delta)^s u(x) - \frac{\lambda}{|x|^{2s}}(x),$$

in $x \in B_1(0)$. Since $w(x) = 0 \leq u(x)$ in $\mathbb{R}^N \subset B_1(0)$ we conclude that

$$u(x) \geq w(x) = A\frac{1}{|x|^\gamma}\Big(\log_+ \frac{1}{|x|}\Big).$$

As in the previous case for $\phi \in C_\prime^\infty(B_1(0))$, by the application of the Picone inequality we find

$$A^{p(\lambda,s)-1} \int_{B_1(0)} \frac{\phi^2}{|x|^{2s}} \Big(\log_+ \frac{1}{|x|}\Big)^{p(\lambda,s)-1} dx \leq \int_{B_1(0)} u^{p-1}\phi^2\, dx \leq \|\phi\|^2_{H_0^s(B_1(0))},$$

which contradicts the Hardy inequality.

(ii) Now we consider the case $\lambda = \Lambda_{N,2s}$. This case is easier. Assume that there exist a positive solution u to problem (11.1.1) with $\lambda = \Lambda_{N,2s}$ and $p \geq p(\Lambda_{N,2s}, s)$. By Theorem 10.4.10, in particular, we must have

$$\int_{B_r(0)} \frac{u^p}{|x|^\gamma}\,dx < \infty, \qquad r > 0.$$

But by Lemma 10.3.8 there exists $C > 0$ such that

$$u(x) \geq \frac{C}{|x|^\gamma}, \qquad \text{for all } x \in B_r(0),$$

and therefore

$$\int_{B_r(0)} \frac{u^p}{|x|^\gamma}\,dx \geq C^p \int_{B_r(0)} \frac{1}{|x|^{(p+1)\gamma}}\,dx,$$

and since for $\Lambda_{N,2s}$, $p(\Lambda_{N,2s}, s) = \frac{N+2s}{N-2s}$ and $\gamma = \frac{N-2s}{2}$, we have

$$(p+1)\gamma \geq \frac{2N}{N-2s}\frac{N-2s}{2} = N,$$

which contradicts the necessary integrability condition. □

Remark 11.5.2. Nonexistence results, using the Caffarelli–Silvestre extension in [105], were proved in [152, Theorem 0.2]. Our proof is more direct, using the natural homogeneities associated to the problem.

11.5.1 Complete blow-up

As a byproduct of the nonexistence result we are able to prove complete blow-up, which is a common fact in all the nonexistence results associated to the Hardy potential. That is, we will see that the solutions u_n of the truncated problems verify that $u_n(x) \to \infty$ when $n \to \infty$ and $x \in \Omega$. Before proving the complete blow-up result, we need the following auxiliary lemma, which is an extension to the fractional semilinear framework of Proposition 6.2.5 (see the original in [87, Lemma 3.2]).

Lemma 11.5.3. *Let $F(x,u) \geq 0$ in $L^\infty(\Omega)$, and let u be the solution of*

$$(P_F) = \begin{cases} (-\Delta)^s u = F(x,u) & \text{in } \Omega, \\ u = 0 & \text{in } \mathbb{R}^N \setminus \Omega. \end{cases}$$

Then,

$$\frac{u(x)}{\delta^s(x)} \geq C \int_\Omega F(x,u)\delta^s(x)\,dx, \qquad x \in \Omega, \tag{11.5.2}$$

where $\delta(x) = \operatorname{dist}(x, \partial\Omega)$ and C is a constant depending only on Ω.

Proof. First of all we will prove (11.5.2) for points that belong to a compact set $K \subset \Omega$. Let $x_0 \in K$. By the mean value inequality in Proposition 11.2.4 (see [287]), we have

$$u(x_0) \geq \int_{\mathbb{R}^N} u(x)\gamma_r(x - x_0)\,dx = \int_{\Omega} u(x)\gamma_r(x - x_0)\,dx > 0,$$

for any $x_0 \in \Omega$, $r \leq \operatorname{dist}(x_0, \partial\Omega)$ and $\gamma_r = (-\Delta)^s \Gamma_r$, where Γ_r is a $C^{1,1}$-paraboloid that matches outside the ball $B(0, r)$ with the fundamental solution. Then there exist a positive constant $c > 0$ and a compact set $K \subseteq \Omega$ such that $u(x_0) > c$ for every $x_0 \in K$. That is,

$$u(x_0) > M \int_{\Omega} u(x)\,dx, \quad x_0 \in K, \tag{11.5.3}$$

where

$$M = c\Big(\int_{\Omega} u(x)\,dx\Big)^{-1}.$$

Consider now the function ξ_1 such that

$$(P_{F=1}) = \begin{cases} (-\Delta)^s \xi_1 = 1 & \text{in } \Omega, \\ \xi_1 = 0 & \text{in } \mathbb{R}^N \setminus \Omega. \end{cases} \tag{11.5.4}$$

Using ξ_1 as a test function in (P_F), by (11.5.3) we obtain

$$u(x_0) \geq M \int_{\Omega} u(x)\,dx = M \int_{\Omega} F(x, u)\xi_1(x), \quad x_0 \in K. \tag{11.5.5}$$

Then, by Hopf's lemma (see [273, Lemma 3.2]),

$$u(x_0) \geq C \int_{\Omega} F(x, u)\delta^s(x)\,dx, \quad x_0 \in K.$$

Moreover, since $c_1 \leq \delta^s(x_0) \leq c_2$ for $x_0 \in K$, we have

$$\frac{u(x_0)}{\delta^s(x_0)} \geq \tilde{C} \int_{\Omega} F(x, u)\delta^s(x)\,dx, \quad x_0 \in K. \tag{11.5.6}$$

Let now w satisfy

$$\begin{cases} (-\Delta)^s w = 0 & \text{in } \Omega \setminus K, \\ w = 0 & \text{in } \mathbb{R}^N \setminus \Omega, \\ w = 1 & \text{in } K. \end{cases}$$

We define

$$v(x) = \frac{u(x)}{\tilde{C}\int_\Omega F(x,u)\delta^s(x)\,dx}, \qquad x \in \mathbb{R}^N.$$

Therefore

$$\begin{cases} (-\Delta)^s v \geq 0 & \text{in } \Omega, \\ v = 0 & \text{in } \mathbb{R}^N \setminus \Omega, \\ v \geq 1 & \text{in } K. \end{cases}$$

Then, by the strong maximum principle of Section 11.2.2, we have $v(x) \geq w(x)$ for $x \in \Omega \setminus K$. Since by Hopf's lemma $w(x) \geq C\delta^s(x)$, it follows that

$$\frac{u(x_0)}{\delta^s(x_0)} \geq C\int_\Omega F(x,u)\delta^s(x)\,dx, \quad x_0 \in \Omega \setminus K. \tag{11.5.7}$$

Hence, by (11.5.6) and (11.5.7), we obtain the desired estimate given in (11.5.2). □

We state now the blow-up result.

Theorem 11.5.4. *Assume that* $0 < \lambda \leq \Lambda_{N,2s}$. *Let* $p \geq p(\lambda, s)$ *and* $\mu > 0$. *Then, there exists complete blow-up of the problem* (P_μ).

Proof. We argue by contradiction. Consider the minimal solution $u_n \geq 0$ to the truncated problem

$$(P_n) = \begin{cases} (-\Delta)^s u_n = \lambda a_n(x) T_n(u_n) + g_n(u_n) + \mu f_n(u_n) & \text{in } \Omega, \\ u_n = 0 & \text{in } \mathbb{R}^N \setminus \Omega, \end{cases}$$

where

$$a_n(x) := T_n\left(\frac{1}{|x|^{2s}}\right), \quad g_n(u) := T_n(u_+^p) \quad \text{and} \quad f_n(u) := T_n(u_+^q),$$

with

$$T_n(x) := \begin{cases} x & \text{if } |x| \leq n, \\ n\frac{x}{|x|} & \text{if } |x| > n. \end{cases}$$

Note that we can affirm that this minimal solution exists because, since

$$\lambda a_n(x) T_n(u_n) + g_n(u_n) + \mu f_n(u_n) \leq C_n,$$

we can consider, for a suitable $c > 0$, $\overline{u} := C_n \xi_1$ and $\underline{u} := c\varphi_1$ as well-ordered super- and subsolutions of (P_n), respectively. Here φ_1 is the nonnegative first eigenfunction of the fractional Laplacian defined in (11.3.3) and ξ_1 is given in (11.5.4).

We suppose that

$$\int_{\Omega} (\lambda a_n(x)T_n(u_n) + g_n(u_n) + \mu f_n(u_n))\delta^s(x)\,dx \le C < \infty, \quad n \in \mathbb{N},$$

with C independent of n. Using ξ_1 given in $(P_{F=1})$ as a test function in problem (P_n), by [273, Proposition 1.1], we obtain

$$\begin{aligned}\int_{\mathbb{R}^N} u_n &= \int_{\mathbb{R}^N} u_n(-\Delta)^s\xi_1 \\ &= \lambda\int_{\Omega} a_nT_n(u_n)\xi_1 + \int_{\Omega} g_n(u_n)\xi_1 + \mu\int_{\Omega} f_n(u_n)\xi_1 \\ &\le C.\end{aligned}$$

Hence, up to a subsequence, $\{u_n\}$ converges in $L^1(\Omega)$ to a nonnegative limit u. Then, since $a_n(x)T_n(u_n) + g_n(u_n) + \mu f_n(u_n)$ increases to $\frac{u}{|x|^{2s}} + u^p + \mu u^q$ in Ω, by the monotone convergence theorem we can pass to the limit in (P_n) obtaining a nonnegative weak solution of the problem

$$\begin{cases} (-\Delta)^s u - \lambda\frac{u}{|x|^{2s}} = u^p + \mu u^q & \text{in } \Omega, \\ u = 0 & \text{in } \mathbb{R}^N \setminus \Omega. \end{cases}$$

But this is a contradiction with the nonexistence results above (see also [151, Theorem 0.2]) and, therefore, we conclude applying Lemma 11.5.3. Indeed, since

$$\frac{u_n(x)}{\delta^s(x)} \ge C\int_{\Omega} (\lambda a_nT_n(u_n) + g_n(u_n) + \mu f_n(u_n))\,\delta^s(x)dx,$$

hence $u_n(x) \to \infty$, as $n \to \infty$, for all $x \in \Omega$. □

11.6 Problems with the Hardy potential and nonlinear terms singular at the boundary

The results in this section have some partial precedents in [1] and are also applicable to the local case.

The aim will be to study the problem

$$\begin{cases} (-\Delta)^s u = \lambda\frac{u}{|x|^{2s}} + \frac{h(x)}{u^\sigma} & \text{in } \Omega, \\ u > 0 & \text{in } \Omega, \\ u = 0 & \text{in } \mathbb{R}^N \setminus \Omega, \end{cases} \tag{11.6.1}$$

where h is a nonnegative function, $\sigma > 0$ and $\lambda \ge 0$. As we pointed out in Remark 10.4.11 it can be easily checked that problem (11.6.1) has no positive solution for $\lambda > \Lambda_{N,2s}$. Hence we will assume $\lambda \le \Lambda_{N,2s}$.

Remark 11.6.1. Call $\mu := -\sigma$. We know by the previous sections that problem (11.6.1) has no positive solution for $\mu > p_+(\lambda)$, where y is defined in (9.4.9). Therefore, finding a solution of (11.6.1) can be seen as proving that there is no lower threshold for the power to solve the semilinear problem.

The main result in this section is the following.

Theorem 11.6.2. *Assume that $\sigma \geq 1$ and $\lambda \leq \Lambda_{N,2s}$. Then, for all $h \in L^1(\Omega)$, problem* (11.6.1) *has a positive weak solution. More precisely,*

1. *if $\sigma = 1$, then $u \in H_0^s(\Omega)$ when $\lambda < \Lambda_{N,2s}$, and $u \in W_0^{s,q}(\Omega)$ for all $q < 2$ if $\lambda = \Lambda_{N,2s}$;*
2. *if $\sigma > 1$, then $u \in H^s_{\mathrm{loc}}(\Omega)$ with $G_k(u) \in H_0^s(\Omega)$ and $T_k^{\frac{\sigma+1}{2}}(u) \in H_0^s(\Omega)$. Moreover if $[\frac{4\sigma}{(\sigma+1)^2} - \frac{\lambda}{\Lambda_{N,2s}}] > 0$, then $u^{\frac{\sigma+1}{2}} \in H_0^s(\Omega)$.*

Proof. Let $h_n := T_n(h)$, the usual truncation of h, and define u_n to be the unique positive solution to the approximated problem

$$\begin{cases} (-\Delta)^s u_n = \lambda \frac{u_n}{|x|^{2s}} + \frac{h_n(x)}{(u_n+\frac{1}{n})^\sigma} & \text{in } \Omega, \\ u_n > 0 & \text{in } \Omega, \\ u_n = 0 & \text{in } (\mathbb{R}^N \setminus \Omega). \end{cases} \tag{11.6.2}$$

The existence follows by minimization and the uniqueness by using the result in Lemma 10.2.23. Since $T_n(h)$ is an increasing function in n, again by Lemma 10.2.23 we conclude that $\{u_n\}_{n\in\mathbb{N}}$ is an increasing function in n. We divide the proof in two cases.

First case: $\sigma = 1$ and $\lambda < \Lambda_{N,2s}$.
Taking u_n as a test function in (11.6.2) and using the Hardy inequality, we obtain

$$\frac{a_{N,s}}{2}\left(1 - \frac{\lambda}{\Lambda_{N,2s}}\right)\|u_n\|^2_{H_0^s(\Omega)} \leq \int_\Omega \frac{h_n u_n}{u_n + \frac{1}{n}}\,dx \leq \int_\Omega h\,dx = C.$$

Thus $\{u_n\}_{n\in\mathbb{N}}$ is bounded in $H_0^s(\Omega)$ and then there exists $u \in H_0^s(\Omega)$ such that, up to a subsequence, $u_n \rightharpoonup u$ weakly in $H_0^s(\Omega)$ and $u_n \uparrow u$ strongly in $L^\eta(\Omega)$ for all $\eta < 2_s^*$.

Since $(-\Delta)^s u_n \geq 0$, using the monotonicity of $\{u_n\}_{n\in\mathbb{N}}$ and the compactness lemma, Lemma 10.2.24, we easily obtain that $u_n \to u$ strongly in $H_0^s(\Omega)$. Hence we conclude that u solves problem (11.6.1).

Second case: $\sigma > 1$ and $\lambda < \Lambda_{N,2s}$.
Using now $G_k(u_n)$ as a test function in (11.6.2) we have

$$a_{N,s} \iint_{D_\Omega} \frac{|G_k(u_n(x)) - G_k(u_n(y))|^2}{|x-y|^{N+2s}}\,dx\,dy - \lambda \int_\Omega \frac{u_n G_k(u_n)}{|x|^{2s}}\,dx \leq \int_\Omega \frac{h_n G_k(u_n)}{(u_n + \frac{1}{n})^\sigma}\,dx$$

and

$$\int_\Omega \frac{h_n G_k(u_n)}{(u_n + \frac{1}{n})^\sigma}\,dx \leq \frac{1}{k^{\sigma-1}} \int_\Omega h\,dx.$$

Moreover, $u_n G_k(u_n) = G_k^2(u_n) + kG_k(u_n)$, and thus

$$a_{N,s} \iint_{D_\Omega} \frac{|G_k(u_n(x)) - G_k(u_n(y))|^2}{|x-y|^{N+2s}}\,dx\,dy - \lambda \int_\Omega \frac{G_k^2(u_n)}{|x|^{2s}}\,dx$$
$$\leq \lambda k \int_\Omega \frac{G_k(u_n)}{|x|^{2s}}\,dx + \frac{1}{k^{\sigma-1}} \int_\Omega h\,dx.$$

Since $\lambda < \Lambda_{N,2s}$, by the Hardy–Sobolev inequality we obtain

$$C \iint_{D_\Omega} \frac{|G_k(u_n(x)) - G_k(u_n(y))|^2}{|x-y|^{N+2s}}\,dx\,dy \leq \lambda k \int_\Omega \frac{G_k(u_n)}{|x|^{2s}}\,dx + C(k,h),$$

and applying the Young and Hardy–Sobolev inequalities in the integral on the right hand side we reach

$$\iint_{D_\Omega} \frac{|G_k(u_n(x)) - G_k(u_n(y))|^2}{|x-y|^{N+2s}}\,dx\,dy \leq C(k,\lambda,\Lambda_{N,2s},h).$$

Therefore $\{G_k(u_n)\}_{n\in\mathbb{N}}$ is uniformly bounded in $H_0^s(\Omega)$, and again by the Hardy–Sobolev inequality,

$$\int_\Omega \frac{G_k^2(u_n(x))}{|x|^{2s}}\,dx \leq C(k,\lambda,\Lambda_{N,2s},h).$$

Then we get

$$\int_\Omega \frac{u_n^2(x)}{|x|^{2s}}\,dx = \int_\Omega \frac{T_k^2(u_n(x))}{|x|^{2s}}\,dx + \int_\Omega \frac{G_k^2(u_n(x))}{|x|^{2s}}\,dx + 2\int_\Omega \frac{T_k(u_n)G_k(u_n)}{|x|^{2s}}\,dx \leq C(k,\lambda,\Lambda_{N,2s},h).$$

Likewise, testing with $T_k^\sigma(u_n)$ in (11.6.2), it follows that

$$a_{N,s} \iint_{D_\Omega} \frac{(T_k^\sigma(u_n(x)) - T_k^\sigma(u_n(y)))(u_n(x) - u_n(y))}{|x-y|^{N+2s}}\,dx\,dy$$
$$\leq \lambda \int_\Omega \frac{u_n T_k^\sigma(u_n)}{|x|^{2s}}\,dx + \int_\Omega \frac{h_n T_k^\sigma(u_n)}{(u_n + \frac{1}{n})^\sigma}\,dx$$
$$\leq k^{\sigma-1}\lambda \int_\Omega \frac{u_n^2}{|x|^{2s}}\,dx + \int_\Omega h_n\,dx \leq C(k,\lambda,\Lambda_{N,2s},h),$$

and applying Lemma 10.2.27 we conclude

$$\iint_{D_\Omega} \frac{(T_k^{\frac{\sigma+1}{2}}(u_n(x)) - T_k^{\frac{\sigma+1}{2}}(u_n(y)))^2}{|x-y|^{N+2s}}\,dx\,dy \leq C(k,\lambda,\Lambda_{N,2s},h,\sigma).$$

Thus $\{T_k^{\frac{\sigma+1}{2}}(u_n)\}_{n\in\mathbb{N}}$ is bounded in $H_0^s(\Omega)$. Furthermore, the strong maximum principle provides that

$$u_n \geq u_1 \geq c(K) > 0, \quad \text{for any compact set } K \subset \Omega.$$

Claim. The sequence $\{T_k(u_n)\}_{n\in\mathbb{N}}$ is bounded in $H^s_{\text{loc}}(\Omega)$.

Since $\{u_n\}_{n\in\mathbb{N}}$ is an increasing sequence, we have $T_k(u_n) \geq T_k(u_1)$, and for all $\Omega' \subset\subset \Omega$, $u_1 \geq C(\Omega')$. Thus,

$$T_k(u_n) \geq \min\{k, C(\Omega')\} =: C_0.$$

For $(x,y) \in \Omega' \times \Omega'$, we define $\alpha_n := \frac{T_k(u_n(x))}{C_0}$ and $\beta_n := \frac{T_k(u_n(y))}{C_0}$. It is clear that $\alpha_n, \beta_n \geq 1$. Therefore the following inequality holds:

$$(\alpha_n - \beta_n)^2 \leq (\alpha_n^{\frac{\sigma+1}{2}} - \beta_n^{\frac{\sigma+1}{2}})^2. \tag{11.6.3}$$

Indeed, if $\alpha_n = \beta_n$ the estimate is trivial. Otherwise, without loss of generality we can assume $\alpha_n > \beta_n \geq 1$. Let $0 < x := \frac{\beta_n}{\alpha_n} < 1$. Since $\sigma > 1$, we easily obtain

$$0 \leq 1 - x \leq 1 - x^{\frac{\sigma+1}{2}},$$

and hence

$$(1-x)^2 \leq (1 - x^{\frac{\sigma+1}{2}})^2.$$

Clearly $\alpha_n^2 < \alpha_n^{\sigma+1}$, and thus

$$\alpha_n^2(1-x)^2 \leq \alpha_n^{\sigma+1}(1 - x^{\frac{\sigma+1}{2}})^2.$$

Recalling the definition of x, (11.6.3) follows.

Finally, by the definition of α_n and β_n, we conclude that for $(x,y) \in \Omega' \times \Omega'$, we have

$$(T_k(u_n(x)) - T_k(u_n(y)))^2 \leq C_0^{1-\sigma}(T_k^{\frac{\sigma+1}{2}}(u_n(x)) - T_k^{\frac{\sigma+1}{2}}(u_n(y)))^2.$$

Thus the claim follows using the boundedness of $\{T_k^{\frac{\sigma+1}{2}}(u_n)\}_{n\in\mathbb{N}}$ in $H_0^s(\Omega)$.

Hence combining the estimates above, we obtain that $\{u_n\}_{n\in\mathbb{N}}$ is bounded in $H^s_{\text{loc}}(\Omega)$ and then, up to a subsequence, there exists $u \in H^s_{\text{loc}}(\Omega)$ such that

$$\begin{aligned} u_n &\rightharpoonup u \quad \text{weakly in } H^s_{\text{loc}}(\Omega), \\ G_k(u_n) &\rightharpoonup G_k(u) \quad \text{weakly in } H_0^s(\Omega) \quad \text{and} \\ T_k^{\frac{\sigma+1}{2}}(u) &\rightharpoonup T_k^{\frac{\sigma+1}{2}}(u) \quad \text{weakly in } H_0^s(\Omega). \end{aligned}$$

Applying the compactness result in Lemma 10.2.25 we obtain that $u_n \to u$ strongly in $H^s_{\mathrm{loc}}(\Omega)$. Thus $u_n \uparrow u$ strongly in $L^r(\Omega)$ for all $1 \le r < 2^*_s$.

Let $\phi \in \mathcal{T}$, where $\mathcal{T}$ was defined in (8.6.2). Testing with ϕ in (11.6.2), it follows that

$$\int_\Omega (-\Delta)^s u_n \phi\, dx = \lambda \int_\Omega \frac{u_n \phi}{|x|^{2s}}\, dx + \int_\Omega \frac{\phi h_n(x)}{(u_n + \frac{1}{n})^\sigma}\, dx. \tag{11.6.4}$$

By the estimates above we reach that when n goes to $+\infty$,

$$\lambda \int_\Omega \frac{u_n \phi}{|x|^{2s}} + \int_\Omega \frac{\phi h_n(x)}{(u_n + \frac{1}{n})^\sigma}\, dx \to \lambda \int_\Omega \frac{u\phi}{|x|^{2s}} + \int_\Omega \frac{\phi h(x)}{u^\sigma}\, dx < +\infty.$$

Moreover, we have

$$\int_\Omega (-\Delta)^s u_n \phi\, dx = \int_\Omega u_n (-\Delta)^s \phi\, dx \to \int_\Omega u(-\Delta)^s \phi\, dx,$$

as $n \to +\infty$. Therefore, passing to the limit in (11.6.4),

$$\int_\Omega (-\Delta)^s u\phi\, dx = \lambda \int_\Omega \frac{u\phi}{|x|^{2s}}\, dx + \int_\Omega \frac{h(x)\phi}{u^\sigma}\, dx,$$

i. e., u is a weak solution.

Finally, for every $\lambda < \Lambda_{N,2s}$ take σ such that $|\frac{4\sigma}{(\sigma+1)^2} - \frac{\lambda}{\Lambda_{N,2s}}| > 0$. By using u_n^σ as a test function in (11.6.2), it follows that

$$a_{N,s} \iint_{D_\Omega} \frac{(u_n(x) - u_n(y))(u_n^\sigma(x) - u_n^\sigma(y))}{|x-y|^{N+2s}}\, dx\, dy \le \lambda \int_\Omega \frac{u_n^{\sigma+1}}{|x|^{2s}}\, dx + \int_\Omega h_n\, dx.$$

By Lemma 10.2.27, we get

$$(u_n(x) - u_n(y))(u_n^\sigma(x) - u_n^\sigma(y)) \ge \frac{4\sigma}{(\sigma+1)^2}\big(u_n^{\frac{\sigma+1}{2}}(x) - u_n^{\frac{\sigma+1}{2}}(y)\big)^2,$$

and hence, by Hardy's inequality,

$$a_{N,s}\Big(\frac{4\sigma}{(\sigma+1)^2} - \frac{\lambda}{\Lambda_{N,2s}}\Big) \iint_{D_\Omega} \frac{(u_n^{\frac{\sigma+1}{2}}(x) - u_n^{\frac{\sigma+1}{2}}(y))^2}{|x-y|^{N+2s}}\, dx\, dy \le \|h\|_{L^1(\Omega)}.$$

Therefore $u^{\frac{\sigma+1}{2}} \in H^s_0(\Omega)$ and this is the sense how the boundary value is reached. □

We now deal with the case $\sigma < 1$. If $h \in L^{(\frac{2^*_s}{1-\sigma s})'}(\Omega)$, the existence of a positive energy solution can be proved proceeding as in the case $\sigma = 1$. However, our goal from now on will be to study the solvability when h has less regularity. Indeed, we have the following result.

Theorem 11.6.3. *Assume $\sigma < 1$, $\lambda < \Lambda_{N,2s}$ and $h \in L^1(\Omega, |x|^{-(1-\sigma)\gamma}\,dx)$, $h \gneq 0$. Then problem* (11.6.1) *has at least a weak solution.*

Proof. Let $\{h_n\}_{n\in\mathbb{N}}$ be such that $h_n \geq 0$ and $h_n \uparrow h$ strongly in $L^1(\Omega)$. Define u_n as the unique positive solution to the approximated problem (11.6.2). Then by setting $v_n := |x|^\gamma u_n$, it follows that v_n satisfies

$$L_\gamma(v_n) = |x|^{-\gamma}\frac{h_n(x)}{(|x|^{-\gamma}v_n + \frac{1}{n})^\sigma} \leq |x|^{-(1-\sigma)\gamma}\frac{h_n}{v_n^\sigma} \quad \text{in } \Omega, \tag{11.6.5}$$

where L_γ was defined in (10.2.3). Using v_n^σ as a test function in (11.6.5) we obtain

$$a_{N,s}\frac{4\sigma}{(\sigma+1)^2}\iint_{D_\Omega}\frac{(v_n^{\frac{\sigma+1}{2}}(x) - v_n^{\frac{\sigma+1}{2}}(y))^2}{|x-y|^{N+2s}}\frac{dx}{|x|^\gamma}\frac{dy}{|y|^\gamma} \leq \int_\Omega \frac{h_n}{|x|^{(1-\sigma)\gamma}}\,dx \leq C,$$

with C independent of n.

Thus we conclude that the sequence $\{v_n^{\frac{\sigma+1}{2}}\}_{n\in\mathbb{N}}$ is bounded in the weighted Sobolev space $W^s_{\gamma,0}(\Omega)$ and hence there exists v_0 such that, up to a subsequence,

$$v_n^{\frac{\sigma+1}{2}} \rightharpoonup v_0^{\frac{\sigma+1}{2}} \quad \text{weakly in } W^s_{\gamma,0}(\Omega)$$

Since $L_\gamma(v_n) \geq 0$, using the fact that $\frac{\sigma+1}{2} < 1$, also $L_\gamma(v_n^{\frac{\sigma+1}{2}}) \geq 0$. Hence by the monotonicity of v_n and Lemma 10.2.24, we obtain

$$v_n^{\frac{\sigma+1}{2}} \to v_0^{\frac{\sigma+1}{2}} \quad \text{strongly in } \phi \in W^s_{\gamma,0}(\Omega).$$

Passing to the limit in (11.6.5), it follows that v_0 solves

$$\begin{cases} L_\gamma(v_0) = |x|^{-(1-\sigma)\gamma}\frac{h}{v_0^\sigma} & \text{in } \Omega, \\ v_0 = 0 & \text{in } \mathbb{R}^N \setminus \Omega, \end{cases}$$

in the weak sense. Defining $u_0 := |x|^{-\gamma}v_0$, then $\lambda\frac{u_0}{|x|^{2s}} \in L^1(\Omega)$ and u_0 is a weak solution of problem (11.6.1). □

To end this section, we consider the problem in the whole space, that is, $\Omega = \mathbb{R}^N$. Then we will work in the space $\dot{H}^s(\mathbb{R}^N)$ defined by the completion of $C_0^\infty(\mathbb{R}^N)$ with respect to the Gagliardo seminorm

$$[\phi]_{\dot{H}^s(\mathbb{R}^N)} = \left(\int_{\mathbb{R}^N}\int_{\mathbb{R}^N}\frac{(\phi(x)-\phi(y))^2}{|x-y|^{N+2s}}\,dx\,dy\right)^{\frac{1}{2}}.$$

We obtain the following existence result.

Theorem 11.6.4. *Consider the problem*

$$\begin{cases} (-\Delta)^s u = \lambda \frac{u}{|x|^{2s}} + \frac{h(x)}{u^\sigma} & \text{in } \mathbb{R}^N, \\ u > 0 & \text{in } \mathbb{R}^N. \end{cases} \tag{11.6.6}$$

Then:

(i) *If $\sigma = 1$, then for all $h \in L^1(\mathbb{R}^N)$, problem (11.6.6) has a solution $u \in \dot{H}^s(\mathbb{R}^N)$.*

(ii) *If $\sigma > 1$, then for all $h \in L^1(\mathbb{R}^N)$, problem (11.6.6) has a weak solution u such that $G_k(u) \in \dot{H}^s(\mathbb{R}^N)$ and $T_k^{\frac{\sigma+1}{2}}(u) \in \dot{H}^s(\mathbb{R}^N)$, for all $k > 0$. Moreover, if*

$$\frac{4\sigma}{(\sigma+1)^2} - \frac{\lambda}{\Lambda_{N,2s}} > 0, \tag{11.6.7}$$

then $u^{\frac{\sigma+1}{2}} \in \dot{H}^s(\mathbb{R}^N)$.

(iii) *If $\sigma < 1$ and $h \in L^m(\mathbb{R}^N)$ with $m = (\frac{2_s^*}{1-\sigma})'$, then problem (11.6.6) has a solution u such that $u \in \dot{H}^s(\mathbb{R}^N)$.*

(iv) *If $\sigma < 1$ and (11.6.7) holds, then for all $h \in L^1(\mathbb{R}^N)$ problem (11.6.6) has a weak solution u such that $u^{\frac{\sigma+1}{2}} \in \dot{H}^s(\mathbb{R}^N)$.*

Proof. Consider u_n to be the unique positive solution to the approximated problem

$$\begin{cases} (-\Delta)^s u_n = \lambda \frac{u_n}{|x|^{2s}+\frac{1}{n}} + \frac{h_n(x)}{(u_n+\frac{1}{n})^\sigma} & \text{in } B_n(0), \\ u_n > 0 & \text{in } B_n(0), \\ u_n = 0 & \text{in } \mathbb{R}^N \setminus B_n(0). \end{cases} \tag{11.6.8}$$

It is clear that u_n is increasing with n.

If $\sigma = 1$, taking u_n as a test function in (11.6.8) and using the Hardy–Sobolev inequality it follows that

$$\frac{a_{N,s}}{2}\left(1 - \frac{\lambda}{\Lambda_{N,2s}}\right)[u_n]^2_{\dot{H}^s(\mathbb{R}^N)} \le C.$$

Hence $\{u_n\}_{n\in\mathbb{N}}$ is uniformly bounded in $\dot{H}^s(\mathbb{R}^N)$ and then, up to a subsequence, $u_n \rightharpoonup u$ weakly in $\dot{H}^s(\mathbb{R}^N)$, where u solves (11.6.6). Using the monotonicity of u_n and a straightforward adaptation of Lemma 10.2.24 we can prove that $u_n \to u$ strongly in $\dot{H}^s(\mathbb{R}^N)$, which proves (i), and (iii) similarly follows.

To prove (ii) we take $G_k(u_n)$ as a test function in (11.6.8) and performing the same computations as in the proof of Theorem 11.6.2 we conclude.

Finally, (iv) follows closely using the arguments in the proof of Theorem 11.6.2. In particular, the existence of $u \in L^1(\Omega)$ is a consequence of the uniform bounds of $\{G_k(u_n)\}_{n\in\mathbb{N}}$ and $\{T_k^{\frac{\sigma+1}{2}}(u_n)\}_{n\in\mathbb{N}}$ in $\dot{H}^s(\mathbb{R}^N)$ and the fact that $\frac{\sigma+1}{2} < 1$. □

Remark 11.6.5. In a similar way to the results in [46] the problem

$$\begin{cases} (-\Delta)^s u = \lambda \frac{u}{|x|^{2s}} + \mu \frac{h(x)}{u^\sigma} + u^p & \text{in } \Omega, \\ u > 0 & \text{in } \Omega, \\ u = 0 & \text{in } \mathbb{R}^N \setminus \Omega \end{cases} \tag{11.6.9}$$

can be analyzed. In fact the existence of a minimal solution for $p < p_+(\lambda)$ and $\lambda \le \Lambda_{N,2s}$ and μ small can be done with minor analytical changes.

Theorem 11.6.6. *Assume that in problem* (11.6.9) *one of the following conditions holds:*
(i) $\sigma < 1$, $p < 2_s^* - 1$ *and* $h \in L^\lambda(\Omega)$ *for some* $\lambda \ge \frac{2_s^*}{2_s^* - (1-\sigma)}$;
(ii) $\sigma \ge 1$, $p < 2_s^* - 1$ *and* $h \in L^{\frac{2_s^*}{2_s^* - (1-\sigma_1)}}(\Omega)$ *for some* $\sigma_1 < 1$;
(iii) $\sigma > 0$, $2_s^* - 1 \le p < p_+(\lambda)$ *and* $h \in L^\infty(\Omega)$.

Then there exists $\mu^* > 0$ *such that for all* $\mu < \mu^*$, *problem* (11.6.9) *has a minimal weak solution* u *and, moreover, for all* $\mu > \mu^*$, *problem* (11.6.9) *has no positive solution.*

The existence of a second positive solution in cases (i) and (ii) for μ small enough is easy to obtain. The result for all $\mu < \mu^*$ by a direct method seems to be an open problem.

11.7 Further comments

11.7.1 Other operators

The results above could be studied also for a larger class of linear operators like those studied in Chapter 8.

Another case that seems very interesting comes from considering the fractional p-Laplacian operator and the corresponding Hardy inequality. See [168]. The techniques to attack this problem may have some mathematical interest for applications.

12 The heat equation with fractional diffusion

12.1 Introduction

This chapter is dedicated to the evolution problem associated to the nonlocal elliptic operators studied in Chapter 8, that is, we will study the following parabolic problem:

$$\begin{cases} u_t + \mathcal{L}u = f & \text{in } Q_T, \\ u(x,t) = 0 & \text{in } (\mathbb{R}^N \setminus \Omega) \times [0,T), \\ u(x,0) = u_0(x) & \text{in } \Omega \times \{0\}, \end{cases} \tag{12.1.1}$$

where $\mathcal{L}$ is given by (8.3.9) with a kernel as in Definition 8.3.2, u_0 and f are functions defined in suitable Lebesgue spaces, $T > 0$ and Ω is a bounded subset of $\mathbb{R}^N$, $N \geq 2$, with Lipschitz boundary. We will study in particular the behavior of the problem when $\mathcal{L} = (-\Delta)^s$, giving a complete proof of the behavior of the kernel and some related results.

This kind of problems appears in different situations. They arise, for instance, in probabilistic processes in which a particle moves randomly in the space subject to a probability that allows long jumps with a polynomial tail. See for instance [98]. This type of problems also appears in fluid mechanics, mathematical economy and, in general, in problems with a fractional-like diffusion. We refer to the introduction in Chapter 8 for motivations and references.

12.2 Finite energy setting

As in the elliptic case, we are interested in the existence, uniqueness and summability of the solution u with respect to the summability of the datum f.

First, we define the meaning of a finite energy solution.

Definition 12.2.1. We say that $u \in L^2(0,T;H_0^s(\Omega)) \cap C([0,T],L^2(\Omega))$ with $u_t \in L^2(0,T;H^{-s}(\Omega))$ is a finite energy solution for the parabolic problem

$$\begin{cases} u_t + \mathcal{L}u = f & \text{in } Q_T, \\ u(x,t) = 0 & \text{in } (\mathbb{R}^N \setminus \Omega) \times [0,T), \\ u(x,0) = u_0(x) & \text{in } \Omega \times \{0\}, \end{cases}$$

with $f \in L^2(0,T;H^{-s}(\Omega))$ and $u_0 \in L^2(\Omega)$ if it satisfies

$$\int_0^T \int_\Omega u_t w \, dx \, dt + \int_0^T \mathcal{E}(u(x,t), w(x,t)) \, dt = \int_0^T \langle f, w \rangle \, dt,$$

for any $w \in L^2(0,T;H_0^s(\Omega))$, where $\mathcal{E}(\cdot,\cdot)$ is the bilinear form defined in (8.3.10).

https://doi.org/10.1515/9783110606270-012

We recall the following classical result.

Theorem 12.2.2. *Let H be a Hilbert space such that*

$$V \overset{\text{dense}}{\hookrightarrow} H \hookrightarrow V'.$$

Let $u \in L^p(a,b;V)$ be such that u_t, defined in the distributional sense, belongs to $L^{p'}(a,b;V')$. Then u belongs to $C([a,b];H)$.

For the proof see, for instance, [133, Chapter XVIII, Section 2, Theorem 1]. The particular case $p = 2$ can be seen in [148, Theorem 3 in Section 5.9].

Remark 12.2.3. Observe that we ask u to belong to $\mathcal{C}([0,T],L^2(\Omega))$ in order to have the identity $u(x,0) = u_0(x)$ in the $L^2(\Omega)$-sense. Anyway, Theorem 12.2.2 guarantees that if $u \in L^2(0,T;H_0^s(\Omega))$ and $u_t \in L^2(0,T;H^{-s}(\Omega))$, then $u \in \mathcal{C}([0,T];L^2(\Omega))$.

The existence and uniqueness of an energy solution to problem (12.1.1) for f in the dual space $L^2(0,T;H^{-s}(\Omega))$ and $u_0 \in L^2(\Omega)$ can be proved by means of a direct abstract Hilbert space approach. For the reader's convenience, we include here the proof inspired by the one of A. N. Milgram (see [242]) for the local case, which is based on a method by Vishik (see [312] and [226]).

Theorem 12.2.4. *Assume that $f \in L^2(0,T;H^{-s}(\Omega))$. Then for any $u_0 \in L^2(\Omega)$ problem* (12.1.1) *has a unique finite energy solution.*

Moreover if f is also a nonnegative function and $u_0 \geq 0$, such a solution is nonnegative too.

Proof. Let $\mathcal{C}_*^\infty(\Omega \times [0,T])$ denote the $\mathcal{C}^\infty(\Omega \times [0,T])$ functions that vanish in $(\mathbb{R}^N \setminus \Omega) \times [0,T]$ and in $\Omega \times \{T\}$. Consider $\phi \in \mathcal{C}_*^\infty(\Omega \times [0,T])$, $u \in L^2(0,T;H_0^s(\Omega))$, and define the operator

$$L_\phi(u) := \int_0^T \int_\Omega -u\phi_t \, dx \, dt + \int_0^T \mathcal{E}(u(x,t),\phi(x,t)) \, dt.$$

Note that u is an energy solution to (P) with $f \in L^2(0,T;H^{-s}(\Omega))$ if and only if

$$L_\phi(u) = \int_0^T \langle f,\phi\rangle \, dt + \int_\Omega u(x,0)\phi(x,0) \, dx.$$

We also define the following inner product:

$$\langle \varphi,\phi\rangle_* = \frac{1}{2}\langle \varphi(x,0),\phi(x,0)\rangle_{L^2(\Omega)} + \int_0^T \mathcal{E}(\varphi(x,t),\phi(x,t)) \, dt, \tag{12.2.1}$$

and denote by $H^*(\Omega\times[0,T])$ the Hilbert space built as the completion of $\mathcal{C}_*^\infty(\Omega\times[0,T])$ with the norm $\|\phi\|_*$ induced by the inner product (12.2.1).

Observe that for any $\varphi \in L^2(0,T;H_0^s(\Omega))$, by the Hölder and Sobolev inequalities, we have

$$|L_\phi(\varphi)| \le c_\phi(\|\varphi\|_{L^2(0,T,L^2(\Omega))} + \lambda^{-1}\|\varphi\|_{L^2(0,T,H_0^s(\Omega))}) \le \tilde{c}_\phi\|\varphi\|_*.$$

Therefore, L_ϕ is a linear continuous functional in $H^*(\Omega\times[0,T])$, and by the Fréchet–Riesz theorem, there exists $\mathcal{T}\phi \in H^*(\Omega\times[0,T])$ such that

$$L_\phi(\varphi) = \langle\varphi, \mathcal{T}\phi\rangle_* \quad \text{for all } \varphi \in H^*(\Omega\times[0,T]).$$

Since $\mathcal{T}$ is a linear operator in $H^*(\Omega\times[0,T])$, it is injective too. Moreover,

$$L_\phi(\phi) = \frac{1}{2}\int_\Omega \phi^2(x,0)\,dx + \int_0^T \mathcal{E}(\phi(x,t),\phi(x,t))\,dt = \|\phi\|_*^2,$$

and consequently, $\langle\phi,\mathcal{T}\phi\rangle_* = \|\phi\|_*^2$. Thus, by the Cauchy–Schwarz inequality,

$$\|\phi\|_*^2 \le \|\phi\|_*\|\mathcal{T}\phi\|_*, \quad \text{i.e.,} \quad \|\phi\|_* \le \|\mathcal{T}\phi\|_*.$$

Therefore, this implies that $\mathcal{T}$ is bijective and its inverse $\mathcal{T}^{-1}$ has a norm less than or equal to 1, and can be extended to the closure M of Rank($\mathcal{T}$).

On the other hand, let us define

$$B_{u_0,f}(\phi) := \int_\Omega u_0\phi(x,0)\,dx + \int_0^T\int_\Omega \phi f\,dx\,dt.$$

Denoting $\phi_0 := \phi(x,0)$, by the Hölder inequality we have

$$|B_{u_0,f}(\phi)| \le \|u_0\|_{L^2(\Omega)}\|\phi_0\|_{L^2(\Omega)} + \|f\|_{L^2(0,T;L^2(\Omega))}\|\phi\|_{L^2(0,T;L^2(\Omega))} \le c_{u_0,f,\lambda}\|\phi\|_*,$$

and thus,

$$|B_{u_0,f}(\mathcal{T}^{-1}\psi)| \le c\|\mathcal{T}^{-1}\psi\|_* \le c\|\psi\|_*.$$

Therefore, by applying the Fréchet–Riesz theorem again, there exists a unique $u \in M$ such that $B_{u_0,f}(\mathcal{T}^{-1}\psi) = \langle\psi, u\rangle_*$ for every $\psi \in M$. Calling $\phi = \mathcal{T}^{-1}\psi$ this means

$$B_{u_0,f}(\phi) = \langle\mathcal{T}\phi, u\rangle_* = L_\phi(u),$$

that is,

$$\int_0^T\int_\Omega -u\phi_t\,dx\,dt + \int_0^T \mathcal{E}(u(x,t),\phi(x,t))\,dt = \int_0^T\int_\Omega f\phi\,dx\,dt + \int_\Omega u(x,0)\phi(x,0)\,dx,$$

where $\phi \in L^2(0,T;H_0^s(\Omega))$ and $\phi_t \in L^2(0,T;H^{-s}(\Omega))$. Finally, by a density argument, one can conclude, integrating by parts, that in fact $u \in L^2(0,T;H_0^s(\Omega))$, $u_t \in L^2(0,T;H^{-s}(\Omega))$ and

$$\int_0^T\int_\Omega u_t\phi\,dx\,dt + \int_0^T \mathcal{E}(u(x,t),\phi(x,t))\,dt = \int_0^T\int_\Omega f\phi\,dx\,dt.$$

Thus $u(x,t)$ is an energy solution of (P). □

12.3 Nonvariational setting: weak solutions

We define the notion of weak solution just asking the regularity needed to give distributional sense to the equation.

Let us consider the following class of test functions:

$$\mathcal{P}(Q_T) = \{\phi(\cdot,t) \in C^1([0,T), \mathcal{C}_0^\beta(\Omega)),\ \text{a solution to } (P)\}, \tag{12.3.1}$$

$$\text{where } (P) = \begin{cases} -\phi_t + \mathcal{L}\phi = \psi & \text{in } Q_T,\\ \phi(x,t) = 0 & \text{in } (\mathbb{R}^N \setminus \Omega)\times[0,T),\\ \phi(x,T) = 0 & \text{in } \Omega,\end{cases}$$

with $\psi \in \mathcal{C}_0^\infty(Q_T)$.

Note that the existence of a regular solution is guaranteed by the result by Felsinger and Kassmann in [169] (for a kernel $\mathcal{K}$ more general than ours).

Definition 12.3.1. We say that $u \in C([0,T];L^1(\Omega))$ is a *weak solution* to (12.1.1) for $f \in L^1(Q_T)$ if for all $\phi \in \mathcal{P}(Q_T)$, $\phi = 0$ on $(\mathbb{R}^n \setminus \Omega)\times[0,T)$ and $\phi(x,T) = 0$ in Ω, we have

$$\int_{Q_T} u\psi\,dx\,dt = \int_{Q_T} f\phi\,dx\,dt + \int_\Omega u_0(x)\phi(x,0)\,dx.$$

Henceforth, let us introduce the following notations: for any measurable v we define

$$A_k(v) = \{(x,t) \in Q_T : v(x,t) > k\},$$
$$\text{and for a. e. } t \in (0,T),\ A_k^t(v) = \{x \in \Omega : v(x,t) > k\}. \tag{12.3.2}$$

Theorem 12.3.2. *Let $f \in L^1(Q_T)$ and $u_0 \in L^1(\Omega)$. Then there exists a unique weak solution to* (12.1.1). *Moreover,*

$$\forall k \geq 0, \quad T_k(u) \in L^2(0,T;H_0^s(\Omega)), \tag{12.3.3}$$

$$u \in L^q(Q_T), \quad \forall q \in \left(1, \frac{N+2s}{N}\right)$$

$$\textit{and} \quad |(-\Delta)^{\frac{s}{2}}u| \in L^r(Q_T),\ \forall r \in \left(1, \frac{N+2s}{N+s}\right). \tag{12.3.4}$$

Remark 12.3.3. Observe that if $s = 1$ the above results correspond to the classical ones for second order parabolic problems (see for instance [67]).

Proof. UNIQUENESS. Let w be a very weak solution of (12.1.1) with $f = 0$ and $u_0 = 0$, i. e.,

$$\begin{cases} w_t + \mathcal{L}w = 0 & \text{in } Q_T, \\ w = 0 & \text{in } (\mathbb{R}^N \setminus \Omega) \times [0, T), \\ w(x, 0) = 0 & \text{in } \Omega; \end{cases}$$

we want to prove that $w \equiv 0$. Consider $F \in C_0^\infty(Q_T)$, and let ϕ_F be the solution of the backward problem

$$\begin{cases} -(\phi_F)_t + \mathcal{L}\phi_F = F & \text{in } Q_T, \\ \phi_F(x, t) = 0 & \text{in } (\mathbb{R}^N \setminus \Omega) \times [0, T), \\ \phi_F(x, T) = 0 & \text{in } \Omega. \end{cases} \tag{12.3.5}$$

Choosing ϕ_F as a test function we deduce that for any $F \in C_0^\infty(Q_T)$,

$$\int_0^T \int_\Omega wF \, dx \, dt = 0,$$

which means $w = 0$ in $\mathcal{D}'(Q_T)$.

EXISTENCE. We obtain the solution as a limit of solutions to approximated problems. Let us consider $u_n \in L^2(0, T; H_0^s(\Omega)) \cap L^\infty(Q_T)$, the solution to the approximated problems

$$\begin{cases} (u_n)_t + \mathcal{L}u_n = f_n & \text{in } \Omega \times (0, T), \\ u_n(x, t) = 0 & \text{in } (\mathbb{R}^N \setminus \Omega) \times (0, T), \\ u_n(x, 0) = u_{n0}(x) & \text{in } \Omega, \end{cases} \tag{12.3.6}$$

where $f_n = T_n(f(x, t))$ and $u_{n0} = T_n(u_0(x))$ are smooth functions.

Step 1. *There exists a positive constant c, depending only on N, Ω and s, such that*

$$\|u_n\|_{L^q(Q_T)} \le c\|f_n\|_{L^1(Q_T)}, \quad \forall q \in \left(1, \frac{N + 2s}{N}\right). \tag{12.3.7}$$

We take $T_k(u_n)$ as a test function in (12.3.6) and, by a standard argument, we deduce that there exists a positive constant c, just depending on Ω and $\|f_n\|_{L^1(Q_T)}$, such that

$$\|u_n\|_{L^\infty(0,T,L^1(\Omega))} \le c \quad \text{and} \quad \|T_k(u_n)\|_{L^2(0,T,H_0^s(\Omega))} \le ck.$$

Observe moreover that by interpolation we have

$$\int_\Omega |T_k(u_n)|^\sigma\,dx \le \Big(\int_\Omega |T_k(u_n)|\,dx\Big)^{(1-\theta)\sigma}\Big(\int_\Omega |T_k(u_n)|^{2_s^*}\,dx\Big)^{\frac{\sigma\theta}{2_s^*}},$$

with $\theta \in (0,1)$ such that $\frac{1}{\sigma} = 1-\theta+\frac{\theta}{2_s^*}$. This and the Sobolev inequality lead to

$$\int_\Omega |T_k(u_n)|^\sigma\,dx \le c\Big(\int_\Omega |T_k(u_n)|\,dx\Big)^{(1-\theta)\sigma}\Big(\int_\Omega |T_k(u_n)|\,dx + \|T_k(u_n)\|_{H_0^s(\Omega)}\Big)^{\sigma\theta},$$

for a $c>0$. Choosing $\theta\sigma = 2$, $\sigma = \frac{2(N+s)}{N}$, after integration over $(0,T)$, we get

$$\int_0^T\int_\Omega |T_k(u_n)|^\sigma\,dx\,dt \le ck.$$

Therefore,

$$k^\sigma\,\mathrm{m}\{(x,t)\in\Omega\times(0,T), |u_n(x,t)|\ge k\} \le C_M k,$$

and then u_n is bounded in the Marcinkiewicz space $\mathcal{M}^{\frac{N+2s}{N}}(Q_T)$. Here $\mathrm{m}(A)$ denotes the Lebesgue measure of the set A in the corresponding Euclidean space.

Step 2. *There exists a positive constant c, depending on q, N, Ω and s, such that*

$$\|(-\Delta)^{\frac{s}{2}}u_n\|_{L^q(Q_T)} \le c\|f_n\|^q_{L^1(Q_T)}, \quad \forall q\in\Big(1,\frac{N+2s}{N+s}\Big). \tag{12.3.8}$$

For any $\lambda>0$, we have

$$\begin{aligned}&\{(x,t)\in Q_T : |(-\Delta)^{\frac{s}{2}}u_n|\ge\lambda\}\\ &\quad=\{(x,t)\in Q_T : |(-\Delta)^{\frac{s}{2}}u_n|\ge\lambda, |u_n|<k\}\cup\{(x,t)\in Q_T : |(-\Delta)^{\frac{s}{2}}u_n|\ge\lambda, |u_n|\ge k\},\end{aligned}$$

so that

$$\{(x,t)\in Q_T : |(-\Delta)^{\frac{s}{2}}u_n|\ge\lambda\}\subset\{(x,t)\in Q_T : |(-\Delta)^{\frac{s}{2}}u_n|\ge\lambda, |u_n|<k\}\cup A_k^t(u_n),$$

where $A_k^t(u_n)$ has been defined in (12.3.2). We have

$$\begin{aligned}&\mathrm{m}(\{(x,t)\in Q_T : |(-\Delta)^{\frac{s}{2}}u_n|\ge\lambda, |u_n|<k\})\\ &\quad\le\frac{1}{\lambda^2}\int_0^T\int_{\{(x,t)\in\Omega_T,|u_n|<k\}}|(-\Delta)^{\frac{s}{2}}u_n|^2\,dx\,dt.\end{aligned}$$

Applying Corollary 8.3.9, we deduce that

$$\begin{aligned} & m(\{(x,t) \in Q_T : |(-\Delta)^{\frac{s}{2}} u_n| \geq \lambda, |u_n| < k\}) \\ & \leq \frac{1}{\lambda^2} \int_0^T \int_{\mathbb{R}^N} |(-\Delta)^{\frac{s}{2}} T_k(u_n)|^2 \, dx \leq \frac{k}{\lambda^2} \|f_n\|_{L^1(Q_T)}, \end{aligned}$$

and then we have for every $k > 0$

$$m(\{(x,t) \in Q_T : |(-\Delta)^{\frac{s}{2}} u_n| \geq \lambda\}) \leq \frac{k}{\lambda^2} \|f_n\|_{L^1(Q_T)} + c\left(\frac{\|f_n\|_{L^1(Q_T)}}{k}\right)^{\frac{N+2s}{N}}.$$

Minimizing in k we find that the minimum is achieved in $k = \lambda^{\frac{N}{N+s}} \|f\|_{L^1(Q_T)}^{\frac{s}{N+s}}$. Then the above inequality becomes

$$m(\{(x,t) \in Q_T : |(-\Delta)^{\frac{s}{2}} u_n| \geq \lambda\}) \leq c\left(\frac{\|f\|_{L^1(Q_T)}}{\lambda}\right)^{\frac{N+2s}{N+s}}. \tag{12.3.9}$$

Step 3. *Passing to the limit.*

We have obtained in step 1 that the sequence $\{u_n\}$ is bounded in the Marcinkiewicz space $\mathcal{M}^{\frac{N+2s}{N}}(Q_T)$. Using the linearity of the operator, we conclude that the sequence u_n a. e. converges to a function u in Q_T. By the embedding of $\mathcal{M}^{\frac{N+2s}{N}}(Q_T)$ in $L^q(Q_T)$ for any $q < \frac{N+2s}{N}$ and the uniqueness of the solution, the whole sequence u_n converges to u in $L^q(Q_T)$. Then, it follows that u is a weak solution to (12.1.1).

As in the elliptic case, from (12.3.9) and the linearity of the operator we conclude that the sequence $(-\Delta)^{\frac{s}{2}} u_n$ converges a. e. in Q_T to $(-\Delta)^{\frac{s}{2}} u$. Therefore, from step 2, using Fatou's lemma, u verifies (12.3.4) □

12.4 *A priori* estimates and summability of the solutions

In this section we deal with the summability of the solutions that we have found in the previous section when the right hand side has extra integrability. More precisely, we will study the optimal summability of the solution in terms of the summability of the data.

Let us represent the summability of the datum $f \in L^r(0,T; L^q(\Omega))$ in a diagram with axes $\frac{1}{q}$ and $\frac{1}{r}$. Since $r, q \in [1, +\infty]$, all possible cases of summability are inside of the square $[0,1] \times [0,1]$ (we use the notation $\frac{1}{\infty} = 0$).

12.4.1 Bounded solutions

The curve $\frac{1}{r} + \frac{N}{2qs} = 1$ will be called the *Aronson–Serrin* (AS) curve to be consistent with the local case. The zone under the AS curve, called *AS zone*, corresponds to the set

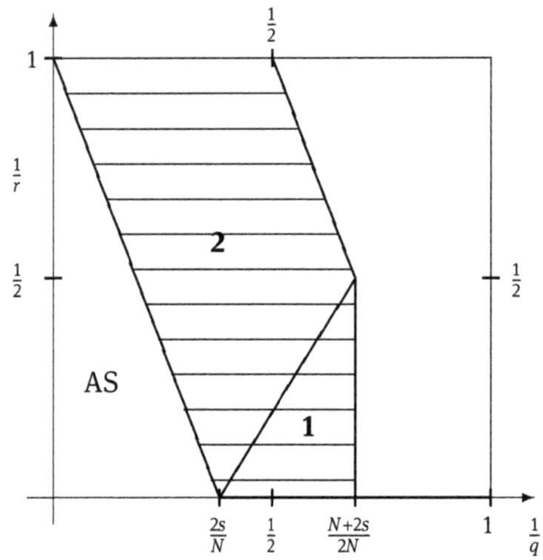

Figure 12.4.1: The *Aronson–Serrin* (AS) zone is delimitated by the line $\frac{1}{r}+\frac{N}{2qs}=1$. Parallel to this AS line, in zone 2, we have the straight line $\frac{N}{2qs}+\frac{1}{r}=1+\frac{N}{4s}$ for $r<2$. Zones 1 and 2 are separated by the line $\frac{1}{r}=\frac{N}{N-2}\frac{1}{q}-\frac{2}{N-2}$ with $r>2$. Finally, zone 1 is delimited on the right by the line $q=\frac{2N}{N+2s}$.

of data for which we find bounded solutions; in the case of second order differential operators (as, for instance, the Laplacian) such a result was obtained by D. G. Aronson and J. Serrin in [39]. The solutions for zones 1 and 2 are not expected to be bounded, but we can determine their summability in some Lebesgue space (for $s=1$, the local case, we recover the regularity results in [81]).

Let us consider the following approximated problems:

$$\begin{cases} u_{nt}+\mathcal{L}u_n=f_n & \text{in } Q_T,\\ u_n(x,t)=0 & \text{in } (\mathbb{R}^N\setminus\Omega)\times[0,T),\\ u_n(x,0)=u_{n0} & \text{in } \Omega\times\{0\}, \end{cases} \tag{12.4.1}$$

where $f_n(x,t)=T_n(f(x,t))$ and $f\geq 0$ in a suitable Lebesgue space.

Observe that since f_n is bounded, the existence of a sequence $\{u_n\}$ of energy solutions to (12.4.1) is guaranteed by Theorem 12.2.4.

The main result in this paragraph is the L^∞-estimate for u_n.

Theorem 12.4.1. *Assume $f\in L^r(0,T;L^q(\Omega))$ with r,q satisfying*

$$\frac{1}{r}+\frac{N}{2qs}<1 \tag{12.4.2}$$

and suppose that $u_{n0}(x)\in L^\infty(\Omega)$. Then there exists a positive constant c such that the unique finite energy solution of (12.4.1) satisfies

$$\|u_n(x,t)\|_{L^\infty(Q_T)}\leq c\|f\|_{L^r(0,T;L^q(\Omega))}.$$

Proof. According to the classical proof by D. G. Aronson and J. Serrin, the idea is to prove an L^∞-bound for u_n in $\Omega \times (0,\tau)$, for a positive small τ (to be fixed), and then iterate such an estimate. Hence, we consider the energy formulation for (12.4.1) in Q_τ and we choose there $G_k(u_n)$ as a test function; hence we get

$$\int_\Omega \varphi_k(u_n(x,\tau))\,dx + \int_0^\tau \mathcal{E}(u_n(x,t), G_k(u_n(x,t)))\,dt = \int_0^\tau \int_\Omega f_n G_k(u_n)\,dx\,dt + \int_\Omega \varphi_k(u_n(x,0))\,dx,$$

where $\varphi_k(s) = \int_0^s G_k(\sigma)\,d\sigma$.

Consider a positive k large enough (let us say $k \geq \|u_{n0}\|_{L^\infty(\Omega)}$, in order to neglect the last term above). We deduce that

$$\|G_k(u_n)\|^2_{L^\infty(0,\tau;L^2(\Omega))} + \|G_k(u_n)\|^2_{L^2(0,\tau;H_0^s(\Omega))} \leq c \int_0^\tau \int_{A_k^t} fG_k(u_n)\,dx\,dt.$$

Note that the term of the right hand side above can be estimated as follows:

$$\int_0^\tau \int_{A_k^t} fG_k(u_n)\,dx\,dt \leq \int_0^\tau \int_{A_k^t} fG_k(u_n)^2\,dx\,dt + \int_0^\tau \int_{A_k^t} f\,dx\,dt. \tag{12.4.3}$$

Let us study each member of the previous inequality. First, we define for any $\chi \in (0,1)$

$$\bar{r} = 2r', \quad \bar{q} = 2q', \quad \hat{r} = \bar{r}(1+\chi), \quad \hat{q} = \bar{q}(1+\chi), \quad \chi = \frac{2\chi_1}{N}, \tag{12.4.4}$$

where $\chi_1 = 1 - \frac{1}{r} - \frac{N}{2qs}$ is strictly positive thanks to (12.4.2). Consequently, $\frac{1}{\hat{r}} + \frac{N}{2\hat{q}s} = \frac{N}{4s}$. Thus by the Hölder and Gagliardo–Nirenberg inequalities (see (8.3.18)), we deduce

$$\begin{aligned}
\int_0^\tau \int_{A_k^t} fG_k(u_n)^2\,dx\,dt &\leq c\mu(k)^{\frac{2\chi}{\hat{r}}} \|G_k(u_n)\|^2_{L^{\hat{r}}(0,\tau;L^{\hat{q}}(A_k^t))} \\
&\leq c\mu(k)^{\frac{2\chi}{\hat{r}}} \left(\int_0^\tau \|G_k(u_n)\|^{(1-\theta)\hat{r}}_{L^2(\Omega)} \|(-\Delta)^{s/2} G_k(u_n)\|^{\hat{r}\theta}_{L^2(\mathbb{R}^N)}\,dt \right)^{\frac{2}{\hat{r}}} \\
&\leq c\mu(k)^{\frac{2\chi}{\hat{r}}} \big[\|G_k(u_n)\|^2_{L^\infty(0,\tau;L^2(\Omega))} + \|(-\Delta)^{s/2} G_k(u_n)\|^2_{L^2(0,\tau;L^2(\mathbb{R}^N))}\big],
\end{aligned}$$

where $\mu(k) = \int_0^\tau |A_k^t|^{\frac{\hat{r}}{\hat{q}}}\,dt$. We also applied the Young inequality and (12.4.4) (i. e., $\hat{r}\theta = 2$).

On the other hand the second term on the right hand side in (12.4.3) can be estimated by means of the Hölder inequality, so that

$$\int_0^\tau \int_{A_k^t} f \, dx\, dt \le c\mu(k)^{\frac{2(1+\chi)}{\hat{r}}}.$$

Denoting

$$|||G_k(u_n)|||^2 = \|G_k(u_n)\|^2_{L^\infty(0,\tau;L^2(\Omega))} + \|(-\Delta)^{s/2}G_k(u_n)\|^2_{L^2(0,\tau;L^2(\mathbb{R}^N))},$$

we deduce

$$|||G_k(u_n)|||^2 \le c[\mu(k)^{\frac{2\chi}{\hat{r}}} |||G_k(u_n)|||^2 + \mu(k)^{\frac{2(1+\chi)}{\hat{r}}}].$$

Note that $\mu(\tau) \le \tau|\Omega|$, so that we can fix τ, independent of u, suitable small in such a way that by using again the Gagliardo–Nirenberg inequality, we deduce

$$\|G_k(u_n)\|^2_{L^{\hat{r}}(0,\tau;L^{\hat{q}}(\Omega))} \le c\mu(k)^{\frac{2(1+\chi)}{\hat{r}}}.$$

Consider $h > k > 0$. Then $A_h \subset A_k$, where A_h is as in (12.3.2), and

$$\mu(h) \le \frac{c}{(h-k)^{\hat{r}}}\mu(k)^{1+\chi}.$$

Applying Lemma 8.4.2, we conclude that there exists a constant d, depending only on $q, r, \|f\|_{L^r(0,T;L^q(\Omega))}$ and s, such that $\mu(d) = 0$, that is,

$$\|u_n\|_{L^\infty(\Omega\times[0,\tau])} \le d.$$

Iterating this procedure in the sets $\Omega \times [\tau, 2\tau], \dots, \Omega \times [j\tau, T]$, where $T - j\tau \le \tau$, we can conclude that

$$\|u_n\|_{L^\infty(Q_T)} \le C \quad \text{uniformly in } n \in \mathbb{N}.$$ □

There are several interesting consequences of the above estimate; here we stress that it directly implies existence of a solution, with the definition of solution changing according to the summability of f.

Corollary 12.4.2. *Let f be a function $f \in L^r(0, T; L^q(\Omega))$ with r, q satisfying*

$$\frac{1}{r} + \frac{N}{2qs} < 1. \tag{12.4.5}$$

Then, the solution obtained as limit of the solutions to (12.4.1) satisfies:

1. $\|u\|_{L^\infty(Q_T)} \le c$;
2. $\|u\|_{L^2(0,T;H_0^s(\Omega))} \le c$; and
3. *moreover,*
   (a) *if* $r \ge 2$, *then* $u_t \in L^2(0,T;H^{-s}(\mathbb{R}^N))$, *and* u *is an energy solution;*
   (b) *if* $1 < r < 2$, *then the equation is satisfied in the following weak sense:*

$$-\int_{Q_T} u\phi_t\,dx\,dt + \int_0^T \mathcal{E}(u,\phi)\,dt = \int_{Q_T} f\phi\,dx\,dt + \int_\Omega u(x,0)\phi(x,0)\,dx,$$

for any $\phi \in L^2(0,T;H_0^s(\Omega)) \cap L^\infty(Q_T) \cap C^1([0,T);L^1(\Omega))$, *with* $\phi(x,T) = 0$.

Proof. We first observe that since the f_n are increasing, by the weak comparison principle (using that the equation is linear) we deduce that the sequence u_n is increasing, too. Thus, the uniform estimate gives (1). Moreover, testing equations (12.4.1) with u_n, part (2) easily follows.

Noting that if $r \ge 2$, then $f \in L^2(0,T;L^{(2_s^*)'}(\Omega))$, we obtain (3a). Finally (3b) is obtained by a simple integration by parts argument in (12.4.1) and passing to the limit. □

Going further into the summability of the solution, we consider the borderline case in which $\frac{1}{r} + \frac{N}{2qs} = 1$. Having in mind what happens in the elliptic case, we prove that solutions are not anymore bounded, but have an exponential summability. Namely, we prove the following result.

Theorem 12.4.3. *Assume that* f *belongs to* $L^r(0,T;L^q(\Omega))$, *with* $r \ge 1$, $q > \frac{N}{2s}$ *such that*

$$\frac{1}{r} + \frac{N}{2qs} = 1, \tag{12.4.6}$$

and let u_n *be the solution of* (12.4.1) *with* $u_{n0}(x) \equiv 0$. *Then there exists an* $\alpha > 0$ *such that*

$$\|e^{\alpha u_n}\|_{L^\infty(0,T;L^2(\Omega))} \le c\|f\|_{L^r(0,T;L^q(\Omega))}. \tag{12.4.7}$$

Consequently, u_n *is bounded in* $L^p(Q_T)$, *for every* $p \ge 1$.

Proof. We only sketch the proof, since it is very similar to the elliptic one that we have proved in Section 8.4 (see Theorem 8.4.3).

Consider the following auxiliary convex function:

$$\Phi_k(\sigma) = \begin{cases} e^{\alpha\sigma} - 1 & \text{if } 0 \le \sigma < k, \\ \alpha e^{\alpha k}(\sigma - k) + e^{\alpha k} - 1 & \text{if } \sigma \ge k \end{cases}$$

for $k > 0$ and for some α to be fixed later. Thus we have

$$\frac{1}{2}\|\Phi_k(u_n)\|^2_{L^\infty(0,T;L^2(\Omega))} + \lambda\|u_n\|^2_{L^2(0,T;H_0^s(\Omega))} \le \int_0^T\int_\Omega \Phi_k'(u_n)\Phi_k(u_n) f\,dx\,dt.$$

Now, observe that for $0 \le u_n \le k$ we have $\Phi_k'(u_n) = \alpha(e^{\alpha u_n} - 1 + 1) = \alpha\Phi_k(u_n) + \alpha$. Hence,

$$\int_0^T \int_\Omega \Phi_k'(u_n)\Phi_k(u_n) f \, dx\, dt \le \alpha \int_0^T \int_{\{u_n \le k\}} \Phi_k(u_n)^2 f + \alpha \int_0^T \int_{\{u_n \le k\}} \Phi_k(u_n) f + \alpha e^{\alpha k} \int_0^T \int_{\{u_n > k\}} \Phi_k(u_n) f.$$

From now on, we follow the same ideas as in the elliptic case, and we get (12.4.7). □

12.4.2 Summability of the solutions outside of the Aronson–Serrin zone

Outside the AS zone, the solutions are not expected to be bounded. We obtain the following summability results for zones 1 and 2 of Figure 12.4.1.

Theorem 12.4.4. *Assume $u_{n0}(x) \equiv 0$ and $f \in L^r(0, T; L^q(\Omega))$ with $r > 1, q > 1$, satisfying*

$$1 < \frac{1}{r} + \frac{N}{2qs} \le 1 + \frac{N}{4s} \quad \text{and} \quad \frac{1}{q} \le \frac{1}{2} + \frac{s}{N} \quad \text{(zones 1 and 2 in Figure 12.4.1).}$$

Then there exists a positive constant c such that the sequence of finite energy solutions of (12.4.1) satisfies

$$\|u_n(x,t)\|_{L^\infty(0,T;L^{2\gamma}(\Omega))} + \|u_n(x,t)\|_{L^{2\gamma}(0,T;L^{2_s^*\gamma})} \le c\|f\|_{L^r(0,T;L^q(\Omega))},$$

where

$$\gamma = \begin{cases} \frac{q(N-2s)}{2(N-2qs)} & \text{if } \frac{1}{r} < \frac{N}{N-2}\frac{1}{q} - \frac{2}{N-2} \quad \text{(zone 1 in Figure 12.4.1),} \\ \frac{qrN}{2(Nr-2qs(r-1))} & \text{if } \frac{1}{r} \ge \frac{N}{N-2}\frac{1}{q} - \frac{2}{N-2} \quad \text{(zone 2 in Figure 12.4.1).} \end{cases} \tag{12.4.8}$$

Proof. Let us consider a positive Lipschitz convex function $\Phi(s)$, with $\Phi(0) = 0$. Note that if $v \in C_0^\infty(\Omega)$,

$$\Phi(v)(-\Delta)^s\Phi(v) \le \Phi'(v)\Phi(v)(-\Delta)^s v. \tag{12.4.9}$$

We point out that since u_n is a bounded energy solution, $\Phi(u_n)$ belongs to the energy space, and then we are allowed to choose it as a test function in problem (12.4.1). Thus by (8.3.15), we deduce that

$$\int_0^T \int_\Omega \Phi(u_n)\Phi(u_n)_t \, dx\, dt + \lambda \int_0^T \|\Phi(u_n)\|^2_{H_0^s(\Omega)} dt \le \int_0^T \int_\Omega \Phi'(u_n)\Phi(u_n) f \, dx\, dt. \tag{12.4.10}$$

Let us apply such an inequality with the choice $\Phi(s) = s^\gamma$, with $\gamma \geq 1$ as in (12.4.8). Thus applying the Hölder and Sobolev inequalities, we obtain

$$\frac{1}{2\gamma} \sup_{t\in[0,T]} \int_\Omega u_n^{2\gamma}\,dx + \lambda \int_0^T \|u_n^\gamma\|^2_{L^{2_s^*}(\Omega)}\,dt$$
$$\leq \gamma \|f_n\|_{L^r(0,T;L^q(\Omega))} \left(\int_0^T \left[\int_\Omega u_n^{(2\gamma-1)q'}\,dx \right]^{\frac{r'}{q'}} dt \right)^{\frac{1}{r'}}. \tag{12.4.11}$$

Denoting

$$A = \left(\int_0^T \|u_n\|^{(2\gamma-1)r'}_{L^{(2\gamma-1)q'}(\Omega)}\,dt \right)^{\frac{1}{r'}},$$

thanks to Sobolev inequality, we get

$$\sup_{[0,T]} \int_\Omega u_n^{2\gamma+1}\,dx \leq 2\gamma A \|f_n\|_{L^r(0,T;L^q(\Omega))}$$

and

$$\int_0^T \left[\int_\Omega u_n^{\gamma 2_s^*}\,dx \right]^{\frac{2}{2_s^*}} dt \leq \frac{S^2}{\lambda} \gamma A \|f_n\|_{L^r(0,T;L^q(\Omega))}.$$

We distinguish now two cases depending on the size of q.

First case: $1 < q < \frac{Nr}{N+2s(r-1)}$, i. e., zone 1 in Figure 12.4.1.

Observe that $1 < q < \frac{Nr}{N+2s(r-1)}$ implies $q < r$ and $\frac{r'}{q'} < \frac{2}{2_s^*}$. Hence, we apply the Hölder inequality to estimate A and we get

$$A \leq T^{\frac{1}{r'} - \frac{2_s^*}{2q'}} \left[\int_0^T \left(\int_\Omega u_n^{(2\gamma-1)q'}\,dx \right)^{\frac{2}{2_s^*}} dt \right]^{\frac{2_s^*}{2q'}}.$$

Thus by the Sobolev inequality (see (8.2.7)), we deduce

$$\int_0^T \left[\int_\Omega u_n^{2_s^*\gamma}\,dx \right]^{\frac{2}{2_s^*}} dt \leq c \left[\int_0^T \left(\int_\Omega u_n^{(2\gamma-1)q'}\,dx \right)^{\frac{2}{2_s^*}} dt \right]^{\frac{2_s^*}{2q'}}.$$

Choose $2_s^*\gamma = (2\gamma - 1)q'$, that is, $\gamma = \frac{1}{2} \cdot \frac{q(N-2s)}{(N-2qs)}$. Note that since $q < \frac{N}{2s}$ it follows that $\frac{2_s^*}{2q'} < 1$. Thus we get

$$\int_0^T \left(\int_\Omega u_n^{(2\gamma_1-1)q'}\,dx \right)^{\frac{2}{2_s^*}} dt \leq c,$$

and therefore, $\|u_n\|_{L^r(0,T;L^{q_s^{**}}(\Omega))} \leq c$, with $r = \frac{q(N-2s)}{(N-2qs)} > q$.

Second case: $\frac{Nr}{N+2(r-1)} \le q < \frac{N}{2s}r'$, i. e., zone 2 in Figure 12.4.1.

In this case we choose $\gamma = \frac{1}{2}\frac{qrN}{Nr-2qs(r-1)}$. Thus by interpolation, we get $\frac{1}{(2\gamma-1)q'} = \frac{1-\theta}{2\gamma} + \frac{\theta}{2_s^*\gamma}$ and consequently

$$\int_0^T \|u_n\|_{L^{(2\gamma-1)q'}(\Omega)}^{r'(2\gamma-1)}\,dt \le c\|u_n\|_{L^\infty(0,T;L^{2\gamma}(\Omega))}^{2\gamma\mu_1} \int_0^T \left(\int_\Omega u_n^{2_s^*\gamma}\,dx\right)^{\frac{2}{2_s^*}\mu_2} dt,$$

where $\mu_1 = \frac{(1-\theta)r'(2\gamma-1)}{2\gamma}$ and $\mu_2 = \frac{\theta r'(2\gamma-1)}{2\gamma}$. Using (12.4.8), we have

$$\int_0^T \|u_n\|_{L^{(2\gamma-1)q'}}^{r'(2\gamma-1)}\,dt \le c\|u_n\|_{L^\infty(0,T;L^{2\gamma}(\Omega))}^{2\gamma\mu_1} \left(\int_0^T \left[\int_\Omega u_n^{2_s^*\gamma}\,dx\right]^{\frac{2}{2_s^*}} dt\right)^{\mu_2},$$

and since $\mu_2 \le 1$, matching the above inequality with (12.4.11) we obtain

$$\int_0^T \left(\int_\Omega u_n^{(2\gamma-1)q'}\,dx\right)^{\frac{r'}{q'}} dt \le c. \qquad \square$$

Remark 12.4.5. Under the assumptions of Theorem 12.4.4, if u is the energy solution of (12.1.1), then it belongs also to the following spaces:

$$L^p(0,T;L^r(\Omega)), \quad p = \frac{4\gamma ms}{N(m-2)}, \quad r = \gamma m, \quad \forall m \in (2, 2_s^*], \tag{12.4.12}$$

where γ is the number defined in Theorem 12.4.4. Moreover uniform estimates hold true in the previous spaces.
In particular, choosing $m = 2_s^*$, we get the best summability exponent in the spacial variable (i. e., $2_s^*\gamma$) together with the lower summability exponent in the time variable (i. e., 2γ) and we obtain

$$u \in L^{2\gamma}(0,T;L^{2_s^*\gamma}(\Omega)).$$

On the other hand, choosing $m = 2\frac{N+2s}{N}$ we get $p = r$ and hence

$$u \in L^\mu(Q), \quad \mu = 2\gamma\frac{N+2s}{N}, \tag{12.4.13}$$

where γ is defined in (12.4.8).

12.4.3 Further summability results

In order to conclude our study of the summability of the solutions, we argue as follows: first we find the summability of the solutions with data in $L^\infty(0,T;L^q(\Omega))$, for $q \in [1, N/2s)$, i. e., the bottom line of Figure 12.4.1.

Once we found that the unique solution of problem (12.4.1) is bounded in the Lebesgue space $L^\infty(0,T;L^{q_s^{**}}(\Omega))$, we argue by interpolation. Indeed, assume to know, for any pairs of exponents (r_1,q_1) and (r_2,q_2), that the solutions u_n of (12.4.1) are bounded in some Lebesgue space.

Recall the following interpolation result for parabolic Sobolev spaces.

Theorem 12.4.6. *Let $r_i,q_i \in (1,\infty]$, $i = 1,2$, and suppose that $v \in L^{r_1}(0,T;L^{q_1}(\Omega)) \cap L^{r_2}(0,T;L^{q_2}(\Omega))$. Then for any $\theta \in (0,1)$, $u \in L^r(0,T;L^q(\Omega))$ with*

$$\frac{1}{r} = \frac{1-\theta}{r_1} + \frac{\theta}{r_2} \quad \textit{and} \quad \frac{1}{q} = \frac{1-\theta}{q_1} + \frac{\theta}{q_2} \tag{12.4.14}$$

and

$$\|v\|_{L^r(0,T;L^q(\Omega))} \le \|v\|^{\theta}_{L^{r_1}(0,T;L^{q_1}(\Omega))}\|v\|^{1-\theta}_{L^{r_2}(0,T;L^{q_2}(\Omega))}.$$

Proof. See [39, Lemma 1]. □

Then, by Theorem 12.4.6, we know the summability of the sequence $\{u_n\}$ for any r and q that satisfy (12.4.14). Roughly speaking, in Figure 12.4.1, given any pairs of coordinates, we can deduce the summability of u_n for any r and q that lay in the segment that joins $(\frac{1}{r_1},\frac{1}{q_1})$ and $(\frac{1}{r_2},\frac{1}{q_2})$.

Therefore, if we know the behavior of u_n in a subset $\mathcal{A}$ of the square $(0,1)\times(0,1)$ of Figure 12.4.1, we know it in its convex envelop $\mathcal{A}^*$. Thus, if we deduce the summability of u_n in the bottom line of Figure 12.4.1, i. e., if such a line belongs to $\mathcal{A}$, then $\mathcal{A}^* \supset (0,1)\times(0,1)$.

We deal, now, with the bottom segment, $r=\infty$ and $1<q<\frac{N}{2s}$. We use the following result.

Lemma 12.4.7. *If v is a positive finite energy solution to*

$$\begin{cases} v_t + \mathcal{L}v = F(x) & \textit{in } Q_T,\\ v(x,t) = 0 & \textit{in } (\mathbb{R}^N\setminus\Omega)\times[0,T),\\ v(x,0)=0 & \textit{in } \Omega, \end{cases} \tag{12.4.15}$$

with $F \in L^{\frac{2N}{N+2s}}(\Omega)$, $F(x)\ge 0$, then $v_t(x,t)\ge 0$ for all (x,t) in Q_T.

Proof. Applying the (weak) maximum principle (i. e., multiplying the equation by $v_n^-(x,t)$ and integrating by parts), we conclude that $v(x,t)\ge 0$, for every (x,t) in Q_T. It is sufficient to prove that $v(x,t)\le v(x,t+\tau)$, for all $\tau\in[0,T-t]$. This immediately follows thanks to the comparison principle for finite energy solutions, since, for $\tau\in[0,T-t]$,

$$v_t(x,t)+\mathcal{L}v(x,t) = F(x) = v_t(x,t+\tau)+\mathcal{L}v(x,t+\tau),$$
$$v(x,0) = 0 \le v(x,t).$$

□

Theorem 12.4.8. *If u_n is a solution to* (12.4.1), *with* $f(x,t) \in L^{\infty}(0,T;L^q(\Omega))$ *and* $u_{n0}(x) \equiv 0$, *then:*
(i) *if* $q > \frac{N}{2s}$, *then* $\|u_n\|_{L^{\infty}(Q_T)} \le c$;
(ii) *if* $1 < q < \frac{N}{2s}$, *then* $\|u_n\|_{L^{\infty}(0,T;L^{q_s^{**}}(\Omega))} \le c$;
(iii) *if* $q = 1$, *then* $\|u_n\|_{L^{\infty}(0,T;L^{\sigma}(\Omega))} \le c$, $\forall \sigma \in (1, \frac{N}{N-2s})$.

Proof. By using Lemma 12.4.7, u_n are subsolutions of the elliptic problem associated to (12.4.1) and applying the results of the elliptic sections, the conclusions follow. □

12.4.4 Parabolic Kato inequality

We have also a parabolic counterpart of the elliptic Kato inequality proved in Theorem 8.3.10. We state it here for completeness, but we only sketch the proof since it follows with minor changes from the previous one.

Theorem 12.4.9 (Parabolic Kato inequality). *Let Ω be a bounded domain in $\mathbb{R}^N$ and φ : $\mathbb{R} \to \mathbb{R}$ a nondecreasing bounded continuous function (except at most in a number of finite points). Then for any weak solution to*

$$u_t + \mathcal{L}u = f \quad \text{in } Q_T,$$

with $f \in L^1(Q_T)$, we have

$$\frac{\partial \Phi(u)}{\partial t} + \mathcal{L}\Phi(u) \le \left(\frac{\partial u}{\partial t} + \mathcal{L}u\right)\varphi(u) \quad \text{in the weak sense,}$$

where

$$\Phi(s) = \int_0^s \varphi(\sigma)\, d\sigma.$$

Proof. If $\varphi(s)$ is smooth, then the proof is direct. Otherwise we argue by approximation and we follow the same line as in the elliptic case. □

Remark 12.4.10. Observe that applying the above result with the choice $\varphi(s) = \text{sign}(s)$, we get

$$\frac{\partial |u|}{\partial t} + \mathcal{L}|u| \le \left(\frac{\partial u}{\partial t} + \mathcal{L}u\right)\text{sign}(u) \quad \text{in the weak sense,}$$

while for $\varphi(u) = \text{sign}^+(u)$, we deduce

$$\frac{\partial u^+}{\partial t} + \mathcal{L}u^+ \le \left(\frac{\partial u}{\partial t} + \mathcal{L}u\right)\text{sign}^+(u) \quad \text{in the weak sense.}$$

12.5 The Cauchy problem for the fractional heat equation

In this section we will consider the Cauchy problem for the fractional heat equation, that is, the problem

$$\begin{aligned} u_t(x,t) + (-\Delta)^s u(x,t) = 0, \quad (x,t) \in \mathbb{R}^N \times (0,\infty),\\ u(x,0) = g(x), \quad x \in \mathbb{R}^N. \end{aligned} \tag{12.5.1}$$

Taking the Fourier transform in the spatial variable, we find

$$\widehat{u}_t + |\xi|^{2s}\widehat{u} = 0, \quad \widehat{u}(\xi,0) = \widehat{g}(\xi). \tag{12.5.2}$$

Then

$$\widehat{u}(x,t) = \widehat{g}(\xi)e^{-|\xi|^{2s}t}, \tag{12.5.3}$$

which is well defined, for instance, if $g \in \mathcal{S}(\mathbb{R}^N)$. By using the inverse Fourier transform we find

$$u(x,t) = \int_{\mathbb{R}^N} P_{2s}(x-y,t)g(y)\,dy, \tag{12.5.4}$$

where

$$P_{2s}(x,t) = (2\pi)^{-N} \int_{\mathbb{R}^N} e^{i\langle x,\xi\rangle} e^{-|\xi|^{2s}t}\,d\xi. \tag{12.5.5}$$

Since for all positive t,

$$e^{-|\xi|^{2s}t} \in L^1(\mathbb{R}^N),$$

$P_{2s}(x,t)$ is a continuous function that goes to zero as $|x| \to \infty$. Also, since $(x_j\xi_j)^k e^{-|\xi|^{2s}t} \in L^1(\mathbb{R}^N), \forall k \in \mathbb{N}$, we have $P_{2s}(x,t) \in \mathcal{C}^\infty(\mathbb{R}^N)$ for all $t > 0$. Note that by scaling,

$$P_{2s}(x,t) = t^{-\frac{N}{2s}} P_{2s}(xt^{-\frac{1}{t^{2s}}},1)$$

and then, with the appropriate normalization constant, it is an approximation to the identity. We will study the behavior of the kernel $P_{2s}(x,1)$. For $s = 1$ it is well known that we find a Gaussian function and for $s = \frac{1}{2}$ we get the Poisson kernel in the half space. For the rest of the values of $s \in (0,1)$ there is no elementary expression. The global behavior of the kernel has been developed mainly from the world of probability theory, more precisely from the theory of stable fractional Markov processes (see for instance the paper by K. Bogdan, A. Stós and P. Sztonyk [84] and the references therein). A first result in dimension $N = 1$ is due to G. Polya in [262] and for $N \geq 1$ due to Blumenthal and Getoor in [63], in both cases using properties of the Bessel functions.

Proposition 12.5.1 (Blumenthal, Getoor, Polya). *Let $\alpha > 0$ and let $P_\alpha(x,1)$ be defined by* (12.5.5) *(with $2s = \alpha$). Then*

$$\lim_{|x|\to\infty} |x|^{N+2\alpha} P_\alpha(x,1) = \alpha 2^{\alpha-1}\left(\frac{1}{\pi}\right)^{\frac{N}{2}+1} \sin\left(\frac{\alpha\pi}{2}\right)\Gamma\left(\frac{N+\alpha}{2}\right)\Gamma\left(\frac{\alpha}{2}\right).$$

A different approach can be found in reference [84], where the arguments are based on the subordination of processes in the sense of Bochner [82]. In a pure analytical language, this means taking $\eta(A)$ such that its Laplace transform verifies

$$\mathcal{L}\eta(A) = \int_0^\infty e^{-Au}\eta(u)\,du = e^{-A^{\alpha/2}},$$

and then estimate $\eta(A)$ via the properties of the Laplace transform; see [142, Theorem 37.1] and also [320, pages 259–268].

However we will perform here a direct approach, with the ideas by Bochner, using only real analysis techniques and avoiding the theory of Bessel functions. So, we will prove the following result.

Theorem 12.5.2. *Let $P_\alpha(x,t)$ be defined by* (12.5.5), *where $\alpha = 2s$, $0 < s < 1$ and $N \geq 2s$. Then, there exists a constant $C > 1$ such that*

$$\frac{1}{C}\frac{1}{(1+|x|^{N+\alpha})} \leq P_\alpha(x,1) \leq \frac{C}{(1+|x|^{N+\alpha})}, \quad \text{for all } x \in \mathbb{R}^N. \tag{12.5.6}$$

To prove Theorem 12.5.2 we will use a celebrated result by S. N. Bernstein (see [59]) about the characterization of *completely monotone functions* via Laplace transform.

To try to make the presentation self-contained we will show the proof of the Bernstein result and then we will recall the Helly selection principle [198], which is a classical tool used in harmonic analysis, probability theory and real analysis. With our current perspective, this can be seen as a Rellich compactness result in the space of *bounded variation functions*. We will formulate the result as we will need it, and later we will give the general version.

Lemma 12.5.3 (The Helly selection principle). *Let $\{\rho_n\}_{n\in\mathbb{N}}$ be a uniformly bounded sequence of non decreasing functions in a given interval $I \subset \mathbb{R}$. Then there exists a subsequence converging pointwise to an increasing function.*

Proof. We apply the Cantor diagonal principle to obtain a subsequence $\{\rho_{n_k}\}_{n_k\in\mathbb{N}} \subset \{\rho_n\}_{n\in\mathbb{N}}$ that converges at any rational $r \in I \cap \mathbb{Q}$. As every ρ_{n_k} is non decreasing, the limit function

$$\rho : I \cap \mathbb{Q} \to \mathbb{R}$$

is also non decreasing. To extend ρ to the whole interval I we proceed as follows: given a point $x \in I \setminus \mathbb{Q}$ we consider an increasing sequence of rational numbers $r_k \to x$ as

$k \to \infty$ and define

$$\rho(x) = \lim_{k\to\infty} \rho(r_k);$$

therefore ρ is well defined in I and again is non decreasing. Hence the discontinuities of ρ are finite jumps. The points of discontinuity of ρ are contained in the set

$$D = \bigcup_1^{\infty} D_n, \quad \text{where } D_n = \left\{x \in I, \rho(x^+) - \rho(x^-) \geq \frac{1}{n}\right\}.$$

Note that since ρ is bounded, each D_n must be finite, and then D is at most countable.

To finish we have to prove that there exists a subsequence that converges pointwise in I. If x is a continuity point of ρ, then for $r_l < x < r_m$ rational numbers we find

$$\rho_{n_k}(r_l) - \rho(r_m) \leq \rho_{n_k}(x) - \rho(x) \leq \rho_{n_k}(r_m) - \rho(r_l).$$

Choosing r_l, r_m rational numbers such that for $\epsilon > 0$ $\rho(r_m) - \rho(r_l) < \epsilon$, we obtain

$$-\epsilon \leq \liminf_{n_k\to\infty}(\rho_{n_k}(x) - \rho(x)) \leq \limsup_{n_k\to\infty}(\rho_{n_k}(x) - \rho(x)),$$

and since ϵ is arbitrary we conclude the proof.

Assume, to finish, that for $x_1 \in D$, by the Bolzano–Weierstrass theorem there exists a subsequence of the previous subsequence that converges in x_1 and since D is numerable, iterating the Cantor process we obtain a subsequence that converges in each $x \in D$. □

Recall that if $I = [a, b]$ is a bounded interval in the real line, a function

$$f : I \to \mathbb{R} \quad \text{is of bounded variation if and only if} \quad f(x) = p(x) - n(x),$$

where p and n are monotone functions (see for instance [275]). As customary we denote by $BV(I)$ the class of bounded variation functions in $I = [a, b]$. For a partition

$$\mathcal{P} = \{a = x_0 < x_1 < \cdots < x_n = b\} \quad \text{we define } v_{\mathcal{P}}(f) = \sum_1^k |f(x_i) - f(x_{i-1})|$$

and the total variation of $f \in BV(I)$ by

$$V_I(f) = \sup\{v_{\mathcal{P}}(f) \mid \mathcal{P} \quad \text{partition of } I\}.$$

We can formulate now the general version of the Helly principle.

Theorem 12.5.4. *Let $\{f_m\}_{n\in\mathbb{N}}$ be a sequence in $BV(I)$ such that:*
(i) *There exists C_1 such that $\sup_I |f_m(x)| < C_1$ for all $m \in \mathbb{N}$.*
(ii) *There exists C_2 such that $V_I(f_m) \leq C_2$ for all $m \in \mathbb{N}$.*

Then there exist a function $f \in BV(I)$ and a subsequence $\{f_{n_k}\}_{n_k\in\mathbb{N}} \subset \{f_n\}_{n\in\mathbb{N}}$ such that

$$\lim_{n_k\to\infty} f_{n_k}(x) = f(x), \quad \textit{for all } x \in I.$$

Proof. We find a couple of sequences $\{p_m\}_{m\in\mathbb{N}}$ and $\{n_m\}_{m\in\mathbb{N}}$ such that $f_n = p_m - n_m$. From Lemma 12.5.3 we find a subsequence

$$\{p_{m_k}\}_{m_k\in\mathbb{N}} \subset \{p_m\}_{m\in\mathbb{N}}$$

and a monotone function p such that

$$p_{m_k}(x) \to p(x) \quad \text{as } m_k \to \infty.$$

Starting now with the sequence $\{n_{m_k}\}_{m_k\in\mathbb{N}}$, again from Lemma 12.5.3 we find a subsequence $\{n_{m_{k_j}}\}_{m_{k_j}\in\mathbb{N}} \subset \{n_{m_k}\}_{m_k\in\mathbb{N}}$ and a monotone function $n(x)$ such that

$$n_{m_{k_j}}(x) \to n(x) \quad \text{as } m_{k_j} \to \infty \text{ for all } x \in I.$$

Therefore, for all $x \in I$,

$$f_{m_{k_j}}(x) = p_{m_{k_j}}(x) - n_{m_{k_j}}(x) \to f(x) = p(x) - n(x), \quad \text{as } m_{k_j} \to \infty,$$

and hence $f \in BV(I)$. □

Some applications of the Helly selection principle to the study of trigonometic series can be found in the classical book by A. Zigmund [322]. More general details about the above result understood as a compactness result can be found in [149].

We return to the study of the result by Bernstein by introducing the precise meaning of *completely monotone function*.

Definition 12.5.5. Let $g : [0,\infty) \to [0,\infty)$ be a continuous function which is additionally in $C^\infty((0,\infty))$. We say that g is completely monotone when

$$\forall n \geq 0,\ \forall t > 0, \quad (-1)^n g^{(n)}(t) \geq 0.$$

In particular if g is completely monotone $g^{(0)} := g \geq 0$ on $[0,\infty)$, $g^{(1)} := g' \leq 0$ on $(0,\infty)$, $g^{(2)} := g'' \geq 0$ on $(0,\infty)$, etc. And moreover, $g(+\infty) := \lim_{t\to+\infty} g(t) = 0$ exists and g is bounded.

Theorem 12.5.6 (Bernstein). *The following statements are equivalent:*
1. *g is completely monotone, that is,*

$$\forall n \geq 0,\ \forall t > 0, \quad (-1)^n g^{(n)}(t) \geq 0;$$

2. *g is the Laplace transform of a nonnegative finite Borel measure μ on $\mathbb{R}_+$:*

$$\forall t \in \mathbb{R}_+, \quad g(t) = \int_0^\infty e^{-xt}\, d\mu(x).$$

Proof. That (2) implies (1) follows from the fact that

$$g^{(n)}(t) = (-1)^n \int_0^\infty x^n e^{-tx}\, d\mu(x).$$

Now we will prove that (1) implies (2). Since $(-1)^n g^{(n)}$ is nonnegative and nonincreasing, we have for any $t > 0$ and $n \geq 1$,

$$|g^{(n)}(t)| = (-1)^n g^{(n)}(t) \leq \frac{2}{t} \int_{t/2}^{t} (-1)^n g^{(n)}(u)\, du = \frac{2}{t} |g^{(n-1)}(t) - g^{(n-1)}(t/2)|.$$

By induction, it follows then that

$$\forall n \geq 1, \quad g^{(n)}(t) = o(t^{-n}), \quad \text{as } t \to +\infty.$$

By integration by parts (the boundary terms vanish thanks to the former result),

$$\begin{aligned} g(x) - g(+\infty) &= -\int_x^\infty g'(t)\, dt \\ &\;\vdots \\ &= \frac{(-1)^{n+1}}{n!} \int_x^\infty (t-x)^n g^{(n+1)}(t)\, dt \\ &= \frac{(-1)^{n+1} n}{n!} \int_{x/n}^\infty (1 - x/(tn))^n (nt)^n g^{(n+1)}(nt)\, dt \\ &= \int_0^\infty \varphi_n(x/t)\, d\left(\frac{(-1)^{n+1} n}{n!} \int_0^t (nu)^n g^{(n+1)}(nu)\, du \right) \\ &= \int_0^\infty \varphi_n(xt)\, d\sigma_n(t), \end{aligned}$$

where $\varphi_n(x) := (1 - x/n)_+^n$ and where

$$\sigma_n(t) := \frac{1}{n!} \int_{1/t}^\infty (-1)^{n+1} n (nu)^n g^{(n+1)}(nu)\, du.$$

Integration by parts again gives

$$\frac{1}{n!} \int_0^\infty x^n |g^{(n+1)}|(x)\, dx = \frac{(-1)^n}{(n-1)!} \int_0^\infty x^{n-1} g^{(n)}(x)\, dx$$

$$\vdots$$

$$= -\int_0^\infty g'(x)\,dx = g(0) - g(+\infty).$$

Therefore, the total variation of σ_n on $[0, +\infty)$ is

$$\frac{1}{n!}\int_0^\infty n(nt)^n |g^{(n+1)}(nt)|\,dt = g(0) - g(+\infty).$$

By Lemma 12.5.3, there exists a subsequence $\sigma_{n_k}(t)$ that converges almost everywhere to a bounded nonnegative nondecreasing function $\sigma(t)$ on $[0, +\infty)$. Since $\varphi_n(x) \to e^{-x}$ uniformly on $[0, \infty)$ as $n \to \infty$, it follows that

$$g(x) - g(+\infty) = \int_0^\infty e^{-tx}\,d\sigma(t).$$

Finally, if we set $\mu := \sigma + g(+\infty)\delta_0$, δ_0 being the Dirac measure concentrated at 0, then we get for all $x \in [0, +\infty)$,

$$g(x) = \int_0^\infty e^{-tx}\,d\mu(t),$$

as we want to prove. □

We will also need the following.

Corollary 12.5.7. *Assume that $g(t)$ is completely monotone and define, for $\xi \in \mathbb{R}^N$, the function $F(\xi) = g(|\xi|^2)$. Then, the inverse Fourier transform of F is given by a positive, radially decreasing function.*

Proof. The proof is a simple consequence of Bernstein's theorem. By hypothesis, there exists a finite Borel measure μ so that

$$F(\xi) = g(|\xi|^2) = \int_0^\infty e^{-|\xi|^2 s}\,d\mu(s).$$

Using the Fubini theorem, we have

$$P(x) = (2\pi)^{-N/2}\int_{\mathbb{R}^N} F(\xi)e^{ix\cdot\xi}\,d\xi = (2\pi)^{-N/2}\int_0^\infty\int_{\mathbb{R}^N} e^{-|\xi|^2 s}e^{ix\cdot\xi}\,d\xi\,d\mu(s)$$

$$= \int_0^\infty \frac{1}{(2s)^{N/2}}e^{-|x|^2/4s}\,d\mu(s). \tag{12.5.7}$$

This clearly gives the result. □

In particular, for the $\eta = \eta_\alpha$, the inverse Laplace Transform of $e^{-A^{\alpha/2}}$

$$P_\alpha(x,1) = (2\pi)^{-N} \int_{\mathbb{R}^N} e^{i\langle x,\xi\rangle - |\xi|^\alpha}\, d\xi = \int_0^\infty \frac{1}{(4\pi s)^{N/2}} e^{-\frac{|x|^2}{4s}} \eta(s)\, ds.$$

So, $P_\alpha(x,1) > 0$ for all $x \in \mathbb{R}^N$. Moreover, the corollary tells us that $P_\alpha(x,1)$ is radially decreasing and that

$$0 < P_\alpha(x,1) \le P_\alpha(0,1), \quad \forall x.$$

Therefore we have

$$P_\alpha(x,t) \le P_\alpha(y,t), \quad \text{if } |x| > |y|.$$

By the scaling

$$P_\alpha(t,x) = t^{-\frac{N}{\alpha}} P_\alpha\left(\frac{x}{t^{\frac{1}{\alpha}}}, 1\right),$$

for $t > s$ we find

$$P_\alpha(x,t) \ge \left(\frac{s}{t}\right)^{\frac{N}{\alpha}} P_\alpha(x,s).$$

Before we go to the proof of Theorem 12.5.2 let us present another result that will be used later.

Corollary 12.5.8. *Assume that $P_\alpha(0,t) \le 1$ and $\delta \ge 2$. Then*

$$P_\alpha\left(\frac{(x-y)}{\delta}, t\right) \ge P_\alpha(x,t) P_\alpha(y,t).$$

Proof. It is clear that the numerical inequality

$$\frac{1}{\delta}|x-y| \le \max\left\{\frac{2}{\delta}|x|, \frac{2}{\delta}|y|\right\} \le \max\{|x|,|y|\}$$

holds. Hence

$$P_\alpha\left(\frac{1}{\delta}|x-y|, t\right) \ge P_\alpha(\max\{|x|,|y|\}, t) \ge P_\alpha(|x|,t) P_\alpha(|y|,t),$$

where the last inequality is a consequence of the radial decrease of the kernel P_α and the hypothesis $P_\alpha(0,t) \le 1$. □

The conclusion of Theorem 12.5.2 is thus a direct consequence of the positivity of the kernel, which the Bernstein theorem provides, and Theorem 12.5.1 by Polya ($N = 1$) and Blumenthal–Getoor ($N \ge 1$).

Let us prove the latter in a straightforward manner.

Proposition 12.5.9. *Given as before $0 < \gamma < 1$, $g(t) = e^{-t^\gamma}$, if we denote by $P_\gamma(x)$ the inverse Fourier transform of $F(\xi) = g(|\xi|^2) = e^{-|\xi|^{2\gamma}}$, then P_γ satisfies the following asymptotic behavior:*

$$\lim_{|x|\to\infty} |x|^{N+2\gamma} P_\gamma(x) = \frac{\gamma}{\Gamma(1-\gamma)} \Gamma\left(\frac{N+2\gamma}{2}\right).$$

Proof. Using that $g^{(n)} \to 0$ as $t \to \infty$, $\forall n = 0, 1, \ldots$, integration by parts and an induction argument we can write

$$\begin{aligned} g(t) &= -\int_t^\infty g'(y)\,dy = \int_t^\infty (y-t)g''(y)\,dy \\ &= \frac{1}{2}\int_t^\infty (y-t)^2 g'''(y)\,dy \cdots = \frac{(-1)^n}{n!}\int_t^\infty (y-t)^n g^{(n+1)}(y)\,dy \\ &= \frac{(-1)^{n+1}}{n!}\int_t^\infty \left(1-\frac{t}{y}\right)^n y^n g^{(n+1)}(y)\,dy \\ &\overset{y=n/u}{=} \frac{(-1)^{n+1}}{n!}\int_0^{n/t} \left(1-\frac{tu}{n}\right)^n \frac{n^{n+1}}{u^{n+2}} g^{(n+1)}\left(\frac{n}{u}\right) du = \int_0^\infty E_n(tu)\eta_n(u)\,du, \end{aligned} \tag{12.5.8}$$

where

$$E_n(s) = \left(1-\frac{s}{n}\right)_+^n$$

and

$$\eta_n(s) = \frac{(-1)^{n+1}}{n!}\frac{n^{n+1}}{s^{n+2}} g^{(n+1)}\left(\frac{n}{s}\right) = \frac{1}{n!}\sum_{k=1}^{n+1} c_k^{n+1}\left(\frac{n}{s}\right)^{k\gamma}\frac{1}{s}g\left(\frac{n}{s}\right). \tag{12.5.9}$$

Let us also set

$$G_n(s) = s^{1+\gamma}\eta_n(s). \tag{12.5.10}$$

Observe that

$$G_n(s) = \left(A_n + \frac{1}{n!}\sum_{k=2}^{n+1} c_k^{n+1}\left(\frac{n}{s}\right)^{(k-1)\gamma}\right) g\left(\frac{n}{s}\right),$$

with $A_n = \frac{c_1^{n+1}}{n!}$ and, therefore, G_n is bounded and

$$\lim_{s\to\infty} G_n(s) = A_n. \tag{12.5.11}$$

We recall, from (12.5.8), that we have for all $n = 1, 2, \dots$

$$g(t) = \int_0^\infty E_n(tu)\eta_n(u)\,du. \tag{12.5.12}$$

Differentiating both sides we get

$$-\gamma t^{\gamma-1} g(t) = -\int_0^\infty \widetilde{E}_n(tu) u\eta_n(u)\,du = -\int_0^\infty \widetilde{E}_n(u)\frac{u}{t^2}\eta_n\left(\frac{u}{t}\right) du,$$

where $\widetilde{E}_n(u) = (1 - \frac{u}{n})_+^{n-1}$. Hence,

$$g(t) = \frac{1}{\gamma}\int_0^\infty \widetilde{E}_n(u) u^{-\gamma} G_n\left(\frac{u}{t}\right) du.$$

Letting $t \to 0^+$ and using the dominated convergence theorem, we get

$$1 = \frac{1}{\gamma} A_n \int_0^\infty \widetilde{E}_n(u) u^{-\gamma}\,du.$$

Therefore,

$$\lim_{n\to\infty} A_n = \gamma\left(\int_0^\infty e^{-u} u^{-\gamma}\,du\right)^{-1} = \frac{\gamma}{\Gamma(1-\gamma)}.$$

Finally, using identity (12.5.12) and the argument in Corollary 12.5.7 we have

$$P_\gamma(x) = \int_0^\infty \frac{1}{u^{N/2}} F_n(|x|/u^{1/2})\eta_n(u)\,du, \tag{12.5.13}$$

where F_n denotes the radial function described by the inverse Fourier transform of $E_n(|\xi|^2)$, that is, $F_n(|x|) = (2\pi)^{-N/2} \int_{\mathbb{R}^N} E_n(|\xi|^2) e^{ix\cdot\xi}\,d\xi$. From the general theory of Bochner–Riesz multiplier operators, each $F_n(t)$ is in $L^1(\mathbb{R}_+, t^{N-1}\,dt)$ for n greater than the critical index $(N-1)/2$ (see [294]). Making the change of variables $u = |x|^2/s$ in (12.5.13) we get

$$\begin{aligned} P_\gamma(x) &= \int_0^\infty \frac{s^{N/2}}{|x|^N} F_n(s^{1/2})\eta_n\left(\frac{|x|^2}{s}\right)\frac{|x|^2}{s^2}\,ds \\ &= \int_0^\infty \frac{s^{\frac{N+2\gamma}{2}}}{|x|^{N+2\gamma}} F_n(s^{1/2}) G_n\left(\frac{|x|^2}{s}\right)\frac{ds}{s}. \end{aligned} \tag{12.5.14}$$

By the dominated convergence theorem again we have

$$\lim_{|x|\to\infty} |x|^{N+2\gamma} P_\gamma(x) = \int_0^\infty s^{\frac{N+2\gamma}{2}} F_n(s^{1/2}) A_n \frac{ds}{s}, \quad \forall n \in \mathbb{N}.$$

Letting $n \to \infty$ and using the above estimates we conclude

$$\lim_{|x|\to\infty} |x|^{N+2\gamma} P_\gamma(x) = \frac{\gamma}{\Gamma(1-\gamma)} \int_0^\infty s^{\frac{N+2\gamma}{2}} e^{-s} \frac{ds}{s} = \frac{\gamma}{\Gamma(1-\gamma)} \Gamma\left(\frac{N+2\gamma}{2}\right). \qquad \square$$

End of the proof of Theorem 12.5.2.

Taking $0 < \epsilon < \frac{1}{2}\frac{\gamma}{\Gamma(1-\gamma)}$, by Proposition 12.5.9 there exists an $R > 0$ such that if $|x| > R$,

$$\gamma\Gamma(1-\gamma) - \epsilon \le (1 + |x|^{N+\alpha}) P_\alpha(x) \le \frac{\gamma}{\Gamma(1-\gamma)} + \epsilon.$$

According to the Bernstein theorem, the kernel $P_\alpha(x,1)$ is positive in $\mathbb{R}^N$, and then we easily conclude the proof. $\square$

12.5.1 Some remarks on regularity of the solution to the Cauchy problem

In this paragraph we summarize some regularity results with respect to the solution of the Cauchy problem in spaces of Hölder continuous functions. We follow the appendix in the paper by L. Caffarelli and A. Figalli [106]. More precisely, we will study the behavior of the fractional heat equation in the following spaces:

– $v \in C^{\alpha,\beta}_{x,t}([t_0, t_1] \times \mathbb{R}^N)$ if

$$\|v\|_{C^{\alpha,\beta}_{x,t}([t_0,t_1]\times\mathbb{R}^N)} := \|v\|_{L^\infty([t_0,t_1]\times\mathbb{R}^N)} + \sup_{[t_0,t_1]\times\mathbb{R}^N} \frac{|v(x,t) - v(x',t')|}{|x-x'|^\alpha + |t-t'|^\beta} < +\infty;$$

– $v \in \operatorname{Lip}_t C^\alpha_x([t_0, t_1] \times \mathbb{R}^N)$ if

$$\|v\|_{\operatorname{Lip}_t C^\alpha_x([t_0,t_1]\times\mathbb{R}^N)} := \|v\|_{L^\infty([t_0,t_1]\times\mathbb{R}^N)} + \sup_{[t_0,t_1]\times\mathbb{R}^N} \frac{|v(x,t) - v(x',t')|}{|x-x'|^\alpha + |t-t'|} < +\infty;$$

– $v \in \log \operatorname{Lip}_t C^\alpha_x([t_0, t_1] \times \mathbb{R}^N)$ if

$$\|v\|_{\log \operatorname{Lip}_t C^\alpha_x([t_0,t_1]\times\mathbb{R}^N)} := \|v\|_{L^\infty([t_0,t_1]\times\mathbb{R}^N)} + \sup_{[t_0,t_1]\times\mathbb{R}^N} \frac{|v(x,t) - v(x',t')|}{|x-x'|^\alpha + |t-t'|(1+\log(|t-t'|))}$$
$$< +\infty.$$

By $v \in C^{\alpha,\beta-0^+}_{x,t}([t_0,t_1]\times\mathbb{R}^N)$ we will understand that $v \in C^{\alpha,\beta-\epsilon}_{x,t}([t_0,t_1]\times\mathbb{R}^N)$ for all $\epsilon > 0$, and similarly in the other spaces.

Note that from the result on the regularity and decay of the kernel $P_{2s}(x,t)$ we find by scaling, for instance, the following estimates:

$$P_{2s}(x,t) \approx \frac{t}{t^{\frac{N+2s}{2s}} + |x|^{N+2s}}, \qquad \left|\frac{\partial}{\partial t}P_{2s}(x,t)\right| \le \frac{1}{t^{\frac{N+2s}{2s}} + |x|^{N+2s}}, \tag{12.5.15}$$

$$\left|\left(\frac{\partial}{\partial t}\right)^2 P_{2s}(x,t)\right| \le \left(\frac{1}{t}\right)\frac{1}{t^{\frac{N+2s}{2s}} + |x|^{N+2s}}, \qquad \left|\nabla_x \frac{\partial}{\partial t}P_{2s}(x,t)\right| \le \left(\frac{1}{|x|}\right)\frac{1}{t^{\frac{N+2s}{2s}} + |x|^{N+2s}}. \tag{12.5.16}$$

We assume that the initial data $v(x,0) = v_0(x) \in C^2(\mathbb{R}^N \cap W^{1,\infty}(\mathbb{R}^N)$ and that $\|D^2 v_0\|_{L^\infty(\mathbb{R}^N)} + \|(-\Delta)^2 v_0\|_{C^{1-s}_x(\mathbb{R}^N)} < \infty$ in order to avoid technicalities at $t = 0$. Then the solution to the linear problem

$$v_t + (-\Delta)^s v = f, \quad \text{if } (x,t) \in \mathbb{R}^N \times [0,T],\ v(x,0) = v_0(x)$$

is given by the convolution with the fractional heat kernel, i. e.,

$$v(x,t) = \int_{\mathbb{R}^N} P_{2s}(x-y,t)v_0(y)\,dy + \int_0^t \int_{\mathbb{R}^N} P_{2s}(x-y,t-\tau)f(y,\tau)\,dy\,d\tau. \tag{12.5.17}$$

As a consequence we obtain

$$-(-\Delta)^s v(x,t) = -\int_{\mathbb{R}^N} P_{2s}(x-y,t)(\Delta)^s v_0(y)\,dy + \int_0^t \int_{\mathbb{R}^N} \frac{\partial}{\partial t}P_{2s}(x-y,t-\tau)[f(y,\tau)-f(x,\tau)]\,dy\,d\tau.$$

Note that the first term on the right hand side is regular according to the regularity of v_0. Then studying the regularity of $(-\Delta)^s v$ is reduced to the following result.

Theorem 12.5.10. *The following inequalities hold:*
1. *If $0 < \beta < 2s$, then*

$$\left\|\int_0^t \int_{\mathbb{R}^N} \frac{\partial}{\partial t}P_{2s}(x-y,t-\tau)[f(y,\tau)-f(x,\tau)]\,dy\,d\tau\right\|_{C^{\beta-0^+,\frac{\beta}{2s}}_{x,t}(\mathbb{R}^N\times[0,T])} \lesssim \|f\|_{L^\infty(0,T],C^\beta_x(\mathbb{R}^N)}. \tag{12.5.18}$$

2. *If $1 > \beta \ge 2s$, then*

$$\left\|\int_0^t \int_{\mathbb{R}^N} \frac{\partial}{\partial t}P_{2s}(x-y,t-\tau)[f(y,\tau)-f(x,\tau)]\,dy\,d\tau\right\|_{\mathrm{logLip}_t C^{\beta-0^+}_x(\mathbb{R}^N\times[0,T])} \lesssim \|f\|_{L^\infty(0,T],C^\beta_x(\mathbb{R}^N)}. \tag{12.5.19}$$

Before proving this result we state the main consequences.

Corollary 12.5.11. *The following estimates hold:*
1. *If* $0 < \beta < 2s$, *then*

$$\|(-\Delta v)\|_{C_{x,t}^{\beta-0^+,\frac{\beta}{2s}}(\mathbb{R}^N\times[0,T])} \lesssim (1 + \|f\|_{L^\infty(0,T],C_x^\beta(\mathbb{R}^N)}). \tag{12.5.20}$$

2. *If* $1 > \beta \geq 2s$, *then*

$$\|(-\Delta v)\|_{\mathrm{logLip}_t C_x^{\beta-0^+}(\mathbb{R}^N\times[0,T])} \lesssim \|f\|_{L^\infty(0,T],C_x^\beta(\mathbb{R}^N)}. \tag{12.5.21}$$

And finally, taking into account the equation, we obtain

$$\|v_t\|_{L^\infty(0,T],C_x^{\beta-0^+}(\mathbb{R}^N)} + \|(-\Delta)^s v\|_{L^\infty(0,T],C_x^{\beta-0^+}(\mathbb{R}^N)} \lesssim (1 + \|f\|_{L^\infty(0,T],C_x^\beta(\mathbb{R}^N)}), \tag{12.5.22}$$

for all $\beta \in (0,1)$.

In order to prove Theorem 12.5.10 we will use the following technical lemma.

Lemma 12.5.12. *Let* $0 < \beta < 1$. *The following estimates hold:*
1. *There exists a constant* $C > 0$ *such that for all* $h \in (0,1]$

$$\int_{\mathbb{R}^N} \frac{\min\{|y|^\beta, 1\}}{h^{\frac{N+2s}{2s}} + |y|^{N+2s}}\, dy \leq C(1 + h^{\frac{\beta}{2s}-1}). \tag{12.5.23}$$

2. *Assume that* $|t-\tau| < 1$. *Then there exists a constant* $C > 0$ *such that for all* $h > 0$,

$$\int_0^t \frac{d\tau}{(t-\tau)^{\frac{N+2s}{2s}} + |h|^{N+2s}} \leq C \min\left\{\frac{1}{h^N}, \frac{1}{|h|^{N+2s}}\right\}. \tag{12.5.24}$$

Proof. We start by proving (12.5.23). We have

$$\begin{aligned}
\int_{\mathbb{R}^N} &\frac{\min\{|y|^\beta, 1\}}{h^{\frac{N+2s}{2s}} + |y|^{N+2s}}\, dy \\
&= \int_{\mathbb{R}^N\setminus B_1} \frac{1}{h^{\frac{N+2s}{2s}} + |y|^{N+2s}}\, dy \\
&\quad + \int_{B_{h^{1/2s}}} \frac{|y|^\beta}{h^{\frac{N+2s}{2s}} + |y|^{N+2s}}\, dy \\
&\quad + \int_{B_1\setminus B_{h^{1/2s}}} \frac{|y|^\beta}{h^{\frac{N+2s}{2s}} + |y|^{N+2s}}\, dy I_1 + I_2 + I_3.
\end{aligned}$$

Then

$$I_1 \leq \int_{\mathbb{R}^N\setminus B_1} \frac{1}{|y|^{N+2s}}\, dy \leq 1,$$

$$I_2 \leq \frac{1}{h^{\frac{N+2s}{2s}}} \int_{B_{h^{1/2s}}} |y|^\beta \, dy \lesssim h^{\frac{\beta}{2s}-1}$$

and

$$I_3 \leq \int_{B_1 \setminus B_{h^{1/2s}}} |y|^{\beta-N-2s} \, dy \lesssim 1 + h^{\frac{\beta}{2s}-1}.$$

To prove (12.5.24), note that the estimate is trivial if $h \geq 1$. For $h \in (0,1]$ the result follows by writing

$$\int_0^t \frac{d\tau}{(t-\tau)^{\frac{N+2s}{2s}} + |h|^{N+2s}} = \int_0^{t-h^{2s}} \frac{d\tau}{(t-\tau)^{\frac{N+2s}{2s}} + |h|^{N+2s}} + \int_{t-h^{2s}}^t \frac{d\tau}{(t-\tau)^{\frac{N+2s}{2s}} + |h|^{N+2s}}.$$

The first integral can be estimated by a constant times $|h|^{-(N+2s)}$ and the second, as above, by

$$\int_{t-h^{2s}}^t \frac{d\tau}{(t-\tau)^{\frac{N+2s}{2s}} + |h|^{N+2s}} \leq \int_{t-h^{2s}}^t \frac{d\tau}{|h|^{N+2s}} \leq \frac{1}{h^N}.$$

□

– *Proof of the regularity with respect to time in* (12.5.20). Assume without loss of generality that $\sigma < t$. We check directly the difference

$$\begin{aligned}
&|(-\Delta)^s v(x,t) - (-\Delta)^s v(x,\sigma)| \\
&= \Bigg| \int_0^t \int_{\mathbb{R}^N} \frac{\partial}{\partial t} P_{2s}(x-y, t-\tau)[f(y,\tau) - f(x,\tau)] \, dy \, d\tau \\
&\quad - \int_0^\sigma \int_{\mathbb{R}^N} \frac{\partial}{\partial t} P_{2s}(x-y, \sigma-\tau)[f(y,\tau) - f(x,\tau)] \, dy \, d\tau \Bigg| \\
&\leq \int_\sigma^t \int_{\mathbb{R}^N} \left| \frac{\partial}{\partial t} P_{2s}(x-y, t-\tau) \right| |f(y,\tau) - f(x,\tau)| \, dy \, d\tau \\
&\quad + \int_0^\sigma \int_{\mathbb{R}^N} \left| \frac{\partial}{\partial t} P_{2s}(x-y, t-\tau) - \frac{\partial}{\partial t} P_{2s}(x-y, \sigma-\tau) \right| \min\{1, |x-y|^\beta\} \, dy \, d\tau \\
&= I_1 + I_2.
\end{aligned}$$

We estimate I_1 by using (12.5.15) and (12.5.23):

$$\begin{aligned}
I_1 &\leq \int_\sigma^t \int_{\mathbb{R}^N} \frac{\min\{1, |x-y|^\beta\}}{(t-\tau)^{\frac{N+2s}{2s}} + |x-y|^{N+2s}} \, dy \, d\tau \leq \int_\sigma^t \left(1 + (\tau-\tau)^{\frac{\beta}{2s}-1}\right) d\tau \\
&\leq (t-\sigma) + (t-\sigma)^{\frac{\beta}{2s}}.
\end{aligned}$$

To estimate I_2 we use (12.5.15), (12.5.16) and (12.5.23):

$$I_2 \le \int_0^\sigma \int_{\mathbb{R}^N} \min\left\{\frac{t-\sigma}{\sigma-\tau}, 1\right\} \frac{1}{(\sigma-\tau)^{\frac{N+2s}{2s}} + |x-y|^{N+2s}} \min\{|x-y|^\beta, 1\}\, dy\, dt$$
$$\lesssim \int_0^{\sigma-(t-\sigma)} \frac{t-\sigma}{\sigma-\tau}\left(1+(\sigma-\tau)^{\frac{\beta}{2s}-1}\right) d\tau + \int_{\sigma-(t-\sigma)}^{\sigma} \left(1+(\sigma-\tau)^{\frac{\beta}{2s}-1}\right) d\tau$$
$$\lesssim (t-\sigma) + (t-\sigma)\log|t-\sigma| + (t-\sigma)^{\frac{\beta}{2s}}.$$

Therefore we conclude the regularity with respect to t.

– *Proof of spatial regularity in* (12.5.20) *and* (12.5.21). We evaluate the variation with respect to the spatial variable with fixed t, namely,

$$\begin{aligned}
&\left|(-\Delta)^s v(x,t) - (-\Delta)^s v(z,t)\right| \\
&= \left| \int_0^t \int_{\mathbb{R}^N} \frac{\partial}{\partial t} P_{2s}(x-y, t-\tau)[f(y,\tau) - f(x,\tau)]\, dy\, d\tau \right. \\
&\quad \left. - \int_0^\sigma \int_{\mathbb{R}^N} \frac{\partial}{\partial t} P_{2s}(z-y, t-\tau)[f(y,\tau) - f(z,\tau)]\, dy\, d\tau \right|.
\end{aligned}$$

We decompose the spatial integral in two complementary sets,

$$A_1 := \left\{|x-z| \le \frac{|x-y|}{2}\right\} \quad \text{and} \quad A_2 := \left\{|x-z| > \frac{|x-y|}{2}\right\}.$$

In A_2, using again (12.5.15), we find

$$\begin{aligned}
&\left|\frac{\partial}{\partial t} P_{2s}(x-y, t-\tau)[f(y,\tau) - f(x,\tau)] - \frac{\partial}{\partial t} P_{2s}(z-y, t-\tau)[f(y,\tau) - f(z,\tau)]\right| \\
&\le \left|\frac{\partial}{\partial t} P_{2s}(x-y, t-\tau) - \frac{\partial}{\partial t} P_{2s}(x-y, \sigma-\tau)\right| \min\{1, |x-y|^\beta\} \\
&\le \frac{|x-y|^\beta}{(t-\tau)^{\frac{N+2s}{2s}} + |x-y|^{N+2s}} + \frac{|z-y|^\beta}{(t-\tau)^{\frac{N+2s}{2s}} + |z-y|^{N+2s}}.
\end{aligned}$$

Therefore, since $A_2 \subset B_{3|x-z|}(x) \cap B_{3|x-z|}(z)$, using again (12.5.24), we obtain

$$\begin{aligned}
&\left| \int_0^t \int_{A_2} \frac{\partial}{\partial t} P_{2s}(x-y, t-\tau)[f(y,\tau) - f(x,\tau)]\, dy\, d\tau \right. \\
&\quad \left. - \int_0^\sigma \int_{A_2} \frac{\partial}{\partial t} P_{2s}(z-y, t-\tau)[f(y,\tau) - f(z,\tau)]\, dy\, d\tau \right|
\end{aligned}$$

$$\lesssim \int_0^t \int_{B_{3|x-z|}(x)} \frac{|x-y|^\beta}{(t-\tau)^{\frac{N+2s}{2s}} + |x-y|^{N+2s}}\, dy\, d\tau$$

$$= \int_{B_{3|x-z|}(x)} |x-y|^\beta \int_0^t \frac{1}{(t-\tau)^{\frac{N+2s}{2s}} + |x-y|^{N+2s}}\, d\tau\, dy$$

$$\lesssim \int_{B_{3|x-z|}(x)} |x-y|^\beta \min\left\{\frac{1}{|x-y|^N}, \frac{1}{|x-y|^{N+2s}}\right\} dy$$

$$\lesssim \int_{B_{3|x-z|}(x)} |x-y|^{\beta-N}\, dy \lesssim |x-z|^\beta.$$

To finish the proof we have to evaluate the integral in A_1. On A_1 the following quantities are comparable:

$$\frac{1}{(t-\tau)^{\frac{N+2s}{2s}} + |x-y|^{N+2s}}, \qquad \frac{1}{(t-\tau)^{\frac{N+2s}{2s}} + |z-y|^{N+2s}}.$$

Hence, the integrand on A can be estimated by

$$\left|\frac{\partial}{\partial t}P_{2s}(x-y,t-\tau) \mid [f(y,\tau)-f(x,\tau)] - \frac{\partial}{\partial t}P_{2s}(z-y,t-\tau)[f(y,\tau)-f(z,\tau)]\right|$$

$$\le |f(y,\tau)-f(x,\tau)|\left|\frac{\partial}{\partial t}P_{2s}(x-y,t-\tau) - \frac{\partial}{\partial t}P_{2s}(z-y,t-\tau)\right|$$

$$+ |f(x,\tau)-f(z,\tau)|\left|\frac{\partial}{\partial t}P_{2s}(x-y,t-\tau)\right| := F(x,y,z,t,\tau).$$

Using (12.5.24), we obtain

$$\int_0^t \int_{A_2} F(x,y,z,t,\tau)\, dy\, d\tau$$

$$\times \int_{A_2}\left(\frac{|x-z|}{|x-y|^{1-\beta}} + |x-z|^\beta\right)\int_0^t \frac{1}{(t-\tau)^{\frac{N+2s}{2s}} + |x-y|^{N+2s}}\, d\tau\, dy$$

$$\lesssim \int_{A_2} |x-z|^\beta \min\left\{\frac{1}{|x-y|^N}, \frac{1}{|x-y|^{N+2s}}\right\} dy$$

$$\le |x-z|^\beta\Bigg(\int_{|x-z|<\frac{|x-y|}{2}\le 1} |x-y|^N\, dy + \int_{\frac{|x-y|}{2}>1} |x-yN-2s\, dy\Bigg)$$

$$\times |x-z|^\beta \log|x-z| \lesssim |x-z|^{\beta-\epsilon} \quad \text{for all } \epsilon > 0.$$

Putting together the previous estimates we conclude the proof of Theorem 12.5.10.

12.6 Regularity in bounded domains: relation between weak solutions and viscosity solutions

The regularity of the solutions to the problem

$$\begin{cases} u_t + (-\Delta)^s u = f(x,t) & \text{in } \mathbb{R}^N \times (0,T), \\ u(x,0) = u_0(x) & \text{if } x \in \mathbb{R}^N \end{cases} \tag{12.6.1}$$

according to the regularity of the data can be seen in [106], where the properties of the fundamental solution play a crucial role. Our aim here is to prove a similar result for the case of a bounded domain. In particular, we want to see that provided that our right hand side is good enough, the solution is regular and can be understood in a pointwise sense. More precisely, we have the following theorem.

Theorem 12.6.1. *Assume that Ω is a $C^{1,1}$-bounded domain in $\mathbb{R}^N$. Let $f \in L^\infty(\Omega \times (0,T))$ be Hölder continuous in space–time and let u be the energy solution to the problem*

$$\begin{cases} u_t + (-\Delta)^s u = f(x,t) & \text{in } \Omega \times (0,T), \\ u(x,t) = 0 & \text{in } (\mathbb{R}^N \setminus \Omega) \times (0,T), \\ u(x,0) = 0 & \text{in } \Omega. \end{cases} \tag{12.6.2}$$

Then, u is a strong solution in the following sense:
(i) $u_t \in C(\Omega \times (0,T))$;
(ii) $u \in C(\Omega \times [0,T))$;
(iii) *the equation holds pointwise, i. e.,*

$$u_t(x,t) + C_{N,s}\, P.V. \int_{\mathbb{R}^N} \frac{u(x,t) - u(y,t)}{|x-y|^{N+2s}}\, dy = f(x,t) \quad \text{for every } (x,t) \in \Omega \times (0,T).$$

The proof of this result is based on the relation between energy and viscosity solutions, and will require the following regularity result [161, Theorem 1.1 and Corollary 4.1].

Theorem 12.6.2. *Assume that Ω is a $C^{1,1}$-bounded domain in $\mathbb{R}^N$ and let $v_0 \in L^2(\Omega)$. Consider v, the unique energy solution to problem*

$$\begin{cases} v_t + (-\Delta)^s v = 0 & \text{in } \Omega \times (0,+\infty), \\ v = 0 & \text{in } \mathbb{R}^N \setminus \Omega \times (0,+\infty), \\ v(x,0) = v_0(x) & \text{in } \Omega. \end{cases}$$

Then, for each $\epsilon > 0$,
(i)

$$\sup_{t>\epsilon} \|v(\cdot,t)\|_{C^s(\mathbb{R}^N)} \le C_1(\epsilon)\|v_0\|_{L^2(\Omega)};$$

(ii) *denoting* $\delta(x) := \operatorname{dist}(x, \partial\Omega)$,

$$\sup_{t>\epsilon}\left\|\frac{v(\cdot,t)}{\delta^s(\cdot)}\right\|_{C^{s-\eta}(\overline{\Omega})} \le C_2(\epsilon)\|v_0\|_{L^2(\Omega)} \quad \textit{for any } \eta > 0.$$

Both constants C_1 *and* C_2 *blow up when* $\epsilon \to 0$. *In addition, for every* $j \in \mathbb{N}$,

$$\sup_{t>\epsilon}\left\|\frac{\partial^j v(\cdot,t)}{\partial t^j}\right\|_{C^s(\mathbb{R}^N)} \le C_j(\epsilon)\|v_0\|_{L^2(\Omega)}.$$

This result in particular provides continuity up to the boundary of the domain, and thus we can formulate an extension result.

Corollary 12.6.3. *Let* v *be as in Theorem* 12.6.2. *Then* v *can be extended as a continuous function to* $\mathbb{R}^N \times (0, +\infty)$.

We can now prove Theorem 12.6.1.

Proof of Theorem 12.6.1. Let $\tau > 0$. By Theorem 12.6.2 and Corollary 12.6.3, the solution v to the problem

$$\begin{cases} v_t(x,\tau,t) + (-\Delta)^s v(x,\tau,t) = 0 & \text{in } \Omega \times (\tau, +\infty), \\ v(x,\tau,t) = 0 & \text{in } \mathbb{R}^N \setminus \Omega \times (\tau, +\infty), \\ v(x,\tau,0) = f(x,\tau) & \text{in } \Omega \end{cases}$$

is continuous up to the boundary and can be extended to the whole space $\mathbb{R}^N \times (\tau, +\infty)$ (we still name it v). Moreover, applying Duhamel's principle, it is easy to see that

$$u(x,t) := \int_0^t v(x,t,\tau)\,d\tau$$

is the extension to $\mathbb{R}^N \times (0, T)$ of the unique solution to (12.6.2).

The goal now is to prove that u is also a viscosity solution. For the definition of this concept and the description of the class of test functions (that we will call $\mathcal{S}$) we refer to [120, Definition 2.3]. Let $(x_0, t_0) \in \Omega \times (0, T)$. Due to the stability of viscosity solutions by uniform limits in compact sets (see [120, Theorem 5.3]), by a mollification argument we can assume that u is regular in a neighborhood of (x_0, t_0) (see for example [282], where this argument is detailed in the elliptic case). Hence, consider a test function $\phi \in S$ verifying

(i) $u(x_0, t_0) = \phi(x_0, t_0)$;
(ii) $u(x,t) < \phi(x,t)$ in some $B_r(x_0) \times (t_0 - \varepsilon, t_0]$;
(iii) $u(x,t) = \phi(x,t)$ in $(\mathbb{R}^N \setminus B_r(x_0)) \times (t_0 - \varepsilon, t_0]$.

Then,

$$\begin{aligned}
\phi_{t^-}&(x_0,t_0)+(-\Delta)^s\phi(x_0,t_0)\\
&=\lim_{h\to 0^+}\frac{\phi(x_0,t_0)-\phi(x_0,t_0-h)}{h}+C_{N,s}\,P.\,V.\int_{\mathbb{R}^N}\frac{\phi(x_0,t_0)-\phi(y,t_0)}{|x_0-y|^{N+2s}}dy\\
&\le\lim_{h\to 0^+}\frac{u(x_0,t_0)-u(x_0,t_0-h)}{h}+C_{N,s}\,P.\,V.\int_{\mathbb{R}^N}\frac{u(x_0,t_0)-u(y,t_0)}{|x_0-y|^{N+2s}}dy\\
&=f(x_0,t_0),
\end{aligned}$$

and we conclude that u is a viscosity subsolution (note that the singular integrals are well defined since we assume u regular). Analogously, it can be seen that u is a viscosity supersolution and therefore we conclude that u is a viscosity solution (12.6.2) and we can use the regularity results of [119, 120, 203]. In particular, by using [203, Corollaries 2.6 and 2.7], for a second member Hölder continuous in space–time, we find that the equation is verified in the pointwise sense. □

Remark 12.6.4. The interior regularity for a general class of operators has been studied in [162]. The study also considers regularity up to the boundary. We recommend the reader to check reference [162] for the details.

12.7 The uniqueness result of Widder type

The heat equation has the hyperplane $t = 0$ as a characteristic surface and therefore the Cauchy problem with initial data on $t = 0$ is not well posed in general. However, there is a clear agreement between the principles of thermodynamics and the model of transfer of heat given by such an equation. This is reflected in the fact that the initial temperature evolves in time as the convolution with a kernel, giving rise to an average, and smoothing out in this way the potential effect of sharp thermal differences (this is in agreement with the entropy effect of the Second Principle of Thermodynamics); also, uniqueness holds true under a positivity assumption on the function (this is in agreement with the so-called Third Principle of Thermodynamics, according to which temperatures are always positive if measured in the Kelvin scale). In this sense D. V. Widder in [317] proved the following classical result:

Assume that $u : \mathbb{R}^N \times [0,T) \subset \overline{\mathbb{R}^N_+} \to \mathbb{R}$ is so that

$$u(x,t)\ge 0,\quad u\in\mathcal{C}(\mathbb{R}^N\times[0,T)),\quad u_t,u_{x_ix_i}\in\mathcal{C}(\mathbb{R}^N\times(0,T)),$$

and satisfies

$$u_t(x,t)-\Delta u(x,t)=0,\quad (x,t)\in\mathbb{R}^N\times(0,T),$$

in the classical sense.

Then

$$u(x,t) = \frac{1}{(4\pi t)^{\frac{N}{2}}} \int\limits_{\mathbb{R}^N} u(y,0) e^{\left(\frac{-|x-y|^2}{4t}\right)} dy.$$

In this chapter we obtain a similar result for the *nonlocal heat equation*, that is, the equation

$$u_t + (-\Delta)^{\alpha/2} u = 0 \quad \text{for } (x,t) \in \mathbb{R}^N \times (0,T),\ 0 < \alpha < 2, \tag{12.7.1}$$

in which the diffusion is given by a power of the Laplacian (we refer to [138] and [294] for basic definitions and properties of the fractional Laplace operator).

It is worth noting that for every $\alpha \in (0,2]$ the operator in (12.7.1) does not satisfy the so-called *Hadamard condition*, that is, the Cauchy problem is ill posed (see [193] or [243, page 290, Theorem 5.3] for more details). As a consequence we face here a problem similar to the one of the heat equation.

In this section we will present the results of the article [48].

As is well known, the operator in (12.7.1) and its inverse are nonlocal. In a sense, equation (12.7.1) is perhaps one of the simplest classical examples of pseudo-differential operators.

In Section 12.5 we have studied the problem

$$u_t + (-\Delta)^{\alpha/2} u = 0 \quad \text{for } (x,t) \in \mathbb{R}^N \times (0,T),\ 0 < \alpha < 2,$$

with initial datum $u(x,0) = u_0(x)$, assuming, say, $u_0 \in C(\mathbb{R}^N) \cap L^\infty(\mathbb{R}^N)$ has a solution obtained by the convolution with the kernel

$$P_t(x) = \frac{1}{t^{N/\alpha}} P\left(\frac{x}{t^{1/\alpha}}\right), \tag{12.7.2}$$

where

$$P(x) := (2\pi)^{-N} \int\limits_{\mathbb{R}^N} e^{ix\cdot\xi - |\xi|^\alpha}\, d\xi. \tag{12.7.3}$$

Note that in this section, for simplicity we will denote by P the kernel P_α of Section 12.5.

Therefore, a solution is given by

$$u(x,t) = \int\limits_{\mathbb{R}^N} P_t(x-y) u_0(y)\, dy. \tag{12.7.4}$$

We look here for a class of solutions of the fractional parabolic equation, such that the sign condition $u(x,t) \geq 0$ implies that u, necessarily, is of the form (12.7.4) with $u_0(x)$ replaced by the trace $u(x,0)$. This is the type of extension that we propose for the classical result of Widder to the nonlocal equation in (12.7.1).

In this section we will describe several interpretations of what *solution to equation* (12.7.1) means, and consider weak, viscosity and strong solutions, in a sense that we introduce next.

Weak solutions

Consider the space

$$\mathcal{L}^{\alpha/2}(\mathbb{R}^N) := \left\{u : \mathbb{R}^N \to \mathbb{R} \text{ measurable: } \int_{\mathbb{R}^N} \frac{|u(x)|}{1+|x|^{N+\alpha}}\,dx < \infty\right\},$$

endowed with the norm

$$\|u\|_{\mathcal{L}^{\alpha/2}(\mathbb{R}^N)} := \int_{\mathbb{R}^N} \frac{|u(x)|}{1+|x|^{N+\alpha}}\,dx.$$

If $u \in \mathcal{L}^{\alpha/2}(\mathbb{R}^N)$, then $(-\Delta)^{\alpha/2}u$ can be defined in the weak sense as a distribution, that is, we can compute the duality product $\langle(-\Delta)^{\alpha/2}u, \varphi\rangle$, for every φ in the Schwartz class.

Definition 12.7.1. We say that $u(x,t)$ is a weak solution of the fractional heat problem

$$\begin{cases} u_t + (-\Delta)^{\alpha/2}u = 0 & \text{for } (x,t) \in \mathbb{R}^N \times (0,T), \\ u(x,0) = u_0(x) & \text{in } \mathbb{R}^N \end{cases}$$

if the following conditions hold:

(i) $u \in L^1([0,T'], \mathcal{L}^{\alpha/2}(\mathbb{R}^N))$ for every $T' < T$;

(ii) $u \in C((0,T), L^1_{\mathrm{loc}}(\mathbb{R}^N))$;

(iii) for every test function $\varphi \in C_0^\infty(\mathbb{R}^N \times [0,T))$ and $0 < T' < T$ one has

$$\begin{aligned} &\int_0^{T'}\int_{\mathbb{R}^N} [-u(x,t)\varphi_t(x,t) + u(x,t)(-\Delta)^{\alpha/2}\varphi(x,t)]\,dx\,dt \\ &\quad = \int_{\mathbb{R}^N} u(x,T')\varphi(x,T')\,dx - \int_{\mathbb{R}^N} u_0(x)\varphi(x,0)\,dx. \end{aligned} \tag{12.7.5}$$

Condition (ii) is imposed so that the right hand side of equality (12.7.5) makes sense for every T'. Note that the continuity would follow in any case from the left hand side of (12.7.5) due to the integrability condition in (i).

Viscosity solutions

Consider $Q = \mathbb{R}^N \times (0,T)$ and

$$\begin{aligned} C_p^{1,2}(Q) = \Big\{ f : Q \to \mathbb{R} \mid f_t \in C(Q),\ f_{x_i,x_j} \in C(Q) \\ \text{and } \sup_{t\in(0,T)} |f(x,t)| \le C(1+|x|)^p \Big\}. \end{aligned}$$

Definition 12.7.2. A function $u \in C(Q)$ is a *viscosity subsolution* (respectively *supersolution)* of

$$u_t + (-\Delta)^{\alpha/2} u = 0 \tag{12.7.6}$$

in Q if for all $(\hat{x}, \hat{t}) \in Q$ and $\varphi \in C_p^{1,2}(Q)$ such that $u - \varphi$ attains a local maximum (minimum) at $(\hat{x}, \hat{t})$ one has

$$\varphi_t(\hat{x}, \hat{t}) + (-\Delta)^{\alpha/2}\varphi(\hat{x}, \hat{t}) \leq 0 \quad \text{(respectively} \geq 0).$$

We say that $u \in C(Q)$ is a *viscosity solution* of (12.7.6) in Q if it is both a viscosity subsolution and a viscosity supersolution. See [202], [119], [120] and [203] for more details about these types of solutions.

Strong solutions

From the results in Section 8.2.1 (see also [286, Proposition 2.1.4]) we know that if $u \in \mathcal{L}^{\alpha/2}(\mathbb{R}^N) \cap C_{\text{loc}}^{\alpha+\gamma}(\mathbb{R}^N)$ (or $C^{1,\alpha+\gamma-1}$ if $\alpha > 1$), for some $\gamma > 0$, then $(-\Delta)^{\alpha/2}u$ is a continuous function and it is well defined by an integral representation as the principal value of

$$(-\Delta)^{\alpha/2}u(x) := C_{N,\alpha/2}\ \text{P. V.} \int_{\mathbb{R}^N} \frac{u(x) - u(y)}{|x - y|^{N+\alpha}}\, dy, \alpha \in (0, 2).$$

See Chapters 8 and 9.

In order to allow a larger class of functions, and following standard procedures in the theory of singular integrals, we will define hereafter the principal value as the two-sided limit

$$\text{P. V.} \int_{\mathbb{R}^N} \frac{u(x) - u(y)}{|x - y|^{N+\alpha}}\, dy = \lim_{\epsilon \to 0} \int_{\{y:\epsilon<|x-y|<1/\epsilon\}} \frac{u(x) - u(y)}{|x - y|^{N+\alpha}}\, dy.$$

We will use the same notation for this extended operator. When $u \in \mathcal{L}^{\alpha/2}(\mathbb{R}^N)$ this definition coincides with the usual one.

Definition 12.7.3. We say that $u(x, t)$ is a strong solution of the fractional heat equation

$$u_t + (-\Delta)^{\alpha/2}u = 0 \quad \text{for } (x, t) \in \mathbb{R}^N \times (0, T)$$

if the following conditions hold:
(i) $u_t \in C(\mathbb{R}^N \times (0, T))$;
(ii) $u \in C(\mathbb{R}^N \times [0, T))$;
(iii) $C_{N,\alpha/2} \int_{|x-y|>\epsilon} \frac{u(x,t) - u(y,t)}{|x - y|^{N+\alpha}}\, dy \to (-\Delta)^{\alpha/2}u(x)$ as $\epsilon \to 0$, in $L_{\text{loc}}^1(\mathbb{R}^N)$;

(iv) the equation is satisfied pointwise for every $(x,t) \in \mathbb{R}^N \times (0,T)$, that is,

$$u_t(x,t) + C_{n,\alpha/2}\,\mathrm{P.V.}\int_{\mathbb{R}^N} \frac{u(x,t)-u(y,t)}{|x-y|^{N+\alpha}}\,dy = 0.$$

Remark 12.7.4. Note that if $u(x,t)$ is a strong solution, then

$$\mathrm{P.V.}\int_{\mathbb{R}^N} \frac{u(x,t)-u(y,t)}{|x-y|^{N+\alpha}}\,dy \in \mathcal{C}(\mathbb{R}^N \times (0,T)).$$

See for details the regularity results in Section 12.5.1.

Observe also that if $u(x,t)$ is a strong solution of the fractional heat equation satisfying $u(x,t) \in L^1([0,T'],\mathcal{L}^{\alpha/2}(\mathbb{R}^N))$ for every $T' < T$, then $u(x,t)$ is a weak solution of the fractional heat equation (as a byproduct of our results, we will see that this holds true for nonnegative strong solutions; see the forthcoming Corollary 12.7.12).

The main result in this section is the following Widder type result.

Theorem 12.7.5. *If $u \geq 0$ is a strong solution of*

$$u_t + (-\Delta)^{\alpha/2}u = 0 \quad \textit{for } (x,t) \in \mathbb{R}^N \times (0,T),\ 0 < \alpha < 2,$$

then

$$u(x,t) = \int_{\mathbb{R}^N} P_t(x-y)u(y,0)\,dy.$$

Note that if for a given polynomial p with real coefficients we consider the differential operator $p(D)$, then we obtain

$$p(D)P_t(x) = (2\pi)^{-N}\int_{\mathbb{R}^N} e^{ix\cdot\xi}p(i\xi)e^{-t|\xi|^\alpha}\,d\xi.$$

Since $e^{-t|\xi|^\alpha}$ is a tempered distribution we deduce that $P_t \in C^\infty(\mathbb{R}^N \times (0,\infty))$. See Section 12.5 (see also, for instance, [143] and [83] for more details).

In particular, and as a consequence of the main result in Theorem 12.7.5, if u is a nonnegative strong solution of the fractional heat equation, then $u \in C^\infty(\mathbb{R}^N \times (0,T))$.

The proof of the main result is organized as follows. In Section 12.7.1 we prove a uniqueness result for weak solutions that, in turn, will be the key step to obtain our representation theorem. In Section 12.7.2 we prove the main result for strong solutions. We start by proving a comparison result that will allow us to show that every positive strong solution $u(x,t)$ is bigger than or equal to the convolution of the trace $u(x,0)$ with the kernel $P_t(x)$. By a scaling argument, we will prove that any positive strong solution is also a weak solution and then, by the uniqueness result of the previous section, we will conclude with the proof of the theorem.

Finally, in the last subsection we establish the pointwise behavior of positive strong solutions that has an interest on its own and that could provide and alternative proof to the representation result. It is appropriate to say that the proof follows a different strategy than the classical for the heat equation.

12.7.1 Uniqueness for weak solutions

To start with, we prove a uniqueness result for weak solutions with vanishing initial condition. The proof is quite complicated and it involves many fine integral estimates. The nonlocal feature of the problem also makes localization and cutoff arguments much harder than in the classical case.

Theorem 12.7.6. *Set $T > 0$, $\alpha \in (0,2)$ and let u be a weak solution of the fractional heat equation*

$$\begin{cases} u_t + (-\Delta)^{\alpha/2} u = 0 & \text{for } (x,t) \in \mathbb{R}^N \times (0,T), \\ u(x,0) = 0 & \text{in } \mathbb{R}^N. \end{cases} \tag{12.7.7}$$

Then $u(x,t) = 0$ for every $t \in (0,T)$ and a. e. $x \in \mathbb{R}^N$.

Proof. We must show that $u(x,t_0) = 0$ for an arbitrary $t_0 \in (0,T)$ and $x \in \mathbb{R}^n$. For this, we fix $R_0 > 0$ and $\theta \in C_0^\infty(B_{R_0})$ and we will prove that

$$\int_{\mathbb{R}^N} u(x,t_0)\theta(x)\,dx = 0.$$

For any $t \in [0,t_0)$, we define

$$\varphi(x,t) := (\theta(\cdot) * P_{t_0-t}(\cdot))(x),$$

where P_t is the kernel defined in (12.7.2) and (12.7.3). By the results in Section 12.5 (see also [210], [134] and [100]) we know that

$$\frac{1}{C}\frac{1}{1+|x|^{N+\alpha}} \le P(x) \le \frac{C}{1+|x|^{N+\alpha}}. \tag{12.7.8}$$

Therefore

$$\hat{\varphi}(\xi,t) = \hat{\theta}(\xi)e^{-(t_0-t)|\xi|^\alpha} = C(\xi)e^{t|\xi|^\alpha}.$$

Since

$$\hat{\varphi}_t(\xi,t) = |\xi|^\alpha \hat{\varphi}(\xi,t),$$

we have

$$\begin{cases} \varphi_t - (-\Delta)^{\alpha/2}\varphi = 0 & \text{for } (x,t) \in \mathbb{R}^N \times [0, t_0), \\ \varphi(x, t_0) = \theta(x) & \text{in } \mathbb{R}^N. \end{cases} \tag{12.7.9}$$

Now we claim that

$$|\varphi(x,t)| = |\theta(x) * P_{t_0-t}(x)| \le \frac{C_1}{1+|x|^{N+\alpha}}, \tag{12.7.10}$$

where C_1 depends of n, α, R_0, $M := \|\theta\|_{L^\infty(B_{R_0})}$ and t_0.

Indeed, considering without loss of generality that $2R_0 > 1$, we will distinguish two cases.

When $|x| \le 2R_0$, using that P_t is a summability kernel in L^1 we have

$$\begin{aligned} \Big|\int_{\mathbb{R}^N} P_{t_0-t}(y)\theta(x-y)\,dy\Big| &\le M \int_{\mathbb{R}^N} P_{t_0-t}(y)\,dy = M \\ &\le M\frac{1+(2R_0)^{N+\alpha}}{1+|x|^{N+\alpha}} \\ &\le \frac{c_1}{1+|x|^{N+\alpha}}, \end{aligned} \tag{12.7.11}$$

where $c_1 = c_1(n, \alpha, R_0, M)$.

Consider now $|x| > 2R_0$. Note that from (12.7.2) and (12.7.8), we obtain

$$P_t(y) \le \frac{C}{t^{\frac{N}{\alpha}}\big(1 + \frac{|y|^{N+\alpha}}{t^{\frac{N+\alpha}{\alpha}}}\big)} \le \frac{Ct}{|y|^{N+\alpha}}. \tag{12.7.12}$$

Then, since for $|x-y| \le R_0$ it follows that $|y| \ge \frac{|x|}{2}$, we have

$$P_{t_0-t}(y) \le \frac{2^{N+\alpha}C(t_0-t)}{|x|^{N+\alpha}}.$$

As a consequence,

$$\begin{aligned} \Big|\int_{\mathbb{R}^N} P_{t_0-t}(y)\theta(x-y)\,dy\Big| &= \Big|\int_{|x-y|\le R_0} P_{t_0-t}(y)\theta(x-y)\,dy\Big| \\ &\le 2^{N+\alpha}CM|B_{R_0}|\frac{t_0-t}{|x|^{N+\alpha}} \\ &\le 2^{N+\alpha}2CM|B_{R_0}|\frac{t_0}{1+|x|^{N+\alpha}} \\ &\le \frac{c_2}{1+|x|^{N+\alpha}}, \end{aligned} \tag{12.7.13}$$

where $c_2 = c_2(N, \alpha, R_0, M, t_0)$ and $|B_{R_0}|$ denotes, as usual, the Lebesgue measure of the ball. Hence, (12.7.10) follows from (12.7.11) and (12.7.13).

Also, applying (12.7.13) to the derivatives of $\theta \in C_0^\infty(B_{R_0})$, we have

$$|\nabla\varphi(x,t)| = |\nabla\theta(x) * P_{t_0-t}(x)| \le \frac{C_2}{1+|x|^{N+\alpha}}, \tag{12.7.14}$$

where C_2 depends also on n, α, R_0, $M' := \|\nabla\theta\|_{L^\infty(B_{R_0})}$ and t_0.

Then, from (12.7.10) and the fact that $u \in L^1([0,t_0], \mathcal{L}^{\alpha/2}(\mathbb{R}^N))$ we deduce that

$$M_2 := \int_0^{t_0}\int_{\mathbb{R}^N} |u(x,t)\varphi(x,t)|\,dx\,dt < \infty. \tag{12.7.15}$$

Let now $\phi \in C^\infty(\mathbb{R})$ be such that

$$\chi_{B_{1/2}} \le \phi \le \chi_{B_1}. \tag{12.7.16}$$

For $R > 2R_0$ we define $\phi_R(x) = \phi(\frac{x}{R})$ and

$$\psi(x,t) := \varphi(x,t)\phi_R(x)$$

as a test function of problem (12.7.7). Also, by the nonlocal Leibniz formula in (8.1.17), one sees that, for any $f, g \in \mathcal{L}^{\alpha/2}(\mathbb{R}^N)$,

$$(-\Delta)^{\alpha/2}(fg)(x) = f(x)(-\Delta)^{\alpha/2}g(x) + g(x)(-\Delta)^{\alpha/2}f(x) - B(f,g)(x),$$

with $B(f,g)$ the bilinear form given by

$$B(f,g)(x) := C_{N,\alpha/2}\int_{\mathbb{R}^N} \frac{(f(x)-f(y))(g(x)-g(y))}{|x-y|^{N+\alpha}}\,dy.$$

Applying this formula (for a fixed t) to the functions φ and ϕ_R and recalling (12.7.9) we obtain

$$(-\Delta)^{\alpha/2}\psi = \varphi(-\Delta)^{\alpha/2}\phi_R + \phi_R\varphi_t - B(\varphi,\phi_R). \tag{12.7.17}$$

Moreover, the function $u\psi$ evaluated at t_0 is $u(x,t_0)\theta(x)\phi_R(x)$, thanks to the terminal time condition in (12.7.9). Thus, considering ψ as a test function in (12.7.7) and using that φ is a solution of problem (12.7.9), we obtain

$$\begin{aligned} &\left|\int_{\mathbb{R}^N} u(x,t_0)\theta(x)\phi_R(x)\,dx\right| \\ &\quad = \left|\int_0^{t_0}\int_{\mathbb{R}^N} [u\varphi_t(x,t)\phi_R(x) - u(-\Delta)^{\alpha/2}(\psi(x,t))]\,dx\,dt\right| \end{aligned} \tag{12.7.18}$$

$$= \left| \int_0^{t_0} \int_{\mathbb{R}^N} [uB(\varphi, \phi_R)(x,t) - u\varphi(x,t)(-\Delta)^{\alpha/2}\phi_R(x)]\, dx\, dt \right|$$

$$\leq \int_0^{t_0} \int_{\mathbb{R}^N} |u\varphi(x,t)| |(-\Delta)^{\alpha/2}\phi_R(x)|\, dx\, dt$$

$$+ \int_0^{t_0} \int_{\mathbb{R}^N} |u(x,t)| |B(\varphi, \phi_R)(x,t)|\, dx\, dt.$$

Then, since θ is supported in B_{R_0} and $R_0 < R/2$, recalling (12.7.16), we conclude

$$\left| \int_{\mathbb{R}^N} u(x,t_0)\theta(x)\, dx \right| \leq \int_0^{t_0} \int_{\mathbb{R}^N} |u\varphi(x,t)| |(-\Delta)^{\alpha/2}\phi_R(x)|\, dx\, dt$$
$$+ \int_0^{t_0} \int_{\mathbb{R}^N} |u(x,t)| |B(\varphi, \phi_R)(x,t)|\, dx\, dt$$
$$=: I_1(R) + C(n,\alpha) I_2(R). \tag{12.7.19}$$

It remains to show that

$$\lim_{R\to\infty} I_1(R) + I_2(R) = 0.$$

Indeed, since

$$|(-\Delta)^{\alpha/2}\phi_R(x)| = R^{-\alpha} \left| ((-\Delta)^{\alpha/2}\phi)\left(\frac{x}{R}\right) \right| \leq C_0 R^{-\alpha},$$

by (12.7.15), it follows that

$$I_1(R) \leq C_0 R^{-\alpha} \int_0^{t_0} \int_{\mathbb{R}^N} |u\varphi(x,t)|\, dx\, dt \leq C_0 M_2 R^{-\alpha}$$

and so

$$\lim_{R\to\infty} I_1(R) = 0. \tag{12.7.20}$$

Now we are going to estimate $I_2(R)$. For this we cover $\mathbb{R}^N \times \mathbb{R}^N$ with six domains suitably described by the radii $R/4$, $R/2$, R and $2R$ and represented (for $N = 1$) in the following picture:

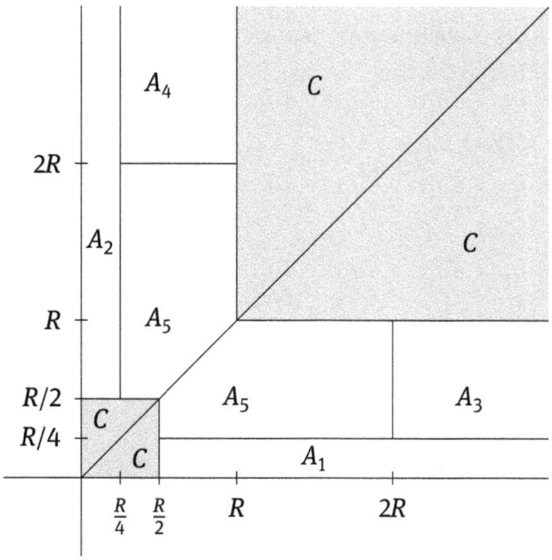

Therefore

$$\mathbb{R}^{2N} = \left(\bigcup_{k=1}^{5} A_k\right) \cup C,$$

where

$$A_1 := \{(x,y) : |x| > R/2, |y| \le R/4\}, \quad A_2 := \{(x,y) : |x| \le R/4, |y| > R/2\},$$
$$A_3 := \{(x,y) : |x| \ge 2R, R/4 < |y| < R\}, \quad A_4 := \{(x,y) : R/4 < |x| < R, |y| \ge 2R\},$$
$$A_5 := \{(x,y) : R/4 < |x| < 2R, R/4 < |y| < 2R\}$$

and

$$C := \{(x,y) : |x| \le R/2, |y| \le R/2\} \cup \{(x,y) : |x| \ge R, |y| \ge R\}.$$

From (12.7.16), we know that $\phi_R(x) - \phi_R(y) = 0$ if $(x,y) \in C$, and so

$$\begin{aligned} I_2(R) &= \int_0^{t_0} \int_{\mathbb{R}^N} |u(x,t)| \int_{\mathbb{R}^N} \frac{|\varphi(x,t) - \varphi(y,t)||\phi_R(x) - \phi_R(y)|}{|x-y|^{N+\alpha}} \, dy\, dx\, dt \\ &\le \sum_{k=1}^{5} I_2^{A_k}(R), \end{aligned} \tag{12.7.21}$$

where

$$I_2^{A_k}(R) = \int_0^{t_0} \int_{A_k} |u(x,t)| \frac{|\varphi(x,t) - \varphi(y,t)||\phi_R(x) - \phi_R(y)|}{|x-y|^{N+\alpha}} \, dy\, dx\, dt,$$

for $k = 1, \dots, 5$.

We are going to estimate each of these five integrals separately. For $(x,y) \in A_1$ we get $|x-y| \geq C|x|$. Moreover by (12.7.10) it follows that

$$|\varphi(x,t)| + |\varphi(y,t)| \leq \frac{C}{1+|y|^{N+\alpha}}.$$

Therefore

$$\begin{aligned} I_2^{A_1}(R) &\leq \int_0^{t_0} \int_{|x|>R/2} \frac{|u(x,t)|}{|x|^{N+\alpha}} \int_{|y|\leq R/4} \frac{C}{1+|y|^{N+\alpha}} \, dy \, dx \, dt \\ &\leq C \int_0^{t_0} \int_{|x|>R/2} \frac{|u(x,t)|}{|x|^{N+\alpha}} \, dx \, dt. \end{aligned} \tag{12.7.22}$$

Following the same ideas, since for $(x,y) \in A_2$ we obtain

$$|\varphi(x,t)| + |\varphi(y,t)| \leq \frac{C}{|x|^{N+\alpha}} \quad \text{and} \quad |x-y| \geq C|y|, \tag{12.7.23}$$

we have

$$\begin{aligned} I_2^{A_2}(R) &\leq \int_0^{t_0} \int_{|x|\leq R/4} \frac{|u(x,t)|}{|x|^{N+\alpha}} \int_{|y|>R/2} \frac{C}{|y|^{N+\alpha}} \, dy \, dx \, dt \\ &\leq CR^{-\alpha} \int_0^{t_0} \int_{|x|\leq R/4} \frac{|u(x,t)|}{|x|^{N+\alpha}} \, dx \, dt. \end{aligned} \tag{12.7.24}$$

Also, since for $(x,y) \in A_3$ (12.7.23) is satisfied, we have

$$I_2^{A_3}(R) \leq CR^{-\alpha} \int_0^{t_0} \int_{|x|\geq 2R} \frac{|u(x,t)|}{|x|^{N+\alpha}} \, dx \, dt. \tag{12.7.25}$$

Similarly, using again the good decay of φ and the fact that $|x-y| \geq C|y|$ for every $(x,y) \in A_4$, we obtain

$$I_2^{A_4}(R) \leq CR^{-\alpha} \int_0^{t_0} \int_{R/4<|x|<R} \frac{|u(x,t)|}{|x|^{N+\alpha}} \, dx \, dt. \tag{12.7.26}$$

Then, using the monotone convergence theorem, by (12.7.15) and the fact that $u \in L^1([0,t_0], \mathcal{L}^{\alpha/2}(\mathbb{R}^N))$, from (12.7.22), (12.7.24), (12.7.25) and (12.7.26) it follows that

$$\lim_{R\to\infty} I_2^{A_1}(R) = \lim_{R\to\infty} I_2^{A_2}(R) = \lim_{R\to\infty} I_2^{A_3}(R) = \lim_{R\to\infty} I_2^{A_4}(R) = 0. \tag{12.7.27}$$

To estimate $I_2^{A_5}(R)$ we will treat separately the cases $\alpha \in [0,1)$ and $\alpha \in [1,2)$.

We start with the case $\alpha \in [0,1)$ and $(x,y) \in A_5$. Since in A_5 the roles of x and y are symmetric, we deduce from (12.7.10) that, in this case,

$$|\varphi(x,t)| + |\varphi(y,t)| \le \frac{C}{|x|^{N+\alpha}}. \tag{12.7.28}$$

Also

$$|\phi_R(x) - \phi_R(y)| \le \frac{C}{R}|x-y|.$$

Thus

$$I_2^{A_5}(R) \le \frac{C}{R}\int_0^{t_0}\int_{\frac{R}{4}\le|x|\le 2R}\frac{|u(x,t)|}{|x|^{N+\alpha}}\int_{\frac{R}{4}\le|y|\le 2R}\frac{1}{|x-y|^{N+\alpha-1}}\,dy\,dx\,dt. \tag{12.7.29}$$

By the change of variables $\tilde{y} := x - y$, it follows from (12.7.29) that

$$\begin{aligned} I_2^{A_5}(R) &\le \frac{C}{R}\int_0^{t_0}\int_{\frac{R}{4}\le|x|\le 2R}\frac{|u(x,t)|}{|x|^{N+\alpha}}\int_{\frac{R}{4}\le|x-\tilde{y}|\le 2R}\frac{1}{|\tilde{y}|^{N+\alpha-1}}\,d\tilde{y}\,dx\,dt \\ &\le \frac{C}{R}\int_0^{t_0}\int_{\frac{R}{4}\le|x|\le 2R}\frac{|u(x,t)|}{|x|^{N+\alpha}}\int_{|\tilde{y}|\le 4R}\frac{1}{|\tilde{y}|^{N+\alpha-1}}\,d\tilde{y}\,dx\,dt \\ &\le CR^{-\alpha}\int_0^{t_0}\int_{\frac{R}{4}\le|x|\le 2R}\frac{|u(x,t)|}{|x|^{N+\alpha}}\,dx\,dt. \end{aligned} \tag{12.7.30}$$

Therefore, using that $u \in L^1([0,t_0], \mathcal{L}^{\alpha/2}(\mathbb{R}^N))$, we conclude

$$\lim_{R\to\infty} I_2^{A_5}(R) = 0, \quad \text{when } \alpha \in [0,1). \tag{12.7.31}$$

We consider now the case $\alpha \in [1,2)$. By (12.7.14), we get

$$|\varphi(x,t) - \varphi(y,t)| \le \frac{C}{1+|z|^{N+\alpha}}|x-y|, \tag{12.7.32}$$

for some z in the segment joining x and y. We take the family

$$Q := \left\{(x,y) \in A_5 : |x-y| \le \frac{R}{100}\right\}.$$

Note that if $(x,y) \in Q$, then every point z lying on the segment from x to y verifies $|z| \ge C|x|$. Hence, (12.7.32) and the previous estimate for ϕ_R gives

$$\begin{aligned} &\int_0^{t_0}\int_{(x,y)\in Q}|u(x,t)|\frac{|(\phi_R(x)-\phi_R(y))(\varphi(x,t)-\varphi(y,t))|}{|x-y|^{N+\alpha}}\,dy\,dx\,dt \\ &\quad\le \int_0^{t_0}\int_{(x,y)\in Q}|u(x,t)|\frac{C}{R|x|^{N+\alpha}|x-y|^{N+\alpha-2}}\,dy\,dx\,dt \end{aligned}$$

$$\leq \frac{C}{R}\int_0^{t_0}\int_{\frac{R}{4}\leq|x|\leq 2R}\frac{|u(x,t)|}{|x|^{N+\alpha}}\int_{\frac{R}{4}\leq|y|\leq 2R}\frac{1}{|x-y|^{N+\alpha-2}}\,dy\,dx\,dt$$

$$\leq \frac{C}{R}\int_0^{t_0}\int_{\frac{R}{4}\leq|x|\leq 2R}\frac{|u(x,t)|}{|x|^{N+\alpha}}\int_{\frac{R}{4}\leq|x-\tilde{y}|\leq 2R}\frac{1}{|\tilde{y}|^{N+\alpha-2}}\,d\tilde{y}\,dx\,dt$$

$$\leq CR^{1-\alpha}\int_0^{t_0}\int_{\frac{R}{4}\leq|x|\leq 2R}\frac{|u(x,t)|}{|x|^{N+\alpha}}\,dx\,dt. \tag{12.7.33}$$

On the other hand, if $(x,y)\in A_5\setminus Q$ we have

$$|x-y| > \frac{R}{100} \geq C|y|. \tag{12.7.34}$$

Then by (12.7.28) and (12.7.34) it follows that

$$\int_0^{t_0}\int_{(x,y)\in A_5\setminus Q}|u(x,t)|\frac{|(\phi_R(x)-\phi_R(y))(\varphi(x,t)-\varphi(y,t))|}{|x-y|^{N+\alpha}}\,dy\,dx\,dt$$

$$\leq \frac{C}{R}\int_0^{t_0}\int_{(x,y)\in A_5\setminus Q}|u(x,t)|\frac{|(\varphi(x,t)-\varphi(y,t))|}{|x-y|^{N+\alpha-1}}\,dy\,dx\,dt$$

$$\leq \frac{C}{R}\int_0^{t_0}\int_{(x,y)\in A_5\setminus Q}\frac{|u(x,t)|}{|x|^{N+\alpha}}\frac{1}{|y|^{N+\alpha-1}}\,dy\,dx\,dt$$

$$\leq \frac{C}{R}\int_0^{t_0}\int_{\frac{R}{4}\leq|x|\leq 2R}\frac{|u(x,t)|}{|x|^{N+\alpha}}\int_{\frac{R}{4}\leq|y|\leq 2R}\frac{1}{|y|^{N+\alpha-1}}\,dy\,dx\,dt$$

$$\leq CR^{-\alpha}\int_0^{t_0}\int_{\frac{R}{4}\leq|x|\leq 2R}\frac{|u(x,t)|}{|x|^{N+\alpha}}\,dx\,dt. \tag{12.7.35}$$

Therefore, from (12.7.33) and (12.7.35)

$$I_2^{A_5}(R)\leq \int_0^{t_0}\int_{(x,y)\in Q}|u(x,t)|\frac{|(\phi_R(x)-\phi_R(y))(\varphi(x,t)-\varphi(y,t))|}{|x-y|^{N+\alpha}}\,dy\,dx\,dt$$

$$+\int_0^{t_0}\int_{(x,y)\in A_5\setminus Q}|u(x,t)|\frac{|(\phi_R(x)-\phi_R(y))(\varphi(x,t)-\varphi(y,t))|}{|x-y|^{N+\alpha}}\,dy\,dx\,dt$$

$$\leq CR^{1-\alpha}\int_0^{t_0}\int_{\frac{R}{4}\leq|x|\leq 2R}\frac{|u(x,t)|}{|x|^{N+\alpha}}\,dx\,dt + CR^{-\alpha}\int_0^{t_0}\int_{\frac{R}{4}\leq|x|\leq 2R}\frac{|u(x,t)|}{|x|^{N+\alpha}}\,dx\,dt. \quad (12.7.36)$$

Then, since $u \in L^1([0,t_0],\mathcal{L}^{\alpha/2}(\mathbb{R}^N))$, using the monotone convergence theorem we obtain

$$\lim_{R\to\infty} I_2^{A_5}(R) = 0, \quad \text{when } \alpha \in [1,2). \quad (12.7.37)$$

That is, by (12.7.31) and (12.7.37), we get

$$\lim_{R\to\infty} |I_2^{A_5}(R)| = 0, \quad \text{whenever } \alpha \in (0,2). \quad (12.7.38)$$

Putting together (12.7.21), (12.7.27) and (12.7.38) it follows that

$$\lim_{R\to\infty} I_2(R) = 0, \quad \text{when } \alpha \in (0,2). \quad (12.7.39)$$

Therefore, from (12.7.19), by (12.7.20) and (12.7.39) we conclude that

$$\int_{\mathbb{R}^N} u(x,t_0)\theta(x)\,dx = 0,$$

for an arbitrary $\theta \in C_0^\infty(B_{R_0})$. □

12.7.2 Uniqueness for strong positive solutions

In this section we will establish the representation of the positive strong solutions of the fractional heat equation as the Poisson integral of the initial value.

Theorem 12.7.7. *Let $(x,t) \in \mathbb{R}^n \times (0,T)$. If $u(x,t) \geq 0$ is a strong solution of the fractional heat equation, then*

$$u(x,t) = \int_{\mathbb{R}^N} P_t(x-y)u(y,0)\,dy.$$

To prove this theorem we will need some previous results that we present as follows.

1. We prove first a maximum principle that allow us to establish that, among all possible positive solutions of the fractional heat equation, the smallest one is given by a formula that involves the convolution with the fractional heat kernel (see Lemma 12.7.9).
2. We use the uniqueness theorem, Theorem 12.7.6.

Lemma 12.7.8 (A maximum principle). *Set* $D_T := \Omega\times(0,T)$ *and let* $v(x,t) \in C(\overline{\Omega}\times[0,T))$ *satisfy, pointwise, the following problem:*

$$\begin{cases} v_t + (-\Delta)^{\alpha/2} v \le 0 & \text{for } (x,t) \in D_T, \\ v(x,t) \le 0 & \text{in } (\mathbb{R}^N \times [0,T)) \setminus D_T. \end{cases} \tag{12.7.40}$$

Then $v \le 0$ *in* $\overline{\Omega} \times [0,T)$.

Proof. Fixing an arbitrary $T' \in (0,T)$, we define

$$v(x_0,t_0) := \max_{\overline{\Omega}\times[0,T']} v(x,t).$$

Our goal is to show that $v(x_0,t_0) \le 0$. The proof is by contradiction, assuming that

$$v(x_0,t_0) > 0. \tag{12.7.41}$$

Then, (x_0,t_0) cannot lie in $(\partial\Omega\times[0,T))\cup(\Omega\times\{0\})$, since $v \le 0$ there, thanks to the boundary conditions in (12.7.40). As a consequence, (x_0,t_0) lies in $\Omega \times (0,T']$ and therefore $v_t(x_0,t_0) = 0$. Therefore the equation in (12.7.40) implies that

$$\begin{aligned} 0 \ge (-\Delta)^{\alpha/2} v(x_0,t_0) &= C_{N,\alpha/2}\, \text{P. V.} \int_{\mathbb{R}^N} \frac{v(x_0,t_0) - v(y,t_0)}{|y-x_0|^{N+\alpha}}\, dy \\ &= C_{N,\alpha/2}\Big(\text{P. V.} \int_{\Omega} \frac{v(x_0,t_0) - v(y,t_0)}{|y-x_0|^{N+\alpha}}\, dy + \int_{\mathbb{R}^N\setminus\Omega} \frac{v(x_0,t_0) - v(y,t_0)}{|y-x_0|^{N+\alpha}}\, dy\Big) \\ &\ge C_{N,\alpha/2}\, \text{P. V.} \int_{\mathbb{R}^N\setminus\Omega} \frac{v(x_0,t_0) - v(y,t_0)}{|y-x_0|^{N+\alpha}}\, dy. \end{aligned}$$

Since $v(y,t_0) \le 0$ for $y \in \mathbb{R}^N \setminus \Omega$, thanks to (12.7.40), the latter integrand is strictly positive, due to (12.7.41), and this is a contradiction. □

Now we are able to prove that strong solutions are upper bounds for the kernel convolutions.

Lemma 12.7.9. *Let* $(x,t) \in \mathbb{R}^n \times (0,T)$. *If* $u(x,t) \ge 0$ *is a strong solution of the fractional heat equation, then*

$$I := \int_{\mathbb{R}^N} P_t(x-y)u(y,0)\, dy \le u(x,t), \tag{12.7.42}$$

where $P_t(x)$ *is the function defined in* (12.7.2).

Proof. First of all, observe that the integral $I = I(x,t)$ exists, for it is given by the integration of two (measurable) positive functions, although we do not know *a priori* that

I is finite. However, this will be a consequence of our result that gives the inequality $I \le u(x,t)$ for every $(x,t) \in \mathbb{R}^N \times [0,T)$. To this aim, we let

$$\phi_R(x) := \begin{cases} 1 & |x| \le R-1, \\ R-|x| & R-1 \le |x| \le R, \\ 0 & |x| > R. \end{cases} \tag{12.7.43}$$

We define

$$v_R(x,t) := \int_{\mathbb{R}^N} P_t(x-y)\phi_R(y)u(y,0)\,dy = (P_t(\cdot) * \phi_R u(\cdot,0))(x).$$

Then

$$\begin{cases} \dfrac{\partial v_R}{\partial t} + (-\Delta)^{\alpha/2} v_R = 0 & \text{for } (x,t) \in \mathbb{R}^N \times (0,T), \\ v_R(x,t) \ge 0 & \text{for } (x,t) \in \mathbb{R}^N \times (0,T), \\ v_R(x,0) = \phi_R(x)u(x,0) & \text{in } \mathbb{R}^N. \end{cases}$$

Let $|x| > R$. As $u(x,t) \in C(\mathbb{R}^N \times [0,T))$ we can define the real number

$$M_R := \sup_{|y|<R} u(y,0) < \infty.$$

By (12.7.12) we have

$$\begin{aligned} 0 \le v_R(x,t) &\le M_R \int_{B_R} P_t(x-y)\,dy \\ &\le CM_R \int_{B_R} \frac{T}{|x-y|^{N+\alpha}}\,dy \\ &\le C(T,M_R) \int_{B_R} \frac{dy}{||x|-R|^{N+\alpha}} \\ &= C(T,M_R,N)\frac{R^N}{||x|-R|^{N+\alpha}}, \end{aligned}$$

for any $(x,t) \in (\mathbb{R}^N \setminus B_R) \times (0,T)$.

Then, for every $\varepsilon > 0$ it follows that

$$0 \le v_R(x,t) \le \varepsilon, \quad \text{for any } |x| \ge \rho,\ t \in (0,T), \tag{12.7.44}$$

where

$$\rho = R + \left(\frac{C(T,M_R,N)R^N}{\varepsilon}\right)^{\frac{1}{N+\alpha}} > 0.$$

Moreover, as $u(x,t) \geq 0$ in $\mathbb{R}^N \times [0,T)$ we obtain

$$0 \leq v_R(x,t) \leq \varepsilon \leq \varepsilon + u(x,t), \quad \text{for any } |x| \geq \rho, t \in (0,T), \tag{12.7.45}$$

and

$$v_R(x,0) = \phi_R(x)u(x,0) \leq u(x,0) \leq \varepsilon + u(x,0), \quad \text{for any } |x| \leq \rho. \tag{12.7.46}$$

Consider the cylinder

$$D_{\rho,T} = B_\rho \times (0,T).$$

We define the function

$$w(x,t) := v_R(x,t) - u(x,t) - \varepsilon.$$

Then, by (12.7.45) and (12.7.46), we get $w(x,t) \leq 0$ in $\mathbb{R}^N \times [0,T) \setminus D_{\rho,T}$. Therefore, since $u(x,t)$ is a strong solution of the fractional heat equation in $\mathbb{R}^N \times [0,T)$, applying Lemma 12.7.8 in $D_{\rho,T}$ to the function $w(x,t)$ we have

$$v_R(x,t) \leq \varepsilon + u(x,t), \quad \text{for } |x| \leq \rho \text{ and } t \in [0,T).$$

Therefore, from (12.7.45), it follows that

$$v_R(x,t) \leq \varepsilon + u(x,t), \quad \text{for } x \in \mathbb{R}^N \text{ and } t \in [0,T).$$

Since ε is fixed but arbitrary, the previous inequality implies that

$$v_R(x,t) \leq u(x,t), \quad \text{for every } x \in \mathbb{R}^N \text{ and } t \in [0,T).$$

Finally, by the monotone convergence theorem, as $\lim_{R\to\infty} \phi_R = 1$, we conclude that

$$0 \leq v(x,t) = \lim_{R\to\infty} v_R(x,t) = \int_{\mathbb{R}^N} P_t(x-y)u(y,0)\,dy \leq u(x,t). \qquad \square$$

By a simple time translation, we obtain from Lemma 12.7.9 the following.

Corollary 12.7.10. *Let $0 < \tau < T$ and $(x,t) \in \mathbb{R}^n \times (0, T-\tau)$. If $u(x,t) \geq 0$ is a strong solution of the fractional heat equation, then*

$$\int_{\mathbb{R}^N} P_t(x-y)u(y,\tau)\,dy \leq u(x,t+\tau), \tag{12.7.47}$$

where $P_t(x)$ is the function defined in (12.7.2). As a consequence, for every $x \in \mathbb{R}^N$ and $t \in [0, T-\tau)$ we have

$$\int_0^{T-t}\int_{\mathbb{R}^N} P_t(x-y)u(y,\tau)\,dy\,d\tau \leq \int_0^{T-t} u(x,t+\tau)\,d\tau. \tag{12.7.48}$$

Moreover we have the following.

Corollary 12.7.11. *Let $(x,t) \in \mathbb{R}^n \times (0,T)$. If $u(x,t) \geq 0$ is a strong solution of the fractional heat equation, then $u(\cdot,t) \in \mathcal{L}^{\alpha/2}(\mathbb{R}^N)$.*

Proof. Let $0 < T' < T$. Taking $t = T - T'$ in (12.7.47), from (12.7.8), we get

$$\frac{T-T'}{C}\int_{\mathbb{R}^N}\frac{u(y,\tau)}{(T-T')^{\frac{N+\alpha}{\alpha}}+|x-y|^{N+\alpha}}\,dy \leq \int_{\mathbb{R}^N} P_{T-T'}(x-y)u(y,\tau)\,dy \leq u(x,T-T'+\tau) < \infty. \tag{12.7.49}$$

Let

$$C(T) := \min\left\{1, \frac{1}{(T-T')^{\frac{N+\alpha}{\alpha}}}\right\}.$$

Then, since

$$\frac{1}{(T-T')^{\frac{N+\alpha}{\alpha}}+|y|^{N+\alpha}} \geq C(T)\frac{1}{1+|y|^{N+\alpha}},$$

taking $x = 0$ in (12.7.49), we conclude that

$$\int_{\mathbb{R}^N}\frac{|u(y,t)|}{(1+|y|^{N+\alpha})}\,dy < \infty.$$

□

We also have the following corollary.

Corollary 12.7.12. *Let $(x,t) \in \mathbb{R}^N \times (0,T)$. If $u(x,t) \geq 0$ is a strong solution of the fractional heat equation, then $u(\cdot,t) \in L^1([0,T'],\mathcal{L}^{\alpha/2}(\mathbb{R}^N))$ for every $0 < T' < T$.*

Proof. Take an arbitrary $0 < T' < T$. By (12.7.48) with $t = T - T'$ we have

$$\int_0^{T'}\int_{\mathbb{R}^N} P_{T-T'}(x-y)u(y,\tau)\,dy\,d\tau \leq \int_0^{T'} u(x,\tau)\,d\tau.$$

Then, as $u \in C(\mathbb{R}^N \times [0,T'])$ we get

$$\int_0^{T'}\int_{\mathbb{R}^N} P_{T-T'}(x-y)u(y,\tau)\,dy\,d\tau < \infty.$$

Therefore proceeding as in the proof of Corollary 12.7.11 we get

$$\int_0^{T'}\int_{\mathbb{R}^N}\frac{u(y,t)}{(1+|y|^{N+\alpha})}\,dy\,dt < \infty.$$

That is, $u \in L^1([0,T'],\mathcal{L}^{\alpha/2}(\mathbb{R}^N))$ for every $0 < T' < T$.

□

Note that Corollary 12.7.12 affirms that if $u(x,t) \geq 0$ is a strong solution of the fractional heat equation, then u is also a weak solution of the same equation.

Now we are able to prove our main result.

Proof of Theorem 12.7.5. By Corollary 12.7.12 we get $u \in L^1([0,T'],\mathcal{L}^{\alpha/2}(\mathbb{R}^N))$ for every $0 < T' < T$. Moreover if we define

$$p(x,t) := \int_{\mathbb{R}^N} P_t(x-y)u(y,0)\,dy,$$

by (12.7.48) with $t = T - T'$, we also have $p \in L^1([0,T'],\mathcal{L}^{\alpha/2}(\mathbb{R}^N))$. Let now the function

$$w(x,t) := u(x,t) - p(x,t) \geq 0.$$

It is clear that w is a strong solution of the fractional heat equation. Moreover, as $w \in L^1([0,T'],\mathcal{L}^{\alpha/2}(\mathbb{R}^N))$, $w(x,t)$ is also a solution in the weak sense with zero initial datum. Therefore applying Theorem 12.7.6 we conclude that $w(x,t) = 0$ for every $(x,t) \in \mathbb{R}^N \times [0,T)$. □

12.7.3 About viscosity solutions

As said at the beginning of this work, it is natural to consider viscosity solutions of the fractional heat equation. Our purpose here is to describe some cases in which a positive viscosity solution has the unique representation in terms of the kernel P_t.

Proposition 12.7.13. *Let $\{u_n\}$ be a sequence of nonnegative, strong solutions of the fractional heat equation converging uniformly over compact sets to a given function u. Then $u \geq 0$ is a viscosity solution of* (12.7.1) *satisfying*

$$\int_{\mathbb{R}^N} P_t(x-y)u(y,0)\,dy \leq u(x,t), \quad (x,t) \in \mathbb{R}^N \times (0,T) \tag{12.7.50}$$

and

$$u \in L^1([0,T'],\mathcal{L}^{\alpha/2}(\mathbb{R}^N)) \quad \textit{for every } 0 < T' < T. \tag{12.7.51}$$

That is, the conclusions of Lemma 12.7.9 *and Corollary* 12.7.12 *are satisfied.*

Proof. By Lemma 12.7.9 it follows that

$$\int_{\mathbb{R}^N} P_t(x-y)u_n(y,0)\,dy \leq u_n(x,t), \quad (x,t) \in \mathbb{R}^N \times (0,T).$$

Applying the Fatou lemma we obtain (12.7.50). Therefore, proceeding as in the proof of Corollary 12.7.12 we conclude (12.7.51). Note also that, by the comparison principle [286,

Corollary 2.1.6], $u_n \geq 0$ is a viscosity solution of (12.7.1) for every $n \in \mathbb{N}$. Therefore, since u is the uniform limit over compact sets of viscosity solutions, u is also a viscosity solution of (12.7.1). □

Note that to conclude the equality in (12.7.50) we would need to know that u is a weak solution.

If we add a monotonicity condition over the sequence $\{u_n\}$, then we obtain the following.

Proposition 12.7.14. *Let $\{u_n\}$ be a monotone sequence of nonnegative, strong solutions of the fractional heat equation converging uniformly over compact sets to a given function u. Then $u \geq 0$ is a strong solution of* (12.7.1) *satisfying*

$$\int_{\mathbb{R}^N} P_t(x-y)u(y,0)\,dy = u(x,t), \quad (x,t) \in \mathbb{R}^N \times (0,T). \tag{12.7.52}$$

Proof. Since, for every $n \in \mathbb{N}$, $u_n \geq 0$ satisfies Theorem 12.7.5, by the monotone convergence theorem we obtain (12.7.52). Clearly this implies that $u \geq 0$ is a strong solution of (12.7.1). □

Remark 12.7.15. It would be interesting to find the biggest class of positive viscosity solutions for which the representation property (12.7.52) holds. To the best of our knowledge, this is an open problem.

12.8 Further results

12.8.1 Some remarks about the fractional Widder theorem

As in the local case (see [318]), given a solution u, one can define the *enthalpy*

$$v(x,t) = \int_0^t u(x,s)\,ds,$$

which is also a solution of the fractional heat equation. Indeed we have the following result.

Lemma 12.8.1. *Let $(x,t) \in \mathbb{R}^n \times [0,T)$. If $u(x,t)$ is a strong solution of the fractional heat equation with vanishing initial condition, then the enthalpy term*

$$v(x,t) = \int_0^t u(x,s)\,ds$$

is also a strong solution of the fractional heat equation. Moreover if u is positive the function v is increasing in t for x fixed and $\alpha/2$-subharmonic as a function of x in the weak sense.

Proof. Let $(x,t) \in \mathbb{R}^N \times [0,T)$. First of all note that $v(x,t)$ satisfies conditions (i)–(iii) of Definition 12.7.3. Therefore, since $u(x,t)$ is a strong solution of the fractional heat equation, by the Fundamental Theorem of Calculus and the Fubini theorem it follows that

$$\begin{aligned}
C_{N,\alpha/2}\,\mathrm{P.V.}\int_{\mathbb{R}^N} \frac{v(x,t)-v(y,t)}{|x-y|^{N+\alpha}}\,dy
&= C_{N,\alpha/2}\,\mathrm{P.V.}\int_{\mathbb{R}^N} \frac{\int_0^t u(x,s)\,ds - \int_0^t u(y,s)\,ds}{|x-y|^{N+\alpha}}\,dy \\
&= C_{N,\alpha/2}\,\mathrm{P.V.}\int_{\mathbb{R}^N}\int_0^t \frac{u(x,s)-u(y,s)}{|x-y|^{N+\alpha}}\,ds\,dy \\
&= \int_0^t (-\Delta)^{\alpha/2} u(x,s)\,ds \\
&= -\int_0^t u_s(s,t)\,ds \\
&= -u(x,t) \\
&= -v_t(x,t),
\end{aligned}$$

for any $(x,t) \in \mathbb{R}^N \times (0,T]$. Then, $v(x,t)$ satisfies the fractional heat equation in the strong sense. Also if $u \geq 0$ from the calculations done before we deduce that

$$-(-\Delta)^{\alpha/2} v(x,t) = v_t(x,t) = u(x,t) \geq 0.$$

So v is $\alpha/2$-subharmonic as a function of x and increasing in t for x fixed. □

Lemma 12.8.1 shows that the enthalpy belongs to the special class of positive strong solutions that are also $\alpha/2$-subharmonic. This class naturally satisfies a polynomial estimate, as shown by the following result.

Proposition 12.8.2. *Let $(x,t) \in \mathbb{R}^n \times [0,T)$. Let $u(x,t) \geq 0$ such that:*
(i) *$u(x,t)$ is an $\alpha/2$-subharmonic function with respect to the variable x;*
(ii) *$u(x,t)$ is a strong solution of the fractional heat equation.*

Then $u(x,t) \leq C(t)(1+|x|^{N+\alpha})$ if $(x,t) \in \mathbb{R}^N \times [0,T)$.

Proof. By hypothesis it is clear that u is an $\alpha/2$-subharmonic function with respect to the variable x for t fixed and therefore u is increasing in time.

Let now $0 < t_1 < T$ and $0 < t_0 < T - t_1$. By Corollary 12.7.10 we have

$$\int_{\mathbb{R}^N} P_t(x-y)u(y,t_1)\,dy \leq u(x,t+t_1), \quad \text{for any } 0 < t < T - t_1.$$

Therefore

$$M_{t_0} := \int_{\mathbb{R}^N} P_{t_0}(y)u(y,t_1)\,dy \le u(0,t_0+t_1) < \infty.$$

Our objective is to show that

$$|u(x,t_1)| \le C(1+|x|^{N+\alpha}). \tag{12.8.1}$$

Once this is done, using that u is increasing in time, we would get

$$0 \le u(x,t) \le u(x,t_1) \le C(1+|x|^{N+\alpha}) \quad \text{for every } (x,t) \in \mathbb{R}^N \times (0,t_1).$$

But, since t_0 and t_1 are fixed but arbitrary, we would conclude that

$$|u(x,t)| \le C(1+|x|^{N+\alpha}).$$

So, we are left to showing that (12.8.1) is true. Note that from Corollary 12.7.11 we have $u \in \mathcal{L}^{\alpha/2}(\mathbb{R}^N)$. Then as u is $\alpha/2$-subharmonic, by Proposition 11.2.4 (see also [286, Proposition 2.2.6]) u also satisfies the following mean value property:

$$u(x_0,t) \le \int_{\mathbb{R}^N} \gamma_\lambda(y-x_0)u(y,t)\,dy \quad \text{for every } x_0 \in \Omega \subseteq \mathbb{R}^N \text{ and } \lambda \le \operatorname{dist}(x_0,\partial\Omega), \tag{12.8.2}$$

where, as in Proposition 11.2.4,

$$\gamma_\lambda(x) := (-\Delta)^{\alpha/2}\Gamma_\lambda(x)$$

and

$$\Gamma_\lambda(x) := \frac{1}{\lambda^{N-\alpha}}\Gamma\left(\frac{x}{\lambda}\right).$$

Here Γ is a $C^{1,1}$-function that coincides with $\Phi(x) := c|x|^{\alpha-N}$ outside the ball B_λ and Γ is a paraboloid inside this ball. Recall that Φ is the fundamental solution of $(-\Delta)^{\alpha/2}$. From (12.8.2) we have

$$\begin{aligned} u(x,t_1)(1+|x|^{N+\alpha})^{-1} &\le (1+|x|^{N+\alpha})^{-1}\int_{\mathbb{R}^N} \gamma_\lambda(y)u(x-y,t_1)\,dy \\ &= (1+|x|^{N+\alpha})^{-1}\int_{\{y:|y|\ge\lambda\}} \gamma_\lambda(y)u(x-y,t_1)\,dy \\ &\quad + (1+|x|^{N+\alpha})^{-1}\int_{\{y:|y|\le\lambda\}} \gamma_\lambda(y)u(x-y,t_1)\,dy \\ &:= I_1(x) + I_2(x). \end{aligned} \tag{12.8.3}$$

Choosing

$$\lambda = |x|/4, \tag{12.8.4}$$

we have $|x - y| \le 5|y|$. Therefore, we obtain

$$\begin{aligned} I_1(x) &\le C(1 + |x|^{N+\alpha})^{-1} \int\limits_{\{y:|y|\ge\lambda\}} \frac{u(x-y,t_1)}{|y|^{N+\alpha}}\,dy \\ &\le C(1 + |x|^{N+\alpha})^{-1} \int\limits_{\mathbb{R}^N} \frac{u(x-y,t_1)}{|x-y|^{N+\alpha}}\,dy \\ &\le C\|u\|_{\mathcal{L}^{\alpha/2}(\mathbb{R}^N)} := C_1. \end{aligned} \tag{12.8.5}$$

Moreover, using that the fractional Laplacian of a paraboloid is bounded, by (12.7.8) and (12.8.4), it follows that

$$\begin{aligned} I_2(x) &\le C(1 + |x|^{N+\alpha})^{-1} \int\limits_{\{y:|y|\le\lambda\}} u(x-y,t_1)\,dy \\ &\le C(1 + |x|^{N+\alpha})^{-1} \int\limits_{\{z:|x-z|\le\lambda\}} u(z,t_1)dz \\ &\le C(1 + |x|^{N+\alpha})^{-1} \int\limits_{\{z:|z|\le 2|x|\}} u(z,t_1)P_1(z)\frac{1}{P_1(z)}dz \\ &\le C(1 + |x|^{N+\alpha})^{-1}(1 + (2|x|)^{N+\alpha}) \int\limits_{\{z:|z|\le 2|x|\}} u(z,t_1)P_1(z)dz \\ &\le CM_1(1 + |x|^{N+\alpha})^{-1}(1 + (2|x|)^{N+\alpha}) \\ &\le C(N,\alpha)u(0,1+t_1) := C_2. \end{aligned} \tag{12.8.6}$$

By (12.8.3), (12.8.5) and (12.8.6), we obtain (12.8.1) and we conclude the proof. □

Remark 12.8.3. It would be interesting to prove that the pointwise behavior in Proposition 12.8.2 and some comparison arguments provide an alternative proof to our main result, namely, the representation of every solution as a convolution with the associated Poisson kernel.

12.8.2 The Fujita exponent for the fractional heat equation

In Chapter 5 we have studied the H. Fujita result in [170] for the heat equation. A similar result for the fractional heat equation was obtained by S. Sugitani in [299]. The estimates on the kernel of the fractional heat equation and the uniqueness of the positive solution allow us to study the behavior of the positive solution to the following semilinear problem:

$$\begin{cases} u_t + (-\Delta)^s u = u^{1+\beta}, & u(x,t) > 0,\ x \in \mathbb{R}^N,\ t > 0 \\ u(x,0) = u_0(x), & x \in \mathbb{R}^N, \end{cases} \tag{12.8.7}$$

where $0 < s < 1$, $\beta > 0$ and $u_0 \in L^\infty(\mathbb{R}^N) \cap \mathcal{C}(\mathbb{R}^N)$, $u_0(x) \gneq 0$.

Given the positive solution to problem (12.8.7), $u(x,t) \geq 0$, we can write by the representation formula

$$u(x,t) = \int_{\mathbb{R}^N} P_{2s}(x-y,t)u_0(y)\,dy + \int_0^t \int_{\mathbb{R}^N} P_{2s}(x-y,t-\tau)u^{1+\beta}(y,\tau)\,dy\,d\tau, \tag{12.8.8}$$

in the times t in which the function u is finite. That is, we consider the normalization

$$\int_{\mathbb{R}^N} P_{2s}(y,1)\,dy = 1.$$

Note that by using Corollary 12.5.8 we can find $t_0 > 0$, $c > 0$ and $\gamma > 0$ such that

$$u(x,t_0) \geq cP_{2s}(x,\gamma). \tag{12.8.9}$$

Indeed, choosing $t_0 > 0$ such that $P_{2s}(0,t_0) \leq 1$, we have

$$P_{2s}(x-y,t_0) = P_{2s}\Big(\frac{2x-2y}{2},t_0\Big) \geq P_{2s}(2x,t_0)P_{2s}(2y,t_0) \geq 2^{-N}P_{2s}\Big(x,\frac{t_0}{2^{2s}}\Big)P_{2s}(2y,t_0),$$

and therefore,

$$u(x,t_0) \geq \Big(2^{-N}\int_{\mathbb{R}^N} P_{2s}(2y,t_0)u_0(y)\,dy\Big)P_{2s}\Big(x,\frac{t_0}{2^{2s}}\Big).$$

That is, (12.8.9) is satisfied for

$$c = \Big(2^{-N}\int_{\mathbb{R}^N} P_{2s}(2y,t_0)u_0(y)\,dy\Big), \quad \gamma = \frac{t_0}{2^{2s}}.$$

We remark that the lower estimate (12.8.9) can be seen as a proof of the *non finite speed of propagation*. Now by using the semigroup property,

$$u(x,t+t_0) = \int_{\mathbb{R}^N} P_{2s}(x-y,t)u(y,t_0)\,dy + \int_0^t \int_{\mathbb{R}^N} P_{2s}(x-y,t-\tau)u^{1+\beta}(y,\tau+t_0)\,dy\,d\tau. \tag{12.8.10}$$

Plugging inequality (12.8.9) in the identity (12.8.10), we obtain

$$\begin{aligned} u(x,t+t_0) &\geq c\int_{\mathbb{R}^N} P_{2s}(x-y,t)P_{2s}(x-y,t)P_{2s}(y,\gamma)\,dy \\ &\quad + \int_0^t d\tau \int_{\mathbb{R}^N} P_{2s}(x-y,t-\tau)u^{1+\beta}(y,\tau+t_0)\,dy \\ &= cP_{2s}(x,t+\gamma) + \int_0^t d\tau \int_{\mathbb{R}^N} P_{2s}(x-y,t-\tau)u^{1+\beta}(y,\tau+t_0)\,dy. \end{aligned} \tag{12.8.11}$$

Call

$$f(t) = \int_{\mathbb{R}^N} P_{2s}(x,t)u(x,t+t_0)\,dx.$$

Multiplying inequality (12.8.11) by $P_{2s}(x,t)$, integrating in the whole space and using Fubini's theorem, we obtain

$$\begin{aligned} f(t) &\equiv \int_{\mathbb{R}^N} P_{2s}(x,t)u(x,t+t_0)\,dx \\ &\geq c\int_{\mathbb{R}^N} P_{2s}(x,\gamma)p(x,t+\tau) + \int_0^t d\tau \int_{\mathbb{R}^N} P_{2s}(x,t) \int_{\mathbb{R}^N} P_{2s}(x-y,t-\tau)u^{1+\beta}(y,\tau+t_0)\,dy\,dx \\ &= cP_{2s}(0,2t+\gamma) + \int_0^t d\tau \int_{\mathbb{R}^N} P_{2s}(y,2t-\tau)u^{1+\beta}(y,\tau+t_0)\,dy \\ &\geq cP_{2s}(0,1)\frac{1}{(2t+\gamma)^{\frac{N}{2s}}} + \int_0^t d\tau\left(\frac{\tau}{2t-\tau}\right)^{\frac{N}{2s}} \int_{\mathbb{R}^N} P_{2s}(y,\tau)u^{1+\beta}(y,\tau+t_0)\,dy. \end{aligned} \tag{12.8.12}$$

Since $d\mu := P_{2s}(x,t)\,dx$ is a probability measure for all t, we can apply the Jensen inequality and conclude the following inequality:

$$f(t) \geq cP_{2s}(0,1)\frac{1}{(2t+\gamma)^{\frac{N}{2s}}} + \int_0^t \left(\frac{\tau}{2t}\right)^{\frac{N}{2s}} f^{1+\beta}(\tau)\,d\tau. \tag{12.8.13}$$

Take $\delta > 0$ and consider $t > \delta$; then multiplying inequality (12.8.13) by $t^{\frac{N}{2s}}$ we find

$$t^{\frac{N}{2s}} f(t) \geq cP_{2s}(0,1)\frac{\delta}{(2\delta+\gamma)^{\frac{N}{2s}}} + \int_0^t \left(\frac{\tau}{2}\right)^{\frac{N}{2s}} f^{1+\beta}(\tau)\,d\tau. \tag{12.8.14}$$

Calling $h(t) = t^{\frac{N}{2s}} f(t)$, inequality (12.8.14) can be written as

$$h(t) \geq cP_{2s}(0,1)\frac{\delta}{(2\delta+\gamma)^{\frac{N}{2s}}} + \int_0^t \left(\frac{\tau}{2}\right)^{\frac{N}{2s}} (h(\tau)\tau^{-\frac{N}{2s}})^{1+\beta}\, d\tau.$$

Considering the function $g(t)$ satisfying

$$g(t) = cP_{2s}(0,1)\frac{\delta}{(2\delta+\gamma)^{\frac{N}{2s}}} + \int_0^t \left(\frac{\tau}{2}\right)^{\frac{N}{2s}} (g(\tau)\tau^{-\frac{N}{2s}})^{1+\beta}\, d\tau,$$

$h(t) \geq g(t)$, and then if g blows up in a finite time, also h and f blow up in finite time. Now $g(t)$ is the solution to the problem

$$\begin{cases} g'(t) = (\frac{t}{2})^{\frac{N}{2s}} (\frac{g(t)}{t^{\frac{N}{2s}}})^{1+\beta}, \\ g(0) = cP_{2s}(0,1)(\frac{\delta}{2\delta+\gamma})^{\frac{N}{2s}} := g_0. \end{cases} \tag{12.8.15}$$

Since $g_0 > 0$, integrating the ordinary differential equation,

$$g(t) = \left(\frac{1}{g_0^{-\beta} - (\frac{\beta}{1-\frac{N}{2s}\beta})t^{\beta N/2s}}\right)^{\frac{1}{\beta}},$$

which blows up in a finite time if $N\beta < 2s$.

As a conclusion we have obtained the following result.

Lemma 12.8.4. *Assume that $\beta < \frac{2s}{N}$. Then if $u(x,t)$ is a positive solution to problem* (12.8.7), *there exists a time T such that*

$$\int_{\mathbb{R}^N} P_{2s}(x,t)u(x,t)\, dx \to \infty \quad \text{as } t \to T.$$

The result can be made even stronger. We have proved the blow-up of the L^1-norm with respect to the probability measure $d\mu := P_{2s}(x,t)\,dx$. But next we will prove that there is a *complete blow-up*, in the sense that for any positive solution to problem (12.8.7), there exists a time T_1 such that $u(x,t) = \infty$ for every $t > T_1$ and every $x \in \mathbb{R}^N$. The complete blow-up is a consequence of the following result.

Lemma 12.8.5. *Assume that $u(x,t) \geq 0$ is a solution to* (12.8.7) *and*

$$f(t) = \int_{\mathbb{R}^N} P_{2s}(x,t)u(x,t)\, dx \to \infty \quad \text{as } t \to T.$$

Then there exists a $T_1 > 0$ such that

$$u(x,t) \to \infty \quad \text{as } t \to T_1, \text{ for all } x \in \mathbb{R}^N.$$

Proof. We can assume that $P_{2s}(0, T_1) \le 1$ and therefore $P_{2s}(0, t) \le 1$ if $t > T_1$. Consider

$$T_1 \le t \le \tau \le \frac{2^3}{2^\beta + 1} t.$$

By using again the homogeneity of the kernel and Corollary 12.5.8, we find

$$\begin{aligned} P_{2s}(x-y, 2^3 t - \tau) &= P_{2s}\left(x-y, \tau\left(\frac{2^3 t - \tau}{\tau}\right)\right) \\ &= \left(\frac{\tau}{2^3 t - \tau}\right)^{\frac{N}{2s}} P_{2s}\left(\left(\frac{\tau}{2^3 t - \tau}\right)(x-y), \tau\right) \\ &\ge \left(\frac{\tau}{2^3 t - \tau}\right)^{\frac{N}{2s}} P_{2s}(x, \tau) P_{2s}(y, \tau). \end{aligned}$$

Then

$$\int_{\mathbb{R}^N} P_{2s}(x-y, 2^3 t - \tau) u(y, \tau)\, dy \ge \left(\frac{\tau}{2^3 t - \tau}\right)^{\frac{N}{2s}} P_{2s}(x, \tau) f(\tau) = \infty.$$

Finally, using the Jensen inequality we conclude that

$$\begin{aligned} u(x, 2^3 t) &\ge \int_t^{\frac{2^3}{2^\beta+1} t} d\tau \int_{\mathbb{R}^N} P_{2s}(x-y, 2^3 t - \tau) u^{1+\beta}(y, \tau)\, dy \\ &\ge \int_t^{\frac{2^3}{2^\beta+1} t} d\tau \left(\int_{\mathbb{R}^N} P_{2s}(x-y, 2^3 t - \tau) u(y, \tau)\, dy\right)^{1+\beta} = \infty. \end{aligned}$$

Summarizing,

$$u(x, t) = \infty, \quad \text{if } t > 2^3 T_1 \text{ for all } x \in \mathbb{R}^N.$$

□

Remark 12.8.6. In [299] the blow-up result is obtained for positive increasing convex functions, $F(u)$, such that for some $\beta \in (0, \frac{2s}{N}]$,

$$\lim_{u \to \infty} \frac{F(u)}{u^{1+\beta}} = \gamma \in (0, \infty).$$

Note that the extreme $\beta = \frac{2s}{N}$ lives in the exponent where *blow-up occurs for all positive functions.*

13 The influence of the Hardy potential on the linear and semilinear fractional heat equations

13.1 Introduction

We will study in this chapter the solvability of the linear problem

$$\begin{cases} u_t + (-\Delta)^s u = \lambda \dfrac{u}{|x|^{2s}} + f & \text{in } \Omega \times (0,T), \\ u(x,t) > 0 & \text{in } \Omega \times (0,T), \\ u(x,t) = 0 & \text{in } (\mathbb{R}^N \setminus \Omega) \times [0,T), \\ u(x,0) = u_0(x) & \text{if } x \in \Omega \end{cases} \tag{13.1.1}$$

and of the semilinear problem

$$\begin{cases} u_t + (-\Delta)^s u = \lambda \dfrac{u}{|x|^{2s}} + u^p + f & \text{in } \Omega \times (0,T), \\ u(x,t) > 0 & \text{in } \Omega \times (0,T), \\ u(x,t) = 0 & \text{in } (\mathbb{R}^N \setminus \Omega) \times [0,T), \\ u(x,0) = u_0(x) & \text{if } x \in \Omega, \end{cases} \tag{13.1.2}$$

where Ω is a $C^{1,1}$-bounded domain in $\mathbb{R}^N$, $N > 2s$, $0 < s < 1$, $p > 1$ and c and λ are positive constants. To avoid the trivial case, we assume $0 \in \Omega$. We suppose that f and u_0 are nonnegative functions satisfying some hypotheses that we will precise later. We recall the integral expression of the fractional Laplacian of order $2s$ obtained in Chapter 8, that is,

$$(-\Delta)^s u(x) := C_{N,s} \, \text{P.V.} \int\limits_{\mathbb{R}^N} \frac{u(x) - u(y)}{|x-y|^{N+2s}} \, dy, s \in (0,1), \tag{13.1.3}$$

where $C_{N,s}$ is defined in (9.3.12).

The local case, $s = 1$, has been studied in Chapter 5, and there problems (13.1.1) and (13.1.2) correspond to the classical heat equation. The linear problem (13.1.1) was first studied by P. Baras and J. Goldstein in [41] and (13.1.2), the semilinear problem, has been studied in [19]. There are many other related papers in the literature.

For $s \in (0,1)$, the fractional setting, there exists also a large literature dealing with the case $\lambda = 0$. We refer to Chapter 12 and to the articles [106, 169] and the references therein for the basic theory of existence. A result on the uniqueness of a positive regular solution to the linear problem in the whole space was studied in Section 12.7 (see also the original article [48]).

The case under consideration in this chapter, $\lambda > 0$ and $s \in (0,1)$, presents many new difficulties; for instance as in the elliptic case, any positive supersolution to problem (13.1.1) is unbounded close to the origin, even for nice data. In the nonlocal framework, similarly to the elliptic problem, the precise rate of growth of the solutions near

https://doi.org/10.1515/9783110606270-013

the origin will be the key to obtain the optimal results. It is worth pointing out here the difference with the local case, where this rate is obtained just by solving an elementary linear differential equation.

The main results in this chapter can be summarized as follows.

- First, to study the local behavior of the solutions we need some sharp local estimates that are based on a *Harnack inequality* for a related problem resulting from the *ground state transformation* by Frank, Lieb and Seiringer [167], studied in Section 9.5. This weak Harnack inequality will give the exact blow-up rate for the positive supersolutions near the t-axis. For the proof of this result, we closely follow the work of Felsinger and Kassmann [155], where the authors develop a weak parabolic Harnack inequality for a general type of nonlocal operators with bounded coefficients. Since the resulting problem after the *ground state transformation* has *singular coefficients*, this result does not apply straightforward to our singular operator, so, based on their scheme, we need to check every step in the Moser protocol (see [244, 246]). The Harnack inequality for singular weights can be useful in related problems.
  The results in this chapter for the linear problem (13.1.1) can be seen as the extension to the fractional setting of those for the heat equation developed by P. Baras and J. A. Goldstein in [41] and studied in Chapter 5. Nevertheless, the proofs that we present are significantly different. More precisely, we obtain the optimal summability required for the data in order to solve the problem and to prove the instantaneous and complete blow-up for $\lambda > \Lambda_{N,2s}$ by using the results of Section 13.3. As a byproduct, we also prove the optimality of the power in the Hardy potential term, i. e., that $p > 1$ is a *supercritical power* for $\frac{u^p}{|x|^{2s}}$. This result in the local framework was obtained by Brezis and Cabré in [87].
- Second, concerning the semilinear problem (13.1.2), the main result that we obtain in this chapter is that for all $0 < \lambda < \Lambda_{N,2s}$ there exists a threshold exponent, $p_+(\lambda, s)$, for the existence of positive solutions. As before, by *threshold* we mean that when we consider an exponent $p > p_+(\lambda, s)$, there are no positive supersolutions even in the weak sense, while if $p < p_+(s, \lambda)$, it is possible to establish a suitable class of nonnegative data for which we can find a positive solution. We will see in particular that the threshold exponent $p_+(s, \lambda)$ is the same as in the elliptic case (see [46, 171]). In fact, this critical power is related to the possibility of finding a supersolution to the elliptic problem in the whole $\mathbb{R}^N$. As in the linear problem the main ingredient is the local estimates of the solutions close to the origin.

The chapter is organized as follows. In Section 7.2.21 we describe the natural functional framework associated to problems (13.1.1) and (13.1.2). We define the two notions of solution we will use along the paper: *weak* solutions and *energy* solutions. Moreover, we prove some interesting comparison principles. In Section 3 we describe the radial solutions of the corresponding homogeneous elliptic problem in $\mathbb{R}^N$. These solutions will allow us to detail the singularity of the supersolutions to problem (13.1.1) near the origin. This is a key point to perform the so-called *ground state transformation* (see [167]), and to obtain the nonexistence results afterwards. Section 13.3 is devoted to obtaining the weak Harnack inequality for the positive supersolutions of the problem resulting from the *ground state transformation* by Frank, Lieb and Seiringer, that introduces the difficulty of dealing with a kernel with *singular coefficients*. The goal of Section 13.4 is to study the linear problem (13.1.1) and some consequences. Finally, in Section 13.5 we consider the semilinear problem (13.1.2). Furthermore, we prove an instantaneous and complete blow-up phenomenon.

13.2 Functional framework: some preliminary results

Along this chapter we will always assume that Ω is a $C^{1,1}$-bounded domain of $\mathbb{R}^N$ such that $0 \in \Omega$. We consider the fractional Sobolev space $H_0^s(\Omega)$ defined in Chapter 8 as

$$H_0^s(\Omega) := \{u \in H^s(\mathbb{R}^N) \text{ with } u = 0 \text{ a. e. in } \mathbb{R}^N \setminus \Omega\},$$

endowed with the norm

$$\|u\|_{H_0^s(\Omega)} := \left(a_{N,s} \iint\limits_{D_\Omega} \frac{|u(x) - u(y)|^2}{|x - y|^{N+2s}}\, dx\, dy \right)^{1/2},$$

where $D_\Omega = \mathbb{R}^{2N} \setminus (\mathcal{C}\Omega \times \mathcal{C}\Omega)$. The pair $(H_0^s(\Omega), \|\cdot\|_{H_0^s(\Omega)})$ yields a Hilbert space. Moreover, as we point out in Chapter 8,

$$(-\Delta)^s : H_0^s(\Omega) \to H^{-s}(\Omega)$$

is a continuous operator, where $(-\Delta)^s$ is defined in (13.1.3).

In what follows we will use the relation between the norm in the space $H_0^s(\Omega)$ and the L^2-norm of the fractional Laplacian (see [138, Proposition 3.6])

$$\|u\|^2_{H_0^s(\Omega)} = \|(-\Delta)^{s/2} u\|^2_{L^2(\mathbb{R}^N)}. \tag{13.2.1}$$

It is easy to check that for u and φ smooth enough, with vanishing conditions outside Ω, we have the following duality product:

$$\int\limits_{\mathbb{R}^N} u(-\Delta)^s \varphi\, dx = a_{N,s} \int\limits_{D_\Omega} \frac{(u(x) - u(y))(\varphi(x) - \varphi(y))}{|x - y|^{N+2s}}\, dx\, dy,$$

which in particular implies the self-adjointness of $(-\Delta)^s$ in $H_0^s(\Omega)$. See Section 8.2.

We will use the Sobolev inequality in Theorem 8.2.7.

The parabolic problems studied in this article are related to the fractional Hardy inequality studied in Section 9.2. Let us recall some relevant facts about the Hardy inequality.

1. The criticality of inequality (9.2.1) is motivated by the coincidence of homogeneity between the fractional Laplacian and the *inverse* 2*s*-*potential*.
2. As we see in Chapter 9, $\Lambda_{N,2s} \to \Lambda_{N,2} := (\frac{N-2}{2})^2$, the classical Hardy constant, when s tends to 1. And by scaling we know that the optimal constant is the same for every domain containing the pole of the Hardy potential.
3. The optimal constant defined in (9.2.2) coincides for every bounded domain Ω containing the pole of the Hardy potential. That is, if $0 \in \Omega$, using (13.2.1) we can rewrite the Hardy inequality (10.1.3) as

$$a_{N,s} \int\limits_{D_\Omega} \frac{|u(x) - u(y)|^2}{|x - y|^{N+2s}} \, dx \, dy \geq \Lambda_{N,2s} \int\limits_{\Omega} \frac{u^2}{|x|^{2s}} \, dx, u \in H_0^s(\Omega). \tag{13.2.2}$$

The optimality of $\Lambda_{N,2s}$ here follows again by a scaling argument.
4. In [167] the fractional Hardy inequality plays an important role in the proof of the *stability of relativistic matter* in a very general setting.

Consider the parabolic problem

$$(P) \begin{cases} u_t + (-\Delta)^s u = f(x,t,u) & \text{in } \Omega \times (0,T), \\ u(x,t) = 0 & \text{in } (\mathbb{R}^N \setminus \Omega) \times [0,T), \\ u(x,0) = u_0(x) & \text{if } x \in \Omega. \end{cases}$$

Let $\alpha, \beta \in (0,1)$. Denote

$$\begin{aligned} \mathcal{T} := \{ \phi : \mathbb{R}^N \times [0,T] \to \mathbb{R}, \text{ s. t. } - \phi_t + (-\Delta)^s \phi = \varphi, \varphi \in L^\infty(\Omega \times (0,T)) \cap \mathcal{C}^{\alpha,\beta}(\Omega \times (0,T)), \\ \phi = 0 \text{ in } (\mathbb{R}^N \setminus \Omega) \times (0,T], \phi(x,T) = 0 \text{ in } \Omega \}. \end{aligned}$$

Note that every $\phi \in \mathcal{T}$ belongs in particular to $L^\infty(\Omega \times (0,T))$; see Section 12.4.1 and [224]. Moreover, according to the results in Section 12.6, $\phi \in \mathcal{T}$ satisfies the equation in a pointwise sense (see [48, Definition 1.3]).

We define the meaning of weak solution.

Definition 13.2.1. Assume $u_0 \in L^1(\Omega)$. We say that $u \in \mathcal{C}([0,T]; L^1(\Omega))$ is a weak supersolution (subsolution) of problem (P) if $f(x,t,u) \in L^1(\Omega \times [0,T))$, $u \geq (\leq) 0$ in $(\mathbb{R}^N \setminus \Omega) \times [0,T)$, $u(x,0) \geq (\leq) u_0(x)$ in Ω, and for all nonnegative $\phi \in \mathcal{T}$ we have

$$\int\limits_0^T \int\limits_\Omega u(-\phi_t + (-\Delta)^s \phi) \, dx \, dt \geq (\leq) \int\limits_0^T \int\limits_\Omega f(x,t,u)\phi \, dx \, dt + \int\limits_\Omega u_0(x)\phi(x,0) \, dx. \tag{13.2.3}$$

If u is a super- and subsolution, then we say that u is a weak solution to (P).

Weak solutions will be considered to formulate the optimal nonexistence results. For existence results, we will consider the classical notion of *finite energy solutions*.

Definition 13.2.2. Assume $u_0(x) \in L^2(\Omega)$. We say that $u \in L^2(0,T;H^s(\mathbb{R}^N))$ with $u_t \in L^2(0,T;H^{-s}(\Omega))$ is a finite energy supersolution (respectively subsolution) of (P) if $f(x,t,u) \in L^2(0,T;H^{-s}(\Omega))$ and it satisfies

$$\int_0^T\int_\Omega u_t\varphi\,dx\,dt + a_{N,s}\int_0^T\int_{D_\Omega}\frac{(u(x,t)-u(y,t))(\varphi(x,t)-\varphi(y,t))}{|x-y|^{N+2s}}\,dx\,dy\,dt$$
$$\geq (\leq)\int_0^T\int_\Omega f(x,t,u)\varphi\,dx\,dt,$$

for any nonnegative $\varphi \in L^2(0,T;H_0^s(\Omega))$, $\varphi = 0$ in $(\mathbb{R}^N\setminus\Omega)\times(0,T)$.

If u is a super- and subsolution, then u is a finite energy solution.

Remark 13.2.3. If $u \in L^2(0,T;H_0^s(\Omega))$ and $u_t \in L^2(0,T;H^{-s}(\Omega))$, then by approximating with smooth functions and taking advantage of the Hilbertian structure of the space, it can be checked that $u \in C([0,T];L^2(\Omega))$.

Note that both definitions can be considered, by scaling, in $\mathbb{R}^N \times [T_1,T_2)$ with $[T_1,T_2) \subset [0,T)$.

The existence and uniqueness of an energy solution to problem (P) when f does not depend on u and is in the dual space $L^2(0,T;H^{-s}(\Omega))$ was obtained in section 12.2 by means of a direct Hilbert space approach due to A. N. Milgram in [242] which are based on a method of Vishik in [312], see too [226]. Such arguments are essentially an extension of the Lax–Milgram theorem to parabolic problems. We recall such result.

Theorem 13.2.4. *Let $f \in L^2(0,T;H^{-s}(\Omega))$. Then problem (P) has a unique finite energy solution.*

See Chapter 12 and [224, Theorem 26] for a detailed proof in this fractional framework.

Remark 13.2.5. Note that by defining

$$L_\phi(u) := \int_0^T\int_\Omega -u\phi_t\,dx\,dt + a_{N,s}\int_0^T\int_{D_\Omega}\frac{(u(x,t)-u(y,t))(\phi(x,t)-\phi(y,t))}{|x-y|^{N+2s}}\,dx\,dy\,dt$$
$$-\lambda\int_0^T\int_\Omega\frac{u\phi}{|x|^{2s}}\,dx\,dt \tag{13.2.4}$$

and

$$\langle\varphi,\phi\rangle_* := \frac{1}{2}\langle\varphi(x,0),\phi(x,0)\rangle_{L^2(\Omega)} + a_{N,s}\left(1-\frac{\lambda}{\Lambda_{N,2s}}\right)\langle\varphi,\phi\rangle_{L^2(0,T;H_0^s(\Omega))},$$

thanks to Hardy's inequality one can reproduce the proof of section 12.2 to ensure the existence and uniqueness of an energy solution to the problem

$$(P_\lambda)\begin{cases} u_t + (-\Delta)^s u - \lambda \frac{u}{|x|^{2s}} = F(x,t) & \text{in } \Omega \times (0,T), \\ u(x,t) = 0 & \text{in } (\mathbb{R}^N \setminus \Omega) \times [0,T), \\ u(x,0) = u_0(x) & \text{if } x \in \Omega, \end{cases}$$

for $F \in L^2(0,T;H^{-s}(\Omega))$, $u_0 \in L^2(\Omega)$ and $\lambda < \Lambda_{N,2s}$. For the case $\lambda = \Lambda_{N,2s}$, consider the Hilbert space $H(\Omega)$ defined as the completion of $\mathcal{C}_0^\infty(\Omega)$ with respect to the norm

$$\|u\|^2_{H(\Omega)} := a_{N,s}\|u\|^2_{H_0^s(\Omega)} - \Lambda_{N,2s}\int_\Omega \frac{u^2}{|x|^{2s}}\,dx. \tag{13.2.5}$$

In [164], the author proves the following improved Hardy inequality:

$$a_{N,s}\|u\|^2_{H_0^s(\Omega)} - \Lambda_{N,2s}\int_\Omega \frac{u^2}{|x|^{2s}}\,dx \geq C(\Omega,q,N,s)\|u\|^2_{W_0^{\tau,2}(\Omega)}, \tag{13.2.6}$$

for all $s/2 < \tau < s$ (see also [171, 22] for alternative proofs without using the Fourier transform).

In Remark 10.2.2 we have defined $H(\Omega)$ as the completion of $\mathcal{C}_0^\infty(\Omega)$ with respect to the norm defined in (13.2.5). Thus we see that $H(\Omega) \subset W_0^{\tau,2}(\Omega)$ and therefore, $H(\Omega)$ is compactly embedded in $L^p(\Omega)$ for all $1 \leq p < 2_s^*$ (see [138, Corollary 7.2]). Therefore, the proof remains the same considering $L_\phi(u)$ as in (13.2.4) (setting $\lambda = \Lambda_{N,2s}$), and defining the scalar product $\langle\cdot,\cdot\rangle_*$ as

$$\langle\varphi,\phi\rangle_* := \frac{1}{2}\langle\varphi(x,0),\phi(x,0)\rangle_{L^2(\Omega)} + \langle\varphi,\phi\rangle_{L^2(0,T;H(\Omega))},$$

where the last term follows from (13.2.5).

In order to study monotonicity approaches, we will need to prove comparison results for both kind of solutions.

Lemma 13.2.6 (Weak comparison principle). *Let $0 \leq \lambda \leq \Lambda_{N,2s}$ and let $u, v \in \mathcal{C}([T_1,T_2]; L^1(\Omega))$ be weak solutions to the problems*

$$\begin{cases} u_t + (-\Delta)^s u - \lambda \frac{u}{|x|^{2s}} = f_1 & \text{in } \Omega \times (T_1,T_2), \\ u = g_1 & \text{in } (\mathbb{R}^N \setminus \Omega) \times [T_1,T_2), \\ u(x,T_1) = h_1(x) & \text{in } \Omega, \end{cases}$$

$$\begin{cases} v_t + (-\Delta)^s v - \lambda \frac{v}{|x|^{2s}} = f_2 & \text{in } \Omega \times (T_1,T_2), \\ v = g_2 & \text{in } (\mathbb{R}^N \setminus \Omega) \times [T_1,T_2), \\ v(x,T_1) = h_2(x) & \text{in } \Omega, \end{cases}$$

respectively, where $f_1, f_2 \in L^1(\Omega \times (T_1, T_2))$, $g_1, g_2 \in L^1((\mathbb{R}^N \setminus \Omega) \times (T_1, T_2))$ *and* $h_1, h_2 \in L^1(\Omega)$.

If $f_1 \le f_2$ *in* $\Omega \times (T_1, T_2)$, $g_1 \le g_2$ *in* $(\mathbb{R}^N \setminus \Omega) \times [T_1, T_2)$ *and* $h_1 \le h_2$ *in* Ω, *then* $u \le v$ *in* $\mathbb{R}^N \times (T_1, T_2)$.

Proof. Define $w = v - u$. Hence, w is a weak solution of

$$\begin{cases} w_t + (-\Delta)^s w - \lambda \frac{w}{|x|^{2s}} = f_2 - f_1 \ge 0 & \text{in } \Omega \times (T_1, T_2), \\ w = g_2 - g_1 \ge 0 & \text{in } (\mathbb{R}^N \setminus \Omega) \times [T_1, T_2), \\ w(x, T_1) = h_2 - h_1 \ge 0 & \text{in } \Omega. \end{cases}$$

Consider now $\Phi \in C_0^\infty(\Omega \times (T_1, T_2))$, $\Phi \ge 0$ and the solution φ_n to the problem

$$\begin{cases} -(\varphi_n)_t + (-\Delta)^s \varphi_n = \lambda \frac{\varphi_{n-1}}{|x|^{2s} + \frac{1}{n}} + \Phi & \text{in } \Omega \times (T_1, T_2), \\ \varphi_n = 0 & \text{in } (\mathbb{R}^N \setminus \Omega) \times [T_1, T_2), \\ \varphi_n(x, T_2) = 0 & \text{in } \Omega, \end{cases} \tag{13.2.7}$$

with

$$\begin{cases} -(\varphi_0)_t + (-\Delta)^s \varphi_0 = \Phi & \text{in } \Omega \times (T_1, T_2), \\ \varphi_0 = 0 & \text{in } (\mathbb{R}^N \setminus \Omega) \times [T_1, T_2), \\ \varphi_0(x, T_2) = 0 & \text{in } \Omega. \end{cases} \tag{13.2.8}$$

Note that the existence of $\{\varphi_n\}_{n\in\mathbb{N}}$ follows from Theorem 13.2.4 after an appropriate change of variable in time. Since φ_n is regular in $\Omega \times [T_1, T_2)$ and bounded in $\mathbb{R}^N \times (T_1, T_2)$ (see Section 12.6), this equation can be understood in a pointwise sense. Moreover, by the *weak comparison principle*, we know that $\varphi_n \ge 0$ and $\varphi_{n-1} \le \varphi_n$ in $\mathbb{R}^N \times [T_1, T_2)$ for all $n \in \mathbb{N}$. Hence, by the definition of weak solutions and using that $w \ge 0$ and $(-\Delta)^s \varphi_n \le 0$ in $(\mathbb{R}^N \setminus \Omega) \times [T_1, T_2)$, it follows that

$$\begin{aligned} \int_{T_1}^{T_2}\int_\Omega w\Phi \, dx\, dt &= \int_{T_1}^{T_2}\int_\Omega w(-\varphi_n)_t \, dx\, dt + \int_{T_1}^{T_2}\int_\Omega w(-\Delta)^s\varphi_n \, dx\, dt - \lambda \int_{T_1}^{T_2}\int_\Omega \frac{w\varphi_{n-1}}{|x|^{2s} + \frac{1}{n}} \, dx\, dt \\ &\ge \int_{T_1}^{T_2}\int_\Omega w(-\varphi_n)_t \, dx\, dt + \int_{T_1}^{T_2}\int_{\mathbb{R}^N} w(-\Delta)^s\varphi_n \, dx\, dt - \lambda \int_{T_1}^{T_2}\int_\Omega \frac{w\varphi_n}{|x|^{2s}} \, dx\, dt \\ &= \int_{T_1}^{T_2}\int_\Omega (f_2 - f_1)\varphi_n \, dx\, dt + \int_\Omega w(x, T_1)\varphi_n(x, T_1)\, dx \ge 0, \end{aligned}$$

for all $\Phi \in C_0^\infty(\Omega \times (T_1, T_2))$, $\Phi \ge 0$. Thus, $w \ge 0$ in $\mathbb{R}^N \times (T_1, T_2)$, and therefore $u \le v$ in $\mathbb{R}^N \times (T_1, T_2)$. □

Corollary 13.2.7 (Uniqueness of weak solutions for the linear problem). *Let us suppose $F \in L^1(\Omega \times (0,T))$. Then problem (P_λ) has at most one weak solution.*

The comparison result for energy solutions can be proved in a standard way, so we skip the proof (see for example [46] for a proof in the elliptic case).

Lemma 13.2.8 (Energy comparison principle). *Let $0 \le \lambda < \Lambda_{N,2s}$ and let $u, v \in L^2(T_1, T_2; H^s(\mathbb{R}^N))$ with $u_t, v_t \in L^2(T_1, T_2; H^{-s}(\Omega))$ be finite energy solutions to the problems*

$$\begin{cases} u_t + (-\Delta)^s u - \lambda \frac{u}{|x|^{2s}} = f_1 & \text{in } \Omega \times (T_1, T_2), \\ u = g_1 & \text{in } (\mathbb{R}^N \setminus \Omega) \times [T_1, T_2), \\ u(x, T_1) = h_1(x) & \text{in } \Omega, \end{cases}$$

$$\begin{cases} v_t + (-\Delta)^s v - \lambda \frac{v}{|x|^{2s}} = f_2 & \text{in } \Omega \times (T_1, T_2), \\ v = g_2 & \text{in } (\mathbb{R}^N \setminus \Omega) \times [T_1, T_2), \\ v(x, T_1) = h_2(x) & \text{in } \Omega, \end{cases}$$

respectively, where $f_1, f_2 \in L^2(T_1, T_2; H^{-s}(\Omega))$, $g_1, g_2 \in L^2(T_1, T_2; L^2(\mathbb{R}^N \setminus \Omega))$ and $h_1, h_2 \in L^2(\Omega)$. If $f_1 \le f_2$ in $\Omega \times (T_1, T_2)$, $g_1 \le g_2$ in $(\mathbb{R}^N \setminus \Omega) \times [T_1, T_2)$ and $h_1 \le h_2$ in Ω, then $u \le v$ in $\mathbb{R}^N \times (T_1, T_2)$.

Remark 13.2.9. Note that if $\lambda = \Lambda_{N,2s}$, we can obtain the same result for $u, v \in L^2(T_1, T_2; H(\Omega))$, where $H(\Omega)$ was defined in (13.2.5), only by exactly repeating this proof.

Finally, consider the problem

$$\begin{cases} u_t + (-\Delta)^s u = 0 & \text{in } \Omega \times (0, T), \\ u(x,t) = 0 & \text{in } (\mathbb{R}^N \setminus \Omega) \times [0, T), \\ u(x, 0) = u_0(x) \gneq 0 & \text{if } x \in \Omega. \end{cases} \tag{13.2.9}$$

We enunciate a *weak Harnack inequality* that we will use along the section (see [155, Theorem 1.1] for a more general setting).

Lemma 13.2.10 (Weak Harnack inequality). *Let $t_0 \in (0,T)$ and $r, \beta > 0$ such that $B_r(0) \times (t_0 - \frac{3}{4}\beta, t_0 + \frac{3}{4}\beta) \subset \Omega \times (0,T)$. If u is a nonnegative supersolution of (13.2.9) in $\Omega \times (0,T)$, then there exists a positive constant $C = C(N, s, r, t_0, \beta)$ such that*

$$\iint_{R^-} u(x,t)\, dx\, dt \le C\Big(\operatorname*{ess\,inf}_{R^+} u\Big),$$

where $R^- := B_r(0) \times (t_0 - \frac{3}{4}\beta, t_0 - \frac{1}{4}\beta)$, $R^+ := B_r(0) \times (t_0 + \frac{1}{4}\beta, t_0 + \frac{3}{4}\beta)$.

As a consequence of this lemma, we can formulate the *strong maximum principle*.

Theorem 13.2.11 (Strong maximum principle). *If u is a nonnegative supersolution of (13.2.9), then $u(x,t) > 0$ in $\Omega \times (0,T)$.*

13.3 Weak Harnack inequality for a weighted problem

According to Lemma 9.4.2 we know that if $\lambda \le \Lambda_{N,2s}$

$$(-\Delta)^s u = \lambda \frac{u}{|x|^{2s}} \quad \text{in } (\mathbb{R}^N \setminus \{0\}) \tag{13.3.1}$$

has radial solutions $v_{\pm\alpha} = |x|^{-\frac{N-2s}{2} \pm \alpha}$ if and only if α is obtained by the identity

$$\lambda = \lambda(\alpha) = \lambda(-\alpha) = \frac{2^{2s}\Gamma(\frac{N+2s+2\alpha}{4})\Gamma(\frac{N+2s-2\alpha}{4})}{\Gamma(\frac{N-2s+2\alpha}{4})\Gamma(\frac{N-2s-2\alpha}{4})}. \tag{13.3.2}$$

In Section 9.5 we have studied the *ground state representation* obtained by Frank, Lieb and Seiringer in [167, Proposition 4.1]. We recall the representation result.

Lemma 13.3.1 (Ground state representation). *Let* $0 < \gamma < \frac{N-2s}{2}$*. If* $\phi \in C_0^\infty(\mathbb{R}^N)$ *and* $\bar{\phi}(x) := |x|^\gamma \phi(x)$*, then*

$$\int_{\mathbb{R}^N} |\xi|^{2s} |\hat{\phi}(\xi)|^2 \, d\xi - (\Lambda_{N,2s} + \Phi_{N,s}(\gamma)) \int_{\mathbb{R}^N} |x|^{-2s} |\phi(x)|^2 \, dx$$
$$= a_{N,s} \iint_{\mathbb{R}^{2N}} \frac{|\bar{\phi}(x) - \bar{\phi}(y)|^2}{|x-y|^{N+2s}} \frac{dx}{|x|^\gamma} \frac{dy}{|y|^\gamma}, \tag{13.3.3}$$

where

$$\Phi_{N,s}(\gamma) := 2^{2s}\left(\frac{\Gamma(\frac{\gamma+2s}{2})\Gamma(\frac{N-\gamma}{2})}{\Gamma(\frac{N-\gamma-2s}{2})\Gamma(\frac{\gamma}{2})} - \frac{\Gamma^2(\frac{N+2s}{4})}{\Gamma^2(\frac{N-2s}{4})} \right). \tag{13.3.4}$$

We will use the monotonicity property of $\Psi_{N,s}$ obtained in Lemma 9.4.2. The result follows from the proof of Lemma 9.4.2 noting that $\lambda(\alpha) = \Psi_{N,s}(\frac{N-2s}{2} - \alpha)$ and γ is defined in (9.4.9).

Thus, for any $\alpha \in (0, \frac{N-2s}{2})$ thanks to Lemma 13.3.1 we know that

$$\int_{\mathbb{R}^N} |\xi|^{2s} |\hat{u}(\xi,t)|^2 \, d\xi - \lambda(\alpha) \int_{\mathbb{R}^N} |x|^{-2s} |u(x,t)|^2 \, dx = a_{N,s} \iint_{\mathbb{R}^{2N}} \frac{|v(x,t) - v(y,t)|^2}{|x-y|^{N+2s}} \frac{dx}{|x|^\gamma} \frac{dy}{|y|^\gamma}, \tag{13.3.5}$$

where $v(x,t) := |x|^\gamma u(x,t)$. Consider now the operator defined as

$$L_\gamma(v(x)) := C_{N,s} \, P.V. \int_{\mathbb{R}^N} (v(x) - v(y)) K(x,y) \, dy,$$

and

$$K(x,y) = \frac{1}{|x|^\gamma} \frac{1}{|y|^\gamma} \frac{1}{|x-y|^{N+2s}}, \quad 0 < \gamma = \frac{N-2s}{2} - \alpha < \frac{N-2s}{2}.$$

Note that by Fubini's theorem it is easy to check that for v and φ smooth enough there holds

$$\int_{\mathbb{R}^N} |x|^\gamma L_\gamma(v(x))\varphi(x) = a_{N,s} \iint_{\mathbb{R}^{2N}} \frac{(v(x)-v(y))(|x|^\gamma\varphi(x)-|y|^\gamma\varphi(y))}{|x-y|^{N+2s}} \frac{dx}{|x|^\gamma} \frac{dy}{|y|^\gamma}$$

and thus, understanding the scalar product in this sense, we can affirm that the Euler–Lagrange equation associated to (13.3.5) is

$$(-\Delta)^s u - \lambda \frac{u}{|x|^{2s}} = |x|^\gamma L_\gamma.$$

The Harnack inequality for this elliptic operator was proved in Section 10.3.

From the *ground state representation*, we conclude that if u is an energy solution of problem (13.1.1) with $0 < \lambda < \Lambda_{N,2s}$, then $v(x,t) := |x|^\gamma u(x,t)$ solves the parabolic equation

$$\begin{cases} |x|^{-2\gamma} v_t + L_\gamma v = |x|^{-\gamma} f(x,t) & \text{in } \Omega \times (0,T), \\ v(x,t) = 0 & \text{in } (\mathbb{R}^N \setminus \Omega) \times [0,T), \\ v(x,0) = v_0(x) := |x|^\gamma u_0(x) & \text{if } x \in \Omega. \end{cases} \tag{13.3.6}$$

Therefore, if we want to analyze the behavior of u near the origin, we may, equivalently, deal with the same question for v. In particular, we will prove that the weighted operator

$$|x|^{-2\gamma} v_t + L_\gamma v$$

satisfies a suitable weak Harnack inequality. In the local case, this kind of result can be obtained as a consequence of some results by Chiarenza–Frasca, Chiarenza–Serapioni and Gutierrez–Wheeden; see [122, 123, 191] and the references therein.

Let us detail first the natural functional framework associated to the new problem (13.3.6). For simplicity we denote

$$d\mu := \frac{dx}{|x|^{2\gamma}} \quad \text{and} \quad d\nu := K(x,y)\, dx\, dy. \tag{13.3.7}$$

Let $\Omega \subseteq \mathbb{R}^N$, $0 \in \Omega$. As in Chapter 10, we define the weighted Sobolev space $W^{s,\gamma}(\Omega)$ as

$$W^{s,\gamma}(\Omega) := \left\{ \phi \in L^2(\Omega, d\mu) : \int_\Omega \int_\Omega (\phi(x)-\phi(y))^2\, d\nu < +\infty \right\}.$$

It is clear that $W^{s,\gamma}(\Omega)$ is a Hilbert space endowed with the norm

$$\|\phi\|_{W^{s,\gamma}(\Omega)} := \left(\int_\Omega |\phi(x)|^2\, d\mu + \int_\Omega \int_\Omega (\phi(x)-\phi(y))^2\, d\nu \right)^{\frac{1}{2}},$$

and we define the space $W_0^{s,\gamma}(\Omega)$ as the completion of $\mathcal{C}_0^\infty(\Omega)$ with respect to this norm. In particular, we denote

$$|||\phi|||_{W^s_{\gamma,0}(\Omega)} := \left(\int_\Omega \int_\Omega (\phi(x) - \phi(y))^2 \, d\nu \right)^{\frac{1}{2}}.$$

If Ω is bounded, the norms $|||\cdot|||_{W_0^{s,\gamma}(\Omega)}$ and $\|\cdot\|_{W^{s,\gamma}(\Omega)}$ are equivalent (see Theorem 10.2.11 for more details). If $\Omega = \mathbb{R}^N$, using the definition of L_γ, we obtain that for all $w_1, w_2 \in W_0^{s,\gamma}(\mathbb{R}^N)$,

$$\langle L_\gamma(w_1), w_2 \rangle = a_{N,s} \int_{\mathbb{R}^N} \int_{\mathbb{R}^N} (w_1(x) - w_1(y))(w_2(x) - w_2(y)) \, d\nu.$$

Let us begin by the following natural definition.

Definition 13.3.2. Let $v \in \mathcal{C}([0,T), L^2(\mathbb{R}^N, d\mu)) \cap L^2(0,T; W^{s,\gamma}(\mathbb{R}^N))$. We say that v is a supersolution to problem (13.3.6) if $v(x,0) \geq v_0(x)$ and for all $\Omega_1 \subset\subset \Omega$, for all $[t_1, t_2] \subset (0,T)$ we have

$$\int_{t_1}^{t_2} \int_{\Omega_1} -\varphi_t v \, d\mu \, dt + a_{N,s} \int_{t_1}^{t_2} \int_{\tilde{D_\Omega}} (v(x,t) - v(y,t))(\varphi(x,t) - \varphi(y,t)) \, d\nu \, dt$$
$$\geq \int_{t_1}^{t_2} \int_{\Omega_1} f\varphi |x|^{-\gamma} \, dx \, dt + \int_{\Omega_1} \varphi(x,t_1) v(x,t_1) \, d\mu - \int_{\Omega_1} \varphi(x,t_2) v(x,t_2) \, d\mu,$$

for any nonnegative $\varphi \in L^2(t_1,t_2; W_0^{s,\gamma}(\Omega_1))$, such that $\varphi_t \in L^2(t_1,t_2; W^{-s,\gamma}(\Omega_1))$, where $\tilde{D_\Omega} := \mathbb{R}^{2N} \setminus (\mathcal{C}\Omega_1 \times \mathcal{C}\Omega_1)$.

The main result of this section is the next theorem.

Theorem 13.3.3 (Weak Harnack inequality). *Assume that f (respectively v_0) ≥ 0 in $\Omega \times (0,T)$ (respectively in Ω). Let $v \in \mathcal{C}([0,T), L^2(\mathbb{R}^N, d\mu)) \cap L^2(0,T; W^{s,\gamma}(\Omega))$ be a supersolution to (13.3.6) with $v \gneqq 0$ in $\mathbb{R}^N \times (0,T)$.*

Then for any $q < 1 + \frac{2s}{N}$, we have

$$\left(\iint_{Q_1} v^q \, d\mu \, dt \right)^{\frac{1}{q}} \leq C \inf_{Q_2} v, \tag{13.3.8}$$

where $Q_1 = B_r(x_0) \times (t_1, t_2)$, $Q_2 = B_r(x_0) \times (t_3, t_4)$ with $0 < t_1 < t_2 < t_3 < t_4 < T$, $B_r(x_0) \subset \Omega$ and $C = C(N, r, t_1, t_2, t_3, t_4) > 0$.

The proof of this result follows the classical arguments by Moser (see [246]) with some necessary adaptations. In the context of the parabolic fractional-like operators the precedent work is the interesting paper for bounded coefficients by Felsinger and Kassmann [155], which we will closely follow here, adapting the proofs to the weighted operator appearing in (13.3.6). The main new feature in the proof below is the use of some radial estimates that allow us to handle the effect of the singular weight. To make the chapter self-contained, we include here the corresponding proofs. First of all, we will need an iteration result, originally proved in [246] and extended by Bombieri and Giusti in [85], to the case of general measures in the elliptic setting (see also [277, Lemma 2.2.6]).

Lemma 13.3.4. *Let $\{U(r)\}_{\theta\le r\le 1}$ be a nondecreasing family of bounded domains $U(r)\subset \mathbb{R}^{N+1}$, and let m, c_0 be positive constants, $\eta\in(0,1)$, $\theta\in[\frac{1}{2},1]$ and $0<p_0\le+\infty$. Let w be a positive, measurable function defined on $U(1)$ satisfying*

$$\|w\|_{L^{p_0}(U(r),\,d\mu\,dt)} \le \left(\frac{c_0}{(R-r)^m |U(1)|_{d\mu\times dt}}\right)^{(\frac{1}{p_0}-\frac{1}{p})} \|w\|_{L^p(U(R),d\mu\,dt)},$$

for all $r,R\in[\theta,1]$, $r<R$ and for all $p\in(0,\min\{\eta p_0,1\})$.

Assume also that

$$\forall s>0: \left|U(1)\cap\{\log w>s\}\right|_{d\mu\times dt} \le \frac{c_0|U(1)|_{d\mu\times dt}}{s}.$$

Then there exists $C(\theta,\eta,c_0,m,p_0)$ such that

$$\left(\iint\limits_{U(\theta)} w^{p_0}\,d\mu\,dt\right)^{\frac{1}{p_0}} \le C\left|U(1)\right|^{\frac{1}{p_0}}_{d\mu\times dt}.$$

Hereafter, we will make use of the following notation. Given $r>0$, we define

$$I_-(r) := (-r^{2s},0),\quad I_+(r) := (0,r^{2s}), \tag{13.3.9}$$

$$Q_-(r) := B_r(0)\times I_-(r),\quad Q_+(r) := B_r(0)\times I_+(r). \tag{13.3.10}$$

The first step to prove Theorem 13.3.3 is to establish the next estimate (see [155, Proposition 3.4]). Note that we just have to consider the case where $B_r(x_0)=B_r(0)$. For simplicity, we will write B_r instead of $B_r(0)$.

Lemma 13.3.5. *Assume that $\frac{1}{2}\le r<R\le 1$ and let $p>0$. Consider $v\gneq 0$, a supersolution to (13.3.6). Then*

$$\left(\iint\limits_{Q_-(r)} v^{-\tau p}\,d\mu\,dt\right)^{\frac{1}{\tau}} \le A\iint\limits_{Q_-(R)} v^{-p}\,d\mu\,dt, \tag{13.3.11}$$

where $\tau := 1+\frac{2s}{N}$ and

$$A := A(N,s,p,r,R,\gamma) = C(N,s,\gamma)(p+1)^2\left(\frac{1}{(R-r)^{2s}}+\frac{1}{R^{2s}-r^{2s}}\right)^{\tau}.$$

Proof. Without loss of generality we can assume that $v \geq \varepsilon > 0$ in $Q_-(R)$ (otherwise we can deal with $v + \varepsilon$ and let $\varepsilon \to 0$ at the end). Let $q > 1$ and $\psi \in W_0^{s,\gamma}(B_R) \cap L^\infty(B_R)$ be a nonnegative radial cutoff function such that $\operatorname{supp}(\psi) \subseteq B_R$ with $r < R$, $0 \leq \psi \leq 1$, $\psi = 1$ in B_r and

$$\frac{(\psi(x) - \psi(y))^2}{|x-y|^2} \leq \frac{C}{(R-r)^2}. \tag{13.3.12}$$

Using $\psi^{q+1}v^{-q}$ as a test function in (13.3.6), it follows that

$$\int_{B_R} \psi^{q+1} v^{-q} v_t \, d\mu + a_{N,s} \int_{\mathbb{R}^N} \int_{\mathbb{R}^N} (v(x,t) - v(y,t))(\psi^{q+1}(x) v^{-q}(x,t) - \psi^{q+1}(y) v^{-q}(y,t)) \, d\nu \geq 0.$$

Hence, using a pointwise inequality proved in [155, Lemma 3.3], it can be deduced that

$$\begin{aligned} &\frac{1}{q-1} \int_{B_R} \psi^{q+1} (v^{1-q})_t \, d\mu + \frac{1}{q-1} a_{N,s} 2 \int_{B_r} \int_{B_r} \big(v^{\frac{1-q}{2}}(x,t) - v^{\frac{1-q}{2}}(y,t)\big)^2 \, d\nu \\ &\quad \leq C q a_{N,s} \int_{\mathbb{R}^N} \int_{\mathbb{R}^N} \Big(\Big(\frac{v(x,t)}{\psi(x)}\Big)^{1-q} + \Big(\frac{v(y,t)}{\psi(y)}\Big)^{1-q} \Big) (\psi(x) - \psi(y))^2 \, d\nu. \end{aligned}$$

Furthermore,

$$\begin{aligned} &\int_{\mathbb{R}^N} \int_{\mathbb{R}^N} \Big(\Big(\frac{v(x,t)}{\psi(x)}\Big)^{1-q} + \Big(\frac{v(y,t)}{\psi(y)}\Big)^{1-q} \Big) (\psi(x) - \psi(y))^2 \, d\nu \\ &\quad \leq 2 \int_{B_R} \int_{B_R} \frac{v^{1-q}(x,t)}{|x|^\gamma} \frac{(\psi(x) - \psi(y))^2 \, dx \, dy}{|y|^\gamma |x-y|^{N+2s}} \\ &\qquad + 4 \int_{\mathbb{R}^N \setminus B_R} \int_{B_R} \frac{v^{1-q}(x,t)}{|x|^\gamma} \frac{(\psi(x) - \psi(y))^2 \, dx \, dy}{|y|^\gamma |x-y|^{N+2s}}. \end{aligned}$$

We set

$$I = \int_{B_R} \int_{B_R} \frac{v^{1-q}(x,t)}{|x|^\gamma} \frac{(\psi(x) - \psi(y))^2 \, dx \, dy}{|y|^\gamma |x-y|^{N+2s}}$$

and

$$J = \int_{\mathbb{R}^N \setminus B_R} \int_{B_R} \frac{v^{1-q}(x,t)}{|x|^\gamma} \frac{(\psi(x) - \psi(y))^2 \, dx \, dy}{|y|^\gamma |x-y|^{N+2s}}.$$

Let us begin by estimating the term J. Since $\frac{1}{|y|^\gamma} \le \frac{1}{|x|^\gamma}$ for $x \in B_R$ and $y \in \mathbb{R}^N \backslash B_R$ and using Fubini's theorem, we reach

$$\begin{aligned} J &\le \int_{B_R} \frac{v^{1-q}(x,t)}{|x|^{2\gamma}} \int_{\{\mathbb{R}^N \backslash B_R\} \cap \{|x-y|>R-r\}} \frac{(\psi(x)-\psi(y))^2\, dy\, dx}{|x-y|^{N+2s}} \\ &\quad + \int_{B_R} \frac{v^{1-q}(x,t)}{|x|^{2\gamma}} \int_{\{\mathbb{R}^N \backslash B_R\} \cap \{|x-y|\le R-r\}} \frac{(\psi(x)-\psi(y))^2\, dy\, dx}{|x-y|^{N+2s}} \\ &\le J_1 + J_2. \end{aligned}$$

Setting $\rho = |x-y|$, we get

$$\begin{aligned} J_1 &\le 4 \int_{B_R} \frac{v^{1-q}(x,t)}{|x|^{2\gamma}} \int_{\{\mathbb{R}^N \backslash B_R\} \cap \{|x-y|>R-r\}} \frac{dy\, dx}{|x-y|^{N+2s}} \\ &\le 4 \int_{B_R} \frac{v^{1-q}(x,t)}{|x|^{2\gamma}}\, dx \int_{R-r}^{\infty} \rho^{-1-2s}\, d\rho \\ &\le \frac{C}{(R-r)^{2s}} \int_{B_R} \frac{v^{1-q}(x,t)}{|x|^{2\gamma}}\, dx. \end{aligned}$$

We estimate now the term J_2. Using (13.3.12), it follows that

$$\begin{aligned} J_2 &\le \frac{C}{(R-r)^2} \int_{B_R} \frac{v^{1-q}(x,t)}{|x|^{2\gamma}} \int_{\{\mathbb{R}^N \backslash B_R\} \cap \{|x-y|\le R-r\}} \frac{|x-y|^2\, dy\, dx}{|x-y|^{N+2s}} \\ &\le \frac{C}{(R-r)^2} \int_{B_R} \frac{v^{1-q}(x,t)}{|x|^{2\gamma}}\, dx \int_0^{R-r} \rho^{1-2s}\, d\rho \\ &\le \frac{C}{(R-r)^{2s}} \int_{B_R} \frac{v^{1-q}(x,t)}{|x|^{2\gamma}}\, dx, \end{aligned}$$

for some constant C possibly changing from line to line. Hence

$$J \le \frac{C}{(R-r)^{2s}} \int_{B_R} v^{1-q}(x,t)\, d\mu.$$

We deal now with I. Using the definition of ψ, we easily get

$$\begin{aligned} I &= \iint_{B_R \times B_R \backslash B_r \times B_r} \frac{v^{1-q}(x,t)}{|x|^\gamma} \frac{(\psi(x)-\psi(y))^2\, dx\, dy}{|y|^\gamma |x-y|^{N+2s}} \\ &= \int_{B_R \backslash B_r} \int_{B_r} \frac{v^{1-q}(x,t)}{|x|^\gamma} \frac{(\psi(x)-\psi(y))^2\, dx\, dy}{|y|^\gamma |x-y|^{N+2s}} + \int_{B_R \backslash B_r} \int_{B_R \backslash B_r} \frac{v^{1-q}(x,t)}{|x|^\gamma} \frac{(\psi(x)-\psi(y))^2\, dx\, dy}{|y|^\gamma |x-y|^{N+2s}} \end{aligned}$$

$$+\int_{B_r}\int_{B_R\setminus B_r} \frac{v^{1-q}(x,t)}{|x|^\gamma} \frac{(\psi(x)-\psi(y))^2\,dx\,dy}{|y|^\gamma|x-y|^{N+2s}}$$
$$= I_1 + I_2 + I_3.$$

Let us begin estimating I_1. Since $(x,y) \in B_r \times B_R\setminus B_r$, we have $|x| \le |y|$, and hence

$$\begin{aligned}
I_1 &\le \int_{B_R\setminus B_r}\int_{B_r} \frac{v^{1-q}(x,t)}{|x|^{2\gamma}} \frac{(\psi(x)-\psi(y))^2\,dx\,dy}{|x-y|^{N+2s}}\\
&\le \int_{B_r} \frac{v^{1-q}(x,t)}{|x|^{2\gamma}} \int_{\{B_R\setminus B_r\}\cap\{|x-y|>R-r\}} \frac{(\psi(x)-\psi(y))^2\,dy\,dx}{|x-y|^{N+2s}}\\
&\quad+\int_{B_r} \frac{v^{1-q}(x,t)}{|x|^{2\gamma}} \int_{\{B_R\setminus B_r\}\cap\{|x-y|\le R-r\}} \frac{(\psi(x)-\psi(y))^2\,dy\,dx}{|x-y|^{N+2s}}\\
&\le I_{11} + I_{12}.
\end{aligned}$$

As in the previous computations, setting $\rho = |x-y|$ and using the fact that ψ is bounded, we conclude that

$$I_{11} \le 4\int_{B_r} \frac{v^{1-q}(x,t)}{|x|^{2\gamma}} \int_{R-r}^{\infty} \rho^{-1-2s}\,d\rho \le \frac{C}{(R-r)^{2s}} \int_{B_r} \frac{v^{1-q}(x,t)}{|x|^{2\gamma}}\,dx.$$

In the same way and using the fact that $\frac{(\psi(x)-\psi(y))^2}{|x-y|^2} \le \frac{C}{(R-r)^2}$, we get

$$I_{12} \le \frac{C}{(R-r)^2} \int_{B_r} \frac{v^{1-q}(x,t)}{|x|^{2\gamma}} \int_0^{R-r} \rho^{1-2s}\,d\rho \le \frac{C}{(R-r)^{2s}} \int_{B_r} \frac{v^{1-q}(x,t)}{|x|^{2\gamma}}\,dx.$$

Therefore,

$$I_1 \le \frac{C}{(R-r)^{2s}} \int_{B_R} \frac{v^{1-q}(x,t)}{|x|^{2\gamma}}\,dx.$$

We deal now with I_2. Since $\frac{1}{2} \le r \le |x|, |y| \le R < 1$, we have

$$\begin{aligned}
I_2 &\le 2^\gamma \int_{B_R\setminus B_r}\int_{B_R\setminus B_r} \frac{v^{1-q}(x,t)}{|x|^{2\gamma}} \frac{(\psi(x)-\psi(y))^2\,dx\,dy}{|x-y|^{N+2s}}\\
&\le \int_{B_R\setminus B_r} \frac{v^{1-q}(x,t)}{|x|^{2\gamma}} \int_{\{B_R\setminus B_r\}\cap\{|x-y|>R-r\}} \frac{(\psi(x)-\psi(y))^2\,dy\,dx}{|x-y|^{N+2s}}\\
&\quad+\int_{B_R\setminus B_r} \frac{v^{1-q}(x,t)}{|x|^{2\gamma}} \int_{\{B_R\setminus B_r\}\cap\{|x-y|\le R-r\}} \frac{(\psi(x)-\psi(y))^2\,dy\,dx}{|x-y|^{N+2s}}\\
&\le \frac{C}{(R-r)^{2s}} \int_{B_R} \frac{v^{1-q}(x,t)}{|x|^{2\gamma}}\,dx,
\end{aligned}$$

by repeating the computations performed for I_1. Let us consider now the term I_3, which is the most complicated. We have

$$\begin{aligned} I_3 &= \int_{r\le|x|\le R} \frac{v^{1-q}(x,t)}{|x|^\gamma}\Big(\int_{|y|\le\frac{|x|}{2}} \frac{(\psi(x)-\psi(y))^2}{|x-y|^{N+2s}|y|^\gamma}\,dy\Big)\,dx \\ &\quad + \int_{r\le|x|\le R} \frac{v^{1-q}(x,t)}{|x|^\gamma}\Big(\int_{\frac{|x|}{2}\le|y|\le r} \frac{(\psi(x)-\psi(y))^2}{|x-y|^{N+2s}|y|^\gamma}\,dy\Big)\,dx \\ &= I_{31}+I_{32}. \end{aligned}$$

If $|y|\le\frac{|x|}{2}$, then $|x-y|\ge\frac{|x|}{2}\ge\frac{r}{2}\ge\frac{1}{4}$, and thus,

$$\begin{aligned} I_{31} &\le C\int_{r\le|x|\le R} \frac{v^{1-q}(x,t)}{|x|^\gamma}\Big(\int_{|y|\le\frac{|x|}{2}} \frac{1}{|y|^\gamma}\,dy\Big)\,dx \\ &\le C\int_{r\le|x|\le R} \frac{v^{1-q}(x,t)}{|x|^\gamma}\Big(\int_0^{\frac{|x|}{2}} \rho^{N-1-\gamma}\,d\rho\Big)\,dx \\ &\le \frac{C}{(R-r)^{2s}}\int_{B_R} \frac{v^{1-q}(x,t)}{|x|^{2\gamma}}\,dx. \end{aligned}$$

To estimate I_{32}, we use the fact that $\frac{1}{|y|^\gamma}\le\frac{2^\gamma}{|x|^\gamma}$, and hence

$$\begin{aligned} I_{32} &\le C\int_{r\le|x|\le R} \frac{v^{1-q}(x,t)}{|x|^{2\gamma}}\Big(\int_{\frac{|x|}{2}\le|y|\le r} \frac{(\psi(x)-\psi(y))^2}{|x-y|^{N+2s}}\,dy\Big)\,dx \\ &\le C\int_{r\le|x|\le R} \frac{v^{1-q}(x,t)}{|x|^{2\gamma}}\Big(\int_{\{\frac{|x|}{2}\le|y|\le r\}\cap\{|x-y|>R-r\}} \frac{(\psi(x)-\psi(y))^2}{|x-y|^{N+2s}}\,dy\Big)\,dx \\ &\quad + C\int_{r\le|x|\le R} \frac{v^{1-q}(x,t)}{|x|^{2\gamma}}\Big(\int_{\{\frac{|x|}{2}\le|y|\le r\}\cap\{|x-y|\le R-r\}} \frac{(\psi(x)-\psi(y))^2}{|x-y|^{N+2s}}\,dy\Big)\,dx. \end{aligned}$$

Using the same computations as in the estimates of I_{11} and I_{12} it follows that

$$I_{32}\le\frac{C}{(R-r)^{2s}}\int_{B_R}\frac{v^{1-q}(x,t)}{|x|^{2\gamma}}\,dx.$$

Combining the estimates above, we get

$$\begin{aligned} &\int_{B_R}\psi^{q+1}(v^{1-q})_t\,d\mu + a_{N,s}\int_{B_r}\int_{B_r}\big(v^{\frac{1-q}{2}}(x,t)-v^{\frac{1-q}{2}}(y,t)\big)^2\,d\nu \\ &\quad\le \frac{Cq^2}{(R-r)^{2s}}a_{N,s}\int_{B_R} v^{1-q}(x,t)\,d\mu. \end{aligned}$$

Set now $\theta(t) := [\min\{\frac{t+R^{2s}}{R^{2s}-r^{2s}}, 1\}]_+$. Then, multiplying the last inequality by θ, integrating in time in $(-R^{2s}, t)$ with $t \in (-r^{2s}, 0)$ and noting that $\theta(t) = 1$ for $t \geq -r^{2s}$ and $|\theta'(t)| \leq \frac{1}{R^{2s}-r^{2s}}$, it follows that

$$\sup_{t\in I_-(r)} \int_{B_r} (v^{1-q})\,d\mu + a_{N,s} \iint_{Q_-(r)} \int_{B_r} (v^{\frac{1-q}{2}}(x,t) - v^{\frac{1-q}{2}}(y,t))^2\,d\nu\,dt$$
$$\leq Cq^2 a_{N,s}\left(\frac{1}{(R-r)^{2s}} + \frac{1}{R^{2s}-r^{2s}}\right) \iint_{Q_-(R)} v^{1-q}(x,t)\,d\mu\,dt. \tag{13.3.13}$$

Recalling that $\tau := 1 + \frac{2s}{N}$, let us define $w := v^{\frac{1-q}{2}}$. Then, applying the Hölder inequality, we have

$$\iint_{Q_-(r)} w^{2\tau}\,d\mu\,dt = \iint_{Q_-(r)} w^2 w^{\frac{4s}{N}}\,d\mu\,dt$$
$$\leq \int_{I_-(r)} \left(\int_{B_r} w^2\,d\mu\right)^{\frac{2s}{N}} \left(\int_{B_r} w^{2_s^*}\,d\mu\right)^{\frac{2}{2_s^*}}\,dt.$$

Since $\gamma > 0$ and $R \leq 1$, we conclude that

$$\iint_{Q_-(r)} w^{2\tau}\,d\mu\,dt \leq \int_{I_-(r)} \left(\int_{B_r} w^2(x,t)\,d\mu\right)^{\frac{2s}{N}} \left(\int_{B_r} \frac{w^{2_s^*}}{|x|^{2_s^*\gamma}}\,dx\right)^{\frac{2}{2_s^*}}\,dt.$$

Now, using the Sobolev inequality obtained in Theorem 10.2.20,

$$\iint_{Q_-(r)} w^{2\tau}\,d\mu\,dt \leq C \sup_{t\in I_-(r)} \left(\int_{B_r} w^2(x,t)\,d\mu\right)^{\frac{2s}{N}}$$
$$\times \left(\iint_{Q_-(r)} \int_{B_r} (w(x,t) - w(y,t))^2\,d\nu\,dt + r^{-2s} \iint_{Q_-(r)} w^2\,d\mu\,dt\right).$$

Applying (13.3.13) twice in this inequality and recalling that $\frac{1}{2} \leq r \leq 1$, it can be checked that

$$\iint_{Q_-(r)} w^{2\tau}\,d\mu\,dt \leq A(q,r,R,N,s)\left(\iint_{Q_-(R)} w^2\,d\mu\,dt\right)^{\tau},$$

where

$$A(q,r,R,N,s) = C(N,s,\gamma)q^2\left(\frac{1}{(R-r)^{2s}} + \frac{1}{R^{2s}-r^{2s}}\right)^{\tau}.$$

Setting $p := q - 1$, we conclude the proof. □

As an application of the previous estimate, we reach a control of $\sup_{Q_-(r)} v^{-1}$. More precisely, we have the following result.

Lemma 13.3.6. *Assume that $\frac{1}{2} \leq r < R \leq 1$ and $p \in (0,1]$. Then, there exists a constant $C = C(N,s,\gamma) > 0$ such that every $v \gneq 0$, supersolution to problem (13.3.6), satisfies*

$$\sup_{Q_-(r)} v^{-1} \leq \left(\frac{C}{\alpha(r,R)}\right)^{\frac{1}{p}} \left(\iint\limits_{Q_-(R)} v^{-p}\, d\mu\, dt\right)^{\frac{1}{p}}, \tag{13.3.14}$$

where

$$\alpha(r,R) = \begin{cases} (R-r)^{N+2s} & \text{if } s \geq \frac{1}{2}, \\ (R^{2s} - r^{2s})^{\frac{N+2s}{2s}} & \text{if } s < \frac{1}{2}. \end{cases}$$

Proof. We consider

$$\mathcal{M}(r,p) := \left(\iint\limits_{Q_-(r)} v^{-p}\, d\mu\, dt\right)^{\frac{1}{p}}.$$

By Lemma 13.3.5, we have

$$\mathcal{M}(r,\tau p) \leq A^{\frac{1}{p}} \mathcal{M}(R,p),$$

where $\tau := 1 + \frac{2s}{N}$. Construct now the sequences $\{r_i\}_{i\in\mathbb{N}}$ and $\{p_i\}_{i\in\mathbb{N}}$ by setting $r_0 := R > r_1 > r_2 > \cdots > r$ and $p_i := p\tau^i$. Using again Lemma 13.3.5, we obtain

$$\mathcal{M}(r,p_{m+1}) \leq \mathcal{M}(r_{m+1},p_{m+1}) \leq A_m^{\frac{1}{\tau^m p}} \mathcal{M}(r_m,p_m),$$

with $A_m := C(p_m+1)^2((r_m - r_{m+1})^{-2s} + (r_m^{2s} - r_{m+1}^{2s})^{-1})$. Iterating this and following the arguments in [155, Theorem 3.5], we conclude the proof. □

We prove now a control of a small positive exponent of u.

Lemma 13.3.7. *Suppose that $\frac{1}{2} \leq r < R \leq 1$ and fix $q \in (0,\tau^{-1}]$, with $\tau := 1 + \frac{2s}{N}$. Then, if $v \gneq 0$ is a supersolution to (13.3.6), we have*

$$\left(\iint\limits_{Q_+(r)} v^{q\tau}\, d\mu\, dt\right)^{\frac{1}{\tau}} \leq \alpha \iint\limits_{Q_+(R)} v^q\, d\mu\, dt, \tag{13.3.15}$$

where

$$\alpha = \alpha(N,s,r,R,\gamma) = C(N,s,\gamma)\left(\frac{1}{(R-r)^{2s}} + \frac{1}{R^{2s} - r^{2s}}\right).$$

Proof. The proof follows similarly to the one of Lemma 13.3.5. We set $a := (1-q) \in [1-\tau^{-1}, 1)$ and $w(x,t) := v^{\frac{1-a}{2}}$. Then, using $v^{-a}\psi^2$ (with ψ defined in the proof of Lemma 13.3.5) as a test function in (13.3.6), we reach

$$\sup_{t\in I_+(r)} \int_{B_r} w^2(x,t)\,d\mu + C(q)a_{N,s} \int_{Q_+(r)} \int_{B_r} (w^2(x,t) - w^2(y,t))^2\,d\nu\,dt$$
$$\leq C(q)a_{N,s}\left(\frac{1}{(R-r)^{2s}} + \frac{1}{R^{2s}-r^{2s}}\right) \int_{Q_+(R)} w^2(x,t)\,d\mu\,dt.$$

Using Theorem 10.2.20 and proceeding as in Lemma 13.3.5, we get (13.3.15). □

Define

$$\mathcal{H}(r,q) = \left(\int_{Q_+(r)} v^q\,d\mu\,dt\right)^{\frac{1}{q}}.$$

From (13.3.15), we get $\mathcal{H}(r,\tau q) \leq \alpha^{\frac{1}{q}}\mathcal{H}(R,q)$. Let us define $q_j := \tau^{-j}$ and

$$r_j := \begin{cases} r + \frac{R-r}{2^j} & \text{if } s \geq \frac{1}{2}, \\ (r^{2s} + \frac{R^{2s}-r^{2s}}{2^j})^{1/2s} & \text{if } s < \frac{1}{2}. \end{cases}$$

By Lemma 13.3.7 for r_n and r_{n-1}, it follows that

$$\mathcal{H}(r_n, q_1\tau) \leq \alpha_n^{\tau}\mathcal{H}(r_{n-1}, q_1), \tag{13.3.16}$$

where $\alpha_n = C(N,s,\gamma)(\frac{1}{(r_{n-1}-r_n)^{2s}} + \frac{1}{r_{n-1}^{2s}-r_n^{2s}})$. By using the definition of r_n and considering that $r \geq \frac{1}{2}$, we get

$$\alpha_n \leq C(N,s,\gamma)\frac{2^{2ns}}{(R-r)^{2s}}.$$

Hence

$$\mathcal{H}(r_n, q_1\tau) \leq \left(C(N,s,\gamma)\frac{2^{2ns}}{(R-r)^{2s}}\right)^{\tau}\mathcal{H}(r_{n-1}, q_1). \tag{13.3.17}$$

Iterating this inequality (see [155, Theorem 3.7]), we reach the next result.

Lemma 13.3.8. *Assume that $\frac{1}{2} \leq r < R \leq 1$ and $q \in (0,\tau^{-1})$, with $\tau := 1 + \frac{2s}{N}$. Then, every supersolution $v \gneq 0$ of problem (13.3.6) satisfies*

$$\iint_{Q_+(r)} v\,d\mu\,dt \leq \left(\frac{C}{|Q_+(1)|_{d\mu\times dt}\overline{\alpha}(r,R)}\right)^{\frac{1-q}{q}}\left(\iint_{Q_+(R)} v^q\,d\mu\,dt\right)^{\frac{1}{q}}, \tag{13.3.18}$$

where $C = C(N, s, \gamma) > 0$ *and*

$$\overline{\alpha}(r,R) = \begin{cases} (R-r)^{\omega_1} & \text{if } s \geq \frac{1}{2}, \\ (R^{2s} - r^{2s})^{\omega_2} & \text{if } s < \frac{1}{2}, \end{cases}$$

with $\omega_1, \omega_2 > 0$ *depending only on s and N.*

In order to apply Lemma 13.3.4 we need to estimate

$$\big|Q_+(1) \cap \{\log v < -m - a\}\big|_{d\mu \times dt} \quad \text{and} \quad \big|Q_-(1) \cap \{\log v > m - a\}\big|_{d\mu \times dt}, \tag{13.3.19}$$

where $m, a > 0$ and v is a supersolution to (13.3.6). To do this, we need the following auxiliary result, whose proof immediately follows from Lemma 4.1 in [155]. See Section 10.3 for the elliptic case.

Lemma 13.3.9. *Let* $I \subset \mathbb{R}$ *and let* $\psi : \mathbb{R}^N \to [0, +\infty)$ *be a continuous function satisfying* $\operatorname{supp}(\psi) = \overline{B_R}$ *for some* $R > 0$ *and* $|||\psi|||_{Y_0^{s,\gamma}(\mathbb{R}^N)} \leq C$. *Then, for* $v : \mathbb{R}^N \times I \to [0, +\infty)$, *the following inequality holds:*

$$\begin{aligned} &\int_{\mathbb{R}^N}\int_{\mathbb{R}^N} (v(x,t) - v(y,t))(-\psi^2(x)v^{-1}(x,t) + \psi^2(y)v^{-1}(y,t))\, d\nu \\ &\quad \geq \int_{B_R}\int_{B_R} \psi(x)\psi(y)\Big(\log\frac{v(y,t)}{\psi(y)} - \log\frac{v(x,t)}{\psi(x)}\Big)^2 d\nu - 3\int_{\mathbb{R}^N}\int_{\mathbb{R}^N} (\psi(x) - \psi(y))^2\, d\nu. \end{aligned}$$

With this result, following the proof of Moser in [244] we can establish the estimates in (13.3.19) (see [155, Proposition 4.2] for more details).

Lemma 13.3.10. *Assume that v is a supersolution to* (13.3.6) *in the cylinder* $Q := B_2 \times (-1, 1)$. *Then there exists a positive constant* $C = C(N, s, \gamma)$ *such that for some constant* $a = a(v)$, *we have*

$$\forall m > 0 : \big|Q_+(1) \cap \{\log v < -m - a\}\big|_{d\mu \times dt} \leq \frac{C|B_1|_{d\mu}}{m} \tag{13.3.20}$$

and

$$\forall m > 0 : \big|Q_-(1) \cap \{\log v > m - a\}\big|_{d\mu \times dt} \leq \frac{C|B_1|_{d\mu}}{m}. \tag{13.3.21}$$

Proof. Suppose that $v \geq \varepsilon > 0$ in Q. Let ψ be such that $\psi^2 = [\min\{\frac{3}{2} - \frac{|x|}{2}, 1\}]_+$, and let us denote $w(x,t) := -\log\frac{v(x,t)}{\psi(x)}$. Using $\frac{\psi^2}{v}$ as a test function in (13.3.6) and noting that $\operatorname{supp}(\psi^2) \subseteq B_{3/2}$, by Lemma 13.3.9 and the fact that $|||\psi|||_{W_0^{s,\gamma}(\mathbb{R}^N)} \leq C$, it follows that

$$\begin{aligned} &\int_{B_{3/2}} \psi^2 w_t\, d\mu + a_{N,s}\int_{B_{3/2}}\int_{B_{3/2}} \psi(x)\psi(y)(w(x,t) - w(y,t))^2\, d\nu \\ &\quad \leq 5a_{N,s}\int_{\mathbb{R}^N}\int_{\mathbb{R}^N} (\psi(x) - \psi(y))^2\, d\nu \leq C. \end{aligned}$$

We set $W(t) := \frac{\int_{B_{3/2}} \psi^2 w(x,t)\, d\mu}{\int_{B_{3/2}} \psi^2\, d\mu}$. Then by the Poincaré type inequality obtained in Theorem 10.2.12, we reach

$$\int_{B_{3/2}} \psi^2 w_t\, d\mu + C \int_{B_{3/2}} (w(x,t) - W(t))^2 \psi(x)^2\, d\mu \le C.$$

Let $(t_1, t_2) \subset (-1, 1)$. Integrating in time the previous inequality, dividing by $\int_{B_{3/2}} \psi^2\, d\mu$ and noting that

$$\int_{B_{3/2}} \psi^2\, d\mu \le 2^{N-2\gamma} |B_1|_{d\mu},$$

one gets

$$\frac{W(t_2) - W(t_1)}{t_2 - t_1} + \frac{C_1}{|B_1|_{d\mu}(t_2 - t_1)} \int_{t_1}^{t_2} \int_{B_1} (w(x,t) - W(t))^2\, d\mu \le C_2.$$

We can suppose that W is differentiable. In the contrary case, it is possible to follow a discretization argument as in [155, Proposition 4.2]. By letting $t_2 \to t_1$, we get

$$W'(t) + \frac{C_1}{|B_1|_{d\mu}} \int_{B_1} (w(x,t) - W(t))^2\, d\mu \le C_2 \quad \text{a.e. in } (-1,1). \tag{13.3.22}$$

Define $\tilde{W}(t) = V(t) - C_2 t$ and $\tilde{w}(x,t) = v(x,t) - C_2 t$. From (13.3.22), it follows that

$$\tilde{W}'(t) + \frac{C_1}{|B_1|_{d\mu}} \int_{B_1} (\tilde{w}(x,t) - \tilde{W}(t))^2\, d\mu \le 0 \quad \text{a.e. in } (-1,1). \tag{13.3.23}$$

Note that from (13.3.23) we deduce $\tilde{W}'(t) \le 0$, and therefore calling $a(v) := W(0)$ we get

$$\tilde{W}(t) \le W(0) =: a(v) \quad \text{for all } t \in (0,1).$$

Let $t \in (0,1)$. Then if we define

$$G_m^+(t) := \{x \in B_1(0) : \tilde{w}(x,t) > m + a\},$$

for $x \in G_m^+(t)$, we have

$$\tilde{w}(x,t) - \tilde{W}(t) \ge m + a - \tilde{W}(t) > 0.$$

Thus

$$\tilde{W}'(t) + \frac{C_1}{|B_1|_{d\mu}} |G_m^+(t)|_{d\mu} (m + a - \tilde{W}(t))^2 \le 0.$$

Hence

$$\frac{-\tilde{W}'(t)}{(m+a-\tilde{W}(t))^2} \geq \frac{C_1}{|B_1|_{d\mu}}|G_m^+(t)|_{d\mu}.$$

Integrating the previous differential inequality for $t \in (0,1)$ and substituting $\tilde{w}$ by its value yields

$$|Q_+(1) \cap \{\log v + C_2 t < -m - a\}|_{d\mu \times dt} \leq \frac{C_1 |B_1|_{d\mu}}{m}.$$

Now

$$\begin{aligned} |Q_+(1) \cap \{\log v < -m - a\}|_{d\mu \times dt} &\leq \left|Q_+(1) \cap \left\{\log v + C_2 t < -\frac{m}{2} - a\right\}\right|_{d\mu \times dt} \\ &\quad + \left|Q_+(1) \cap \left\{C_2 t > \frac{m}{2}\right\}\right|_{d\mu \times dt} \\ &\leq \frac{C|B_1|_{d\mu}}{m}, \end{aligned}$$

which finishes the proof of (13.3.20). Estimate (13.3.21) follows using the same approach. □

We are now able to prove the weighted weak Harnack inequality.

Proof of Theorem 13.3.3. Roughly speaking, the key to prove this result will be to define appropriate functions and parameters so that we can deduce the result from Lemma 13.3.4. We divide the proof in two cases. Let $0 < r < 1$ such that $B_r \subset \Omega$.

1. Assume first that $s \geq \frac{1}{2}$.
   We set $\theta_1 = \theta_2 = \frac{1}{2}$ and define $U_1(r) = B_r \times (1 - r^{2s}, 1)$, $U_2(r) = B_r \times (-1, -1 + r^{2s})$. In the same way we consider $U_1(1) = Q_+(1)$ and $U_2(1) = Q_-(1)$.
   Let $w_1 := e^{-a} v^{-1}$, $w_2 := e^a v$, where $a = a(v)$ is defined in Lemma 13.3.10. From this result we obtain

$$|Q_+(1) \cap \{\log w_1 > m\}|_{d\mu \times dt} \leq \frac{C|B_1|_{d\mu}}{m}$$

and

$$|Q_-(1) \cap \{\log w_2 > m\}|_{d\mu \times dt} \leq \frac{C|B_1|_{d\mu}}{m}.$$

Using Lemma 13.3.6, it follows that $(w_1, U_1(r))$ satisfies the conditions of Lemma 13.3.4 with $p_0 = \infty$ and η any positive constant. Moreover, by Lemma 13.3.8, $(w_2, U_2(r))$ satisfies the same conditions with $p_0 = 1$ and $\eta = \frac{N}{N+2s} < 1$. Hence we conclude that

$$\sup_{U_1(\frac{1}{2})} w_1 \leq C \quad \text{and} \quad \|w_2\|_{L^1(U_2(\frac{1}{2}), d\mu)} \leq \tilde{C}.$$

Using these estimates and the definitions of w_1 and w_2, we get

$$\|v\|_{L^1(U_2(\frac{1}{2}),d\mu)} \leq C \inf_{U_1(\frac{1}{2})} v,$$

and the result follows in this case.

2. If $0 < s < \frac{1}{2}$, we have to change the domains by setting $\theta_1 = \theta_2 = (\frac{1}{2})^{2s}$ and $U_1(r) = B_{r^{\frac{1}{2s}}} \times (1-r,1)$, $U_2(r) = B_{r^{\frac{1}{2s}}} \times (-1,-1+r)$. Then the same arguments as in the previous case allow us to conclude the proof. □

From the weighted weak Harnack inequality, we immediately deduce the next corollary.

Corollary 13.3.11. *Let $\lambda \leq \Lambda_{N,2s}$. Assume that u is a nonnegative function such that $u \not\equiv 0$, $u \in L^1(\Omega\times(0,T))$ and $\frac{u}{|x|^{2s}} \in L^1(\Omega\times(0,T))$. If u satisfies $u_t+(-\Delta)^s u-\lambda\frac{u}{|x|^{2s}} \geq 0$ in the weak sense in $\Omega\times(0,T)$, then there exist $r_1 > 0$, $t_2 > t_1 > 0$ and a constant $C = C(N,r_1,t_1,t_2)$ such that for each cylinder $B_r(0)\times(t_1,t_2) \subset\subset \Omega\times(0,T)$, $0 < r < r_1$,*

$$u \geq C|x|^{-\frac{N-2s}{2}+\alpha} \quad \text{in } B_r(0)\times(t_1,t_2),$$

where α is the singularity of the homogeneous problem given in Lemma 9.4.3. In particular, for r conveniently small we can assume that $u > 1$ in $B_r(0)\times(t_1,t_2)$.

Finally, to end this section, we can establish a boundedness condition on the solutions of (13.3.6).

Proposition 13.3.12. *Let $v \in C([0,T);L^2(\Omega,d\mu)) \cap L^2(0,T;W_0^{s,\gamma}(\Omega))$ be a solution to (13.3.6) with $u_0 \in L^\infty(\Omega)$. If $f \in L^r(0,T;L^q(\Omega))$ with $r, q > 1$ and $\frac{1}{r}+\frac{N}{2qs} < 1$, then $v \in L^\infty(\Omega\times(0,T))$.*

Proof. The proof follows the same idea of the classical result by D. G. Aronson and J. Serrin in [39] as in Theorem 12.4.1. We test with $G_k(v) \in W_0^{s,\gamma}(\Omega)$ in (13.3.6), and defining

$$|||G_k(v)|||^2 := \|G_k(v)\|^2_{L^\infty(0,T;L^2(\Omega,d\mu))} + \|G_k(v)\|^2_{L^2(0,T;W_0^{s,\gamma}(\Omega))},$$

the result is obtained in a similar way as Theorem 12.4.1. See too [224, Theorem 29]. The presence of the singular term is handled in a straightforward way, so we skip the details. □

Remark 13.3.13. Apart from the integral version we proved, if the solution is bounded we can prove the strong Harnack inequality by classical arguments as in the elliptic case. We skip the details because we are not using this inequality in the applications.

Corollary 13.3.14. *If u is a solution to (13.1.1) with a sufficient regular datum f, then $u \leq C|x|^{-\frac{N-2s}{2}+\alpha}$ in $\Omega\times(0,T)$.*

13.4 The linear problem: dependence on the spectral parameter λ

Along this section we will study the problem

$$\begin{cases} u_t + (-\Delta)^s u = \lambda \frac{u}{|x|^{2s}} + g(x,t) & \text{in } \Omega \times (0,T), \\ u(x,t) = 0 & \text{in } (\mathbb{R}^N \setminus \Omega) \times [0,T), \\ u(x,0) = u_0(x) \gneq 0 & \text{if } x \in \Omega, \quad 0 \in \Omega, \end{cases} \tag{13.4.1}$$

where $g(x,t)$ is a nonnegative function. The goal will be to establish some necessary and sufficient conditions on g and u_0 in order to find solutions of this problem. These results correspond to the ones obtained by P. Baras and J. A. Goldstein for the heat equation in the presence of the inverse square potential (see [41]).

First, we deal with the necessary summability conditions on g and u_0.

Theorem 13.4.1. *Let $0 < \lambda \le \Lambda_{N,2s}$. Assume that $\tilde{u}$ is a positive weak supersolution to problem (13.4.1). Then g and u_0 must satisfy*

$$\int_{t_1}^{t_2} \int_{B_r(0)} |x|^{-\gamma} g \, dx \, dt < +\infty, \qquad \int_{B_r(0)} |x|^{-\gamma} u_0 \, dx < +\infty$$

for any cylinder $B_r(0) \times (t_1, t_2) \subset\subset \Omega \times (0,T)$, where γ was defined in (9.4.9).

Proof. Fix $\varepsilon > 0$. Let us consider φ_n, the positive solution to

$$\begin{cases} -(\varphi_n)_t + (-\Delta)^s \varphi_n = \lambda \frac{\varphi_{n-1}}{|x|^{2s} + \frac{1}{n}} + 1 & \text{in } \Omega \times (-\varepsilon, T), \\ \varphi_n = 0 & \text{in } (\mathbb{R}^N \setminus \Omega) \times (-\varepsilon, T], \\ \varphi_n(x,T) = 0 & \text{in } \Omega, \end{cases} \tag{13.4.2}$$

with

$$\begin{cases} -(\varphi_0)_t + (-\Delta)^s \varphi_0 = 1 & \text{in } \Omega \times (-\varepsilon, T), \\ \varphi_0 = 0 & \text{in } (\mathbb{R}^N \setminus \Omega) \times (-\varepsilon, T], \\ \varphi_0(x,T) = 0 & \text{in } \Omega. \end{cases} \tag{13.4.3}$$

Note that φ_0 is a strong solution in $\Omega \times (-\varepsilon, T]$, and therefore every φ_n is a strong solution too. Furthermore, $\varphi_{n-1} \le \varphi_n \le \varphi$, where φ is the positive solution to

$$\begin{cases} -\varphi_t + (-\Delta)^s \varphi - \lambda \frac{\varphi}{|x|^{2s}} = 1 & \text{in } \Omega \times (-\varepsilon, T), \\ \varphi = 0 & \text{in } (\mathbb{R}^N \setminus \Omega) \times (-\varepsilon, T], \\ \varphi(x,T) = 0 & \text{in } \Omega. \end{cases}$$

As a consequence of Corollary 13.3.11, for any cylinder $C_{r_1,t_1,t_2} := B_r(0) \times (t_1,t_2) \subset\subset \Omega \times (-\varepsilon, T)$, $0 < r < r_1$, there exists a constant $A = A(N, s, C_{r_1,t_1,t_2})$ such that

$$\varphi(x,t) \ge \frac{A}{|x|^\gamma}, \quad (x,t) \in C_{r_1,t_1,t_2}, \quad \gamma = \frac{N-2s}{2} - \alpha. \tag{13.4.4}$$

Since φ_n is regular and bounded we can use it as a test function in (13.4.1); hence we get

$$\begin{aligned}
&\int_0^T\int_\Omega g\varphi_n\,dx\,dt + \int_\Omega u_0\varphi_n(x,0)\,dx \\
&\le -\int_0^T\int_\Omega \tilde{u}(\varphi_n)_t\,dx\,dt + \int_0^T\int_{\mathbb{R}^N} \tilde{u}(-\Delta)^s\varphi_n\,dx\,dt - \lambda\int_0^T\int_\Omega \frac{\tilde{u}\varphi_n}{|x|^{2s}}\,dx\,dt \\
&\le -\int_0^T\int_\Omega \tilde{u}(\varphi_n)_t\,dx\,dt + \int_0^T\int_\Omega \tilde{u}(-\Delta)^s\varphi_n\,dx\,dt - \lambda\int_0^T\int_\Omega \frac{\tilde{u}\varphi_{n-1}}{|x|^{2s}+\frac{1}{n}}\,dx\,dt \\
&= \int_0^T\int_\Omega \tilde{u}\,dx\,dt = C < +\infty,
\end{aligned}$$

where we have used the fact that $\tilde{u}$ is a weak supersolution (see Definition 13.2.1). Since both integrals on the left hand side are positive, in particular each one is uniformly bounded. Thus, $\{g\varphi_n\}$ is an increasing sequence uniformly bounded in $L^1(\Omega\times(0,T))$, and applying the monotone convergence theorem and (13.4.4) we get

$$\begin{aligned}
C\int_{t_1}^{t_2}\int_{B_r(0)} |x|^{-\frac{N-2s}{2}+\alpha} g\,dx\,dt &\le \int_{t_1}^{t_2}\int_{B_r(0)} g\varphi\,dx\,dt \\
&\le \int_0^T\int_\Omega g\varphi\,dx\,dt = \lim_{n\to\infty}\int_0^T\int_\Omega g\varphi_n\,dx\,dt < +\infty.
\end{aligned}$$

Likewise, $\{u_0\varphi_n(x,0)\}$ is also an increasing sequence, uniformly bounded in $L^1(\Omega)$, and thus, choosing t_1 and t_2 so that $0\in(t_1,t_2)\subsetneq(-\varepsilon,T)$ as above, we conclude

$$\tilde{C}\int_{B_r(0)} |x|^{-\frac{N-2s}{2}+\alpha} u_0(x)\,dx \le \int_\Omega u_0\varphi(x,0)\,dx < +\infty. \qquad \square$$

Conversely, we would like to find the optimal summability conditions on g and u_0 to prove existence of a weak solution. In this direction, note that if $g\in L^2(0,T;H^{-s}(\Omega))$ and $u_0\in L^2(\Omega)$, by Remark 13.2.5 we can ensure the existence of an energy solution of problem (13.4.1) when $\lambda<\Lambda_{N,2s}$, and in $H(\Omega)$ (see Remark 10.2.2 and (13.2.5)) for $\lambda=\Lambda_{N,2s}$. A sharper result, for a more general class of data, is the following.

Theorem 13.4.2. *Assume that $0<\lambda\le\Lambda_{N,2s}$ and that g and u_0 satisfy*

$$\int_\Omega \frac{u_0}{|x|^\gamma}\,dx < +\infty, \qquad \int_0^T\int_\Omega \frac{g}{|x|^\gamma}\,dx\,dt < +\infty,$$

where γ is defined in (9.4.9). Then problem (13.4.1) has a positive weak solution.

Proof. Consider the approximated problems

$$\begin{cases} u_{nt} + (-\Delta)^s u_n = \lambda \frac{u_{n-1}}{|x|^{2s}+\frac{1}{n}} + g_n & \text{in } \Omega \times (0,T), \\ u_n(x,t) > 0 & \text{in } \Omega \times (0,T), \\ u_n(x,t) = 0 & \text{in } (\mathbb{R}^N \setminus \Omega) \times [0,T), \\ u_n(x,0) = T_n(u_0(x)) & \text{if } x \in \Omega, \end{cases} \tag{13.4.5}$$

where

$$\begin{cases} u_{0t} + (-\Delta)^s u_0 = g_1 & \text{in } \Omega \times (0,T), \\ u_0(x,t) > 0 & \text{in } \Omega \times (0,T), \\ u_0(x,t) = 0 & \text{in } (\mathbb{R}^N \setminus \Omega) \times [0,T), \\ u_0(x,0) = T_1(u_0(x)) & \text{if } x \in \Omega, \end{cases} \tag{13.4.6}$$

with $g_n = T_n(g)$ and as customary,

$$T_n(g) = \begin{cases} g & \text{if } |g| \le n, \\ \frac{g}{|g|} n & \text{if } |g| > n. \end{cases}$$

By Lemma 13.2.6, it follows that $u_0 \le u_1 \le \cdots \le u_{n-1} \le u_n$ in $\mathbb{R}^N \times (0,T)$. Note that since the right hand sides of these problems are bounded, every u_n is actually an energy solution.

Consider φ, the solution of the problem

$$\begin{cases} -\varphi_t + (-\Delta)^s \varphi - \lambda \frac{\varphi}{|x|^{2s}} = 1 & \text{in } \Omega \times (-\varepsilon, T), \\ \varphi > 0 & \text{in } \Omega \times (-\varepsilon, T), \\ \varphi = 0 & \text{on } (\mathbb{R}^N \setminus \Omega) \times (-\varepsilon, T], \\ \varphi(x,T) = C & \text{in } \Omega, \end{cases} \tag{13.4.7}$$

where $C > 0$. As a consequence of the weak Harnack inequality, Theorem 13.3.3 and Proposition 13.3.12, for any cylinder $B_r(0) \times [t_1, t_2] \subset \Omega \times (-\varepsilon, T)$ we find $c_1, c_2 > 0$ such that

$$\frac{c_1}{|x|^\gamma} \le \varphi(x,t) \le \frac{c_2}{|x|^\gamma}. \tag{13.4.8}$$

Since φ also belongs to $L^2(0,T; H_0^s(\Omega))$, we can use it as a test function in (13.4.5). Thus,

$$\int_0^T \int_\Omega (u_n)_t \varphi \, dx \, dt + \int_0^T \int_{\mathbb{R}^N} u_n (-\Delta)^s \varphi \, dx \, dt = \lambda \int_0^T \int_\Omega \frac{u_{n-1}\varphi}{|x|^{2s} + \frac{1}{n}} \, dx \, dt + \int_0^T \int_\Omega g_n \varphi \, dx \, dt$$

$$\leq \lambda \int_0^T \int_\Omega \frac{u_n \varphi}{|x|^{2s}}\,dx\,dt + \int_0^T \int_\Omega g_n \varphi\,dx\,dt.$$

Integrating in time and applying (13.4.7) and (13.4.8), we conclude that

$$\begin{aligned} C \int_\Omega u_n(x,T)\,dx + \int_0^T \int_\Omega u_n\,dx\,dt &\leq \int_0^T \int_\Omega g_n \varphi\,dx\,dt + \int_\Omega T_n(u_0(x))\varphi(x,0)\,dx \\ &\leq \int_0^T \int_\Omega g\varphi\,dx\,dt + \int_\Omega u_0(x)\varphi(x,0)\,dx \\ &\leq C \int_0^T \int_\Omega \frac{g}{|x|^\gamma}\,dx\,dt + C \int_\Omega \frac{u_0(x)}{|x|^\gamma}\,dx < +\infty, \end{aligned}$$

by hypothesis. Hence, since the sequence $\{u_n\}_{n\in\mathbb{N}}$ is increasing, we can define $u := \lim_{n\to\infty} u_n$ and conclude that $u \in L^1(\Omega\times(0,T))$ by applying the monotone convergence theorem.

Note that using the same computations as above and integrating in $\Omega \times [0,t]$ with $t \leq T$, by considering the estimates on $\{u_n\}_{n\in\mathbb{N}}$, we reach

$$\sup_{t\in[0,T]} \int_\Omega u_n(x,t)\,dx + \int_0^T \int_\Omega u_n\,dx\,dt \leq C \quad \text{for all } n. \tag{13.4.9}$$

Fix $T_1 > T$ and define $\tilde{\varphi}$ as the unique solution to the problem

$$\begin{cases} -\tilde{\varphi}_t + (-\Delta)^s \tilde{\varphi} = 1 & \text{in } \Omega \times (-\varepsilon, T_1), \\ \tilde{\varphi} > 0 & \text{in } \Omega \times (-\varepsilon, T_1), \\ \tilde{\varphi} = 0 & \text{on } (\mathbb{R}^N \setminus \Omega) \times (-\varepsilon, T_1], \\ \varphi(x, T_1) = 0 & \text{in } \Omega. \end{cases} \tag{13.4.10}$$

It is clear that $\tilde{\varphi} \in L^\infty(\Omega\times(-\varepsilon,T_1))$ and $\tilde{\varphi}(x,t) \geq \bar{C} > 0$ for all $(x,t) \in B_r(0)\times[0,T]$, where $B_r(0) \subset\subset \Omega$. Now, using $\tilde{\varphi}$ as a test function in (13.4.5) and integrating in $\Omega \times (0,T)$, it follows that

$$\int_\Omega u_n(x,T)\tilde{\varphi}(x,T)\,dx\,dt + \int_0^T \int_\Omega u_n\,dx\,dt \geq \lambda \int_0^T \int_\Omega \frac{u_{n-1}\tilde{\varphi}}{|x|^{2s} + \frac{1}{n}}\,dx\,dt.$$

Thus

$$\lambda \int_0^T \int_\Omega \frac{u_{n-1}\tilde{\varphi}}{|x|^{2s} + \frac{1}{n}}\,dx\,dt \leq C \sup_{\{t\in[0,T]\}} \int_\Omega u_n(x,t)\,dx + \int_0^T \int_\Omega u_n\,dx\,dt \leq C \quad \text{for all } n.$$

Hence

$$\int_0^T\int_\Omega \frac{u_{n-1}}{|x|^{2s}+\frac{1}{n}}\,dx\,dt = \int_0^T\int_{B_r(0)} \frac{u_{n-1}}{|x|^{2s}+\frac{1}{n}}\,dx\,dt + \int_0^T\int_{\Omega\setminus B_r(0)} \frac{u_{n-1}}{|x|^{2s}+\frac{1}{n}}\,dx\,dt$$
$$\leq C\int_0^T\int_\Omega \frac{u_{n-1}\tilde{\varphi}}{|x|^{2s}+\frac{1}{n}}\,dx\,dt + C\int_0^T\int_\Omega u_{n-1}\,dx\,dt \leq C.$$

Therefore, by the monotone convergence theorem we find

$$\frac{u_{n-1}}{|x|^{2s}+\frac{1}{n}} + g_n \uparrow \frac{u}{|x|^{2s}} + g \quad \text{strongly in } L^1(\Omega\times(0,T)).$$

To conclude that u is a weak solution to problem (13.4.1), it remains to check that $u \in \mathcal{C}([0,T);L^1(\Omega))$. We claim that $\{u_n\}_{n\in\mathbb{N}}$ is a Cauchy sequence in $\mathcal{C}([0,T];L^1(\Omega))$, and hence the result follows. In order to prove this claim, we closely follow the arguments in [267].

For $n,m \in \mathbb{N}$ such that $n \geq m$, denote $u_{n,m} := u_n - u_m$ and $g_{n,m} := g_n - g_m$. Clearly, $u_{n,m}, g_{n,m} \geq 0$. We set

$$C_{n,m} := \lambda\int_0^t\int_\Omega \frac{u_{n,m}}{|x|^{2s}}\,dx\,d\tau + \int_0^t\int_\Omega g_{n,m}T_1(u_{n,m})\,dx\,d\tau,$$

and then $C_{n,m} \to 0$ as $n,m \to \infty$.

By the definition of the approximated problems in (13.4.5) and the linearity of the operator, for $t \leq T$,

$$\int_0^t\int_\Omega (u_{n,m})_t T_1(u_{n,m})\,dx\,d\tau + \int_0^t\int_\Omega (-\Delta)^s(u_{n,m})T_1(u_{n,m})\,dx\,d\tau \leq C_{n,m}.$$

Since $u_{n,m} \in L^2(0,T;H_0^s(\Omega))$, it follows that (see Chapter 12 for details)

$$\int_0^t\int_\Omega (-\Delta)^s(u_{n,m})T_1(u_{n,m})\,dx\,d\tau \geq \int_0^t \|T_1(u_{n,m})\|^2_{H_0^s(\Omega)}\,d\tau \geq 0,$$

and therefore,

$$\int_0^t\int_\Omega (u_{n,m})_t T_1(u_{n,m})\,dx\,d\tau \leq C_{n,m}.$$

Let us define

$$\Psi(s) := \int_0^s T_1(\sigma)\,d\sigma.$$

Since $u_n \in \mathcal{C}([0,T];L^2(\Omega))$, we have

$$\int_0^t \int_\Omega (u_{n,m})_t T_1(u_{n,m})\,dx\,d\tau = \int_\Omega (\Psi(u_{n,m})(t) - \Psi(u_{n,m})(0))\,dx.$$

Thus

$$\int_\Omega \Psi(u_{n,m})(t)\,dx \le C_{n,m} + \int_\Omega \Psi(u_{n,m})(0)\,dx.$$

Taking into account that $\Psi(u_{n,m})(0) = \Psi(T_n(u_0) - T_m(u_0))$ and by noting that $\Psi(s) \le |s|$ and $T_n(u_0) - T_m(u_0) \to 0$ strongly in $L^1(\Omega)$ as $n, m \to \infty$, we obtain

$$\int_\Omega \Psi(u_{n,m})(t)\,dx \to 0 \text{ as } n, m \to \infty, \quad \text{uniformly in } t.$$

Now, since

$$\int_{|u_{n,m}|<1} \frac{|u_{n,m}|^2(t)}{2}\,dx + \int_{|u_{n,m}|>1} |u_{n,m}|(t)\,dx \le \int_\Omega \Psi(u_{n,m})(t)\,dx,$$

we conclude that $u_{n,m} \to 0$ uniformly in t.

Thus $\{u_n\}_{n\in\mathbb{N}}$ is a Cauchy sequence in $\mathcal{C}([0,T];L^1(\Omega))$ and passing to the limit in the weak formulation of the approximated problems, one obtains that u is a positive weak solution of problem (13.4.1) in $\Omega \times (0,T)$. □

Next, we see that $\Lambda_{N,2s}$ provides a real obstruction on λ.

Proposition 13.4.3. *If $\lambda > \Lambda_{N,2s}$, problem (13.4.1) has no positive weak supersolution.*

Proof. Consider $\tilde{u}$ as a weak supersolution to the problem

$$\begin{cases} u_t + (-\Delta)^s u - \Lambda_{N,2s}\frac{u}{|x|^{2s}} = (\lambda - \Lambda_{N,2s})\frac{u}{|x|^{2s}} + g & \text{in } \Omega \times (0,T), \\ u(x,t) > 0 & \text{in } \Omega \times (0,T), \\ u(x,t) = 0 & \text{in } (\mathbb{R}^N \setminus \Omega) \times [0,T). \end{cases} \tag{13.4.11}$$

Since in the left hand side the constant is $\Lambda_{N,2s}$, we are in the case $\alpha = 0$, and by Theorem 13.4.1, necessarily

$$\left((\lambda - \Lambda_{N,2s})\frac{\tilde{u}}{|x|^{2s}} + g\right)|x|^{-\frac{N-2s}{2}} \in L^1(B_r(0) \times (t_1,t_2)),$$

for any $B_r(0) \times (t_1,t_2) \subset\subset \Omega \times (0,T)$ small enough. In particular, this implies

$$(\lambda - \Lambda_{N,2s})\frac{\tilde{u}}{|x|^{2s}}|x|^{-\frac{N-2s}{2}} \in L^1(B_r(0) \times (t_1,t_2)),$$

and hence, applying Corollary 13.3.11 again,

$$(\lambda - \Lambda_{N,2s})|x|^{-N} \in L^1(B_r(0) \times (t_1, t_2)),$$

which is a contradiction. Therefore, there does not exist a positive supersolution if $\lambda > \Lambda_{N,2s}$. □

Remark 13.4.4. The previous nonexistence result implies that for $\lambda > \Lambda_{N,2s}$ an instantaneous and complete blow-up phenomenon occurs. The proof is a simple adaptation of Theorem 13.5.4, where this result is proved for a more involved semilinear problem.

Furthermore, we can state a nonexistence result that shows the optimality of the power $p = 1$ in the singular term $\frac{u^p}{|x|^{2s}}$. The proof for this nonlocal problem closely follows the classical case, due to H. Brezis and X. Cabré (see [87]).

Theorem 13.4.5. *Let $p > 1$, and let $u \geq 0$ satisfy*

$$u_t + (-\Delta)^s u \geq \frac{u^p}{|x|^{2s}} \quad \text{in } \Omega \times (0, T),$$

in the weak sense. Then $u \equiv 0$.

Proof. Consider a cylinder $B_\tau(0) \times (t_1, t_2)$. If $u \gneq 0$, by the maximum principle (Theorem 13.2.11), we know that there exists $\varepsilon > 0$ so that

$$u \geq \varepsilon \quad \text{in } B_\tau(0) \times (t_1, t_2).$$

Let us define

$$\phi(s) := \begin{cases} \frac{1}{(p-1)\varepsilon^{p-1}} - \frac{1}{(p-1)s^{p-1}} & \text{if } s \geq \varepsilon, \\ \frac{1}{\varepsilon^p}(s - \varepsilon) & \text{if } s < \varepsilon. \end{cases}$$

Note that $0 \leq \phi < +\infty$ in $[\varepsilon, +\infty)$, $\phi(\varepsilon) = 0$, $\phi'(\varepsilon) = 0$, and ϕ is a C^1 function satisfying $\phi'(s) = \frac{1}{s^p}$ for $s \geq \varepsilon$. Moreover, since ϕ is concave, it follows that $(-\Delta)^s(\phi(u)) \geq \phi'(u)(-\Delta)^s u$ and thus,

$$(\phi(u))_t + (-\Delta)^s(\phi(u)) \geq \phi'(u)(u_t + (-\Delta)^s u) \geq \frac{1}{|x|^{2s}} \quad \text{in } B_\tau(0) \times (t_1, t_2),$$

with $u \geq \varepsilon$.

Without loss of generality we can assume that $\tau = 1$. Define $w(x, t) := (t - t_1)\vartheta(x)$, where

$$\vartheta(x) := \begin{cases} \log(\frac{1}{|x|}) & \text{if } |x| < 1, \\ 0 & \text{if } |x| \geq 1. \end{cases}$$

Then $w(x, t_1) = 0$ in $B_1(0)$ and $w(x, t) = 0$ in $\mathbb{R}^N \backslash B_1(0)$.

We claim that

$$w_t + (-\Delta)^s w \leq \frac{C}{|x|^{2s}} \quad \text{in } B_1(0) \times (t_1, t_2),$$

where $C \equiv C(t_1, t_2) > 0$. Indeed, we have

$$w_t + (-\Delta)^s w = \vartheta(x) + (t - t_1)(-\Delta)^s \vartheta.$$

It is clear that $\vartheta(x) \leq \frac{C_1}{|x|^{2s}}$ in $B_1(0) \times (t_1, t_2)$, and hence to prove the claim we have to show that

$$(-\Delta)^s \vartheta(x) \leq \frac{C_2}{|x|^{2s}} \quad \text{for all } x \in B_1(0).$$

In fact,

$$\begin{aligned}
(-\Delta)^s \vartheta(x) &= a_{N,s} \int\limits_{\mathbb{R}^N} \frac{(\vartheta(x) - \vartheta(y))}{|x - y|^{N+2s}}\, dy \\
&= a_{N,s} \int\limits_{\{|y|<1\}} \frac{(\vartheta(x) - \vartheta(y))}{|x - y|^{N+2s}}\, dy + a_{N,s} \int\limits_{\{|y|>1\}} \frac{\vartheta(x)}{|x - y|^{N+2s}}\, dy \\
&= I_1(x) + I_2(x).
\end{aligned}$$

We follow the arguments in [158] to estimate the integrals above. By setting $r := |x|$ and $\rho := |y|$, we have $x = rx'$, $y = \rho y'$, where $|x'| = |y'| = 1$. We avoid the constant $a_{N,s}$ in the next computations. Thus,

$$I_1(x) = \int_0^1 \log\Big(\frac{\rho}{r}\Big) \rho^{N-1} \Big(\int\limits_{|y'|=1} \frac{dH^{N-1}(y')}{|rx' - \rho y'|^{N+2s}} \Big)\, d\rho.$$

Calling $\sigma := \frac{\rho}{r}$,

$$I_1(x) = \frac{D_1(|x|)}{|x|^{2s}},$$

where

$$D_1(r) = \int_0^{\frac{1}{r}} \log(\sigma) \sigma^{N-1} K(\sigma)\, d\sigma,$$

and

$$K(\sigma) := \int\limits_{|y'|=1} \frac{dH^{N-1}(y')}{|x' - \sigma y'|^{N+2s}} = 2 \frac{\pi^{\frac{N-1}{2}}}{\Gamma(\frac{N-1}{2})} \int_0^{\pi} \frac{\sin^{N-2}(\eta)}{(1 - 2\sigma \cos(\eta) + \sigma^2)^{\frac{N+2s}{2}}}\, d\eta.$$

In the same way we have

$$I_2(x) = \frac{D_2(|x|)}{|x|^{2s}}, \quad \text{where } D_2(r) = -\log(r)\int_{\frac{1}{r}}^{+\infty} \sigma^{N-1}K(\sigma)\,d\sigma.$$

Combining the estimates above, we get

$$(-\Delta)^s \vartheta(x) = \frac{D(|x|)}{|x|^{2s}}$$

with

$$D(r) := D_1(r) + D_2(r) = \int_0^{\frac{1}{r}} \log(\sigma)\sigma^{N-1}K(\sigma)\,d\sigma - \log(r)\int_{\frac{1}{r}}^{+\infty} \sigma^{N-1}K(\sigma)\,d\sigma.$$

Note that $K(\sigma) \le C|1-\sigma|^{-1-2s}$ as $\sigma \to 1$ and $K(\frac{1}{\sigma}) = \sigma^{N+2s}K(\sigma)$ for all $\sigma > 0$.

If $s \le \frac{1}{2}$, then using the behavior of K at $+\infty$ we can easily prove that $|D_1(r)| + |D_2(r)| \le C$ for all $r \le 1$. To study the general case we need to do some sharp computations. Since

$$D(r) \le \int_0^{+\infty} \log(\sigma)\sigma^{N-1}K(\sigma)\,d\sigma = \bar{D},$$

to finish we have to show that $|\bar{D}| < \infty$. Note that

$$\bar{D} = \int_0^1 \log(\sigma)\sigma^{N-1}K(\sigma)\,d\sigma + \int_1^{+\infty} \log(\sigma)\sigma^{N-1}K(\sigma)\,d\sigma.$$

Putting $\theta := \frac{1}{\sigma}$ in the first integral and using the fact that $K(\frac{1}{\theta}) = \theta^{N+2s}K(\theta)$, it follows that

$$\bar{D} = \int_1^{+\infty} K(\sigma)\log(\sigma)(\sigma^{N-1} - \sigma^{2s-1})\,d\sigma.$$

Now, due to the behavior of K near 1 and ∞ we obtain $0 < \bar{D} < \infty$. Thus

$$(-\Delta)^s \vartheta(x) \le \frac{\bar{D}}{|x|^{2s}}.$$

Hence

$$w_t + (-\Delta)^s w \le \frac{C}{|x|^{2s}} \quad \text{in } B_1(0) \times (t_1, t_2),$$

where $C = C_1 + (t_2 - t_1)\bar{D}$ and the claim follows.

Fixing $\varepsilon > 0$ such that $\varepsilon C \ll 1$, we obtain

$$(\phi(u) - \varepsilon w)_t + (-\Delta)^s(\phi(u) - \varepsilon w) \geq 0 \quad \text{in } B_1(0) \times (t_1, t_2)$$

in the weak sense. Since $(\phi(u) - \varepsilon w)(x, t_1) \geq 0$ in $B_1(0)$ and $(\phi(u) - \varepsilon w)(x, t_1) \geq 0$ in $\mathbb{R}^N \backslash B_1(0) \times (t_1, t_2)$, the comparison principle implies that $\phi(u) - \varepsilon w \geq 0$ in $B_\tau(0) \times (t_1, t_2)$. Now, w is unbounded and this reaches a contradiction with the fact that ϕ is, however, a bounded function. The proof is finished. □

Thus, as a straightforward consequence we obtain the following result.

Corollary 13.4.6. *Let $g \in L^1(\Omega \times (0, T))$, $g \geq 0$, $\lambda > 0$ and $p > 1$. Therefore, the problem*

$$\begin{cases} u_t + (-\Delta)^s u = \lambda \frac{u^p}{|x|^{2s}} + g & \text{in } \Omega \times (0, T), \\ u(x, t) = 0 & \text{in } (\mathbb{R}^N \setminus \Omega) \times [0, T), \\ u(x, 0) = u_0(x) \gneq 0 & \text{if } x \in \Omega \end{cases}$$

has no positive weak solution.

13.5 Existence and nonexistence results for a semilinear problem

The goal of this section is to study how the addition of a semilinear term of the form u^p, with $p > 1$, interferes with the solvability of the previous problems. As in the classical heat equation (see [19]), the relevant feature is that, as in the elliptic case, for every $0 < \lambda < \Lambda_{N,2s}$ there exists a threshold for the existence, $p_+(\lambda, s)$, that depends on the spectral parameter. Indeed, we will consider the problem

$$\begin{cases} u_t + (-\Delta)^s u = \lambda \frac{u}{|x|^{2s}} + u^p + f & \text{in } \Omega \times (0, T), \\ u(x, t) > 0 & \text{in } \Omega \times (0, T), \\ u(x, t) = 0 & \text{in } (\mathbb{R}^N \setminus \Omega) \times [0, T), \\ u(x, 0) = u_0(x) & \text{if } x \in \Omega, \end{cases} \tag{13.5.1}$$

with $p > 1$, $0 < \lambda < \Lambda_{N,2s}$, $f(x, t) \geq 0$ and $u_0(x) \geq 0$. By weak or energy solutions of this problem, we mean solutions in the sense of Definition 13.2.1 and Definition 13.2.2 by fixing $F = \lambda \frac{u}{|x|^{2s}} + u^p + cf$.

We will prove that there exists such critical exponent $p_+(\lambda, s)$ so that one can prove existence of solution for problem (13.5.1) when $1 < p < p_+(\lambda, s)$, and nonexistence for $p > p_+(\lambda, s)$. Following the same ideas as in Chapter 11, one can expect $p_+(\lambda, s)$ to depend on s and λ, and in particular to satisfy

$$p_+(\lambda, s) = 1 + \frac{2s}{\frac{N-2s}{2} - \alpha} = 1 + \frac{2s}{\gamma}.$$

Note that if $\lambda = \Lambda_{N,2s}$, namely, $\alpha = 0$, then $p_+(\lambda, s) = 2_s^* - 1$, and if $\lambda = 0$, i. e., $\alpha = \frac{N-2s}{2}$, then $p_+(\lambda, s) = \infty$.

We will need some auxiliary results that allow us to build a solution whenever we have a supersolution. To prove existence of a weak solution to (13.5.1) with L^1-data from a weak supersolution, we will consider the *solution obtained as limit of solutions of approximated problems* (see for instance [129] in the local parabolic operators case).

Lemma 13.5.1. *If $\bar{u} \in C([0,T]; L^1(\Omega))$ is a weak positive supersolution to the equation in* (13.5.1) *with $\lambda \le \Lambda_{N,2s}$ and $f \in L^1(\Omega \times (0,T))$, then there exists a positive minimal weak solution to problem* (13.5.1) *obtained as limit of solutions of approximated problems.*

Proof. If $\bar{u}$ is a positive supersolution to (13.5.1) with $\lambda \le \Lambda_{N,2s}$, we construct a sequence $\{u_n\}_{n\in\mathbb{N}}$ starting with

$$\begin{cases} u_{0t} + (-\Delta)^s u_0 = T_1(f) & \text{in } \Omega \times (0,T), \\ u_0(x,t) > 0 & \text{in } \Omega \times (0,T), \\ u_0(x,t) = 0 & \text{in } (\mathbb{R}^N \setminus \Omega) \times [0,T), \\ u_0(x,0) = T_1(u_0(x)) & \text{if } x \in \Omega. \end{cases} \tag{13.5.2}$$

By the weak comparison principle (Lemma 13.2.6), it follows that $u_0 \le \bar{u}$ in $\mathbb{R}^N \times (0,T)$. Define in a recurrence way

$$\begin{cases} u_{nt} + (-\Delta)^s u_n = \lambda \frac{u_{n-1}}{|x|^{2s} + \frac{1}{n}} + u_{n-1}^p + T_n(f) & \text{in } \Omega \times (0,T), \\ u_n(x,t) > 0 & \text{in } \Omega \times (0,T), \\ u_n(x,t) = 0 & \text{in } (\mathbb{R}^N \setminus \Omega) \times [0,T), \\ u_n(x,0) = T_n(u_0(x)) & \text{if } x \in \Omega. \end{cases} \tag{13.5.3}$$

We have $\{u_n\}_{n\in\mathbb{N}} \subset C([0,T]; L^1(\Omega)) \cap L^p([0,T]; L^p(\Omega))$ (see [224]). As before, it follows that $u_0 \le \cdots \le u_{n-1} \le u_n \le \bar{u}$ in $\mathbb{R}^N \times (0,T)$, so we obtain the pointwise limit $u := \lim u_n$ that verifies $u \le \bar{u}$ and

$$\begin{cases} u_t + (-\Delta)^s u = \lambda \frac{u}{|x|^{2s}} + u^p + f & \text{in } \Omega \times (0,T), \\ u(x,t) > 0 & \text{in } \Omega \times (0,T), \\ u(x,t) = 0 & \text{in } (\mathbb{R}^N \setminus \Omega) \times [0,T), \\ u(x,0) = u_0(x) & \text{if } x \in \Omega \end{cases} \tag{13.5.4}$$

in the weak sense. The fact that $u \in C([0,T]; L^1(\Omega))$ follows as in the proof of Theorem 13.4.2. Finally, we prove the minimality of the solution found above. Assume that v is another positive weak solution of problem (13.5.4). Then, in particular it is a positive weak supersolution. By the weak comparison principle we can prove by recurrence that the approximated solution verifies that $u_n \le v$ in any $(x,t) \in \Omega \times (0,T)$ and for all $n \in \mathbb{N}$. Therefore $u \le v$. □

Likewise, if the supersolution belongs to the energy space, the minimal solution obtained by approximations as above is also an energy solution.

Lemma 13.5.2. *If $\bar{u} \in L^2(0,T;H_0^s(\Omega))$ with $\bar{u}_t \in L^2(0,T;H^{-s}(\Omega))$ is a positive finite energy supersolution to* (13.5.1) *with $\lambda \le \Lambda_{N,2s}$ and $f \in L^2(0,T;H^{-s}(\Omega))$, then there exists a positive minimal energy solution to problem* (13.5.1) *obtained as limit of solutions of the approximated problems.*

Proof. Proceeding as in the proof of Lemma 13.5.1 and using the comparison principle for energy solutions (Lemma 13.2.8), we can build a sequence $\{u_n\}_{n\in\mathbb{N}}$ of energy solutions of the approximated problems (13.5.3), so that

$$u_0 \le u_1 \le \cdots \le u_n \le \cdots \le \bar{u} \quad \text{in } \mathbb{R}^N \times (0,T).$$

Hence, by the monotone convergence theorem we can define $u := \lim_{n\to\infty} u_n \le \bar{u}$. Moreover, applying the energy formulation of u_n,

$$\begin{aligned}
\|u_n\|^2_{L^2(0,T;H_0^s(\Omega))} &= \int_0^T \|u_n(\cdot,t)\|^2_{H_0^s(\Omega)}\,dt = \int_0^T\int_Q \frac{(u_n(x,t)-u_n(y,t))^2}{|x-y|^{N+2s}}\,dx\,dy\,dt \\
&= \frac{1}{a_{N,s}}\Big\{\int_0^T\int_\Omega \Big(\lambda\frac{u_n^2}{|x|^{2s}} + u_n^{p+1} + T_n(f)u_n\Big)\,dx\,dt - \int_0^T\int_\Omega (u_n)_t u_n\,dx\,dt\Big\} \\
&= \frac{1}{a_{N,s}}\Big\{\int_0^T\int_\Omega \Big(\lambda\frac{u_n^2}{|x|^{2s}} + u_n^{p+1} + T_n(f)u_n\Big)\,dx\,dt - \frac12\int_\Omega u_n(x,T)^2\,dx \\
&\quad + \frac12\int_\Omega u_n(x,0)^2\,dx\Big\} \\
&\le \frac{1}{a_{N,s}}\Big\{\int_0^T\int_\Omega \Big(\lambda\frac{\bar{u}^2}{|x|^{2s}} + \bar{u}^{p+1} + f\bar{u}\Big)\,dx\,dt + \frac12\int_\Omega \bar{u}(x,0)^2\,dx\Big\} \\
&\le C.
\end{aligned}$$

Thus, up to a subsequence, we know that $u_n \rightharpoonup u$ in $L^2(0,T;H_0^s(\Omega))$. Since for every $0 \le t \le T$,

$$\begin{aligned}
\|(u_n)_t\|_{H^{-s}(\Omega)} &= \sup_{\|\varphi\|_{H_0^s(\Omega)}\le 1}\Big|\int_\Omega (u_n)_t\varphi\,dx\Big| \\
&\le a_{N,s}\sup_{\|\varphi\|_{H_0^s(\Omega)}\le 1}\Big|\iint_Q \frac{(u_n(x)-u_n(y))(\varphi(x)-\varphi(y))}{|x-y|^{N+2s}}\,dx\,dy\Big| \\
&\quad + \sup_{\|\varphi\|_{H_0^s(\Omega)}\le 1}\Big\{\Big|\int_\Omega u_n^p\varphi\,dx + \lambda\int_\Omega \frac{u_n\varphi}{|x|^{2s}}\,dx + \int_\Omega T_n(f)\varphi\Big\}
\end{aligned}$$

$$\leq \sup_{\|\varphi\|_{H_0^s(\Omega)}\leq 1} \left\{ \|u_n\|_{H_0^s(\Omega)}\|\varphi\|_{H_0^s(\Omega)} + \int_\Omega \bar{u}^p \varphi\,dx + \lambda \int_\Omega \frac{\bar{u}\varphi}{|x|^{2s}}\,dx + \int_\Omega f\varphi\,dx \right\}$$

$$\leq C(\|u_n\|_{H_0^s(\Omega)} + 1 + \|f\|_{L^2(\Omega)}).$$

Hence,

$$\int_0^T \|(u_n)_t\|^2_{H^{-s}(\Omega)}\,dt \leq C(\|u_n\|^2_{L^2(0,T;H_0^s(\Omega))} + 1 + \|f\|^2_{L^2(0,T;L^2(\Omega))}) \leq C,$$

and therefore, up to a subsequence, $(u_n)_t \rightharpoonup u_t$ in $L^2(0,T;H^{-s}(\Omega))$, and we can pass to the limit to conclude that u is a finite energy solution to (13.5.1).

The minimality follows as in the case of weak solutions. □

13.5.1 Nonexistence results for $p > p_+(\lambda, s)$, instantaneous and complete blow-up

Assume first that p is greater than the threshold exponent $p_+(\lambda, s)$. Then we can formulate the nonexistence result as follows.

Theorem 13.5.3. *Let $\lambda \leq \Lambda_{N,2s}$ and suppose that $(f, u_0) \in L^1(\Omega \times (0,T)) \times L^1(\Omega)$. If $p > p_+(\lambda, s)$, then problem* (13.5.1) *has no positive weak supersolution. In the case where $f \equiv 0$, the unique nonnegative supersolution is $u \equiv 0$.*

Proof. Without loss of generality, we can assume $f \in L^\infty(\Omega \times (0,T))$. We argue by contradiction. Assume that $\tilde{u}$ is a positive weak supersolution of (13.5.1). Then $\tilde{u}_t + (-\Delta)^s \tilde{u} - \lambda \frac{\tilde{u}}{|x|^2} \geq 0$ in $\Omega \times (0,T)$ in the weak sense.

Since $\tilde{u}$ is also a weak supersolution in any $B_R(0) \times (T_1, T_2) \subset\subset \Omega \times (0,T)$, by Lemma 13.5.1, the problem

$$\begin{cases} u_t + (-\Delta)^s u = \lambda \frac{u}{|x|^{2s}} + u^p + f & \text{in } B_R(0) \times (T_1, T_2), \\ u(x,t) > 0 & \text{in } B_R(0) \times (T_1, T_2), \\ u(x,t) = 0 & \text{in } (\mathbb{R}^N \setminus B_R(0)) \times [T_1, T_2), \\ u(x,T_1) = \tilde{u}(x,T_1) & \text{if } x \in B_R(0) \end{cases} \tag{13.5.5}$$

has a solution u obtained by approximation of truncated problems in $B_R(0) \times (T_1, T_2)$. In particular $u = \lim u_n$, with $u_n \in L^\infty(B_R(0) \times (T_1, T_2))$ being the energy solution to (13.5.3) in $B_R(0) \times (T_1, T_2)$.

Note that $u_t + (-\Delta)^s u - \lambda \frac{u}{|x|^{2s}} \geq 0$ in $B_{r_1}(0) \times (T_1, T_2)$ in the weak sense, and therefore, by Corollary 13.3.11, for any cylinder $B_r(0) \times (t_1, t_2)$, with $0 < r < r_1 < R$, $0 < T_1 < t_1 < t_2 < T_2 \leq T$, there exists a constant $C = C(N, r_1, t_1, t_2)$ such that $u \geq C|x|^{-\frac{N-2s}{2}+\alpha}$ and then for r small enough $u > 1$ in $B_r(0) \times (T_1, T_2)$.

In particular, since $u \in L^1(\Omega \times (0,T))$, using the fact that $u \geq 1$ in $B_r(0) \times (T_1, T_2)$, we reach $\log(u) \in L^p(B_r(0) \times (t_1, t_2))$, for all $p \geq 1$. By a suitable scaling, we can assume that the cylinder is $B_r(0) \times (0, \tau)$.

Let $\phi \in C_0^\infty(B_r(0))$. Then using $\frac{|\phi|^2}{u_n}$ as a test function in the approximated problems (13.5.3) and applying the Picone (Theorem 8.5.1) and Sobolev (Theorem 8.2.7) inequalities,

$$\int_0^\tau \int_{B_r(0)} u_n^{p-1}\phi^2\,dx\,dt \leq \int_0^\tau \int_{B_r(0)} \frac{|\phi|^2}{u_n} u_{nt}\,dx\,dt + \int_0^\tau \int_{B_r(0)} (-\Delta)^s u_n \frac{|\phi|^2}{u_n}\,dx\,dt$$
$$\leq \int_{B_r(0)} |\log u_n(x,\tau)|\phi^2\,dx + C'(N,s,\tau)\|\phi\|^2_{H_0^s(\Omega)}.$$

Therefore, passing to the limit as n tends to infinity and considering that $u \geq C|x|^{-\frac{N-2s}{2}+\alpha}$ in $B_r(0) \times (0,\tau)$, we obtain

$$\int_{B_r(0)} |\log u(x,\tau)|\phi^2\,dx + C'(N,s,\tau)\|\phi\|^2_{H_0^s(\Omega)} \geq \int_0^\tau \int_{B_r(0)} u^{p-1}\phi^2\,dx\,dt$$
$$\geq C\int_0^\tau \int_{B_r(0)} \frac{\phi^2}{|x|^{(p-1)(\frac{N-2s}{2}-\alpha)}}\,dx\,dt.$$

Using the Hölder and Sobolev inequalities, it follows that

$$\int_{B_r(0)} |\log(u(x,\tau))||\phi|^2\,dx \leq \Big(\int_{B_r(0)} |\phi|^{2_s^*}\,dx\Big)^{\frac{2}{2_s^*}} \Big(\int_{B_r(0)} |\log u(x,\tau)|^{\frac{N}{2s}}\,dx\Big)^{\frac{2s}{N}}$$
$$\leq \Big(\int_{B_r(0)} |\log u(x,\tau)|^{\frac{N}{2s}}\,dx\Big)^{\frac{2s}{N}} \mathcal{S}\|\phi\|^2_{H_0^s(\Omega)},$$

where $\mathcal{S}$ is the optimal constant in the Sobolev embedding. Thus we have

$$\Big[C'(N,s,\tau) + \Big(\int_{B_r(0)} |\log u(x,\tau)|^{\frac{N}{2s}}\,dx\Big)^{\frac{2s}{N}} \mathcal{S}\Big]\|\phi\|^2_{H_0^s(\Omega)} \geq C\int_{B_r(0)} \frac{\phi^2}{|x|^{(p-1)(\frac{N-2s}{2}-\alpha)}}\,dx.$$

Since $p > p_+(\lambda, s)$, for a cylinder small enough, $(p-1)(\frac{N-2s}{2} - \alpha) > 2s$ and we obtain a contradiction with the Hardy inequality. □

The previous nonexistence result is very strong in the sense that a complete and instantaneous blow-up phenomenon occurs. That is, if u_n is the solution to the approximated problems (13.5.3), then $u_n(x,t) \to \infty$ as $n \to \infty$, where (x,t) is an arbitrary point in $\Omega \times (0,T)$.

Theorem 13.5.4. *Let u_n be a solution to problem* (13.5.3) *with* $p > p_+(\lambda, s)$. *Then* $u_n(x_0, t_0) \to \infty$, *for all* $(x_0, t_0) \in \Omega \times (0, T)$.

Proof. Without loss of generality, we can assume that $\lambda \le \Lambda_{N,2s}$. The existence of a positive solution to problem (13.5.3) is clear and, due to comparison principle, we know that $u_n \le u_{n+1}$ for all $n \in \mathbb{N}$.

Suppose by contradiction that there exists $(x_0, t_0) \in \Omega \times (0, T)$ such that

$$u_n(x_0, t_0) \to C_0 < \infty \quad \text{as } n \to \infty.$$

By using the Harnack inequality (see Lemma 13.2.10), there exists $s_0 > 0$ and a positive constant $C = C(N, s_0, t_0, \beta)$ such that

$$\iint_{R_0^-} u_n(x,t)\,dx\,dt \le C \operatorname{ess\,inf}_{R_0^+} u_n \le C,$$

where $R_0^- = B_{s_0}(x_0) \times (t_0 - \frac{3}{4}\beta, t_0 - \frac{1}{4}\beta)$ and $R_0^+ = B_{s_0}(x_0) \times (t_0 + \frac{1}{4}\beta, t_0 + \frac{3}{4}\beta)$.

Without loss of generality, we can suppose $x_0 = 0$. Otherwise, we can find a finite sequence of points $\{x_i\}_{i=0}^M$, ending with $x_M = 0$, and of radii $\{s_i\}_{i=0}^M$ such that $B_{s_i}(x_i) \subset \Omega$, $B_{s_i}(x_i) \cap B_{s_{i+1}}(x_{i+1}) \neq \emptyset$, for all $i = 0, \dots, M$, and, by the Hanarck inequality,

$$\iint_{R_i^-} u_n(x,t)\,dx\,dt \le C \operatorname{ess\,inf}_{R_i^+} u_n,$$

where $R_i^- = B_{s_i}(x_i) \times (t_i - \frac{3}{4}\beta, t_i - \frac{1}{4}\beta)$ and $R_i^+ = B_{s_i}(x_i) \times (t_i + \frac{1}{4}\beta, t_i + \frac{3}{4}\beta)$, $t_i \in (0, T)$, and β is small enough so that $t_i - \frac{3}{4}\beta > 0$ and $t_i + \frac{3}{4}\beta < T$ for all $i = 0, \dots, M$. Let us choose now $t_i = t_{i-1} - \beta$ for $i = 1, \dots, M$. Note that in this case

$$\left(t_i + \frac{1}{4}\beta, t_i + \frac{3}{4}\beta\right) = \left(t_{i-1} - \frac{3}{4}\beta, t_{i-1} - \frac{1}{4}\beta\right),$$

and in particular, $R_i^+ \cap R_{i-1}^- \neq \emptyset$. Thus,

$$\begin{aligned}
\iint_{R_M^-} u_n(x,t)\,dx\,dt &\le \operatorname{ess\,inf}_{R_M^+} u_n(x,t) \le \operatorname{ess\,inf}_{R_M^+ \cap R_{M-1}^-} u_n(x,t) \\
&\le \frac{1}{|R_M^+ \cap R_{M-1}^-|} \iint_{R_M^+ \cap R_{M-1}^-} u_n(x,t)\,dx\,dt \\
&\le \frac{1}{|R_M^+ \cap R_{M-1}^-|} \iint_{R_{M-1}^-} u_n(x,t)\,dx\,dt \\
&\le \dots \le C \iint_{R_0^-} u_n(x,t)\,dx\,dt \le \tilde{C}.
\end{aligned}$$

Therefore, supposing $x_0 = 0$, by the monotone convergence theorem there exists $u \geq 0$ such that $u_n \uparrow u$ strongly in $L^1(B_r(0) \times (t_1, t_2))$. Let now φ be the solution to the problem

$$\begin{cases} -\varphi_t + (-\Delta)^s \varphi = \chi_{B_r(0)\times[t_1,t_2]} & \text{in } \Omega \times (0, T), \\ \varphi(x,t) > 0 & \text{in } \Omega \times (0, T), \\ \varphi(x,t) = 0 & \text{in } (\mathbb{R}^N \setminus \Omega) \times [0, T), \\ \varphi(x,T) = 0 & \text{in } \Omega. \end{cases} \tag{13.5.6}$$

Note that due to the regularity of the right hand sides of problems (13.5.3) and (13.5.6), both u_n and φ are in the energy space, and thus both can be used as test functions in the energy formulation of the problems. Indeed, considering first u_n as test function in (13.5.6) and then, after integrating by parts, φ in (13.5.3), and defining $\eta := \inf_{B_r(0)\times(t_1,t_2)} \varphi(x,t)$, we have

$$\begin{aligned} C &\geq \int_{t_1}^{t_2} \int_{B_r(0)} u_n(x,t)\, dx\, dt \\ &\geq \lambda \int_0^T \int_\Omega \frac{u_{n-1}}{|x|^{2s} + \frac{1}{n}} \varphi\, dx\, dt + \int_0^T \int_\Omega u_{n-1}^p \varphi\, dx\, dt + \int_0^T \int_\Omega T_n(f) \varphi\, dx\, dt \\ &\geq \lambda\eta \int_{t_1}^{t_2} \int_{B_r(0)} \frac{u_{n-1}}{|x|^{2s} + \frac{1}{n}}\, dx\, dt + \eta \int_{t_1}^{t_2} \int_{B_r(0)} u_{n-1}^p\, dx\, dt + \eta \int_{t_1}^{t_2} \int_{B_r(0)} T_n(f)\, dx\, dt. \end{aligned}$$

By the monotone convergence theorem,

$$\begin{aligned} u_{n-1}^p &\to u^p \text{ in } L^1(B_r(0) \times (t_1, t_2)), \\ \frac{u_{n-1}}{|x|^{2s}+\frac{1}{n}} &\nearrow \frac{u}{|x|^{2s}} \text{ in } L^1(B_r(0) \times (t_1, t_2)), \\ T_n(f) &\to f \text{ in } L^1(B_r(0) \times (t_1, t_2)). \end{aligned}$$

Thus it follows that u is a weak supersolution to (13.1.1) in $B_r(0)\times(t_1, t_2)$, a contradiction with Theorem 13.5.3. □

13.5.2 Existence results for $1 < p < p_+(\lambda, s)$

The goal now is to consider the complementary interval of powers, $1 < p < p_+(\lambda, s)$, and to prove that under some suitable hypotheses on f and u_0, problem (13.5.1) has a positive solution. We will consider here the case $f \equiv 0$. For the case $f \not\equiv 0$, see Remark 13.5.6.

First of all, note that if $0 < \lambda \le \Lambda_{N,2s}$ and $1 < p < p_+(\lambda, s)$, the stationary problem

$$(-\Delta)^s u = \lambda \frac{u}{|x|^{2s}} + u^p \text{ in } \Omega, \quad u > 0 \text{ in } \Omega, \quad u = 0 \text{ in } \mathbb{R}^N \setminus \Omega \tag{13.5.7}$$

has a positive supersolution w, depending on the following cases.

(A) If $0 < \lambda < \Lambda_{N,2s}$, in Section 11.3 (see also [46, Proposition 2.3]), we find a positive solution to the problem

$$(-\Delta)^s u = \lambda \frac{u}{|x|^{2s}} + u^p + \mu u^q \text{ in } \Omega, \quad u > 0 \text{ in } \Omega, \quad u = 0 \text{ in } \mathbb{R}^N \setminus \Omega, \tag{13.5.8}$$

for μ small enough, $0 < q < 1$ and $1 < p < p_+(\lambda, s)$. In particular, for every $\mu \ge 0$ this solution is a supersolution of (13.5.7). Note that for $1 < p \le 2_s^* - 1$ this supersolution is in the energy space, and for $2_s^* - 1 < p < p_+(\lambda, s)$, it is a weak positive supersolution.

(B) If $\lambda = \Lambda_{N,2s}$, then $p_+(\lambda, s) = 2_s^* - 1$. Thus, instead of $H_0^s(\Omega)$, we consider the Hilbert space $H(\Omega)$ defined in (13.2.5). Since $H(\Omega)$ is compactly embedded in $L^p(\Omega)$ for all $1 \le p < 2_s^*$, classical variational methods in the space $H(\Omega)$ allow us to prove the existence of a positive solution w to the stationary problem (13.5.7).

Theorem 13.5.5. *Assume that $0 < \lambda \le \Lambda_{N,2s}$ and $1 < p < p_+(\lambda, s)$. Suppose that $u_0(x) \le \overline{w}$, where $\overline{w}$ is a supersolution to the stationary problem*

$$(-\Delta)^s w = \lambda \frac{w}{|x|^{2s}} + w^p \textit{ in } \Omega, \quad w(x) > 0 \textit{ in } \Omega, \quad w(x) = 0 \textit{ on } \mathbb{R}^N \setminus \Omega.$$

Then for all $T > 0$, the problem

$$\begin{cases} u_t + (-\Delta)^s u = \lambda \frac{u}{|x|^{2s}} + u^p & \textit{in } \Omega \times (0, T), \\ u(x, t) = 0 & \textit{in } (\mathbb{R}^N \setminus \Omega) \times [0, T), \\ u(x, 0) = u_0(x) & \textit{if } x \in \Omega \end{cases} \tag{13.5.9}$$

has a global positive solution. If $\overline{w}$ is a weak supersolution, the solution will also be weak, and similarly, if $\overline{w}$ is an energy supersolution, problem (13.5.9) *will have an energy solution.*

Proof. Since $\overline{w}(x) \ge u_0(x)$ for all $x \in \Omega$, $\overline{w}$ is a positive supersolution to problem (13.5.9). Hence, we conclude just by applying Lemma 13.5.1, when $\overline{w}$ is a weak supersolution, or Lemma 13.5.2, if $\overline{w}$ is an energy supersolution. □

Remark 13.5.6.

(I) With the results above we find the optimality of the power $p_+(\lambda, s)$, which was our main aim. Nevertheless, it could be interesting to know the optimal class of data for which there exists a solution and the regularity of such solutions according to the regularity of the data. In this direction, considering $g = u^p + cf$ in problem (13.4.1), Theorem 13.4.1 establishes that, necessarily,

$$\int_{t_1}^{t_2}\int_{B_r(0)} |x|^{-\frac{N-2s}{2}+\alpha} g\,dx\,dt < +\infty \quad \text{and} \quad \int_{B_r(0)} |x|^{-\frac{N-2s}{2}+\alpha} u_0\,dx < +\infty,$$

if we expect to find a solution of problem (13.5.1).

(II) In the presence of a source term $f \gneqq 0$, if $f(x,t) \le \frac{c_0(t)}{|x|^{2s}}$ with $c_0(t)$ bounded and *sufficiently small*, then the computation above allows us to prove the existence of a supersolution. Then the existence of solution to problem (13.1.1) follows for all $p < p_+(\lambda, s)$.

(III) The asymptotic behavior of the solution remains essentially open.

13.6 Further results and comments

13.6.1 Fujita exponent depending on λ

We have analyzed the Fujita phenomenon for the fractional heat equation in Section 12.8.2; the Fujita exponent happens to be $F(0,s) = 1 + \frac{2s}{N}$. Similarly to the local case (see Section 5.6), the Hardy potential produces a deviation to the right of the Fujita exponent obtained in Section 12.8.2. Precisely, with the parameters defined in Section 9.4, the following result holds.

Theorem 13.6.1. *Suppose that* $1 < p < F(\lambda, s) := 1 + \frac{2s}{N-\gamma}$, *where* $\gamma = \frac{N-2s}{2} - \alpha$, *and let* u *be a positive solution to problem (7.5.2). Then there exists* $T^* := T^*(u_0)$ *such that*

$$\lim_{t\to T^*} \int_{\mathbb{R}^N} |x|^{-\gamma} u(x,t)\,dx = \infty.$$

See [25] for the proof. Note also that $F(0,s) < F(\lambda,s) < p_+(\lambda,s)$.

13.6.2 Other operators

As in the elliptic case, the results above could be extended for a larger class of linear operators, like those studied in Chapter 8.

Another case that could have some interest would be the elliptic part of the fractional p-Laplacian operator taking into consideration the corresponding Hardy inequality. See [168]. The techniques to be used, such as Harnack's inequality, may have a genuine mathematical interest. These aspects have not been considered in this book.

Bibliography

[1] B. Abdellaoui, A. Attar, *Quasilinear elliptic problem with Hardy potential and singular term*. Commun. Pure Appl. Anal. 12 (2013) no. 3, 1363–1380.

[2] B. Abdellaoui, A. Attar, R. Bentifour, *On the fractional p-laplacian equations with weights and general datum*. Adv. Nonlinear Anal. 8 (2019) no. 1, 144–174.

[3] B. Abdellaoui, R. Bentifour, *Caffarelli-Kohn-Nirenberg type inequalities of fractional order and applications*. J. Funct. Anal. 272 (2017) no. 10, 3998–4029.

[4] B. Abdellaoui, E. Colorado, I. Peral, *Existence and nonexistence results for a class of linear and semilinear parabolic equations related to some Caffarelli–Kohn–Nirenberg inequalities*. J. Eur. Math. Soc. 6 (2004) no. 1, 119–148.

[5] B. Abdellaoui, E. Colorado, I. Peral, *Some remarks on elliptic equations with singular potentials and mixed boundary conditions*. Adv. Nonlinear Stud. 4 (2004), 501–531.

[6] B. Abdellaoui, E. Colorado, I. Peral, *Some improved Caffarelli-Kohn-Nirenberg inequalities*. Calc. Var. Partial Differ. Equ. 23 (2005) no. 3, 327–345.

[7] B. Abdellaoui, E. Colorado, I. Peral, *Effect of the boundary conditions in the behavior of the optimal constant of some Caffarelli-Kohn-Nirenberg inequalities: applications to critical nonlinear elliptic problems*. Adv. Differ. Equ. 11 (2006) no. 6, 667–720.

[8] B. Abdellaoui, A. Dall' Aglio, I. Peral, *Some remarks on elliptic problems with critical growth in the gradient*. J. Differ. Equ. 222 (2006) no. 1, 21–62.

[9] B. Abdellaoui, A. Dall' Aglio, I. Peral, *Regularity and nonuniqueness results for nonlinear parabolic problems having natural growth in the gradient*. J. Math. Pures Appl. 90 (2008), 242–269.

[10] B. Abdellaoui, V. Felli, I. Peral, *Existence and nonexistence results for quasilinear elliptic equations involving the p-laplacian*. Boll. Unione Mat. Ital. (8) 9-B (2006), 445–484.

[11] B. Abdellaoui, M. Medina, I. Peral, A. Primo, *Optimal results for the fractional heat equation involving the Hardy potential*. Nonlinear Anal. 140 (2016), 166–207.

[12] B. Abdellaoui, M. Medina, I. Peral, A. Primo, *A note on the effect of the Hardy potential in some Calderón-Zygmund properties for the fractional Laplacian*. J. Differ. Equ. 260 (2016) no. 11, 8160–8206.

[13] B. Abdellaoui, I. Peral, *Some results for semilinear elliptic equations with critical potential*. Proc. R. Soc. Edinb., Sect. A 132 (2002) no. 1, 1–24.

[14] B. Abdellaoui, I. Peral, *Existence and nonexistence results for quasilinear elliptic equations involving the p-Laplacian with a critical potential*. Ann. Mat. Pura Appl. (4) 182 (2003) no. 3, 247–270.

[15] B. Abdellaoui, I. Peral, *On quasilinear elliptic equations related to some Caffarelli-Kohn-Nirenberg inequalities*. Commun. Pure Appl. Anal. 2 (2003) no. 4, 539–566.

[16] B. Abdellaoui, I. Peral, *Competition reaction-absorption in some elliptic and parabolic problems related to the Caffarelli-Kohn-Nirenberg inequalities*. J. Math. Anal. Appl. 314 (2006) no. 2, 590–617.

[17] B. Abdellaoui, I. Peral, *The equation $-\Delta u - \lambda \frac{u}{|x|^2} = |\nabla u|^p + cf(x)$: the optimal power*. Ann. Sc. Norm. Super. Pisa, Cl. Sci. (5) 6 (2007) no. 1, 159–183.

[18] B. Abdellaoui, I. Peral, *The effect of Harnack inequality on the existence and nonexistence results for quasi-linear parabolic equations related to Caffarelli-Kohn-Nirenberg inequalities*. Nonlinear Differ. Equ. Appl. 14 (2007) no. 3–4, 335–360.

[19] B. Abdellaoui, I. Peral, A. Primo, *Elliptic problems with a Hardy potential and critical growth in the gradient: non-resonance and blow-up results*. J. Differ. Equ. 239 (2007) no. 2, 386–416.

https://doi.org/10.1515/9783110606270-014

[20] B. Abdellaoui, I. Peral, A. Primo, *Breaking of resonance and regularizing effect of a first order quasi-linear term in some elliptic equations*. Ann. Inst. Henri Poincaré, Anal. Non Linéaire 25, (2008) 969–985.

[21] B. Abdellaoui, I. Peral, A. Primo, *Influence of the Hardy potential in a semilinear heat equation*. Proc. R. Soc. Edinb., Sect. A, Math. 139 (2009) no. 5, 897–926.

[22] B. Abdellaoui, I. Peral, A. Primo, *Strong regularizing effect of a gradient term in the heat equation with the Hardy potential*. J. Funct. Anal. 258 (2010), 1247–1272.

[23] B. Abdellaoui, I. Peral, A. Primo, *Optimal results for parabolic problems arising in some physical models with natural growth in the gradient respect to a Hardy potential*. Adv. Math. 225 (2010), 2967–3021.

[24] B. Abdellaoui, I. Peral, A. Primo, *A remark on the fractional Hardy inequality with a remainder term*. C. R. Acad. Sci. Paris, Ser. I 352 (2014) 299–303.

[25] B. Abdellaoui, I. Peral, A. Primo, *A note on Fujita exponent in fractional heat equation involving the Hardy potential*. Math. Eng. 2 (2020) no. 4, 639–656.

[26] B. Abdellaoui, I. Peral, A. Primo, F. Soria, *Fractional KPZ equations with critical growth in the gradient respect to Hardy potential*. Nonlinear Anal. 201 (2020), 111942. https://doi.org/10.1016/j.na.2020.111942.

[27] M. Abramowitz, I. A. Stegun, *Handbook of Mathematical Functions with Formulas, Graphs and Mathematical Tables*. Dover Publications, New York, 1992.

[28] R. A. Adams, *Sobolev Spaces*. Academic Press, New York, 1975.

[29] Adimurthi, N. Chaudhuri, M. Ramaswamy, *An improved Hardy-Sobolev inequality and its application*. Proc. Am. Math. Soc. 130 (2002) no. 2, 489–505.

[30] J. A. Aguilar, I. Peral *Global behaviour of the Cauchy problem for some critical nonlinear parabolic equations*. SIAM J. Math. Anal. 31 (2000) no. 6, 1270–1294.

[31] N. Alaa, M. Pierre, *Weak solutions of some quasilinear elliptic equations with data measures*. SIAM J. Math. Anal. 24 (1993) no. 1, 23–35.

[32] S. Alama, *Semilinear elliptic equations with sublinear indefinite nonlinearities*. Adv. Differ. Equ. 4 (1999), 813–842.

[33] G.Alberti, G.Bellettini, *A nonlocal anisotropic model for phase transitions. I. The optimal profile problem*. Math. Ann. 310 (1998), 527–560.

[34] W. Allegretto, Y. X. Huang, *A Picone's identity for the p − Laplacian and applications*. Nonlinear Anal. 32 (1998) no. 7, 819–830.

[35] A. Ambrosetti, H. Brezis, G. Cerami, *Combined effects of concave and convex nonlinearities in some elliptic problems*. J. Funct. Anal. 122 (1994) no. 2, 519–543.

[36] A. Ambrosetti, J. Garcia, I. Peral, *Multiplicity of solutions for semilinear and quasilinear elliptic problems*. J. Funct. Anal. 137 (1996), 219–242.

[37] A. Ambrosetti, P. H. Rabinowitz, *Dual variational methods in critical point theory and applications*. J. Funct. Anal. 14 (1973), 349–381.

[38] D. Applebaum, *LéVy Processes and Stochastic Calculus*, Cambridge Studies in Advanced Mathematics, 116. Second edition. Cambridge University Press, Cambridge, 2009.

[39] D. G. Aronson, J. Serrin, *Local behavior of solutions of quasilinear parabolic equations*. Arch. Ration. Mech. Anal. 25 (1967), 81–122.

[40] E. Artin, *The gamma function*, in Rosen, Michael (Ed.) Exposition by Emil Artin: A Selection; History of Mathematics 30. American Mathematical Society, Providence, RI, 2006.

[41] P. Baras, J. A. Goldstein, *The heat equation with a singular potential*. Trans. Am. Math. Soc. 284 (1984) no. 1, 121–139.

[42] P. Baras, M. Pierre, *Problèmes paraboliques semilineéaires avec données mesures*. Appl. Anal. 18 (1984), 111–149.

[43] M. T. Barlow, R. F. Bass, Z.-Q. Chen, M. Kassmann, *Non-local Dirichlet forms and symmetric jump processes*. Trans. Am. Math. Soc. 361 (2009), 1963–1999.

[44] B. Barrios, E. Colorado, R. Servadei, F. Soria, *A critical fractional equation with concave-convex nonlinearities*. Ann. Inst. Henri Poincaré, Anal. Non Linéaire 32 (2015) no. 4, 875–900.

[45] B. Barrios, I. De Bonis, M. Medina, I. Peral, *Semilinear problems for the fractional Laplacian with a singular nonlinearity*. Open Math. 13 (2015), 390–407.

[46] B. Barrios, M. Medina, I. Peral, *Some remarks on the solvability of non-local elliptic problems with the Hardy potential*. Commun. Contemp. Math. 16 (2014) no. 4, 1350046.

[47] B.Barrios, L. Montoro, I. Peral, F. Soria, *Neumann conditions for the higher order s-fractional Laplacian, $(-\Delta)^s$ with $s > 1$*. Nonlinear Anal. 193 (2020) 111368. https://doi.org/10.1016/j.na.2018.10.012.

[48] B. Barrios, I. Peral, F. Soria, E. Valdinoci, *A Widder's type theorem for the heat equation with nonlocal diffusion*. Arch. Ration. Mech. Anal. 213 (2014) no. 2, 629–650.

[49] B. Barrios, I. Peral, S. Vita, *Some Remarks About the Summability of Nonlocal Nonlinear Problems*. Adv. Nonlinear Anal. 4 (2015) no. 2, 91–107.

[50] R. Beals, *On the boundedness of pseudo-differential operators*. Commun. Partial Differ. Equ. 2 (1977) no. 10, 1063–1070.

[51] R. Beals, *A general calculus of pseudodifferential operators*. Duke Math. J. 42 (1975), 1–42.

[52] R. Beals, C. Fefferman, *On local solvability of linear partial differential equations*. Ann. Math. 97 (1973) no 2, 482–498.

[53] J. Bebernes and D. Eberly, *Mathematical Problems in Combustion Theory*. Springer, New York, 1989.

[54] W. Beckner, *Pitt's inequality and the uncertainty principle*. Proc. Am. Math. Soc. 123 (1995) no. 6, 1897–1905.

[55] A. Ben-Artzi, P. Souplet, F. B. Weissler, *The local theory for viscous Hamilton-Jacobi equations in Lebesgue spaces*. J. Math. Pures Appl. (9) 81 (2002) no. 4, 343–378.

[56] P. Bénilan, L. Boccardo, T. Gallouët, R. Gariepy, M. Pierre, J. L. Vazquez, *An L^1-theory of existence and uniqueness of solutions of nonlinear elliptic equations*. Ann. Sc. Norm. Super. Pisa, Cl. Sci. (4) 22 (1995) no. 2, 241–273.

[57] H. Berestycki, S. Kamin, G. Sivashinsky, *Metastability in a flame front evolution equation*. Interfaces Free Bound. 3 (2001) no. 4, 361–392.

[58] H. Berestycki, L. Nirenberg, S. R. S. Varadhan, *The principal eigenvalue and maximum principle for second-order elliptic operators in general domains*. Commun. Pure Appl. Math. 47 (1994) no. 1, 47–92.

[59] S. N. Bernstein *Sur les fonctions absolument monotones*. Acta Math. 52 (1928), 1–66. https://doi.org:10.1007/BF02592679.

[60] D. Blanchard, *Truncations and monotonicity methods for parabolic equations*. Nonlinear Anal. 21 (1993) no. 10, 725–743.

[61] D. Blanchard, F. Murat, *Renormalised solutions of nonlinear parabolic problems with L^1 data: existence and uniqueness*. Proc. R. Soc. Edinb., Sect. A 127 (1997) no. 6, 1137–1152.

[62] D. Blanchard, A. Porretta, *Nonlinear parabolic equations with natural growth terms and measure initial data*. Ann. Sc. Norm. Super. Pisa, Cl. Sci. (4) 30 (2001) no. 3–4, 583–622.

[63] R. M. Blumenthal, R. K. Getoor, *Some theorems on stable processes*. Trans. Am. Math. Soc. 95 (1960) 263–273.

[64] L. Boccardo, G. Croce, *Esistenza e regolaritá di soluzioni di alcuni problemi ellitici*. Quaderni dell'UMI 51. Bologna, 2010.

[65] L. Boccardo, M. Escobedo, I.Peral, *A remark on elliptic problems involving critical exponent*. Nonlinear Anal. 24 (1995) no. 11, 1639–1648.

[66] L. Boccardo, A. Dall'Aglio, T. Gallouët, L. Orsina, *Nonlinear parabolic equations with measure data.* J. Funct. Anal. 147 (1997) no. 1, 237–258.

[67] L. Boccardo, A. Dall'Aglio, T. Gallouët, L. Orsina, *Existence and regularity results for some nonlinear parabolic equations.* Adv. Math. Sci. Appl. 9 (1999) no. 2, 1017–1031.

[68] L. Boccardo, T. Gallouët, *Strongly nonlinear elliptic equations having natural growth terms and L^1 data.* Nonlinear Anal. 19 (1992) no. 6, 573–579.

[69] L. Boccardo, T. Gallouët, *Nonlinear elliptic equations with right-hand side measures.* Commun. Partial Differ. Equ. 17 (1992), 641–655.

[70] L. Boccardo, T. Gallouet, F. Murat, *A unified presentation of two existence results for problems with natural growth.* Res. Notes Math. 296 (1993), 127–137.

[71] L. Boccardo, T. Gallouët, L. Orsina, *Existence and nonexistence of solutions for some nonlinear elliptic equations.* J. Anal. Math. 73 (1997), 203–223.

[72] L. Boccardo, D. Giachetti, *Existence results via regularity for some nonlinear elliptic problems.* Commun. Partial Differ. Equ. 14 (1989), 663–680.

[73] L. Boccardo, F. Murat, *Almost everywhere convergence of the gradients of solutions to elliptic and parabolic equations.* Nonlinear Anal. 19 (1992) no. 6, 581–597.

[74] L. Boccardo, F. Murat, J. P. Puel, *Existence de solutions non bornées pour certaines équations quasi-linéaires.* Port. Math. 41 (1982) no. 1–4, 507–534.

[75] L. Boccardo, F. Murat, J. P. Puel, *Existence de solutions faibles pour des équations elliptiques quasi-linéaires à croissance quadratique*, in Nonlinear Partial Differential Equations and Their Applications. Collège de France Seminar, Vol. IV (Paris, 1981/1982). Res. Notes in Math. 84, pp. 19–73. Pitman, Boston, Mass.-London, 1983.

[76] L. Boccardo, F. Murat, J. P. Puel, *Résultats d'existence pour certains problèmes elliptiques quasilinéaires.* Ann. Sc. Norm. Super. Pisa, Cl. Sci. (4) 11 (1984) no. 2, 213–235.

[77] L. Boccardo, F. Murat, J. P. Puel, *Existence results for some quasilinear parabolic equations.* Nonlinear Anal. 13 (1989) no. 4, 373–392.

[78] L. Boccardo, F. Murat, J. P. Puel, *L^∞ estimate for some nonlinear elliptic partial differential equations and application to an existence result.* SIAM J. Math. Anal. 23 (1992) no. 2, 326–333.

[79] L. Boccardo, L. Orsina, *Semilinear elliptic equations with singular nonlinearities.* Calc. Var. 37 (2010), 363–380.

[80] L. Boccardo, L. Orsina, I. Peral, *A remark on existence and optimal summability of solutions of elliptic problems involving Hardy potential.* Discrete Contin. Dyn. Syst. 16 (2006) no. 3, 513–523.

[81] L. Boccardo, M. M. Porzio, A. Primo, *Summability and existence results for nonlinear parabolic elliptic equations.* Nonlinear Anal. 71 (2009) no. 3–4, 978–990.

[82] S. Bochner, *Diffusion equations and stochastic processes.* Proc. Natl. Acad. Sci. USA 38 (1949), 368–370.

[83] K. Bogdan, T. Jakubowski, *Estimates of the heat kernel of fractional Laplacian perturbed by gradient operators.* Commun. Math. Phys. 271 (2007) no. 1, 179–198.

[84] K. Bogdan, A. Stós, P. Sztonyk, *Harnack inequality for stable processesin in d-sets.* Stud. Math. 158 (2003) no. 2, 163–198.

[85] E. Bombieri, E. Giusti, *Harnack's inequality for elliptic differential equations on minimal surfaces.* Invent. Math. 15 (1972), 24–46.

[86] H. Brezis, *Functional Analysis, Sobolev Spaces and Partial Differential Equations.* Springer, Berlin, 2011.

[87] H. Brezis, X. Cabré, *Some simple nonlinear PDE's without solutions.* Boll. Unione Mat. Ital. Sez. B Artic. Ric. Mat. (8) 1 (1998) no. 2, 223–262.

[88] H. Brezis, T. Cazenave, Y. Martel, A. Ramiandrisoa, *Blow up for $u_t - \Delta u = g(u)$ revisited*. Adv. Differ. Equ. 1 (1996) no. 1, 73–90.

[89] H. Brezis, L. Dupaigne, A. Tesei, *On a semilinear elliptic equation with inverse-square potential*. Sel. Math. New Ser. 11 (2005) no. 1, 1–7.

[90] H. Brezis, S. Kamin, *Sublinear elliptic equations in R^n*. Manuscr. Math. 74 (1992) no. 1, 87–106.

[91] H. Brezis, E. H. Lieb, *A relation between pointwise convergence of functions and convergence of functionals*. Proc. Am. Math. Soc. 88 (1983) no. 3, 486–490.

[92] H. Brezis, P. Mironescu, *Gagliardo-Nirenberg, composition and products in fractional Sobolev spaces*. J. Evol. Equ. 1 (2001) no. 4, 387–404.

[93] H. Brezis, P. Mironescu, *Gagliardo-Nirenberg inequalities and non-inequalities: the full story*. Ann. Inst. Henri Poincaré, Anal. Non Linéaire 35 (2018) no. 5, 1355–1376.

[94] H. Brezis, L. Nirenberg *Positive solutions of nonlinear elliptic equations involving critical Sobolev exponents*. Commun. Pure Appl. Math. XXXVI (1983), 437–477.

[95] H. Brezis, L. Oswald, *Remarks on sublinear elliptic equations*. Nonlinear Anal. 10 (1986) no. 1, 55–64.

[96] H. Brezis, A. C Ponce, *Kato's inequality when Δu is a measure*. C. R. Math. Acad. Sci. Paris 338 (2004) no. 8, 599–604.

[97] H. Brezis, J. L. Vázquez, *Blow-up solutions of some nonlinear elliptic problems*. Rev. Mat. Univ. Complut. Madr. 10 (1997) no. 2, 442–469.

[98] C. Bucur, E. Valdinoci, *Nonlocal Diffusion and Applications*. Lecture Notes of the Unione Matematica Italiana 20. Springer Unione Matematica Italiana, Bologna, 2016.

[99] X. Cabré, G. Poggesi *Stable solutions to some elliptic problems: minimal cones, the Allen-Cahn equation, and blow-up solutions*. In Geometry of PDEs and related problems, CIME Found. Subser., Lecture Notes in Math. 2220, pp. 1–45, Springer.

[100] X. Cabré, J. M. Roquejoffre, *The influence of fractional diffusion in Fisher-KPP equations*. Commun. Math. Phys. 320 (2013) no. 3, 679–722.

[101] X. Cabré, J. Sola-Morales, *Layer solutions in a half-space for boundary reactions*. Commun. Pure Appl. Math. 58 (2005), 1678–1732.

[102] X. Cabré, A. Figalli, X. Ros-Oton, J. Serra *Stable solutions to semilinear elliptic equations are smooth up to dimension 9*. Acta Math. 224 (2020) no. 2, 187–252.

[103] L. Caffarelli, *Further regularity for the Signorini problem*. Commun. Partial Differ. Equ. 4 (1979), 1067–1075.

[104] L. Caffarelli, J. M. Roquejoffre, Y. Sire, *Variational problems in free boundaries for the fractional Laplacian*. J. Eur. Math. Soc. 12 (2010), 1151–1179.

[105] L. Caffarelli, L. Silvestre, *An extension problem related to the fractional Laplacian*. Commun. Partial Differ. Equ. 32 (2007) no. 7–9, 1245–1260.

[106] L. Caffarelli, A. Figalli, *Regularity of solutions to the parabolic fractional obstacle problem*. J. Reine Angew. Math. 680 (2013), 191–233.

[107] L. Caffarelli, R. Kohn, L. Nirenberg, *Partial regularity of suitable weak solutions of the Navier-Stokes equations*. Commun. Pure Appl. Math. 35 (1982) no. 6, 771–831.

[108] L. Caffarelli, R. Kohn, L. Nirenberg, *First order interpolation inequalities with weights*. Compos. Math. 53 (1984) no. 3, 259–275.

[109] L. Caffarelli, A. Vasseur, *Drift diffusion equations with fractional diffusion and the quasi-geostrophic equation*. Ann. Math. 171 (2010), 1903–1930.

[110] A. P. Calderon, A. Zygmund, *On the existence of certain singular integrals*. Acta Math. 88 (1952), 85–139.

[111] A. P. Calderón, A. Zygmund, *On singular integrals*. Am. J. Math. 78 (1956), 289–309.

[112] A. P. Calderón, A. Zygmund, *Singular integral operators and differential equations*. Am. J. Math. 79 (1957), 901–921.

[113] A. P. Calderón, R. Vaillancourt, *A class of bounded pseudo-differential operators*. Proc. Natl. Acad. Sci. USA 69 (1972), 1185–1187.

[114] S. Campanato, *Sistemi ellittici in forma divergenza. Regolaritá all'interno*. Quaderni Scuola Normale Superiore di Pisa, Pisa 1980.

[115] K. M. Case, *Singular potentials*. Phys. Rev. 80 (1950) no. 1, 797–806.

[116] F. Catrina, Z. Q. Wang, *On the Caffarelli-Kohn-Nirenberg inequalities: sharp constants, existence (and nonexistence), and symmetry of extremal functions*. Commun. Pure Appl. Math. 54 (2001) no. 2, 229–258.

[117] R. R. Coifman, G. Weiss, *Analyse Harmonique Non-Conmutative sur Certains Espaces Homogenes*. Lectures Notes in Mathematics 242. Springer-Verlag, Berlin, 1971.

[118] S. Chandrasekar, *An Introduction to the Study of Stellar Structure*. Dover Publ. Inc. New York, 1985.

[119] H. Chang-Lara, G. Dávila, *Regularity for solutions of non local parabolic equations*. Calc. Var. Partial Differ. Equ. 49 (2014) no. 1–2, 139–172.

[120] H. Chang-Lara, G. Dávila, *Regularity for solutions of nonlocal parabolic equations II*. J. Differ. Equ. 256 (2014) no. 1, 130–156.

[121] F. Charro, E. Colorado, I. Peral *Multiplicity of solutions to uniformly elliptic fully nonlinear equations with concave-convex right-hand side*. J. Differ. Equ. 246 (2009), 4221–4248.

[122] F. M. Chiarenza, M. Frasca, *Boundedness for the solutions of a degenerate parabolic equation*. Appl. Anal. 17 (1984) no. 4, 243–261.

[123] F. M. Chiarenza, R. P. Serapioni, *A Harnack inequality for degenerate parabolic equations*. Commun. Partial Differ. Equ. 9 (1984) no. 8, 719–749.

[124] M. Cozzi, *Regularity results and Harnack inequalities for minimizers and solutions of nonlocal problems: a unified approach via fractional de giorgi classes*. J. Funct. Anal. 272 (2017), 4762–4837.

[125] W. Craig, M. Groves, *Hamiltonian long-wave approximations to the water-wave problem*. Wave Motion 19 (1994), 367–389.

[126] W. Craig, U. Schanz, C. Sulem, *The modulational regime of three-dimensional water waves and the Davey-Stewartson system*. Ann. Inst. Henri Poincaré, Anal. Non Linéaire, 5 (1997), 615–667.

[127] M. G. Crandall, P. L. Lions, *Viscosity solutions of Hamilton-Jacobi equations*. Trans. Am. Math. Soc. 277 (1983) no. 1, 1–42.

[128] M. Cwikel, *Weak type estimates for singular values and the number of bound states of Schrödinger operators*. Ann. Math. 106 (1977), 93–102.

[129] A. Dall' Aglio, *Approximated solutions of equations with L^1 data. Application to the H-convergence of quasi-linear parabolic equations*. Ann. Mat. Pura Appl. (4) 170 (1996), 207–240.

[130] A. Dall' Aglio, D. Giachetti, I.Peral, *Results on parabolic equations related to some Caffarelli-Kohn-Nirenberg inequalities*. SIAM J. Math. Anal. 36 (2004/2005) no. 3, 691–716.

[131] A. Dall'Aglio, D. Giachetti, S. Segura, *Nonlinear parabolic problems with a very general quadratic gradient term*. Differ. Integral Equ. 20 (2007) no 4, 361–396.

[132] G. Dal Maso, F. Murat, L. Orsina, A. Prignet, *Renormalized solutions of elliptic equations with general measure data*. Ann. Sc. Norm. Super. Pisa, Cl. Sci. (4) 28 (1999) no. 4, 741–808.

[133] R. Dautray, J. L. Lions, *Analyse mathématique et calcul numérique pour les sciences et les techniques*. Tome 1, 2 & 3. Collection du Commissariat à l'Énergie Atomique: Série Scientifique. Masson, Paris, 1984.

[134] E. Di Benedetto, *Degenerate Parabolic Equations*. Universitext. Springer-Verlag, New York, 1993.
[135] A. Di Castro, T. Kuusi, G. Palatucci, *Nonlocal Harnack inequalities*. J. Funct. Anal. 267 (2014) no. 6, 1807–1836.
[136] A. Di Castro, T. Kuusi, G. Palatucci, *Local behavior of fractional p-minimizers*. Ann. Inst. Henri Poincaré, Anal. Non Linéaire 33 (2016) no. 5, 1279–1299.
[137] E. De Giorgi, *Sulla differenziabilità e l'analiticità delle extremali degli integrali multipli regolari*. Mem. Accad. Sci. Torino, Cl. Sci. Fis. Mat. Nat. (3) 3 (1957), 25–43.
[138] E. Di Nezza, G. Palatucci, E. Valdinoci, *Hitchhiker's guide to the fractional Sobolev spaces*. Bull. Sci. Math. 136 (2012) no. 5, 521–573.
[139] S. Dipierro, E. Valdinoci, *A density property for fractional weighted Sobolev spaces*. Atti Accad. Naz. Lincei, Rend. Lincei, Mat. Appl. 26 (2015) no. 4, 397–422.
[140] S. Dipierro, L. Montoro, I. Peral, B. Sciunzi, *Qualitative properties of positive solutions to nonlocal critical problems involving the Hardy-Leray potential*. Calc. Var. Partial Differ. Equ. 55 (2016) no. 4, 99.
[141] S Dipierro, X Ros-Oton, E Valdinoci, *Nonlocal problems with Neumann boundary conditions*. Rev. Mat. Iberoam. 33 (2017), 377–416.
[142] G. Doetsch, *Introduction to the Theory and Applications of the Laplace Transformation*. Springer, New York, 1974.
[143] J. Droniou, T. Gallouët, J. Vovelle, *Global solution and smoothing effect for a non-local regularization of a hyperbolic equation*. J. Evol. Equ. 3 (2003) no. 3, 499–521.
[144] J. Duoandikoetxea, *Fourier Analysis*. Graduate Studies in Mathematics 29. American Mathematical Society, Providence, RI, 2001.
[145] L. Dupaigne, *A nonlinear elliptic PDE with the inverse square potential*. J. Anal. Math. 86 (2002), 359–398.
[146] B. Dyda, M. Kassmann, *On weighted poincaré inequalities*. Ann. Acad. Sci. Fenn., Math. 38 (2013) no. 2, 721–726.
[147] I. Ekeland, *On the variational principle*. J. Math. Anal. Appl. 47 (1974), 324–353.
[148] L. C. Evans, *Partial Differential Equations*. Graduate Studies in Mathematics 19. American Mathematical Society, Providence, RI, 1998.
[149] L. C. Evans, R. F. Gariepy, *Measure Theory and Fine Properties of Functions*. CRC Press, Boca Raton, Fl. USA, 1991.
[150] E. B. Fabes, C. E. Kenig, R. P. Serapioni, *The local regularity of solutions of degenerate elliptic equations*. Commun. Partial Differ. Equ. 7 (1982) no. 1, 77–116.
[151] M. M. Fall, *Semilinear elliptic equations for the fractional Laplacian with Hardy potential*. Preprint. arXiv:1109.5530v4 [math.AP].
[152] M. M. Fall, *Nonexistence of distributional supersolutions of a semilinear elliptic equation with Hardy potential*. J. Funct. Anal. 264 (2013) no. 3, 661–690.
[153] M. M. Fall, T. Weth, *Nonexistence results for a class of fractional elliptic boundary value problems*. J. Funct. Anal. 263 (2012) no. 8, 2205–2227.
[154] V. Felli, M. Schneider, *Perturbation results of critical elliptic equations of Caffarelli-Kohn-Nirenberg type*. J. Differ. Equ. 191 (2003), 121–142.
[155] M. Felsinger, M. Kassmann, *Local regularity for parabolic nonlocal operators*. Commun. Partial Differ. Equ. 38 (2013), 1539–1573.
[156] V. Ferone, F. Murat, *Quasilinear problems having quadratic growth in the gradient: an existence result when the source term is small*, in Equations aux dérivées partielles et applications, pp. 497–515. Gauthier-Villars, Ed. Sci. Méd. Elsevier, Paris, 1998.
[157] V. Ferone, F. Murat, *Nonlinear problems having natural growth in the gradient: an existence result when the source terms are small*. Nonlinear Anal. 42 (2000) no. 7, 1309–1326.

[158] F. Ferrari, I. E. Verbitsky, *Radial fractional Laplace operators and Hessian inequalities.* J. Differ. Equ. 253 (2012) no. 1, 244–272.

[159] C. L. Fefferman, *The uncertainty principle.* Bull. Am. Math. Soc. 9 (1983) no. 2, 129–206.

[160] C. Fefferman, D. H. Phong, *On positivity of pseudo-differential operators.* Proc. Natl. Acad. Sci. USA 75 (1978) no. 10, 4673–4674.

[161] X. Fernandez-Real, X. Ros-Oton *Boundary regularity for the fractional heat equation.* Rev. Acad. Cienc. Ser. A Mat. 110 (2016), 49–64.

[162] X. Fernandez-Real, X. Ros-Oton *Regularity theory for general stable operators: parabolic equations.* J. Funct. Anal. 272 (2017), 4165–4221.

[163] A. Fiscella, P. Pucci, *On certain nonlocal Hardy-Sobolev critical elliptic Dirichlet problems.* Adv. Differ. Equ. 21 (2016) no. 5–6, 571–599.

[164] R. Frank, *A simple proof of Hardy-Lieb-Thirring inequalities.* Commun. Math. Phys. 290 (2009), 789–800.

[165] R. Frank, *Sobolev inequalities and uncertainty principles in Mathematical Physics. Part I.* http://www.math.caltech.edu/~rlfrank/sobweb1.pdf.

[166] R. Frank, *Cwikel's theorem and the CLR inequal.* J. Spectr. Theory 4 (2014) no. 1, 1–21.

[167] R. Frank, E. Lieb, R. Seiringer, *Hardy-Lieb-Thirring inequalities for fractional Schrödinger operators.* J. Am. Math. Soc. 21 (2007) no. 4, 925–950.

[168] R. Frank, R. Seiringer, *Non linear ground state representations and sharp Hardy inequalities.* J. Funct. Anal.. 255 (2008), 3407–3430.

[169] D. A. Frank-Kamenenetskii, *Diffusion and Heat Transfer in Chemical Kinetics.* Plenum Press, New York, 1969.

[170] H. Fujita, *On the nonlinear equations* $\Delta u + \exp u = 0$ *and* $v_t = \Delta u + \exp u$. Bull. Am. Math. Soc. 75 (1969), 132–135.

[171] H. Fujita, *On the blowing up of solutions of the Cauchy problem for* $u_t = \Delta u + u^{1+\alpha}$. J. Fac. Sci., Univ. Tokyo, Sect. I 13 (1966), 109–124.

[172] M. Fukushima, *On an* L^p*-estimate of resolvents of Markov processes.* Publ. RIMS, Kyoto Univ. 13 (1977), 277–284.

[173] E. Gagliardo, *Ulteriori proprietà di alcune classi di funzioni in più variabili.* Ric. Mat. 8 (1959), 24–51.

[174] V. A. Galaktionov, S. P. Kurdyumov, A. P. Mikhailov, A. A. Samarskii, *On unbounded solutions of the Cauchy problem for the parabolic equation* $u_t = \nabla(u^\sigma \nabla u) + u^\beta$. *(Russian) Dokl. Akad. Nauk SSSR 252, no. 6 (1980), 1362–1364. English translation*: Sov. Phys. Dokl. 25 (1980), 458–459.

[175] V. A. Galaktionov, H. A. Levine, *A general approach to critical Fujita exponents in nonlinear parabolic problems.* Nonlinear Anal. 34 (1998) no. 7, 1005–1027.

[176] T. Gallouet, F. Mignot, J. P. Puel, *Quelques resultats sur le probleme* $-\Delta u = \exp u$. C. R. Acad. Sci. 307 (1988), 289–292.

[177] J. García Azorero, I. Peral, *Multiplicity of solutions for elliptic problems with critical exponents or with a non-symmetric term.* Trans. Am. Math. Soc. 323 (1991) no. 2, 877–895.

[178] J. García Azorero, I. Peral, *On the existence of a second positive solution for a quasilinear elliptic problem involving the critical Sovolev exponent.* Indiana Univ. Math. J. 43 (1994) no. 3, 941–957.

[179] J. García Azorero, I. Peral, *Hardy inequalities and some critical elliptic and parabolic problems.* J. Differ. Equ. 144 (1998) no. 2, 441–476.

[180] J. García Azorero, J. J. Manfredi, I. Peral, *Sobolev versus Hölder minimizers with applications to quasilinear elliptic equations.* Commun. Contemp. Math. 2 (2000) no. 3, 385–404.

[181] J. García Azorero, I. Peral, A. Primo, *A borderline case in elliptic problems involving weights of Caffarelli-Kohn-Nirenberg type*. Nonlinear Anal. 67 (2007) no. 6, 1878–1894.

[182] I. M. Gelfand, *Some problems in the theory of quasilinear equations (Section 15 by G.I. Barenblatt)*. Am. Math. Soc. Transl. (2) 29 (1963), 295–381.

[183] N. Ghoussoub, D. Preiss *A general mountain pass principle for locating and classifying critical points*. Ann. Inst. Henri Poincaré, Anal. Non Linéaire 6 (1989) no. 5, 321–330.

[184] N. Ghoussoub, C. Yuan, *Multiple solution for quasi-linear PDEs involving the critical Sobolev and Hardy exponents*. Trans. Am. Math. Soc. 352 (2000) no. 12, 5703–5743.

[185] M. Giaquinta, E. Giusti, *On the regularity of the minima of variational integrals*. Acta Math. 148 (1982), 31–46.

[186] G. Gilbarg, N. S. Trudinger, *Elliptic Partial Differential Equations of Second Order*. Reprint of the 1998 edition. Classics in Mathematics. Springer-Verlag, Berlin, 2001.

[187] B. Gilding, M. Guedda, R. Kersner, *The Cauchy problem for $u_t = \Delta u + |\nabla u|^q$*. J. Math. Anal. Appl. 284 (2003) no. 2, 733–755.

[188] E. Giusti *Metodi diretti nel Calcolo delle variationi*. Unione Matematica Italiana, Bologna, 1994.

[189] L. Grafakos, *Classical and Modern Fourier Analysis*. Pearson Education, Prentice Hall, New Jersey, 2004.

[190] C. E. Gutiérrez, R. L. Wheeden, *Harnack's inequality for degenerate parabolic equations*. Commun. Partial Differ. Equ. 16 (1991) no. 4–5, 745–770.

[191] C. E. Gutiérrez, R. L. Wheeden, *Mean value Harnack inequalities for degenerate parabolic equations*. Colloq. Math. 60–61 (1990) no. 1, 157–194.

[192] M. de Guzmán, *Differentiation of Integrals in $\mathbb{R}^N$*. Lectures Notes in Mathematics 481. Springer-Verlag, Berlin 1975.

[193] J. Hadamard, *Lectures on Cauchy's Problem in Linear Partial Differential Equations*. Dover Publications, New York, 1953.

[194] K. Hansson, V. G. Maz'ya, I. E. Verbitsky, *Criteria of solvability for multidimensional Riccati equations*. Ark. Mat. 37 (1999) no. 1, 87–120.

[195] G. H. Hardy, *Note on a Theorem of Hilbert*. Math. Z. 6 (1920), 314–317.

[196] G. Hardy, J. E. Littlewood, G. Polya, *Inequalities*. Reprint of the 1952 edition. Cambridge Mathematical Library. Cambridge University Press, Cambridge, 1988.

[197] J. Heinonen, T. Kilpelainen, M. O. Martio, *Nonlinear Potential Theory of Degenerate Elliptic Equations*. Oxford Mathematical Monographs. Oxford Science Publications. The Clarendon Press, Oxford University Press, New York, 1993.

[198] E. Helly, *Über lineare Funktionaloperationen*. Sitzungsber. Naturwiss. Kl. Akad. Wiss. Wien 121 (1912) 265–297.

[199] I. W. Herbst, *Spectral theory of the operator $(p^2 + m^2)^{1/2} - Ze^2/r$*. Commun. Math. Phys. 53 (1977), 285–294.

[200] D. Hilbert, *Sur les problèmes futurs des Mathématiques*. Compte Rendu au Deuxième Congrès International des Mathématiciens. Paris, 1900.

[201] L. Hörmander, *The analysis of linear partial differential operators. III. Pseudo-differential operators*. Classics in Mathematics. Springer, Berlin, 2007.

[202] C. Imbert, *A non-local regularization of first order Hamilton-Jacobi equations*. J. Differ. Equ. 211 (2005) no. 1, 218–246.

[203] T. Jin, J. Xiong, *Schauder estimates for solutions of linear parabolic integro-differential equations*, Ann. Inst. Henri Poincaré, Anal. Non Linéaire 33 (2016) no. 5, 1375–1407.

[204] D. D. Joseph, T. S. Lundgren, *Quasilinear Dirichlet problems driven by positive sources*. Arch. Ration. Mech. Anal. 49 (1972/1973), 241–269.

[205] M. Kardar, G. Parisi, Y. C. Zhang, *Dynamic scaling of growing interfaces*. Phys. Rev. Lett. 56 (1986), 889–892.

[206] K. H. Karlsen, F. Petitta, S. Ulusoy, *A duality approach to the fractional Laplacian with measure data*. Publ. Mat. 55 (2011), 151–161.

[207] M. Kassmann, *A priori estimates for integro-differential operators with measurable kernels*. Calc. Var. 34 (2009), 1–21.

[208] M. Kassmann, *Harnack inequality: an Introduction*. Bound. Value Probl. (2007) 81415.

[209] T. Kato, *Schrödinger operators with singular potentials*. Proceedings of the International Symposium on Partial Differential Equations and the Geometry of Normed Linear Spaces (Jerusalem, 1972). Isr. J. Math. 13 (1972), 135–148.

[210] T. Kawanago, *Existence and behaviour of solutions for* $u_t = \Delta(u^m) + u^l$. Adv. Math. Sci. Appl. 7 (1997) no. 1, 367–400.

[211] J. L. Kazdan, F. W. Warner, *Curvature functions for compact 2-manifolds*. Ann. Math. 99 (1974), 14–47.

[212] J. L. Kazdan, R. J. Kramer, *Invariant criteria for existence of solutions to second-order quasilinear elliptic equations*. Commun. Pure Appl. Math. 31 (1978) no. 5, 619–645.

[213] J. Kinnunen, N. Shanmugalingam, *Regularity of quasi-minimizers on metric spaces*. Manuscr. Math. 105 (2001), 401–423.

[214] L. Korkut, M. Pašić, D. Žubrinić, *Some qualitative properties of solutions of quasilinear elliptic equations and applications*. J. Differ. Equ. 170 (2001), 247–280.

[215] J. Korvenpaa, T. Kuusi, G. Palatucci, *The obstacle problem for nonlinear integro-differential operators*. Calc. Var. Partial Differ. Equ. 55 (2016) no. 3, 63.

[216] N. V. Krylov, M. V. Safonov, *Certain properties of solutions of parabolic equations with measurable coefficients (Russian)*. Izv. Akad. Nauk SSSR 40 (1980), 161–175.

[217] H. Kumano-go, *Pseudodifferential Operators*. MIT Press, Cambridge, Mass.-London, 1981.

[218] T. Kuusi, G. Mingione, Y. Sire, *Nonlocal equations with measure data*. Commun. Math. Phys. 337 (2015) no. 3, 1317–1368.

[219] M. Kwasnicki, *Ten equivalent definitions of the fractional Laplace operator*. Fract. Calc. Appl. Anal. 20 (2017) no. 1, 7–51.

[220] O. A. Ladyženskaja, V. A. Solonnikov, N. N. Ural'ceva, *Linear and Quasilinear Equations of Parabolic Type (Russian). Translated from the Russian by S. Smith*. Translations of Mathematical Monographs 23. American Mathematical Society, Providence, R.I., 1967.

[221] R. Landes, V. Mustonen, *On parabolic initial-boundary value problems with critical growth for the gradient*. Ann. Inst Henri Poincaré, Anal. Non Linénaire 11 (1994) no. 2, 135–158.

[222] R. Landes, *On the existence of weak solutions for quasilinear parabolic initial-boundary value problems*. Proc. Roy. Soc. Edinburgh Sect. A 89 (1981) no. 3–4, 217–237.

[223] N. Landkof, *Foundations of Modern Potential Theory*. Die Grundlehren der mathematischen Wissenschaften 180. Springer-Verlag, New York-Heidelberg, 1972.

[224] T. Leonori, I. Peral, A. Primo, F. Soria, *Basic estimates for solutions of a class of nonlocal elliptic and parabolic equations*. Discrete Contin. Dyn. Syst., Ser. A 35 (2015) no. 12, 6031–6068.

[225] J. Leray, *Sur le mouvement d'un liquide visqueux emplissant l'espace*. Acta Math. 63 (1934) no. 1, 193–248.

[226] J. Leray, J. L. Lions, *Quelques résultats de Višik sur les problèmes elliptiques nonlinéaires par les méthodes de Minty-Browder*. Bull. Soc. Math. Fr. 93 (1965), 97–107.

[227] H. Levine, *The role of critical exponents in blowup theorems*. SIAM Rev. 32 (1990) no. 2, 262–288.

[228] C. K. Li, M. A. Aguirre, *The distributional products on spheres and Pizetti's formula*. J. Comput. Appl. Math. 235 (2011), 1482–1489.

[229] E. H. Lieb, *Bounds on the eigenvalues of the Laplace and Schrödinger operators*. Bull. Am. Math. Soc. 82 (1976), 751–752.

[230] E. Lieb, *Sharp constants in the Hardy-Littlewood-Sobolev and related inequalities*. Ann. Math. 118 (1983), 349–374.

[231] E. H. Lieb, M. Loss, *Analysis*. Graduate Studies in Mathematics 14. Second edition. American Mathematical Society, Providence, RI, 2001.

[232] P. L. Lions, *Generalized Solutions of Hamilton-Jacobi Equations*. Research Notes in Mathematics 69. Pitman (Advanced Publishing Program), Boston, Mass.-London, 1982.

[233] P. L. Lions, *The concentration-compactness principle in the calculus of variations. The limit case, part 1*. Rev. Mat. Iberoam. 1 (1985) no. 1, 145–201.

[234] P. L. Lions, *The concentration-compactness principle in the calculus of variations. The limit case, part 2*. Rev. Mat. Iberoam. 1 (1985), no. 2, 45–121.

[235] A. Majda, E. Tabak, *A two-dimensional model for quasigeostrophic flow: comparison with the two-dimensional Euler flow*. Physica D 98 (1996), 515–522.

[236] J. Maly, W. P. Ziemer, *Fine Regularity of Solutions of Elliptic Partial Differential Equations*. Mathematical Surveys and Monographs 51. American Mathematical Society, Providence, RI, 1997.

[237] I. Markina, *Potential theory: the origin and applications*. https://folk.uib.no/ima083/courses_files/potential.pdf.

[238] J. Mawhin, M. Willem, *Critical Point Theory and Hamiltonian Systems*. Springer-Verlag, Berlin 1989.

[239] V. G. Maz'ja, *Sobolev Spaces*. Springer Verlag, Berlin, 1985.

[240] F. Mignot, J. P. Puel, *Sur une classe de probleme non lineaire avec non linearite positive, croissante, convexe*. Commun. Partial Differ. Equ. 5 (1980) no. 8, 791–836.

[241] F. Mignot, J. P. Puel, *Solution radiale singuliere de* $-\Delta u = \lambda e^u$. C. R. Acad. Sci., Sér. 1 307 (1988), 379–382.

[242] A. N. Milgram, in L. Bers, F. John, M. Schechter (Eds.) Supplement II in Partial Differential Equations: Vol III in Lectures in Applied Mathematics, pp. 229. Interscience New York, 1964.

[243] S. Mizohata, *The Theory of Partial Differential Equations*. Cambridge University Press, London 1973.

[244] J. Moser, *A Harnack inequality for parabolic differential equations*. Commun. Pure Appl. Math. 17 (1964), 101–134.

[245] J. Moser, *A new proof of De Giorgi's theorem concerning the regularity problem for elliptic differential equations*. Commun. Pure Appl. Math. 13 (1960), 457–468.

[246] J. Moser, *On a pointwise estimate for parabolic differential equations*. Commun. Pure Appl. Math. 17 (1971), 101–134.

[247] N. F. Mott, H. S. W. Massey, *Theory of Atomic Collisions*. Clarendon Press, Oxford, 1949.

[248] F. Murat, *L'injection du cone positif de* H^{-1} *dans* $W^{-1,q}$ *est compacte pour tout* $q < 2$. J. Math. Pures Appl. 60 (1981), 309–322.

[249] F. Murat, A. Porretta, *Stability properties, existence, and nonexistence of renormalized solutions for elliptic equations with measure data*. Commun. Partial Differ. Equ. 27 (2002) no. 11–12, 2267–2310.

[250] J. Nash, *Continuity of solutions of parabolic and elliptic equations*. Am. J. Math. 80 (1958), 931–954.

[251] L. Nirenberg, *On elliptic partial differential equations*. Ann. Sc. Norm. Super. Pisa (3) 13 (1959), 115–162.

[252] L. Nirenberg, *Pseudo-differential operators*, in 1970 Global Analysis. Proc. Sympos. Pure Math. Amer. Math. Soc. Vol. XVI, Berkeley, Calif., pp. 149–167. Providence, R.I. (1968).

[253] L. Nirenberg, F. Trèves, *On local solvability of linear partial differential equations. II. Sufficient conditions*. Commun. Pure Appl. Math. 23 (1970), 459–509.
[254] L. Nirenberg, F. Trèves, *Conditions nécessaires de résolubilité locale des équations pseudo-différentielles*. C. R. Acad. Sci., Ser. A-B 269 (1969), A774–A777.
[255] L. Oswald, *Isolated positive singularities for a nonlinear heat equation*. Houst. J. Math. 14 (1988) no. 4, 543–572.
[256] I. Peral, *Multiplicity of solutions for the p-Laplacian*, in Second School of Nonlinear Functional Analysis and Applications to Differential Equations 21 April–9 May 1997 International Center for Theoretical Physics, Lecture Notes, Trieste (Italy).
[257] I. Peral, J. L. Vazquez, *On the stability or instability of the singular solution of the semilinear heat equation with exponential reaction term*. Arch. Ration. Mech. Anal. 129 (1995) no. 3, 201–224.
[258] F. Petitta, *Renormalized solutions of nonlinear parabolic equations with general measure data*. Ann. Mat. Pura Appl. 187 (2008) no. 4, 563–604.
[259] M. Picone, *Sui valori eccezionali di un parametro da cui dipende un'equazione differenziale lineare ordinaria del second'ordine*. Ann. Sc. Norm. Super. Pisa, Cl. Sci. 11 (1910), 1–144.
[260] P. Pizzetti, *Sulla media dei valore che una funzione dei punti della spazio assume alla superticie di una sfera*. Rend. Lincei, Ser. V XVIII (1909), 182–185.
[261] S. I. Pohozaev, *Eigenfunctions of the equation* $\Delta u + \lambda f(u) = 0$. Sov. Math. Dokl. 6 (1965), 1408–1411 (English translation Dokl. Akad Nauk. SSRR 165 (1965), 33–36).
[262] G. Polya, *On the zeros of an integral function represented by Fourier's integral*. Messenger Math. 52 (1923), 185–188.
[263] G. Polya, G. Szegö, *Isoperimetric Inequalities in Mathematical Physics*. Princeton University Press, 1951.
[264] A. C. Ponce, *Elliptic PDEs, Measures and Capacities. From the Poisson Equation to Nonlinear Thomas-Fermi Problems*. Tracts in Mathematics 23. European Mathematical Society, 2016.
[265] A. Porretta, *Existence results for nonlinear parabolic equations via strong convergence of truncations*. Ann. Mat. Pura Appl. (4) 177 (1999), 143–172.
[266] A. Porretta, *Nonlinear equations with natural growth terms and measure data* in 2002-Fez Conference on Partial Differential Equations 09, pp. 183–202, 2002. http://ejde.math.swt.edu or http://ejde.math.unt.edu.
[267] A. Prignet, *Existence and uniqueness of "entropy" solutions of parabolic problems with* L^1 *data*. Nonlinear Anal. 28 (1997) no. 12, 1943–1954.
[268] P. Pucci, J. Serrin, *A general variational identity*. Indiana Univ. Math. J. 35 (1985) no 3, 681–703.
[269] P. Quittner, P. Souplet, *Superlinear Parabolic Problems. Blow-up, Global Existence and Steady States*. Birkhäuser Advanced Texts: Basler Lehrböcher. Birkhäuser Verlag, Basel, 2007.
[270] M. Reed, B. Simon, *Methods of Modern Mathematical Physics. II. Fourier Analysis, Self-Adjointness*. Academic Press, New York-London, 1975.
[271] F. Riesz, *Sur une inégalité intégrale*. J. Lond. Math. Soc. 5 (1930), 162–168.
[272] M. Riesz, *Intégrales de Riemann-Liouville et potenciels*. Acta Sci. Math. 9 (1938), 1–42.
[273] X. Ros-Oton, J. Serra, *The Dirichlet problem for the fractional Laplacian: regularity up to the boundary*. J. Math. Pures Appl. 101 (2014), 275–302.
[274] X. Ros-Oton, J. Serra, *The Pohozaev identity for the fractional Laplacian*. Arch. Ration. Mech. Anal. 213 (2014), 587–628.
[275] H. L. Royden, *Real Analysis*. Macmillan Publishing Co., New York, 1968.
[276] G. V. Rozenblum, *Distribution of the discrete spectrum of singular differential operators*. Sov. Math. Dokl. 13 (1972), 245–249 and Soviet Math. (Iz. VUZ) 20 (1976), 63–71.

[277] L. SALOFF-COSTE, *Aspects of Sobolev-type inequalities*. London Mathematical Society Lecture Note Series, vol. 289. Cambridge University Press, Cambridge, 2002

[278] O. SAVIN, E. VALDINOCI, *Elliptic PDEs with fibered nonlinearities*. J. Geom. Anal. 19 (2009), 420–432.

[279] R. SERVADEI, E. VALDINOCCI, *Mountain pass solutions for non-local elliptic operators*. J. Math. Anal. Appl. 389 (2012) no. 2, 887–898.

[280] R. SERVADEI, E. VALDINOCI, *The Brezis-Nirenberg result for the fractional Laplacian*. Trans. Am. Math. Soc. 367 (2015) no. 1, 67–102.

[281] R. SERVADEI, E. VALDINOCI, *Variational methods for non-local operators of elliptic type*. Discrete Contin. Dyn. Syst. 33 (2013) no. 5, 2105–2137.

[282] R. SERVADEI, E. VALDINOCI, *Weak and viscosity solutions of the fractional Laplace equation*. Publ. Mat.. 58 (2014), 133–154.

[283] R. SERVADEI, E. VALDINOCI, *A Brezis-Nirenberg result for non-local critical equations in low dimension*. Commun. Pure Appl. Anal. 12 (2013) no. 6, 2445–2464.

[284] X. SHANG, J.ZHANG, R. YIN, *Existence of positive solutions to fractional elliptic problems with Hardy potential and critical growth*. Math. Mech. Appl. Sci. 42 (2019), 115–136.

[285] A. SIGNORINI, *Questioni di elasticitá non linearizzata e semilinearizzata*. Rend. Mat. Appl. 18 (1959), 95–139.

[286] L. SILVESTRE, PhD thesis. The University of Texas at Austin, 2005.

[287] L. SILVESTRE, *Regularity of the obstacle problem for a fractional power of the Laplace operator*. Commun. Pure Appl. Math. 60 (2007) no. 1, 67–112.

[288] L. SILVESTRE, *Hölder estimates for solutions of integro-differential equations like the fractional Laplace*. Indiana Univ. Math. J. 55 (2006), 1155–1174.

[289] J. SIMONS, *Minimal varieties in Riemannian manifolds*. Ann. Math. 88 (1968) no. 1, 62–105.

[290] Y. SIRE, E. VALDINOCI, *Fractional Laplacian phase transitions and boundary reactions: a geometric inequality and a symmetry result*. J. Funct. Anal. 256 (2009), 1842–1864.

[291] G. STAMPACCHIA, *Le problème de Dirichlet pour les équations elliptiques du second ordre à coefficients discontinus*. Ann. Inst. Fourier (Grenoble) 15 (1965) no. 1, 189–258.

[292] G. STAMPACCHIA, *Èquations elliptiques du second ordre à coefficients discontinus*. Séminaire de Mathématiques Supérieures, 16 (Été, 1965). Les Presses de l'Université de Montréal, Montreal. 1966.

[293] E. M. STEIN, *Singular Integrals and Differentiability Properties of Functions*. Princeton Mathematical Series 30. Princeton University Press, Princeton, N.J., 1970.

[294] E. M. STEIN, *Harmonic analysis: real-variable methods, orthogonality, and oscillatory integrals*, in Monographs in Harmonic Analysis, III. Princeton Mathematical Series 43. Princeton University Press, Princeton, NJ, 1993.

[295] E. M. STEIN, G. WEISS, *Fractional integrals on n-dimensional Euclidean space*. J. Math. Mech. 7 (1958), 503–514.

[296] E. M. STEIN, G. WEISS, *Introduction to Fourier Analysis on Euclidean Spaces*. Princeton University Press, New Yersey, 1975.

[297] J. STOKER, *Water Waves: The Mathematical Theory with Applications*. Pure and Applied Mathematics IV. Interscience Publishers, Inc., New York, 1957.

[298] M. STRUWE, *Variational Methods. Applications to Nonlinear Partial Differential Equations and Hamiltonian Systems*. Springer-Verlag, Berlin. 2000.

[299] S. SUGITANI, *On nonexistence of global solutions for some nonlinear integral equations*. Osaka J. Math. 12 (1975), 45–51.

[300] C. A. SWANSON, *Remarks on Picone's identity and related identities*. Atti Accad. Naz. Lincei Mem. Cl. Sci. Fis. Mat. Natur. Sez. Ia (8) 11 (1972/1973) no. 1, 3–15.

[301] G. Talenti, *Inequalities in Rearrangement Invariant Function Spaces*. Nonlinear Analysis, Function Spaces and Applications 5 (Prague, 1994), pp. 177–230. Prometheus, Prague, 1994.
[302] G. Talenti, *Best constant in Sobolev inequality*. Ann. Mat. Pura Appl. (4) 110 (1976), 353–372.
[303] L. Tartar, *Imbedding theorems of Sobolev spaces into Lorentz spaces*. Boll. Unione Mat. Ital. Sez. B Artic. Ric. Mat. (8) 1 (1998), 479–500.
[304] S. Terracini, *On positive entire solutions to a class of equations with a singular coefficient and critical exponent*. Adv. Differ. Equ. 1 (1996) no. 2, 241–264.
[305] J. Toland, *The Peierls-Nabarro and Benjamin-Ono equations*. J. Funct. Anal. 145 (1997), 136–150.
[306] F. Treves, *Basic Linear Partial Differential Equations*. Academic Press, New York, 1975.
[307] F. Treves *Introduction to Pseudodifferential and Fourier Integral Operators. Vol. 1. Pseudodifferential Operators*. The University Series in Mathematics. Plenum Press, New York-London, 1980.
[308] N. Trudinger, *Remarks concerning the conformal deformation of Riemannian structures on compact manifolds*. Ann. Sc. Norm. Super. Pisa. 22 (1968), 265–274.
[309] E. Valdinoci, *From the long jump random walk to the fractional Laplacian*. Bol. Soc. Esp. Mat. Apl. 49 (2009), 33–44.
[310] J. L. Vázquez, *A strong maximum principle for some quasilinear elliptic equations*. Appl. Math. Optim. 12 (1984) no. 3, 191–202.
[311] J. L. Vázquez, E. Zuazua, *The Hardy inequality and the asymptotic behaviour of the heat equation with an inverse-square potential*. J. Funct. Anal. 173 (2000) no. 1, 103–153.
[312] M. I. Vishik, *Mixed boundary problems and an approximate method for their solution (Russian)*. Dokl. Akad. Nauk SSSR 97 (1954), 193–196.
[313] C. Wang, S. Zheng, *Critical Fujita exponents of degenerate and singular parabolic equations*. Proc. R. Soc. Edinb., Sect. A 136 (2006) no. 2, 415–430.
[314] Z. Q. Wang, M. Willem, *Caffarelli-Kohn-Nirenberg inequalities with remainder terms*. J. Funct. Anal. 203 (2003) no. 2, 550–568.
[315] E. T. Whittaker, G. N. Watson, *A Course of Modern Analysis*. 4th edition, Cambridge Univ Press, 1996.
[316] F. B. Weissler, *Existence and nonexistence of global solutions for a semilinear heat equation*. Isr. J. Math. 38 (1981) no. 1–2, 29–40.
[317] D. V. Widder, *Positive temperatures on an infinite rod*. Trans. Am. Math. Soc. 55 (1944) no. 1, 85–95.
[318] D. V. Widder, *The Heat Equation*. Academic Press, New York, 1975.
[319] D. Yafaev, *Sharp constants in the Hardy-Rellich inequalities*. J. Funct. Anal. 168 (1999) no 1, 121–144.
[320] K. Yosida, Functional Analysis. 4th edition. Springer Verlag, Berlin 1974.
[321] Y. B. Zeldovich, G. I. Barenblatt, V. B. Librovich, G. M. Makhviladze, *The Mathematical Theory of Combustion and Explosions*. Consultants Bureau, New York and London, 1985.
[322] A. Zygmund *Trigonometric Series: Vols. I, II*. Second edition. Cambridge University Press, London-New York 1968.

Alphabetical Index

www.ingramcontent.com/pod-product-compliance
Ingram Content Group UK Ltd.
Pitfield, Milton Keynes, MK11 3LW, UK
UKHW051029041026
15470UKWH00016B/1816